Industrial Environmental Management

Industrial Environmental Management

Engineering, Science, and Policy

Tapas K. Das

WILEY

This edition first published 2020

Registered Office
John Wiley & Sons, Inc., 111 River Street, Hoboken, NJ 07030, USA

Editorial Office
111 River Street, Hoboken, NJ 07030, USA

For details of our global editorial offices, customer services, and more information about Wiley products visit us at www.wiley.com.

Wiley also publishes its books in a variety of electronic formats and by print-on-demand. Some content that appears in standard print versions of this book may not be available in other formats.

Library of Congress Cataloging-in-Publication Data

Names: Das, Tapas K., author.
Title: Industrial environmental management : engineering, science, and policy / Tapas K. Das.
Description: First edition. | Hoboken, NJ : Wiley, 2020. | Includes bibliographical references and index.
Identifiers: LCCN 2019035347 (print) | LCCN 2019035348 (ebook) | ISBN 9781119591580 (hardback) | ISBN 9781119591559 (adobe pdf) | ISBN 9781119591566 (epub)
Subjects: LCSH: Industrial management–Environmental aspects. | Industrial engineering–Environmental aspects. | Environmental management.
Classification: LCC HD30.255 .D37 2020 (print) | LCC HD30.255 (ebook) | DDC 658.4/083–dc23
LC record available at https://lccn.loc.gov/2019035347
LC ebook record available at https://lccn.loc.gov/2019035348

Cover Design: Wiley
Cover Images: Medicine abstract background © Zoezoe33/Shutterstock, Abstract plant © Viktoriya/Shutterstock

Set in 9.5/12.5pt STIXTwoText by SPi Global, Pondicherry, India

Printed in the United States of America

V10017117_012320

To our current and future students.

Contents

About the Author

Tapas K. Das, PhD, PE, BCEE, is a chemical and environmental engineer; a Fellow member of the American Institute of Engineers; Past Chair of the AIChE's Environmental Division; former Chair of the Air Pollution Control Committee of the American Academy of Environmental Engineers and Board of Trustee with the Academy; American Academy Who's Who in Environmental Engineering from 2002 to the present; former environmental engineer at the Washington Department of Ecology; as an adjunct faculty member, Dr. Das has been teaching several undergraduate and graduate courses in Civil, Environmental, and Mechanical Engineering programs at Saint Martin's University School of Engineering, Washington; formerly assistant professor in the College of Natural Resources and Paper Science and Engineering at the University of Wisconsin, Stevens Point; and recipient of the Professor S. K. Sharma Medal and CHEMCON Distinguished Speaker Award for 2007 given by the Development Organization for Sustainable Transformation (DOST), Indian Institute of Chemical Engineers. Dr. Das holds a BS in Chemical Engineering from Jadavpur University in Kolkata, India, and PhD from Bradford University, Bradford, England. Dr. Das was a postdoctoral fellow at London's Imperial College of Science, Technology, and Medicine and a visiting scientist at Princeton University. He has wide practical and theoretical experience in various areas, including air toxics and aerosols, industrial wastewater treatment for water reuse, solid waste management and combustion, profitable process pollution prevention, reuse, recycle, redesign, sustainable engineering, and sustainability. Dr. Das is a registered professional engineer in the state of Washington. Dr. Das is the author of the book *Toward Zero Discharge: Innovative Methodology and Technologies for Process Pollution Prevention* (Wiley, 2005).

Preface

Recently, I taught a similar course titled "Industrial Environmental Management" to senior engineering students and found out that there isn't a single textbook available to cover the depth and breadth of this subject matter. That alone motivated me to write this textbook with real-world examples, challenging problems, and solutions provided for each chapter.

Tomorrow's and today's sustainable products and processes require engineers to carefully consider environmental, economic, social factors, while using sustainable feedstocks, renewable energy, water, chemicals, and materials in creating their design. Some quantitative tools for incorporating sustainability concepts into engineering designs and performing metrics are highlighted in the text; sustainable engineering and its principles introduce these tools and show how to apply them in lean manufacturing. In general, engineers and managers working in manufacturing industries find valuable and up-to-date information about lean manufacturing, Six Sigma, workers' health and safety issues and environmental regulations, monitoring, reporting, and compliance. Also, consulting engineers will find useful information about sustainable design principles and methodology, plus best available control technologies for environmental remediation in cost-effective ways.

This book is dedicated to undergraduate and graduate students. This book is designed to be a textbook that is prepared primarily for junior-level and senior-level students in multidisciplinary engineering fields including, but not limited to, aerospace, chemical, civil, environmental, industrial and manufacturing, materials science and engineering, mechanical, paper science and engineering, petroleum engineering, and business management. The subject matters covered in this textbook will be suitable for offering a course in multiple engineering disciplines within colleges and schools of engineering programs. This book has 10 chapters. I have written the text for Chapters 1 through 8 for students in clear and simple language. Theories, real-world problems, and applications are embedded throughout these first eight chapters so that students can check their understanding before continuing on to new sections. Chapters 9 and 10 are more suitable for a graduate-level course in sustainable engineering, sustainable manufacturing, or related topics.

4 June 2019

Tapas K. Das
Olympia, WA, USA

Acknowledgments

This book publication wouldn't have been successful without the helping hands of many individuals. I would like to thank John Berg, Clint Bowman, Jae Chung, Meghasree Dey, Dibyendu Narayan Ghosh, Linn Hergert, Clint Lamoreaux, Joseph Mailhot, Robert Peters, Katherine Porter, Selma Thagard, Sandra Tully, and staff members at the Timberland Regional Library in the City of Lacey, Washington, who helped to prepare the manuscript, proofread the text, provided reference materials, figures, graphics for the book, and made the book more readable for students. Also, I want to acknowledge my teachers, professors, and my classmates in India, my doctoral and postdoctoral advisors, mentors, and colleagues in England and United States for their encouragement, noble efforts, dedication and integrity to their professions, and exemplary lifestyle.

Finally, I would like to express my sincere appreciation to Bob Esposito, Associate Publisher at Wiley for accepting the idea of this textbook; Beryl Mesiadhas, Senior Project Editor; Devi Ignasi, Production Editor; Michael Leventhal; and the entire editorial and publishing team at Wiley.

About the Companion Website

This book is accompanied by a companion website:

www.wiley.com/go/Das/IEM_1e

1

Why Industrial Environmental Management?

1.1 Introduction

This introductory chapter addresses why industrial environmental management is important! Environmental management is a very crucial part of human well-being that needs to be deeply considered. Formulated design seeks to steer the development process to take advantage of opportunities, avoid hazards, mitigate problems, and prepare people for unavoidable difficulties by improving adaptability and resilience. It is a process concerned with human–environment interactions, and seeks to identify: what is environmentally desirable; what are physical, economic, social, and technological constraints to achieving that process; and what are the most feasible options. Actually there can be no concise universal definition of environmental management; however, it can be briefly summarized as supporting sustainable development; demanding multidisciplinary and interdisciplinary or even holistic approaches; it should integrate and reconcile different development viewpoints, co-ordinate science, engineering, technology, social, policy making and planning; state proactive processes; timescales and concerns ranging from local to global issues; and one stresses stewardship rather than exploitation while dealing with a world affected by humans.

In other perspectives, environmental management can be explained as methods of ways when dealing with issues due to the importance of the need to improve environmental stewardship by integrating ecology, policy making, planning, and social development. The goals include sustaining and (if possible) improving existing resources; preventing and overcoming environmental problems; establishing limits; founding and nurturing institutions that effectively support environmental research, monitoring and management resources; warning of threats and identifying positive change opportunities; (where possible) improving quality of life; and finally, identifying new technology or policies that are useful. And moreover, actually environmental management may be subdivided into a number of fields, including the following:

- Environmental economics
- Sustainable development issues
- Environmental assessment, modeling, forecasting, and "hand-casting"
- Corporate environmental management activities
- Pollution recognition and control
- Environmental enforcement and legislation
- Environmental and development institutions and ethics
- Environmental management systems and quality issues
- Environmental planning and management
- Assessment of stakeholders involved in environmental management
- Environmental perceptions and education
- Community participation
- Natural resources management
- Environmental rehabilitation
- Environmental politics
- Environment aid and institution building

Generally, the environmental managers must ensure there is optimum balance between environmental protection and allowing human liberty. Then, the question is "how it can be done?" Basically, there are several steps in environmental management implementation. First, we need to identify goals and define problems, then determine appropriate actions, which will be continued as draw-up plan. Next, implement the plan, which will be followed by ongoing development management. After that monitor and evaluate the situation. If there are adjustments needed, do it. Then finally when the exact model and standards of environmental management are generated, continuous development will be conducted.

Industrial Environmental Management: Engineering, Science, and Policy, First Edition. Tapas K. Das.

Companion website: www.wiley.com/go/Das/IEM_1e

The overall objective of environmental management is improved human life quality. It involves the mobilization of resources and the use of government to administer the use of both natural and economic goods and services. It is based on the principles of ecology. It uses systems analysis and conflict resolution to distribute the costs and benefits of development activities throughout the affected populations and seeks to protect the activities of development from natural hazards. Conflict identification is one of the most important tasks in environmental management planning and the resolution of conflicts is a fundamental part of what makes up "environmentally sound development. ISO (International Organization for Standardization) 14000 is a family of standards related to environmental management that exists to help organizations (i) minimize how their operations (processes, etc.) negatively affect the environment (i.e. cause adverse changes to air, water, or land); (ii) comply with applicable laws, regulations, and other environmentally oriented requirements; and (iii) continually improve in the above.

1.1.1 ISO in Brief

ISO is the International Organization for Standardization. It has a membership of 160 national standards institutes from countries large and small, industrialized, developing and in transition, in all regions of the world. ISO's portfolio of more than 18 000 standards provides practical tools for all 3 dimensions of sustainable development: economic, environment, and societal.

ISO standards for business, government, and society as a whole make a positive contribution to the world we live in. They ensure vital features such as quality, ecology, safety, economy, reliability, compatibility, interoperability, conformity, efficiency, and effectiveness. They facilitate trade, spread knowledge, and share technological advances and good management practice. ISO develops only those standards that are required by the market. This work is carried out by experts on loan from the industrial, technical, and business sectors which subsequently put them to use. These experts may be joined by others with relevant knowledge, such as representatives of governmental agencies, testing laboratories, consumer association and academia, and by nongovernmental or other stakeholder organizations that have a specific interest in the issues addressed in the standards. Published under the designation of International Standards, ISO standards represent an international consensus on the state of the art in the technology or good practice concerned.

1.1.2 ISO and the Environment

ISO has a multifaceted approach to meeting the needs of all stakeholders from business, industry, governmental authorities, and nongovernmental organizations, as well as consumers, in the field of the environment.

1) ISO develops standards that help organizations to take a proactive approach to managing environmental issues: the ISO 14000 family of environmental management standards can be implemented in any type of organization in either public or private sectors – from companies to administrations to public utilities.
2) ISO helps to meet the challenge of climate change with standards for greenhouse gas accounting, verification and emissions trading, and for measuring the carbon footprint of products.
3) ISO develops standard documents to facilitate the fusion of business and environmental goals by encouraging the inclusion of environmental aspects in product design.
4) ISO offers a wide-ranging portfolio of standards for sampling and test methods to deal with specific environmental challenges. It has developed some 570 International Standards for the monitoring of such aspects as the quality of air, water and the soil, as well as noise, radiation, and for controlling the transport of dangerous goods. They also serve in a number of countries as the technical basis for environmental regulations.

1.1.3 Benefits

ISO 14001 was developed primarily to assist companies with a framework for better management control that can result in reducing their environmental impacts. In addition to improvements in performance, organizations can reap a number of economic benefits, including higher conformance with legislative and regulatory requirements (Sheldon 1997) by adopting the ISO standard. By minimizing the risk of regulatory and environmental liability fines and improving an organization's efficiency (Delmas 2009), benefits can include a reduction in waste, consumption of resources, and operating costs. Secondly, as an internationally recognized standard, businesses operating in multiple locations across the globe can leverage their conformance to ISO 14001, eliminating the need for multiple registrations or certifications (Hutchens 2010). Thirdly, there has been a push in the last decade by consumers for companies to adopt better internal controls, making the incorporation of ISO 14001 a smart approach for the long-term viability of businesses. This can provide them with a competitive advantage against companies that do not adopt the standard (Potoski and Prakash 2005). This in turn can have a positive impact on a company's asset value (Van der Veldt 1997). It can lead to improved public perceptions of the business, placing them in a better position to operate in the international marketplace (Potoski and Prakash 2005; Sheldon 1997). The use of ISO 14001 can demonstrate an innovative and forward-thinking approach to customers and prospective employees. It can increase a business's access to new customers and business partners. In some markets it can potentially reduce public liability insurance costs. It can serve to reduce trade barriers between registered businesses (Van der Veldt 1997). There is growing interest in including certification to ISO 14001 in tenders for public–private partnerships for infrastructure renewal. Evidence of value in terms of environmental quality and benefit to the taxpayer has been shown in highway projects in Canada.

ISO 14001 addresses not only the environmental aspects of an organization's processes but also those of its products and services. Therefore, ISO/TC 207 has developed additional tools to assist in addressing such aspects. Life-cycle assessment (LCA) is a tool for identifying and evaluating the environmental aspects of products and services from the "cradle to the grave": from the extraction of resource inputs to the eventual disposal of the product or its waste. The ISO 14040 standards give guidelines on the principles and conduct of LCA studies that provide an organization with information on how to reduce the overall environmental impact of its products and services. ISO 14064 parts 1, 2, and 3 are international greenhouse gas (GHG) accounting and verification standards which provide a set of clear and verifiable requirements to support organizations and proponents of GHG emission reduction projects.

1.2 Environmental Management in Industries

Today many industries and companies have recognized the importance of proper environmental management and have switched over from traditional end-of-pipe solutions to the integration of environment management in overall management process of the industry. A few major driving forces for such changes are stringent legislation; demand for better work environment for employees; customers' demands; company's image; the growing pressure from all stakeholders regarding the environmental, economical, and social responsibilities. Environmental considerations are no longer regarded on ad-hoc basis, rather these considerations form the part of industries' everyday reality. Still, there is a lack of holistic approach where environment management is a natural part of overall management system. While the environmental departments are busy with generating reports and petitions for external purposes, the top management is not making use of the competence already present within the organization. The main objective is to focus on the application of various environmental tools and methods so that there could be a shift from reactive regulatory approach to proactive environmental decision making; there could be full support of top management on the environmental information system; and the environmental issues could be prioritized and the implementation could be accelerated. Ultimately, the integration of the environmental responsibility with the environmental systems and allocation of the resources needed shall lead to implementation of the environmental strategies and it can contribute to both improvements in the environmental performance and in increasing long-term profitability of the industry.

1.2.1 Environmental Challenges

Our avid interest in environmental sustainability and environmental management issues can be traced directly to awareness that as the world's population continues to expand and to consume natural resources, humanity faces shortages that threaten quality of life in developed areas and elsewhere on the Earth, life itself. In attempts to find solutions to these problems, we have created an ever growing inventory of manufactured goods, chemicals, drugs, ostensibly to improve the quality of life that has in fact contributed to the pollution of our environment. "Pollution prevention," an environmental buzz word since the 1990s, encompasses designing processes that generate no waste to plants that emit only harmless compounds such as pure water.

Zero defect and zero effect (ZDZE) is different from pollution prevention in that it converts raw materials into useful products or valuable resources that have "no defect" in manufactured products and "zero effect" has no adverse effect on health and environment. In this book, the meanings of "Zero Effect," "Zero Discharge," or "Zero Emissions" are complimentary and all terms are used interchangeably (Das 2005). Within the *ZDZE* paradigm the goal of resource extraction, refining, or commodity production is approached in much the same way that the mining, iron and steel, pulp and paper, petroleum, energy, automobiles, petrochemical, pharmaceutical, fertilizer, agricultural, and chemical industries go about processing raw materials. Sometimes the conversion of wastes or by-products into resources having value to another industry is more efficient than the implementation of pollution prevention techniques – that is industrial ecology (also see Chapter 9).

In this book, we will focus on the best management practices, best available industrial manufacturing processes, techniques, and technologies that treat raw materials into no-defect products, as well as innovative and emerging processes that have best potential for achieving the highest standards in pollution prevention at the plant and industry levels, leading to *no defect and zero effect* (NDZE) – a common goal toward industrial environmental management. To move toward *NDZE* via *process pollution prevention* (P3) and *profitable pollution prevention* (P3), industries must use processes that deploy materials and energy efficiently enough to neutralize and control contaminants in the waste stream. The ultimate goal is to remove pollutants from the waste streams and convert them into products or feeds for other processes. Logically then, P3 refers to industrial manufacturing processes by which materials and energy are efficiently utilized to achieve the end product(s) that have "no defect," while reducing or eliminating the creation of pollutants or waste at the source that is "zero discharge or zero effect." The primary goal is to educate the engineering students to prepare them as current and future generation engineers who will learn and practice sustainable engineering and who will be our champion stewards in industrial environmental management as needed caretakers of the Earth.

1.3 Waste as Pollution

A waste is defined as an unwanted by-product or damaged, defective, or superfluous material of a manufacturing process. Most often, in its current state, it has or is perceived to have no value. It may or may not be harmful or toxic if released to the environment. Pollution is any release of waste to environment (i.e. any routine or accidental emission, effluent, spill, discharge, or disposal to the air, land, or water) that contaminates or degrades the environment.

1.4 Defining Pollution Prevention

In this book, we define pollution prevention fairly broadly as any action that prevents the release of harmful materials to the environment. This definition manifests itself in the form of a pollution prevention hierarchy, with safe disposal forms at the base of the pyramid and minimizing the generation of waste at the source at the peak (Figure 1.1).

In contrast, the U.S. Environmental Protection Agency (USEPA) (1992) definition of pollution prevention recognizes only source reduction and conservation, which encompasses only the upper two tiers in the hierarchy – minimize generation and minimize introduction. The USEPA describes the seven-level hierarchy of Figure 1.1 as "environment management options." The European Community, on the other hand, includes the entire hierarchy in its definition of pollution prevention. The tiers in the pollution prevention hierarchy are broadly described as follows.

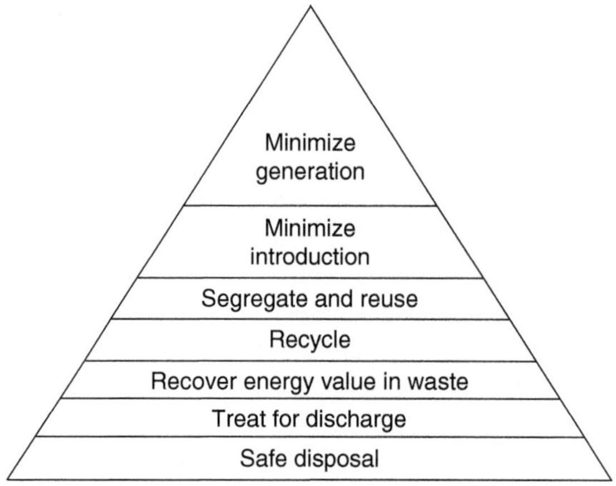

Figure 1.1 Pollution prevention hierarchy.

- Minimize generation: Reduce to a minimum the formation of nonsalable by-products in chemical reaction steps and waste constituents (such as tars, fines, etc.) in all chemical and physical separation steps.
- Minimize introduction: Cut down as much as possible on the amounts of process materials that pass through the system unreacted or are transformed to make waste. This implies minimizing the introduction of materials that are not essential ingredients in making the final product. For examples, plant designers can decide not to use water as a solvent when one of the reactants, intermediates, or products could serve the same function, or they can add air as an oxygen source, heat sink, diluent, or conveying gas instead of large volumes of nitrogen.
- Segregate and reuse: Avoid combining waste streams together with no consideration to the impact on toxicity or the cost of treatment. It may make sense to segregate a low-volume, high-toxicity wastewater stream from high-volume, low-toxicity wastewater streams. Examine each waste stream at the source and identify any that might be reused in the process or transformed or reclassified as valuable coproducts.
- Recycle: Many manufacturing facilities, especially chemical plants, have internal recycle streams that are considered part of the process. In addition, however, it is necessary to recycle externally such materials as polyester film and bottles, Tyvek envelopes, paper, and spent solvents.

- Recover energy value in waste: As a last resort, spent organic liquids, gaseous streams containing volatile organic compounds (VOCs), and hydrogen gas can be burned for their fuel value. Often the value of energy and resources required to make the original compounds is much greater than that which can be recovered by burning the waste streams for their fuel value (also see Appendix G).
- Treat for discharge: Before any waste stream is discharged to the environment, measures should be taken to lower its toxicity, turbidity, global warming potential, pathogen content, and so on. Examples include, but not limited to, biological wastewater treatment, carbon adsorption, filtration, and chemical oxidation.
- Safe disposal: Render waste streams completely harmless so that they do not adversely impact the environment. In this book, we define this as total conversion of waste constituents to carbon dioxide, water, and nontoxic minerals. An example would be posttreatment of a wastewater treatment plant effluent in a private wetland. So-called secure landfills do not fall within this category unless the waste is totally encapsulated in granite.

1.4.1 Resource Efficiency

Resource efficiency reflects the understanding that current, global, economic growth, and development cannot be sustained with the current manufacturing, production, and consumption patterns. Globally, we are extracting more resources to produce goods than the planet can replenish. Resource efficiency is the reduction of the environmental impact from the production and consumption of these goods, from final raw material extraction to last use and disposal. This process of resource efficiency can address sustainability (see Chapter 10).

In this book, we will focus on the upper three options of the industrial pollution prevention hierarchy; that is, recovering the energy value in waste, treating for discharge, and arranging for safe disposal. To improve this bottom line, however, businesses should address the upper three tiers first: that is minimize generation, minimize introduction, and segregate and reuse. This is where the real opportunity exists for reducing the volume of wastes to be treated. The volume of the waste stream, in turn, has a strong influence on treatment cost and applicability. Thus, useful technologies such as membrane processes in highly flexible separation techniques for water, solvent and solute recovery, or condensation of VOCs from air are not economic at large volumetric flow rates. The focus has shifted from *end-of-the-pipe* solutions to more fundamental structural changes in industrial manufacturing processes.

1.5 The *ZDZE* Paradigm

No defect and zero effect, or something very close to it, is the ultimate goal of P3, while the processes themselves are the tools and pathways to achieve it. Thus, industries were to be reorganized into "clusters" in which the wastes or by-products of each industrial process' were fully matched with others industries' input requirements; the integrated process would produce only clean product, perfectly matching with given specifications, while no waste of any kind. As described in Chapter 8 Section 1.5, this solution is being applied in scattered areas throughout the world, from modern industrial nations such as Sweden to developing countries such as Bangladesh.

Traditionally, pollution control technology processed a "waste" until it was benign enough for discharge into the environment. This was achieved through dilution, destruction, separation, or concentration. Within the ZDZE paradigm, many of these processes will still be applied, but as mentioned earlier, the goal will be resource extraction, refining, or commodity production, not simply removal of waste from the premises. Engineering firms will need to develop conversion technologies that create "designer wastes" to meet input specifications of other industries.

1.6 Zero Discharge Industries

While there are several practical definitions of *zero effect or zero discharge (ZD)* manufacturing, a ZD system is most commonly understood to be one that discharges no waste from a processing and manufacturing site. In such a manufacturing facility (see Figure 1.2) an absolute minimum amount of waste, "ideally zero," is generated and leaves the plant. The only inputs to the facility are the raw materials needed to make salable products and energy. The only outputs are salable products and any by-products as feedstocks to another plant. The wastes (in air, water, or as solid) or by-products generated during manufacturing process are recovered using various technologies (Das 2005).

The materials and energy recovered from waste streams either are reused in the plant or are sold to another plant as feedstock. It is in practice, as well as in theory, possible to isolate some industrial facilities almost completely from the environment by recycling all wastes into materials that can then be manufactured into consumer products. An example of such a facility is a coal-fired power plant. An electron beam–ammonia conversion unit adds ammonia to the effluent gases, which it then irradiates electronically, producing ammonium nitrate and ammonium sulfate that are sold as feedstock to fertilizer manufacturing. The details are given in Section 9.2.3.

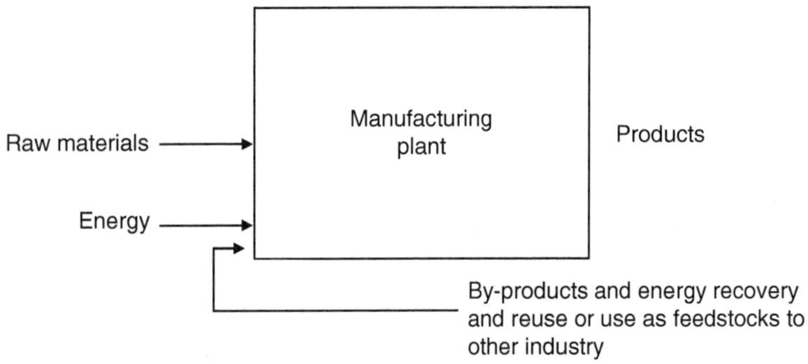

Figure 1.2 "Zero" waste manufacturing facility.

The concept of Zero Emissions was inspired by the business programs of zero defects (total quality management), zero inventory (just-in-time), and zero accidents (*workplace safety*), and while its driver is improved business performance, the environmental benefits are also significant.

The basic premise of Zero Emissions is converting wastes from one industry to the material input of another industry. The application and development of Zero Emissions systems will be the purview of industry, specifically manufacturers and consulting engineering firms. To a degree, the move toward Zero Emissions is already happening with pollution prevention, waste minimization, and design for the environment. While these systems require further improvement, industries employing them have seen the benefits already.

But it is important to understand that some manufacturing processes inherently produce wastes, even after all reasonable efforts at pollution prevention. Thus, in some cases the use of a conversion technology may be more appropriate than a program of pollution prevention: many industrial wastes can be processed to render them viable as material inputs to another industry or to part of an industrial cluster of several connected industries.

1.7 Sustainability, Industrial Ecology, and Zero Discharge (Emissions)

The concept of *Zero Discharge or Zero Emissions* is the key to sustainable development but is by itself a subset of industrial ecology (also see Section 9.2) (see Figure 1.3).

Sustainability can be defined as follows: "Sustainable development" is the challenge of meeting human needs for natural resources, industrial products, energy, food, transportation, shelter, and effective waste management while conserving and protecting environmental quality and the natural resource base essential for future development; and "the development that meets the needs of the present without compromising the ability of future generations to meet their own needs" (World Commission on Environment and Development 1987). Sustainability is a worthy vision, but inherently ambiguous, and inescapably expressed in value-laden terms subject to differing ideological interpretations. Accordingly, while the concept provides a useful direction, it is almost impossible to put into operation. Standing alone, it can guide neither technology development nor policy formulation.

Industrial ecology is "an approach to the design of industrial products and processes that evaluated these activities through the dual perspectives of product competitiveness and environmental interactions." The field has been developed, largely academically, over the last 10 years to be "the means by which humanity can deliberately and rationally approach and maintain a desirable carrying capacity" (Graedel and Allenby 1995). It can be thought of as the science of sustainability for industrial systems – the multidisciplinary study of industrial and economic systems and their linkages with fundamental natural systems (also see Section 9.2).

Even though it is still under development, industrial ecology can provide the theoretical scientific basis upon which understanding, and reasoned improvement, of current practices can be based. It incorporates, among other things, research involving energy supply and use, new materials, new technologies, basic sciences, economics, law, management, and social sciences. It encompasses concurrent engineering, *design for the environment (DFE)*, dematerialization, pollution prevention, waste conversion, waste exchange, waste minimization, and recycling, with *Zero Emissions* as an important subset. Industrial ecology can be a policy tool, but it is neither a policy nor a planning system.

The goal of Zero Emissions is to restructure manufacturing so that there are no wastes. "Zero emissions" is thus applied "industrial ecology" at the manufacturing/service level, and indeed the terms are often used interchangeably. Note that the emphasis is on the manufacturing level, and not firm or industry level: whereas both "firm" and

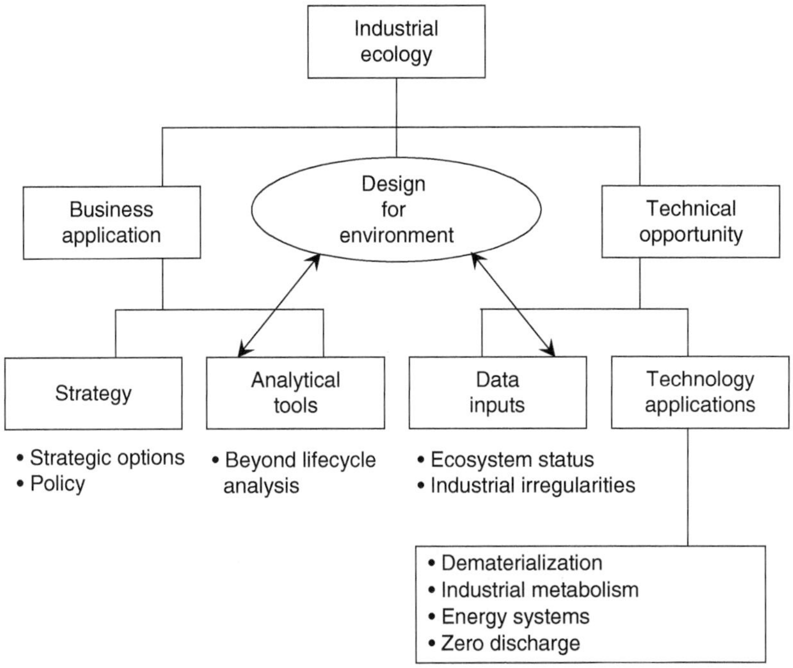

Figure 1.3 Zero discharge (emissions) is a subset of industrial ecology (see Section 9.2). Design for the environment and dematerialization are discussed later in the chapter. Industrial metabolism compromises the energy and value-yielding process essential to economic development.

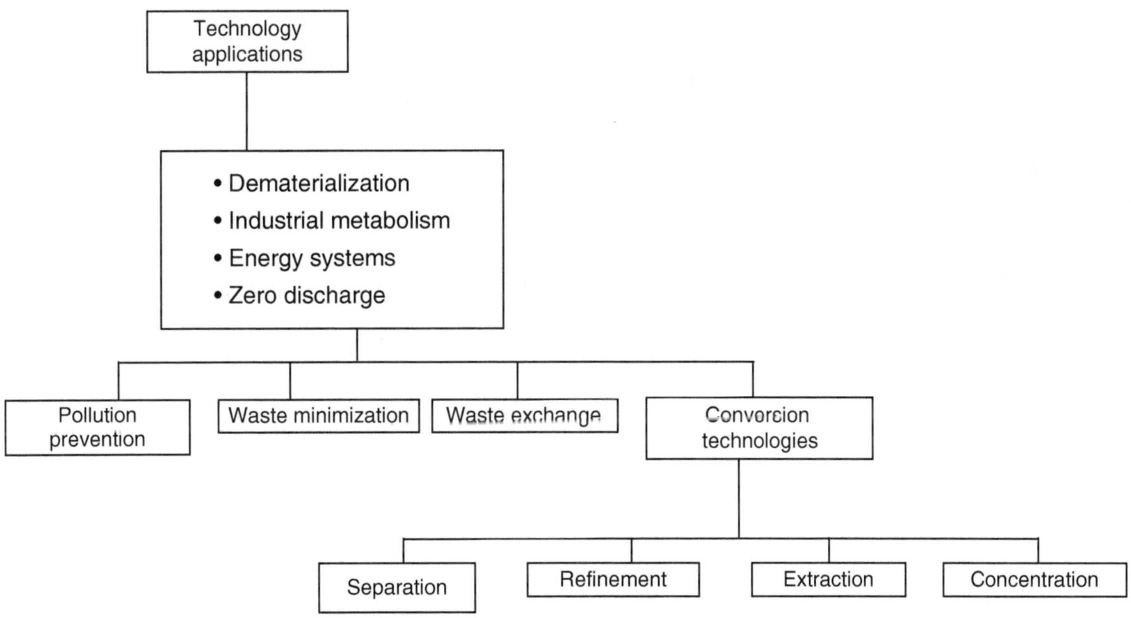

Figure 1.4 Zero discharge is supported by an array of tools and methodologies.

"industry" imply singular facilities or sectors, wastes are converted most successfully when several facilities are linked in an industrial cluster. As can be seen in Figure 1.4, pollution prevention and waste minimization are part of Zero Emissions and are technologies to be explored and where possible, optimized. Since it is not always feasible to prevent the generation of wastes early in the industrial cycle, the practice has arisen of trading them after they have been generated. Thus, critical components of ZD are waste exchange and the conversion of wastes so that they are viable inputs in other sectors. This interaction of systems widens the focus of the waste management effort.

1.8 Why Zero Discharge Is Critical to Sustainability

To understand why *Zero Discharge* is a critical component of sustainability, it is important to recognize that one principle of sustainability is the efficient and wise use of resources, especially with regard to limiting the amount and type of resource extraction and subsequent pollution loadings. To see how these are related, it may help to think of a cycle with three parts: sources, systems, and sinks.

- The sources include raw materials such as minerals, water, topsoil, and fossil fuels. On this planet, these are limited but have huge external reserves.
- The systems are our ability to manipulate energy to turn source materials into finished products. Economic and industrial systems are limited only by imagination.
- The sinks are the global waste bins. The ultimate long-term sink is the deep trenches of the oceans; short-term sinks are biosystems such as the atmosphere, rivers, wetlands, and the land. The ability of the sinks to handle wastes is limited; most show adverse effects of pollutant loading in just a few years.

The most restricting rate-limiting component in this three-part model is the sinks. Currently, the depletion of resources is rarely the driving force for resource substitution; instead, change is driven by process innovation to beat the competition, or regulatory intervention imposed from outside. But given the limitations of the sinks, the pressure for modification of an industrial system will increasingly come from the need to reduce loadings to environmental sinks.

To achieve sustainability, the Earth's environment must be protected in multiple ways. For example, planners must aim to minimize or eliminate anthropogenic changes to climate, net increases in acidification, loses of topsoil, and withdrawals of fossil water; moreover, biodiversity must be preserved, and buildup of toxic metals and other nonbiodegradable toxics in soils or sediments must be stopped. *Zero discharge* technologies attempt to accomplish these goals by reducing resource extraction and loadings to sinks. The objective is a closed loop in the economic subsystem, so that wastes inevitably created by human activities do not escape to contaminate the environment. *Zero Discharge/Emissions* proponent Gunter Pauli, the founder and former director of *Zero Emission Research Institute*, also notes that in the effort to eliminate waste, *Zero Emissions* "is nothing more than a persistent drive to cut costs." Waste is a form of inefficiency, and an "economic system cannot be considered efficient, or ultimately competitive, if it generates waste" (Pauli 1996).

Zero Discharge (or Emissions) leaves behind the linear "cradle to grave" concept of materials use (Figure 1.5) and embraces a cyclical, "cradle to cradle" vision (Figure 1.6), in which wastes become value-added inputs and the raw materials for other production cycles. This is how natural systems dispose of waste, and according to Pauli, the only way to achieve sustainability.

Meanwhile, the biological sinks are not increasing in capacity. Existing industries will keep operating and generating wastes – some of these wastes, as will be discussed later, containing richer concentrations of recoverable materials than virgin ones. In the interim, there will be a demand for technologies to manage and convert today's wastes into usable feedstocks. Chemical process design engineers and consulting firms will provide focal services to meet this demand through technology development, system integration, and facility operation.

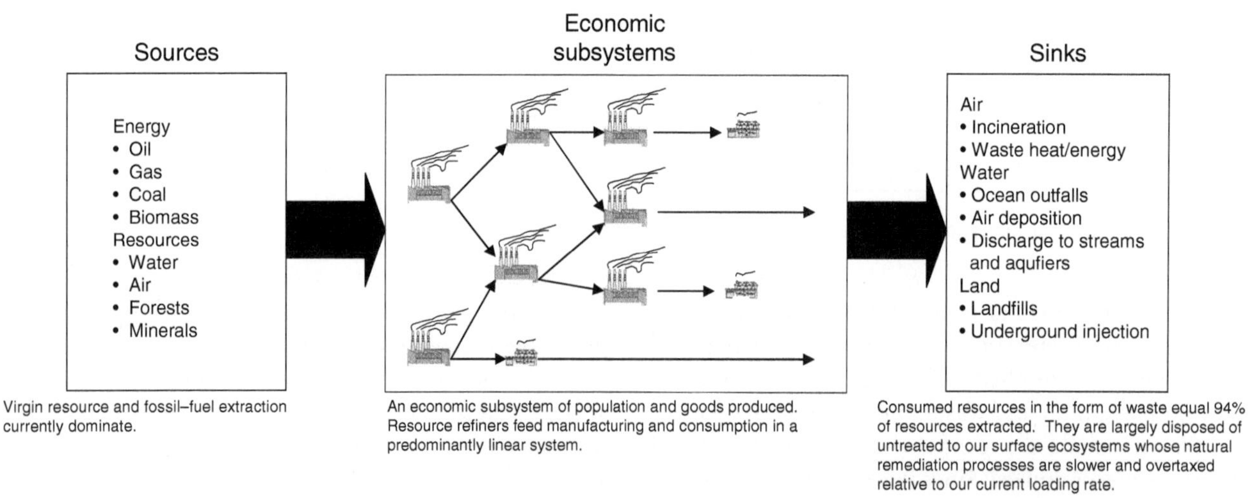

Figure 1.5 The interrelationship of sources, systems, and sinks for a linear (cradle to grave) materials use pattern.

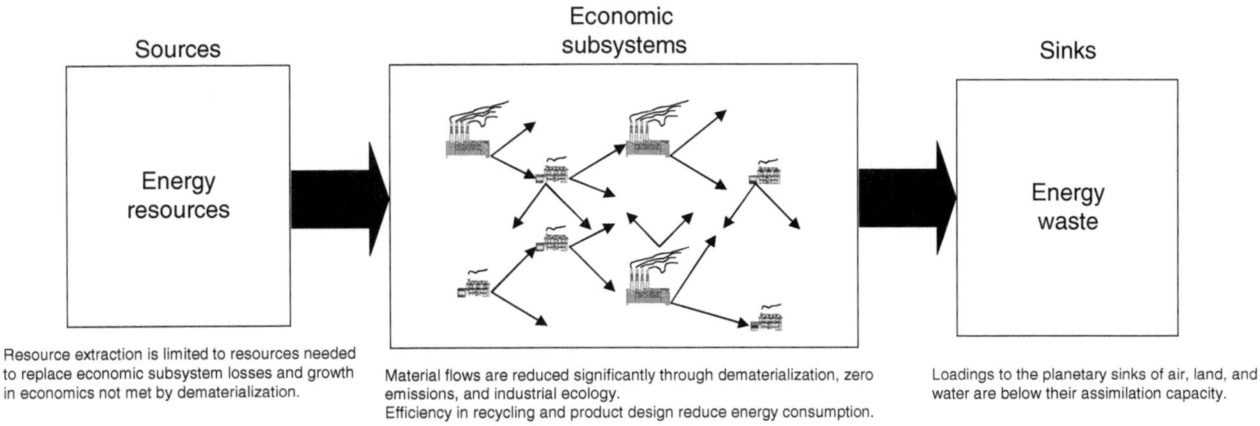

Figure 1.6 The interrelationship of sources, systems, and sinks for a cyclic (Zero Emissions) materials use pattern.

1.9 The New Role of Process Engineers and Engineering Firms

Chemical process and product design engineers, environmental engineers, and consulting engineering firms can play a pivotal role as industries move toward the *Zero Emissions* or *Zero Discharge* paradigm, especially firms whose traditional niche has been to treat waste so that it is benign and acceptable for discharge. The role for these engineers in the twenty-first century is to transform the effluent of one process to serve as the raw material for another process. The new role is not simply facilitating waste exchange; rather, the new jobs include the following:

- Assessing material flows through the economy and the use of raw materials, water, and energy
- Designing databases with a wider set of information about material flows and manufacturing processes
- Working with design firms to understand the production processes of the industries that produce the wastes
- Designing conversion processes
- Identifying purchasers for converted wastes
- Designing material transfer systems to carry wastes to industries that will use them as feedstock
- Identifying industrial clusters and understanding how to fit diverse industries into a successful industrial cluster
- Designing eco-industrial parks and negotiating arrangements that are commercially sound and profitable, yet based on good personal relationships; voluntary, and yet in close collaboration with regulatory agencies

What this means for engineering firms is the need for a broader set of engineering skills and services. As can be seen, consulting engineering firms will find that achieving Zero Emissions entails expertise in areas that have not been part of engineering curriculum, or the professional engineer's exam.

Zero Emissions engineers need to be not only well trained in design for the environment, concurrent engineering, and industrial engineering but also be able to think and design outside the traditional boundaries of the factory to work in terms of industrial clusters.

Many of the skills and services enumerated above would be applicable to the development of an agro-industrial cluster such as the one in Namibia, described in Mini-Case Study 1.1.

Mini-Case Study 1.1 Beer to Mushrooms: Focusing on the Productivity of Raw Materials

In the town of Tsumeb in the African desert, Namibian nationals will be implementing Zero Emissions technology at a brewery inaugurated in January of 1997. The three main inputs into beer – grain, water, and energy – are scarce commodities in a developing nation. Brewing uses only 8–10% of the nutrients in grain, consumes 10 L of water for every liter of beer produced, and generally requires imported coal, an expensive and polluting energy source.

The lignin-cellulose component of spent grain, which makes up 70–80% of its bulk, is indigestible to cattle, but it is easily broken down by the enzymes of mushrooms. It takes 4 T of spent grain to produce 1 T of mushrooms, which are a potentially lucrative cash crop for export, because most southern African nations currently import mushrooms. The protein content of the spent grain – up to 26% – is used by earthworms, which in turn are fed to chickens and pigs. Processing the waste from the animals in a digester could supply all of the vapor energy required for brewing. Brewery wastewater is high in nutrients but is too alkaline for crops. However, it can be used to grow spirulina, which generate up to 70% protein.

The brewery's thermal waste could heat greenhouses or the brewery. These interrelated industries will form an optimal industrial cluster for increasing the productivity of the brewery's raw materials in ways that also produce food for humans that is high in nutritional value.

1.10 Zero Discharge (Emissions) Methodology

Over the last few years, the members of *zero discharge communities* and industries have developed a five-step methodology for implementing Zero Emissions. Pauli's *Breakthroughs* (1996) provides a far more comprehensive approach that extends well beyond the manufacturing site. The summary provided here emphasizes the use and impact of a ZD approach at the manufacturing level.

1.10.1 Analyze Throughput

The first step toward achieving *Zero Discharge* and/or *Zero Emissions* is an in-depth review of the industry to see if total throughput is possible. This means determining whether all material inputs can be found in the final product – if there are no wastes, all inputs must have ended up in the product. One of the few industries where this can occur is cement manufacturing. In the Mini-Case Study 1.1, however, only a small fraction of the nutrients in the grain ends up in beer.

If throughput is not total, the next step is to determine whether the products manufactured can be easily reintegrated into the ecosystem without additional costs for processing, energy, or transportation. However, since this is rarely possible, most industries will not achieve Zero Discharge unilaterally.

Process and product design engineers will probably find their first opportunities by meeting with their traditional clients, analyzing each client's throughput, and looking for opportunities for pollution prevention and waste minimization that the client's in-house experts may not have seen. The analysis would include evaluating products and services presently being produced, processes and materials used, and management of environmental issues including energy efficiency, as well as clarifying the full scope of emissions.

1.10.2 Inventory Inputs and Outputs

Once the initial analysis has determined that total throughput is not possible, and that wastes will be generated, the next step is to assess the industry's inputs and outputs, and to inventory all the outputs ("wastes"). A diagram of the inputs and outputs of a system like that of Figures 1.2, 1.5, and 1.6 are then used to compile basic overview of the company's resources and needs. From this information, design engineers and process specialists can attempt to modify the manufacturing process so that it can become a *Zero Emissions* system.

Extracting raw materials and processing them imposes significant environmental burdens. An analysis of the industrial metabolism of the product (i.e. its input, materials use, and life cycle expectancy) will help determine the path of least environmental impact. Some materials choices will yield better throughput or by-products that are more suited for use as an input for another industry.

Additional audits and inventories may be needed to determine manufacturing efficiency by percentage of input wasted, to quantify amounts of waste landfill by type of material, to account for amounts of materials collected for recycling, and to identify major emissions of waste heat and the site and amounts of wastewater discharges. Analysis of these outputs may reveal the most effective ways to reuse these outputs and help to determine which industries could use the wastes as raw materials. For example, at the Namibian brewery, spent grain, excess heat, and wastewater all have potential uses in producing food items.

1.10.3 Build Industrial Clusters

In sectors that cannot achieve *Zero Emissions* unilaterally, it may be necessary to build industrial clusters. The input–output analysis leads directly into development of clusters of industries that can use each other's outputs. Developing effective clusters calls for executives look beyond single industries and make innovative connections among seemingly unrelated potential partners in new industrial clusters. Companies are loathe to implement such changes, however. In addition to concerns about antitrust regulations, and the need to rely on single vendors for supply, there is fear that relinquishing information about waste stream composition will allow their competitors to deduce proprietary secrets.

Also critical is the geographic location of the client's potential partners, as transportation is a key factor in optimizing waste exchange and use of conversion technologies. The most obvious link in the search for industrial cluster partners will be obtaining industrial input data for other industrial sectors and determining if the client's waste flows could serve (in some converted form) as a material input to another sector. The second place to look for candidate industrial clusters is the historical records of waste exchanges. These material flows will demonstrate which materials being discarded by a sector are of a volume and quality desirable to another sector.

Industries that buy process wastes are taking in nonvirgin material of a grade that may fall short of the purchasers' specification. This is an opportunity for materials blending. For example, plastics can be recycled to make a lumber-like product, but the grade of such recycled products is not always acceptable as a direct input. If a contaminated waste flow is not of sufficient volume, however, blending in virgin plastics can bring both quality and volume up to manufacturing specifications.

Once the potential partners have been identified, the industrial cluster should be designed and developed. Kalundborg (Grann 1994) is an excellent example of a cluster that includes heavy industries, while Tsumeb's cluster is based primarily around food production and processing. Elsewhere in the world, industrial cluster is yet to be developed.

India and China are industrializing nations that have abundant and cheap supplies of coal, but burning it to generate electricity produces CO_2, SO_x, and NO_x. The key to sustainability for industrializing countries will therefore be development of industrial clusters that link energy, agriculture, and sewage treatment, in the fundamental format for *Zero Emission* communities. The most effective incentive to develop such clusters is economics, and, unlike conventional SO_x and NO_x treatments, the system for the electron beam/ammonia conversation of these pollutant has a financial payback of 10–15 years. Section 9.2.3 treats this technology in more detail.

1.10.4 Develop Conversion Technologies

The easiest connections for industrial clusters are through a simple, direct waste exchange. The next easiest route is to develop an intermediary process that will take one industry's current waste stream, convert it to a usable form, and transfer it to a purchasing industry. Now we consider briefly the pivotal function of conversion technologies as illustrated by the problems encountered in the paper recycling industry as it works toward attaining *Zero Emissions* in the United States (see also Chapters 7 and 9).

Paper recycling is quintessentially "green." But current processes used to de-ink paper remove only 70–80% of the ink particles, leaving recycled papers an unattractive gray. The wastes are a toxic mix of ink, short fibers, coating chemicals, and paper fillers that requires both primary and secondary treatment before disposal. De-inking is both inefficient and expensive, and results in a product that is often higher priced and lower quality.

Under the auspices of ZD industries, a conversion technology is being developed that results in 100% removal of ink and 3 viable outputs. The recaptured ink could be reused in printing or for making pencils (as is already done with ink from photocopiers). The long fibers could be made into paper again or used in cardboard. The remaining sludgy mixture of short fibers and residues could be dried and used as acoustic insulation inside building walls or as ceiling tiles. The sludge could be used to make shock-absorbent packaging such as egg cartons or replacements for corrugated cardboard.

The industrial cluster built around paper recycling thus includes recapturing ink, making new paper, and making building and packaging materials. Canada, Latvia, and Italy have tested this conversion technology, the steam explosion system. Many cities, states, and national governments, however, require that recycled paper be used in newspapers, and where these regulations are in force, a system that produces a grade of paper better than that needed by newspapers is not likely to be implemented.

1.10.5 Designer Wastes

So-called designer wastes can serve as direct feedstock to another sector, or, if properly processed through a conversion technology, as processed feedstock. If the beer industry for example, used a sugar-based cleanser instead of a caustic cleanser for its bottle-washing process, the discharge water could serve as a direct feed to fish ponds, without needing any conversion. Yet other solutions will require installation assistance in adjusting or adapting material flows within the client's processes so that its waste output is in an acceptable form.

Building industrial clusters may involve working upstream within a facility to modify its production process so that the waste is produced in an acceptable "designer" form for conversion. Or, perhaps downstream, working with the purchaser industry, to help it modify its processes so that it can accept the converted waste.

This means that the involved engineering consultants must become familiar with the products and materials provided by suppliers to the producer and the materials needed by the purchaser of the designer wastes. All parties must have access to a database of specific information about the level of impurities acceptable in each final product. For example, companies that produce basic material inputs (such as plastics, oils, lubricants, and papers) deal in high volumes and some manufacturing processes can tolerate impurities. These companies are already experienced in their own refining processes and may also be knowledgeable about the manufacturing requirements of their customers, but their expertise may not extend to particular impurities present in the producer's waste stream. The database can provide the information necessary to complete the connections.

1.10.6 Reinvent Regulatory Policies

Experienced professionals are probably already aware of how government policies inadvertently inhibit creativity in reuse of wastes. These policies can also inhibit formation of effective industrial clusters. For example, breweries are regulated as an industry and the facilities are generally located in industrial areas. However, to make efficient use of their discharge water in aquaculture, breweries should be located in agricultural zones. Similarly, regulations aimed at providing a market for recycled newsprint need to be revised so that technologies that produce a better grade of paper can flourish – and allow more complete reuse of all the other by-products associated with complete de-inking.

Ironically, *ZD* goals are inhibited in the United States by regulations pursuant to the Resource Conservation and Recovery Act of 1976 (RCRA). Transferring "wastes" among the members of industrial clusters is often prohibited by RCRA because its regulatory net entangles all wastes, whether hazardous or not. Its broad scope has the unintended consequences of creating disincentives to invest in recovery technologies and blocking progress toward pollution prevention and recycling. RCRA waste classifications can put kinks in potential closed-loop systems. Indeed, as Robert Herman (1989) put it, "the essence of the environmental crisis is not nearly so much bad actors as the whole, often contradictory structure of incentives in the economy." Regulatory policies need to be reinvented to foster development of breakthrough conversion technologies and encourage cross-sector markets for designer wastes.

1.11 Making the Transition

The shift toward a *ZD* culture, especially in a world dominated by industrial ecology, will see the development of new products, services, and industries. Our global economic system depends on extracting massive quantities of materials from the environment – after extraction and processing, the "annual accumulation of active materials embodied in durables, after some allowance for discard and demolition, is probably not more than six percent of the total. The other 94 percent is converted into waste residuals as fast as it is extracted" (Allen and Shonnard 2012; Allenby and Richards 1994; Ayres et al. 1996). In the United States, this means more than 10 T of "active" mass (excluding fresh water) per person each year.

Of this mass, roughly 75% is mineral and nonrenewable, and 25% is from biological sources. Of the biological materials, none of the food or fuel becomes part of durable goods and even most timber is burned as fuel or made into pulp and paper products that are disposed of. Of the mineral materials, about 80% of the mass of the ores is unwanted impurities, and of the final products, a large portion is processed into consumables and throwaways. Only in the case of nonmetallic minerals is as much as 50% of the mass embodied in durable goods such as cement and ceramics (Allenby and Richards 1994). All of this translates into an estimated more than 12 billion T of industrial wastes annually in the United States (Allen and Rosselot 1997).

In addition to materials lost as waste residuals during extraction and processing, finished goods are dissipated/lost because they are present in concentrations too small to be economically recoverable. Many products are inherently dissipative, and lost with a single normal use. These include packaging, lubricants, solvents, flocculants, antifreezes, detergents, soaps, bleaches, dyes, paints, paper, cosmetics, pharmaceuticals, fertilizers, pesticides, herbicides, and germicides. From one-half to as much as seven-eighths of the toxic heavy metals including lead, cadmium, chromium, cobalt, in insecticides (arsenic) and in wood preservatives, fungicides, catalysts, and plastic stabilizers are dispersed into the environment beyond economic recoverability. Other materials are lost to uses that are not inherently dissipative but are so in effect because of the difficulty of recycling. Allenby and Richards (1994) point out that the total elimination of manufacturing wastes probably is an unattainable goal because it would require, in addition to technological advances not yet in place, 100% cooperation by consumers.

1.11.1 Recycling of Materials and Reuse of Products

A critical element of an interim strategy is enhanced recovery. This can be approached from two directions: reuse of products and recycling of materials. Reuse of products includes return, reconditioning, and remanufacturing. The energy required for reuse and recycling is one of the key factors determining recoverability of a product. The closer the recovered product is to the form it needs to be in for recycling, the less energy is required to make that transformation. From the standpoint of economic development it is worth pointing out that the reconditioning or remanufacturing cycle is relatively less costly; it requires roughly half the energy and twice the labor per physical unit of output.

Recycling materials means closing the loop between the supply of post-consumer waste and the demand for resources for production. Recycling of materials will be the business of the Zero Emissions engineer; reuse of products will also involve the *Zero Emissions* engineer, but it will have lots of front-end work from another professional, the concurrent engineer. Concurrent engineering, which incorporates aspects of industrial engineering, product design, and product manufacturing, is an integrated approach that seeks to optimize materials, assembly, and factory operation. These engineers examine the broader context of a product, including technology for managing the environmental impacts of its transport, intended use, recyclability, and disposal, as well as the environmental consequences of the extraction of the raw material used in its production.

The ultimate fate of all materials is thus dissipation, being discarded, or recycling and recovery. With 94% of materials extracted from the environment being converted to wastes, current levels of recovery are clearly not sufficient. Recovery of materials from wastes will reduce the extent of resource extraction (but will not slow the speed of material flows through the economy). Aluminum and lead are two resources currently being heavily recycled, but evidence shows that there is potential for a lot more resources to be recovered from wastes.

Sherwood plots are diagrams that permit the graphic comparison of concentrations of materials in nature against their commodity cost. The sample plot in Figure 1.7 shows that the price for a commodity depends on its concentration in nature before extraction and refining. Figure 1.8 is a similar plot for metal price (2004) as a function of dilution (concentration) of metals in commercial ores; the relationship illustrates the concept that the more dilute a material is in its native ore, the more expensive it will be to purify into a commodity material (Johnson et al. 2007).

Together, the Sherwood plots demonstrate the recovery potential of materials. The elements plotted above the line in Figure 1.7 should be vigorously recycled because they are present in individual by-products in relatively high concentrations. Lead, zinc, copper, nickel, mercury, arsenic, silver, selenium, antimony, and thallium are more economical to recover from waste than from nature. Extensive waste trading could significantly reduce the quantity of material requiring disposal because resource extraction uses from wastes, not virgin feedstocks (Allen and Behmanesh 1994).

1.11.2 Dematerilization

One critical component of the industrial ecology paradigm is dematerialization. Dematerialization means using less material to make products that perform the same function as predecessors. Sometimes this means smaller or lighter products, but other aspects can include increasing the lifetime of a product or its efficiency. The net effect is a reduction in overall resource extraction. Dematerialization is thus a way to increase the percentage of active materials embodied in durables, and to reduce the percentage that is left as waste residuals.

However, dematerialization has limits in achieving *Zero Emissions*. We may also need to think of rematerialization – products that may or may not have a lighter weight in their final form, but whose production, use, and subsequent conversion or recyclability fits within the Zero Emissions paradigm. This is demonstrated by the ease and benefit of recycling older model cars versus the newer ones. Nonetheless, economists and engineers point out that optimizing for environmental protection alone means some loss in safety, efficiency, durability, convenience, attractiveness, and price.

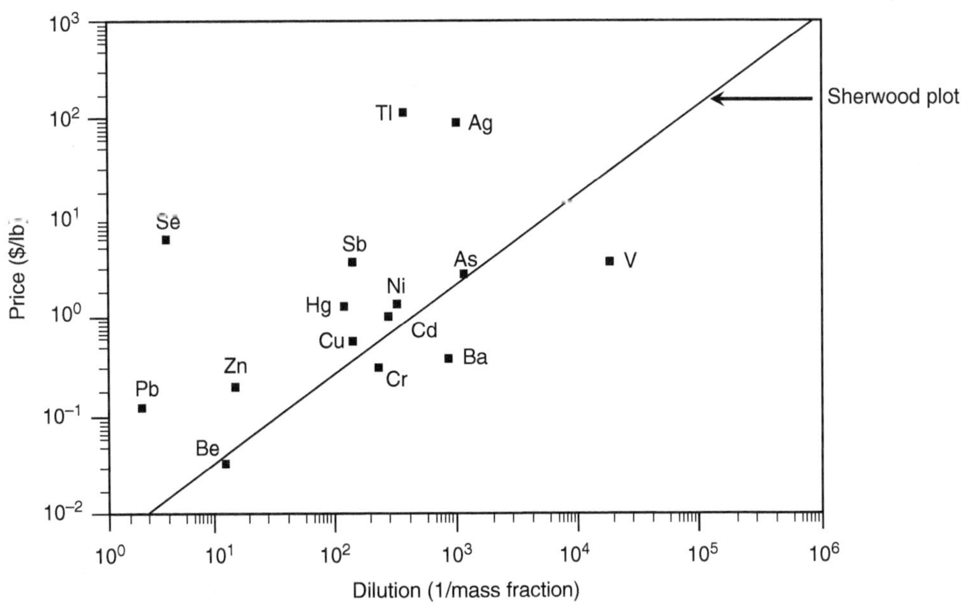

Figure 1.7 Metals-specific Sherwood plot for waste streams: minimum concentration of metal wastes undergoing recycling versus metal prices. *Source:* From Johnson et al. (2007). American Chemical Society.

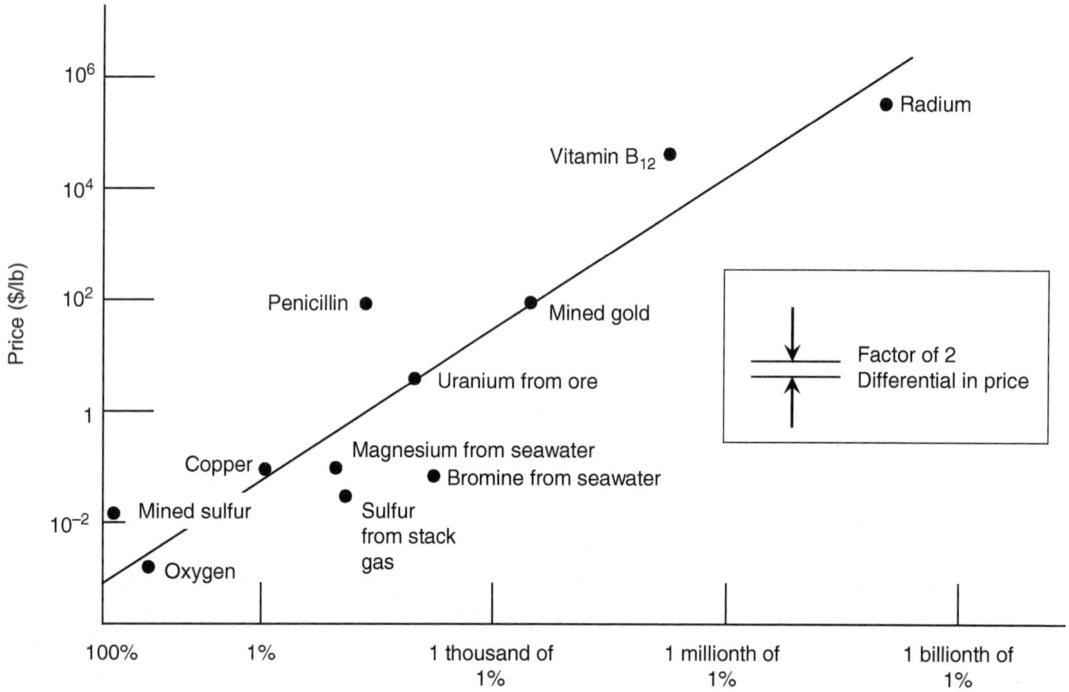

Figure 1.8 A Sherwood diagram showing the correlation between the selling price of materials and their degree of dilution in the matrix from which they are separated.

1.11.3 Investment Recovery

During the transition to Zero Emissions, another early need will be for companies to fill the "decomposer niche," a term for a specialized form of recycling developed by Raymond Cote at Dalhousie University in Ottawa (1995). Just as decomposer organisms turn dead animals and vegetable matter into forms that can become food for other animals and plants, decomposer niche companies will "consume" otherwise unusable wastes by processing them into usable feedstock or disassembling equipment and marketing reusable components and materials.

The work of decomposers also can be visualized as investment recovery. Taking a systematic approach to ending waste, investment recovery is a traditional service, according to Cote, "an integrated business process that identified, for redeployment, recycling, or remarketing, nonproductive assets generated in the normal course of business." These assets include idle, obsolete, unused, or inoperable equipment, machinery, or facilities; excess raw materials, operating inventories, and supplies; construction debris; equipment and fixtures in facilities scheduled for demolition; off-grade, out of specification, or discontinued products; and process waste (Cote 1995, 2003).

The goal of investment recovery is to develop strategies and procedures to recapture the highest value from all surplus assets in a company or community. It seeks to reduce operating and disposal costs, prevent disposal of assets as waste, and find markets for redistributing the by-products for increased economic value. One of the operating paradigms for an investment recovery firm would be integration of its functions into a comprehensive strategy for an eco-industrial park. Firms that specialize in this work base their fees on a retainer plus a percent of savings and/or revenues if there is an incentive.

1.11.4 New Technologies and Materials

During transitional stages, existing industries can be identified as potential members of a cluster if minimal design engineering can make them compatible and most of the transfers of materials are occurring in a more basic commodity form, rather than as "designer wastes." Once Zero Emissions has been incorporated at the drawing board level, facilities can be planned to work together in clusters so that the by-products of each enterprise meet the feedstock specifications of other industries in the cluster. This will require innovations in materials and methods alike.

1.11.4.1 New, Less Toxic Chemicals and Materials

Examples of new materials include biopolymers, fiber-reinforced composites, high-performance ceramics, and extra-strength concrete. To meet basic environmental requirements, new materials will be biodegradable, nonpolluting, recyclable or convertible; made from renewable

Mini-Case Study 1.2 DaimlerChrysler's ZD Wastewater Treatment Plant in Mexico

DaimlerChrysler's production complex in Toluca, Mexico, home of the Chrysler PT Cruiser, has received much attention not only because of its in-demand product but also because of its state-of-the-art *ZD* wastewater treatment plant (WWTP). Located 37 miles north of Mexico City, Toluca suffered for years from a worsening water shortage due to urban sprawl, regional drought, and increased industrial activity. The city is one of the leading producers of beverages, textiles, and automobiles in Mexico, as well as a center for food processing.

DaimlerChrysler, one of Mexico's largest manufacturers, mindful of the mounting strain on the world's natural resources, has consistently sought ways to decrease operational waste, reduce costs, and increase process efficiencies. Upon locating in Mexico, the automaker began to study the region's rapidly dropping aquifer, hoping to minimize the stress on this valuable resource, yet keep its operations in compliance with the federal government's water quality standards.

In 1999, the company hit upon a solution. It would build its own $17 million WWTP that would treat sanitary and manufacturing-process water generated by the facility's four separate plants – engine, transmission, stamping, and assembly. And to make this WWTP truly state of the art, a comprehensive zero liquid discharge (ZLD) system would be installed. By using a ZLD system, the Toluca complex would avoid further depleting the local aquifer, the environmentally friendly and cost-efficient system would discharge no process water, but rather would recycle it to use throughout the facility. It was projected that implementing a ZLD solution and thus reusing water could extend the facility's life without disrupting production and causing costly overhauls.

resources; have low energy requirements in production and use; and provide a final product with greater strength and durability and lower weight and volume.

1.11.4.2 Improved Processes

Consultants can help managers cut costs and create new values by instituting real-time monitoring and eliminating inefficiencies in the use of resources all along a product's life cycle. These inefficiencies include incomplete utilization of material and energy resources, poor process controls, product defects, waste storage costs, discarded packaging, costs passed on to consumers for pollution or low energy efficiency, and the ultimate loss of resources through disposal and dissipative use. Poor resource productivity also can entail costs for waste disposal and regulatory penalties. Methods that are less energy-intensive and more labor-intensive are more sustainable environmentally and socially.

1.11.5 New Mindset

As a result of changes in materials and processes, engineering professionals will have to expand the purview of their design parameters. A larger, more integrated design for facilities and manufacturing processes is called for. This is where design for environment comes in: DFE examines the life cycle of the product and considers not only its primary use but the environmental consequences of its production, assembly, testing, servicing, and recycling. In designing an eco-industrial park, for example, manufacturing processes would be linked to material flows and to energy flows. As a result, the design period and overall costs would be higher. Return on investment, however, would be much shorter (see Sections 7.3 and 7.3.1).

1.11.5.1 System Design

DaimlerChrysler put in operation ZLD systems of two kinds. The first uses reverse osmosis (RO) to produce a concentrate of total dissolved solids (TDS), which is sent to a large evaporator and eventually on to a lagoon or solar evaporator pond. Used in dry, arid areas of low elevation, this system is frequently found in the WWTPs of Northern Mexico's automotive facilities. The other system, used at the Toluca facility, softens and removes silica from the RO concentrate through microfiltration before sending the water on to another RO unit where it is further concentrated. Water is then returned and blended with the water from the first stage water, where the concentrate is sent on to either an evaporator or a crystallizer to dry TDS to powder and eliminate the need to dispose of liquid.

In essence, industrial plants with ZLD installations can expect to recover nearly 100% of water that would otherwise be discharged to the environment as wastewater. At the Toluca facility, the WWTP recovers 95% or more of the water used for processing, with a recovery rate of up to 237 500 gallons per day (gpd). In actuality, the ZLD installation at the Toluca facility is two separate systems: a sanitary water system that biologically treats wastewater from the complex's restrooms, showers, cafeterias, and other domestic areas, and a manufacturing-process water system that chemically treats wastewater mixed with heavy metals and paint from the assembly plant. The latter also treats wastewater containing emulsified and soluble oils from the facility's stamping, transmission, and engine plants.

In the sanitary water system, domestic water is collected and sent through a screening mechanism before moving on to the biological treatment system's equalization tank, ensuring a constant, even flow of water through the system. This water is then passed through jet aeration sequential batch reactors that treat the water with microorganisms and air to reduce the biological oxygen demand (BOD) and chemical oxygen demands (COD), as well as suspended solids. The complex uses the 150 000–200 000 gpd of disinfected water to irrigate its landscape. The microorganisms and solids recovered from the batch reactors are then sent through a sludge digester and eventually a filter press that eliminates the water. While the dewatered sludge is used as fertilizer, the filtered water re-enters the system.

Wastewater from the Toluca facility's three machining plants is directed through the manufacturing-process system where it is first chemically treated, passing through a filtering screen. In a separate tank, chemicals are used to de-emulsify the free-floating oils that comprise most of the waste. Afterward, the oils are removed and stored in another tank before disposal. The process water from the machining plants is then mixed with water from the assembly plant that contains residue from the spray painting, phosphating, E-coating, and body-wash operations. Upon being mixed with a combination of ferric chloride, lime and magnesium oxide, metal pollutants and silica are rendered insoluble and turned into sludge that is removed and sent to a landfill. Then, to further lower the proportion of unwanted organic compounds, the water is pumped to a biological system that reduces the BOD to 20–30 ppm.

1.11.5.2 Results

Since installing the wastewater recovery system, the Toluca facility has noted several benefits, including decreased production and operation costs, reduced aquifer use, better environmental friendliness, and greater employee safety (Zacerkowny 2002). Moreover, the integrated system helps preserve the environment, is safe for employees to work with, and provides almost 7000 jobs to local residents. The Toluca industrial complex uses approximately 250 000 gpd of water, recovering more than 95% of its processing water. The ZLD system allows the facility to treat more than 550 000 gpd, significantly reducing the amount of water that must be drawn from the local aquifer. Using treated water might also extend the life of the facility's equipment, as the salt content of the processed industrial water is much lower than that of the aquifer.

1.11.6 In the Full ZD (Emission) Paradigm

Designing ZD systems requires an expansion of the focus and outputs of the traditional design engineer. Concurrent engineers need to incorporate design for the environment. Industrial engineers need to think in terms of industrial clusters. Environmental engineers need to understand upstream processes better so that they can develop designer wastes. Environmental engineers also need to revamp their processes to begin mimicking resource refining.

ZD engineering firms are expected to be working with design for environment engineers, concurrent engineers, and industrial engineers. They should all be seeking to design wastes, conversion processes, and industrial clusters. Setting the stage for overall product design, the industrial ecology approach assists companies in looking beyond the product to its functionality over its life cycle. Services and products should be designed and delivered differently as the following six strategic elements of industrial ecology are applied:

- Selection of materials with desired properties at the outset
- Use of "just in time" materials
- Substitution of processes to eliminate toxic feedstock
- Modification of processes to contain, remove, and treat toxics in waste streams
- Engineering of a robust and reliable process
- Consideration of durability and end of life recyclability

ZD solutions that use conversion technologies should be developed, designed, built, and marketed by the appropriate professionals who understand not only the industrial clusters and the processes involved but also the upstream and downstream requirements.

1.11.6.1 Opening New Opportunities

As the ZD mission gains currency, new opportunities are revealed – to provide cost-saving new design applications, to design new product lines, and to win new customers. These opportunities include the following:

- Finding cost savings and new revenues in existing operations. Initial cost savings at existing operations should come from pollution prevention and waste minimization, which may already have been optimized. However, new revenues will come from identifying a viable market for the waste stream after it has been converted.
- Entering new markets for existing goods and services. New markets will be entered when cluster partners are identified; for example, brewery specialists will expand into agricultural sectors. The market for handling "designer wastes" is expected to grow significantly,

especially for firms specializing in reprocessing. Producers of a wide range of materials processing equipment such as grinding, sifting, sorting, purifying, separating, and packaging will find new markets. However, in some cases the conversion process will be handled by an intermediary company that will alter the wastes mechanically, chemically, or biologically to meet customer specifications.

- Developing new technologies, processes, and materials. Many of the pollution control firms will begin partnering with the upstream commodity producers (e.g. petroleum, chemical, and mining companies) and learning their refining techniques.
- Supporting the organizational changes, and technical and information needs of a Zero Emissions-based economy. This will be a business opportunity primarily for those offering skills in informational and organizational systems.
- Integrating technologies and methods into innovative new systems. As Zero Emissions expands, professionals from different sectors will connect to benefit from each other's skills and experience.
- Developing the infrastructure for eco-industrial parks. Requires equipment to channel the flow of materials, water, or heat between plants and communities. Civil engineering firms that specialize in urban design and infrastructure systems should find great opportunities in providing the integrated system designs.

1.11.6.2 Providing Return on Investment

A simple economic metric, return on investment (ROI) is a quick measure of when an item will pay for itself. The time between the initiation of an investment and the achievement of ROI is called the payback period. Pollution control technology, which is not traditionally viewed as an investment that is able to generate a return on investment, is usually measured in terms of lowest available cost to meet regulatory guidelines (see also Chapter 7). If wastes are viewed as materials, however, as in ZD, the whole picture changes. Some have joked about giving all materials produced in a facility a product name and an advertising budget and/discontinuing any "product" that does not sell.

Engineers have historically been compensated based on overall project cost. Incentives need to be shifted to designs that reduce material and energy flows. Compensation based on energy efficiency is being implemented in some projects and it is successful because energy efficiency can be measured in a single unit (joules of energy saved). While material flows are not as easy to quantify in a single unit, waste recovery systems can provide an excellent return on investment, as illustrated in Mini-Case Study 1.3.

Mini-Case Study 1.3 Recovery of Wastes from Palm Oil Extraction Yields High Return on Investment

Recovery of wastes from agro-industries is an extremely promising aspect of Zero Emissions. This project focuses on recovering all of the solid, liquid, gaseous, and thermal wastes from the Golden Hope Plantation in Malaysia, the largest oil palm plantation in the world. With the commitment of Meta Epsi, a large engineering group with substantial interests in palm oil plantations, operation of the pilot project for the total use of palm oil biomass commenced in the summer of 1996.

The pilot project uses steam explosion to provide for conversion of biomass into recoverable fibers, with a goal of reusing the spent seeds, bunches, leaves, and trunks that Golden Hope used to pay to have disposed of. It costs \$50 Malaysian (M\$50) to produce 1 T of commercially usable fiber, which can be sold for approximately M\$350. Products made from the fiber include MDF board, stuffing for car seats, and bedding for medical use. Just one of the mills built to process waste fiber generates pre-tax profits of approximately M\$12.5 million (about \$3 million) (Malaysian Ringgit ~ \$0.24).

1.12 Constraints and Challenges

The implementation of Zero Emissions faces constraints and challenges, as well as new opportunities. For example, the use of dissipative materials poses a design challenge: If solvents and flocculants are no longer to be used, what would it be replaced? Chemical manufacturers need to work with design engineers to arrive at an understanding of the constraints of separation technologies so that manufacturing any material without emissions is difficult, but working with chemicals is particularly challenging because of the need to develop nontoxic materials that are also biodegradable. Two possible solutions are biological.

- Biopolymers are an outgrowth of chemurgy, the division of applied chemistry that deals with industrial utilization of organic raw materials, especially from agro-business. These substances, complex molecules formed in biological systems, can replace toxic, dissipative materials currently used as adhesives, absorbents, lubricants, soil conditioners, cosmetics, drug delivery vehicles, and textile dissipative. Substitutes for toxic materials and mechanical processes to substitute for dissipative materials are aspects of the same principle.
- Enzymes are natural catalysts that speed up chemical reactions without being consumed in the process. They function best in mild conditions, so their use requires up

to one-third less energy than many synthetic chemicals; paradoxically, this lower need for energy can be an obstacle in a system that still rewards large-scale energy use with reduced rates. Enzymes are especially useful in systems designed to reduce or eliminate dissipative losses.

There is also a need for a taxonomy of environmental technologies that clarifies opportunities for fast developing, generic processes to address such recurring problem as process large streams of contaminated water from various processes and oxidation in air. Chemical engineering and related professions ought to be able to make rapid advances in such areas.

Many of the industries in the investment recovery or "decomposer niche" are hard put to compete against large-scale facilities that produce materials from virgin materials. More recently, however, economies of scale for resource extractors and processors, along with cheap energy supplies, have been introduced almost everywhere in the world. For example, economies of scale have enabled chemical companies to produce plastics at a price that other manufacturers, as well as the individual consumer, can afford.

1.12.1 The Challenges in Industrial Environmental Management

We need to educate and train current generation engineers, managers, business owners, and policy makers with the skills and knowledge they need to be our champion stewards of environmental management and expert on lean manufacturing of major industries demonstrating ZDZE operations. Engineers have always faced design constraints. Historically, these constraints were the laws of physics, availability of materials, and energy. Modern engineers still face the limitation of the laws of physics but have been granted larger amounts of energy and a wide variety of materials. Modern society has added design parameters that include safety, durability, convenience, regulatory compliance, attractiveness, and price.

A true engineer does not view regulation as an obstruction, but rather as a design constraint like efficiency and durability. The goal has always been to develop the optimal design within given constraints, whether they are the laws of nature or society. Unfortunately, the actions of our modern society have placed undue burdens on nature. Nature's ability to absorb excessive amounts of pollutants and stressors while still providing critical services of acceptable water quality, clean air, food, and biodiversity is limited.

Industrial ecology is Western society's response to meeting the challenge of sustainable development. To manufacturers falls the challenge of attaining Zero Emissions. They in turn pass this directive to their engineers. To engineers, the advance of technology has meant increasing degrees of freedom with regard to design. The collective body of knowledge and our harnessing of materials and energy has been the source of these freedoms. Safety was the first man-made design constraint that society imposed under the name of social good. Engineers responded to meet the challenge. Now society recognizes the need to impose a design parameter of Zero Emissions. Engineers will meet this challenge and accept it as they have the laws of physics – as a given.

1.12.2 Codes of Ethics in Engineering

Codes of ethics state the moral responsibilities of engineers as seen by the profession, and as represented by a professional society. Because they express the profession's collective commitment to ethics, codes are enormously important, not only in stressing engineers' responsibilities but also the freedom to exercise them.

Codes of ethics play at least eight essential roles: serving and protecting the public, providing guidance, offering inspiration, establishing shared standards, supporting responsible professionals, contributing to education, deterring wrongdoing, and strengthening a profession's image (also see Appendix J).

1.13 The Structure of the Book

1.13.1 What Is in the Book?

Chapter 1: This chapter addresses why industrial environmental management is important! Environmental management is a very crucial part of human well-being that needs to be deeply considered. Formulated design seeks to steer the development process to take advantage of opportunities, avoid hazards, mitigate problems, and prepare people for unavoidable difficulties by improving adaptability and resilience. It is a process concerned with human–environment interactions, and seeks to identify: what is environmentally desirable; what are physical, economic, social and technological constraints to achieving that process; and what are the most feasible options. Actually, there can be no concise universal definition of environmental management; however, it can be briefly summarized as supporting sustainable development; demanding multidisciplinary and interdisciplinary or even holistic approaches; it should integrate and reconcile different development viewpoints, co-ordinate science, engineering, technology, social, policy making, and planning; state proactive processes; timescales and concerns ranging from local to global issues; and one stresses stewardship rather than exploitation while dealing with a world affected by humans.

In other perspectives, environmental management can be explained as methods of ways when dealing issues due to the importance of the need to improve environmental stewardship by integrating ecology, policy making, planning, and social development. The goals include sustaining and (if possible) improving existing resources; preventing and overcoming environmental problems; establishing limits; founding and nurturing institutions that effectively support environmental research, monitoring, and management resources; warning of threats and identifying positive change opportunities; (where possible) improving quality of life; and finally, identifying new technology or policies that are useful.

Chapter 2: The objective of this chapter is to introduce the genesis of world environmental problems and to provide an overview of the history behind present environmental laws and regulations of pollution in various countries and continents.

Engineers in all disciplines practice a profession that must obey rules governing their professional conduct and ethics. One important set of rules that all engineers should be aware of is *environmental statues*, which are laws enacted by US Congress and governments of other countries around the world.

Environmental law, also known as *environmental and natural resources law*, is a collective term describing the network of treaties, statutes, regulations, common, and customary laws addressing the effects of human activity on the natural environment. The core environmental law regimes address environmental pollution. A related but distinct set of regulatory regimes, now strongly influenced by environmental legal principles, focus on the management of specific natural resources, such as forests, minerals, or fisheries. Other areas, such as environmental impact assessment, may not fit neatly into either category but are important components of environmental law.

Chapter 3: This chapter provides a summary of industrial wastewater sources, wastewater characteristics, wastewater treatment, reuse and discharge, industrial sources of air pollutions, inventories, air pollution control, solid waste and hazardous waste characteristics, treatments, and management.

Industrial waste is the waste produced by industrial activity which includes any material that is rendered unusable during a manufacturing process such as that of factories, industries, mills, and mining operations. Mass manufacturing has existed since the start of the Industrial Revolution. Some examples of industrial wastes are discussed including (but not limited to) chemical solvents, paints, sandpaper, paper products, industrial by-products, metals, plastics, and radioactive wastes.

Toxic waste, chemical waste, industrial solid waste, and municipal solid waste are designations of industrial wastes. Sewage treatment plants can treat some industrial wastes, i.e. those consisting of conventional pollutants such as biochemical oxygen demand, COD, suspended solid, and total suspended solid. Industrial wastes containing toxic pollutants require specialized treatment systems.

Chapter 4: This chapter describes and deals with various important aspects of selecting the best remedial control technologies for pollutants, managing wastes, monitoring, sampling industrial water, air, and solid and hazardous materials, modes of sample collections, sample analyses by various analytical, physical–chemical methods approved by governmental agency, as required quality control and quality assurance, properly conducted laboratory auditing, testing, monitoring, permitting, report keeping, reporting, and compliance with local, state, and federal governments for discharging wastewater, emitting air, pollutants, and safely disposing of solid and hazardous materials.

Chapter 5: This chapter addresses risk assessment, which is an organized process used to describe and estimate the likelihood of adverse health and environmental impacts from exposures to chemicals released to air, water, and land. Risk assessment is also a systematic, analytical method used to determine the probability of adverse effects. A common application of risk assessment methods is to evaluate human health and ecological impacts of chemical releases into the environment. Information collected from environmental monitoring or modeling is incorporated into models of worker activity and exposure forms conclusion about the likelihood of adverse effects are formulated. As such, risk assessment is an important tool for making decisions with environmental and public health consequences, along with economic, societal, technological, and political consequences of proposed actions. This chapter addresses the assessment of risks to human health as well as ecological risks and, briefly, ecological risk management. In addition a major section is devoted to industrial and manufacturing process safety, federal and state occupational safety laws and regulations, and management occupational health.

Chapter 6: This chapter describes the wastes produced by industrial activities, which include materials that are rendered unusable during manufacturing processes such as that of factories, industries, mills, and mining operations. This wastefulness has existed since the start of the Industrial Revolution. Some examples of industrial wastes and sources are chemicals and allied products, solvents, pigments, sludge, metals, ash, paints, furniture and fixtures, paper and allied products, plastics, rubber, leather, textile mill products, petroleum refining and related industries, electronic equipment and components, industrial by-products, metals, radioactive wastes, miscellaneous manufacturing industries, and the list goes on. Hazardous or toxic wastes, chemical waste, industrial solid waste, and municipal solid waste are also designations of industrial wastes.

More than 12 billion T of industrial wastes are generated annually in the United States alone. This is roughly equivalent to more than 40 T of waste for every man, woman, and a child in the United States. The sheer magnitude of these numbers is cause for big environmental concern and drives us to identify the characteristics of the wastes, the various industrial operations that are generating the waste, the manner in which the waste are being managed, and the industrial pollution prevention policy and strategies. The first portion of this chapter is devoted to the pollution prevention hierarchy. Next there is an overview of how LCA tools can be applied to choose best available technologies (BACT) to minimize the waste at various stages of manufacturing processes of products. Finally, a few case studies on industrial competitive processes and products applying LCA tools are reviewed; and hence, also selections of BACT to demonstrate hierarch pollution prevention (P^2) and environmental performance strategies.

Chapter 7: The role of economics in pollution prevention is of tantamount important, even as important as the ability to identify technologies changes to the process, new and emerging technologies, ZD technologies', technologies for biobased engineered chemicals, products, renewable energy sources, and associated costs. This chapter shows some methods that can be used to assess the costs of implementing pollution prevention technologies and making cost comparisons to evaluate the cost-effectiveness of various operations. The concept of best available control technologies is introduced and we analyze the costs and benefits of manufacturing biobased products. The topics treated illustrate that biobased new development can lead to sustainable economic progress and a healthier planet.

Sustainable development is about creating a business climate in which better goods and services are produced using less energy and materials with no or less waste and pollution. Natural steps and systems are a model for thinking about how to produce, consume, and live in sustainable cycles: nature produces little or no waste, relies on free and abundant energy from sun, and uses renewable resources. In this chapter, we focus on a framework that integrates environmental, social, and economic interests into effective chemical and allied business strategies.

Chapter 8: In this chapter our major focus is on lean manufacturing of various products while applying techniques and methodologies to achieve zero defects in products, and significantly eliminate waste and discharges to environment from the manufacturing process.

The quality of products, processes, and services has become a major decision factor in most industries and businesses today. Regardless of whether the consumer is an individual, a corporation, a military defense program, or a retail store, the consumer is making purchase decisions, he or she is likely to consider quality to be equal in importance to cost and schedule. Consequently, quality improvement has become a major concern to industries and businesses.

Quality means fitness for use with *zero defects or zero effects* in environmentally conscious manufacturing. For example, you or I may purchase automobiles that we expect to be free of manufacturing defects and that should provide reliable and economical transportation, a retailer buys finished goods with the expectation that they are properly packaged and arranged for easy storage and display, and a manufacturer buys raw material and expects to process it with no rework or scrap. In other words, all consumers expect that the products and services they buy will meet their requirements. These requirements define fitness for use.

Quality or fitness for use is determined through the interaction of quality of design and quality of conformance. Quality of design for environment and other aspects is defined by the different grades or levels of performance, reliability, serviceability, and function that are the result of deliberate engineering and environmental management decisions. By quality of conformance, we mean systematic reduction of variability and elimination of defects until every unit, batch, and product produced is identical in physical and chemical properties (*zero defect and zero effect*).

Chapter 9: The rate of industrial hazardous waste generation in the United States is approximately 750 million T/Y. Once these materials are designated as hazardous, the costs of managing, treating, storing, and disposing of them increase dramatically. This chapter describes some specific industrial waste minimization processes and technologies that have been successfully operating and provides other methodologies including industrial ecology, eco-industrial park, *manufacturing process intensification*, and *integration*. The wastes (in air, water, or as solid) or by-products generated during manufacturing process are recovered. The materials and energy recovered from waste streams either are reused in the plant or are sold to another plant as feedstock. It is possible in practice, as well as in theory, to isolate some industrial facilities almost completely from the environment by recycling all wastes into materials that can then be manufactured into consumer products. An example of such a facility is a coal-fired power plant. An electron beam–ammonia conversion unit adds ammonia to the effluent gases, which then irradiates electronically, producing ammonium nitrate and ammonium sulfate that are sold as feedstock to fertilizer manufacturing; there is enhanced recovery of mercury from flue gas by adsorption and mercury recovery is complete. The details of these two processes are given as case studies later in Sections 9.2 and 9.4. Also, two separate case studies have been presented that highlight a profitable "waste-to-energy" recovery generating electricity and heat, and making chemicals and energy from gasification of black liquor as by-products of pulping process.

Our goal is to modify industrial processes so that services and manufactured goods can be produced without waste. But it is important to understand that some manufacturing processes inherently produce wastes, even after all reasonable efforts at pollution prevention. Thus, in some cases the use of a conversion technology may be more appropriate than a program of pollution prevention: many industrial wastes can be processed to render them viable as material inputs to another industry or to part of an industrial cluster of several connected industries – as part of the movement of "industrial ecology."

Chapter 10: Engineers play an important role in global sustainable manufacturing and development by designing production systems for materials: minerals, chemicals, energy, water, electricity generation and distribution, transportation, buildings plus other structures, and consumer products. These designs have impact on the environment, economics, and societal benefits at scales that vary from local to global and temporal scales that vary from minutes to decades. As engineers create designs, they do not only evaluate their designs at multiple use and sustainability index (scale), they also embed their designs in complex systems.

The field of transportation provides an illustration of the multiple layers of systems in which engineers create designs. Among the most visible products designed by engineers are automobiles. Engineers design engines, and improvements to the design of a fossil fuel–powered engine for an automobile can increase fuel efficiency and reduce environmental impacts of emissions associated with burning fuels, while simultaneously reducing the cost of operating the vehicle. The size, power, and fuel efficiency of the engine must be balanced with the weight of the vehicle. The use of materials and fuels by automobiles are embedded in complex fuel and material supply systems. Developing systems to recycle the materials that make up the automobile at the end of its useful life might improve the environmental and economic performance of global materials flows. Use of alternative power sources, such as electricity or biofuels can impact flows of fuels, which, in turn, might impact global flows of materials such as water. Finally, the design of cities that reduce the need for personal transportation could dramatically reduce the environmental impacts of transportation systems and would also transform social structures.

In this chapter, *sustainable design* is very much emphasized and lays a foundation. Sustainable engineering is the design, commercialization, and use of processes and products that are feasible and economical while minimizing both the generation of pollution at the source and the risk to human health and the environment. The discipline embraces the concept that decisions to protect human health and the environment can have the greatest impact and cost effectiveness when applied early in the "design and development phase of a process or product." Sustainable engineering transforms existing engineering disciplines and practices to those that promote sustainability. This new discipline incorporates the development and implementation of technologically and economically viable products, processes, and systems that promote human welfare while protecting human health and elevating the protection of the biosphere as a criterion in engineering solutions. To fully implement sustainable green engineering solutions, engineers use numerous principles and tools that are described in this chapter.

Problems

1.1 State why industrial environmental management is a critically important part of human well-being and sustainable development?

1.2 Define waste as pollution.

1.3 Explain the difference between pollution prevention and minimization of waste.

1.4 What is the key point of the Pollution Prevention Act of 1990?

1.5 Define sustainability. Explain why the concept of "Zero Discharge and Zero Waste" in manufacturing are the keys to sustainability.

1.6 What are the basic environmental challenges that must be met to meet world demands for (a) clean air, (b) clean water, and (c) arable land?

1.7 Using Zero Waste, Zero Defect and Zero Effect, and Zero Discharge, (a) what conversion technologies need to be developed? [Hint: Check out Figures 1.7

and 1.8.] (b) What are the investment management points of view when manipulating the profit margins? (c) What are the constraints and challenges that must be met to meet regulatory requirements?

1.8 Fifteen fields of environmental management are introduced in this chapter. Answer the following: (a) Choose seven fields that work closely together and explain the commonality that the eight remaining fields have in common. (b) How is consensus methodology achieved between groups? (c) Which field of environmental management requires global support for human health?

1.9 Does ISO support industrial environmental management system, and if it does, how?

1.10 What are the key elements of an environmentally conscious manufacturing strategy?

1.11 As chemical process and product design engineers, as well as construction, industrial, and environmental engineers, how you can play a pivotal role as industries move toward the *Zero Emissions or Zero Discharge* paradigm.

References

Allen, D.T. and Behmanish, N. (1994). *Wastes as raw materials*. In: *The Greening of Ecosystems* (eds. B. Allenby and D. Richards). Washington, DC: National Academy Press.

Allen, D.T. and Rosselot, K.S. (1997). *Pollution Prevention for Chemical Processes*, 19–32. New York, NY, Chapter 2: Wiley.

Allen, D.T. and Shonnard, D.R. (2012). *Sustainable Engineering: Concepts, Design, and Case Studies*. Upper Saddle River, NJ: Prentice Hall.

Allenby, B.R. and Richards, D.J. (1994). *The Greening of Industrial Ecosystems*. Washington, DC: National Academy Press.

Ayres, R.U., Ayres, L.W., and Ayres, L. (1996). *Industrial Ecology: Towards Closing the Materials Cycle*. Cheltenham, UK: Edward Elgar.

Cote, R.P. (1995). Supporting pillars for industrial ecosystems. *Journal of Cleaner Production* 5 (1–2): 67–74.

Cote, R.P. (2003). *A Primer on Industrial Ecosystems: A Strategy for Sustainable Industrial Development*, 1–34. Halifax, NS: Eco-Efficiency Center, School for Resources and Environmental Studies, Dalhousie University.

Das, T.K. (2005). *Toward Zero Discharge: Innovative Methodology and Technologies for Process Pollution Prevention*. Hoboken, NJ: Wiley.

Delmas, M. (2009). Erratum to "Stakeholders and Competitive Advantage: The Case of ISO 14001". *Production and Operations Management*. 13 (4): 398.

Graedel, T.E. and Allenby, B.R. (1995). *Industrial Ecology*. Englewood Cliffs, NJ: Prentice Hall.

Grann, H. (1994). The industrial Symbiosis at Kalundborg, Denmark. Paper presented at the National Academy of Engineering's Conference on Industrial Ecology, Irvine, CA (9–13 May 1994).

Herman, R. (1989). *Technology and Environment*. Washington, DC: National Academy Press.

Hutchens, S. (2010). Using ISO 9001 or ISO 14001 to gain a competitive advantages. *Intertake*. www.intertek-sc.com (accessed 21 September 2019).

Johnson, J., Harper, E.M., Lifset, R., and Graedel, T.E. (2007). Dining at the periodic table: metals concentrations as they relate to recycling. *Environmental Science and Technology* 41: 1759–1765.

Pauli, G. (1996). *Breakthroughs*. Haslemere, Surrey: Epsilon Press.

Potoski, M. and Prakash, A. (2005). Green clubs and voluntary governance: ISO 14001 and firms' regulatory compliance. *American Journal of Political Science* 49 (2): 235.

Sheldon, C. (1997). *ISO 14001 and Beyond: Environmental Management Systems in the Real World*. New York: Prentice Hall.

USEPA (1992). *Facility Pollution Prevention Guide*, EPA/600/R-92/088. Washington, DC: USEPA, Office of Research and Development.

Van der Veldt, D. (1997). Case studies of ISO 14001: a new business guide for global environmental protection. *Environmental Quality Management* 7 (1): 1–19.

World Commission on Environment and Development (1987). *Our Common Future*. Oxford University Press, Oxford, UK.

Zacerkowny, O. (2002). Not a drop leaves the plant. *Pollution Engineering* 34: 19–22.

2

Genesis of Environmental Problem Worldwide

International Environmental Regulations

2.1 Introduction

Engineers in all disciplines practice a profession of those who must obey rules governing their professional conduct. One important set of rules that all engineers should be aware of is *environmental statues*, which are laws enacted by US Congress and governments of other countries around the world.

Environmental law, also known as *environmental and natural resources law*, is a collective term describing the network of treaties, statutes, regulations, common, and customary laws addressing the effects of human activity on the natural environment. The core environmental law regimes address environmental pollution. A related but distinct set of regulatory regimes, now strongly influenced by environmental legal principles, focus on the management of specific natural resources, such as forests, minerals, or fisheries. Other areas, such as environmental impact assessment, may not fit neatly into either category but are nonetheless important components of environmental law.

The objective of this chapter is to provide an overview of history behind present environmental laws and regulations and environmental pollution in various countries and continents. Much of the materials presented in this chapter has been adopted from various sources that include the following: United States Code (USC); Code of Federal Regulations (CFR); United States Environmental Protection Agency (USEPA); United Nations Environmental Program (UNEP); Centre for International Environmental Law; Environmental Law Handbook by Sullivan and Adams (1997); Environmental Regulatory Calculations Handbook by Stander and Theodore (2008); Environmental Law by Upadhyay (2012); Environmental Law and Policy in India: Cases, Materials and Statues by Divan and Rosencranz (2002); Craig, Selected Environmental Law Statutes, West Academic Publishing Co. (2016); Lynch (1995); Google; and Wikipedia. Most of these sources can be found online.

Environmental law came into existence as a result of confrontation with the serious problems concerning environment, prior, during, and post Industrial Revolution. In response to environmental problems, laws seek to protect and promote environment. It is designed to prevent, control, and regulate environmental pollution.

Environmental protection and its preservation is today's concern of all. Present environmental conditions in many parts of the world clearly indicate that human activities on the Earth are interconnected. Today society's interaction with nature is so extensive that environment questions have assumed all proportions affecting humanity at large. Environmental destruction and pollution have seriously threatened human life, health, and livelihood. Thus, there has been a thrust on the protection of environment all across the world. If the quality of life is to be assured to present and future generations, they should be saved from the environmental failures. Nature's gifts to humanity in the form of flora and fauna have to be preserved in natural form. The proper balance of the world's eco system is in urgent need. The only answer to tackle this problem is sustainable development. The purpose of environmental law is to protect and preserve our water, air, Earth, and atmosphere from pollution. But the law alone cannot tackle the problem of pollution. There has to be awareness of the problem and sustained efforts are required to manage it.

The situation may be clarified by noting that with the development of science and technology and with ever-increasing world population came tremendous changes in the human environment. Such changes upset the eco-laws, shifted the balance between human life and the environment. In such a situation, it became necessary to regulate human behaviors and social transactions with new laws, designed to cope with the changing conditions and values.

Industrial Environmental Management: Engineering, Science, and Policy, First Edition. Tapas K. Das.

Companion website: www.wiley.com/go/Das/IEM_1e

Accordingly, a new branch of law called environmental law, developed in order to face the myriad challenges of such system.

The need of environmental protection is a big issue and ranks high among people's priorities. Also the issue of environmental protection is big in terms of the size of the problems on hand and the measures required to solve them. Some of the major problems, for example, are population growth (Figure 2.1), global warming (Figure 2.2), depletion of ozone layer, acid rain, toxic waste, and deforestration.

Long-term global population growth is difficult to predict. The United Nations and the U.S. Census Bureau both give different estimates – according to the UN, the world population reached about 7.31 billion in late 2015 (United Nations Secretariat 2015). Figure 2.1 shows a trend that the human population has been growing rapidly since the mid-1800s, which goes hand in hand with the degradation of the environment's natural equilibrium.

Population increase over the last two decades, at least in the United States, has also been accompanied by a shift to an increase in urban areas from rural areas (Wallach 2005), which concentrates the demand for water into certain areas, and puts stress on the fresh water supply from industrial and human contaminants (Water 2011). Urbanization causes overcrowding and increasingly unsanitary living conditions, especially in developing countries, which in turn exposes an increasing number of people to disease. About 79% of the world's population is in developing countries, which lack access to sanitary water and sewer systems, giving rise to disease and deaths from contaminated water and increased numbers of disease-carrying insects (Powell 2009).

Figure 2.2 shows recent behavior of average global temperature anomaly (land and ocean combined). As can be seen in Figure 2.2, the bulk of the warming has occurred in two periods – from 1910 through 1940 and from 1980 to the present. Line plot of monthly mean global surface temperature anomaly, with the base period 1951–1980. The black line shows meteorological stations only; circles are the land–ocean temperature index, as described in Hansen et al. (2010). The land–ocean temperature index uses sea surface temperatures.

These problems are global issues which require an appropriate global response. The question of environmental protection is big issue in terms of the range of problems and issues, namely air pollution, water pollution, safe drinking water, pesticides, noise pollution, waste disposal, radio activity and waste, and conservation of forests and wild life. The list is exhaustive. Moreover, the issue in the field of environment is big in terms of the knowledge and skills required to understand a particular issue to be solved. Accordingly, it can be said that environmental issues range "from a local street

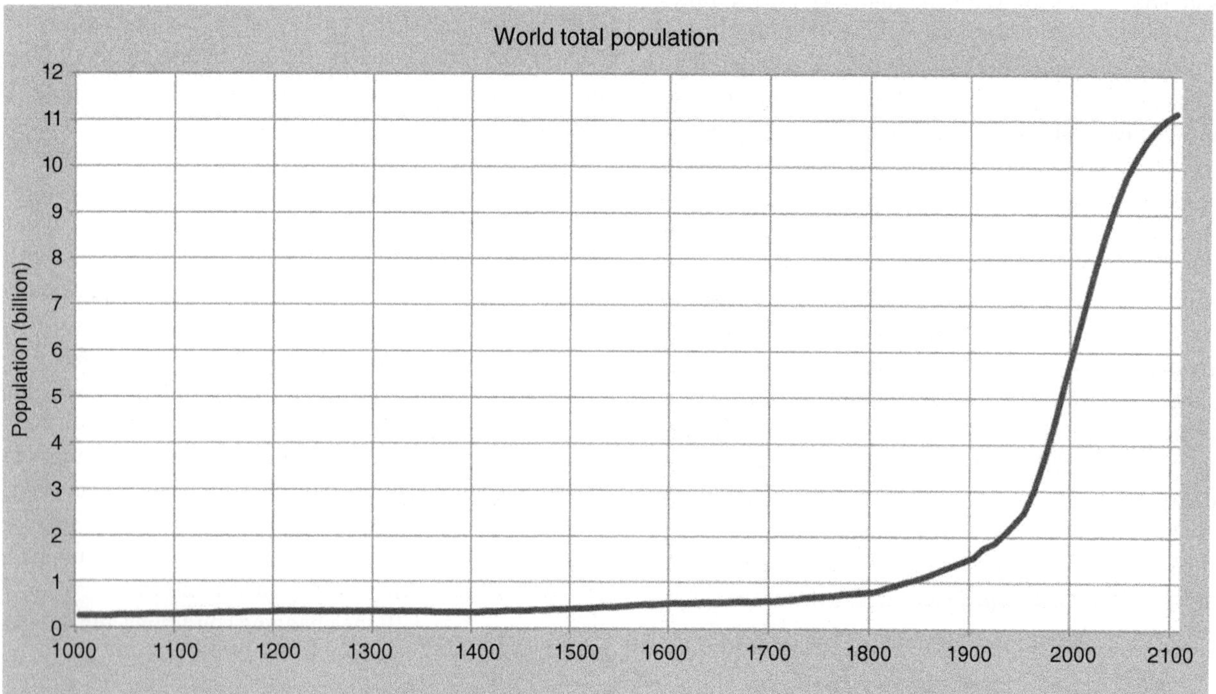

Figure 2.1 World's population growth over 100 years. *Source:* United Nations, Department of Economic and Social Affairs, Population Division (2017).

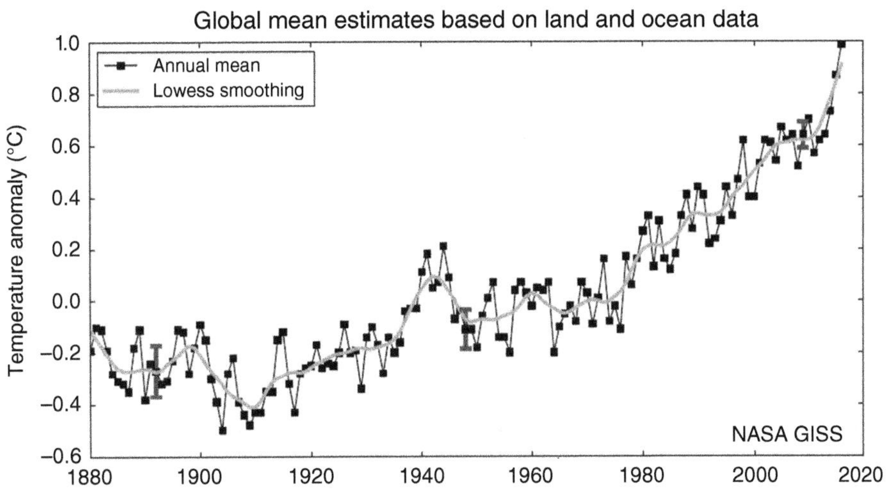

Figure 2.2 Earth's temperature rise since industrialization over 100 years. *Source:* http://data.giss.nasa.gov/gistemp.graphs.

corner to the stratosphere." It is thus clear that the humanity faces overwhelming environmental problems. Law is invoked to protect humanity and our eco-system by solving environmental problems as they strike directly at our most intimate links to the biosphere, which is a thin shell of life – only about five miles thick – covering the planet like the skin of an apple.

The range of environmental protection has worldwide coverage and is not confined to an isolated area or nation. The problem of environment pollution is global and concerns all countries irrespective of their size, level of development, or ideology. Notwithstanding political division of the world into national units, oceanic world is an interconnected whole and winds that blow over the countries are also one. Environment is a universal phenomenon pervading the whole world at large. If the nuclear test is carried out in one part of the world, the fall out may be carried by winds to any other part of the world and such fall out of irresponsible disposal of use radioactivity from a remote energy planet in one country may turn out to have greater adverse effect on the neighboring countries than the danger of a full-fledged war.

2.1.1 Environmental History

Environmental history is the study of human interaction with the natural world over time. In contrast to other historical disciplines, it emphasizes the active role nature plays in influencing human affairs. Environmental historians study how humans both shape their environment and are shaped by it.

Environmental history emerged in the United States out of the environmental movement of the 1960s and 1970s, and much of its impetus still stems from present-day global environmental concerns (Chakrabarti 2006). The field was founded on conservation issues but has broadened in scope to include more general social and scientific history and may deal with cities, population, or sustainable development. As all history occurs in the natural world, environmental history tends to focus on particular time-scales, geographic regions, or key themes. It is also a strongly multidisciplinary subject that draws widely on both the humanities and natural science.

The subject matter of environmental history can be divided into three main components (Chakrabarti 2007). The first, nature itself and its change over time, includes the physical impact of humans on the Earth's land, water, atmosphere, and biosphere. The second category, how humans use nature, includes the environmental consequences of increasing population, more effective technology, and changing patterns of production and consumption. Other key themes are the transition from nomadic hunter-gatherer communities to settled agriculture in the Neolithic Revolution, the effects of colonial expansion and settlements, and the environmental and human consequences of the industrial and technological revolutions (Cronon 1995). Finally, environmental historians study how people think about nature – the way attitudes, beliefs, and values influence interaction with nature, especially in the form of myths, religion, and science.

2.2 Genesis of the Environmental Problem

Environmental problems have bedeviled humanity since the first person discovered fire. The earliest humans appear to have inhabited a variety of locales within a tropical and semitropical belt stretching from Ethiopia to southern Africa about 1.9 million years ago. These first humans provided for themselves by a combination of gathering food and hunting animals. Humans, for the majority of their two million years' existence, lived in this manner. The steady development and dispersion of these humans was largely due to an increase in brain size. This led to the ability to think abstractly, which was vital in the development of technology and ability to speak. This in turn led to cooperation and more elaborate social organization. The ability to use and communicate and the developed technology to overcome the hostile environment ultimately led to the expansion of these first human settlements (Ponting 1991; Sander and Theodore 2008).

By about 10 000 years ago humans had spread over every continent, living in small mobile groups. A minority of these groups lived in close harmony with the environment and did minimum damage. Evidence has been found where groups tried to conserve resources in an attempt to maintain subsistence for long periods of time. In some cases, restrictions on hunting a particular species at a time, a certain period of time of the year, or only in a certain area every few years helped to maintain population levels of certain animals (Goudie 1981). The Cree in Canada used a form of rotational hunting, only returning to an area after a considerable length of time, which allowed animal populations to recover. But the majority of these groups exploited the environment and animals inhabiting it. In Colorado, bison were often hunted by stampeding them off a cliff, ending up with about 200 corpses, most of which could not be used. On Hawaii, within a 1000 years of human settlement, 39 species of land birds had become extinct (Ponting 1991). In Australia, over the last 100 000 years, 86% of the large animals have become extinct. The large number of species lost was largely due to tendency for hunters to concentrate on one species to the exclusion of others. The main reason why these groups avoided further damage to nature was the fact that their numbers were so small that the pressure they exerted on the environment was limited.

The major shift in human evolution took place between 10 000 and 12 000 years ago. Humans learned how to domesticate animals and cultivate plants and in so doing made a transition from nomadic hunter-gatherer to rooted agriculturist. The global population at this time was about four million people, which was about the maximum that could readily be supported by a gathering and hunting way of life (Ponting 1991). The increasing difficulty in obtaining food is believed to be a major contributor to this sudden change. The farmer changed the landscape of the planet and was far more destructive than the hunter. While farming fostered the rise of cities and civilizations, it also led to practices that denuded the land of its nutrients and water-holding capacity. Great civilizations flourished and then disappeared as once-fertile land, after generations of over-farming and erosion, was transformed into barren wasteland.

However, many years later different dimensions of the problem of environmental protection and its management have taken a serious turn in the present era. This is so because humans' interaction with the natural environment is so extensive that the environment question has assumed proportions affecting all humanity. The dominant factors which are responsible to environmental deterioration throughout the world are rapid industrialization, urbanization, population explosion (Figure 2.1), poverty, overexploitation of resources, depletion of traditional resources of energy and raw materials, and research for new sources of energy and raw materials.

As a result of this, there is air pollution, water pollution, noise pollution, land pollution, and so on. The list is virtually endless. Moreover, the seriousness of environmental problem may be judged in terms of knowledge and skills required to understand a particular issues demanding solution. In order to achieve sustainable development, environmental protection constitutes an integral part of development process and it cannot be considered in isolation. Peace, development, and environment are interdependent and indivisible (Rio Declaration, UNESCO 1992). In the same continuation, Earth Summit of 1992 in Rio de Janerio, through Rio Declaration and agenda 21 has further concretized the concept of environmental protection and sustainable development essential to survival of human race. Today we are confronted with a perpetuation of disparities between and within nations, a worsening of poverty among developing countries, hunger, ill health due to malnutrition, poor sanitation and lack of safe drinking water and proper health care, illiteracy, and continuing deterioration of the ecosystem on which we depend for our well-being. However, integration of environment and development concerns and greater attention to them will lead to the fulfillment of basic needs, improved living standards for all, better protected and managed ecosystems, and a safer, more prosperous future.

2.3 Causes of Pollution and Environmental Degradation

Causes of pollution and environmental degradation are of two types:

1) Natural causes
2) Man-made causes

2.3.1 Natural Causes

Drought, flood, tsunami, cyclone, hurricane, twister, torrents, earthquake, molten lava of volcano, and epidemic are the main natural causes/factors which cause environmental pollution. Since these are natural-caused events and man has no control over them, they are known as natural causes.

2.3.2 Man-Made Causes

There are four main man-made causes:

1) Population growth
2) Poverty
3) Urbanization
4) Industrialization

2.3.3 Population Growth

"The Earth is finite and world population is finite." Every new born consumes plenty of natural and nonnatural products, which are also ultimately provided after utilizing natural resources. Thus, every birth increases the consumption of natural resources. But the fact is a finite world can support only a finite population. In other words, natural resources shrink as people multiply. The current world population is 7.73 billion as of August 2019 according to the most recent United Nations estimates elaborated by Worldometers. The world's population has grown almost sixfold in this century (Figure 2.1). India alone has about 17% population of the world land area. This rise in urban population is at a very high rate. It indicates an increasing demand for fuel, food, pollution free clean water and air, space to live in, and healthy conditions. Increasing population in urban areas has created the problem of land, air, and water pollution, and unsanitary conditions that all cumulatively adversely affecting the quality of life. Some big cities are rated as choked cities due to polluting industries around them. Continuous rise in urban population has enhanced the density of population in various areas which has also created various social, physical, and psychological problems for people. These high-density populated areas have also resulted in deforestation and disappearance of vegetation cover, which is only 11% of the total area against 33% which is essential. Increasing population also results in poverty which is also a cause of pollution.

2.3.4 Poverty

Poverty contributes equally to both – population growth and environmental pollution. Poverty has been defined as the inability of an individual or household to attain a minimum standard of living (World Commission on Health and Environment 1992). The poor people usually have low life expectancy, high infant mortality, and higher incidence of disablement. Unhygienic, unsanitary, and poor health conditions are due to poverty. The impoverishment of the poor is accompanied by simultaneous and systematic erosion of the basic means of their subsistence, the environment, with its life-supporting natural resources – land, forest, water, and air. Poverty reduces people's capacity to use resources in a sustainable way, which also intensifies pressure on the environment in underdeveloped and developing countries. It is now been aptly observed that the poor and poverty are linked to the environment (Our Common Future 1987).

2.3.5 Urbanization

Rapid and unplanned urbanization has also contributed to environmental pollution and degradation of human environment. This is a direct result of rapid population growth and unending migration of the poor from small towns and villages to urban centers of developing countries. The burning of coal and wood in concentrated areas made the cities the primary sources of pollution. Environmental factors have been given too little consideration in the thinking on urbanization in many cities of underdeveloped and developing countries. Yet they are extremely important factors to be considered with increasing urbanization. The levels of water and smog pollution are already high in many cities in underdeveloped and developing nations.

2.4 Industrialization and Urbanization in the United States

Early in the nineteenth century, an awesome new force was gathering strength in Europe. The term "Industrial Revolution" was coined by the French as a metaphor of the affinity between technology and the great political revolution of modern times. Soon exported to the United States, the Industrial Revolution swept away any visions of America being an agrarian society. The steam engine, the railroad, the mechanical thresher, and hundreds of other ingenious artifacts that increased man's ability to transform the natural world and put it to use would soon be puffing and clattering and roaring in all corners of the land. The new machines swiftly accelerated the consumption of raw materials from the nation's farms, forests, and mines.

2.4.1 Mini Case Studies

Industrialization and urbanization, as a result of Industrial Revolution, began long before the late nineteenth and early twentieth centuries, but it accelerated greatly during this period because of technological innovations, social changes, and a political system increasingly apt to favor economic growth beyond any other concern. Before 1880, industrialization depended upon a prescribed division of labor – breaking most jobs up into smaller tasks, and assigning the same people to repeat one task indefinitely. After 1880, industrialization depended much more on mechanization – the replacement of people with machines – to increase production and maximize profits. The development of the modern electrical grid, starting in the early 1880s, facilitated such technological advances. Henry Ford's assembly line and the rise of mass production after the turn of the twentieth century only strengthened this effect. As a result, the total manufacturing output of the United States was 28 times greater in 1929 than it was 1859. Adjust that number for the growth in population over the same period, and it still multiplied seven times over (Wright 1941).

Cities in America date back to the beginning of the colonial period, but the tendency for new industrial factories to be located in or near urban areas meant that cities grew much faster during the late nineteenth century than ever before. This trend was most apparent in large cities like New York, which expanded from approximately half a million to around 3.5 million people between 1850 and 1900, and Philadelphia, which increased in size from slightly more than 100 000 inhabitants to more than 1.2 million people over the same period. During the last half of the late nineteenth century, Chicago proved to be the fastest growing city in the world. Overall, 15.3% of Americans lived in cities in 1850. By 1900, that percentage had increased to 39.7, and kept growing. The 1920 Census revealed that more Americans lived in cities than the countryside for the first time (Rees 2013).

Not every city in the country developed as fast as the largest cities did. Important regional differences existed in urbanization because of differences in the nature of industrial growth. The largest cities in the Northeast were manufacturing powerhouses that contained everything, from large factories building railroad locomotives to small shops producing textiles in people's apartments. The Northeast also gave rise to smaller cities that concentrated on particular industries, like Rochester, New York, which specialized in men's clothing, boots, and shoes. Following on a tradition of manufacturing from earlier in the century, New Bedford and Fall River, Massachusetts, increased in size because of their cotton textile factories. Other cities, like Elizabeth, New Jersey, grew as by-products of the expansion of their larger neighbors.

Chicago, the largest city in the Midwest, made its name processing natural resources from the Western frontier before those resources traveled eastward as finished products. Grain and lumber – two industries that had been crucial for Chicago's early growth – relied on Chicago for marketing and storage. With perfection of the refrigerated railroad car, meat processing became such an enormous industry that the vast majority of the meat that Americans ate was processed in the stockyards on that city's south side. (That activity would disperse again, after the turn of the twentieth century, to other cities like Fort Worth and Kansas City.) Smaller cities in America's industrial heartland would grow around other manufacturing pursuits like steel in Youngstown, and machine tools and cash registers in Dayton, Ohio (Warner 1995).

The South had lagged behind the rest of the country since before the Civil War. As a result, many advocates for outside investment in this region expanded their activities after the war. They were somewhat successful. While the rate of industrialization (and therefore urbanization) picked up in the South during the late nineteenth and early twentieth centuries, it still has not fully caught up with the rest of the country. Birmingham, Alabama, for example, founded in 1871, flourished as a center for iron and steel manufacturing during the 1880s, when two railroads first linked that city to the region's mineral resources (Misa 1999). The growth of cotton mills in the "upcountry" section of the Carolinas began during the 1870s. After the turn of the twentieth century, this region became an important center of activity for the textile industry, in large part because of the available cheap, nonunion labor.

What separates this period from earlier periods in urban and industrial history is that this was the first time in American history that cities had moved to the center of American life. Cities were places where most of the new factories were built. Waves of immigrants settled in cities because that is where the job openings in industrial factories were. Cities were also places where the effects of industrialization, especially the increased inequality of wealth, were most visible. That means that the problems of cities became the problems of America.

2.4.2 The Electrical Grid and Improvements in Transportation

One of the reasons that later industrialization progressed at such a greater pace than before was the improvement in power sources. The early Industrial Revolution depended upon steam engines and waterpower. The earliest engines were large and prohibitively expensive for all but the largest firms. Water wheels were a possibility for smaller concerns, but they could not perform nearly as much work as later power sources could. Between 1869 and 1929, total available horsepower in the United States increased from 2.3 to 43 million units. In factories, the greatest part of that growth came from a huge increase in the use of electricity (Wright 1941).

Although factories had grown larger and more efficient over the entire nineteenth century, they grew particularly large after 1880, as the power to run them became cheaper, cleaner, and more convenient to acquire. Starting in the late 1870s, Thomas Edison turned the attention of his extensive laboratory toward harnessing electricity to create affordable electric light. This achievement depended not only upon the creation of an efficient, inexpensive, incandescent light bulb but also on the creation of an electrical system to power it – everything from generators, to electrical wires, to switches. Without a precedent for any of these things, the Edison Electric Company and many related subsidiaries (later gathered together under the umbrella of General Electric) had to manufacture just about everything to make the grid operate. "Since capital is timid, I will raise and supply it," explained Edison to one of his investors. "The issue is factories or death!" (Jonnes 2004). Other companies soon followed, because creating the central stations and the grid that eventually powered just about everything was so obviously lucrative.

Symbolizing the importance of capital to Edison's efforts, the first person to have his home successfully wired for electricity was the banker J. P. Morgan, in 1882. Despite setbacks, his experience with electric light encouraged him to invest further in Edison's efforts. Edison built the first central generating station in New York City later that same year. The first area of Manhattan that Edison wired was a neighborhood filled with the homes and workplaces of those who operated the financial institutions he hoped to convince to invest in his enterprises, as well as two major newspapers that would publicize his achievements. By 1902, there were 2250 electrical generating stations in the United States. By 1920, that number grew to just short of 4000. Electricity spread from large cities to small cities and eventually out into rural areas by the 1920s (Cowan 1985; Cowan and Hersch 2017).

This kind of growth required substantial improvement beyond Edison's initial vision of an electrical system. The effects of a reliable electric grid on the cities where it first appeared were numerous, ranging from less coal smoke in the air to new sounds produced by various electrical creations – everything from streetcars to arc lights. Early arc lights were so bright people thought they could stop crime and vice by exposing the people who perpetrated these crimes. In smaller cities, obtaining electric light was a sign of modernization, which implied future growth. Modern light in urban workplaces made office work easier by lessening strain on the eyes. As electric light companies moved in, the much-hated urban gas companies lost a considerable amount of economic power. Since people preferred electric light to gas, it became increasingly popular, as the grid expanded and the costs dropped. Electric light even changed the way people lived inside their houses. For example, children could now be trusted to put themselves to bed since there was no longer a fire risk from the open flames that were once needed to get to bed in the dark.

Nevertheless, the growing electrical grid created new urban dangers. High-voltage electrical wires strung above ground joined other wires from telephones, telegraphs – even stock tickers – posed a new urban "wire menace." Many came down in bad weather. They were a hazard for electric company employees and pedestrians alike. "The overhead system is a standing menace to health and life," reported one medical journal in 1888 (Freeberg 2014). In 1889, a fire caused by overheated electrical wires ignited a building full of dry goods and burned down much of downtown Boston.

The most noteworthy effect of high-quality, affordable lighting was the widespread practice of running factories 24 hours a day – which made them much more productive without any improvements in the technology of production. Replacing putrid gas lamps also made the smell of factories better for the workmen who worked there. As the electrical grid became more reliable, electric motors gradually began to replace steam engines as the source of power in manufacturing. Using small electric motors as a source of power freed factories from having to be located near water sources to feed boilers and made it possible for them to be smaller too.

Between 1880 and 1900, factories tended to adopt electric lighting but kept using earlier sources of power for their operation. Electric power for factory operations came quickly between 1900 and 1930. Both these developments (along with the large supply of immigrant workers) contributed to the industrialization of cities. The electrification of industrial facilities of all kinds proceeded quickly during the first two decades of the twentieth century. Businesses got wired for electricity much faster than cities because they could make the most use of what started out as a relatively expensive service.

Because factories were concentrated in or near cities, it was a lot cheaper to wire them than it was to wire farms or even smaller cities away from electrical generating stations. Many of the new factories built during this later period appeared outside city limits, another new development. Electrification allowed managers to automate jobs once done by hand labor, thereby eliminating inefficiency, gaining greater control over the production process, and boosting overall productivity. New devices like time clocks and even new modes of production like the assembly line also depended upon electric power.

The advent of cheap and readily available electricity had a particularly important effect upon the physical layout of American cities during this period. Frank Sprague, an electrical engineer who had once worked for Thomas Edison, designed the first electric streetcar system for Richmond, Virginia, in 1888. Such systems supplanted horse-drawn carriages, making it possible for people to travel further and faster than they would have otherwise. This gave rise to a burst of suburbanization, a spate of new towns on the outskirts of American cities where wealthy and middle-class people could move to escape from the difficulties of modern urban life but still be close enough to enjoy many of its advantages.

The new suburbanites often traveled to and from work via new electric streetcars. The electrical equipment manufacturer Westinghouse was one of the major manufacturers of vehicles powered by an overhead wire. Electric streetcars had the advantage over horses of not leaving manure or of dying in the streets. Streetcars were more popular during weekends than during the week as working-class people took advantage of low fares to explore new neighborhoods or to visit amusement parks, like Coney Island, generally built at the end of these lines.

In the same way that employers and city planners depended upon streetcars to move people, manufacturers became more dependent upon railroads, after 1880, to move their finished products. Railroad track mileage grew greatly after the Civil War, connecting cities and leading to the growth of new factories in places that were convenient to the necessary resources to make marketable goods. Eventually, mass distribution was a prerequisite to benefit from all that increased productivity. For all these reasons, separating the causes and effects of industrialization and urbanization is practically impossible.

Throughout the nineteenth century, factories usually had to be built near shipping ports or railroad stops because these were the easiest way to get factory products out to markets around the world. As more railroad tracks were built late in the nineteenth century, it became easier to locate factories outside of downtowns. Streetcars helped fill up the empty space downtown where factories would have gone. They made it easier to live further away from work and still commute to the heart of downtown, thereby making it possible for other kinds of businesses to locate there. One example would be the large urban department store, a phenomenon that predates 1880, but grew into its own after that date. Such stores like Wanamaker's in Philadelphia or Marshall Field's in Chicago bought the products of industrialization in bulk and sold them at a discounted price to workers who might have had trouble getting access to them any other way.

2.4.3 Structural Steel and Skyscrapers

While retail emporiums could be blocks long and only a few stories tall, other business rented space in thinner buildings built much higher. By the late 1880s, structures that had once been built with iron began to be built with a structural steel – a new, stronger kind of steel. The practice had begun in Chicago, championed by the architect Louis Sullivan, who designed the first skyscrapers there. A skyscraper, Sullivan wrote, "must be every inch a tall and soaring thing, rising in sheer exultation that from bottom to top it is a unit without a single dissenting line" (Alexiou 2013). That kind of design required a skeleton of structural steel upon which other substances like brick or granite could hang. Even then, such skyscrapers had to be tapered; otherwise, the weight from the top floors could make the whole structure collapse.

Creating structural steel for skyscrapers required entirely different production methods than had been required to make Bessemer steel (which had been used primarily for railroad rails). Quantity and speed were the main requirements of producing Bessemer steel. Structural steel required a more carefully made product. The demands of structural steel encouraged steelmakers like Andrew Carnegie to redesign entire factories, most notably replacing older Bessemer converters with the open-hearth process. This new kind of steelmaking not only produced higher quality steel but also required fewer skilled workers. This encouraged Carnegie's company to lock out its union workforce at Homestead, Pennsylvania, in 1892, so that it could save money by employing cheap replacement workers.

The other innovation that made skyscrapers possible was the electric elevator. Elisha Graves Otis designed the first reliable elevator in 1857. With electric power, it became possible to rise 60 stories in a matter of seconds. Before the elevator, rental spaces in commercial buildings cost more on lower floors because people didn't want to have to walk up stairs to get to the top. With elevators, tenants willing paid a premium in order to get better views out of their windows. Without elevators, nobody would have bothered to erect a building taller than five stories (Misa 1999, 2016).

The construction of skyscrapers was itself a terrific example of the industrial age coordination of labor and materials distribution. Steel skeletons meant that the unornamented higher sections of a building could be worked on even before the inevitable elaborate ornamental fringes on the lower part of the building were finished. This saved both time and money. When New York got so crowded that there was no space to store raw materials, the appearance of those materials would be carefully choreographed, and they would be taken directly off of flatbed trucks and placed in their exact positions near the tops of new buildings. Around the turn of the twentieth century, a major skyscraper could be built in as little as one year. The faster a building could be built, the faster an owner could collect rents and begin to earn back construction expenses.

The great benefit of skyscrapers was the ability to compress economic activity into smaller areas. "The skyscraper," explained one New Yorker in 1897, "gathers into a single edifice an extraordinary number of activities, which otherwise would be widely separated. Each building is an almost complete city, often comprising within its walls, banks and insurance offices, post office and telegraph office, business exchanges restaurants, clubrooms and shops." These same miniature cities also included numerous retail outlets, where the products of industrialized manufacturing could be purchased (Rees 2013). Shorter distances between these locations accelerated the pace of economic activity, which promoted further economic growth. However, large projects (like the many skyscrapers associated with the building of New York's Grand Central Station) eliminated or at least obscured urban industrial areas.

Unburdened by the need to pay federal income tax, industrial titans from across the United States displayed their massive wealth by building lavish mansions along New York's Fifth Avenue during the 1890s. By the 1920s, the value of land in Manhattan grew so fast because of its possible use for skyscrapers that second-generation industrial families sold their mansions, since they no longer wanted to pay huge property taxes on them. Blocks of what was known as "Vanderbilt Alley," named after the children of the steamship and railroad pioneer who had built mansions in the same area, were replaced by skyscrapers and high-end retail emporiums.

The same basic principles of skyscraper production – build it quick and large, and pack it with people – motivated the way that builders produced other kinds of urban domiciles. "Today, three-fourths of [New York City's] people live in the tenements," wrote the reformer Jacob Riis in his 1890 classic, *How the Other Half Lives*, "and the nineteenth-century drift of the population to the cities is sending ever-increasing multitudes to crowd them" (Riis 1914). The best-known tenement house design of this period was the dumbbell tenement of about five or six stories tall. They came about as the result of a design contest but were generally so crowded that they did more harm than good to the people who lived in them. Four families might live on a single floor with only two bathrooms between them. Designed to let light and air into central courtyards (which explains why they were shaped like a dumbbell from above), stacked up back-to-back, one against the other they did neither. Widely copied, New York City actually outlawed this design for new buildings in 1901 – but the old structures remained.

Apartment houses made it easier to pack people into small urban areas and therefore live closer to where they worked. Wealthy people could buy space and separation from one's neighbors, while those middle-class people who could not afford to live in suburbs lost the space they had before urbanization accelerated. To counter these unequal tendencies, New Yorkers developed the idea of the cooperative, where many people bought a single building and managed it themselves. Lavish apartments became alternatives for mansions once Manhattan real estate became too expensive for all except those with huge fortunes.

2.4.4 The Assembly Line

The farther away that people lived from central business districts, the more they needed efficient transportation. Streetcars helped, to an extent, but passenger lines that centered on downtown neighborhoods left large areas that could be occupied with housing for a growing working population, provided that these residents had their own way to get around. "I will build a car for the great multitude," declared Henry Ford in 1908. "It will be so low in price that no man making a good salary will be unable to own one" (Watts 2005). That car was the Model T, and it revolutionized both auto-making and the American landscape. It also revolutionized the entire concept of American production. Ford did not worry about whether his cars would have a market. He would make a market for his cars by producing them so cheaply that nearly every American could afford one.

Ford could achieve both quality and a low price at scale because of the assembly line. This particular conceptual breakthrough owed much to the "disassembly lines" that had been pioneered in the meatpacking industry during the previous century. In the same way that a single carcass was picked apart by men with specialized jobs as it moved along a line, mounted upon a hook, Ford arranged his new factory at Highland Park so that men with highly specialized assignments could build an automobile much faster than before. The assembly line moved work to the men rather than forcing men to move to the work, thereby saving valuable time and energy. It also extended the concept of the division of labor to its logical extreme so that workers would perform only one function in a much larger assembly process all day, every day. The applicability of these principles to the manufacturing of just about everything is what made

Ford such an important figure in the history of industrialization. Mass production became possible for all kinds of things that had once seemed far removed from the automobile.

Ford built Model T's at three different facilities over the entire history of that vehicle. He improved his production methods over time (which included introducing and improving upon the assembly line) so that he could produce them more cheaply and efficiently. Efficiency depended on speed, and speed depended upon the exact place in the factory where those machines were placed. Because Ford made only one car, he could employ single-purpose machine tools of extraordinarily high quality. The company also used lots of other automated manufacturing equipment, like gravity slides and conveyors, to get parts of the car from one place to another in its increasingly large, increasingly mechanized factories.

Because the assembly line moved the work to the men rather than the men to the work, the company could control the speed of the entire operation. Like earlier manufacturers, Ford depended upon standardized, identical parts to produce more cars for less, but the assembly line also made it possible to conserve labor – not by mechanizing jobs that had once been done by hand, but by mechanizing work processes and paying employees just to feed and tend to those machines. This was not fun work to do. "The chain system you have is a slave driver!" wrote an anonymous housewife based on her husband's experience working on the assembly line. "My God! Mr. Ford. My husband has come home and thrown himself down and won't eat his supper – so done out! Can't it be remedied"? Ford instituted an unprecedented wage of $5/day to keep workers on his assembly line, but this reward did not make the work any easier (Hounshell 1984).

Before Ford came along, cars were boutique goods that only rich people could afford to operate. After Ford introduced the assembly line (actually a series of assembly lines for every part of the car), labor productivity improved to such a degree that mass production became possible. Perhaps more important than mass production was mass consumption, since continual productivity improvements meant that Ford could lower the price of the Model T every year, while simultaneously making small but significant changes that steadily improved the quality of the car. Mass production eliminated choice, since Ford produced no other car, but Ford built variations of the Model T, like the runabout with the same chassis, and owners retro-fitted their Model T's for everything from camping to farming.

The increased number of automobiles on city streets further congested already congested downtown areas. Streetcars got blocked. Pedestrians died in gruesome traffic accidents. One of the basic requirements of having so many new cars on the roads was to improve the quality and quantity of roads. Local city planners tended to attack such problems on a case-by-case basis, laying pavement on well-traveled roads and widening them when appropriate. New traffic rules, such as the first one-way streets, appeared in an effort to alleviate these kinds of problems. Traffic control towers and traffic lights – the mechanical solution to a problem inspired by industrialization – also appeared for the first time during this era.

Cities grew when industries grew during this era. Since people had to live near where they worked (and few people lived in skyscrapers), many builders built out into undeveloped areas. If a city had annexed much of the land around it previous to these economic expansions (like Detroit), those areas became parts of a larger city. If they hadn't, much of this growth occurred in new suburbs (like Philadelphia). Chicago was so confident of further growth during this period that it built streetcar lines into vacant fields. To meet rising demand for housing, homebuilders applied industrial principles to building – using standardized parts that were themselves the result of mass production techniques. By the 1920s, buying precut mail order houses became big business.

2.4.5 The Origins of Mass Production

After 1880, mechanization made factories even more productive thanks to technological improvements. This can be traced back to Thomas Edison's labs in New Jersey, where he practiced systematic invention to exploit the great commercial opportunities that modern life created. The electrical and chemical industries formed the vanguard for the blending of science and the useful arts during this era. By the 1920s, engineers had been formally integrated into the management hierarchies of countless American industries.

Reorganization of production merged with technological improvement had made mass production possible long before Ford developed the assembly line. James Bonsack's cigarette rolling machine, for example, patented in 1881, could produce 70 000 cigarettes in a single 10-hour day. By the end of that decade, it could produce 120 000 cigarettes in a day (Chandler 1977). When James "Buck" Duke bought exclusive rights to this machine in 1885, it became the basis of his American Tobacco Company, which quickly controlled most of the industry.

By the 1920s, mass production had arrived in industries that produced goods that were much more expensive than cigarettes. Ford's principles of mass production spread quickly throughout the manufacturing sector, to products of all kinds, because Henry Ford was so open about the way he designed his factories. Among the other manufacturers that used Ford's principles during the 1920s were the makers of home appliances, like refrigerators and radios. General Electric, for example, built an $18 million assembly line for its Monitor Top refrigerator and sold 1 million refrigerators just 4 years after its introduction in 1927 (Cowan 1985).

Even craft-dominated industries like furniture making came to depend upon mass production to make their products more available to the masses. People who moved from farms to cities desperately needed furniture for their new urban residences, but in industrial towns like Grand Rapids, Michigan, they could not afford pieces made by craftsman. New mass-produced models made with minimal carving and overlays, based on stylish patterns, found a market all over the country. It helped that companies like Bassett, founded in Virginia in 1902, discouraged their workers from forming unions, just like Ford did. An unorganized workforce made it easier for industrialists to impose changes in the production process without resistance from employees.

The changeover from the Model T to the Model A, in 1927, demonstrated the limits of industrialized mass production. The Model A was incredibly expensive, and Ford had to shut his main plant for months to retool the production line for his new models. While the new car sold well initially, sales dropped precipitously as the Depression deepened. "Mass production is not simply large-scale production," wrote the department store magnate Edward Filene, in 1932. "It is large-scale production based upon a clear understanding that increased production demands increased buying" (Hounshell 1984). Mass buying became difficult when people had little money with which to buy the products of industrialization. Urban building slowed precipitously during the Depression too. Since cities were the focal points of industrialization, urban citizens suffered disproportionately when production waned. Of course, when the United States sank into the economic downturn of the Great Depression, both urban and industrial growth decreased sharply.

2.5 Important Technological Developments

The commencement of the Industrial Revolution is closely linked to a small number of innovations, beginning in the second half of the eighteenth century (Bond et al. 2003). By the 1830s the following gains had been made in important technologies:

- Textiles – Mechanized cotton spinning powered by steam or water increased the output of a worker by a factor of 500. The power loom increased the output of a worker by a factor of over 40 (Ayres 1989). The cotton gin increased productivity of removing seed from cotton by a factor of 50 (Wickham 1916). Large gains in productivity also occurred in spinning and weaving of wool and linen, but they were not as great as in cotton (Beckert 2014; Landes 1969).
- Steam power – The efficiency of steam engines increased so that they used between one-fifth and one-tenth as much fuel. The adaptation of stationary steam engines to rotary motion made them suitable for industrial uses (Landes 1969). The high-pressure engine had a high power to weight ratio, making it suitable for transportation. Steam power underwent a rapid expansion after 1800.
- Iron making – The substitution of coke for charcoal greatly lowered the fuel cost for pig iron and wrought iron production (Landes 1969). Using coke also allowed larger blast furnaces, resulting in economies of scale (Landes 1969; Rosen 2012). The cast iron blowing cylinder was first used in 1760. It was later improved by making it double acting, which allowed higher blast furnace temperatures. The puddling process produced a structural grade iron at a lower cost than the finery forge (Landes 1969). The rolling mill was 15 times faster than hammering wrought iron. Hot blast greatly increased fuel efficiency in iron production in the following decades.
- Invention of machine tools – The first machine tools were invented. These included the screw cutting lathe, cylinder boring machine, and the milling machine (Hounshell 1984).

As the nineteenth century was drawing to a close, three very special individuals made their entrance on the US national stage. Gifford Pinchot, John Muir, and Theodore Roosevelt were to write the first pages of modern environmental history in the United States, which in turn led to the birth of the modern environmental movement early in the twentieth century. However, pollution and environmental degradation was a fact of life across most of America during the first half of the twentieth century, and phrases such as "the smell of money," "good, clean soot," "God bless it," "it's our life-blood," and "an index to local activity and enterprise" were used to describe air pollution.

At this point of time, muscle and animal power were replaced with electricity, internal-combustion engines, and nuclear reactors. At the same time, industry was consuming natural resources at an incredible rate. All of these events began to escalate at a dangerous rate after World War II. Soon after, in the late summer of 1962, a marine biologist named Rachael Carson, author of *Silent Spring*, the best-selling book about ocean life, opened the eyes of the world to the dangers of attacking the environment (Carson 1962). It was perhaps at this point that America began calling in earnest for reform of the destruction of nature and constraints on environment laws that addressed these issues. It all began in 1970 with the birth of the EPA. For additional literature regarding Early History of the American Environmental Movement and American technology, the interested readers are referred to the book by Philip Shabecoff, titled *A Fierce Green Fire* (1993). This outstanding book, as well as Ponting's *A Green History of the World* (1991), and Ruth Cowan's book, titled *A Social History of American Technology* (1985), is a "must" for anyone who works in or has interests in the environment.

Industry is the axis to gear up the economy of a modern society – known as the indispensable motor of growth and development. On the other hand, it has been identified as a major source of environmental degradation and pollution. Therefore, development without destruction and environmental sustainability and sustainable development are the urgent needs of our time. The problem we are facing is how to strike a balance between the benefits of rising living standard and its cost in terms of deterioration of the physical environment and quality of life. In the past, the danger of polluting air, water, and land was not fully recognized, but now there is no doubt that it is a matter of great concern.

Famous Minamata Disease in Japan (1956), Flixborough (1974), Love Canal (1978), Three Mile Island incident (1979), Bhopal gas tragedy (1984), Chernobyl Atomic Reactor accident (1986), and Tennessee Valley Authority Kingston Coal Power Plant Toxic Ash Spill in Emory River (2008) have reminded us that industrialization has posed a serious threat not only to humans but also to animals, aquatic life, and vegetation cover. On one hand, industrialization has helped us to raise the standard and quality of life, but on the other it has deteriorated our environment. Thus, pollutants fate and transport in environment through human activities, e.g. acid rain, smog, global warming, ocean acidification, wild fires, cancer, are worst possible forms of pollution which is a direct result of industrialization.

Industries degrade the environment and pollute in the following ways:

a) Use of natural resources by industries, as it destroys nature and affects natural environment. Wheat, rice, barley, corn, cotton, trees, plywood, rubber, sugar cane, iron, coal, oil, natural gas, etc. are all natural resources for food processing, packing, paper, clothes, and other finished products. Thus, increasing needs of industries have resulted in over exploitation and stress on natural resources.
b) Residues and by-products of industries are released in water, air, and land with or without any treatment which pollutes the water, air, and land, affecting the air quality, aquatic life, and ground water.
c) Fossil fuel used by industries like coal, kerosene, diesel, and nuclear energy pollutes the air in the form of smokes, soot, small particulate matter, smog, ozone, and radioactive wastes.
d) Noise is also a major by-product of industries that cause noise pollution to human health.
e) Industrial wastes, particularly hazardous wastes and radioactive wastes, have become a major environmental pollution problem.

2.6 Industrial Disasters

2.6.1 Bhopal: The World's Worst Industrial Tragedy

Thirty three years ago, on the night of 2 December 1984, an accident at the Union Carbide pesticide plant in Bhopal, India, released at least 30 T of a highly toxic gas called methyl isocyanate (MIC), as well as a number of other poisonous gases. The pesticide plant was surrounded by shanty towns, leading to more than 600 000 people being exposed to the deadly gas cloud that night. The gases stayed low to the ground, causing victims' throats and eyes to burn, inducing nausea, and many deaths. Estimates of the death toll vary from as few as 3 800 to as many as 16 000, but government figures now refer to an estimate of 15 000 killed over the years. Toxic material remains, and 30 years later, many of those who were exposed to the gas have given birth to physically and mentally disabled children. For decades, survivors have been fighting to have the site cleaned up, but they say the efforts were slowed when Michigan-based Dow Chemical took over Union Carbide in 2001. Human rights groups say that thousands of tons of hazardous waste remain buried underground, and the government has conceded the area is contaminated. There has, however, been no long-term epidemiological research which conclusively proves that birth defects are directly related to the drinking of the contaminated water.

2.6.1.1 What Happened that Evening!

Due to lack of environmental regulations, enforcement and compliance, maintenance and operation safety, the plant in Bhopal where the disaster happened started to produce "Carbaryl" in 1977. Carbaryl is mainly used as an insecticide. At first, the production was 2500 T/Y. This was no problem, as the plant had been designed for an output of 5000 T. At the beginning of the 1980s, Carbaryl did not sell very well. For this reason, the owners of the plant started to cut costs. This included employing fewer people, doing maintenance less frequently, and using parts that were made of lower-grade steel. Closing the plant was being considered as well. When the disaster happened, there was no production at the plant because there was a surplus on the market.

The disaster happened because water entered a tank containing *MIC*. This caused a chemical reaction which resulted in the buildup of much carbon dioxide, among other things. The resulting reaction increased the temperature inside the tank to reach over 200 °C (392 °F). The pressure was more than the tank was built to withstand. The tank had valves to control the pressure. These were triggered in an emergency, which reduced the pressure. As a result, large amounts of toxic gases were released into the environment. The pipes were rusty. The rust in the iron pipes made the reaction faster. All the contents of the tank were released within a period of about two hours. The water had entered the tank because of a sequence of events. The tank had been maintained badly. When cleaning work was done, water could enter the tank. The leakage of MIC gas from Union Carbide Corporation, Bhopal, gave impetus to the development of environmental law and principles of quantum of compensation (*Union Carbide v. Union of India* 1989).

2.6.1.2 Taj Mahal Acid Rain Attack

Yellowing of a historical monument, the Taj Mahal at Agra, was attacked by acid gases due to emissions of oxides of sulfur (SO_x) from foundries, coal-fired power plants, chemical and hazardous industries, and oil refinery. The sulfur dioxide emitted from these industries, combined with atmospheric oxygen in presence of moisture and sun, formed sulfuric acid called "acid rain" affecting the marble of the Taj Mahal (Mehta 1987).

2.6.1.3 River Ganges and River Yamuna

The industries which made the water of the holy River Ganges and a river of the south Chennai toxic were found to be tanneries (Mehta 1988; Vellore Citizen 1996). In the Ganges pollution case, tanneries discharged untreated effluents in the river, and near Kanpur the water of Ganges was found to be highly toxic. In the other case, the Pallar River of the state of Tamil Nadu became highly polluted because tanneries discharged chemicals used in treating leather, which resulted in nonavailability of potable water. Recently, the Supreme Court of India ordered the closure of industries or to shift them from the territory of the State of Delhi as their untreated effluent and sludge was polluting the holy River Yamuna (Hindustan Times 2000; Times of India 2000).

2.6.1.4 Flixborough

On Saturday, 1 June 1974, the Nypro (UK) site at Flixborough was severely damaged by a large explosion. Twenty-eight workers were killed and a further 36 suffered injuries. It is recognized that the number of casualties would have been more if the incident had occurred on a weekday, as the main office block was not occupied. Offsite consequences resulted in 53 reported injuries. Property in the surrounding area was damaged to a varying degree.

The chemical plant was designed to produce 70 000 T/Y of caprolactam, a raw material for the production of nylon. The process used cyclohexane as a feed and oxidized it to cyclohexanol in the presence of air within a series of six catalytic reactors. Under process conditions, cyclohexane vaporizes immediately upon mixed depressurization, forming a cloud of flammable cyclohexane vapor mixed with air. Reactor 5 was found to have a small crack in the stainless steel structure in the series using a 20 in. pipe, even though the reactors are normally connected using 28 in. pipe. The temporary section of piping was not properly supported and it ruptured upon pressurization, releasing an estimated 30 T of cyclohexane in a large cloud. An unknown ignition source caused the cloud to explode, leveling the entire plant facility. The resulting fire in the plant burned for over 10 days. The accident could have been prevented by following proper safety design and operating procedures, including reducing the inventory of flammable liquids onsite (CCPS 1993; Crowl and Louver 1990).

2.6.1.5 Love Canal Tragedy

Quite simply, Love Canal is one of the most appalling environmental tragedies in American history. But that's not the most disturbing fact. What is worse is that it cannot be regarded as an isolated event. It could happen again – anywhere in the United States – unless we move expeditiously to prevent it.

It is a cruel irony that Love Canal was originally meant to be a dream community. That vision belonged to the man for whom the three-block tract of land on the eastern edge of Niagara Falls, New York, was named – William T. Love. Love felt that by digging a short canal between the upper and lower Niagara Rivers, power could be generated cheaply to fuel the industry and homes of his would-be model city.

But despite considerable backing, Love's project was unable to endure the one-two punch of fluctuations in the economy and Nikola Tesla's discovery of how to economically transmit electricity over great distances by means of an alternating current. By 1910, the dream was shattered. All that was left to commemorate Love's hope was a partial ditch where construction of the canal had begun. In the

1920s the seeds of a genuine nightmare were planted. The canal was turned into a municipal and industrial chemical dumpsite.

Landfills can of course be an environmentally acceptable method of hazardous waste disposal, assuming they are properly sited, managed, and regulated. Love Canal will always remain a perfect historical example of how not to run such an operation. In 1953, the Hooker Chemical Company, then the owners and operators of the property, covered the canal with earth and sold it to the city for one dollar. It was a bad buy. In the late 1950s, about 100 homes and a school were built at the site. Perhaps it wasn't William T. Love's model city, but it was a solid, working-class community. On the first day of August 1978, the lead paragraph of a front-page story in the New York Times read: "Niagara Falls, N.Y. – Twenty five years after the Hooker Chemical Company stopped using the Love Canal here as an industrial dump, 82 different compounds, 11 of them suspected carcinogens, have been percolating upward through the soil, their drum containers rotting and leaching their contents into the backyards and basements of 100 homes and a public school built on the banks of the canal."

In an article prepared for the February 1978 *EPA Journal*, I wrote that, regarding chemical dumpsites in general, "even though some of these landfills have been closed down, they may stand like ticking time bombs." Just months later, Love Canal exploded. The explosion was triggered by a record amount of rainfall. Shortly thereafter, the leaching began.

Corroding waste-disposal drums could be seen breaking up through the grounds of backyards. Trees and gardens were turning black and dying. One entire swimming pool had been popped up from its foundation, afloat now on a small sea of chemicals. Puddles of noxious substances were pointed out to me by the residents. Some of these puddles were in their yards, some were in their basements, others yet were on the school grounds. Everywhere the air had a faint, choking smell. Children returned from play with burns on their hands and faces.

And then there were the birth defects. The New York State Health Department is continuing an investigation into a disturbingly high rate of miscarriages, along with five birth-defect cases detected thus far in the area. The father of one the children with birth defects said, "I heard someone from the press saying that there were *only* five cases of birth defects here," he told me. "When you go back to your people at EPA, please don't use the phrase '*only* five cases'. People must realize that this is a tiny community. Five birth defect cases here is terrifying."

A large percentage of people in Love Canal are also being closely observed because of detected high white-blood-cell counts, a possible precursor of leukemia. When the citizens of Love Canal were finally evacuated from their homes and their neighborhood, pregnant women and infants were deliberately among the first to be taken out.

"We knew they put chemicals into the canal and filled it over," said one woman, a long-time resident of the Canal area, "but we had no idea the chemicals would invade our homes. We're worried sick about the grandchildren and their children." Two of this woman's four grandchildren have birth defects. The children were born and raised in the Love Canal community. A granddaughter was born deaf with a cleft palate, an extra row of teeth, and slight retardation. A grandson was born with an eye defect.

Of the chemicals which comprise the brew seeping through the ground and into homes at Love Canal, one of the most prevalent is benzene – a known human carcinogen, and one detected in high concentrations. But the residents characterize things more simply. "I've got this slop everywhere," said another man who lives at Love Canal. His daughter also suffers from a congenital defect.

On 7 August, New York Governor Hugh Carey announced to the residents of the Canal that the State Government would purchase the homes affected by chemicals. On that same day, President Carter approved emergency financial aid for the Love Canal area (the first emergency funds ever to be approved for something other than a "natural" disaster), and the U.S. Senate approved a "sense of Congress" amendment saying that Federal aid should be forthcoming to relieve the serious environmental disaster which had occurred.

By the month's end, 98 families had already been evacuated. Another 46 had found temporary housing. Soon after, all families would be gone from the most contaminated areas – a total of 221 families have moved or agreed to be moved. State figures show more than 200 purchase offers for homes have been made, totaling nearly $7 million.

A plan is being set in motion now to implement technical procedures designed to meet the seemingly impossible job of detoxifying the Canal area. The plan calls for a trench system to drain chemicals from the Canal. It is a difficult procedure, and we are keeping our fingers crossed that it will yield some degree of success. I have been very pleased with the high degree of cooperation in this case among local, State, and Federal governments, and with the swiftness by which the Congress and the President have acted to make funds available.

But this is not really where the story ends. Quite the contrary. We suspect that there are hundreds of such chemical dumpsites across United States. Unlike Love Canal, few are situated so close to human settlements. But without a doubt, many of these old dumpsites are time bombs with burning fuses – their contents slowly leaching out. And the next victim cold be a water supply, or a sensitive wetland.

The presence of various types of toxic substances in our environment has become increasingly widespread – a fact that President Carter has called "one of the grimmest discoveries of the modern era."

Chemical sales in the United States now exceed a mind-boggling $112 billion/year, with as many as 70 000 chemical substances in commerce. Love Canal can now be added to a growing list of environmental disasters involving toxics, ranging from industrial workers stricken by nervous disorders and cancers to the discovery of toxic materials in the milk of nursing mothers.

Through the national environmental program it administers, the EPA is attempting to draw a chain of Congressional acts around the toxics problem. The Clean Air and Water Acts, the Safe Drinking Water Act, the Pesticide Act, the Resource Conservation and Recovery Act, the Toxic Substances Control Act – each is an essential link.

Under the Resource Conservation and Recovery Act, EPA is making grants available to States to help them establish programs to assure the safe handling and disposal of hazardous wastes. As guidance for such programs, we are working to make sure that State inventories of industrial waste disposal sites include full assessments of any potential dangers created by these sites.

Also, USEPA recently proposed a system to ensure that more than 35 million T of hazardous wastes produced in the United States each year, including most chemical wastes, are disposed of safely. Hazardous wastes will be controlled from point of generation to their ultimate disposal, and dangerous practices now resulting in serious threats to health and environment will not be allowed.

Although we are taking these aggressive strides to make sure that hazardous waste is safely managed, there remains the question of liability regarding accidents occurring from wastes disposed of previously. This is a missing link. But no doubt this question will be addressed effectively in the future. Regarding the missing link of liability, if health-related dangers are detected, what are we as a people willing to spend to correct the situation? How much risk are we willing to accept? Who's going to pick up the tab? One of the chief problems we are up against is that ownership of these sites frequently shifts over the years, making liability difficult to determine in cases of an accident. And no secure mechanisms are in effect for determining such liability.

It is within our power to exercise intelligent and effective controls designed to significantly cut such environmental risks. A tragedy, unfortunately, has now called upon us to decide on the overall level of commitment we desire for defusing future Love Canals. And it is not forgotten that no one has paid more dearly already than the residents of Love Canal.

2.6.1.6 Tennessee Valley Authority Kingston Coal Power Plant Toxic Ash Sludge Spill

On 22 December 2008, a retention pond wall collapsed at Tennessee Valley Authority's (TVA) Kingston plant in Harriman, Tennessee, releasing a combination of water and fly ash that flooded 12 homes, spilled into nearby Watts Bar Lake, contaminated the Emory River, and caused a train wreck. Officials said 4–6 ft of material escaped from the pond to cover an estimated 400 acres of adjacent land. A train bringing coal to the plant became stuck when it was unable to stop before reaching the flooded tracks (White 2008). Hundreds of fish were floating dead downstream from the plant. Water tests showed elevated levels of lead and thallium (Knoxville News Sentinel 2008a, b).

Originally, TVA estimated that 1.7 million cubic yards of waste had burst through the storage facility. Company officials said the pond had contained a total of about 2.6 million cubic yards of sludge. However, the company revised its estimates on 26 December, when it released an aerial survey showing that 5.4 million cubic yards (1.09 billion gal) of fly ash was released from the storage facility (Knoxville News Sentinel 2008a). Several days later, the estimate was increased to over 1 billion gal spilled (CNN 2008). The size of the spill was larger than the amount TVA claimed to have been in the pond before the accident, a discrepancy that TVA was unable to explain (New York Times 2008). The TVA spill was 100 times larger than the Exxon Valdez spill in Alaska, which released 10.9 million gal of crude oil (Encyclopedia of the Earth 2018), and it was expected to take weeks and cost tens of millions of dollars to clean it (Knoxville News Sentinel 2008c). According to the TVA, rain totaling 6 in. in 10 days and 12 °F temperatures were factors that contributed to the failure of the earthen embankment (Valley Precipitation 2008).

The 40-acre pond was used to contain ash created by the coal-burning plant (White 2008). The water and ash that were released in the accident were filled with toxic substances. Each year coal preparation creates waste containing an estimated 13 T of mercury, 3236 T of arsenic, 189 T of beryllium, 251 T of cadmium, and 2754 T of nickel, and 1098 T of selenium (Associated Press 2008; Valley Precipitation 2008).

2.6.1.7 Cuyahoga River Fire

The Cuyahoga River is in the United States, located in Northeast Ohio, that feeds into Lake Erie. The river is famous for having been so polluted that it "caught fire" in 1969 (Figure 2.3). It was the disaster that ignited an environmental revolution. On that day, 22 June 1969, the Cuyahoga River burst into flames in Cleveland when sparks from a passing train set fire to oil-soaked debris

Figure 2.3 Cuyahoga River caught fire.

floating on the water's surface. By 1969, the Cuyahoga River was not a unique experience in the United States. A river flowing into Baltimore, Maryland, caught fire on 8 June 1906 (CPD 1926). In Philadelphia, the Schuykill burned in the 1950s (Kernan 1958). The Buffalo River in upper New York state burned in the 1960s (UPI 1984). The Rouge River in Dearborn, Michigan, repeatedly burned (US 1974).

So, why is the Cuyahoga River fire a seminal event in the history of water pollution control in the United States? Because it was a catalyst for change in federal government's role in water pollution control. Although the federal government had powerful tools to control water pollution, for example, the River and Harbors Act of 1899 and the Water Quality Act of 1965. States and cities were left to fend for themselves. The flaming Cuyahoga became a figurehead for America's mounting environmental issues and sparked wide-ranging reforms, including the passage of the Clean Water Act (CWA) (1972) and the creation of federal and state environmental protection agencies.

But the episode itself did not quite live up to its billing. It was not the first fire, or even the worst, on the Cuyahoga, which had lit up at least a dozen other times before. And industrial dumping was already improving by the time of the 1969 blaze. The reality is that the 1969 Cuyahoga fire was not a symbol of how bad conditions on the nation's rivers could become, but how bad they had once been. The 1969 fire was not the first time an industrial river in the United States had caught on fire, but the last. The event helped to spur the environmental movement in the United States (Adler 2003).

2.6.1.8 The Great Smog of London

Great Smog of 1952 was a severe air-pollution event that affected the British capital of London in early December 1952. A period of cold weather, combined with an anticyclone and windless conditions, collected airborne pollutants – mostly arising from the use of coal – to form a thick layer of smog over the city. It lasted from Friday, 5 December to Tuesday, 9 December 1952 and then dispersed quickly when the weather changed.

Figure 2.4 Nelson Tower showing the poor visibility.

Figure 2.5 Source of pollution from Battersea Coal Power Plant, London.

It caused major disruption by reducing visibility and even penetrating indoor areas, far more severe than previous smog events experienced in the past, called "pea-soupers." Government medical reports in the following weeks, however, estimated that up until 8 December, 4 000 people had died as a direct result of the smog and 100 000 more were made ill by the smog's effects on the human respiratory tract. More recent research suggests that the total number of fatalities was considerably greater, about 12 000 (Figures 2.4 and 2.5) (The Great Smog of 1952 2014).

The Prime Minister at that time, Winston Churchill, was adamant that it would pass, simply dismissing it as a "weather event." London had suffered since the thirteenth century from poor air quality (Brimblecombe 1976), which worsened in the 1600s (The Observer 2002), but the Great Smog is known to be the worst air-pollution event in the history of the United Kingdom, and the most significant in terms of its effect on environmental research, government regulation, and public awareness of the relationship between air quality and health (Bell et al. 2004; The Observer 2002). It led to several changes in practices and regulations, including the Clean Air Act 1956.

2.7 Environmental Law

Environmental law, also known as environmental and natural resources law, is a collective term describing the network of treaties, statutes, regulations, common, and customary laws addressing the effects of human activity on the natural environment. The core environmental law regimes address environmental pollution. A related but distinct set of regulatory regimes, now strongly influenced by environmental legal principles, focus on the management of specific natural resources, such as forests, minerals, or fisheries. Other areas, such as environmental impact assessment, may not fit neatly into either category but are nonetheless important components of environmental law.

In many parts of Eastern Europe, the former Soviet Union, and the developing countries, in Asia, Africa, and South America, pollution conditions persist today. Global and regional environmental issues are increasingly the subject of international law. The acronym ISO stands for *International Organization for Standardization*. It is a worldwide program that was founded in 1947 to promote the development of international manufacturing, trade, and communication standards. ISP membership includes over 100 countries.

2.7.1 History of Environmental Law

Early examples of legal enactments designed to consciously preserve the environment, for its own sake or human enjoyment, are found throughout history. In the common law, the primary protection was found in the law of nuisance, but this only allowed for private actions for damages or injunctions if there was harm to land. Thus, smells emanating from pig sties, (*Aldred's Case* 1610), strict liability against dumping rubbish (*R v. Stephens* 1866), or damage from exploding dams (*Rylands v. Fletcher*, 1868). Private enforcement, however, was limited and found to be woefully inadequate to deal with major environmental threats, particularly threats to common resources. During the "Great Stink of 1858, the dumping of sewerage into the River Thames began to smell so ghastly in the summer heat that Parliament had to be evacuated. Ironically, the Metropolitan Commission of Sewers Act 1848 had allowed the Metropolitan Commission for Sewers to close cesspits around the city in an attempt to "clean up," but this simply led people to pollute the river. In 19 days, Parliament passed a further Act to build the London sewerage system. London also suffered from terrible air pollution, and this culminated in the "Great Smog" of 1952, which in turn triggered its own legislative response: the Clean Air Act 1956. The basic regulatory structure was to set limits on emissions for households and business (particularly burning coal) while an inspectorate would enforce compliance.

Notwithstanding early analogs, the concept of "environmental law" as a separate and distinct body of law is a twentieth-century development (Lazarus 2006). The recognition that the natural environment was fragile and in need of special legal protections, the translation of that recognition into legal structures, the development of those structures into a larger body of "environmental law," and the strong influence of environmental law on natural resource laws did not occur until about the 1960s. At that time, numerous influences – including a growing awareness of the unity and fragility of the biosphere; increased public concern over the impact of industrial activity on natural resources and human health; the increasing strength of the regulatory state; and more broadly the advent and success of environmentalism as a political movement -coalesced to produce a huge new body of law in a relatively short period of time. While the modern history of environmental law is one of continuing controversy, by the end of the twentieth century environmental law had been established as a component of the legal landscape in all developed nations of the world, many developing ones, and the larger project of international law.

2.8 Pollution Control Laws

2.8.1 Air Quality Law

Industrial air pollution, now regulated by air quality law:

Air quality laws govern the emission of air pollutants into the atmosphere. A specialized subset of air quality laws regulate the quality of air inside buildings. Air quality laws are often designed specifically to protect human health by limiting or eliminating airborne pollutant concentrations. Other initiatives are designed to address broader ecological problems, such as limitations on chemicals that affect the ozone layer, and emissions trading programs to address acid rain or climate change. Regulatory efforts include identifying and categorizing air pollutants, setting limits on acceptable emissions levels, and dictating necessary or appropriate mitigation technologies.

2.8.2 Water Quality Law

A typical stormwater outfall, subject to water quality law:

Water quality laws govern the release of pollutants into water resources, including surface water, ground water, and stored drinking water. Some water quality laws, such as drinking water regulations, may be designed solely with reference to human health. Many others, including restrictions on the alteration of the chemical, physical, radiological, and biological characteristics of water resources, may also reflect efforts to protect aquatic ecosystems more broadly. Regulatory efforts may include identifying and categorizing water pollutants, dictating acceptable pollutant concentrations in water resources, and limiting pollutant discharges from effluent sources. Regulatory areas include sewage treatment and disposal, industrial and agricultural waste water management, and control of surface runoff from construction sites and urban environments.

2.8.3 Waste Management Law

A municipal landfill, operated pursuant to waste management law:

Waste management laws govern the transport, treatment, storage, and disposal of all manner of waste, including municipal solid waste, hazardous waste, and nuclear waste, among many other types. Waste laws are generally designed to minimize or eliminate the uncontrolled dispersal of waste materials into the environment in a manner that may cause ecological or biological harm and include laws designed to reduce the generation of waste and promote or mandate waste recycling. Regulatory efforts include identifying and categorizing waste types and mandating transport, treatment, storage, and disposal practices.

2.8.4 Contaminant Cleanup Law

Oil spill emergency response, governed by environmental cleanup law:

Environmental cleanup laws govern the removal of pollution or contaminants from environmental media, such as soil, sediment, surface water, or ground water. Unlike pollution control laws, cleanup laws are designed to respond after-the-fact to environmental contamination, and consequently must often define not only the necessary response actions but also the parties who may be responsible for undertaking (or paying for) such actions. Regulatory requirements may include rules for emergency response, liability allocation, site assessment, remedial investigation, feasibility studies, remedial action, post-remedial monitoring, and site reuse.

2.8.5 Chemical Safety Laws

Chemical safety laws govern the use of chemicals in human activities, particularly man-made chemicals in modern industrial applications. As contrasted with media-oriented environmental laws (e.g. air or water quality laws), chemical control laws seek to manage the (potential) pollutants themselves. Regulatory efforts include banning specific chemical constituents in consumer products (e.g. Bisphenol A in plastic bottles) and regulating pesticides.

2.8.6 Water Resources Law

An irrigation ditch, operated in accordance with water resources law:

Water resources laws govern the ownership and use of water resources, including surface water and ground water. Regulatory areas may include water conservation, use restrictions, and ownership regimes.

2.8.7 Mineral Resources Law

Mineral resource laws cover several basic topics, including the ownership of the mineral resource and who can work them. Mining is also affected by various regulations regarding the health and safety of miners, as well as the environmental impact of mining.

2.8.8 Forest Resources Law

A timber operation, regulated by forestry law:

Forestry laws govern activities in designated forest lands, most commonly with respect to forest management and timber harvesting. Ancillary laws may regulate forest land acquisition and prescribed burn practices. Forest management laws generally adopt management policies, such as multiple use and sustained yield, by which public forest resources are to be managed. Governmental agencies are generally responsible for planning and implementing forestry laws on public forest lands and may be involved in forest inventory, planning, and conservation, and oversight of timber sales. Broader initiatives may seek to slow or reverse deforestation.

2.8.9 Wildlife and Plants Protection Laws

Wildlife laws govern the potential impact of human activity on wild animals, whether directly on individuals or populations, or indirectly via habitat degradation. Similar laws may operate to protect plant species. Such laws may be enacted entirely to protect biodiversity or as a means for protecting species deemed important for other reasons. Regulatory efforts may include the creation of special conservation statuses, prohibitions on killing, harming, or disturbing protected species, efforts to induce and support species recovery, establishment of wildlife refuges to support conservation, and prohibitions on trafficking in species or animal parts to combat poaching.

2.8.10 Fish and Game Laws

Fish and game laws regulate the right to pursue and take or kill certain kinds of fish and wild animal (game). Such laws may restrict the days to harvest fish or game, the number of animals caught per person, the species harvested, or the weapons or fishing gear used. Such laws may seek to balance dueling needs for preservation and harvest and to manage both environment and populations of fish and game. Game laws can provide a legal structure to collect license fees and other money which is used to fund conservation efforts as well as to obtain harvest information used in wildlife management practice.

2.8.11 Principles

Environmental law has developed in response to emerging awareness of and concern over issues impacting the entire world. While laws have developed piecemeal and for a variety of reasons, some effort has gone into identifying key concepts and guiding principles common to environmental law as a whole (UNEP 1992). The principles discussed below are not an exhaustive list and are not universally recognized or accepted. Nonetheless, they represent important principles for the understanding of environmental law around the world.

2.9 Resource Sustainability

2.9.1 Environmental Impact Assessment

Environmental impact assessment (EA) is the assessment of the environmental consequences both positive and negative of a plan, policy, program, or actual projects prior to the decision to move forward with the proposed action. In this context, the term "environmental impact assessment" is usually used when applied to actual projects by individuals or companies and the term "strategic environmental assessment" (SEA) applies to policies, plans, and programs most often proposed by organization of state (Eccleston 2017). Environmental assessments may be governed by rules of administrative procedure regarding public participation and documentation of decision making and may be subject to judicial review.

2.9.2 Sustainable Development

Defined by the United Nations Environment Program (UNEP 1986, 1992, 2001, 2006, 2013) as "development that meets the needs of the present without compromising the ability of future generations to meet their own needs," sustainable development may be considered together with the concepts of "integration" (development cannot be considered in isolation from sustainability) and "interdependence" (social and economic development, and environmental protection, are interdependent). Laws mandating EPA and requiring or encouraging development to minimize environmental impacts may be assessed against this principle.

The modern concept of sustainable development was a topic of discussion at the 1972 United Nations Conference on the Human Environment (Stockholm Conference) and the driving force behind the 1983 World Commission on Environment and Development (WCED, or Bruntland Commission). In 1992, the first UN Earth Summit resulted in the Rio Declaration, Principle 3 of which reads: "The right to development must be fulfilled so as to equitably meet developmental and environmental needs of present and future generations." Sustainable development has been a core concept of international environmental discussion ever since, including at the World Summit on Sustainable Development (Earth Summit 2002), and the United Nations Conference on Sustainable Development (Earth Summit 2012).

2.9.3 Equity

Further information: Intergenerational equity

Defined by UNEP to include intergenerational equity – "the right of future generations to enjoy a fair level of the common patrimony" – and intragenerational equity – "the right of all people within the current generation to fair access to the current generation's entitlement to the Earth's natural resources" – environmental equity considers the present generation under an obligation to account for long-term impacts of activities, and to act to sustain the global environment and resource base for future generations. Pollution control and resource management laws may be assessed against this principle.

2.9.4 Transboundary Responsibility

Defined in the international law context as an obligation to protect one's own environment, and to prevent damage to neighboring environments, UNEP considers transboundary responsibility at the international level as a potential limitation on the rights of the sovereign state. Laws that act to limit externalities imposed upon human health and the environment may be assessed against this principle.

2.9.5 Public Participation and Transparency

Identified as essential conditions for "accountable governments ··· industrial concerns," and organizations generally, public participation and transparency are presented by UNEP as requiring "effective protection of the human right to hold and express opinions and to seek, receive and impart ideas ... a right of access to appropriate, comprehensible and timely information held by governments and industrial concerns on economic and social policies regarding the sustainable use of natural resources and the protection of the environment, without imposing undue financial burdens upon the applicants and with adequate protection of privacy and business confidentiality," and "effective judicial and administrative proceedings." These principles are present in EPA, laws requiring publication and access to relevant environmental data, and administrative procedure.

2.9.6 Precautionary principle

One of the most commonly encountered and controversial principles of environmental law, the Rio Declaration formulated the precautionary principle as follows:

In order to protect the environment, the precautionary approach shall be widely applied by States according to their capabilities. Where there are threats of serious or irreversible damage, lack of full scientific certainty shall not be used as a reason for postponing cost-effective measures to prevent environmental degradation.

The principle may play a role in any debate over the need for environmental regulation.

2.9.7 Prevention

The concept of prevention ... can perhaps better be considered an overarching aim that gives rise to a multitude of legal mechanisms, including prior assessment of environmental harm, licensing or authorization that set out the conditions for operation and the consequences for violation of the conditions, as well as the adoption of strategies and policies. Emission limits and other product or process standards, the use of best available techniques and similar techniques can all be seen as applications of the concept of prevention.

2.10 Polluter Pays Principle

The polluter pays principle stands for the idea that "the environmental costs of economic activities, including the cost of preventing potential harm, should be internalized rather than imposed upon society at large." All issues related to responsibility for cost for environmental remediation and compliance with pollution control regulations involve this principle.

2.11 Theory/Environmental Law Debate

Environmental law is a continuing source of controversy. Debates over the necessity, fairness, and cost of environmental regulation are ongoing, as well as regarding the appropriateness of regulations vs. market solutions to achieve even agreed-upon ends.

Allegations of scientific uncertainty fuel the ongoing debate over greenhouse gas regulation and are a major factor in debates over whether to ban particular pesticides. In cases where the science is well-settled, it is not unusual to find that corporations intentionally hide or distort the facts, or sow confusion (Oreskes and Conway 2010).

It is very common for regulated industry to argue against environmental regulation on the basis of cost (Pizer and Kopp 2003). Difficulties arise in performing cost-benefit analysis of environmental issues. It is difficult to quantify the value of an environmental value such as a healthy ecosystem, clean air, or species diversity. Many environmentalists' response to pitting economy vs. ecology is summed up by former Senator and founder of Earth Day Gaylord Nelson: "The economy is a wholly owned subsidiary of the environment, not the other way around" (Nelson et al. 2002). Furthermore, environmental issues are seen by many as having an ethical or moral dimension, which would transcend financial cost. Even so, there are some efforts underway to systemically recognize environmental costs and assets, and account for them properly in economic terms.

While affected industries spark controversy in fighting regulation, there are also many environmentalists and public interest groups who believe that current regulations are inadequate, and advocate for stronger protection (Hiss 2014; Stein and Beckel 2004). Environmental law conferences – such as the annual Public Interest Environmental Law Conference in Eugene, Oregon – typically have this focus, also connecting environmental law with class, race, and other issues.

An additional debate is to what extent environmental laws are fair to all regulated parties. For instance, researchers Preston Teeter and Jorgen Sandberg highlight how smaller organizations can often incur disproportionately larger costs as a result of environmental regulations, which can ultimately create an additional barrier to entry for new firms, thus stifling competition and innovation (Teeter and Sandberg 2017).

2.11.1 Environmental Impact Statement and NEPA Process

An environmental impact statement (EIS), under US environmental law, is a document required by the National Environmental Policy Act (NEPA) for certain actions "significantly affecting the quality of the human environment" (NEPA 1969). An EIS is a tool for decision making. It describes the positive and negative environmental effects of a proposed action, and it usually also lists one or more alternative actions that may be chosen instead of the action described in the EIS. Several US state governments require that a document similar to an EIS be submitted to the state for certain actions.

2.11.2 Purpose of NEPA

The purpose of the NEPA is to promote informed decision making by federal agencies by making "detailed information concerning significant environmental impacts" available to both agency leaders and the public (*Robertson v. Methow Valley Citizens Council* 1989). The NEPA was the first piece of legislation that created a comprehensive method to assess potential and existing environmental risks at once. It also encourages communication and cooperation between all the actors involved in environmental decisions, including government officials, private businesses, and citizens (Felleman 2013).

In particular, an EIS acts as an enforcement mechanism to ensure that the federal government adheres to the goals and policies outlined in the NEPA. An EIS should be created in a timely manner as soon as the agency is planning development or is presented with a proposal for development. The statement should use an interdisciplinary approach so that it accurately assesses both the physical and social impacts of the proposed development (EIS 2010). In many instances an action may be deemed subject to NEPA's EIS requirement even though the action is not specifically sponsored by a federal agency. These factors may include actions that receive federal funding, federal licensing, or authorization, or that are subject to federal control (Eccleston 2008).

An EIS typically has four sections (Eccleston 2014):

1) An introduction including a statement of the purpose and need of the proposed action.
2) A description of the affected environment.
3) A range of alternatives to the proposed action. Alternatives are considered the "heart" of the EIS.
   An analysis of the environmental impacts of each of the possible alternatives. This section covers topics such as the following:
   - Impacts to threatened or endangered species
   - Air and water quality impacts
   - Impacts to historic and cultural sites, particularly sites of significant importance to indigenous peoples.
   - Social and economic impacts to local communities, often including consideration of attributes such as impacts to available housing stock, economic impacts to businesses, property values, aesthetics, and noise within the affected area.
   - Cost analysis for each alternative, including costs to mitigate expected impacts, to determine if the proposed action is a prudent use of taxpayer dollars.

2.12 International Law

It is well known that the United Nations Conference in Stockholm on the human environment is a landmark milestone at the international arena for the protection of the deteriorating environment. The conference laid emphasis on the need that man's capabilities to transform his surroundings must be wisely used. Wrong and unwise use can do incalculable harm to human beings and the human environment. It was suggested by the Conference that developing countries must direct their efforts toward balancing their priorities with the need to check increasing population. Moreover, the conference identified the areas and laid down the principles on which the nations should take up and enact laws for protecting environment. These principles have been incorporated in the Stockholm Declaration (British Institute of International and Comparative Environmental Law 1992; Caldwell 1996; Koivurova 2014; Muralikrishna and Manickam 2017). In this process, there are national and international dimensions of environmental law.

Global and regional environmental issues are increasingly the subject of international law. Debates over environmental concerns implicate core principles of international law and have been the subject of numerous international agreements and declarations (see Appendix B).

Customary international law is an important source of international environmental law. These are the norms and rules that countries follow as a matter of custom and they are so prevalent that they bind all states in the world. When a principle becomes customary law is not clear cut and many arguments are put forward by states not wishing to be bound. Examples of customary international law relevant to the environment include the duty to warn other states promptly about icons of an environmental nature and environmental damages to which another state or states may be exposed, and Principle 21 of the Stockholm Declaration.

Numerous legally binding international agreements encompass a wide variety of issue areas, from terrestrial, marine, and atmospheric pollution through to wildlife and biodiversity protection. International environmental agreements are generally multilateral (or sometimes bilateral) treaties (a.k.a. convention, agreement, protocol, etc.). Protocols are subsidiary agreements built from a primary treaty. They exist in many areas of international law but are especially useful in the environmental field, where they may be used to regularly incorporate recent scientific knowledge. They also permit countries to reach agreement on a framework that would be contentious if every detail were to be agreed upon in advance. The most widely known protocol in international environmental law is the Kyoto Protocol, which followed from the United Nations Framework Convention on Climate Change.

While the bodies that proposed, argued, agreed upon, and ultimately adopted existing international agreements vary according to each agreement, certain conferences, including 1972s United Nations Conference on the Human Environment, 1983s World Commission on Environment and Development, 1992s United Nations Conference on Environment and Development, and 2002s World Summit on Sustainable Development have been particularly important. Multilateral environmental agreements sometimes create an International Organization, Institution, or Body responsible for implementing the agreement. Major examples are the Convention on International Trade in Endangered Species of Wild Fauna and Flora (CITES) and the International Union for Conservation of Nature.

International environmental law also includes the opinions of international courts and tribunals. While there are few and they have limited authority, the decisions carry much weight with legal commentators and are quite influential on the development of international environmental law. One of the biggest challenges in international decisions is to determine an adequate compensation for environmental damages (Hardman Reis 2011). The courts include the International Court of Justice (ICJ), the international Tribunal for the Law of the Sea, the European Court of Justice, European Court of Human Rights, and other regional treaty tribunals.

2.12.1 Africa

According to the International Network for Environmental Compliance and Enforcement, the major environmental issues in Africa are "drought and flooding, air pollution, deforestation, loss of biodiversity, freshwater availability, degradation of soil and vegetation, and widespread poverty." The USEPA is focused on the "growing urban and industrial pollution, water quality, electronic waste, and indoor air from cook stoves." They hope to provide enough aid on concerns regarding pollution before their impacts contaminate the African environment as well as the global environment. By doing so, they intend to "protect human health, particularly vulnerable populations such as children and the poor" (EPA 2012). In order to accomplish these goals in Africa, EPA programs are focused on strengthening the ability to enforce environmental laws as well as public compliance to them. Other programs work on developing stronger environmental laws, regulations, and standards.

2.12.2 Asia

The Asian Environmental Compliance and Enforcement Network (AECEN) is an agreement between 16 Asian countries dedicated to improving cooperation with environmental laws in Asia. These countries include Cambodia, China, Indonesia, India, Maldives, Japan, Korea, Malaysia, Nepal, Philippines, Pakistan, Singapore, Sri Lanka, Thailand, Vietnam, and Lao PDR (AECEN 2018).

2.12.3 European Union

The European Union issues secondary legislation on environmental issues that are valid throughout the EU (so-called regulations) and many directives that must be implemented into national legislation from the 28 member states (national states). Examples are the Regulation (EC) No. 338/97 on the implementation of CITES; or the Natura 2000 network the centerpiece for nature and biodiversity policy, encompassing the bird directive (79/409/EEC/ changed to 2009/147/EC) and the habitats directive (92/43/EEC), which are made up of multiple SACs (Special Areas of Conservation, linked to the habitats directive) and SPAs (Special Protected Areas, linked to the bird directive) throughout Europe.

EU legislation is ruled in Article 249 Treaty for the Functioning of the European Union. Topics for common EU legislation are as follows:

- Climate change
- Air pollution
- Water protection and management
- Waste management
- Soil protection
- Protection of nature, species, and biodiversity
- Noise pollution
- Cooperation for the environment with third countries (other than EU member states)
- Civil protection

2.12.4 Middle East

The USEPA is working with countries in the Middle East to improve "environmental governance, water pollution and water security, clean fuels and vehicles, public participation, and pollution prevention."

2.12.5 Oceania

The main concerns on environmental issues in the Oceanic Region are "illegal releases of air and water pollutants, illegal logging/timber trade, illegal shipment of hazardous wastes, including e-waste and ships slated for destruction, and insufficient institutional structure/lack of enforcement capacity." The Secretariat of the Pacific Regional Environmental Program (SPREP) is an international organization between Australia, the Cook Islands, Fiji, France, Kiribati, Marshall Islands, Nauru, New Zealand, Niue, Palau, Papua New Guinea, Samoa, Solomon Island, Tonga, Tuvalu, United States, and Vanuatu. The SPREP was established in order to provide assistance in improving and protecting the environment as well as to assure sustainable development for future generations (SPREP n.d.; Taylor et al. 2013).

2.12.6 Australia

The Environment Protection and Biodiversity Conservation Act 1999 is the centerpiece of environmental legislation in the Australian Government. It sets up the "legal framework to protect and manage nationally and internationally important flora, fauna, ecological communities and heritage places" (EPBC 1999). It also focuses on protecting world heritage properties, national heritage properties, wetlands of international importance, nationally threatened species and ecological communities, migratory species, Commonwealth marine areas, Great Barrier Reef Marine Park, and the environment surrounding nuclear activities (EPBC 1999). *Commonwealth v. Tasmania* (1983), also known as the "Tasmanian Dam Case," is the most influential case for Australian environmental law (*Commonwealth v Tasmania* 1983).

2.12.7 Brazil

The Brazilian government created the Ministry of Environment in 1992 in order to develop better strategies of protecting the environment, use natural resources sustainably, and enforce public environmental policies. The Ministry of Environment has authority over policies involving environment, water resources, preservation, and environmental programs involving the Amazon.

2.12.8 Canada

The Department of the Environment Act establishes the Department of the Environment in the Canadian government as well as the position Minister of the Environment. Their duties include "the preservation and enhancement of the quality of the natural environment, including water, air and soil quality; renewable resources, including migratory birds and other nondomestic flora and fauna; water; meteorology" (Department of the Environment Act 1985/2009). The Environmental Protection Act is the main piece of Canadian environmental legislation that was put into place on 31 March 2000. The Act focuses on "respecting pollution prevention and the protection of the environment and human health in order to contribute to sustainable development." Other principle federal statutes include the Canadian Environmental Assessment Act and the Species at Risk Act. When provincial and federal legislation are in conflict, federal legislation takes precedence, that being said individual provinces can have their own legislation such as Ontario's Environmental Bill of Rights and Clean Water Act (1985).

2.12.9 China

According to the USEPA, "China has been working with great determination in recent years to develop, implement, and enforce a solid environmental law framework. Chinese officials face critical challenges in effectively implementing the laws, clarifying the roles of their national and provincial governments, and strengthening the operation of their legal system" (EPA Collaboration with China 2017). Explosive economic and industrial growth in China has led to significant environmental degradation, and China is currently in the process of developing more stringent legal controls (McElwee 2011). The harmonization of Chinese society and the natural environment is billed as a rising policy priority (NRDC 2014; Pettit 2014; Stern 2013; Wang 2013).

2.12.10 Ecuador

With the enactment of the 2008 Constitution, Ecuador became the first country in the world to codify the Rights of Nature. The Constitution, specifically Articles 10 and 71–74, recognizes the inalienable rights of ecosystems to exist and flourish, gives people the authority to petition on the behalf of ecosystems, and requires the government to remedy violations of these rights. The rights approach is a break away from traditional environmental regulatory systems, which regard nature as property and legalize and manage degradation of the environment rather than prevent it (CELDF 2017).

The Rights of Nature articles in Ecuador's constitution are part of a reaction to a combination of political, economic, and social phenomena. Ecuador's abusive past with the oil industry, most famously the class-action litigation against Chevron, and the failure of an extraction-based economy and neoliberal reforms to bring economic prosperity to the region has resulted in the election of a New Leftist regime, led by President Rafael Correa, and sparked a demand for new approaches to development. In conjunction with this need, the principle of "Buen Vivir," or good living – focused on social, environmental, and spiritual wealth versus material wealth – gained popularity among citizens and was incorporated into the new constitution (Gudynas 2011).

The influence of indigenous groups, from whom the concept of "Buen Vivir" originates, in the forming of the constitutional ideals also facilitated the incorporation of the Rights of Nature as a basic tenet of their culture and conceptualization of "Buen Vivir" (Becker 2011).

2.12.11 Egypt

The Environmental Protection Law outlines the responsibilities of the Egyptian government to "preparation of draft legislation and decrees pertinent to environmental management, collection of data both nationally and internationally on the state of the environment, preparation of periodical reports and studies on the state of the environment, formulation of the national plan and its projects, preparation of environmental profiles for new and urban areas, and setting of standards to be used in planning for their development, and preparation of an annual report on the state of the environment to be prepared to the President" (Ministry of Environment Egyptian Environmental Affairs 2009).

2.12.12 Germany

Since 15 November 1994, environmental protection has been enshrined as an objective of the state in Article 20a of the German Basic Law. Constitutional status has thus been afforded to environmental protection and its objectives. All state bodies – in particular the legislature – are required to be "mindful also of [their] responsibility toward future generations" and to protect the environment (Seider 2010):

- Law on Conservation and Environmental Care (Gesetz über Naturschutz und Landschaftspflege – Bundesnatur schutzgesetz – BNatSchG)
- Law on Protection for Environmental Harms due to Air Pollution, Noise, etc. (Gesetz zum Schutz vor schädlichen Umwelteinwirkungen durch Luftverunreinigungen, Geräusche, Erschütterungen und ähnliche Vorgänge – Bundes-Immissionsschutzgesetz – BImSchG)
- Regulation on Drinking Water Quality (Trinkwasservero rdnung – TrinkwV)
- Regulation on Soil Protection (Bundesbodenschutzgeset z – BBSchG)
- Regulation on Waste Management (Kreislaufwirtschafts gesetz – KrwG)
- Regulation on Water Usage (Wasserhaushaltsgesetz – WHG)

2.12.12.1 Environmental Rules for Doing Business in Germany: Legal Requirements

The environmental laws at the federal and state level are generally implemented by the Länder. The highest national authority for environmental matters is the Federal Ministry for the Environment, Nature Conservation and Nuclear Safety. The 16 Länder also have their own environment ministries.

The Federal Ministry for the Environment and Environment ministries of the Länder collates all acts and regulations within its area of competence. It is broken down into the following fields: General environmental protection; Waste management; Laws on chemicals; Renewable energy/climate protection; Water conservancy; Emission protection; Nuclear safety/radiological protection; Nature and landscape conservation; Chemicals Act (Act on protection against hazardous substances); Chemicals Prohibition Regulation; Hazardous Substances Regulation; and Chemicals Penalty Regulation.

2.12.13 India

In India, environmental law is governed by the Environment Protection Act (EPA, India), 1986 (The Environmental Act 1986; Upadhyay 2012). This act is enforced by the Central Pollution Control Board (CPCB) and the numerous State Pollution Control Boards (SPCBs). Apart from this, there are also individual legislations specifically enacted for the protection of water, air, wildlife, etc. Such legislations include the following:

- The Water (Prevention and Control of Pollution) Act, 1974
- The Water (Prevention and Control of Pollution) Cess Act, 1977
- The Forest (Conservation) Act, 1980
- The Air (Prevention and Control of Pollution) Act, 1981
- Air (Prevention and Control of Pollution) (Union Territories) Rules, 1983
- The Biological Diversity Act, 2002 and the Wild Life Protection Act, 1972
- Batteries (Management and Handling) Rules, 2001
- Recycled Plastics, Plastics Manufacture and Usage Rules, 1999
- The National Green Tribunal established under the National Green Tribunal Act of 2010 has jurisdiction over all environmental cases dealing with a substantial environmental question and acts covered under the Water (Prevention and Control of Pollution) Act, 1974.

- Water (Prevention and Control of Pollution) Cess Rules, 1978
- Ganga Action Plan, 1986
- The Forest (Conservation) Act, 1980
- The Public Liability Insurance Act, 1991 and the Biological Diversity Act, 2002. The acts covered under Indian Wild Life Protection Act 1972 do not fall within the jurisdiction of the National Green Tribunal. Appeals can be filed in the Hon'ble Supreme Court of India (The Indian Wildlife Protection Act 1972)
- Basel Convention on Control of Transboundary Movements on Hazardous Wastes and Their Disposal, 1989 and Its Protocols
- Hazardous Wastes (Management and Handling) Amendment Rules, 2003 (Malik and Malik 2015)

2.13 The Legal and Regulatory Framework for Environmental Protection in India

2.13.1 Introduction

Over the years, together with a spreading of environmental consciousness, there has been a change in the traditionally held perception that there is a trade-off between environmental quality and economic growth as people have come to believe that the two are necessarily complementary. The current focus on environment is not new environmental considerations have been an integral part of the Indian culture. The need for conservation and sustainable use of natural resources has been expressed in Indian scriptures, more than 3000 years old and is reflected in the constitutional, legislative, and policy framework as also in the international commitments of the country.

Even before India's independence in 1947, several environmental legislations existed but the real impetus for bringing about a well-developed framework came only after the UN Conference on the Human Environment (Stockholm, 1972). Under the influence of this declaration, the National Council for Environmental Policy and Planning within the Department of Science and Technology was set up in 1972. This Council later evolved into a full-fledged Ministry of Environment and Forests (MOEF) in 1985 which today is the apex administrative body in India for regulating and ensuring environmental protection. After the Stockholm Conference, in 1976, constitutional sanction was given to environmental concerns through the 42nd Amendment, which incorporated them into the Directive Principles of State Policy and Fundamental Rights and Duties.

Since the 1970s an extensive network of environmental legislation has grown in the country. The MOEF and the pollution control boards (i.e. CPCB and SPCBs) together form the regulatory and administrative core of the sector.

A policy framework has also been developed to complement the legislative provisions. The Policy Statement for Abatement of Pollution and the National Conservation Strategy and Policy Statement on Environment and Development were brought out by the MOEF in 1992, to develop and promote initiatives for the protection and improvement of the environment. The Environmental Action Program (EAP) was formulated in 1993 with the objective of improving environmental services and integrating environmental considerations in to development programs.

Other measures have also been taken by the government to protect and preserve the environment. Several sector-specific policies have evolved, which are discussed at length in the concerned sections. This attempts to highlight only legislative initiatives toward the protection of the environment.

2.13.2 Legislation for Environmental Protection in India

2.13.2.1 Water

Water quality standards especially those for drinking water are set by the Indian Council of Medical Research. These bear close resemblance to WHO standards. The discharge of industrial effluents is regulated by the Indian Standard Codes, and recently water quality standards for coastal water marine outfalls have also been specified. In addition to the general standards, certain specific standards have been developed for effluent discharges from industries, such as iron and steel, aluminum, pulp and paper, oil refineries, petrochemicals, and thermal power plants. Legislation to control water pollution is listed below:

- Water (Prevention and Control of Pollution) Act, 1974

  This Act represented India's first attempts to comprehensively deal with environmental issues. The Act prohibits the discharge of pollutants into water bodies beyond a given standard, and lays down penalties for noncompliance. The Act was amended in 1988 to conform closely to the provisions of the USEPA in 1986. It set up the CPCB which lays down standards for the prevention and control of water pollution. At the State level, the SPCBs function under the direction of the CPCB and the state government.
- Water (Prevention and Control of Pollution) Cess Act, 1977

  This Act provides for a levy and collection of a cess on water consumed by industries and local authorities.

It aims at augmenting the resources of the central and state boards for prevention and control of water pollution. Following this Act, *The Water (Prevention and Control of Pollution) Cess Rules* were formulated in 1978 for defining standards and indications for the kind of and location of meters that every consumer of water is required to install.

2.13.2.2 Air

- Air (Prevention and Control of Pollution) Act, 1981

  To counter the problems associated with air pollution, ambient air quality standards were established under the 1981 Act. The Act provides means for the control and abatement of air pollution. The Act seeks to combat air pollution by prohibiting the use of polluting fuels and substances as well as by regulating appliances that give rise to air pollution. Under the Act, establishing or operating of any industrial plant in the pollution control area requires consent from state boards. The boards are also expected to test the air in air pollution control areas, inspect pollution control equipment, and manufacturing processes.

National Ambient Air Quality Standards (NAAQS) for major pollutants were notified by the CPCB in April 1994. These are deemed to be levels of air quality necessary with an adequate margin of safety, to protect public health, vegetation, and property (CPCB 1995 cited in Gupta 1999). The NAAQS prescribe specific standards for industrial, residential, rural, and other sensitive areas. Industry-specific emission standards have also been developed for iron and steel plants, cement plants, fertilizer plants, oil refineries, and the aluminum industry. The ambient quality standards prescribed in India are similar to those prevailing in many developed and developing countries.

To empower the central and state pollution boards to meet grave emergencies, the *Air (Prevention and Control of Pollution) Amendment Act*, 1987, was enacted. The boards were authorized to take immediate measures to tackle such emergencies and recover the expenses incurred from the offenders. The power to cancel consent for nonfulfillment of the conditions prescribed has also been emphasized in the Air Act Amendment.

The Air (Prevention and Control of Pollution) Rules formulated in 1982 defined the procedures for conducting meetings of the boards, the powers of the presiding officers, decision making, the quorum, manner in which the records of the meeting were to be set, etc. They also prescribed the manner and the purpose of seeking assistance from specialists and the fee to be paid to them.

Complementing the above acts is the *Atomic Energy Act* of 1982, which was introduced to deal with radioactive waste. In 1988, the *Motor Vehicles Act* was enacted to regulate vehicular traffic, besides ensuring proper packaging, labeling, and transportation of the hazardous wastes. Various aspects of vehicular pollution have also been notified under the USEPA of 1986. Mass emission standards were notified in 1990, which were made more stringent in 1996. In 2000 these standards were revised yet again and for the first time separate obligations for vehicle owners, manufacturers, and enforcing agencies were stipulated. In addition, fairly stringent Euro I and II emission norms were notified by the Supreme Court on 29 April 1999 for the city of Delhi. The notification made it mandatory for car manufacturers to conform to the Euro I and Euro II norms by May 1999 and April 2000, respectively, for new noncommercial vehicle sold in Delhi.

2.13.2.3 Forests and Wildlife

- The Wildlife (Protection) Act, 1972, Amendment 1991

  The WPA (Wildlife Protection Act), 1972, provides for protection to listed species of flora and fauna and establishes a network of ecologically important protected areas. The WPA empowers the central and state governments to declare any area a wildlife sanctuary, national park, or closed area. There is a blanket ban on carrying out any industrial activity inside these protected areas. It provides for authorities to administer and implement the Act; regulate the hunting of wild animals; protect specified plants, sanctuaries, national parks, and closed areas; restrict trade or commerce in wild animals or animal articles; and miscellaneous matters. The Act prohibits hunting of animals except with permission of authorized officer when an animal has become dangerous to human life or property or so disabled or diseased as to be beyond recovery (WWF-India, 1999). The near-total prohibition on hunting was made more effective by the Amendment Act of 1991.

The Forest (Conservation) Act, 1980 This Act was adopted to protect and conserve forests. The Act restricts the powers of the state in respect of de-reservation of forests and use of forestland for non-forest purposes (the term non-forest purpose includes clearing any forestland for cultivation of cash crops, plantation crops, horticulture, or any purpose other than re-afforestation).

2.13.3 General

2.13.3.1 Environment (Protection) Act, 1986 (USEPA)

This Act is an umbrella legislation designed to provide a framework for the coordination of central and state authorities established under the Water (Prevention and Control) Act, 1974 and Air (Prevention and Control) Act, 1981.

Under this Act, the central government is empowered to take measures necessary to protect and improve the quality of the environment by setting standards for emissions and discharges, regulating the location of industries, management of hazardous wastes, and protection of public health and welfare.

From time to time the central government issues notifications under the USEPA for the protection of ecologically sensitive areas or issues guidelines for matters under the USEPA.

The following are some notifications issued under this Act:

- *Doon Valley Notification* (1989), which prohibits the setting up of an industry in which the daily consumption of coal/fuel is more than 24 million T/day in the Doon Valley.
- *Coastal Regulation Zone Notification* (1991), which regulates activities along coastal stretches. As per this notification, dumping ash or any other waste in the CRZ is prohibited. The thermal power plants (only foreshore facilities for transport of raw materials, facilities for intake of cooling water and outfall for discharge of treated waste water/cooling water) require clearance from the MOEF.
- *Dhanu Taluka Notification* (1991), under which the district of Dhanu Taluka has been declared an ecologically fragile region and setting up power plants in its vicinity is prohibited.
- *Revdanda Creek Notification (1989)*, which prohibits setting up industries in the belt around the Revdanda Creek as per the rules laid down in the notification.

  The Environmental Impact Assessment of Development Projects Notification (1994 and as amended in 1997). As per this notification:
  - All projects listed under Schedule I require environmental clearance from the MOEF.
  - Projects under the delicensed category of the New Industrial Policy also require clearance from the MOEF.
  - All developmental projects whether or not under the Schedule I, if located in fragile regions must obtain MOEF clearance.
  - Industrial projects with investments above ₹500 million must obtain MOEF clearance and are further required to obtain a LOI (Letter Of Intent) from the Ministry of Industry, and an NOC (No Objection Certificate) from the SPCB and the State Forest Department if the location involves forestland. Once the NOC is obtained, the LOI is converted into an industrial license by the state authority.
  - The notification also stipulated procedural requirements for the establishment and operation of new power plants. As per this notification, two-stage clearance for site-specific projects such as pithead thermal power plants and valley projects is required. Site clearance is given in the first stage and final environmental clearance in the second. A public hearing has been made mandatory for projects covered by this notification. This is an important step in providing transparency and a greater role to local communities.
- *Ash Content Notification* (1997), required the use of beneficiated coal with ash content not exceeding 34% with effect from June 2001. This applies to all thermal plants located beyond 1000 km from the pithead and any thermal plant located in an urban area or sensitive area irrespective of the distance from the pithead except any pithead power plant.
- *Taj Trapezium Notification* (1998), provided that no power plant could be set up within the geographical limit of the Taj Trapezium assigned by the Taj Trapezium Zone Pollution (Prevention and Control) Authority.

  Disposal of Fly Ash Notification (1999), the main objective of which is to conserve the topsoil, protect the environment, and prevent the dumping and disposal of fly ash discharged from lignite-based power plants. The salient feature of this notification is that no person within a radius of 50 km from a coal-or lignite-based power plant shall manufacture clay bricks or tiles without mixing at least 25% of ash with soil on a weight-to-weight basis. For the thermal power plants the utilization of the fly ash would be as follows:
  - Every coal-or lignite-based power plant shall make available ash for at least ten years from the date of publication of the above notification without any payment or any other consideration, for the purpose of manufacturing ash-based products such as cement, concrete blocks, bricks, panels, or any other material or for construction of roads, embankments, dams, dykes, or for any other construction activity.
  - Every coal- or lignite-based thermal power plant commissioned subject to environmental clearance conditions stipulating the submission of an action plan for full utilization of fly ash shall, within a period of nine years from the publication of this notification, phase out the dumping and disposal of fly ash on land in accordance with the plan.

Rules for the Manufacture, Use, Import, Export and Storage of Hazardous Micro-organisms/Genetically Engineered Organisms or Cell were introduced in 1989 with the view to protect the environment, nature, and health in connection with gene technology and micro-organisms, under the Environmental Protection Act, 1986. In 1991, the government further decided to institute a national label scheme for environmentally friendly products called the

"ECOMARK." The scheme attempts to provide incentives to manufactures and importers to reduce adverse environmental impacts, reward genuine initiatives by companies, and improve the quality of the environment and sustainability of available resources. Besides the above attempts, notifications pertaining *to Recycled Plastics Manufacture and Usage Rules, 1999* were also incorporated under the Environment (Protection) Act of 1986.

2.13.3.2 The Environment (Protection) Rules, 1986

These rules lay down the procedures for setting standards of emission or discharge of environmental pollutants. The Rules prescribe the parameters for the Central Government, under which it can issue orders of prohibition and restrictions on the location and operation of industries in different areas. The Rules lay down the procedure for taking samples, serving notice, submitting samples for analysis and laboratory reports. The functions of the laboratories are also described under the Rules along with the qualifications of the concerned analysts.

2.13.3.3 The National Environment Appellate Authority Act, 1997

This Act provided for the establishment of a National Environment Appellate Authority to hear appeals with respect to restriction of areas in which any industry operation or process or class of industries, operations or processes could not carry out or would be allowed to carry out subject to certain safeguards under the Environment Protection Act, 1986.

In addition to these, various acts specific to the coal sector have been enacted. The first attempts in this direction can be traced back to the *Mines Act, 1952*, which promoted health and safety standards in coal mines. Later the *Coal Mines (Conservation and Development) Act* (1974) came up for conservation of coal during mining operations. For conservation and development of oil and natural gas resources, a similar legislation was enacted in 1959.

2.13.4 Hazardous Wastes

There are several legislations that directly or indirectly deal with hazardous waste. The relevant legislations are the Factories Act, 1948, the Public Liability Insurance Act, 1991, the National Environment Tribunal Act, 1995, and some notifications under the Environmental Protection Act of 1986. A brief description of each of these is given in the following.

Under the USEPA 1986, the MOEF has issued several notifications to tackle the problem of hazardous waste management. These include the following:

- *Hazardous Wastes (Management and Handling) Rules, 1989*, which brought out a guide for manufacture, storage, and import of hazardous chemicals and for management of hazardous wastes.
- *Biomedical Waste (Management and Handling) Rules, 1998*, were formulated along parallel lines, for proper disposal, segregation, transport, etc. of infectious wastes.
- *Municipal Wastes (Management and Handling) Rules, 2000*, whose aim was to enable municipalities to dispose municipal solid waste in a scientific manner.
- *Hazardous Wastes (Management and Handling) Amendment Rules, 2000*, a recent notification issued with the view to providing guidelines for the import and export of hazardous waste in the country.

2.13.4.1 Factories Act, 1948 and Its Amendment in 1987

The Factories Act, 1948 was a postindependence statute that explicitly showed concern for the environment. The primary aim of the 1948 Act has been to ensure the welfare of workers not only in their working conditions in the factories but also their employment benefits. While ensuring the safety and health of the workers, the Act contributes to environmental protection. The Act contains a comprehensive list of 29 categories of industries involving hazardous processes, which are defined as a process or activity where unless special care is taken, raw materials used therein or the intermediate or the finished products, by-products, wastes, or effluents would

- cause material impairment to health of the persons engaged
- result in the pollution of the general environment

2.13.4.2 Public Liability Insurance Act (PLIA), 1991

The Act covers accidents involving hazardous substances and insurance coverage for these. Where death or injury results from an accident, this Act makes the owner liable to provide relief as is specified in the Schedule of the Act. The PLIA was amended in 1992, and the Central Government was authorized to establish the Environmental Relief Fund, for making relief payments.

2.13.4.3 National Environment Tribunal Act, 1995

The Act provided strict liability for damages arising out of any accident occurring while handling any hazardous substance and for the establishment of a National Environment Tribunal for effective and expeditious disposal of cases arising from such accident, with a view to give relief and compensation for damages to persons, property, and the environment and for the matters connected therewith or incidental thereto.

2.13.5 International Agreements on Environmental Issues

India is signatory to a number of multilateral environment agreements (MEA) and conventions. An overview of some of the major MEAs and India's obligations under these is presented below. This issue is discussed in the Section 2.9.4 in Chapter 2.

2.13.5.1 Convention on International Trade in Endangered Species (CITES), of Wild Fauna and Flora, 1973

The aim of CITES is to control or prevent international commercial trade in endangered species or products derived from them. CITES does not seek to directly protect endangered species or curtail development practices that destroy their habitats. Rather, it seeks to reduce the economic incentive to poach endangered species and destroy their habitat by closing off the international market. India became a party to the CITES in 1976. International trade in all wild flora and fauna in general and species covered under CITES is regulated jointly through the provisions of The Wildlife (Protection) Act 1972, the Import/Export policy of Government of India, and the Customs Act 1962 (Bajaj 1996).

2.13.5.2 Montreal Protocol on Substances that Deplete the Ozone Layer (to the Vienna Convention for the Protection of the Ozone Layer), 1987

The Montreal Protocol to the Vienna Convention on Substances that deplete the Ozone Layer, came into force in 1989. The protocol set targets for reducing the consumption and production of a range of ozone depleting substances (ODS). In a major innovation, the Protocol recognized that all nations should not be treated equally. The agreement acknowledges that certain countries have contributed to ozone depletion more than others. It also recognizes that a nation's obligation to reduce current emissions should reflect its technological and financial ability to do so. Because of this, the agreement sets more stringent standards and accelerated phase-out timetables to countries that have contributed most to ozone depletion (Divan and Rosencranz 2002).

India acceded to the Montreal Protocol along with its London Amendment in September 1992. The MOEF has established an Ozone Cell and a steering committee on the Montreal Protocol to facilitate implementation of the India Country Program, for phasing out ODS production by 2010.

To meet India's commitments under the Montreal Protocol, the Government of India has also taken certain policy decisions:

- Goods required to implement ODS phase-out projects funded by the Multilateral Fund are fully exempt from duties. This benefit has been also extended to new investments with non-ODS technologies.
- Commercial banks are prohibited from financing or refinancing investments with ODS technologies.

The Gazette of India on 19 July 2000 notified rules for regulation of ODS phase-out called the *Ozone Depleting Substances (Regulation and Control) Rules, 2000*. They were notified under the Environment (Protection) Act, 1986. These rules were drafted by the MOEF following consultations with industries and related government departments.

2.13.5.3 Basel Convention on Transboundary Movement of Hazardous Wastes, 1989

Basel Convention, which entered into force in 1992, has three key objectives:

1) To reduce transboundary movements of hazardous wastes
2) To minimize the creation of such wastes
3) To prohibit their shipment to countries lacking the capacity to dispose hazardous wastes in an environmentally sound manner

India ratified the Basel Convention in 1992, shortly after it came into force. The Indian Hazardous Wastes Management Rules Act, 1989 encompasses some of the Basel provisions related to the notification of import and export of hazardous waste, illegal traffic, and liability.

2.13.5.4 UN Framework Convention on Climate Change (UNFCCC), 1992

The primary goals of the UNFCCC were to stabilize greenhouse gas (GHG) emissions at levels that would prevent dangerous anthropogenic interference with the global climate. The convention embraced the principle of common but differentiated responsibilities which has guided the adoption of a regulatory structure.

India signed the agreement in June 1992, which was ratified in November 1993. As per the convention, the reduction/limitation requirements apply only to developed countries. The only reporting obligation for developing countries relates to the construction of a GHG inventory. India has initiated the preparation of its First National Communication (base year 1994) that includes an inventory of GHG sources and sinks, potential vulnerability to climate change, adaptation measures, and other steps being taken in the country to address climate change. The further details on UNFCC and the Kyoto Protocol are provided in Chapter 5.

2.13.5.5 Convention on Biological Diversity, 1992

The Convention on Biological Diversity (CBD) is a legally binding, framework treaty that has been ratified until now by 180 countries. The CBD has three main thrust areas: conservation of biodiversity, sustainable use of biological resources and equitable sharing of benefits arising from their sustainable use.

The Convention on Biological Diversity came into force in 1993. Many biodiversity issues are addressed in the convention, including habitat preservation, intellectual property rights, biosafety, and indigenous people's rights.

India's initiatives under the Convention are detailed in Section 2.13.2.3. These include the promulgation of the Wildlife (Protection) Act of 1972, amended in 1991, and participation in several international conventions such as CITES.

2.13.5.6 UN Convention on Desertification, 1994

Delegates to the 1992 UN Conference on Environment and Development (UNCED) recommended establishment of an intergovernmental negotiating committee for the elaboration of an international convention to combat desertification in countries experiencing serious drought and/or desertification. The UN General Assembly established such a committee in 1992 that later helped formulation of Convention on Desertification in 1994.

The convention is distinctive as it endorses and employs a bottom-up approach to international environmental cooperation. Under the terms of the convention, activities related to the control and alleviation of desertification and its effects are to be closely linked to the needs and participation of local land-users and nongovernmental organizations. Seven countries in the South Asian region are signatories to the Convention, which aims at tackling desertification through national, regional, and sub-regional action programs. The Regional Action Program has six Thematic Program Networks (TPNs) for the Asian region, each headed by a country task manager. India hosts the network on agroforestry and soil conservation. For details refer to Section 2.12.13.

2.13.5.7 International Tropical Timber Agreement and the International Tropical Timber Organization (ITTO), 1983, 1994

The ITTO established by the International Tropical Timber Agreement (ITTA), 1983, came into force in 1985 and became operational in 1987. The ITTO facilitates discussion, consultation, and international cooperation on issues relating to the international trade and utilization of tropical timber and the sustainable management of its resource base. The successor agreement to the ITTA (1983) was negotiated in 1994 and came into force on 1 January 1997. The organization has 57 member countries. India ratified the ITTA in 1996.

2.13.6 An Assessment of the Legal and Regulatory Framework for Environmental Protection in India

The extent of the environmental legislation network is evident from the above discussion, but the enforcement of the laws has been a matter of concern. One commonly cited reason is the prevailing command and control nature of the environmental regime. Coupled with this is the prevalence of all or nothing approach of the law; they do not consider the extent of violation. Fines are levied on a flat basis and in addition there are no incentives to lower the discharges below prescribed levels.

Some initiatives have addressed these issues in the recent past. The Government of India came out with a Policy Statement for Abatement of Pollution in 1992, before the Rio conference, which declared that market-based approaches would be considered in controlling pollution. It stated that economic instruments will be investigated to encourage the shift from curative to preventive measures, internalize the costs of pollution and conserve resources, particularly water. In 1995, MOEF constituted a task force to evaluate market-based instruments, which strongly advocated their use for the abatement of industrial pollution. Various economic incentives have been used to supplement the command-and-control policies. Depreciation allowances, exemptions from excise or customs duty payment, and arrangement of soft loans for the adoption of clean technologies are instances of such incentives. Another aspect that is evident is the shift in the focus from end-of-pipe treatment of pollution to treatment at source. The role of remote sensing and geographical information systems in natural resource management and environmental protection has also gained importance over time. India has made commendable advances in the use of remote sensing for natural resource management and in integrating environmental and development at the policy planning and management levels.

An important recent development is the rise of judicial activism in the enforcement of environmental legislation. This is reflected in the growth of environment-related public litigation cases that have led the courts to take major steps such as ordering the shutdown of polluting factories.

Agenda 21 highlights the need for integration of environmental concerns at all stages of policy, planning, and decision-making processes, including the use of an effective legal and regulatory framework, economic instruments, and other incentives. These very principles were fundamental to guiding environmental protection in the country well before Rio and will be reinforced, drawing on India's own experiences and those of other countries.

2.13.7 Emerging Environmental Challenges

India's economic development propelled by rapid industrial growth and urbanization is causing severe environmental problems that have local, regional, and global significance. Deforestation, soil erosion, water pollution, and land degradation continue to worsen and are hindering economic development in rural India, while the rapid industrialization and urbanization in India's booming metropolizes are straining the limits of municipal services and causing serious environmental problems.

More than 20 cities in India have populations of over one million, and some of them, including New Delhi, Mumbai, Chennai, and Kolkata, are among the world's most polluted. Assuming continued economic liberalization and increased urbanization, the damage to environment and health could be enormous if precautionary measures are not taken. The challenge, therefore, is to maintain the quality of air, water, and land and protect the environment by reconciling environmental, social, and economic imperatives.

Air quality data in India's major cities indicate that ambient levels of air pollutants exceed both the World Health Organization and Indian standards, particularly for particulate matter. Of the total air pollution load nationwide, vehicular sources contribute 64%, thermal power plants 16%, industries 13%, and the domestic sector 7%. Environmental effects from growing fossil fuel use can only worsen as India seeks to meet the energy needs of its growing economy. It is estimated that over 96% of India's total demand for commercial energy is met by fossil fuel with coal contributing 60% and petroleum products providing the remaining 36%.

India's rivers and streams suffer from high levels of pollution from waste generated primarily from industrial processes and municipal activities. Untreated sewage and nonindustrial wastes account for four times as much pollution as industrial effluents. While it is estimated that 75% of the wastewater generated is from municipal sources, industrial waste from large- and medium-sized plants contributes to over 50% of the total pollution loads. In major cities, less than 5% of the total waste is collected and less than 25% of this treated.

To address these environmental challenges in coordination with the state governments, the central government has identified and targeted 17 highly polluting industries and 24 environmental problem areas. The chemical and engineering industries are at the top of the government's list since they are the major contributors to air, water, and waste pollution. These industries include integrated iron and steel plants, nonferrous metallurgical units, pharmaceutical and petrochemical complexes, fertilizers and pesticide plants, thermal power plants, textiles, pulp and paper, tanneries, and chloralkali units.

The Government of India has established an environmental legal and institutional system to meet these challenges within the overall framework of India's development agenda and international principles and norms. Recently, the Government put forward the National Environment Policy which provides a guide to action in regulatory reform, environmental conservation, and enactment of legislation by government agencies at all levels.

2.13.7.1 Japan

The Basic Environmental Law is the basic structure of Japan's environmental policies replacing the Basic Law for Environmental Pollution Control and the Nature Conservation Law. The updated law aims to address "global environmental problems, urban pollution by everyday life, loss of accessible natural environment in urban areas and degrading environmental protection capacity in forests and farmlands" (Govt. of Japan and Ministry of the Environment 1967).

The three basic environmental principles that the Basic Environmental Law follows are "the blessings of the environment should be enjoyed by the present generation and succeeded to the future generations, a sustainable society should be created where environmental loads by human activities are minimized, and Japan should contribute actively to global environmental conservation through international cooperation." From these principles, the Japanese government have established policies such as "environmental consideration in policy formulation, establishment of the Basic Environment Plan which describes the directions of long-term environmental policy, environmental impact assessment for development projects, economic measures to encourage activities for reducing environmental load, improvement of social infrastructure such as sewerage system, transport facilities etc., promotion of environmental activities by corporations, citizens and NGOs, environmental education, and provision of information, promotion of science and technology" (Govt. of Japan, Ministry of the Environment 1967).

2.13.7.2 New Zealand

The Ministry for the Environment and Office of the Parliamentary Commissioner for the Environment were established by the Environment Act 1986. These positions are responsible for advising the Minister on all areas of environmental legislation. A common theme of New Zealand's environmental legislation is sustainably managing natural and physical resources, fisheries, and forests. The Resource Management Act 1991 is the main piece of environmental legislation that outlines the government's strategy to managing the "environment, including air, water soil, biodiversity, the coastal environment, noise, subdivision, and land use planning in general" (Govt. of New Zealand, Environmental Act 1986; Wells 1984).

2.13.7.3 Russia

The Ministry of Natural Resources and Environment of the Russian Federation makes regulation regarding "conservation of natural resources, including the subsoil, water bodies, forests located in designated conservation areas, fauna and their habitat, in the field of hunting, hydrometeorology and related areas, environmental monitoring and pollution control, including radiation monitoring and control, and functions of public environmental policy making and implementation and statutory regulation" (Ministry of Natural Resources and Environment of the Russian Federation 1986).

2.13.7.4 South Africa

South African environmental law describes the legal rules in South Africa relating to the social, economic, philosophical, and jurisprudential issues raised by attempts to protect and conserve the environment in South Africa. South African environmental law encompasses natural resource conservation and utilization, as well as land-use planning and development. Issues of enforcement are also considered, together with the international dimension, which has shaped much of the direction of environmental law in South Africa. The role of the country's Constitution, crucial to any understanding of the application of environmental law, also is examined. The National Environmental Management Act provides the underlying framework for environmental law (South African National Environmental Management Act 1998).

2.13.7.5 United Kingdom

United Kingdom environmental law concerns the protection of the environment in the United Kingdom. Environmental law is increasingly a European and an international issue, due to the cross-border issues of air and water pollution, and man-made climate change.

History In the common law, the primary protection was found in the law of nuisance, but this only allowed for private actions for damages or injunctions if there was harm to land. Thus, smells emanating from pig sties (*Aldred's Case* 1610), strict liability against dumping rubbish (*R v. Stephens* 1866), or damage from exploding dams (*Rylands v. Fletcher* 1898). Private enforcement, however, was limited and found to be woefully inadequate to deal with major environmental threats, particularly threats to common resources.

- 1306, Edward I briefly banned coal fires in London.
- John Evelyn, *Fumifugium* (1661) argued for burning fragrant wood instead of mineral coal, which he believed would reduce coughing.
- *Ballad of Gresham College* (1661) describes how the smoke "does our lungs and spirits choke, Our hanging spoil, and rust our iron."
- In 1800, 1 million T of coal were burned in London, and 15 million across the United Kingdom.
- Smoke Nuisance Abatement (Metropolis) Act 1853.
- John Snow in 1854 discovered that the water pump on Broad Street, Soho, was responsible for 616 cholera deaths because it was contaminated by an old cesspit leaking fecal bacteria. Germ theory of disease began to replace miasma theory that had lingered since the Black Death.

During the "Great Stink" of 1858, the dumping of sewerage into the River Thames began to smell so ghastly in the summer heat that Parliament had to be evacuated. Ironically, the Metropolitan Commission of Sewers Act 1848 had allowed the Metropolitan Commission for Sewers to close cesspits around the city in an attempt to "clean up," but this simply led people to pollute the river. In 19 days, Parliament passed a further Act to build the London sewerage system.

- Alkali Act 1863 and Alkali Act 1874, amended 1906
- WS Jevons, *The Coal Question; An Inquiry Concerning the Progress of the Nation, and the Probable Exhaustion of Our Coal Mines* (1865)
- Ground Game Act 1880, Night Poaching Act 1828, Game Act 1831, game preservation
- James Johnston (socialist politician), president of the Smoke Abatement League, international conference in 1911

London also suffered from terrible air pollution, and this culminated in the "Great Smog" of 1952, which in turn triggered its own legislative response: the Clean Air Act 1956. The basic regulatory structure was to set limits on emissions for households and business (particularly burning coal), while an inspectorate would enforce compliance. It required zones for smokeless fuel to be burned and relocated power stations.

- Clean Air Act 1968 required tall chimneys to disperse pollution.

2.14 United States Environmental Law

United States environmental law concerns legal standards to protect human health and improve the natural environment of the United States. While subject to criticism at home and abroad on issues of protection, enforcement, and over-regulation, the country remains an important source of environmental legal expertise and experience.

2.14.1 Scope

The United States Congress has enacted federal statutes intended to address pollution control and remediation, including for example the Clean Air Act (air pollution), the Clean Water Act (water pollution), and the Comprehensive Environmental Response, Compensation, and Liability Act (CERCLA, or Superfund) (contaminated site cleanup). There are also federal laws governing natural resources use and biodiversity which are strongly influenced by environmental principles, including the Endangered Species Act, National Forest Management Act, and Coastal Zone Management Act. The National Environmental Policy Act, governing environmental impact review in actions undertaken or approved by the US federal government, may implicate all of these areas.

Federalism in the United States has played a role in the shape of national environmental legislation. Many federal environmental laws employ cooperative federalism mechanisms – many federal regulatory programs are administered in coordination with the US states. Furthermore, the states generally have enacted their own laws to cover areas not preempted by federal law. This includes areas where Congress had acted in limited fashion (e.g. state site cleanup laws to handle sites outside Superfund) and where Congress has left regulation primarily to the states (e.g. water resources law).

2.14.2 History

The history of environmental law in the United States can be traced back to early roots in common law doctrines, for example, the law of nuisance and the public trust doctrine. The first environmental statute was the Rivers and Harbors Act of 1899, which has been largely superseded by the Clean Water Act of 1972 (CWA). However, most current major environmental statutes, such as the federal statutes listed previously, were passed in the time spanning the late 1960s through the early 1980s. Prior to the passage of these statutes, most federal environmental laws were not nearly as comprehensive.

Silent Spring, a 1962 book by Rachel Carson, is frequently credited as launching the environmental movement in the United States. The book documented the effects of pesticides, especially DDT, on birds and other wildlife (Carson 1962; Hynes 1989). Among the most significant environmental disasters of the 1960s was the 1969 Santa Barbara oil spill, which generated considerable public outrage as Congress was considering several major pieces of environmental legislation. (*See* Environmental movement in the United States, Clean Water Act (1972), USEPA.)

One lawsuit that has been widely recognized as one of the earliest environmental cases is *Scenic Hudson Preservation Conference v. Federal Power Commission*, decided in 1965 by the Second Circuit Court of Appeals, prior to passage of the major federal environmental statutes. The case helped halt the construction of a power plant on Storm King Mountain in New York State. The case has been described as giving birth to environmental litigation and helping create the legal doctrine of standing to bring environmental claims (Scenic Hudson Inc. 1963). The Scenic Hudson case also is said to have helped inspire the passage of the NEPA, and the creation of such environmental advocacy groups as the Natural Resources Defense Council.

2.14.3 Legal Sources

Laws from every stratum of the laws of the United States pertain to environmental issues. Congress has passed a number of landmark environmental regulatory regimes, but many other federal laws are equally important, if less comprehensive. Concurrently, the legislatures of the 50 states have passed innumerable comparable sets of laws. These state and federal systems are foliated with layer upon layer of administrative regulation. Meanwhile, the US judicial system reviews not only the legislative enactments but also the administrative decisions of the many agencies dealing with environmental issues. Where the statutes and regulations end, the common law begins (Superfund Regulations 2017).

2.14.4 Federal Regulation

Consistent with the federal statutes that they administer, US federal agencies promulgate regulations in the CFR that fill out the broad programs enacted by Congress.

Primary among these is Title 40 of the CFR, containing the regulations of the EPA. Other important CFR sections include Title 10 (energy), Title 18 (Conservation of Power and Water Resources), Title 21 (Food and Drugs), Title 33 (Navigable Waters), Title 36 (Parks, Forests and Public Property), Title 43 (Public Lands: Interior), and Title 50 (Wildlife and Fisheries).

2.14.5 Judicial Decisions

The federal and state judiciaries have played an important role in the development of environmental law in the United States, in many cases resolving significant controversy regarding the application of federal environmental laws in favor of environmental interests. The decisions of the Supreme Court in cases such as *Calvert Cliffs Coordinating Committee v. U.S. Atomic Energy Commission* (broadly reading the procedural requirements of NEPA), *Tennessee Valley Authority v. Hill* (broadly reading the Endangered Species Act), and, much more recently 14 May 2015, *Massachusetts v. EPA* (requiring EPA to reconsider regulation of greenhouse gases under the Clean Air Act) have had policy impacts far beyond the facts of the particular case.

2.14.6 Common Law

The common law of tort is an important tool for the resolution of environmental disputes that fall beyond the confines of regulated activity. Prior to the modern proliferation of environmental regulation, the doctrines of nuisance (public or private), trespass, negligence, and strict liability apportioned harm and assigned liability for activities that today would be considered pollution and likely governed by regulatory regimes. These doctrines remain relevant, and most recently have been used by plaintiffs seeking to impose liability for the consequences of global climate change (Lehman and Phelps 2004).

The common law also continues to play a leading role in American water law, in the doctrines of riparian rights and prior appropriation.

2.14.7 Administration

In the United States, responsibilities for the administration of environmental laws are divided between numerous federal and state agencies with varying, overlapping, and sometimes conflicting missions. EPA is the most well-known federal agency, with jurisdiction over many of the country's national air, water, and waste and hazardous substance programs (USEPA 2017a, b). Other federal agencies, such as the U.S. Fish and Wildlife Service and National Park Service pursue primarily conservation missions (US Fish & Wildlife Service 2018; US National Park Service 2018), while still others, such as the United States Forest Service and the Bureau of Land Management, tend to focus more on beneficial use of natural resources (US Bureau of Land Management 2018; US Forest Service 2018).

Federal agencies operate within the limits of federal jurisdiction. For example, EPA's jurisdiction under the CWA is limited to "waters of the United States." In many cases, federal laws allow for more stringent regulation by states, and of transfer of certain federally mandated responsibilities from federal to state control. US state governments, therefore, administering state law adopted under state police powers or federal law by delegation, uniformly include environmental agencies (USEPA Heath and Environmental Agencies 2018). The extent to which state environmental laws are based on or depart from federal law varies from jurisdiction to jurisdiction.

Thus, while a permit to fill nonfederal wetlands might require a permit from a single state agency, larger and more complex endeavors – for example, the construction of a coal-fired power plant – might require approvals from numerous federal and state agencies.

2.14.8 Enforcement

In the United States, violations of environmental laws are generally civil offenses, resulting in monetary penalties and, perhaps, civil sanctions such as injunction. Many environmental laws also provide for criminal penalties for egregious violations. Some federal laws, such as the CWA, also allow a US citizen to file a lawsuit against a violator, if the government has failed to take enforcement action (USEPA Clean Water Act 1972).

Environmental agencies often include separate enforcement offices, with duties including monitoring permitted activities, performing compliance inspections, issuing citations, and prosecuting wrongdoing (civilly or criminally, depending on the violation). EPA's Office of Enforcement and Compliance Assurance is one such agency. Others, such as the United States Park Police, carry out more traditional law enforcement activities.

Adjudicatory proceedings for environmental violations are often handled by the agencies themselves under the structures of administrative law. In some cases, appeals are also handled internally (e.g. EPA's Environmental Appeals Board). Generally, final agency determinations may subsequently be appealed to the appropriate court.

2.14.9 Education and Training

Environmental law courses are offered as elective courses in the second and third years of JD study at many American law schools. Curricula vary: an introductory course might focus on the "big five" federal statutes – National

Environmental Policy Act (NEPA), Clean Air Act, CWA, CERCLA (Superfund), and Resource Conservation and Recovery Act (or, alternatively, the Federal Insecticide, Fungicide, and Rodenticide Act) – and may be offered in conjunction with a natural resources law course. Smaller seminars may be offered on more focused topics. Some US law schools also offer an LLM or JSD specialization in environmental law. Additionally, several law schools host legal clinics that focus on environmental law, providing students with an opportunity to learn about environmental law in the context of real world disputes involving actual clients (Babich 2004). U.S. News & World Report has consistently ranked Vermont Law School, Lewis & Clark Law School, and Pace University School of Law as the top three Environmental Law programs in the United States, with Lewis & Clark and Vermont frequently trading the top spot.

Many American law schools host student-published law journals. The environmental law reviews at Yale, Harvard, Stanford, Columbia, NYU, and Lewis & Clark Law School are regularly the most-cited such publications (http://lawlib.wlu.edu/LJ).

International environmental lawyers often receive specialized training in the form of an LLM degree at US institutions, after having a first law degree – often in another country from where they got their first law degree.

2.14.10 Vietnam

Vietnam is currently working with the USEPA on dioxin remediation and technical assistance in order to lower methane emissions. In March 2002, the United States and Vietnam signed the US–Vietnam Memorandum of Understanding on Research on Human Health and the Environmental Effects of Agent Orange/Dioxin (Vietnam Environment Administration, Ministry of Natural Resources and Environment 2002).

2.15 ISO 9000 and 14000

ISO, the International Standards Organization, is a worldwide program that was founded in 1947 to promote the development of international manufacturing, trade, and communication standards. The initial focus of body with representatives from all industrialized nations. ISO membership includes over 100 countries. The American National Standards Institute (ANSI) is the US counterpart and representative to ISO. ISO essentially receives input from government, industry, and other interested parties before developing a standard. All standards developed by ISO are voluntary; thus, there are no legal requirements to force countries to adopt them. However, countries and industries often adopt ISO standards as requirements for doing and maintaining business.

ISO develops standards in all industries except those related to electrical and electronic engineering. Standards in those areas are developed by the Geneva-based International Electrotechnical Commission, which has members from many countries, including the United States.

The purpose and goal of ISO is to improve the climate for international trade by "leveling the playing field." The concept is that by encouraging uniform practices around the world, barriers to trade will be reduced. If the management processes of companies in any other country, then international trade would be made simpler.

2.15.1 Green Accounting Practices and Other Quality Manufacturing and Business Management Paradigms

Readers familiar with various quality industrial management paradigms might have noticed that green accounting and capital budgeting practices are frequently compatible with general strategies for improving manufacturing and business management. These various strategies tend to work together to form a general philosophy of quality improvement; companies and industries that are accustomed to tracking and improving the productivity of labor and capital are just now realizing that it benefits them to do the same for energy and other resources. In quality management, many companies seek external certification of their management systems through ISO or similar organizations at a national level. More recently, ISO standards have been set for management practices.

ISO 9000 lays down guidelines of how to establish and operate an efficient quality assurance system, covering most aspects of a business and its procedures, and specifies that such procedures must be documented in a quality assurance manual (ISO 9000 2018a). How the system is implemented, managed, and periodically reviewed to ensure compliance and continued effectiveness also has to be clearly documented. Customers of any ISO 9000-approved company should feel assured that they are buying from an organization that exercises tight control over its whole business, and the end product will be consistent with the declared specifications. ISO 9000 is a recognized business management standards for quality systems or assurances, and certification is fairly common now. To be certified, companies and businesses must show that they have the required quality management system (QMS) in place (see also Section 8.1).

ISO 14000 has also developed standards for environmental management. ISO environmental management standards are similar to ISO 9000 quality management standards except that they focus on environmental management, of which total cost accounting is a component. Note that ISO 14000 certification is based on whether or not a company has systems in place for managing environmental responsibilities, but not on environment performance. It does NOT require compliance with the regulations of the country in which the company is located. In some countries, it is possible that regulations may be more stringent than the standard. It seems likely, however, that in some companies achieving certification of adherence to the standard would improve the quality of environmental practices in that country. If, as expected, many countries adopt laws that require imported products to have been produced by companies certified to be adhering to ISO 140000, then environmental practices will almost certainly be improved worldwide (ISO 14000 2018b).

ISO 14000 describes in considerable detail what a company must do without prescribing how it must or can be accomplished. Examples of the some components of the ISO 14000 environmental management systems are as follows:

1) Environmental management principles
2) Environmental labeling
3) Environmental performance evaluation
4) Life cycle assessment
5) Principles of environmental auditing
6) Terms and definitions

The environmental management system (EMS) of ISO 14001 is part of the general management system that includes organizational structure, planning activities, responsibilities, practices, procedures, and resources for developing, implementing, achieving, reviewing, and maintaining the environmental policy of an organization. It is a structured process for the achievement of continual improvement related to environmental matters. The facility has the flexibility to define its boundaries and many choose to carry out this standard with respect to the entire organization or to focus the EMS on specific operating units or activities of the organization.

The EMS enables an organization to identify the significant environmental impacts that may have arisen or that may arise from the organization's past, existing, or planned activities, products, or services. It helps the organization to identify relevant environment, legislative, and regulatory requirements that may be imposed on it. Finally, the EMS helps in planning, monitoring, auditing, corrective action, and review activities to assure compliance with established policy and allows a company to be proactive in terms of meeting anticipated new standards and compliance objectives.

The following are the advantages and disadvantages of the ISO 14000 series:

Advantages

1) The ISO 14000 standards provide industry with a structure for managing their environmental problems, which presumably will lead to better environmental performance.
2) It facilitates trade and minimizes trade barriers by harmonization of difference national standards. As a consequence, multiple inspections, certifications, and other conflicting requirements could be reduced.
3) It expands possible market opportunities.
4) In developing countries, ISO 14000 can be used as a way to enhance regulatory systems that are either nonexistent or weak in their environmental performance requirements.
5) A number of potential cost savings can be expected, including
   - increased overall operating efficiency and higher productivity
   - minimized liability claims and risk
   - improved compliance record (avoided fines and penalties)
   - lower insurance rates

Disadvantages

1) Implementation of ISO 14000 standards can be a tedious and expensive process.
2) ISO 14000 standards can indirectly create a technical trade barrier to both small business and developing countries due to limited knowledge and resources (e.g. complexity of the process and high cost of implementation, lack of registration and accreditation infrastructure, etc.)
3) ISO 14000 standards are voluntary. However, some countries may make ISO 14000 standards a regulatory requirement that can potentially lead to a trade barrier for foreign countries who cannot comply with the standards.
4) Certification/registration issues, including
   - the role of self-declaration versus third-party auditing
   - accreditation of the registrars
   - competence of ISO 14000 auditors
   - harmonization and worldwide recognition of ISO 14000 registration

Auditing a facility for certification involves several steps. Proper planning and management are very essential for effective auditing. The (lead) auditor must prepare an audit plan to ensure a smooth audit process. The audit plan must, in general, remain flexible so that any changes to the audit that are found necessary during the actual audit process can be made without compromising the audit.

An audit plan must include the following 10 items:

1) A stated scope and objective(s) for the audit. This includes the reason for conducting the audit, the information required, and the expectation of the audit.
2) Specification of the place, the facility, the date of the audit, and the number of days required to perform the audit.
3) Identification of high-priority items, the facility's and/or organization's EMS.
4) Identification of key personnel who will be involved in the auditing process.
5) Identification of standards and procedures (ISO 14001) that will be used to determine the conformance of various EMS elements.
6) Identification of audit team members, including their special skills, experience, and audit background.
7) Specification of opening and closing meeting times.
8) Specification of confidentiality requirements during the audit process.
9) Specification of the format of the audit report, the language, distribution requirement, and the expected date of issue of the final report.
10) Identification of safety and related issues associated with entry and inspection of various portions of the facility, along with other equipment required to conduct an effective and efficient audit.

Example 2.1 Read the following eight statements regarding the ISO 14000 series of standards. Carefully justify your answer as True (T) or False (F).

1) ISO 14000 standards are based on a principle assumption that better environmental management will lead to better environment performance, increased efficiency, and a greater return on investment. The standards do not explicitly indicate how to achieve these goals, nor prescribe what environmental performance standards an industry must achieve.
2) ISO 14000 standards are regulatory standards developed by the International Organization for Standardization (ISO).
3) ISO 14000 standards are market driven and therefore are based on voluntary involvement of all interests in the marketplace.
4) The adoption of ISO 14000 is a one-time commitment. The company, however, needs to renew the certificate yearly.
5) A main driving force of ISO 14000 standards is the need of the EPA for the replacement of an obsolete regulatory system.
6) A minimum education requirement for an auditor with five years "appropriate work experience" is a high school diploma or equivalent.
7) Companies can only demonstrate compliance through third-party registration.
8) A single ISO certificate can cover several sites or facilities or portions of sites or facilities within a single company.

Solution

1) *True.* ISO 14000 standards are process standards, not performance standards. They do not prescribe to a company what environmental performance they must achieve. They provide a building block for a system to achieve environmental goals. As a consequence, these standards will lead a company to cost saving through better performance of the environmental aspects of an organization's operations.
2) *False.* ISO 14000 standards are international, voluntary standards developed by Technical Committee 207 (TC 207) of the ISO.
3) *True.* ISO 14000 standards are market driven and therefore are based on voluntary involvement of all interests in the marketplace.
4) *False.* The adoption of ISO 14000 standards is a continual commitment. Top management must establish the company's environmental policy and make a commitment to continual improvement and prevention of pollution and to comply is certified; the certificate is normally valid for three years. This may vary depending upon the certification body. The certification body must conduct surveillance audits no less frequently than once a year and carry out a full audit after three years.
5) *False.* USEPA has been participating actively in the standards development process. At present there is no indication of adoption of the ISO 14000 standards as a possible regulatory requirement. The driving force of the ISO 14000 series of standards is mainly from the private sector.
6) *True.* General qualification criteria for environmental management system auditors include education and work experience. Auditors should have completed at least a secondary education or equivalent with five years appropriate working experience or a college degree with two years appropriate work experience.
7) *False.* Companies can demonstrate compliance through either a self-declaration or third-party registration.
8) *True.* Under ISO 14000 certification, a single certificate can cover a specific site of a company, a specific facility, several facilities, or portions of sites or facilities. For example, one ISO 14001 certificate might encompass four different sites of a company in four different states and a portion of a site in a fifth state if these sites are audited at the same time against the same standard.

Example 2.2 A third-party audit is conducted in a pulp and paper mill facility. You are one of the auditors assigned to verify written procedures that are in place and thus to verify the effectiveness of the facility's environmental management system. One of the personnel you are interviewing is the facility's environmental manager. You asked the manager to verify that she has access to all of the regulations and laws that are applicable to the plant. The manager signals you to follow her to a library in the next room. She stops in the library and proudly points toward the four bookcases occupied with texts on environmental laws.

You are impressed but you continue to ask the environmental manager another question. "Can you show me the procedure you use to evaluate regulatory compliance within your facility?" The manager immediately responds by saying that she does not need a procedure for compliance evaluation because she is intimately familiar with what documentation needs to be reviewed when a compliance issue arises. Discuss the above situation with regard to conformance and nonconformance.

Solution

This is a major nonconformance because the facility lacks a procedure to evaluate its compliance with applicable regulations. The ISO 14001 standard, Section 4.2.2, states that "That organization shall establish and maintain a procedure to identify and have access to legal and other requirements to which the organization subscribes directly applicable to the environment aspects of its activities products or services." Thus, the facility is in nonconformance with Section 4.2 and Clause 4.2.2.

The issuance of nonconformance does not mean that the facility cannot be certified. Since this is a major nonconformance, however, the facility must take corrective action and show that the procedure is in place and effective in order to come into conformance with ISO 14001 standards. Having all this in place, the facility may request that the registrar reconsider its certification.

2.16 Current Environmental Regulatory Development in the United States: From End-of-Pipe Laws and Regulations to Pollution Prevention

2.16.1 Introduction

It was 1970, a cornerstone year for the US modern environmental policy. The National Environmental Policy Act (NEPA) was enacted on 1 January 1970. NEPA was not based on specific legislation; instead, it referred in general manner to environmental and quality of life concerns. The Council for Environmental Quality (CEQ), established by NEPA, was one of the councils mandated to implement legislation. 22 April 1970 brought Earth Day, where thousands of demonstrators gathered all around the nation. NEPA and Earth Day were the beginning of a long, seemingly never ending debate over environmental issues.

The US Administration at that time became preoccupied with not only trying to pass more extensive environmental legislation but also implementing the laws. The White House Commission on Executive Reorganization proposed in the Reorganizational Plan of 1970 that a single, independent agency be established, separate from the CEQ. The plan was sent to Congress by President Nixon on 9 July 1970, and a new USEPA began operation on 2 December 1970. The EPA was officially born.

In many ways, the EPA is so broad. The EPA is charged to protect the nation's air, water, and land. The EPA works with the states and local governments to develop and implement comprehensive environmental programs. The key Federal laws such as the Clean Air Act (CAA); the Clean Water Act (CWA); the Safe Drinking Water Act (SDWA); the Resource Conservation and Recovery Act (RCRA); Toxic Substance Control Act (TSCA); the Comprehensive Environmental Response, Compensation and Liability Act (CERCLA – SUPERFUND); the Occupational Safety and Health Act (OSHA); and the Pollution Prevention Act (PPA), all mandate involvement by 50 states and local governments in the detail of implementations.

2.17 Greenhouse Gases

USEPA decided to regulate greenhouse gases as air pollutants under the CAA. Three major development have occurred (USEPA 2011). First, in the Supreme Court case of *Massachusetts v. EPA*, the court ruled in a 5-to-4 decision that the EPA has the statutory authority to regulate CO_2 and other greenhouse gases as air pollutants under the CAA (Cornell University 2007). Second, on 22 September 2009, the EPA administrator signed the Final Mandatory Reporting of Greenhouse Gases Rule. Under this rule large emission sources and suppliers are required to report greenhouse gas emissions (2009a; 2009b). The intention of the rule is to collect accurate data for future policy decision making on climate change mitigation. Third, in December 2009 the EPA made a finding that greenhouse gases endanger human health and welfare, in response to the (CO_2), methane (CH_4), nitrous oxide (N_2O), hydrofluorocarbons (HFCs), perfluorocarbons (PFCs), and sulfur hexafluoride (SF_6). This endangerment finding is a requisite to the EPA developing emission standards for greenhouse gases.

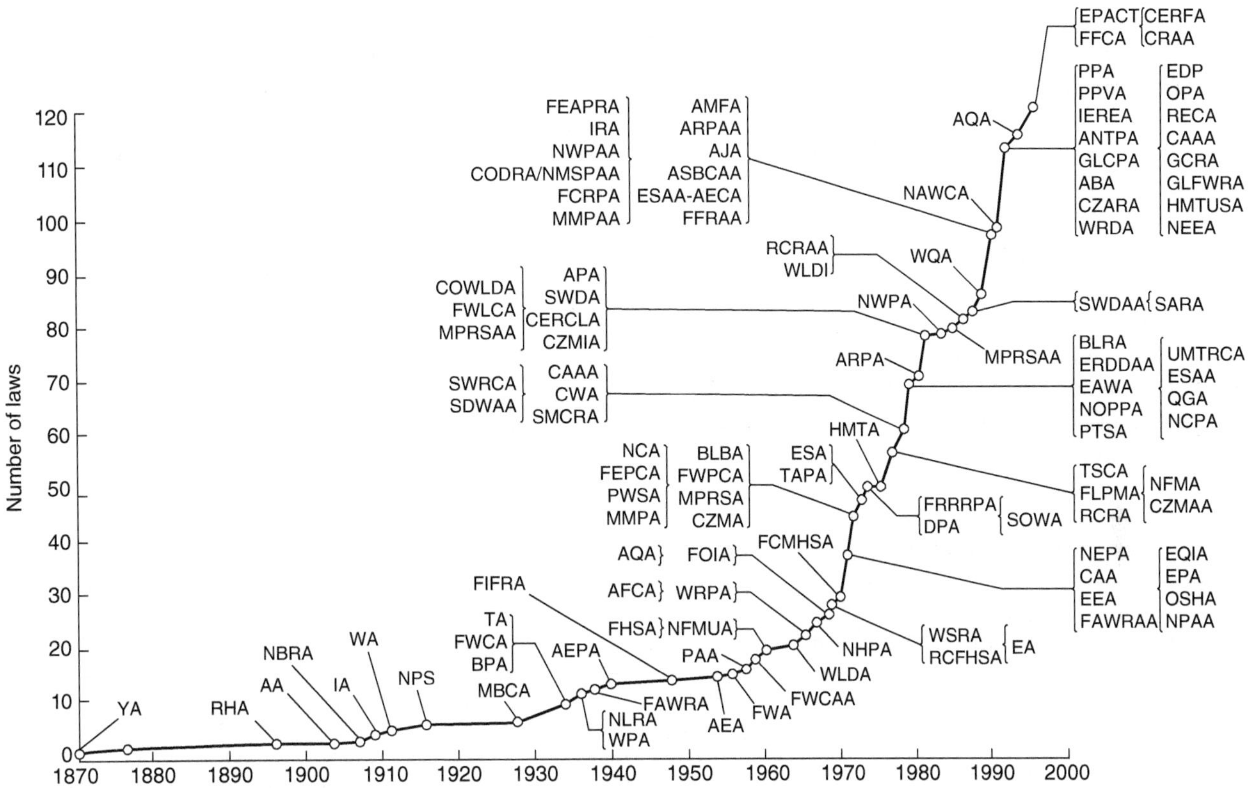

Figure 2.6 Cumulative growth in federal environmental laws and amendments in the United States. *Source:* Adopted from Allen and Shonnard (2002) and Sullivan and Adams (1997).

There are approximately 20 major US federal statues, hundreds of states and local ordinances, thousands of federal and state regulations and even more federal and state court cases and administrative adjustments, etc. that deal with environmental issues. Taken together, they make up the field of environmental law, which has seen remarkable growth in the last 30 some years, as shown in Figure 2.6 (Allen and Shonnard 2002).

All engineers, particularly in the branches of chemical, civil, environmental, mechanical, metallurgy, mining, and nuclear should be familiar with environmental laws and regulations because they affect the operations of many processes and professional responsibilities involved in their respective fields. Environmental regulations and the common law system of environmental law require actions by affected entities. For examples, the CAA (an environmental statute) requires facilities which emit pollutants from a stack (point source) into an air-shed to apply for a *Prevention of Significant Deterioration* (PSD) permit; and the CWA requires facilities that discharge pollutants from a pipe or an outfall (point source) into navigable waters to apply for a *National Pollutant Discharge Elimination System* permit. In many companies, engineers are responsible for applying and obtaining these permits. The common law created by judicial decision also encourages engineers to act responsibly when permitting their professional duties because environmental laws and regulations do not cover every conceivable environmental wrong. Engineers need to be aware of potential legal liability resulting from violation of environmental laws and regulations to protect their companies and themselves from legal and administrative actions.

2.17.1 Nine Prominent Federal Environmental Statues

This section provides some of the key provisions of nine federal environmental statutes that all engineers, in particular chemical engineers, should know. Taken together, these laws regulate chemicals throughout their life cycle, from creation and production to use and disposal.

The nine laws are as follows:

1) The Toxic Substances Control Act (TSCA), 1976 (regulating testing and necessary use restrictions on chemical substances)
2) The Federal Insecticide, Fungicide, and Rodenticide Act (FIFRA), 1972 (the manufacture and use of pesticides)

3) The Occupational Safety and Health Act (OSHA), 1970 (to protect health and safety in the workplace)
4) The Clean Air Act (CAA), 1970 (to protect and enhance the quality of the Nation's air resources)
5) The Clean Water Act (CWA), 1972 (to restore and maintain the physical, chemical, and biological integrity of the Nation's water resources)
6) The Resource Conservation and Recovery Act (RCRA), 1976 (primarily the regulation of hazardous and non-hazardous waste treatment, storage, and safe disposal)
7) The Comprehensive Environmental Response, Compensation, and Liability Act (CERCLA, also known as Superfund), 1980 (the cleanup of abandoned and inactive hazardous waste sites)
8) The Emergency Planning and Community Right-to-Know Act (EPCRA), 1986 (responding to chemical emergencies and reporting of toxic chemical usage)
9) The Pollution Prevention Act (PPA), 1990 (a proactive approach to reducing environmental impact using pollution prevention hierarchy: minimize generation, minimize introduction, segregation, reuse, and recycle, recover energy value in waste, treat for discharge, and safe disposal)

A summary of these prominent federal environmental statues is provided in Table 2.1. The most important regulatory backgrounds for each statute are slated along with a listing of some key provisions for chemical processing facilities.

Table 2.1 Summary table for selected environmental laws.

Environmental statute	Date enacted	Background	Key provisions
Regulation of Chemical Manufacturing The Toxic Substances Control Act (TSCA)	1976	Highly toxic substances, such as polychlorinated biphenyls (PCBs), began appearing in the environment and in food supplies. This prompted the federal government to create a program to assess the risks of chemicals before they are introduced into commerce.	Chemical manufacturers, importers, or processors, must submit a report detailing chemical and processing information for each chemical. Extensive testing by companies may be required for chemicals of concern. For newly created chemicals, a Premanufacturing Notice must be submitted.
The Federal Insecticide, Fungicide, and Rodenticide Act (FIFRA)	Enacted, 1947 Amended, 1972	Because all pesticides are toxic to plants and animals, they may pose an unacceptable risk to human health and the environment. FIFRA is a federal regulatory program whose purpose is to assess the risks of pesticides and to control their usage so that any exposure that may result poses an acceptable level of risk.	Before any pesticide can be distributed or sold in the United States, it must be registered with the EPA. The data is difficult and expensive to develop and must prove that the chemical is effective and safe to humans and the environment. Labels must be placed on pesticide products that indicate approved uses and restrictions.
The Occupational Safety and Health Act (OSH Act)	1970	The agency that oversees the implementation of the OSH Act is the Occupational Safety and Health Administration (OSHA). All private facilities having more than 10 employees must comply with the OSH Act requirements.	Companies must adhere to all OSHA health standards (exposure limits to chemicals) and safety standards (physical hazards from equipment). The OSH Act's Hazard Communication Standard requires companies to develop hazard assessment data (material safety data sheet), label chemical substances, and inform and train employees in the safe use of chemicals.
Regulation of Discharges to the Air, Water, and Soil Clean Air Act (CAA)	1970	The CAA is intended to control the discharge of air pollution by establishing uniform ambient air quality standards that are in some instances health based and in others technology based. The CAA also addresses specific air pollution problems such as hazardous air pollutants, stratospheric ozone depletion, and acid rain.	The CAA established the National Ambient Air Quality Standards (NAAQS) for maximum concentrations in ambient air of CO, Pb, NO_2, O_3, particulate matter, and SO_2. States must develop source-specific emission limits to achieve the NAAQS. States issue air emission permits to facilities. Stricter requirements established for hazardous air pollutants and for new sources.

Table 2.1 (Continued)

Environmental statute	Date enacted	Background	Key provisions
Clean Water Act (CWA)	1972	The CWA is the first comprehensive federal program designed to reduce pollutant discharges into the nation's waterways ("zero discharge" goal). Another goal of the CWA is to make water bodies safe for swimming, fishing, and other forms of recreation ("swimmable" goal). This act is considered largely successful because significant improvements have been made in the quality of the nation's waterways since its enactment.	The CWA established the National Pollutant Discharge Elimination System permit program that requires any point source of pollution to obtain a permit. Permits contain either effluent limits or require the installation of specific pollutant treatment. Permit holders must monitor discharges, collect data, and keep records of the pollutant levels of their effluents. Industrial sources that discharge into sewers must comply with EPA pretreatment standards by applying the best available control technology.
Resource Conservation and Recovery Act (RCRA)	1976	The RCRA was enacted to regulate the "cradle-to-grave" generation, transport, and disposal of both nonhazardous and hazardous wastes to land, encourage recycling, and promote the development of alternative energy sources based on solid waste materials.	Generators must maintain records of the quantity of hazardous waste generated, where the waste was sent for treatment, storage, or disposal, and file this data in biennial reports to the EPA. Transporters and disposal facilities must adhere to similar requirements for record keeping as well as for monitoring the environment.
Clean-Up, Emergency Panning, and Pollution Prevention The Comprehensive Environmental Response, Compensation, and Liability Act (CERCLA)	1980	CERCLA began a process of identifying and cleaning up the many sites of uncontrolled hazardous waste disposal at abandoned sites, industrial complexes, and federal facilities. EPA is responsible for creating a list of the most hazardous sites of contamination, which is termed the National Priority List (NPL). It was amended by the Superfund Amendments and Reauthorization Act (SARA) of 1986.	After a site is listed in the NPL, EPA identifies potentially responsible parties (PRPs) and notifies them of their potential CERCLA liability, which is strict, joint and several, and retroactive. PRPs are (i) present or (ii) past owners of hazardous waste disposal facilities, (iii) generators of hazardous waste, and (iv) transporters of hazardous waste.
The Emergency Planning and Community Right to Know Act (EPCRA)	1986	Title III of (SARA) contains a separate piece of legislation called the (EPCRA). There are two main goals of EPCRA: (i) to have states create local emergency units that must develop plans to respond to chemical release emergencies and (ii) to require EPA to compile an inventory of toxic chemical releases to the air, water, and soil from manufacturing facilities.	Facilities must work with state and local entities to develop emergency response plans in case of an accidental release. Affected facilities must report annually to EPA data on the maximum amount of the toxic substance on-site in the previous year, the treatment and disposal methods used, and the amounts released to the environment or transferred off-site for treatment and/or disposal.
Pollution Prevention Act (PPA)	1990	The act established pollution prevention as the nation's primary pollution management strategy with emphasis on source reduction. Established a Pollution Prevention Information Clearinghouse whose goal is to compile source reduction information and make it available to the public.	The only mandatory provision of the PPA requires owners and operators of facilities that are required to file a Form R under the SARA Title III to report to the EPA information regarding the source reduction and recycling efforts that the facility has undertaken during the previous year.
Total Maximum Daily Load (TMDL) (2000) Section 303(d) of the CWA	2000	Requires states to develop prioritized lists of polluted or threatened water bodies and to establish the maximum amount of pollutant (TMDL) that a water body can receive and still meet water quality standard.	A TMDL is the sum of (i) the individual waste-load allocations (WLAs) for point sources (industrial and municipal), (ii) load allocations for nonpoint sources, (iii) natural background levels, and (iv) a margin of safety (USEPA 2000).

Source: From https://www.epa.gov/sites/production/···/ch3-green-engineering-textbook_508.pdf and USEPA (2000).

Examples (Multiple Choice)

Example 2.3 National Ambient Air Quality Standards (NAAQS) are not promulgated by EPA for which of the following pollutants?

A Nitrogen dioxide (NO_2)
B Sulfur dioxide (SO_2)
C Mercury
D Ozone

Example 2.4 Primary NAAQS are established to

A protect children from toxic air pollutants
B protect the public welfare, including protection against decreased visibility, damage to animals, crops, vegetation, and buildings
C protect the public health, including the health of "sensitive" populations such as asthmatics, children, and the elderly
D both B and C

Example 2.5 A waste is determined to be characteristically toxic under the Resource Conservation and Recovery Act (RCRA) if

A the waste contains specific compounds at greater than threshold concentrations
B it is determined that chemicals in the waste are bioaccumulative, toxic, and persistent in the environment
C leachate from the waste contains specific compounds at greater than threshold concentrations
D exposure to the waste causes adverse health effects in laboratory animals

Solution

The leachability of specific compounds at greater than threshold concentration determines whether a waste is characteristically RCRA hazardous for toxicity.

Example 2.6 Under Superfund legislation, a company can be held liable for all costs associated with remediation of a contaminated site.

A Only if the company disposed of wastes at the site illegally
B Only if the company is solely responsible for contaminated at the site
C Only if the company disposed of waste at the site illegally and is solely responsible for contamination at the site
D If the company disposed of wastes at the site legally and contributed only a small fraction of wastes disposed of at the site

Solution

The strict, joint, and several provisions of the Superfund legislation mean that a company might be held responsible for all of the cleanup costs associated with a site, even if the company's activities were legal and contributed only a small fraction of the contamination at the site.

Example 2.7 ISO 14000 refers to a set of voluntary standards that include

A worker safety standards
B specific techniques for monitoring and sampling
C standards for environmental management systems
D concentrations of pollutants that are allowed after site remediation is completed

Solution

ISO 14000 includes standards for environmental management systems.

Example 2.8 A parts manufacturer uses a solvent degreaser for cleaning parts. The solvent used in the degreaser is a suspected carcinogen. The manufacturer reduces the amount of solvent it disposes of by filtering dirty solvent and returning it to the degreaser. This pollution prevention strategy is known as

A source reduction
B in-process recycling
C out-of-process recycling
D end-of-pipe treatment

Solution

Source reduction would be a reduction in the amount of solvent used; in this case, no changes are made to eliminate or reduce the use of the solvent. An example of out-of-process recycle would be to send the solvent off-site to be recycled. End-of-pipe treatment would consist of treating the solvent to make it less toxic before disposing of it. In this case, the solvent is filtered and returned to the same equipment that created the dirty solvent. Therefore, this is in-process recycling.

Example 2.9 Mixed wastes are wastes that

A contain radioactive wastes regulated under the Atomic Energy Act (AEA) and hazardous wastes regulated under the Resource Conservation and Recovery Act (RCRA)
B are comprised of both solid and liquids
C include both nonhazardous and hazardous wastes regulated under RCRA
D contain municipal solid wastes and hazardous wastes regulated under RCRA

Solution

Mixed waste is defined as a waste mixture that contains both radioactive materials subject to the AEA and a hazardous waste component regulated under the RCRA.

Problems

2.1 Develop a short essay on the water law policies in a country of your choice.

2.2 Develop a short essay on the water law policies in a country in Africa.

2.3 Develop a short essay on the water law policies in a country in Europe.

2.4 Develop a short essay on the water law policies in a country in Asia.

2.5 Provide terms and definitions for "environmental management."
Defining the following terms:
E Pollution prevention
F Source reduction
G In-process vs on-site vs off-site recycling
H Waste treatment
I Disposal
J Direct release

2.6 Analysis of Federal Environmental Statutes.

2.7 The Federal Insecticide, Fungicide, and Rodenticide Act (FIFRA) of 1972.

2.8 Enforcement

2.9 The Occupational Safety and Health Act (OSH Act) of 1970.

2.10 Clean Air Act (CAA) of 1970.

2.11 The Clean Water Act (CWA) of 1972.

2.12 Resource Conservation and Recovery Act (RCRA) of 1976.

2.13 The Comprehensive Environmental Response, Compensation, and Liability Act (CERCLA) of 1980.

2.14 The Emergency Planning and Community Right to Know Act (EPCRA).

2.15 Pollution Prevention Act of 1990.

2.16 U.S. Supreme Court Decision on the Clean Air Act and Greenhouse Gases.

2.17 Superfund Site Investigation Go to the EPA Superfund website.

2.18 Premanufacturing Notice from the TSCA Go to the EPA website.

2.19 Worker Protection Standard under FIFRA.

2.20 Green Jobs and Occupational Safety Issues.

2.21 Programs of the Clean Air Act.

2.22 National Pollutant Discharge Elimination System of the Clean Water Act.

2.23 National Hazardous Waste Biennial Report of RCRA.

2.24 "Quality Improvement": Define the ISO 14000 for an EMS and environmental policy pertinent to product QMS and industrial waste minimization.

References

Adler, J.H. (2003). Fables of the Cuyahoga reconstructing a history of environmental protection. *Fordham Environmental Law Journal*, Case Western Reserve University 14: 95–98, 103–104.

AECEN (Asian Environmental Compliance and Enforcement Network) (2018). https://aric.adb.org/initiative/asian-environmental-compliance-and-enforcement-network (accessed 5 January 2018).

Aldred's Case (1610). 9 Co Rep 57b; (1610) 77 ER 816.

Alexiou, A.S. (2013). *The Flatiron: The New York City Landmark and the Incomparable City that Arose with It*. New York, NY: Thomas Dunne.

Allen, D.T. and Shonnard, D.R. (2002). *Green Engineering: Environmentally Conscious Design of Chemical Processes*. Upper Saddle River, NJ: Prentice-Hall.

Associated Press (2008). Group: stronger warnings needed in Tenn. ash spill, 28 December 2008.

Ayres, R. (1989). Technological Transformations and Long Waves. *IIASA Report, IIASA, Luxemburg, Austria, RR-89-001*.

Babich, A. (2004). The Apolitical Clinic. Law School, Tulane University, Faculty Notebook.

Bajaj, R. (1996). *Convention on International Trade in Endangered Species (CITES) and the Wildlife Trade in India*. New Delhi, India: Center for Environmental Law.

Becker, M. (2011). Correa, indigenous movements, and the writing of a new constitution in Ecuador. *Latin American Perspectives* 38 (1): 47–62.

Beckert, S. (2014). *Empire of Cotton: A Global History*. New York: Vintage Books Division Penguin Random House.

Bell, M.L., Davis, D.L., and Fletcher, T. (2004). *A retrospective assessment of mortality from the London Smog Episode of 1952: the role of influenza and pollution. Environmental Health Perspectives* 112 (1): 6–8.

Bond, E., Gingerich, S., Antonsen, O. A., Purcell, L, and Macklem, E. (2003). *The Industrial Revolution – Innovations*. Industrialrevolution.sea.ca. (accessed 30 January 2011).

Brimblecombe, P. (1976). Attitudes and responses towards air pollution in Medieval England. *Journal of the Air Pollution Control Association* 26 (10): 941–945.

British Institute of International and Comparative Environmental Law (1992). *British Documents of International Environmental and Comparative Law*. Cambridge University Press.

Caldwell, L.K. (1996). *International Environmental Policy: From the Twentieth to the Twenty-First Century*, 3e. Durham, NC and London: Duke University Press.

Canada Water Act, R.S.C. (1985). (accessed 5 January 2018).

Carson, R. (1962). *Silent Spring*. Boston, MA: Houghton-Mifflin Company. https://laws-lois.justice.gc.ca/eng/acts/c-11/page-1.html

Center for Chemical Process Safet (CCPS) (1993). *Guidelines for Engineering Design of Process Safety*. New York, NY: American Institute of Chemical Engineers.

Chakrabarti, R. (ed.) (2006). *Does Environmental History Matter: Sustenance and the Sciences*. Kolkata, India: Readers Service.

Chakrabarti, R. (ed.) (2007). *Situating Environmental History*. New Delhi: Manohar.

Chandler, A.D. (1977). *The Visible Hand: The Managerial Revolution in American Business*. Cambridge, MA: Harvard University Press.

Clean Water Act (CWA) (1972). Federal Water Pollution Control Act Amendment of 1972. USEPA.

CNN (2008). Tennessee sludge spill estimate grows to 1 billion gallons, 26 December 2008.

Commonwealth v Tasmania Dam Case (1983). Tasmania Dam Case. http://www.envlaw.com.au/tasmanian-dam-case (accessed 5 January 2018).

Community Environmental Legal Defense Fund (CELDF). (2017). Community Rights. http://www.celdf.org-community-rights (accessed 6 January 2018).

Cornell University, Law School Legal Information Institute (2007). Supreme Court of the United States, *Massachusetts et al., Petitioners v. Environmental Protection Agency et al.* http://www.law.cornell.edu/supct/pdf/05-1120PZO (accessed 4 January 2018).

Cowan, R.S. (1985). *A Social History of American Technology*, 1e. New York, NY: Oxford University Press.

Cowan, R.S. and Hersch, M. (2017). *A Social History of American Technology*, 2e. New York, NY: Oxford University Press.

CPD (1926). River fire rolls under Baltimore, *Cleveland Plain Dealer*, 9 June 1926.

Craig, R. (2016). *Selected Environmental Law Statutes*. St. Paul, MN: West Academic Publishing.

Cronon, W. (ed.) (1995). *Uncommon Ground: Toward Reinventing Nature*. New York, NY: W.W. Norton.

Crowl, D.A. and Louver, J.F. (1990). *Chemical Process Safety: Fundamentals with Applications*. Englewood Cliff, NJ: Prentice Hall PTR.

Department of the Environmental Act (1985/2009). Government of Canada. Retried 5 January 2018.

Divan, S. and Rosencranz, A. (2002). *Environmental Law and Policy in India: Cases, Materials and Statues*, 2e. New Delhi, India: Oxford University Press.

Eccleston, C.H. (2008). *NEPA and Environmental Planning: Tools, Techniques, and Approaches for Practitioners*. Boca Raton, FL: CRC Press.

Eccleston, C.H. (2014). *The EIS Book: Managing and Preparing Environmental Impact Statements*. Boca Raton, FL: CRC Press.

Eccleston, C.H. (2017). *Environmental Impact Assessment: A Guide to Best Professional Practices*. Boca Raton, FL: CRC Press.

Encyclopedia of the Earth (2018). https://www.google.com/search?q=encyclopedia+of+the+earth+2018&source=univ&tbm=shop&tbo=u&sa=X&ved=0ahUKEwiq2bO85qzkAhUO7J4KHeFOBe8QsxgILg&biw=1366&bih=651 (accessed 4 January 2018).

Environment Protection and Biodiversity Conservation Act 1999 (EPBC) (1999). Australian Government, Department of Environment and Energy. http://www.environment.gov.au/epbc (accessed 5 January 2018).

Environmental Impact Statement (2010). Part 1502, U.S. Council on Environmental Quality.

EPA (2012). African International Programs.

EPA (2017). Collaborate with China. https://www.epa.gov/international-cooperation/epa-collaboration-china (accessed 5 January 2018).

Felleman, J. (2013), Environmental Impact Assessment, *The Encyclopedia of Earth*.

Freeberg, E. (2014). *The Age of Edison: Electric Light and the Invention of Modern America*, 1e. New York: Penguin Books.

Goudie, A. (1981). *The Human Impact: Man's Role in Environmental Change*. Cambridge, MA: The MIT Press.

Govt. of Japan, Ministry of the Environment (1967). The Basic Environmental Law. https://www.env.gov.jp/en/laws/policy.html (accessed 6 January 2018).

Govt. of New Zealand, Ministry of the Environment (1986). Environment Act.

Gudynas, E. (2011). Buen Vivir: Today's tomorrow. *Development* 54 (4): 441–447.

Gupta, S.P. (1999). Forest Projects Committee Annual Report, CPCB, India.

Hansen, J., Ruedy, R., Sato, M., and Lo, K. (2010). "Global surface temperature change," *Reviews of Geophysics*, V-48, Issue RG4004, p. 1–29.

Hardman Reis, T. (2011). *Compensation for Environmental Damages Under International Law*. Hague, The Netherlands: Kluwer Law International.

Hindustan Times, News Item (2000). A.Q.F.M. Yamuna vs. Central Pollution Control Board.

Hiss, T. (2014). Can the world really set aside half of the planet for wildlife? *Smithsonian Magazine*, September.

Hounshell, D.A. (1984). *From the American System to Mass Production, 1800-1932: The Development of Manufacturing Technology in the United States*. Baltimore, MD: Johns Hopkins University Press.

Hynes, P.H. (1989). *Recurring Silent Spring*. New York, NY: Pergamon Press.

International Organization for Standardization (ISO) (2018a). ISO 9000 quality management. www.iso.org (accessed 22 November 2018).

International Organization for Standardization (ISO) (2018b). ISO 14000 family – environmental management. www.iso.org (accessed 23 Novemebr 2018).

Jonnes, J. (2004). *Empires of Light: Edison, Tesla, Westinghouse and the Race to Electrify the World*. New York: Random House Trade Paperback.

Kernan (1958). Kenan v American Dredging Co. 355 US 426, 427.

Knoxville News Sentinel (2008a). Ash Spill: TVA Triples Amount of Sludge Released, 26 December 2008.

Knoxville News Sentinel (2008b). Lead and Thallium Taint Water near TVA Pond Breach, 26 December 2008.

Knoxville News Sentinel (2008c). The Cleanup: Weeks, Millions Needed to Fix Impact from TVA Pond Breach, 27 December 2008.

Koivurova, T. (2014). *Introduction to International Environmental Law*. London/New York: Routledge/Taylor & Francis Group.

Landes, D.S. (1969). *The Unbound Prometheus*, 40. Cambridge: Press Syndicate of the University of Cambridge.

Lazarus, R.J. (2006). *The Making of Environmental Law*. Chicago, IL: Chicago University Press.

Lehman, J. and Phelps, S. (eds.) (2004). *West's Encyclopedia of American Law*, 2e. Detroit: Gale.

Lynch, H. (1995). *A Chemical Engineer's Guide to Environmental Law and Regulation*. Ann Arbor, MI: National Pollution Prevention Center for Higher Education, University of Michigan.

Malik, S. and Malik, S. (eds.) (2015). *Supreme Court on Environment Law*. Lucknow: Eastern Book Company.

McElwee, C.R. (2011). *Environmental Law in China: Mitigating Risk and Ensuring Compliance*. New York: Oxford University Press.

Mehta, M.C. (1987). M.C. Mehta vs. Union of India, 2 SCC 540.

Mehta, M.C. (1988). M.C. Mehta vs. Union of India, 1 SCC 471.

Ministry of Environment Egyptian Environmental Affairs (2009). Environmental protection law. http://www.eeaa.gov.eg/en-us/laws/envlaw.aspx (accessed 6 January 2018).

Ministry of Natural Resources and Environment of the Russian Federation (1986). The Russian Government. www.mnr.gov.ru (accessed 6 January 2018).

Misa, T.J. (1999). *Nation of Steel: The Making of Modern America, 1865–1925*, New e. Baltimore, MD: Johns Hopkins University Press.

Misa, T.J. (2016). *Nation of Steel*, New e. Baltimore, MD: Johns Hopkins University Press.

Muralikrishna, V.I. and Manickam, V. (2017). *Environmental Management: Science and Engineering for Industry*, 1e. Oxford: Butterworth-Heinemann, Elsevier.

National Resources Defense Council (NRDC) (2014). Environmental Law in China. https://act.nrdc.org/donate/donate-monthly?source=MRNRDCc3FR&gclid=EAIaIQobChMIiqTQ6-as5AIVWCCtBh3mUQixEAAYASAAEgIJevD_BwE (accessed 5 January 2018).

Nelson, G., Campbell, S., and Wozniak, P. (2002). *Beyond Earth Day: Fulfilling the Promise*. Madison, WI: University of Wisconsin Press.

NEPA (1969). The National Environmental Policy Act of 1969, as amended, 42 USC Sections 4321-4347 (enacted 1970-01-01) from Council on Environmental Quality NEPAnet.

New York Times (2008). Tennessee ash flood larger than initial estimate, 26 December 2008.

Oreskes, N. and Conway, E.M. (2010). *Merchants of Doubt*. The Christian Science Monitor.

Our Common Future (1987). Brundtland Report of the World Commission on Environment and Development.

Pettit, D. (2014). China's New Environmental Law and the U.S. Clean Air Act.

Pizer, W.A. and Kopp, R. (2003). *Calculating the Costs of Environmental Regulation*, Resources for the Future, Paper 03-06. Washington, DC: Resources for the Future.

Ponting, A.C. (1991). *Green History of the World*. New York, NY: St. Martin's Press.

Powell, F. (2009). Environmental Degradation and Human Disease. Lecture. SlideBoom. 2009. Web. (accessed 14 November 2011).

R v Stephens (1866). LR 1 QB 702.

Rees, J. (2013). *Industrialization and the Transformation of American Life: A Brief Introduction*, 1e. London, New York: Routledge.
Riis, J.A. (1914). *How the Other Half Lives*. Dover Publication.
Rio Declaration, UNESCO (1992). Rio Declaration on Environment and Development, The United Nations, Rio de Janerio, Brazil, 3–14 June 1992.
Robertson v. Methow Valley Citizens Council (1989). 490 U.S. 332, 349.
Rosen, W. (2012). *The Most Powerful Idea in the World: A Story of Steam, Industry and Invention*, 149. University of Chicago Press ISBN: 978-0-226-72634-2.
Ryland v Fletcher (1898). UKHL 1. *Halsbury's Laws of England*, vol. 78. *"The rule in Rylands v Fletcher"* paragraph 148 (5th Ed.)
Scenic Hudson, Inc. (1963). Scenic Hudson collection: records relating to the Storm King Case, 1963–1981. Archives and Special Collections, Marist College, Poughkeepsie, NY.
Secretariat of the Pacific Regional Environmental Program (SPREP) (2012). The 15th Meeting of The Noumea Convention Convenes In Apia. www.sprep.org (accessed 5 January 2018).
Seider, S. (2010). *The German Environmental Constitutional Law: The Basic Law Book*. Government Institutes, Inc.
Shabecoff, P.A. (1993). *A Fierce Green Fire: The American Environmental Movement*. HarperCollins Canada.
South Africa National Environmental Management Act (1998). https://www.gov.za/national-environmental-management-act (accessed 6 January 2018).
Stander, L. and Theodore, L. (2008). *Environmental Regulatory Calculations Handbook*. Hoboken, NJ: Wiley.
Stein, J. and Beckel, M. (2004). A guide to environmental non-profits. *Mother Jones*, March/April Issue.
Stern, R.E. (2013). *Environmental Litigation in China: A Study in Political Ambivalence*. Cambridge University Press.
Sullivan, T.F.P. and Adams, T.L. (1997). *Environmental Law Handbook*. Rockville, MD: Government Institutes.
Superfund Regulations (2017). U.S. Environmental Protection Agency, January 19.
Taylor, P., Stroud, L., and Peteru, C. (2013). *Multilateral Environmental Agreement Negotiator's Handbook: Pacific Region*. Samoa/New Zealand: Secretariat of the Pacific Regional Environment Program/New Zealand Centre for Environmental Law, University of Auckland.
Teeter, P. and Sandberg, J. (2017). Constraining or enabling green capability development? How policy uncertainty affects organizational responses to flexible environmental regulations. *British Journal of Management* 28 (4): 649–665.
The Environment (Protection) Act (1986). http://www.envfor.nic.in/legis/wildlife (accessed 6 January 2018).
The Great Smog of 1952 (2014). Metoffice.gov.uk. Archived from the original. (accessed 12 October 2014).
The Indian Wildlife Protection Act (1972). www.envfor.nic.in (accessed 6 January 2018).
The Observer (2002). *Great Smog is history, but foul air still kills*, 24 November 2002.
Times of India (2000). Supreme Court axe falls on the Delhi Polluting Units, 8 December, p. 1.
US (1974). United States v. Ashland Oil and Transport Co., 504 F.2d 1317, 1326 (6th Circuit).
US Bureau of Land management (2018). About Us. www.blm.gov (accessed 8 January 2018).
Union Carbide vs. Union of India (1989).
United Nations, Department of Economic and Social Affairs, Population Division (2017). World population prespectives: the 2017 revisions. http://eas.un.org/unpd/wpp (accessed 22 January 2018).
UNEP (1992). Training Manual on International Environmental Law: emerging principles and concepts. International Environmental Law, Derived from the 1972 Stockholm Conference, the 1992 Rio Declaration, and more recent developments.
United Nations Secretariat (2015). World Population Projects: Population Division of the Department of Economic and Social Affairs of the United Nations Secretariat. (accessed in 2015).
United States (1972). Clean Water Act, sec. 505, 33 U.S.C.
Upadhyay, J.J.R. (2012). *Environmental Law*, 3e. Allahabad, India: Central Law Agency.
UPI (1984). Significant progress on water pollution reported. *New York Times*, 12 February 1984.
US Fish & Wildlife Service (2018). www.fws.gov (accessed 8 January 2018).
US Forest Service (2018). https://www.fs.fed.us/about-agency (accessed 8 January 2018).
US National Park Service (2018). *About Us*. Washington, DC. https://www.nps.gov/state/dc/index.htm (accessed 8 January 2018).
USEPA (2000). *Total Maximum Daily Load (TMDL)*, EPA 841-F-00-009. Washington, DC: USEPA.
USEPA (2009a). Final Mandatory Reporting of Greenhouse Gases Rule. http://www.epa.gov/climatechange/emissions/ghgrulemaking.html (accessed 17 January 2018).
USEPA (2009b). Endangerment and cause or contribute findings for greenhouse gases under the Clean Air Act. http://www.epa.gov/climatechange/endangerment.html (accessed 17 January 2018).
USEPA (2011). *PSD and Title V Permitting Guidance for Greenhouse Gases*. Research Triangle Park, NC: Office of Air Quality Planning and Standards.
USEPA (2017a). Laws and regulations.
USEPA (2017b). Regulatory information by topic.
USEPA (2018). Health and Environmental Agencies of U.S. States and Territories. https://www.epa.gov/home/health-

and-environmental-agencies-us-states-and-territories (accessed 8 January 2018).

Valley Precipitation (2008). TVA website (accessed 28 December 2008).

Vellore Citizen (1996). Vellore Citizens' Welfare Forum vs. Union of India, 5 SCC 647.

Vietnam Environment Administration (2002). Ministry of Natural Resources and Environment.

Wallach, B. (2005). *Understanding the Cultural Landscape.* New York: Guilford.

Wang, A. (2013). The search for sustainable legitimacy: environmental law and bureaucracy in China. *Harvard Environmental Law Review* 37: 365.

Warner, S.B. (1995). *The Urban Wilderness: A History of the American City.* Berkeley: University of California Press.

Water (2011). Climate Institute. Web. (accessed 3 November 2011).

Watts, S. (2005). *The People's Tycoon: Henry Ford and the American Century.* New York: Random House.

Wells, N.E. (1984). *A Guide to Environmental Law in New Zealand.* Wellington Brooker & Friend Ltd.

White, C. (2008) Dike bursts, floods 12 homes, spills into Watts Bar Lake, *Knoxville News Sentinel,* 22 December 2008.

Wickham, R.J. (1916). *English and American Tool Builders.* New Haven, CT: Yale University Press, LCCN 16011753. Reprinted by McGraw-Hill, New York and London, 1926 (LCCN 27-24075); and by Lindsay Publications, Inc., Bradley, IL.

World Health Organization (WHO) Commission on Health and Environment (1992). Our Planet, Our Health. Report of the WHO Commission, Geneva.

Wright, C.W. (1941). *Economic History of the United States,* 1e, 1941. New York: McGraw Hill.

UNEP (2001) UNEP Manual, 12–19.

UNEP (1986) The Environment (Protection) Act, UNEP Manual, 20–23.

UNEP (2006). Training Manual on International Environmental Law, UNEP Manual, 24–28.

UNEP (2013). UNEP Manual, 58.

UNEP (1992). Declaration of the United Nations Conference on the Human Environment, UNEP Manual, 58, Rio Declaration Principle 16.

3

Industrial Pollution Sources, Its Characterization, Estimation, and Treatment

3.1 Introduction

This chapter provides a summary of industrial wastewater sources, wastewater characteristics, wastewater treatment, reuse and discharge, industrial sources of air pollutions, inventories, air pollution control, solid waste and hazardous waste characteristics, treatments, and management.

Industrial waste is the waste produced by industrial activity which includes any material that is rendered useless during a manufacturing process such as that of factories, industries, mills, and mining operations. It has existed since the start of the Industrial Revolution (Pink 2006). Some examples of industrial wastes are chemical solvents, paints, sandpaper, paper products, industrial by-products, metals, plastics, and radioactive wastes.

Toxic waste, chemical waste, industrial solid waste, and municipal solid waste are designations of industrial wastes. Sewage treatment plants can treat some industrial wastes, i.e. those consisting of conventional pollutants such as biochemical oxygen demand (BOD), chemical oxygen demand (COD), suspended solid (SS), and total suspended solid (TSS). Industrial wastes containing toxic pollutants require specialized treatment systems (United States Code, Clean Water Act, Section 402(p) 33 U.S. Code §1342(p) 1999).

3.2 Wastewater Sources

3.2.1 Point Source

Point source water pollution refers to contaminants that enter a waterway from a single, identifiable source, such as a pipe or ditch. Examples of sources in this category include discharges from a factory, a sewage treatment plant or a publicly owned treatment works (POTWs), or a city storm drain. The US Clean Water Act (CWA) defines point source for regulatory enforcement purposes (United States Code Clean Water Act Section 502 (14) 33 U.S.S 1362(14) 1999). The CWA definition of point source was amended in 1987 to include municipal storm sewer systems, as well as industrial storm water, such as from construction sites.

3.2.2 Nonpoint Source

Nonpoint source (NPS) pollution refers to diffuse contamination that does not originate from a single discrete source. NPS pollution is often the cumulative effect of small amounts of contaminants gathered from a large area. A common example is the leaching out of nitrogen compounds from fertilized agricultural lands (Moss 2008). Nutrient runoff in storm water from "sheet flow" over an agricultural field or a forest is also cited as examples of NPS pollution.

Contaminated storm water washed off of parking lots, roads and highways, called urban runoff, is sometimes included under the category of NPS pollution. However, because this runoff is typically channeled into storm drain systems and discharged through pipes to local surface waters, it becomes a point source.

Fugitive emissions are also NPS pollution in that the emissions of gases or vapors take place from pressurized equipment due to leaks and other unintended or irregular releases of gases, mostly from industrial activities. As well as the economic cost of lost commodities, fugitive emissions contribute to air pollution and climate change.

3.3 Wastewater Characteristics

Prior to about 1940, most municipal wastewater was generated from domestic sources. After 1940, as industrial development in the United States grew significantly, increasing amounts of industrial wastewater have been and continue to be discharged into municipal collection systems. The amounts of heavy metals and synthesized organic compounds generated by industrial activities have increased; some 10000 new organic compounds are added each year. Many of these compounds are now found in the wastewaters.

Industrial Environmental Management: Engineering, Science, and Policy, First Edition. Tapas K. Das.

Companion website: www.wiley.com/go/Das/IEM_1e

As technological changes take place in manufacturing, changes also occur in the compounds discharged and the resulting wastewater characteristics. Numerous compounds generated from industrial processes are difficult and costly to treat by conventional wastewater treatment processes. Therefore, effective industrial pretreatment becomes an essential part of an overall water quality management program. Enforcement of an industrial pretreatment program is a daunting task; some of the regulated pollutants still escape to the municipal wastewater collection system and must be treated. In the future with the objective of pollution prevention, every effort should be made by industrial discharges to assess the environmental impacts of any new compounds that may enter the wastewater stream before being approved for use. If a compound cannot be treated effectively with existing technology, it should not be used.

The wastewater from industries varies greatly in both flow and concentration of pollutants. So, it is impossible to assign fixed values to their constituents. In general, industrial wastewaters may contain suspended, colloidal, and dissolved (mineral and organic) solids. In addition, they may be either excessively acidic or alkaline and may contain high or low concentrations of colored matter. These wastes may contain inert, organic, or toxic materials and possibly pathogenic bacteria. These wastes may be discharged into the sewer system provided they have no adverse effect on treatment efficiency or undesirable effects on the sewer system. It may be necessary to pretreat (Section 3.5.2) the wastes prior to release to the municipal system or it is necessary to make a full treatment when the wastes will be discharged directly to surface or ground waters.

The physical and chemical characterization presented below is valid for most wastewaters, both industrial and municipal.

3.3.1 Physical Characteristics

The principal physical characteristics of wastewater include solids content, color, odor, and temperature.

3.3.2 Total Suspended Solids

The total solids in a wastewater consist of the insoluble or total suspended solids and the soluble compounds dissolved in water. The suspended solids content is found by drying and weighing the residue removed by the filtering of the sample. When this residue is ignited the volatile solids are burned off. Volatile solids are presumed to be organic matter, although some organic matter will not burn and some inorganic salts break down at high temperatures. The organic matter consists mainly of proteins, carbohydrates and fats. Between 40 and 65% of the solids in an average wastewater are suspended. Settleable solids, expressed as milligrams per liter (mg/l), are those that are removed by sedimentation. Usually about 60% of the suspended solids in a wastewater are settleable (Crites and Tchbanoglous 1998). Solids may be classified in another way as well: those that are volatilized at a high temperature (600 °C) and those that are not. The former are known as volatile solids, the latter as fixed solids. Usually, volatile solids are organic.

3.3.3 Color

Color is a qualitative characteristic that can be used to assess the general condition of wastewater. Wastewater that is light brown in color is less than six hours old, while a light-to-medium grey color is characteristic of wastewaters that have undergone some degree of decomposition or that have been in the collection system for some time. Lastly, if the color is dark grey or black, the wastewater is typically septic, having undergone extensive bacterial decomposition under anaerobic conditions. The blackening of wastewater is often due to the formation of various sulphides, particularly ferrous sulphide. This results when hydrogen sulphide produced under anaerobic conditions combines with divalent metal, such as iron, which may be present. Color is measured by comparison with standards.

3.3.4 Odor

The determination of odor has become increasingly important, as the general public has become more concerned with the proper operation of wastewater treatment facilities. The odor of fresh wastewater is usually not offensive, but a variety of odorous compounds are released when wastewater is decomposed biologically under anaerobic conditions. The different unpleasant odors produced by certain industrial wastewater are presented in Table 3.1.

3.3.5 Temperature

The temperature of wastewater is commonly higher than that of the water supply because warm municipal water has been added. The measurement of temperature is important because most wastewater-treatment schemes include biological processes that are temperature dependent. The temperature of wastewater will vary from season to season and also with geographic location. In cold regions, the temperature will vary from about 7 to 18 °C, while in warmer regions the temperatures vary from 13 to 24 °C (Crites and Tchobanoglous 1998).

Table 3.1 Unpleasant odors in some industries.

Industries	Origin of odors
Cement works, Lime Kilns	Dibutyl amines, mercaptans, dibutylsulfide, hydrogen sulfide, sulfur dioxide
Food industries	Acetic acid, acetaldehyde
Food industries (fish)	Butyl amine, mercaptans, dimethyl sulfide, amines
Pharmaceutical industries	Fermentation by-product produces
Pulp and paper industries (kraft)	Total reduced sulfur compounds, TRS: (hydrogen sulfide, mercaptans, methyl disulfide, dimethyl disulfide), sulfur dioxide
Rubber industries	Sulfides, mercaptans
Textile industries[a]	Phenyl mercaptan, phenolic compounds
Tomato cannery	Acetic acid, acetaldehyde, thiophenol

Source: From Eckenfelder (2000).
[a] Yohe and Rich (1995).

3.4 Chemical Characteristics

3.4.1 Inorganic Chemicals

The principal chemical tests include free ammonia, inorganic nitrogen as nitrate, nitrite, organic phosphorus, and inorganic phosphorus. Nitrogen and phosphorus are important because these two nutrients are responsible for the growth of aquatic plants. Other tests such as chloride, sulphate, pH, and alkalinity are performed to assess the suitability of reusing treated wastewater and in controlling the various treatment processes (Rouessac and Rouessac 2007).

Trace elements, which include some heavy metals, are not determined routinely, but trace elements may be a factor in the biological treatment of wastewater. All living organisms require varying amounts of some trace elements such as iron, copper, zinc, and cobalt for proper growth. Heavy metals can also produce toxic effects; therefore, determination of the amounts of heavy metals is especially important where the further use of treated effluent or sludge is to be evaluated. Many metals are also classified as priority pollutants such as arsenic, cadmium, chromium, mercury, etc.

Measurements of gases such as hydrogen sulphide, oxygen, methane, and carbon dioxide are made to help the system to operate. The presence of hydrogen sulphide needs to be determined not only because it is an odorous and very toxic gas but also because it can affect the maintenance of long sewers on flat slopes, since it can cause corrosion. Measurements of dissolved oxygen are made in order to monitor and control aerobic biological treatment processes. Methane and carbon dioxide measurements are used in connection with the operation of anaerobic digesters.

3.4.2 Organic Chemicals

Over the years, a number of different tests have been developed to determine the organic content of wastewaters. In general, the tests may be divided into those used to measure gross concentrations of organic matter greater than about 1 mg/l and those used to measure trace concentrations in the range of 10^{-12}–10^{-3} mg/l. Laboratory methods commonly used today to measure gross amounts of organic matter (>1 mg/l) in wastewater include (i) BOD, (ii) COD, and (iii) total organic carbon (TOC). Trace organics in the range of 10^{-12}–10^{-3} mg/l are determined using instrumental methods including gas mass spectroscopy and chromatography. Specific organic compounds are determined to assess the presence of priority pollutants (Metcalf & Eddy 2003). The BOD, COD, and TOC tests are gross measures of organic content and as such do not reflect the response of the wastewater to various types of biological treatment technologies.

3.4.3 Volatile Organic Compounds

Volatile organic compounds (VOCs), such as benzene, toluene, xylenes, trichloroethane, dichloromethane, and trichloroethylene (TCE), are common soil pollutants in industrialized and commercialized areas. One of the more common sources of these contaminants is leaking underground storage tanks. Improperly discarded solvents and landfills, built before the introduction of current stringent regulations, are also significant sources of soil VOCs. Many of organic substances are classified as priority pollutants such as PCBs, polycyclic aromatic, acetaldehyde, formaldehyde, 1,3-butadiene, 1,2-dichloroethane, dichloromethane, hexachlorobenzene, etc. In Table 3.2, a list of typical inorganic and organic substances present in industrial effluents is presented.

3.4.4 Heavy Metal Discharges

Several industries discharge heavy metals, it can be seen that of all of the heavy metals, chromium is the most widely used and discharged to the environment from different sources. As shown in Figure 3.1, many of the pollutants entering aquatic ecosystems (e.g. mercury lead, pesticides, and herbicides) are very toxic to living organisms. They can lower reproductive success, prevent proper growth and development, and even cause death.

Table 3.2 Substances present in industrial effluents.

Substances	Present in wastewaters from
Acetic acid	Acetate rayon, beet root manufacture
Acids	Chemical manufacture, mines, textiles manufacture
Alkalies	Cotton and straw kiering, wool scouring
Ammonia	Gas and coke and chemical manufacture
Arsenic	Wood treatment, galvanizing process
Benzene	Hydraulic fracking
Cadmium	Plating
Chromium	Plating, chrome tanning, alum anodizing
Citric acid	Soft drinks and citrus fruit processing
Copper	Copper plating, copper pickling
Cyanides	Gas manufacture, plating, metal cleaning
Fats, oils, grease	Wool scouring, laundries, textile industry
Fluorides	Scrubbing of flue gases, glass etching
Formaldehyde	Synthetic resins and penicillin manufacture
Free chlorine	Laundries, paper mills, textile bleaching
Hydrocarbons	Petrochemical and rubber factories
Free chlorine	Laundries, paper mills, textile bleaching
Mercaptans	Oil refining, pulp
Nickel	Plating
Nitro compounds	Explosives and chemical works
Organic acids	Distilleries and fermentation plants
Phenols	Gas and coke manufacture, chemical plants
Starch	Food processing, textile industries
Sugars	Dairies, breweries, sweet industry
Sulfides	Textile industry, tanneries, gas manufacture, fracking
Sulfites	Pulp processing, viscose film manufacture
Tannic acid	Tanning, sawmills
Tartaric acid	Dyeing, wine, leather, chemical manufacture
Toluene, VOC	Hydraulic fracking

Source: From Bond and Straub (1974).

However, chromium is not the metal that is most dangerous to living organisms. Much more toxic are cadmium, lead, and mercury. These have a tremendous affinity for sulfur and disrupt enzyme function by forming bonds with sulfur groups in enzymes. Protein carboxylic acid ($-CO_2H$) and amino ($-NH_2$) groups are also chemically bound by heavy metals. Cadmium, copper, lead, and mercury ions bind to cell membranes, hindering transport processes through the cell wall. Heavy metals may also precipitate phosphate biocompounds or catalyze their decomposition.

Figure 3.1 Discharge of untreated industrial wastewater to a river.

3.4.5 Some Inorganic Pollutants of Concern

Cyanide ion, CN-, is probably the most important of the various inorganic species in wastewater. Cyanide, a deadly poisonous substance, exists in water as HCN which is a weak acid. The cyanide ion has a strong affinity for many metal ions, forming relatively less toxic ferrocyanide, $Fe(CN)_6^{4-}$, with iron (II), for example. Volatile HCN is very toxic and has been used in gas chamber executions in the United States. Cyanide is widely used in industry, especially for metal cleaning and electroplating. It is also one of the main gas and coke scrubber effluent pollutants from gas works and coke ovens. Cyanide is widely used in certain mineral processing operations.

Ammonia is the initial product of the decay of nitrogenous organic wastes, and its presence frequently indicates the presence of such wastes. It is a normal constituent of some sources of groundwater and is sometimes added to drinking water to remove the taste and odor of free chlorine. Since the pK_a (the negative log of the acid ionization constant) of the ammonium ion, NH_4^+, is 9.26, most ammonia in water is present as NH_4^+ rather than NH_3.

Hydrogen sulphide, H_2S, is a product of the anaerobic decay of organic matter containing sulfur. It is also produced in the anaerobic reduction of sulphate by microorganisms and is developed as a gaseous pollutant from geothermal waters. Wastes from chemical plants, paper mills, textile mills, and tanneries may also contain H_2S. Nitrite ion, NO^{2-}, occur in water as an intermediate oxidation state of nitrogen. Nitrite is added to some industrial processes to inhibit corrosion; it is rarely found in drinking water at levels over 0.1 mg/l. Sulphite ion, SO_3^{2-}, is found in some industrial wastewaters. Sodium sulphite is commonly added to boiler feed-waters as an oxygen scavenger:

$$2SO_3^{2-} + O_2 \rightarrow 2SO_4^{2-}$$

3.4.6 Organic Pollutants of Concern

Effluent from industrial sources contains a wide variety of pollutants, including organic pollutants. Primary and secondary sewage treatment processes remove some of these pollutants, particularly oxygen-demanding substances, oil, grease, and solids. Others, such as refractory (degradation-resistant) organics (organochlorides, nitro compounds, etc.), salts, and heavy metals, are not efficiently removed. Soaps, detergents, and associated chemicals are potential sources of organic pollutants. Most of the environmental problems currently attributed to detergents do not arise from the surface-active agents which basically improve the wetting qualities of water. The greatest concern among environmental pollutants has been caused by polyphosphates added to complex calcium functioning as a builder.

Bio-refractory organics are poorly biodegradable substances, prominent among which are aromatic or chlorinated hydrocarbons (benzene, bornyl alcohol, bromobenzene, chloroform, camphor, dinitrotoluene, nitrobenzene, styrene, etc.). Many of these compounds have also been found in drinking water. Water contaminated with these compounds must be treated using physical and chemical methods, including air stripping, solvent extraction, ozonation, and carbon adsorption.

First discovered as environmental pollutants in 1966, polychlorinated biphenyls (PCB compounds) have been found throughout the world in water, sediments, and in bird and fish tissues. They are made by substituting between 1 and 10 Cl atoms onto the biphenyl aromatic structure. This substitution can produce 209 different compounds (Rouessac and Rouessac 2007).

3.4.7 Thermal Pollution

Considerable time has elapsed since the scientific community and regulatory agencies officially recognized that the addition of large quantities of heat to a recipient possesses the potential of causing ecological harm. The really significant heat loads result from the discharge of condenser cooling water from the ever-increasing number of steam electrical generating plants and equivalent-sized nuclear power reactors. Large numbers of power plants currently require approximately 50% more cooling water for a given temperature rise than that required of fossil-fuel plants of an equal size. The degree of thermal pollution depends on thermal efficiency, which is determined by the amount of heat rejected into the cooling water. Thermodynamically, heat should be added at the highest possible temperature and rejected at the lowest possible temperature if the greatest amount of effect is to be gained and the best thermal efficiency realized. The current and generally accepted maximum operating conditions for conventional thermal stations are about 500 °C and 24 MPa, with a corresponding heat rate of 2.5 kWh, 1.0 kWh resulting in power production and 1.5 kWh being wasted. Plants have been designed for 680 °C and 34 MPa; however, metallurgical problems have kept operating conditions at lower levels.

Nuclear power plants operate at temperatures from 250 to 300 °C and pressures of up to 7 MPa, resulting in a heat rate of approximately 3.1 kWh. Thus, for nuclear plants, 1.0 kWh may be used for useful production, whereas 2.1 kWh is wasted. Most steam-powered electrical generating plants are operated at varying load factors, and consequently the heated discharges demonstrate wide variation with time. Thus, the biota is not only subjected to increased or decreased temperature, but also to a sudden or "shock," temperature change. Increased temperature will cause remarkable reduction in the self-purification capacity of a receiving water body and cause the growth of undesirable algae (Krenkel and Novotny 1980). The addition of heated water to the receiving water can be considered equivalent to the addition of sewage or other organic waste material, since both pollutants may cause a reduction in the oxygen resources of the receiving waters. Also, elevated temperatures in the receiving water could cause undesirable algae bloom (Hauser 2018; USEPA 2018).

3.5 Industrial Wastewater Variation

3.5.1 Pollution Load and Concentration

In most industries, wastewater effluents result from the following water uses:

1) Sanitary wastewater (from washing, drinking, etc.)
2) Cooling (from disposing of excess heat to the environment)
3) Process wastewater (including both water used for making and washing products and for removal and transport of waste and by-products)
4) Cleaning (including wastewater from cleaning and maintenance of industrial areas)

Excluding the large volumes of cooling water discharged by the electric power industry, the wastewater production from urban areas is about evenly divided between industrial and municipal sources. Therefore, the use of water by industry can significantly affect the water quality of receiving waters. The level of wastewater loading from industrial sources varies markedly with the water quality objectives enforced by the regulatory agencies. There are many possible in-plant changes, process modifications, and water-saving measures through which industrial wastewater loads can be significantly reduced. Up to 90% of recent wastewater reductions have been achieved by industries employing such methods as recirculation, operation modifications, effluent reuse, or more efficient operation. As a rule, treatment of an industrial effluent is much more expensive without water-saving measures than the total cost of in-plant modifications and residual effluent treatment. Industrial wastewater effluents are usually highly variable, with quantity and quality variations brought about by bath discharges, operation start-ups and shut-downs, working-hour distribution, and so on. A long-term detailed survey is usually necessary before a conclusion on the pollution impact from an industry can be reached. Because of the wide variety of industries and levels of pollutants, we can present a snapshot view of the characteristics. The values of typical concentration of conventional pollutants (BOD_5, COD, TSS) and pH for different industrial effluents are given in Table 3.3. A similar sampling for nonconventional pollutants is given in Table 3.4.

3.5.2 Industrial Pretreatment

Industrial wastewater may contain pollutants which cannot be removed by conventional sewage treatment. Also, variable flow of industrial waste associated with production cycles may upset the population dynamics of biological

Table 3.3 Comparative strengths of industrial wastewaters for conventional pollutants.

Type of waste	BOD5 (mg/l)	COD (mg/l)	TSS (mg/l)	pH
Apparel				
Cotton	200–1000	400–1800	200	8–12
Wool scouring	2000–5000	2000–5000[a]	3000–30000	9–11
Wool composite	1	—	100	9–10
Tannery	1000–2000	2000–4000	2000–3000	11–12
Laundry	1600	2700	250–500	8–9
Food				
Brewery	850	1700	90	4–8
Distillery	7	10	Low	—
Dairy	600–1000	—	200–400	<7
Agriculture				
Citrus	2000	—	7000	Acid
Pea	570	—	130	<7
Slaughterhouse	1500–2500	—	400–1000	7–8
Potato processing	2000	3500	2500	11–13
Sugar beet	450–2000	600–3000	800–1500	7–8
Farm	1000–2000	—	1500–3000	7.5–8.5
Poultry	500–800	600–1050	450–800	6.5–9
Industries				
Pulp; sulfite	1400–1700	—	Variable	
Pulp; kraft	100–350	170–600	75–300	7–9.5
Paperboard	100–450	300–1400	40–100	
Strawboard	950	—	1350	
Coke oven	780	1650[a]	70	7–11
Oil refinery	100–500	150–800	130–600	2–6

[a] COD as $KMnO_4$ mg O_2/l.

Table 3.4 Examples of industrial wastewater concentrations for nonconventional pollutants.

Industry	Pollutant	Concentration (mg/l)
Coke by-product (steel mill)	Ammonia (as N)	200
	Organic nitrogen (as N)	100
	Phenol	2000
Metal plating	Chromium (VI)	3–550
Nylon polymer	COD	23000
	TOC	8800
Synthetic textile	COD	3300
	Nitrogen	40
Meat processing and packing	COD	2100
	Nitrogen	150
	Phosphorus	16
	Grease	500
	Temperature	28 °C
Plywood-plant glue waste	COD	2000
	Phenol	200–2000
	Phosphorus (as PO_4)	9–15
Chlorophenolic manufacture	Chloride	27000
	Phenol	140

treatment units, such as the activated sludge process. Thus, industrial wastewaters can pose serious hazardous to municipal systems because the collection and treatment systems have not been designed to carry or treat them.

Highly regulated industrial effluent usually receives at least pretreatment, if not full treatment, at the factories themselves to reduce the pollutant load, before discharge to the sewer. Using proven treatment technologies and manufacturing practices that promote recycling, industries can remove or eliminate pollutants before discharge of wastewater. This practice is called industrial wastewater treatment or pretreatment.

The pretreatment program is a component of the USEPA (United States Environmental Protection Agency) national NPDES (National Pollutant Discharge Elimination System) program. It is a cooperative effort of federal, state, and local environmental regulatory agencies established to protect water quality. Similar to how USEPA authorizes the NPDES permit program to state, tribal, and territorial governments to perform permitting, administrative, and enforcement tasks for discharges to surface waters (NPDES program), EPA and authorized NPDES state pretreatment programs approve local municipalities to perform permitting, administrative, and enforcement tasks for discharges into the municipalities' POTWs. A treatment works (as defined by CWA section 212) is owned by a state or municipality [as defined by CWA section 502(4)]. This definition includes any devices or systems used in the storage, treatment, recycling, and reclamation of municipal sewage or industrial wastes of a liquid nature. It also includes sewers, pipes, or other conveyances only if they convey wastewater to a POTW treatment plant.

The national pretreatment program identifies specific discharge standards and requirements that apply to sources of nondomestic wastewater discharged to a POTW. By reducing or eliminating waste at the industries ("source reduction"), fewer toxic pollutants are discharged to and treated by the POTW, providing benefits to both the POTWs and the industrial users. Oversight of the program is by the POTW, state, or the EPA. The control authority issues permits for discharge of industrial wastewaters to the POTW. The permits contain numerical limits for controlled pollutants and requirements for sampling, flow measurements, laboratory testing, and reporting.

3.6 Industrial Wastestream Variables

3.6.1 Dilute Solutions

The discharges from continuous manufacturing processes are normally dilute solutions of compatible and sometimes nonconventional pollutants. They may be discharged to the industry's pretreatment system or directly to the POTW without any pretreatment. Manufacturing processes such as plating bath rinses, raw food cleaning, and crude oil dewatering are all examples of dilute solutions of pollutants that may be discharged directly to a POTW sanitary sewer. If a problem occurs in the manufacturing process, a probable result is that the quality of wastewater will change; it may be more laden with pollutants. Some wastestreams from utility services, such as cooling tower and boiler blowdown, are continuous and represent the discharge of dilute solutions.

Another low-strength wastewater is storm water runoff from chemical handling and storage areas. Products which may have spilled on the industry's grounds are washed off during a rainstorm or during the spring thaw. The pollutant concentration is usually too dilute to require pretreatment before discharge to the sewer, but it exceeds the discharge standards for discharge to surface waters. While the strength of the storm runoff may be low, the volume that must be treated in addition to normal flow to the pretreatment system or to the POTW can cause hydraulic capacity problems. Excessive flows can be diverted to storage reservoirs or basins and then gradually discharged to the pretreatment system. A great deal of attention is presently focused on cleaning up groundwater sources that have been contaminated by leaking underground storage tanks. Cleanup projects of this nature typically involve large quantities of wastes that may contain high concentrations of solvents, fuels, heavy metals, and pesticides. Because of the public attention surrounding groundwater cleanup projects, pretreatment of the contaminated water is almost always required and the result is usually a "high-quality" industrial wastewater.

3.6.2 Concentrated Solutions

Typically, concentrated solutions are batch-generated and the frequency of generation is usually not daily but weekly, monthly, annually, or even longer. These solutions are process chemicals or products that cannot be reconditioned or reused in the same manufacturing process. Concentrated solutions such as spent plating baths, acids, alkalies, static drag out solutions, and reject product may have concentrations of pollutants hundreds or thousands of times higher than the discharge limits of the POTW or higher than can be adequately treated by the pretreatment system if discharged all at once. Time has to be taken to examine and understand each manufacturing process, then identify these concentrated solutions and take the necessary steps to prevent damage to the treatment facilities.

Some wastes may be considered concentrated by the POTW but not by the industry. For example, the 10% sulfuric acid solution used for pickling parts is considered "dilute" by comparison to the 98% or 505 stock solution that the industry uses to make up the pickling solution. When this solution is spent or can no longer be used as a pickling solution, proper treatment and disposal are required. From the industrial manufacturer's point of view, the solution is spent and no longer concentrated. However, from a wastewater treatment point of view, the solution is concentrated since it contains high concentrations of acid ($pH < 1.0$) and heavy metals (1000 mg/l) compared to the normal pH of 1.0–4.0 and heavy metal concentrations of less than 100 mg/l (IWT 1999). Another source of concentrated solutions is the wastewater from equipment cleanup. While the amount of material in the process chemical bath may be considered dilute by industry standards, it forms a concentrated wastestream when discharged during the cleanup of manufacturing equipment. Cleanup wastestreams contain a high concentration of the product during the first washing of the tank, pipe, or pump. This discharge of concentrated waste is followed by successive rinses which contain less and less pollutants. If cleanup flow concentrations are not equalized, the cleanup cycle can cause problems in the industrial waste treatment system (IWTS). Spills of process chemicals to the floor, if not contained, can flow directly to the floor drain and the pretreatment or sewer system. The adverse effects on the pretreatment system and POTW are the same as those of any other concentrated solutions. This is why chemical containment areas must not have drains.

3.7 Concentration vs. Mass of the Pollution

An understanding of the concentration and the mass of a pollutant in an industrial waste is needed to determine the effects on the industry's pretreatment system, the POTW collection, treatment, and disposal systems, and the sampling of the industry's discharge. The concentration of a substance in wastewater is normally expressed as milligrams per liter and is a measurement of the mass per unit of volume. The mass of a substance is normally expressed in pounds or kilograms and is a weight measurement. A mass discharge rate is a measurement of weight per unit time and is usually expressed as pounds or kilograms per day. Many of the electroplating and all of the metal finishing categorical standards are written in concentrations, whereas most of the other categorical standards are written as mass discharge rate standards. The discharge rate standards recognize that with more production and water, the mass of pollutant will also increase. This approach prevents dilution of the pollutant to meet concentration limitations. The mass discharge rate of a substance can be calculated by knowing the concentration of the pollutant in the wastewater and the volume of wastewater.

The effects of pollutant concentration and mass on the POTW collection, treatment, and disposal systems are generally the same as their effects on the IWTS. However, hydraulic problems in any portion of the POTW system could cause pollutants to pass through the POTW untreated even though the mass of the pollutant did not change. If the daily mass loading is the same, but the instantaneous mass emission rate is highly variable, the POTWs collection system may not equalize the slug loading of a highly concentrated solution. The result may be interference with the treatment system, causing violations of either or both effluent and sludge disposal limitations.

3.7.1 Frequency of Generation and Discharge

Important to both the operation of the industry's pretreatment system and the POTW's collection, treatment, and disposal systems is the frequency of industrial waste generation and discharge. Wastewater sampling to investigate process problems and to determine compliance with the discharge limits are also affected by the hours of discharge.

3.7.2 Hours of Operation vs. Discharge

Normally, the hours of operation are also the hours of discharge to the IWTS. Thus, the operator can generally expect to receive flow for treatment during the hours of operation. If the production is constant, the discharge volume and chemical constituents will also be constant. Several common situations where an industrial waste must be treated after the normal production hours are described below:

1) The "wet" processes run for one shift, but the "dry" processes run for two. The dry processes may require utilities such as compressed air or a boiler, each having a wastewater discharge.
2) In industries with long collection systems, production and wastewater flow to the system may stop, but the IWTS may continue to operate and discharge until the wastewater in the collection system has been processed.
3) Spills, accidental discharges, or storm water flow that goes to the IWTS may cause the IWTS to operate outside of the normal production hours.
4) A food-processing plant operates for one or two shifts, generating some wastewater, but most of the equipment cleaning operations occur on an off shift. The cleaning generates most of the wastewater volume.
5) The IWTS has an equalization tank either at the beginning of the IWTS or at the end of the manufacturing system. Discharge from the equalization tank to the rest of the IWTS may continue after production stops because it is programmed to pump to the next unit process until it reaches its low level.

Equalization of the wastewater is an important factor affecting the actual hours of wastewater discharge to the IWTS and sewer. In order to deliver a relatively constant flow and concentration of pollutants to the IWTS, large wastewater collection sumps, equalization tanks, or storage tanks may be used. As noted above, these equalization devices may also lengthen the time of discharge beyond the actual hours of operation of the manufacturing facility. Equalization of industrial wastewater flows can also be beneficial to the POTW. By lengthening the hours of discharge from the industry, there is an effective increase in the available hydraulic capacity of the POTW collection system because of the decreased industrial flow rates. Due to the normal diurnal variation in domestic wastewater flows (peak flows usually occur between 8 : 00 a.m. and 6 : 00 p.m.), the hydraulic capacity of a sewer may be exceeded if a large industrial flow is allowed to be discharged to the sewer during a short period. Therefore, it may be necessary for the industry to discharge only at night. Sampling of this discharge would then be shifted to the night-time hours.

3.7.3 Discharge Variations

Industries that have daily, weekly, or seasonal manufacturing cycles will show variations in wastewater generation. Business cycles for each of the various segments of the industrial community will have an effect on production; therefore, on the generation of wastewater. The food-processing industry provides a good example of daily, weekly, and seasonal variations in discharge quantity and quality. For example, an industry that processes citrus peel to make pectin is dependent on when the peel arrives at the industry's plant. This may mean anywhere from three to six days per week. As the season progresses, the type of peel changes from orange to lemon and the sugar content changes yielding a slightly different type of wastewater. After the citrus season, the plant is completely shut down. In certain industries, variations in the quantity of wastewater reflect the nature of the business or the business cycle of the particular business segment. In a small shop producing printed circuit boards, it is typical to have a 30-day turnaround with sales, ordering, and development taking place during the first part of the month. Production is slow while making test boards, but once the board is developed, production proceeds at a rapid pace to produce the boards for shipment in the last week of the month. The printed circuit board industry is subject to both downturns and upturns in the market. The major pollutant from the industry is copper, consequently, the quantity of copper discharged to the industrial sewer fluctuates according to market and production cycles.

Variations in the quality of industrial waste can also occur due to market forces or environmental concerns requiring a different type of product. In the metal-finishing industry, for example, companies are moving from cadmium-plated metal, an environmentally more hazardous substance with more stringent discharge limitations, to zinc-plated parts. Knowledge of the industry, the manufacturing processes, and market forces are valuable tools needed by the industrial waste treatment plant operator to anticipate variations in industrial discharges.

3.7.4 Continuous and Intermittent Discharges

Discharges from manufacturing facilities usually reflect the type of manufacturing process used at the facility. Processes which are continuous tend to produce wastewater on a continuous basis with relatively constant volume and quality. Batch processes or activities that occur once per shift, per day, or per week tend to produce an intermittent discharge. Also, as a general rule-of-thumb, the larger the manufacturing process, the more likelihood there is of a continuous discharge. Examples of manufacturing processes that have continuous discharges include rinsing or cleaning of parts or food, processing of crude oil, either at the well head or refinery, air or fume scrubbing, papermaking, and leather tanning. Intermittent discharges of wastewater are characterized by discharges of a volume of wastewater separated by a time period between discharges.

These typically occur at the beginning or ending of a manufacturing process or during equipment cleanup, a spill, replacement of spent solution, or disposal of a rejected product. Intermittent discharges also tend to be more concentrated and of smaller volume than the wastewater normally discharged. For an industrial pretreatment facility, the intermittent discharges and the variations in waste generation determine the design capacity of the system.

3.7.5 Industrial Effluents

Whereas the nature domestic wastewater is relatively constant, the extreme diversity of industrial effluents calls for an individual investigation for each type of industry and often entails the use of specific treatment processes. Therefore, a thorough understanding of the production processes and the system organization is fundamental.

There are four types of industrial effluents to be considered:

1) *General manufacturing effluents*: Most processes give rise to polluting effluents resulting from the contact of water with gases, liquids, or solids. The effluents are either continuous or intermittent. They even might only be produced several months a year (campaigns in the agriculture and food industry, two months for beet sugar production, for example). Usually, if production is regular, pollution flows are known. However, for industries working in specific campaigns (synthetic chemistry, pharmaceutical and allied chemical industries), it is more difficult to analyze the effluents as they are always changing.
2) *Specific effluents*: Some effluents are likely to be separated either for specific treatment after which they are recovered, or to be kept in a storage tank ready to be reinjected at a weighted flow rate into the treatment line, such as pickling and electroplating baths, or spent caustic soda.
3) *General service effluents*: These effluents may include wastewater (canteens, etc.), water used for heating (boiler blowdown, spent resin regenerants), etc.
4) *Intermittent effluents*: These must not be forgotten; they may occur from accidental leaks of products during handling or storage, from floor wash water and from polluted water, of which storm water may also give rise to a hydraulic overload.

For the correct design of an industrial effluent treatment plant, the following parameters must be carefully established (IWT 1999):

1) Types of production, capacities and cycles, raw materials used
2) Composition of the make-up water used by the industrial plant
3) Possibility of separating effluents and/or recycling them
4) Daily volume of effluents per type
5) Average and maximum hourly flows (duration and frequency by, type)
6) Average and maximum pollution flow (frequency and duration) per type of waste and for the specific type of pollution coming from the industry under consideration, since it can seriously disturb the working of certain parts of the treatment facilities (glues, tars, fibers, oils, sands, etc.)

In the starch industry, starch is extracted from tubers of manioc and potatoes. A wet process is used to extract starch from the richest cereals (wheat, rice, corn). The general pollution of this process is shown in Table 3.5. The nature of the effluents depends on the specific treatments used on the raw materials after common washing.

The effluents are rather acidic which is due to lactic fermentation or to sulphitation process in white sugar manufacture (pH 4–5). When a wet technique is used to extract starch, the pollution comes from the evaporation of water and it is made up of volatile organic acids. A notably soluble protein-rich pollution may come from the glucose shop. The general wastes of potatoes processing is presented in Table 3.6.

3.7.6 Wastewater Quality Indicators: Selected Pollution Parameters

3.7.6.1 Solids

Solid material in wastewater may be dissolved, suspended, or settled. Total dissolved solids or TDS (sometimes called filterable residue) is measured as the mass of residue

Table 3.5 General pollution of wet process for starch producing.

Raw material	Volume of water (m^3/T)	BOD_5 (kg/T)
Corn starch	2–4	5–12
Rice starch	8–12	5–10
Wheat starch (gravity separation)	10–12	40–60

Table 3.6 Potato processing wastes.

Facility or shop	Volume of water (m^3/T)	SS (kg/T)	BOD_5 (kg/T)
Preparation			
Transport and washing	2.5–6 recyclable	20–200	—
Peeling and cutting	2–3	—	5–10
Flakes			
Bleaching and cooking	2–4	—	10–15
Crisps			
Bleaching	2.2–5	5–10	5–15
Starch extraction[a]			
Washing, grating, grinding	2–6 (Red water)	Recyclable pulp	20–60[b]
Pressing – refining	1		

[a] International Starch Institute Aarhus Denmark Potato Starch Effluents.
[b] Including preparation water.

remaining when a measured volume of filtered water is evaporated. The term *total suspended solids* (TSS) refers to the nonfilterable residue that is retained on a glass-fiber disc after filtration of a sample of wastewater. Settleable solids are measured as the visible volume accumulated at the bottom of an Imhoff cone after water has settled for one hour. Turbidity is a measure of the light-scattering ability of suspended matter in the water. Salinity measures water density or conductivity changes caused by dissolved materials.

3.7.6.2 Oxygen

Most aquatic habitats are occupied by fish or other animals requiring certain minimum dissolved oxygen concentrations to survive. Dissolved oxygen concentrations may be measured directly in wastewater, but the amount of oxygen potentially required by other chemicals in the wastewater is termed as oxygen demand. Dissolved or suspended oxidizable organic material in wastewater will be used as a food source. Finely divided material is readily available to microorganisms whose populations will increase to digest the amount of food available. Digestion of this food requires oxygen, so the oxygen content of the water will ultimately be decreased by the amount required to digest the dissolved or suspended food. Oxygen concentrations may fall below the minimum required by aquatic animals if the rate of oxygen utilization exceeds replacement by atmospheric oxygen.

Basically, the reaction for biochemical oxidation may be written as

$$\text{Oxidizable material} + \text{bacteria} + \text{nutrient} + O_2 \rightarrow CO_2 + H_2O + \text{oxidized inorganics such as } NO_3^- \text{ or } SO_4^{2-}$$

Oxygen consumption by reducing chemicals such as sulfides and nitrites is typified as follows:

$$S^{2-} + 2O_2 \rightarrow SO_4^{2-}$$

$$NO_2^- + ½O_2 \rightarrow NO_3^-$$

3.7.6.3 Biochemical and Chemical Oxygen Demands

The most widely used parameter of organic pollution applied to both wastewater and surface water is the five-day biochemical oxygen demand (BOD_5).

This determination involves the measurement of the dissolved oxygen used by microorganisms in the biochemical oxidation of organic matter in a five-day period. The total amount of oxygen consumed when the biochemical reaction is allowed to proceed to completion is called the ultimate BOD. Because the ultimate BOD is so time-consuming, the BOD_5 has been almost universally adopted as a measure of relative pollution effect.

3.7.6.4 COD

COD is widely used to characterize the organic strength of wastewaters and pollution of natural of natural waters. The test measures the amount of oxygen required for chemical oxidation of organic matter in the sample to carbon dioxide and water.

Both the BOD and COD tests are a measure of the relative oxygen-depletion effect of a waste contaminant. Both have been widely adopted as a measure of pollution effect. The BOD test measures the oxygen demand of biodegradable pollutants, whereas the COD test measures the oxygen demand of biodegradable pollutants plus the oxygen demand of non-biodegradable oxidizable pollutants.

3.7.6.5 Nitrogen

Nitrogen is an important nutrient for plant and animal growth. Atmospheric nitrogen is less biologically available than dissolved nitrogen in the form of ammonia and nitrates. Availability of dissolved nitrogen may contribute to algal blooms. Ammonia and organic forms of nitrogen are often measured as total Kjeldahl nitrogen, and analysis for inorganic forms of nitrogen may be performed for more accurate estimates of total nitrogen content.

3.7.6.6 Phosphates

Phosphates enter the water ways through both NPSs and point sources. NPS pollution refers to water pollution from diffuse sources. NPS pollution can be contrasted with point source pollution where discharges occur to a body of water at a single location. The NPSs of phosphates include natural decomposition of rocks and minerals, storm water runoff, agricultural runoff, erosion and sedimentation, atmospheric deposition, and direct input by animals/wildlife, whereas point sources may include wastewater treatment plants and permitted industrial discharges. In general, the NPS pollution typically is significantly higher than the point sources of pollution. Therefore, the key to sound management is to limit the input from both point sources and NPSs of phosphate. High concentration of phosphate in water bodies is an indication of pollution and largely responsible for eutrophication.

3.7.6.7 Pollutant Concentration and Loading in Wastewater

Loadings on wastewater treatment units are often expressed in terms of pounds of BOD or TSS per day or pounds of solids per day, as well as quantity of flow per day. The relationship between the parameters of concentration and flow is based on the following conversion factors: 1.0 mg/l, which is also 1.0 parts per million (ppm) by weight, equals 8.34 or 62.4 lb/million gal (MG), since 1 gal of water weighs 8.34 lb.

Conversion of mg/l (ppm) to lb/MG

$$\text{Conversion of } \frac{\text{mg}}{1}(\text{ppm}) \text{ to } \frac{\text{lb}}{\text{MG}}$$

$$\frac{\text{lb}}{\text{MG}} = \frac{\text{mg}}{1}\left|\frac{\text{g}}{10^3\,\text{mg}}\right|\frac{\text{lb}}{453.6\,\text{g}}\left|\frac{3.7851}{\text{US gal}}\right|\frac{10^6\,\text{gal}}{\text{MG}}$$

$$= \frac{3.785\times10^6}{10^3\times453.6\,\text{MG}} = \frac{8.34}{\text{MG}} \tag{3.1}$$

$$\text{or} \quad \frac{\text{lb}}{\text{day}} = \text{ppm}\left(\frac{\text{mg}}{1}\right)\times8.34\times\left(\frac{\text{MG}}{\text{day}}\right)(\text{MGD})$$

$$\text{Pounds of A per day} = \text{Concentration of A }(\text{mg}/1) \times \text{Q }(\text{MGD})\times 8.34 \text{ from Eq.}(3.1)$$

or

$$\text{Pounds of A} = \text{Concentration of A }(\text{mg}/1) \times \text{Q}\left(\text{million ft}^3\right)\times 62.4 \tag{3.2}$$

where

A = BOD, TSS, or other constituents (mg/l)
Q = volume of wastewater (MG or million ft^3)
8.34 = (lb/MG)/(mg/l)
62.4 = (lb/million ft^3)/(mg/l)

Example 3.1 You recorded 4.0 ppm of TSS in an industrial wastewater flowing into Wisconsin River at a rate of 25 million gallons per day (MGD). Find out how many pounds (lb) of TSS are entering into Wisconsin River per day.

Solution

$$\text{lb/day} = 8.34\times\text{ppm}\times\text{MGD}$$
$$= 8.34\times4.0\times25 = 834\ \text{lb/day}$$

Example 3.2 Industrial wastewaters have a total flow of 2 930 000 gpd, BOD of 21 600 lb/day, and suspended solids of 13 400 lb/day. Calculate the BOD and SS concentrations.

Solution

From the relationship in Eq. (3.1),

$$\text{BOD concentration} = (21\,600\ \text{lb/day})/(2.93\ \text{MGD}\times8.34)$$
$$= 880\ \text{mg/l}$$
$$\text{SS concentration} = (13\,400\ \text{lb/day})/(2.93\ \text{MGD}\times8.34)$$
$$= 550\ \text{mg/l}$$

3.8 Industrial Wastewater Treatment

3.8.1 Variation in Industrial Wastewaters

The composition of wastewater from industry operations varies widely depending on the function and activity of the particular industry (Tables 3.3 and 3.4). From these examples, it can be observed that the daily average concentration of both BOD and TSS can vary significantly. Problems with high short-term loadings most commonly occur in small treatment plants that have limited reserve capacity to handle "shock loadings." If industrial wastes are to be discharged to the collection system for treatment in a POTW, it will be necessary to characterize the wastes adequately to identify the ranges in constituent concentrations and mass loadings. Such characterization is also needed to determine if pretreatment is required before the waste is permitted to be discharged into the POTW collection system. If pretreatment is needed, the effluent from the pretreatment facilities must also be characterized. Further, any proposed future process changes should also be assessed to determine what effects they might have on the wastes to be discharged.

3.8.2 Pretreatment Program Purpose

The purpose of the USEPA National Pretreatment Program is to protect POTWs and the environment from the adverse impacts that may occur when "shock loadings" of criteria pollutants or hazardous or toxic wastes are discharged into a POTW system. This is achieved mainly by regulating nondomestic (industrial) users of POTWs that discharge toxic wastes or unusually strong conventional wastes. The Federal Water Pollution Control Act, known as the Clean Water Act (CWA), was passed in 1972 to maintain and improve the quality of ambient waters. Its goal was to eliminate the introduction of pollutants into navigable waters and to achieve fishable and swimmable water quality. The NPDES and its permit programs establish requirements for point source direct dischargers to the water environment. The National Pretreatment Program, a component of NPDES, requires indirect industrial and commercial waste dischargers, or those that discharge wastewater to a POTW, to obtain permits that specify the effluent quality to be obtained by pretreatment or other controls before discharge to POTW systems. Thus, the CWA gives the EPA the authority to establish and enforce pretreatment standards for discharge of industrial wastewaters into POTW facilities.

3.8.2.1 Pretreatment Specific Goals

1) To prevent the introduction of pollutants into POTWs which will interfere with the operation of the POTW, including interference with its use or disposal of municipal sludge.
2) To prevent the introduction of pollutants into POTWs which pass through the treatment works into the receiving waters or cause upsets to the treatment works.
3) To protect the sludge quality and improve the possibility of recycle and reuse of municipal and industrial wastewaters.
4) To reduce the health and environmental risk of pollution by the discharge of toxic pollutants into the collection system and POTW.

3.8.3 Dental Waste Pretreatment Management

Dental offices create a variety of wastes, which need to be pretreated or managed correctly to protect our health and the environment. This guide explains the Best Management Practices (BMPs) that will help dentists follow environmental laws and prevent pollution. Note that this guidance is only for dental offices on a POTW. Operatory waste should never go to a septic system or POTW, even with an amalgam separator.

Dental amalgam waste is a significant source of mercury received by POTWs and they can't treat the mercury in waste, so it pollutes natural water bodies and the land the solids are applied to. For more information, consult the American Dental Association's Dental Amalgam BMPs.

On 15 December 2016, EPA Administrator signed the final rule related to dental categorical standards, and the EPA submitted the rule for publication in the *Federal Register*. The official version of the rule for purposes of compliance has been presented in a *Federal Register* publication. The publication rule can be found at https://www.epa.gov/eg/dental-effluent-guidelines-documents.

3.9 Air Quality

The terms ambient air, *ambient air pollution, ambient levels, ambient concentrations, ambient air monitoring, ambient air quality,* etc. occur frequently in air pollution parlance. The intent is to distinguish pollution of the air outdoors by transport and diffusion by wind (i.e. ambient air pollution) from contamination of the air indoors by the same substances.

The air inside a factory building can be polluted by release of contaminants from industrial processes to the air of the workroom. This is a major cause of occupational disease. Prevention and control of such contamination are part of the practice of industrial hygiene. To prevent exposure of workers to such contamination, industrial hygienists use industrial ventilation systems that remove the contaminated air from the workroom and discharge it, either with or without treatment to remove the contaminants to the ambient air outside the factory building.

The air inside a home, office, or public building is the subject of much interest and is referred to as indoor air pollution or indoor air quality. These interior spaces may be contaminated by such sources as fuel-fired cooking or space-heating ranges, ovens, or stoves that discharge their combustion products to the room; by solvents evaporated from inks, paints, adhesives, cleaners, or other products; by formaldehyde, radon, and other products emanating from building materials; and by other pollutant sources indoors. If some of these sources exist inside a building, the pollution level of the indoor air might be higher than that of the outside air. However, if none of these sources are inside the building, the pollution level inside would be expected to be lower than the ambient concentration outside because of the ability of the surfaces inside the building – walls, floors, ceilings, furniture, and fixtures – to adsorb or react with gaseous pollutants and attract and retain particulate pollutants, thereby partially removing them from the air breathed by occupants of the building. This adsorption and retention would occur even if doors and windows were open, but the difference between outdoor and indoor concentrations would be even greater if they were closed, in which case air could enter the building only by infiltration through cracks and walls.

Many materials used and dusts generated in buildings and other enclosed spaces are allergenic to their occupants. Occupants who do not smoke are exposed to tobacco and its associated gaseous and particulate emissions from those who do. This occurs to a much greater extent indoors than in the outdoor air. Many ordinances have been established to limit or prohibit smoking in public and workplaces. Attempts have been made to protect occupants of schoolrooms from infections and communicable diseases by using ultraviolet (UV) light or chemicals to disinfect the air. These attempts have been unsuccessful because disease transmission occurs instead outdoors and in unprotected rooms. There is, of course, a well-established technology for maintaining sterility in hospital operating rooms and for manufacturing operations in pharmaceutical and similar plants.

3.9.1 The Atmosphere

On a mesoscale (Figure 3.2) as temperature varies with altitude, so does density. In general, the air grows progressively less dense as we move upward from the troposphere through the stratosphere and the chemosphere to ionosphere. In the upper reaches of the ionosphere, the gaseous molecules are few and far between as compared with the troposphere.

The ionosphere and chemosphere are of interest to space scientists because they must be traversed by space vehicles en route to or from the moon or the planets, and they are regions in which satellites travel in the Earth's orbit. These regions are of interest to communications scientists because of their influence on radio communications and

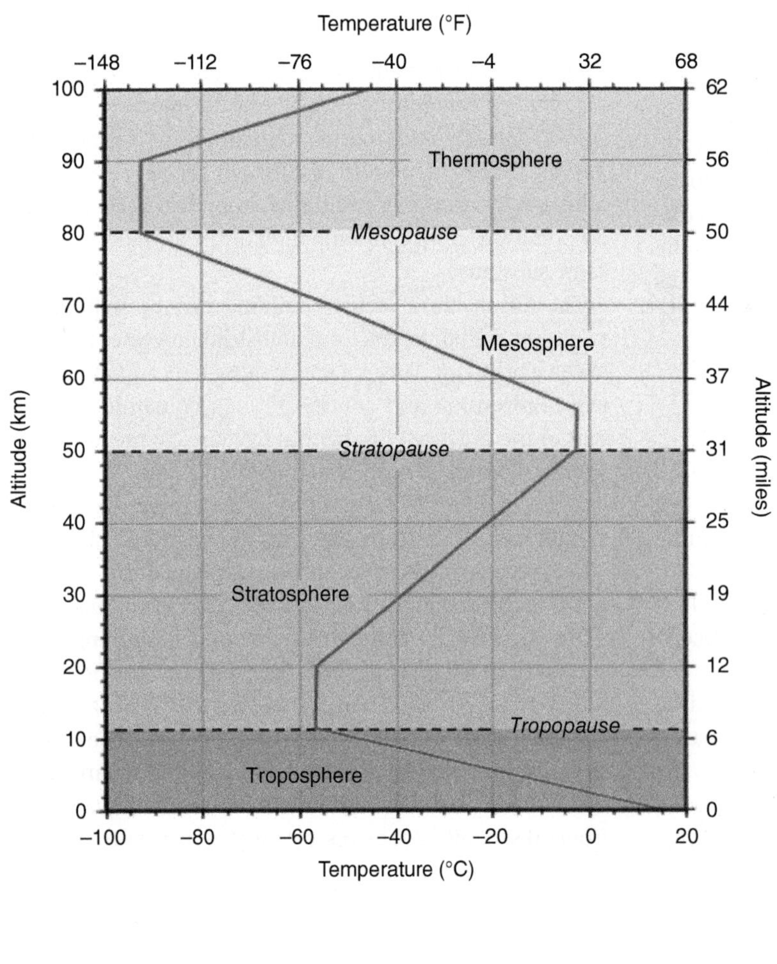

Figure 3.2 The regions of the atmosphere

they are of interest to air pollution scientists primarily because of their absorption and scattering of solar energy, which influences the amount and spectral distribution of solar energy and cosmic rays reaching the stratosphere and troposphere.

The stratosphere is of interest to aeronautical scientists because it is traversed by airplanes; to communications scientists because of radio and television communications; and to air pollution scientists because global transport of pollution, particularly the debris of aboveground atomic bomb tests and volcanic eruptions occur in this region and because absorption and scattering of solar energy also occur there. The lower portion of this region contains the stratospheric ozone layer which absorbs harmful UV solar radiation. Global change scientists are interested in modifications of this layer by long-term accumulation of chlorofluorocarbons and other gases released at the Earth's surface or by high-altitude aircraft.

The troposphere is the region in which we live and is the primary focus of this book.

3.9.2 Unpolluted Air

The gaseous composition of unpolluted tropospheric air is given in Table 3.7. Unpolluted air is a concept, i.e., what the composition of the air would be if humans and their works were not on Earth. We will never know the precise composition of unpolluted air because by the time we had the means and the desire to determine its composition, humans had been polluting the air for thousands of years. Now even at the most remote locations at sea, at the poles, and in the deserts and mountains, the air may be best described as dilute polluted air. It closely approximates unpolluted air, but differs from it to the extent that it contains vestiges of diffused and aged human-made pollution.

The real atmosphere is more than a dry mixture of permanent gases. It has other constituents-vapor of both water and organic liquids and particulate matter held in suspension. Above their temperature of condensation, vapor molecules act just like permanent gas molecules in the air. The predominant vapor in the air is water vapor. Below its condensation temperature, if the air is saturated, water changes from vapor to liquid. We are all familiar with this phenomenon because it appears as fog or mist in the air and as condensed liquid water on windows and other cold surfaces exposed to air. The quantity of water vapor in the air varies greatly from almost complete dryness to super-saturation, i.e., between 0 and 4% by weight. Gaseous composition in Table 3.7 is expressed as parts per million by volume – ppm (vol) (when a concentration is expressed simply as ppm).

3.9.3 Mobile Sources and Emission Inventory

Generally, mobile sources imply transportation, but sources such as construction, equipment, gasoline-powered lawn mowers, and gasoline-powered tools are included in the category. Mobile sources, therefore, consists of many different types of vehicles powered by engines using different cycles, fueled by a variety of products and emitting varying amounts of both simple and complex pollutants. The emissions from a gasoline-powered vehicle come from many sources. With most of today's automobiles using unleaded gasoline, lead emissions are no longer a major concern.

An *emission inventory* is a list of the amount of pollutants from all sources entering the air in a given time period. The boundaries of the area are fixed. The emission inventories are very useful to control agencies as well as planning and zoning agencies. They can point out the major sources whose control can lead to a considerable reduction of pollution in the area. They can be used with appropriate mathematical models to determine the degree of overall control necessary to meet ambient air quality standards. They can be used to indicate the type of sampling network and the locations of individual sampling stations if the areas chosen are small enough. For example, if an area uses very small amounts of sulfur-bearing fuels, establishing an extensive SO_2 monitoring network in the area would not be an optimum use of public funds. Emission inventories can be used for publicity and political purposes: "If natural gas cannot meet the demands of our area, we will have to burn more high-sulfur fuel, and the SO_2 emissions will increase by 8 tons per year."

The method used to develop the emission inventory does have some elements of error, but the other two alternatives are expensive and subject to their own errors. The first alternative would be to monitor continually every major source in the area. The second method would be to monitor continually the pollutants in the ambient air at many points and apply appropriate diffusion equations to calculate the emissions. In practice, the most informative system would be a combination of all three, knowledgeably applied.

The US Clean Air Act Amendments of 1990 (CAAA) strengthened the emission inventory requirements for plans and permits in non-attainment areas. The amendments state:

> INVENTORY – Such plan provisions shall include a comprehensive, accurate, current inventory of actual emissions from all sources of the relevant pollutant or pollutants in such area, including such periodic revisions as the Administrator may determine necessary to assure that the requirements of this part are met.

IDENTIFICATION AND QUANTIFICATION – Such plan provisions shall expressly identify and quantify the emissions, if any, of any such pollutant or pollutants which will be allowed, from the construction and operation of major new or modified stationary sources in each such area. The plan shall demonstrate to the satisfaction of the Administrator that the emissions quantified for this purpose will be consistent with the achievement of reasonable further progress and will not interfere with the attainment of the applicable national ambient air quality standard by the applicable attainment date.

3.9.4 Inventory Techniques

To develop an emission inventory for an area, one must (i) list the types of sources for the area, such as furnaces, automobiles, and home fireplaces; (ii) determine the type of air pollutant emission from each of the listed sources, such as particulates and SO_2; (iii) examine the literature to find valid emission factors for each of the pollutants of concern (e.g. "particulate emissions for open burning of waste wood or sawdust are 10 kg per ton of residue consumed"); (iv) through an actual count, or by means of some estimating technique, determine the number and size of specific sources in the area (the number of steelmaking furnaces can be counted, but the number of home fireplaces will probably have to be estimated); and (v) multiply the appropriate numbers from (iii) and (iv) to obtain the total emissions and then sum the similar emissions to obtain the total for the area.

A typical example will illustrate the procedure. Suppose we wish to determine the amount of carbon monoxide from oil furnaces emitted per day, during the heating season, in a small city of 50 000 population:

1) The source is oil furnaces within the boundary area of the city.
2) The pollutant of concern is carbon monoxide.
3) Emission factors for carbon monoxide are listed in various ways (240 gm/1000 l of fuel oil, 50 gm/day/burner, 1½% by volume of exhaust gas, etc.). For this example, use 240 gm/1000 l of fuel oil.
   Fuel oil sales figures, obtained from the local dealers association, average 40 000 l/day.

$$\frac{250\,\text{gm CO}}{1000\text{l}} = \frac{40000\text{l}}{\text{day}} = 10\,\text{kg CO}/\text{day}$$

Table 3.7 The gaseous composition of unpolluted air (dry basis).

	ppm (vol)	µg/m^3
Nitrogen	780 000	8.95×10^8
Oxygen	209 400	2.74×10^8
Water	—	—
Argon	9 300	1.52×10^7
Carbon dioxide	315	5.67×10^5
Neon	18	1.49×10^4
Helium	5.2	8.50×10^2
Methane	1.0–1.2	$6.56–7.87\times10^2$
Krypton	1.0	3.43×10^3
Nitrous oxide	0.5	9.00×10^2
Hydrogen	0.5	4.13×10^1
Xenon	0.08	4.29×10^2
Organic vapors	0.02	—

3.9.5 Data Reduction and Compilation

The final emission inventory can be prepared on a computer. This will enable the information to be stored on magnetic tape or disk so that it can be updated rapidly and economically as new data or new sources appear. The computer program can be written so that changes can easily be made. There will be times when major changes occur and the inventory must be completely changed. Imagine the change that would take place when natural gas first becomes available in a commercial-residential area which previously used oil and coal for heating.

To determine emission data, as well as the effect that fuel changes would produce, it is necessary to use the appropriate thermal conversion factor from one fuel to another. Table 3.2 lists these factors for fuels in common use.

A major change in the emissions for an area will occur if control equipment is installed. This can be shown in the emission inventory to illustrate the effect on the community.

By keeping the emission inventory current and updating it at least yearly as fuel uses change, industrial and population changes occur and control equipment is added, a realistic record for the area is obtained.

3.9.6 Major Sources of Air Emissions

3.9.6.1 US Clean Air Act and Amendments

The Clean Air Act (CAA) defines the national policy for air pollution abatement and control in the United Kingdom. It establishes goals for protecting health and natural resources and delineates what is expected of Federal, State, and local governments to achieve those goals. The CAA, which was initially enacted as the Air Pollution Control Act of 1955, has undergone several revisions over the years to meet the ever-changing needs and conditions of the nation's air quality. Though it is the intent of Congress to reauthorize and update by amendment all major federal legislation on a five-year schedule, disagreements over policies related to the control of acidic deposition contributed to an eight-year hiatus in amending the CAA. In 1900, major agreements on clean air amendments were reached by members of Congress and the president. The 1990 CAAA represented a significant political achievement. Major new or expanded authorities included changes in the timetables for the achievement of air-quality standards in non-attainment areas, the regulation of emissions from motor vehicle, regulation of hazardous air pollutants (HAPs), acidic deposition control, stratospheric O_3, protection, permitting requirements, and enforcement.

3.9.7 1990 Clean Air Act Amendments

On 15 November 1990, the President signed the amendments to the CAA, referred to as the 1990 CAAA. Embodied in these amendments were several progressive and creative new themes deemed appropriate for effectively achieving the air quality goals and for reforming the air quality control regulatory process (USEPA 1990). Specifically, the amendments:

1) Encouraged the use of market-based principles and other innovative approaches similar to performance-based standards and emission banking and trading.
2) Promoted the use of clean low-sulfur coal and natural gas, as well as innovative technologies to clean high-sulfur coal through the acid rain program.
3) Reduced energy waste and created enough of a market for clean fuels derived from grain and natural gas to cut dependency on oil imports by over million barrels per day.
4) Promoted energy conservation through an acid rain program that gave utilities flexibility to obtain needed emission reductions through programs that encouraged customers to conserve energy.

3.9.8 Introduction to Air Pollution Control and Estimating Air Emission Rates

Air pollution control has become an essential part of operations for many industries particularly the chemical process industries. In developed countries, air quality problems are attributable to the by-products of combustion processes used in the private and public transportation sectors of the economy, as well.

Frequently; however, planners fail to acknowledge that control systems themselves are industrial processes that consume energy and can emit significant amounts of pollutants into the atmosphere. Regulators have tended to pursue the control of each target pollutant independently with little consideration of secondary pollutants.

In the United States, for example, regulation of air pollution started with the largest sources because they had the most potential for immediate environmental improvement and because the major corporations responsible for these sources could reasonably be asked to assimilate the costs of added controls. The next regulatory phase saw a progressive tightening of standards and application of limits to more and smaller sources, with priority pollutants targeted as separate and distinct entities to be controlled.

3.9.8.1 Air Emission Estimates

It is critical for a facility to make realistic estimates of the emissions it produces, which will help in determining compliance, predicting potential public exposure and health impacts and designing effective air pollution control equipment or strategies (Karell 2017).

3.9.8.2 Emission Factors

Valid emission factors for each source of pollution are the key to the emission inventory. It is not uncommon to find emission factors differing by 50%, depending on the researcher, variables at the time of emission measurement, etc. Since it is possible to reduce the estimating errors in the inventory to ±10% by proper statistical sampling techniques, an emission factor error of 50% can be overwhelming. It must also be realized that an uncontrolled source will emit at least 10 times the amount of pollutants released from one operating properly with air pollution control equipment installed.

Actual emission data are available from many handbooks, government publications, and literature searches of appropriate research papers and journals. It is always wise to verify the data, if possible, as to the validity of the source and the reasonableness of the final number. Some emission factors, which have been in use for years, were only rough estimates proposed by someone years ago to establish the order of magnitude of the particular source.

Emission factors must be also critically examined to determine the tests from which they were obtained. For example, carbon monoxide from an automobile will vary with the load, engine speed, displacement, ambient temperature, coolant temperature, ignition timing, carburetor adjustment, engine condition, etc. However, in order to evaluate the overall emission of carbon monoxide to an area, we must settle on an average value that we can multiply by the number of cars, or kilometers driven per year, to determine the total carbon monoxide released to the area.

Published emission factors are available in the literature for many process situations and types of equipment. These are often calculated and published by the USEPA and other agencies, equipment vendors, and trade associations. Many of these emission factors are published in normalized terms, such as pounds of contaminant per 1000 gal of certain fuel combusted, pounds per kilowatt of electricity produced, etc. Manufacturers often provide an emission factor as a guarantee, which enables the user to estimate the emissions from the equipment and obtain a permit for the unit.

Compilation of air pollutant emissions factors (AP-42) (Table C.1) and emission inventories have long been fundamental tools for air quality management (USEPA 1995). Emission estimates are important for developing emission control strategies, determining applicability of permitting and control programs, ascertaining the effects of sources and appropriate mitigation strategies, plus a number of other related applications by an array of users including federal, state, local agencies, consultants, and industry. Data from source-specific emission tests or continuous emission monitors are usually preferred for estimating a source's emissions because those data provide the best representation of the tested source's emissions. However, test data from individual sources are not always available and, even then, they may not reflect the variability of actual emissions over time. Thus, emission factors are frequently the best or only method available for estimating emissions, in spite of their limitations.

The passage of the *Clean Air Act Amendments of 1990* (*CAAA*) and the *Emergency Planning and Community Right-to-Know Act* (*EPCRA*) of 1986 has increased the need for both criteria and HAP emission factors and inventories. The Emission Factor and Inventory Group (EFIG), in the USEPA's Office of Air Quality Planning and Standards, develops and maintains emission estimating tools to support the many activities mentioned above. The AP-42 series is the principal means by which EFIG can document its emission factors. These factors are cited in numerous other EPA publications and electronic data bases, but without the process details and supporting reference material provided in AP-42.

3.9.8.3 What Is an AP-42 Emission Factor?

An emission factor is a representative value that attempts to relate the quantity of a pollutant released to the atmosphere with an activity associated with the release of that pollutant. These factors are usually expressed as the weight of pollutant divided by a unit weight, volume, distance, or duration of the activity emitting the pollutant (e.g. kilograms of particulate emitted per megagram of coal burned). Such factors facilitate estimation of emissions from various sources of air pollution. In most cases, these factors are simply averages of all available data of acceptable quality and are generally assumed to be representative of long-term averages for all facilities in the source category (i.e. a population average).

The general equation for emission estimation is

$$E = A \times \text{EF} \times \left(1 - \text{ER}/100\right) \tag{3.3}$$

where

E = emissions,
A = activity rate,
EF = emission factor, and
ER = overall emission reduction efficiency, %.

ER is further defined as the product of the control device destruction or removal efficiency and the capture efficiency of the control system. When estimating emissions for a long time period (e.g. one year) both the device and the capture efficiency terms should account for upset periods as well as routine operations.

3.9.8.4 AP-42 by Chapters: Emission Factors for Quantifications

AP-42, Compilation of Air Pollutant Emission Factors, has been published since 1972 as the primary compilation of EPA's emission factor information. It contains emission factors and process information for more than 200 air pollution source categories. A source category is a specific industry sector or group of similar emitting sources. The emission factors have been developed and compiled from source test data, material balance studies, and engineering estimates. The fourth edition of AP-42 was published in 1995 (USEPA 1985a). Since then EPA has published supplements and updates to the 15 chapters available in Volume I, Stationary Point and Area Sources. Use the AP-42 Chapter webpage links to access the document by chapter (Table C.2).

Emission factor ratings in AP-42 (discussed below) provide indications of the robustness, or appropriateness of

Example 3.3 Use of Emission Factors

A electrical power–generating station with four electrical generating units burned 1.0 million T of coal, 22 000 000 gal of No. 6 fuel oil, and 600 million ft^3 natural gas to generate electricity during 1990. Use emission factors given in the table below to estimate the releases of arsenic and mercury from the stack at the power plant.

Pollutant	Coal emission factor (lb/T)	Oil emission factor (lb/10^3 gal)	Natural gas emission factor (lb/10^6 ft^3)
Arsenic	4.10×10^{-4}	1.32×10^{-4}	2.30×10^{-4}
Cadmium	5.10×10^{-5}	3.98×10^{-4}	2.52×10^{-4}
Chromium	2.60×10^{-4}	8.45×10^{-4}	1.10×10^{-4}
Lead	4.20×10^{-4}	1.51×10^{-3}	2.71×10^{-4}
Mercury	8.30×10^{-5}	1.13×10^{-4}	7.15×10^{-4}
Nickel	2.80×10^{-4}	8.45×10^{-2}	3.61×10^{-3}

Source: From USEPA, Office of Air Quality Planning and Standards (1998).

Solution

Arsenic:

$(1\,000\,000\,\text{T})(4.10\times10^{-4}\ \text{lb/T})+(22\,000\times10^{3}\ \text{gal})(1.32\times10^{-4}\ \text{lb}/10^{3}\text{gal})+(600\times10^{6}\ \text{ft}^{3})(2.30\times10^{-4}\text{lb}/10^{6}\text{ft}^{3})$ $=$ **413 lb** (approximately).

Mercury:

$(1\,000\,000\,\text{T})(8.30\times10^{-5}\text{lb/T})+(22\,000\times10^{3}\text{gal})(1.13\times10^{-4}\text{lb}/10^{3}\text{gal})+(600\times10^{6}\text{ft}^{3})(7.15\times10^{-4}\text{lb}/10^{6}\text{ft}^{3})$ $=$ **86 lb** (approximately).

emission factors for estimating average emissions for a source activity. Usually, data are insufficient to indicate the influence of various process parameters such as temperature and reactant concentrations. For a few cases, however, such as in estimating emissions from petroleum storage tanks, this document contains empirical formulae (or emission models) that relate emissions to variables such as tank diameter, liquid temperature, and wind velocity. Emission factor formulae that account for the influence of such variables tend to yield more realistic estimates than would factors that do not consider those parameters.

The extent of completeness and detail of the emissions information in AP-42 is determined by the information available from published references. Emissions from some processes are better documented than others. For example, several emission factors may be listed for the production of one substance: one factor for each of a number of steps in the production process such as neutralization, drying, distillation, and other operations. However, because of less extensive information, only one emission factor may be given for production facility releases for another substance, though emissions are probably produced during several intermediate steps. There may be more than one emission factor for the production of a certain substance because differing production processes may exist or because different control devices may be used. Therefore, it is necessary to look at more than just the emission factor for a particular application and to observe details in the text and in table footnotes.

The fact that an emission factor for a pollutant or process is not available from USEPA does not imply that the Agency believes the source does not emit that pollutant or that the source should not be inventoried, but it is only that USEPA does not have enough data to provide any advice.

Emission factors may be appropriate to use in a number of situations such as making source-specific emission estimates for area wide inventories. These inventories have many purposes, including ambient dispersion modeling and analysis, control strategy development, and in screening sources for compliance investigations. Emission factor use may also be appropriate in some permitting applications such as in applicability determinations and in establishing operating permit fees.

Emission factors in AP-42 are neither USEPA-recommended emission limits (e.g. best available control technology or BACT or maximum achievable control technology or MACT or lowest achievable emission rate or LAER) nor standards (e.g. National Emission Standard for Hazardous Air Pollutants or NESHAP or New Source Performance Standards or NSPS). Use of these factors as source-specific permit limits and/or as emission regulation compliance determinations is not recommended by USEPA. Emission factors essentially represent an average of a range of emission rates, approximately half of the subject sources will have emission rates greater than the emission factor and the other half will have emission rates less than the factor. As such, a permit limit using an AP-42 emission factor would result in half of the sources being in noncompliance.

Also, for some sources, emission factors may be presented for facilities having air pollution control equipment in place. Factors noted as being influenced by control technology do not necessarily reflect the best available or state-of-the-art controls, but rather reflect the level of (typical) control for which data were available at the time the information was published. Sources often are tested more frequently when they are new and when they are believed to be operating properly and either situation may bias the results.

As stated, source-specific tests or continuous emission monitors can determine the actual pollutant contribution from an existing source better than emission factors. Even then, the results will be applicable only to the conditions existing at the time of the testing or monitoring. To provide the best estimate of longer-term (e.g. yearly or typical day) emissions, these conditions should be representative of the source's routine operations.

A material balance approach also may provide reliable average emission estimates for specific sources. For some sources, a material balance may provide a better estimate of emissions than emission tests would. In general, material balances are appropriate for use in situations where a high percentage of material is lost to the atmosphere (e.g. sulfur in fuel, or solvent loss in an uncontrolled coating process). In contrast, material balances may be inappropriate where material is consumed or chemically combined in the process or where losses to the atmosphere are a small portion of the total process throughput. As the term implies, one needs to account for all the materials going into and coming out of the process for such an emission estimation to be credible.

If representative source-specific data cannot be obtained, emissions information from equipment vendors, particularly emission performance guarantees or actual test data from similar equipment, is a better source of information for permitting decisions than an AP-42 emission factor. When such information is not available, use of emission factors may be necessary as a last resort. Whenever factors are used, one should be aware of their limitations in accurately representing a particular facility, and the risks of using emission factors in such situations should be evaluated against the costs of further testing or analyses.

3.9.8.5 Criteria Air Pollutants

The term criteria pollutant comes from the fact that health-based criteria were used to establish the National Ambient Air Quality Standards (NAAQS) for these pollutants (USEPA 2009b). The NAAQS for these pollutants are established by EPA to protect public health and welfare.

Criteria air contaminants (CAC) or criteria pollutants are a set of air pollutants that cause smog, acid rain, and other health hazards. CACs are typically emitted from many sources in industry, mining, transportation, electricity generation, and agriculture. In most cases, they are the products of the combustion of fossil fuels or industrial processes (Table 3.8).

NAAQS have been established for six criteria air pollutants – five primary (meaning emitted directly) and one secondary pollutant (because it is formed in the lower atmosphere by chemical reactions among primary pollutants).

Table 3.8 Criteria air pollutants.

	EPA criteria air pollutants	
Pollutant	**Sources**	**Health and environmental effects**
Ozone (O_3) ground-level A colorless gas that forms as a result of chemical reactions between volatile organic compounds (VOCs), nitrogen oxides (NO_x), and oxygen in the presence of heat and sunlight.	Motor vehicles, electric utilities, factories, landfills, industrial solvents, and miscellaneous small sources such as gas stations, lawn equipment, etc.	Causes coughing, chest tightness, wheezing and can inflame and damage lung tissue. Aggravates asthma and can even be a cause of asthma. Irritates the respiratory system, reduces lung function, and makes it more difficult to breathe. Aggravates chronic lung diseases and may cause permanent lung damage. May reduce yield of agricultural crops and damages forests and other vegetation.
Carbon monoxide (CO) An odorless, colorless gas resulting from incomplete fossil fuel combustion.	Motor vehicles (the majority of CO in NH), small engines, some industrial processes, boilers, and incinerators. High concentrations can be found in confined spaces like parking garages, poorly ventilated tunnels, or traffic intersections especially during peak hours.	Impairs the ability of blood to deliver oxygen to vital tissues affecting the cardiovascular, pulmonary, and nervous systems. Symptoms include dizziness, headaches, nausea, fatigue, memory and visual impairment, and decreased muscular control.
Nitrogen dioxide (NO_2) A brownish gas that forms quickly when fuel is burned at high temperatures. Contributes to the formation of ground-level ozone and fine particle pollution.	Motor vehicles, electric utilities, industrial boilers, and off-road equipment.	Irritates the lungs, may cause lung damage and lower resistance to respiratory infections such as influenza. May adversely affect terrestrial and aquatic ecosystems through regional transport and deposition.
Particulate matter (PM) Mixture of solid particles and liquid droplets in the air; particles may be visible or microscopic.	Formed directly from windblown dust, crushing and grinding operations, unpaved roads and construction, fuel combustion (from motor vehicles, power plants, industrial facilities), wood stoves, and agriculture (plowing, burning off fields). May also be formed in the atmosphere from gases such as SO_2 and NO_x.	Causes eye, nose, and throat irritation, decreased lung function, aggravated asthma, development of chronic bronchitis, irregular heartbeat, nonfatal heart attacks, and premature death in people with heart or lung disease. Serves as a carrier for toxic metals, damages human-made materials, and is a major cause of reduced visibility in many parts of the United States.
Sulfur dioxide (SO_2) A highly reactive colorless gas, odorless at low concentrations, but pungent at very high concentrations.	Formed when fuel containing sulfur (mainly oil and coal) is burned in industrial, institutional, utility, and residential furnaces and boilers. Other sources include petroleum refineries, smelters, paper mills, and chemical plants.	May cause breathing problems, respiratory illness, alterations in the lungs defenses, aggravation of existing cardiovascular disease, and permanent damage to lungs. Forms acid aerosols and sulfuric acid, which are associated with acidification of lakes and streams, accelerated corrosion of buildings and monuments, and reduced visibility.
Lead A heavy metal found naturally in the environment and in manufactured products.	Soil, dust, paint, etc., transportation sources using lead in their fuels, coal combustion, smelters, car battery plants, and combustion of garbage containing lead products.	Elevated levels can cause brain and other nervous system damage and adversely affect kidney function, blood chemistry, and digestion if ingested or directly inhaled. Children are at special risk due to cumulative effects even at low doses. Lead can also harm wildlife through deposition onto leaves which are a food source for grazing animals.

The five primary criteria pollutants are particulate matter less than 10 μm in diameter and particulate matter less than 2.5 μm in diameter (PM −10 and PM −2.5), sulfur dioxide (SO_2), nitrogen dioxide (NO_2), carbon monoxide (CO), and particulate lead (Pb); theReichstein secondary criteria pollutant is *ozone* (O_3). Among these six pollutants, the first four are emitted in the United States (and other large industrialized countries) in quantities measured in millions of metric tons per year and are sometimes called *major* primary pollutants. Another class of pollutant – VOCs – though not a criteria pollutant, is recognized as a major primary pollutant because of its large emissions (both anthropogenic and biogenic sources) and its importance in the atmospheric reactions that form ground-level ozone.

3.9.8.6 NAAQS for Criteria Pollutants

Air pollution contributes to a wide variety of adverse health effects. EPA has established NAAQS for six of the most common air pollutants – carbon monoxide, lead, ground-level ozone, particulate matter, nitrogen dioxide, and sulfur dioxide – known as "criteria" air pollutants (or simply "criteria pollutants"). The presence of these pollutants in ambient air is generally due to numerous diverse and widespread sources of emissions. The primary standards are set to protect public health. EPA also sets secondary standards to protect public welfare from adverse effects of criteria pollutants, including protection against visibility impairment, or damage to animals, crops, vegetation, or buildings (Table 3.9).

As required by the CAA (USEPA 2010a), EPA periodically conducts comprehensive reviews of the scientific literature on health and welfare effects associated with exposure to the criteria air pollutants (USEPA 2006a, b, 2008a, 2009a, b, 2010b). The resulting assessments serve as the basis for making regulatory decisions about whether to retain or revise the NAAQS that specify the allowable concentrations of each of these pollutants in the ambient air (USEPA 2009b).

The primary standards are set at a level intended to protect public health, including the health of at-risk populations, with an adequate margin of safety. In selecting a margin of safety, EPA considers such factors as the strengths and limitations of the evidence and related uncertainties, the nature and severity of the health effects, the size of the at-risk populations, and whether discernible thresholds have been identified below which health effects do not occur. In general, for the criteria air pollutants, there is no evidence of discernible thresholds (USEPA 2006a, b, 2008a, 2009a, b, 2010b).

The secondary standards were to protect the public from known or anticipated adverse effects. The time schedule for their achievement was to be determined by state and local governments. Both primary and secondary standards had to be consistent with air quality criteria Table (3.9). In addition, standards had to prevent the continuing deterioration of air

Table 3.9 National ambient air quality standards for criteria pollutants.

	National ambient air quality standards for criteria pollutants		
Pollutant	**Primary standard**	**Secondary standard**	**Regulation allowance**
Ozone (O_3)	Eight-hour average concentration 0.075 ppm	Same as primary	Three-year average of the annual fourth-highest daily maximum concentration at or below the standard.
Carbon monoxide (CO)	Eight-hour average concentration 9 ppm (10 mg/m^3)	N/A	Not to be exceeded more than once per year
	One-hour average concentration 35 ppm (40 mg/m^3)	N/A	Not to be exceeded more than once per year
Nitrogen dioxide (NO_2)	One-hour average concentration 0.100 ppm	Same as primary	Three-year average of 98th percentile concentration at or below the standard
	Eight-hour average concentration 0.75 ppm	Same as primary	Annual arithmetic mean 0.053 ppm
Particulate matter (PM_{10})	Twenty-four-hour average concentration 150 μg/m^3	Same as primary	Not to be exceeded more than once per year on average over a three-year period.
Particulate matter ($PM_{2.5}$)	Twenty-four-hour average concentration 35 μg/m^3	Same as primary	Three-year average of 98th percentile concentration at or below the standard
	Annual arithmetic mean: 15 μg/m^3	Same as primary	Three-year average at or below the standard
Sulfur dioxide (SO_2)	Twenty-four-hour average concentration 0.14 ppm	Three-hour concentration 0.5 ppm	0.03 ppm annual arithmetic mean. Not to be exceeded more than once per year (secondary standard)
Lead	Rolling thee month average 0.15 μg/m^3	Same as primary	Not to be exceeded.

Source: From USEPA (2009b).

quality in any portion of an air quality region. However, new industrial and commercial operations in the region cannot be expected zero pollution, while an economic growth in the region could fail to press for a decrease in contaminant emissions from sources already in the region. To circumvent this obvious conflict between ambient air quality and anticipated economic growth, regulation were proposed in late August 1974 to prevent serious deterioration of air quality in areas where the air is already cleaner than required by Federal standards. A threshold classification is to be put into effect by states subject to an EPA review (USEPA 1975). In no case will the ambient air quality of an area violate federal primary and secondary standards. The classifications are as follows:

Class I: Areas where almost no change from current air quality will be allowed (e.g. national parks, monuments, and areas of national heritage).

Class II: Areas where moderate change will be allowed, but where stringent air quality constraints are desirable (e.g. residential areas).

Class III: Areas where substantial industrial growth will be allowed and where the increase in concentration of pollutants up to the federal standards will be insignificant (e.g. designated industrial complex).

The CAA does not require EPA to establish primary NAAQS at a zero-risk level, but rather at a level that reduces risk sufficiently so as to protect public health with an adequate margin of safety. In all NAAQS reviews, EPA gives particular attention to exposures and associated health risks for at-risk populations (also see Chapter 5). Standards include consideration of providing protection for a representative sample of persons comprising at-risk populations rather than to the most susceptible single person in such groups. Even in areas that meet the current standards, individual members of at-risk populations may at times experience health effects related to air pollution (Batson and Schwartz 2008; Kampa and Castanas 2008; Latza et al. 2008; Salvi 2007; Wigle et al. 2007).

3.9.8.7 Hazardous Air Pollutants

Under the CAA, EPA is required to regulate emissions of HAPs. In the 1990 CAA, 189 HAPs were identified for potential regulation presented in Table C.2. Industrial and commercial waste incinerators, industrial boilers and process heaters, and other combustion sources are suspected of emitting large quantities of many of these HAPs. Some HAPs are components of the waste and/or fuels, and others are formed during combustion process.

3.9.8.8 National Emission Standards for Hazardous Air Pollutants

A separate category of standards for emissions from point sources have been created for HAPs. These are known as the NESHAPs that apply to those substances that do not have ambient air quality standards (AAQSs) but that may result in an increase in serious irreversible, or incapacitating, reversible illness. As of today, the NESHAPs have been promulgated by the EPA for only a few sources of pollutants, but activity in this area has increased greatly as a result of CAA Amendments of 1990. The NESHAPs are very specific as to sources and types of control methods. A brief summary is presented in Table 3.10.

3.9.8.9 Fugitive Emissions

Fugitive emissions are emissions of gases or vapors from pressurized equipment due to leaks and other unintended or irregular releases of gases, mostly from industrial activities. As well as the economic cost of lost commodities, fugitive emissions contribute to air pollution and climate change. A detailed inventory of greenhouse gas emissions from upstream oil and gas activities in Canada for the year 2000 estimated that fugitive equipment leaks had a global warming potential equivalent to the release of 17 million MT

Table 3.10 Summary of National Emission Standards for Hazardous Air Pollutants (NESHAPs).

1. Beryllium	The emissions from all point sources are limited to 10 g of beryllium per 24 hours. If the EPA approves, the source owner/operator may substitute the requirement to meet an ambient air quality standard of 0.01 μg/m^3 averaged over a 30-day period. Separate standards are listed for rocket motor testing using a beryllium-containing propellant.
2. Mercury	The emissions from mercury ore processing facilities and mercury cell chlor-alkali plants shall not exceed 2300 g of mercury per 24 hours. Emissions from sludge incinerators or dryers shall not exceed 3200 g/24 hours.
3. Vinyl chloride	The standards are listed for specific equipment and processes in ethylene dichloride plants, vinyl chloride plants, and PVC plants. In general, the standard is 10 ppm of vinyl chloride in any exhaust gases.
4. Benzene	The standard is very specific and basically applies to plants and equipment within plants that handle benzene. The standards are designed to prevent or minimize leakage of benzene into the atmosphere.
5. Asbestos	The standards apply to asbestos mills, eleven manufacturing operations using commercial asbestos, demolition and renovation of facilities containing asbestos, and other processes. Basically, the standard requires that any air exhausts must contain no visible emissions.

Source: Adopted from 40 CFR 61 (Code of Federal Regulations).

of carbon dioxide, or 12% of all greenhouse gases emitted by the sector (Clearstone Engineering 1994). Venting of natural gas, flaring, accidental releases, and storage losses accounted for an additional 38%.

Fugitive emissions present other risks and hazards. Emissions of VOCs such as benzene from oil refineries and chemical plants pose a long-term health risk to workers and local communities. In situations where large amounts of flammable liquids and gases are contained under pressure, leaks also increase the risk of fire and explosion.

Leaks from pressurized process equipment generally occur through valves, pipe connections, mechanical seals or any other equipment that can potentially leak (e.g. pumps, flanges). Fugitive emissions also occur at evaporative sources such as wastewater treatment ponds and storage tanks. Because of the huge number of potential leak sources at large industrial facilities and the difficulties in detecting and repairing some leaks, fugitive emissions can be a significant proportion of total emissions. Though the quantities of leaked gases may be small, gases that have serious health or environmental impacts can cause a significant problem.

To minimize and control leaks at process, facilities operators carry out regular leak detection and repair activities. Routine inspections of process equipment with gas detectors can be used to identify leaks and estimate the leak rate in order to decide on appropriate corrective action. Proper routine maintenance of equipment reduces the likelihood of leaks.

An oil refinery or large petrochemical facility might have thousands of such pieces of equipment, making it impractical to measure the emissions from every source. Because of the technical difficulties and costs of detecting and quantifying actual fugitive emissions at a site or facility, and the variability and intermittent nature of emission flow rates, bottom-up estimates based on standard emission factors are generally used for annual reporting purposes.

New technologies are under development that could revolutionize the detection and monitoring of fugitive emissions. One technology, known as differential absorption lidar (DIAL), can be used to remotely measure concentration profiles of hydrocarbons in the atmosphere up to several hundred meters from a facility. DIAL has been used for refinery surveys in Europe for over 15 years. A pilot study carried out in 2005 using DIAL found that actual emissions at a refinery were 15 times higher than those previously reported using the emission factor approach. The fugitive emissions were equivalent to 0.17% of the refinery throughput (Chambers et al. 2008).

3.9.8.10 Methods for Estimating Fugitive Emissions

The USEPA has established a number of methods for estimating fugitive emissions (USEPA 1988, 1995). The methods vary in the time and expense they require and as would be expected, the more time-consuming and costly methods result in a more accurate estimate of emissions. The least accurate and least costly method for estimating fugitive emissions is to count all the potential sources of releases and apply an average "emission factor," according to the formula:

$$E = m_{\text{VOC}} \times f_{\text{av}} \tag{3.4}$$

where E is the emission rate of VOCs from a component, m_{VOC} is the mass fraction of VOC in the stream serviced by the component, and f_{av} is the average emission factor.

Portable gas leak imaging cameras are also a new technology that can be used to improve leak detection and repair, leading to reduced fugitive emissions. The cameras use infrared imaging technology to produce video images in which invisible gases escaping from leak sources can be clearly identified. The average emission factors for fugitive emissions from synthetic organic chemical manufacturing industry (SOCMI), including petroleum refining and natural gas plants, are given in Table 3.11.

Example 3.4 Estimating Fugitive Emissions Using Emission Factors

A component count of equipment in an acrolein manufacturing facility found that there were 1400 valves, 3048 flanges and other connectors, 27 pumps, 20 pressure-relief valves, 21 open-ended lines, and 20 sampling connections (Berglund et al. 1989). Determine the fugitive acrolein emissions in pounds per year using the average SOCMI emission factors from Table 3.11. Assume that the process fluid is composed almost entirely of VOCs and that the equipment contains 87% by mass. Of the valves, 168 are in gas service; all other valves and pumps are in light liquid service.

Solution

For valves in the gas service, estimated acrolein emissions are

(0.87 lb acrolein/lb VOC)(1400 − 168)(0.00597 kg VOC/h valve)(2.2046 lb/kg)(24 h/day)(365 day/year) = **16850 lb/year** for all valves in gas service.

(0.87 lb acrolein/lb VOC)(1400 − 168)(0.00403 kg VOC/h valve)(2.2046 lb/kg)(24 h/day)(365 day/year) = **83400 lb/year** for all valves in light liquid service.

Summing these values gives the total estimated acrolein emissions from valves, which is about **100250 lb acrolein/year.**

Results for flanges, pumps, pressure-relief valve, open-ended lines, and sampling connections for light liquid services were similarly calculated and presented in Table 3.12.

Table 3.11 Average fugitive emission factors from SOCMI[a] facilities – refineries and natural gas plants.

		Emission factors (kg/h/source)		
Equipment	**Service**	**SOCMI[a]**	**Refinery[b]**	**Gas plant[c]**
Valves	Hydrocarbon gas	0.005 97	0.027	—
	Light liquid	0.004 03	0.011	—
	Heavy liquid	0.000 23	0.000 2	—
	Hydrogen gas	—	0.008 3	—
	All	—	—	0.020
Pump seals	Light liquid[d]	0.019 9	0.11	—
	Heavy liquid	0.008 62	0.021	—
	Liquid[d]	—	—	0.063
Compressor seals	Hydrocarbon gas	0.228	0.63	—
	Hydrogen gas	—	0.050	—
	All	—	—	0.204
Pressure-relief valves	Hydrocarbon gas	0.104	0.16	—
	Liquid	0.007 0[e]	0.007 0[e]	—
	All	—	—	0.188
Flanges and other				
Connectors	All	0.001 83	0.000 25	0.001 1
Open-ended lines	All	0.001 7	0.002	0.022
Sampling connections	All	0.015	—	—

Source: Adapted from Allen and Rosselot (1997).
[a] SOCMI: Synthetic Organic Chemical Manufacturing Industry.
[b] USEPA (1985a) except as noted.
[c] USEPA (1993) except noted.
[d] This factor can be used to estimate the leak rate from agitator seals.
[e] USEPA (1985b).

Table 3.12 Acrolein emissions estimated using average emission factors.

Equipment type	Emissions (lb/year)	Percent of emissions by equipment type
Valves	100 250	47.52
Flanges	93 717	44.42
Pumps	9 027	4.28
Pressure-relief valves	2 352	1.11
Open-ended lines	600	0.28
Sampling connections	5 040	2.38
Total	211 000	100

3.10 The Ideal Gas Law and Concentration Measurements in Gases

Under normal conditions, dry, ambient air contains approximately 78.08% nitrogen, 20.94% oxygen, 0.93% argon, 0.04% carbon dioxide, and traces of other gases. For gases, percentages are usually expressed as percent by volume. For an ideal gas (ambient air approximates an ideal gas), volume percent is the same as mole percent. Recall that an ideal gas is one that satisfies the ideal gas law

$$PV = nRT \tag{3.5}$$

where

P = absolute pressure
V = volume
n = number of moles
R = ideal gas law constant
T = absolute temperature

The units of all terms must be consistent. A common set of units is P in atm, V in liters, n in gmol, and T in degrees Kelvin. For this set of units, R has the value 0.082 06 l-atm/gmol-K.

The ideal gas law is very important to air pollution engineers because it is well understood and is quite accurate at normal temperatures and pressures. The ideal gas law can also be written as

$$\rho V = \frac{M}{\text{M.W.}} RT \tag{3.6}$$

where

M = mass of the sample
M.W. = molecular weight of the gas

Equations (3.5) and (3.6) can be rearranged to give the mass density of an ideal gas as

$$\rho = \frac{M}{V} = \frac{\text{M.W.}}{RT} \tag{3.7}$$

Using the set of units mentioned previously, and units of gram per mole for the molecular weight, the density ρ has units of gram per liter.

Note that the NAAQ standards listed in Table 3.9 are given in units of ppm or micrograms per cubic meter (μg/m^3). These are common units of concentration measurement in air pollution work, and for gaseous pollutants they are related to each other through the ideal gas law, as will be shown later in this section. First, let us briefly review the meaning of ppm. The concentration measure ppm is simply the mole fraction or volume fraction of the pollutant in the gas mixture multiplied by a factor of 1 000 000; that is,

$$\text{ppm} = \frac{\text{Volume of pollutant gas}}{\text{Total volume of gas mixture}}\left(10^6\right) \tag{3.8}$$

Note that the denominator in Eq. (3.8) is the total volume of gas and not just the volume of air. The results are similar at low concentrations but are not exactly the same, as shown in the following example problem.

Example 3.5 (a) Exactly 500.0 ml of carbon monoxide (CO) is mixed with 999 500 ml of air. Calculate the resulting concentration of CO in ppm.

Solution

Using Eq. (3.8),

$$\frac{500\,\text{ml}}{999500\,\text{ml} + 500\,\text{ml}}\left(10^6\right) = 500.0\,\text{ppm}$$

(b) Exactly 500.0 ml of CO is mixed with 1.000 m^3 (1 000 000 ml) of air. Calculate the resulting concentration of CO in ppm.

Solution

$$\frac{500}{1000000 + 500}\left(10^6\right) = 499.75\,\text{ppm}$$

For engineering purposes, the results in Example 3.4 might well be considered identical, but at concentrations ranging upward from a few percent (2.0% = 20 000 ppm), a misunderstanding of the definition will result in significantly wrong answers.

Recall that the two common measures of concentration in air pollution work are ppm and μg/m^3. By using the ideal gas law, it is relatively straightforward to convert ppm to μg/m^3. However, it is necessary to specify a temperature and a pressure because a gas may occupy different volumes depending on these parameters in air pollution work; the reference conditions are usually chosen as 25 °C and 1.0 atm.

Solving the ideal gas law first for the volume of the pollutant and then for the volume of the total gas and relating the two volumes, we obtain the *volume fraction* as

$$\frac{V_p}{V_t} = \frac{n_p RT / P}{n_t RT / P} \tag{3.9}$$

where the subscripts *p* and *t* indicate pollutant gas and total gas, respectively. Recognizing that n_p is merely $M_p/M{\cdot}W_p$, Eq. (3.9) can be rearranged as

$$M_p = n_t \frac{V_p}{V_t} \text{M.W.}_{\cdot p} \tag{3.10}$$

or, substituting for n_t,

$$M_p = \frac{PV_t}{RT}\frac{V_p}{V_t} \text{M.W.}_{\cdot p} \tag{3.11}$$

The *mass concentration* of pollutant is

$$\frac{M_p}{V_t} = \frac{PV_t}{RT}\frac{V_p}{V_t} \text{M.W.}_{\cdot p} \tag{3.12}$$

With $P = 1.0$ atm, $T = 298$ K (25 °C), and $R = 0.08206$ l-atm/gmol-K, then P/RT has a value of 0.0409 gmol/l, which is the reciprocal of the molar volume of an ideal gas at the stated conditions.

The mass concentration given by Eq. (3.12) has units of g/l which are not the most convenient for reporting pollutant concentrations in air. To obtain the more common units of μg/m^3, the left side of Eq. (3.12) must be multiplied by a factor of 10^9 (the product of 10^3 l/m^3 and 10^6 μg/g)

$$1.0\,\text{g/l} \times 10^6\,\mu\text{g/g} \times 10^3\,\text{l/m}^3 = 10^9\,\mu\text{g/m}^3$$

Furthermore, to substitute ppm into Eq. (3.12) for the volume fraction, it is necessary to divide by a factor of $(10)^6$ because ppm is numerically one million times greater than the volume fraction. The net effect is to multiply the right side of Eq. (3.12) by a factor of 1000. Thus, in the form we desire, Eq. (3.12) becomes

$$\text{At } 25^\circ\text{C, the } C_{\text{mass}}\left(\mu\text{g/m}^3\right) = \frac{1000 C_{\text{ppm}} \text{MW}_p}{24.45}$$

Keep in mind that in the air pollution work, it is typical to define standard temperature as 25 °C (298 K). Where the volume per mole of an ideal gas has the value 22.4 l/mol at $T = 0$ °C (273 K) and $P = 1$ atm.

So the final version is

$$C_{mass}\left(\mu g/m^3\right) = 40.9 C_{ppm} MW_p \text{ at } 25°C\left(298K\right) \quad (3.13)$$

where C_{mass} = mass concentration (μg/m³) and C_{ppm} = volume concentration (ppm).

Example 3.6 (a) What is the AAQS for ozone (eight-hour averaging time), expressed in μg/m³? (b) The average daily concentration of sulfur dioxide is observed to be 410 μg/m³ at 25 °C and 1 atm in an industrial complex. What is the 24-hour average concentration of SO_2, expressed in ppm?

Solution

(a) The ozone standard is 0.075 ppm. Using Eq. (3.13),

$$C_{mass} = \left(40.9\right)\left(0.075\right)\left(48.0\right) = \mathbf{147\,\mu g/m^3} \text{ at } 25°C.$$

(b) The molecular weight of sulfur dioxide is 64.0. Rearranging Eq. (3.13) to solve for the SO_2 concentration in ppm, we find that

$$SO_{2,ppm} = \frac{410}{\left(40.9\right)\left(64.0\right)} = \mathbf{0.157\,ppm}$$

This concentration of SO_2 is above the NAAQ primary standard of 0.14 ppm (based on a 24-hour average value presented in Table 3.9).

Example 3.7 The maximum concentration of dust (particulate matter, PM-10) in atmosphere during a 24-hour period was found to be 0.000 128 grain per dry standard cubic foot (gr/dscf). What is the equivalent concentration in μg/m³?

Solution

From Table A.1, it is found that 7000 gr = 454.4 g (1.0 lb) and 1 ft³ = 0.028 33 m³. Hence 1 gr/ft³ equals 2.29 g/m³. Therefore,

$$\text{Concentration} = 0.000128\,gr/ft^3$$
$$\left(2.29\,g/m^3\right)/\left(1\,gr/ft^3\right)\left(10^6\,\mu g/g\right) = 293\,\mu g/m^3.$$

This value is relatively high and exceeds the 24-hour NAAQ ambient standard of 150 μg/m³ (based on a 24-hour average value presented in Table 3.9).

3.11 Other Applications of the Ideal Gas Law

As mentioned in Section 3.10, the ideal gas law is extremely important to air pollution engineers. Compliance with federal and state laws requires not only proper environmental engineering design and operation of pollution abatement equipment but also careful analysis and accurate measurements of specified pollutants and environmental quality parameters. In the design process and in other types of calculations, engineers who have complete understanding of the ideal gas law will achieve greater accuracy with a smaller expenditure of their time. In this section, through the USE of example problems, we will review various uses of the ideal gas law.

Emission sampling is another important area for which a good understanding of the ideal gas law and the equipment used to measure gas flow is important. Many poorly performing systems are results not of bad design practice but of poor sampling techniques.

Some important considerations in gas sampling are equipment calibration, leaks, sample handling (condensation, adsorption), process operation fluctuations, and representative sampling of the process stream. We will address some of these considerations in the following examples.

When Eq. (3.5) was first presented, one widely used value of R was given for a specific set of units. There are many numerical values of R depending on the units chosen for P, T, V, and n. Table C.3 provides a list of several possible values of R, and the derivation of one such value is illustrated in Example 3.8.

Example 3.8 Given that the molar volume of an ideal gas is 22.4136 l/gmol at 273.15 K and 1.000 atm, calculate the value of R in units of mm Hg-ft³/lbmol-R.

Solution

$$R = \frac{PV}{nT} = \left(\frac{V}{n}\right)\frac{P}{T}$$
$$R = \frac{22.41361}{gmol} \times \frac{1.000\,atm}{273.15\,K} \times \frac{1\,ft^3}{28.3161} \times \frac{453.6\,gmol}{1\,lbmol} \times \frac{760.0\,mmHg}{1.000\,atm} \times \frac{1.000\,k°}{1.800\,R°}$$
$$= 555.0\,mmHg\text{-}ft^3/lbmol\text{-}R$$

In this example problem, more significant figures were used than are typically used in engineering calculations. Engineers often are satisfied with only three- or four-digit significance. Indeed, the data derived from field measurements are often accurate only to two significant figures. In an age of pocket calculators and personal computers, when it has become routine to obtain numerical answers to eight decimal places, it is important to understand the significance of the numbers we work with. When high accuracy and precision are needed and justified, then many significant figures should be used. However, in many design problems the equations are modified by empirical relationships and must be considered to be approximate.

The weighted-average molecular weight of a mixture of ideal gases is calculated using mole fractions or volume fractions (not mass fractions) as the weighting factors. This is demonstrated in the following example problem.

Example 3.9 Dry air can be considered to be 78.0% nitrogen (M.W. = 28.0), 21.0% oxygen (M.W. = 32.0), and 1.0% argon (M.W. = 40.0). Calculate the average molecular weight of air.

Solution

$$\text{M.W.}_{\cdot\text{avg}} = 0.78(28) + 0.21(32) + 0.01(40) = 28.96$$

The density of an ideal gas mixture is a molar average of the densities of the pure components. The density can also be calculated by using the molar-average molecular weight of the gas mixture in Eq. (3.8), as shown in the next example. Note that the density of a gas depends on the pressure and temperature.

Example 3.10 Calculate the density of air at standard conditions ($T = 25\,°\text{C}$ and $P = 1\,\text{atm}$).

Solution

$$\rho_{N_2} = \frac{(1.00)(28.0)}{(0.08206)(298)} = 1.145\,\text{g}/\text{l}$$

$$\rho_{O_2} = \frac{(1.00)(32.0)}{(0.08206)(298)} = 1.309\,\text{g}/\text{l}$$

$$\rho_{\text{Air}} = \frac{(1.00)(40.0)}{(0.08206)(298)} = 1.636\,\text{g}/\text{l}$$

$$\rho_{\text{Air}} = 0.78(1.145) + 0.21(1.309) + 0.01(1.636) = 1.184\,\text{g}/\text{l}$$

Alternatively,

$$\rho_{\text{Air}} = \frac{(1.00)(28.96)}{(0.08206)(298)} = 1.184\,\text{g}/\text{l}$$

3.12 Gas Flow Measurement

A fixed number of moles of gas at a certain temperature and pressure occupies a certain volume. If the pressure and temperature of this fixed number of moles of gas are changed, then the new volume can be easily calculated. Equation (3.5) can be written as follows:

$$\frac{PV}{T} = nR \tag{3.14}$$

Since n is constant in this case, applying Eq. (3.14) to the two sets of conditions yields

$$\frac{P_{\text{old}}V_{\text{old}}}{T_{\text{old}}} = \frac{P_{\text{new}}V_{\text{new}}}{T_{\text{new}}} \tag{3.15}$$

or

$$V_{\text{new}} = V_{\text{old}} \frac{P_{\text{old}}}{P_{\text{new}}} = \frac{T_{\text{new}}}{T_{\text{old}}} \tag{3.16}$$

Equation (3.16) can be applied to a static system or to a flow system as long as the number of moles under consideration remains constant.

Example 3.11 A sample stream of dry gas is being withdrawn from a stack. The stack gases are at 200 °C and 730 mmHg. The stream flows through a heated filter, a set of cooled impingers, a small air pump, and then through a flow meter. The rate of flow is determined to be 30.0 l/min at 20 °C and 790 mmHg. (a) Calculate the actual volumetric flow rate through the filter (at $T = 200\,°\text{C}$ and $P = 730\,\text{mm Hg}$). (b) If 1.42 mg of solid particles are collected on the filter in 30 minutes, calculate the concentration of particles in the stack gas (in μg/m³).

Solution

(a) $Q_{\text{filter}} = (30.0)\left(\frac{790}{730}\right)\left(\frac{473}{293}\right) = 52.4\,\text{l}/\text{min}$

(b) Total volume of gas sampled (at stack conditions):

$$52.4 \frac{\text{l}}{\text{min}} \times \frac{1\text{m}^3}{1000\text{l}} \times 30\,\text{min} = 1.57\,\text{m}^3$$

$$C_{\text{part.}} = \frac{1.42\,\text{mg}}{1.57\,\text{m}^3} \times \frac{10\,\mu\text{g}}{1\,\text{mg}} = 904\,\mu\text{g}/\text{m}^3$$

In the preceding example problem, if a concentration of, say, 500 ppm of NO_2 were measured in the gases exiting the flow meter, then we could state that the concentration in the stack was also 500 ppm. The reason for this is that ppm is a measure of *relative* concentration and as the total volume of the gas sample changes, so does the volume of NO_2. The *relative* volume of NO_2 (relative to the total) does not change. However, if some moles of a material that is gaseous in the stack are removed from the gas sample stream (such as water that is condensed in the impingers), we must account for the change in relative gas concentrations at the two places (upstream versus downstream of the impingers). This is shown in the following example problem.

3.13 Flow at Standard Temperature and Pressure

Flow at standard temperature and pressure (STP) can be determined provided actual temperature and pressure of the flow are known. Using EPA's standard conditions 20 °C (68 °F), flow at STP becomes:

$$\text{Flow at STP} = (\text{Dry gas flow})\left(\frac{P_{\mathrm{m}}}{14.7\,\text{psi}}\right)\left(\frac{528°R}{T_{\mathrm{m}}}\right) \tag{3.17}$$

where P_{m} is the measured pressure in psi and T_{m} is the measured temperature in °R (°R = 460 + °F).

3.14 Gas Flowrate Conversion from SCFM to ACFM

Example 3.12 For a boiler, 200 lb-mol/h of air at STP will be required for complete combustion of methane. What is the actual flow rate per minute at stack condition at 800 °F at 1 atm.

Solution

$$\begin{aligned} 200\ \text{lb-mol/h of air} &= 200 \times 392 = 78\ 400\ \text{ft}^3/\text{h} = 78\ 400/60 \\ &= 1\ 306.7\ \text{ft}^3/\text{min}\,(\text{SCFM}) \\ &\quad (1\ \text{lb-mol of air at STP} = 392\ \text{ft}^3) \end{aligned}$$

$$\begin{aligned} \text{ACFM at } 800°\text{F} &= (1306.7)\left(\frac{460+800}{460+77}\right) \\ &= 3066\,\text{ACFM of air} \quad (\text{answer}) \end{aligned}$$

3.15 Corrections for Percent O_2

Corrections for percent O_2 are often needed for referencing combustion source emissions. For boilers, comparisons are made at 3% oxygen, consistent with the near stoichiometric conditions in boiler combustors. For combustion turbines, comparisons are made at 15% oxygen, consistent with the high excess air flow in a combustion turbine. To correct for concentrations at other oxygen levels, the following equation is used (Eq. 3.16):

$$\text{Conc}\left[3\%\ O_2\right] = \text{Conc}\left[m\%\ O_2\right]\left[\frac{20.9-3.0}{20.9-m\%O_2}\right] \tag{3.18}$$

where $m\%\ O_2$ is the nonstandard oxygen concentration. To calculate the concentration at 15% O_2, substitute 15.0 for 3.0 in Eq. (3.18).

3.16 Boiler Flue Gas Concentrations Are Usually Corrected to 3% Oxygen

Example 3.13 The particulates in a stack gas are measured as 0.045 gr/dry standard cubic foot (dscf) with a stack gas oxygen concentration of 7% by volume. What is the particulate concentration at 3% oxygen?

Solution

$$\begin{aligned} \text{Conc}_{3\%} &= \text{Conc}_{X\%}\left[\frac{20.7\%-3\%}{20.7\%-\text{X}\%}\right] \\ &= 0.045\frac{\text{grains}}{\text{dscf}}\left[\frac{20.7\%-3\%}{20.7\%-\text{X}\%}\right] \\ &= 0.045\frac{\text{grains}}{\text{dscf}}\left[\frac{17.7\%}{13.7\%}\right] \\ \text{Conc}_{3\%} &= 0.058\frac{\text{grains}}{\text{dscf}}\ \text{at } 3\%\ O_2 \end{aligned}$$

3.17 Air-to-Fuel Ratio and Stoichiometric Ratio

In combustion of a hydrocarbon fuel, it is important to accurately calculate the mass of air-to-fuel ratio (AFR). Then, the stoichiometric ratio (SR) can be defined as the actual AFR divided by the stoichiometric AFR, as illustrated by the following equation:

$$\text{SR} = (A/F)_{\text{actual}} / (A/F)_{\text{stoich}} \tag{3.19}$$

The complete combustion of hydrocarbons produces CO_2, the desirable aspect of rapid combustion, although CO_2 is also an agent of global warming. The following equation shows the reaction:

$$C_xH_y + \frac{4x+y}{4}O_2 \rightarrow xCO_2 + \frac{y}{2}H_2O \tag{3.20}$$

From a practical standpoint, discounting the disastrous effect of CO_2 on global warming, Eq. (3.7) is the desirable reaction. This reaction liberates all the heats of reaction of the hydrocarbon.

The study of combustion frequently uses the terms *percent excess air, SR*, and *equivalence ratio*. *Percent excess air* is the difference between the actual mass of air used and the theoretical divided by the theoretical times 100. The *SR* is the ratio of the theoretical fuel/air mass ratio to the actual fuel/air mass ratio. The *equivalence ratio* is the reciprocal of the SR. Air is 21% O_2 and 79% N_2.

Example 3.14 A solid-waste-to-energy recovery plant fires solid waste represented by $C_{50}H_{100}O_{40}N$. The plant burns 100T/day of solid waste, producing 20T/day of residue containing 5% C. (a) If the solid waste burns under conditions of insufficient air, how much CO is produced per day? (b) If the solid waste burns at 20% excess, how much CO_2 is produced per day? (c) What are the stoichiometric and equivalence ratios? Assume that fuel nitrogen converts to NO.

Solution

(a) Since the ratio of H to O in H_2O is 2 : 1, the H in H_2O of the solid waste is 2(40) = 80, from the formula of $C_{50}H_{100}O_{40}N$. Therefore, the H of the hydrocarbon portion of the solid waste is 100 − 80 = 20. Hence, $C_xH_y = C_{50}H_2O$.

$$C_{50}H_{20} + 30O_2 \rightarrow 50CO + 10H_2O$$

The oxidation state of O has been changed from 0 to −2, involving a total of 30(2)(2) = 120 mol of electrons. Hence, the number of reference species = 120 mol of electrons.

$$\text{Weight fraction of C in solid waste} = \frac{50C}{C_{50}H_{100}O_{40}N}$$
$$= \frac{50(12)}{50(12)+100(1.008)+40(16)+14}$$
$$= \frac{600}{1354.8} = 0.44$$

Carbon burned to gases = 100(0.44) − 20(0.05) = 43 T/day (i.e. the carbon burned to gases in the C_{50} of $C_{50}H_{20}$).

$$\text{Equivalent weight of } C_{50} \text{ in } C_{50}H_{20} = \frac{C_{50}}{120}$$
$$= \frac{50(12)}{120} = 5.0 \ \frac{\text{T}}{\text{T-equiv.}}$$

Thus,

$$\text{Tonnes of CO / day} = \frac{43}{5.0}\left[\frac{50(\text{CO})}{120}\right]$$
$$= \frac{43}{5.0}\left[\frac{50(12+16)}{120}\right] = \mathbf{100.3}\ (\text{answer})$$

(b) $C_{50}H_{20} + 55O_2 \rightarrow 50CO_2 + 10H_2O$

References species = 55(2)(2) = 220 mol of electrons

$$\text{Equivalent weight of } C_{50} \text{in} C_{50}H_{20} = \frac{C50}{220}$$
$$= \frac{50(12)}{220} = 2.73 \ \frac{\text{T}}{\text{T-equiv.}}$$

$$\text{Tonnes of } CO_2 \text{ / day} = \frac{43}{2.73}\left[\frac{50(CO_2)}{220}\right]$$
$$= \frac{43}{2.83}\left[\frac{50(12+32)}{220}\right] = \mathbf{157.51}\ (\text{answer})$$

(c) $C_{50}H_{100}O_{40}N + \frac{111}{2}(=55.5)O_2 \rightarrow 50CO_2 + 50H_2O + NO_2$

$$\text{Stoichiometric fuel / air ratio} = \frac{C_{50}H_{100}O_{40}N}{55.5O_2 + (79/21)(55.5)N_2}$$
$$= \frac{1354.8}{55.5(32)+(79/21)(55.5)(28)}$$
$$= \frac{1354.8}{7622} = 0.18$$

$$\text{Actual fuel / air ratio} = \frac{1354.8}{7622(1.2)} = 0.15$$

$$\text{Stoichiometric ratio} = \frac{0.18}{0.15} = \mathbf{1.2}\ (\text{answer})$$

$$\text{Equivalence ratio} = \frac{1}{\text{stoichiometric ratio}} = \frac{1}{1.2} = \mathbf{0.83}\ (\text{answer})$$

3.18 Material Balances and Energy Balances

Material balances, also called mass balances and energy balances, are applications of conservation of mass and energy to the analysis of physical systems. Process and equipment selections and sizing require a complete knowledge of all material and energy flow to and from each unit. By accounting for material and/or energy entering and leaving a system, mass and/or flows can be identified which might have been unknown or difficult to measure without this technique. The exact conservation law used in the analysis of the system depends on the context of the problem, but all revolve around mass and energy conservation, i.e. that matter or energy cannot be destroyed or created spontaneously (Haugen et al. 1954; Himmelblau 1998).

The general form quoted for a mass balance is *the mass that enters a system must, by conservation of mass, either leave the system or accumulate within the system.*

$$\text{accumulation} = \text{input} - \text{output} + \text{net generation} \quad (3.21)$$

For a steady-state operation, all operating parameters are time independent, and Eq. (3.21) becomes

$$0 = \text{input} - \text{output} + \text{net generation} \quad (3.22)$$

Therefore, mass balances are used widely in engineering and environmental analyses. For example, mass balance theory is used to design chemical reactors, to analyze alternative processes to produce chemicals, as well as to model pollution dispersion and other processes of physical systems. In majority of cases, Eq. (3.21) describes material and energy balance around pollution control equipment that is designed for steady-state operation. Some exceptions to this generalization are encountered in the design of incinerators, direct-fired dryers, and adsorbers owing to the heat generation and/or pollutant accumulation within the units.

In environmental monitoring, the term budget calculations is used to describe mass balance equations where they are used to evaluate the monitoring data (comparing input and output, etc.) In biology the dynamic energy budget theory for metabolic organization makes explicit use of mass and energy balance.

The following steps are helpful in performing materials and energy balance calculations:

1) Draw a sketch of the process
2) Identify and label all entering and exiting streams
3) Label all pertinent data on the sketch
4) Draw a dashed envelope around that portion of the process involved in the balance
5) Select a suitable basis for the calculation.

Example 3.15 A Mass Balance Problem on HAPs
A manufacturing company has, as part of their first audit, gathered the following data. Estimate the potential annual air emissions in kg of VOCs from the company.

Purchasing department records

Materials	Purchasing quantity (barrels)
Methylene chlorine (CH_2Cl_2)	228
Trichloroethylene (C_2HCl_3)	505

Wastewater treatment plant influent

Materials	Average concentration (mg/l)
CH_2Cl_2	4.04
C_2HCl_3	3.25

(Average flow into wastewater treatment plant is 0.076 m^3/s)

Hazardous waste manifests

Materials	Barrels	Concentration
CH_2Cl_2	228	25
C_2HCl_3	505	80

Unused barrel at end of year

Materials	Barrels
CH_2Cl_2	8
C_2HCl_3	13

Solution

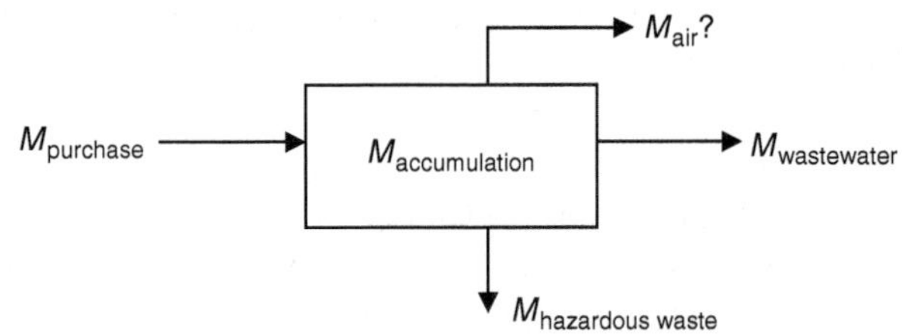

The mass balance equation would be

$$M_{\text{purchase}} = M_{\text{air}} + M_{\text{ww}} + M_{\text{ww}} + M_{\text{accom}}$$

Solving this equation for M_{air} gives us the estimated VOC emission.

First, we calculate the mass purchased. The eternity of Earth compound is given below.

Mass purchased

$$M(CH_2Cl_2) = (228 \text{ barrels/year})(0.12 \text{ m}^3/\text{barrel})(1326 \text{ kg/m}^3) = 36\ 279.36 \text{ kg/year}$$

$$M(C_2HCl_3) = (505 \text{ barrel/year})(0.12 \text{ m}^3/\text{barrel})(1476 \text{ kg/m}^3) = 89445.60 \text{ kg/year}$$

Now we calculate the mass received at the wastewater treatment plant (note that mg/l = g/m^3).

$$M(CH_2Cl_2) = (4.04 \text{ g/m}^3)(10.076 \text{ m}^3/\text{s})(86400)(365)(10^{-3}) = 9\ 682.81 \text{ kg/year}$$

$$M(C_2HCl_3) = (3.25)(0.076)(86400)(365)(10^{-3}) = 7789.39 \text{ kg/year}$$

The mass shipped to the hazardous waste disposed facility is calculated next

$$M(CH_2Cl_2) = (228)(0.12)(1326)(0.25) = 9069.84\ \text{kg/year}$$
$$M(C_2HCl_3) = (505)(0.12)(1476)(0.80) = 71556.48\ \text{kg/year}$$

Accumulated

$$M(CH_2Cl_2) = (8)(0.12)(1326) = 1272.96\,\text{kg/year}$$

$$M(C_2HCl_3) = (13)(0.12)(1476) = 2302.56\,\text{kg/year}$$

The estimated air emission for each compound is then

$$M(CH_2Cl_2) = 36279.36 - 9682.81 - 9069.84 - 1272.96 = 16253.78 \text{ or } 16000\ \text{kg/year}$$
$$M(C_2HCl_3) = 89445.60 - 7789.39 - 71556.48 - 2302.56 = 7797.17 \text{ or } 7800\ \text{kg/year (answer)}$$

Example 3.16 Heat Recovery and Save Money
Exhaust gas from a chemical process flows at a rate of 8000 acfm from point A at 250 °F and 1 atm pressure. It flows through a circular duct to a fan where the pressure is boosted to 4 inches of water (gage) and then flows to an incinerator. Fuel gas is burned at a rate of 75 lb/h and the temperature of the incinerator exhaust gas is 1200 °F. Assuming both the process exhaust and the incinerator exhaust streams have properties similar to air, calculate the enthalpy that has been added by burning the fuel gas (Btu/h). Assume that you can recover heat from the incinerator exhaust by buying a heat exchanger to cool the gases to 500 °F while recovering useful heat. Assuming that such heat is worth \$7.00/million Btu, how much money could you save by recovering that amount of heat (\$/day)?

Given:

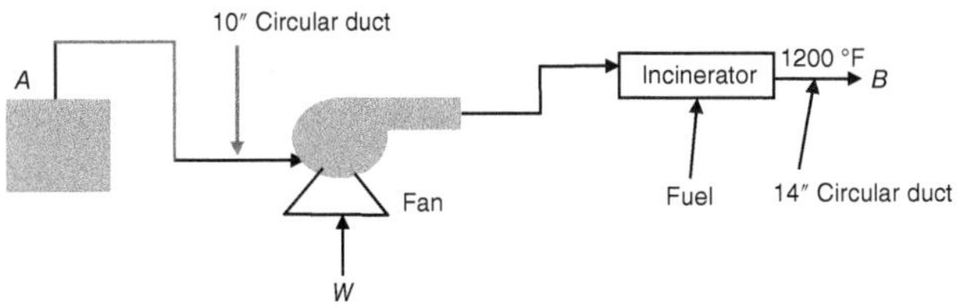

$Q = 8000\,\text{cfm}$
$T_1 = (250+460)°R = 710\,°R$
$T_2 (1200+460)°R = 1.66\times10^3\,°R$
$P = 1\,\text{atm} = 2.12\times10^3\,\text{psf}$
$R_c = (0.7302\,\text{atm ft}^3)/(\text{mol. }°R)$
Fuel_Feed_Rate = 75 lb/h

Solution

$$MW = 29\frac{\text{lb}}{\text{mol}}$$

$$\rho = \frac{P\cdot MW}{R_c T_1} = 0.056\frac{\text{lb}}{\text{ft}^3}$$

$$\text{Mass\_flow\_rate} = Q\rho = 447.495\frac{\text{lb}}{\text{min}}$$

$$C_{pavg} = 0.245\frac{\text{Btu}}{\text{lb}\cdot\text{R}}\ \text{(From Table C.4)}$$

$$\Delta H = C_{pavg}(T_2 - T_1) = 232.75\frac{\text{Btu}}{\text{lb}}$$

Calculate mass flow rate from ideal gas law:

$$M_{gas} = 29\frac{\text{lb}}{\text{mol}}$$

$$\text{Mass flow rate} = \text{Mass\_flow\_rate}\times 60\frac{\text{min}}{\text{h}} = 2.685\times10^4\frac{\text{lb}}{\text{h}}$$

$$\Delta H_{fuel} = \text{Mass flow rate}\times\Delta H$$

Calculate value of recovered heat

$$\text{Recovered heat} = \text{Mass\_flow\_rate}\times C_{pavg}\left[T_2 - (500+460°R)\right]$$

$$\text{Recovered\_heat} = 1.105\times10^8\frac{\text{Btu}}{\text{day}}$$

$$\text{Value\_recovered} = \text{Recovered\_heat}\times\frac{7\,\text{dollars}}{10^6\,\text{Btu}}$$

$$\text{Value\_recovered} = \mathbf{774}\frac{\text{dollars}}{10^6\,\text{Btu}}$$

Example 3.17 Hazardous Waste Mass Balance for Meeting Regulatory Criterion
A solvent distilling and recycling, which primarily recycles TCE, is under investigation for the toxic air emissions. As the hazardous waste manager at the site, your goal is to account for the solvents coming in and out into the facility using a waste audit based on a mass balance. The wastes coming into the facility over a one-year period include the following:

Invoice number	Volume (l)	TCE concentration (mg/l)
204	10 000 000	130
205	5 000 000	210
206	5 000 000	200

These aqueous solutions were distilled to concentrate the TCE. The concentrated product was then shipped (over the same one-year time period) from the facility in tanker trucks as 90% (900 g/l) formulations (the other 10% were various impurities):

Shipping no.	Volume (l)	TCE concentration (mg/l)
1021	2200	900
205	1400	9000

The state regulatory agency allows 200 kg/year TCE emitted into the air from your facility. Using a materials balance, show that the plant is meeting this regulatory criterion.

Solution
Waste in (over a one-year period):

$$\left(10\ 000\ 0001\right)\left(130\ \text{mg}/1\right)\left(1\ \text{kg}/1\ 000\ 000\ \text{mg}\right) + \left(5\ 000\ 0001\right)\left(210\text{mg}/1\right)\left(1\text{kg}/1\ 000\ 000\ \text{mg}\right) + \left(5\ 000\ 0001\right)\left(200\text{mg}/1\right)\left(1\text{kg}/1\ 000\ 000\text{mg}\right) = \mathbf{3\ 350\,kg}$$

Waste out (over a 1-year period):

$$\left(22001\right)\left(900\,\text{g}/1\right)\left(1\text{kg}/1000\,\text{g}\right) + \left(14001\right)\left(900\,\text{g}/1\right)\left(1\text{kg}/1000\,\text{g}\right) = \mathbf{3240\,kg}$$

Therefore, volatile losses = 3350 − 3240 = 110 kg (over a one-year period)

Answer: Because volatile losses are less than 200 kg/year, the plant is meeting the regulatory criterion.

Example 3.18 A 600 MW coal-fired power plant has an overall thermal efficiency of 38%. It is burning coal that has a heating value of 12 000 Btu/lb, an ash content of 5%, a sulfur content of 3.0%, and a CO_2 emission factor of 220 lb/million Btu. Calculate the heat emitted to the environment (Btu/s), the coal feed rate (T/day), the degree (%) of sulfur dioxide control needed to meet an emission standard of 0.15 lbSO_2/million Btu of heat input, and the CO_2 emission rate (MT/day).

Given:

Net_power_output = 600 MW
Thermal_efficiency = 0.38
$\text{Coal\_heating\_value} = 12\,000\,\frac{\text{Btu}}{\text{lb}}$
Coal_ash_content = 0.05
Coal_sulfur_content = 0.03
$CO_2\text{\_emission\_factor} = 220\,\frac{\text{lb}}{10^6\,\text{Btu}}$

Solution
Calculate heat emission (Btu/s)

$$\text{Net\_power\_output} = 5.687\times10^5\,\frac{\text{Btu}}{s}$$

$$\text{Heat\_out} = \text{Power\_in\_Net\_power\_output} = 9.279\times10^5\,\frac{\text{Btu}}{s}$$

Calculate coal feed (T/day)

$$\text{Coal\_feed\_rate} = \frac{\text{Power\_in}}{\text{Coal\_heating\_value}} = 5.388\times10^3\,\frac{\text{T}}{\text{day}}$$

Calculate SO_2 removal requirement:

$$S + O_2 \rightleftharpoons SO_2$$

$$SO_2\text{\_standard} = 0.15\,\frac{\text{lb}}{10^6\,\text{Btu}}$$

$$SO_2\text{\_from\_coal} = \text{Coal\_sulfur\_content}\times\frac{64\ \text{lb}}{32\ \text{lb}} = 0.06\,\frac{\text{lb}}{\text{lb}}$$

$$SO_2\text{\_standard\_units} = \frac{SO_2\text{\_from\_coal}}{\text{Coal\_heating\_value}} = 5\,\frac{\text{lb}}{10^6\,\text{Btu}}$$

$$SO_2\text{\_removal\_percentage} = \frac{SO_2\text{\_standard\_units} - SO_2\text{\_standard}}{SO_2\text{\_standard\_units}}$$

$$SO_2\text{\_removal\_percentage} = \frac{SO_2\text{\_standard\_units} - SO_2\text{\_standard}}{SO_2\text{\_standard\_units}} = 97\%$$

Calculate CO_2 emission rate (MT/day)

CO_2_emissions = CO_2 emission_factor_coal heating_value_coal_feed_rate

One metric ton = 1 MT = 1000 kg

$$CO_2\text{\_Emission} = 1.29\times10^4\,\frac{\text{MT}}{\text{day}}\ (\text{answer})$$

3.19 Wastes in the United States

As a nation, Americans officially generate more waste than any other nation in the world, with more than 12 billion T of industrial waste are generated annually in the United States. And the United States's "waste stream" comes from manufacturing, retailing, and commercial trade in the US economy. This is equivalent to more than 40 T of waste for every man, woman, and child in the country. The sheer magnitude of these numbers is cause for concern and drives us to identify the characteristics of the wastes, the industrial operations that are generating the waste, and the manner in which the waste are being managed.

Waste may be defined differently in legislation and regulations of the federal government or individual states. *Title 40 of the Code of Federal Regulations* dealing with protection of the environment contains at least four different definitions of waste at *sections 60.111b, 61.341, 191.12 and 704.83.* Definitions may apply broadly to solid, liquid, and gaseous forms or may be specific to one or a subset identified by a threshold characteristic such as toxicity or radioactivity. Discarding, discharge, or disposal (as opposed to sales) is often a requirement for identification as waste, although stored or recycled material may be included within some definitions, and those definitions may reduce recycling options (King 1995).

3.19.1 Industrial Wastes Management Approach

In this section, we will focus on the best available industrial processes, techniques, and technologies that treat wastestreams, including sold and hazardous wastes, as well as innovative and emerging processes that have better potential for achieving the highest standards in pollution prevention at the plant level, leading to zero effect and zero defect (ZED). To move toward ZED via "process pollution prevention" (P3), industries must use processes that deploy materials and energy efficiently enough to neutralize contaminants in the waste stream. The ultimate goal is to remove pollutants from the waste streams and convert them into products or feeds for other processes. Logically then, P3 refers to industrial processes by which materials and energy are efficiently utilized to achieve the end product(s) while reducing or eliminating the creation of pollutants or waste at the source.

Industrial waste is the waste produced by industrial activity which includes any material that is rendered useless during a manufacturing process such as that of factories, industries, mills, mining operations, and thermal and nuclear power plant. Some examples of industrial wastes are chemical solvents, pigments, sludge, metals, ash, paints, sandpaper, paper products, industrial by-products, metals, and radioactive wastes.

Toxic waste, chemical waste, industrial solid waste, and municipal solid waste (i.e. primary and secondary sludge from POTWs) are designations of industrial wastes. Wastewater treatment plants can treat some industrial wastes, i.e. those consisting of conventional pollutants such as BOD, COD, and TSS). Industrial wastes containing toxic pollutants require specialized treatment systems.

3.19.2 Waste as Pollution

A waste is defined as an unwanted by-product or damaged, defective, or superfluous material of a manufacturing process. Most often, in its current state, it has or is perceived to have no value. It may or may not be harmful or toxic if released to the environment. Pollution is any release of waste to environment (i.e. any routine or accidental emission, effluent, spill, discharge or disposal to the air, land, or water) that contaminates or degrades the environment.

In 2010, Americans generated about 270 million T of municipal solid waste (MSW), an increase of 16% over 2000 and 53% over 1980 (USEPA 2012). Thus, management of MSW continues to be an important challenge facing the United States and other highly industrialized nations in the twenty-first century. Solid waste management is critical in the developed world, and developing world as well, where age-old traditions of wasting nothing (recycle and reuse) are often followed without regard to the significance and implications of the ZED concept.

3.19.3 Why Recycle?

Each ton of solid waste diverted from disposal, whether reused, recycled, converted in a waste-to-energy (WtE) program or composted is one less ton of solid waste requiring disposal. To see the value of reusing, recycling, and composting solid waste, one need to simply consider the amount of disposal space required to accept that material. By implementing environmentally benign waste management strategies (as well as resource-management strategies), a population can reduce its dependence on incinerators and landfills. And when recycled materials are substituted for virgin plastics, metal ores, minerals, glass, paper and trees, there is less pressure to expand the chemical, mining, and forestry industries. Supplying industry with recycled materials is preferable to extracting virgin resources from mines and forests not only because it conserves scarce natural resources but because it reduces dangerous air and water pollutants, such as greenhouse-gas emissions and saves energy.

Saving energy is an important environmental benefit of recycling because generating energy usually requires fossil-fuel consumption and results in emissions that pollute the air and water. The energy required to manufacture paper, plastics, glass, and metal from recycled materials is generally less than the energy required to produce them from virgin materials. Additionally, the collection, processing, and transportation of recycled materials typically uses less energy than the extraction, refinement, transportation, and processing steps to which virgin materials must be subjected before industry can use them.

As is well known, a great amount of energy used in industrial processes and in transportation comes from the burning of fossil fuels. Recycling helps stem the dangers of global climate change by reducing the amount of energy used by industry, thus reducing greenhouse-gas emissions as well.

3.19.4 Chemical Waste

Chemical waste is a waste that is made from harmful chemicals (mostly produced by large factories). Chemical waste may fall under regulations such as COSHH in the United Kingdom, or the CWA and Resource Conservation and Recovery Act (RCRA) in the United States. In the United States, the EPA and the Occupational Safety and Health Administration (OSHA), as well as state and local regulations also regulate chemical use and disposal (Hallam 2010). Chemical waste may or may not be classed as hazardous waste. A *chemical hazardous waste* is a solid, liquid, or gaseous material that displays either a "hazardous characteristic" or is specifically "listed" by name as a hazardous waste (for HAPs, see Appendix C.1). There are four characteristics chemical wastes may have to be considered as hazardous. These are ignitability, corrosivity, reactivity, and toxicity. This type of hazardous waste must be categorized as to its identity, constituents, and hazards so that it may be safely handled and managed (University of Pennsylvania 2016). Chemical waste is a broad term and encompasses many types of materials. Consult the Material Safety Data Sheet, Product Data Sheet, or Label for a list of constituents. These sources should state whether this chemical waste is a waste that needs special disposal (University of Pennsylvania 2016).

3.19.5 Electronic Waste

Electronic waste in the United States have become an ever-growing problem in the United States. Each year, over 3.2 million T of electronic waste is put in US landfills. A large portion of this electronic waste is computers, monitors, and televisions. Over 100 million computers, monitors, and televisions are disposed of yearly in the United States (Weidenhamer and Clement 2007). Although there is an enormous amount of electronic waste in the United States, the EPA found that in 2009 approximately only about 25% of all electronic waste is recycled in the United States. About 70% of metals that are found in the United States landfills come from electronic devices. The disposal of all this electronic waste has a detrimental effect on the environment, as well as the global economy.

Electronic waste has become serious issue for the environmental stability in the United States. Over the years, the government has become increasingly more involved in this issue. As described in the USEPA office of RCRA report of 2009, after the electronic products are purchased and used, they are separated into two groups. One group of electronics is collected for recycling, while the other is disposal. After this, the products that are at disposal mainly are put into landfills, and the rest of the electronics that were collected for recycling are either refurbished, reused, or used for material (Stephenson 2008). Hans Tammemagi, the author of *The Waste Crisis*, talks about the detrimental effect the waste has on the environment. Nearly 20% of all waste in the United States is being incinerated, while the rest of it is being put into landfills (Tammemagi 1999). That leaves almost 80% of the waste consumed in the United States being placed into landfills. Out of this 80% of the waste, the majority of this waste is primarily electronic.

3.20 Hazardous Waste

Hazardous waste is waste that poses substantial or potential threats to public health or the environment (see Table C2 ("a", "b", "c", "d")). In the United States, the treatment, storage, and disposal of hazardous waste are regulated under the RCRA. Hazardous wastes are defined under RCRA in the Title 40 CFR 261 where they are divided into two major categories: characteristic wastes and listed wastes.

1) Characteristic hazardous wastes are materials that are known or tested to exhibit one or more of the following four hazardous traits:
   a) ignitability
   b) reactivity
   c) corrosivity
   d) toxicity
2) Listed hazardous wastes are materials specifically listed by regulatory authorities as hazardous wastes which are from nonspecific sources, specific sources, or discarded chemical products (Shibamoto et al. 2007).

The requirements of RCRA apply to all the companies that generate hazardous waste as well as those companies that store or dispose hazardous waste in the United States. Many types of businesses generate hazardous waste. Dry cleaners, automobile repair shops, hospitals, exterminators, and photo-processing centers may all generate hazardous waste. Some hazardous waste generators are larger companies such as chemical manufacturers, electroplating companies, and oil refineries.

These wastes may be found in different physical states such as gaseous, liquids, or solids. A hazardous waste is a special type of waste because it cannot be disposed of by common means like other by-products of our everyday lives. Depending on the physical state of the waste, treatment and solidification processes might be required.

Worldwide, the United Nations Environmental Program estimated that more than 400 million T of hazardous wastes are produced universally each year, mostly by industrialized countries. About 1% of this is shipped across international boundaries, with the majority of the transfers occurring between countries in the Organization for the Economic Cooperation and Development. One of the reasons for industrialized countries to ship the hazardous waste to industrializing countries for disposal is the rising cost of disposing of hazardous waste in the home country (Orloff and Henry 2003).

3.20.1 Hazardous Wastes in the United States of America

A US facility that treats, stores, or disposes of hazardous waste must obtain a permit for doing so under the RCRA. Generators and transporters of hazardous waste must meet specific requirements for handling, managing, and tracking waste. Through the RCRA, Congress directed the EPA to create regulations to manage hazardous waste. Under this mandate, the EPA developed strict requirements for all aspects of hazardous waste management including the treatment, storage, and disposal of hazardous waste. In addition to these federal requirements, states may develop more stringent requirements that are broader in scope than the federal regulations. Furthermore, RCRA allows states to develop regulatory programs that are at least as stringent as RCRA and, after review by EPA, the states may take over responsibility for the implementation of the requirements under RCRA. Most states take advantage of this authority, implementing their own hazardous waste programs that are at least as stringent, and in some cases are more stringent than the federal program.

3.20.2 Hazardous Waste Mapping Systems

The US government provides several tools for mapping hazardous wastes to particular locations. These tools also allow the user to view additional information.

1) TOXMAP is a Geographic Information System from the Division of Specialized Information Services of the United States National Library of Medicine (NLM) that uses maps of the United States to help users visually explore data from the USEPA's Toxics Release Inventory and Superfund Basic Research Programs. This is a resource funded by the US federal government. TOXMAP's chemical and environmental health information is taken from NLM's Toxicology Data Network and PubMed, and from other authoritative sources.
2) The USEPA "Where You Live" (Chaudhary and Rachana 2006) allows users to select a region from a map to find information about Superfund sites in that region.

3.20.3 Universal Wastes

Universal wastes are a special category of hazardous wastes that (in the United States):

- generally pose a lower threat relative to other hazardous wastes that are ubiquitous and produced in very large quantities by a large number of generators.

Some of the most common "universal wastes" are fluorescent light bulbs, some specialty batteries (e.g. lithium- or lead-containing batteries), cathode ray tubes, and mercury-containing devices.

Universal wastes are subject to somewhat less stringent regulatory requirements. Small quantity generators of universal wastes may be classified as "conditionally exempt small quantity generators" which release them from some of the regulatory requirements for the handling and storage of hazardous wastes. Universal wastes must still be disposed of properly at a regional or county landfill facility.

3.20.4 Final Disposal of Hazardous Waste

Historically, some hazardous wastes were disposed of in regular landfills. This resulted in unfavorable amounts of hazardous materials seeping into the ground. These chemicals eventually entered to natural hydrologic systems. Many landfills now require countermeasures against groundwater contamination. For example, a barrier has to be installed along the foundation of the landfill to contain the hazardous substances that may remain in the disposed waste. Currently, hazardous wastes must often be stabilized and solidified in order to enter a landfill and must undergo different treatments in order to stabilize and dispose them. Most flammable materials can be recycled into industrial fuel. Some materials with hazardous constituents can be recycled, such as lead acid batteries.

3.20.5 Recycling

Some hazardous wastes can be recycled into new products. Examples may include lead-acid batteries or electronic circuit boards. When heavy metals in these types of ashes go through the proper treatment, they could bind to other pollutants and convert them into easier-to-dispose solids, or they could be used as pavement filling. Such treatments reduce the level of threat of harmful chemicals, like fly and bottom, while also recycling the safe product.

3.20.6 Portland Cement

Another commonly used treatment is cement-based solidification and stabilization. Cement is used because it can treat a range of hazardous wastes by improving physical characteristics and decreasing the toxicity and transmission of contaminants. The cement produced is categorized into five different divisions, depending on its strength and components. This process of converting sludge into cement might include the addition of pH adjustment agents, phosphates, or sulfur reagents to reduce the settling or curing time, increase the compressive strength, or reduce the leach ability of contaminants.

3.21 Incineration, Destruction, and WtE

Hazardous waste may be "destroyed." For example, by incinerating it at a high temperature, flammable wastes can sometimes be burned as energy sources. For example, many cement kilns burn hazardous wastes like used oils or solvents. Today, incineration treatments not only reduce the amount of hazardous waste but also generate energy from the gases released in the process. It is known that this particular waste treatment releases toxic gases produced by the combustion of by-product or other materials which can affect the environment. However, current technology has developed more efficient incinerator units that control these emissions to a point where this treatment is considered a more beneficial option. There are different types of incinerators which vary depending on the characteristics of the waste. Starved air incineration is another method used to treat hazardous wastes. Just like in common incineration, burning occurs, however, controlling the amount of oxygen allowed proves to be significant to reduce the amount of harmful by-products produced. Starved air incineration is an improvement of the traditional incinerators in terms of air pollution. Through using this technology, it is possible to control the combustion rate of the waste and therefore reduce the air pollutants produced in the process (see a Case Study on WtE in Sections 7.13.1.1 and 7.13.3).

3.22 Hazardous Waste Landfill (Sequestering, Isolation, etc.)

Hazardous waste may be sequestered in an hazardous waste landfill or permanent disposal facility. "In terms of hazardous waste, a landfill is defined as a disposal facility or part of a facility where hazardous waste is placed or on land and which is not a pile, a land treatment facility, a surface impoundment, an underground injection well, a salt dome formation, a salt bed formation, an underground mine, a cave, or a corrective action management unit" (Title 40 CFR 260.10).

3.22.1 Pyrolysis

Some hazardous waste types may be eliminated using pyrolysis in an ultrahigh temperature electrical arc, in inert conditions to avoid combustion. This treatment method may be preferable to high-temperature incineration in some circumstances such as in the destruction of concentrated organic waste types, including PCBs, pesticides, and other persistent organic pollutants (RCRA, USEPA, Title 40 CFR 261).

3.23 Radioactive Waste

Radioactive waste is waste that contains radioactive material. Radioactive waste is usually a by-product of nuclear power generation and other applications of nuclear fission or nuclear technology such as research and medicine. Radioactive waste is hazardous to all forms of life and the environment, and it is regulated by government agencies in order to protect human health and the environment.

Radioactivity naturally decays over time; therefore, radioactive waste has to be isolated and confined in appropriate disposal facilities for a sufficient period until it no longer poses a threat. The time radioactive waste must be stored for depends on the type of waste and radioactive isotopes. Current approaches to managing radioactive waste have been segregation and storage for short-lived waste, near-surface disposal for low and some intermediate level waste, and deep burial or partitioning/transmutation for the high-level waste.

A summary of the amounts of radioactive waste and management approaches for most developed countries are presented and reviewed periodically as part of the International Atomic Energy Agency (IAEA) Joint Convention on the Safety of Spent Fuel Management and on the Safety of Radioactive Waste Management (IAEA 2007).

3.23.1 Sources

Radioactive waste comes from a number of sources. In countries with nuclear power plants, nuclear armament or nuclear fuel treatment plants, the majority of waste originates from the nuclear fuel cycle and nuclear weapons reprocessing. Other sources include medical and industrial wastes as well as naturally occurring radioactive materials (NORM) that can be concentrated as a result of the processing or consumption of coal, oil, and gas and some minerals as discussed below.

3.23.2 Nuclear Fuel Cycle

3.23.2.1 Front End

Waste from the front end of the nuclear fuel cycle is usually alpha-emitting waste from the extraction of uranium. It often contains radium and its decay products.

Uranium dioxide (UO_2) concentrate from mining is a thousand or so times as radioactive as the granite used in buildings. It is refined from yellowcake (U_3O_8), then converted to uranium hexafluoride gas (UF_6). As a gas, it undergoes enrichment to increase the U-235 content from 0.7% to about 4.4% low enriched uranium (LEU). It is then turned into a hard ceramic oxide (UO_2) for assembly as reactor fuel elements (Cochran 1999).

The main by-product of enrichment is depleted uranium (DU), principally the U-238 isotope, with a U-235 content of approximately 0.3%. It is stored either as UF_6 or as U_3O_8. Some is used in applications where its extremely high density makes it valuable such as antitank shells, and on at least one occasion even a sailboat keel (IHS Jane's 360). It is also used with plutonium for making mixed oxide fuel and to dilute, or downblend, highly enriched uranium from weapons stockpiles which is now being redirected to become reactor fuel.

3.23.2.2 Back End

The back-end of the nuclear fuel cycle, mostly spent fuel rods, contains fission products that emit beta and gamma radiation, and actinides that emit alpha particles, such as uranium-234 (half-life 245 000 years), neptunium-237 (2.144 million years), plutonium-238 (87.7 years), and americium-241 (432 years), and even sometimes some neutron emitters such as californium (half-life of 898 years for Cf-251). These isotopes are formed in nuclear reactors.

It is important to distinguish the processing of uranium to make fuel from the reprocessing of used fuel. Used fuel contains the highly radioactive products of fission (see Section 3.25.2). Many of these are neutron absorbers, called neutron poisons in this context. These eventually build up to a level where they absorb so many neutrons that the chain reaction stops, even with the control rods completely removed. At that point, the fuel has to be replaced in the reactor with fresh fuel, even though there is still a substantial quantity of uranium-235 and plutonium present. In the United States, this used fuel is usually "stored," while in other countries such as Russia, the United Kingdom, France, Japan, and India, the fuel is reprocessed to remove the fission products, and the fuel can then be reused, thus cutting costs, reducing health risks, saving time, and in general being far safer (Regulation of TENORM 2012). The fission products removed from the fuel are a concentrated form of high-level waste as are the chemicals used in the process. While these countries reprocess the fuel carrying out single plutonium cycles, India is the only country known to be planning multiple plutonium recycling schemes.

3.23.2.3 Nuclear Industry

Industrial source waste can contain alpha, beta, neutron, or gamma emitters. Gamma emitters are used in radiography, while neutron emitting sources are used in a range of applications, such as oil well logging (Nuclear Logging 2009).

Annual release of uranium and thorium radioisotopes from coal combustion, predicted by ORNL, cumulatively amount to 2.9 million T over the 1937–2040 period, from the combustion of an estimated 637 billion T of coal worldwide (Gabbard 1993).

Substances containing natural radioactivity are known as NORM. After human processing that exposes or concentrates this natural radioactivity (such as mining bringing coal to the surface or burning it to produce concentrated ash), it becomes technologically enhanced NORM (TENORM) (USEPA, June 6 2006c). A lot of this waste is alpha particle-emitting matter from the decay chains of uranium and thorium. The main source of radiation in the human body is potassium-40 (°K), typically 17 mg in the body at a time and 0.4 mg/day intake (Radio activity in nature, Idaho State University). Most rocks, due to their components, have a low level of radioactivity. Usually ranging from 1 to 13 mSv annually depending on location, average radiation exposure from natural radioisotopes is 2.0 mSv/person a year worldwide (UNSCEAR 2008). This makes up the majority of typical total dosage (with mean annual exposure from other sources amounting to 0.6 mSv from medical tests averaged over the whole populace, 0.4 mSv from cosmic rays, 0.005 mSv from the legacy of past atmospheric nuclear testing, 0.005 mSv occupational exposure, 0.002 mSv from the Chernobyl disaster, and 0.0002 mSv from the nuclear fuel cycle) (UNSCEAR 2008).

TENORM is not regulated as restrictively as nuclear reactor waste, though there are no significant differences in the radiological risks of these materials (Regulation of TENORM, Aug 1 2012).

3.24 Coal

Coal contains a small amount of radioactive uranium, barium, thorium, and potassium but, in the case of pure coal, this is significantly less than the average concentration of those elements in the Earth's crust. The surrounding strata, if shale or mudstone, often contain slightly more than average and this may also be reflected in the ash content of "dirty" coals (Gabbard 1993; Cosmic Origin of Uranium 2006). The more active ash minerals become concentrated in the fly ash precisely because they do not burn well (Gabbard 1993). The radioactivity of fly ash is about the same as black shale and is less than phosphate rocks, but it is more of a concern because a small amount of the fly ash ends up in the atmosphere where it can be inhaled (U.S. Geological Survey 1997). According to U.S. NCRP reports, population exposure is from 1000-MWe power plants amounts to 490 person-rem/year for coal power plants, and thus is 100 times as great as nuclear power plants (4.8 person-rem/year). (The exposure from the complete nuclear fuel cycle from mining to waste disposal is 136 person-rem/year; the corresponding value for coal use from mining to waste disposal is "probably unknown") (Gabbard 1993).

3.24.1 Oil and Gas

Residues from the oil and gas industry often contain radium and its decay products. The sulfate scale from an oil well can be very radium rich, while the water, oil, and gas from a well often contain radon. The radon decays to form solid radioisotopes which form coatings on the inside of pipe-work. In an oil-processing plant, the area of the plant where propane is processed is often one of the more contaminated areas of the plant as radon has a similar boiling point to propane.

3.25 Low-Level Waste

Low-level waste (LLW) is generated from hospitals and industry, as well as the nuclear fuel cycle. LLWs include paper, rags, tools, clothing, filters, and other materials which contain small amounts of mostly short-lived radioactivity. Materials that originate from any region of an Active Area are commonly designated as LLW as a precautionary measure even if there is only a remote possibility of being contaminated with radioactive materials. Such LLW typically exhibits no higher radioactivity than one would expect from the same material disposed of in a non-active area, such as a normal office block.

Some high-activity LLW requires shielding during handling and transport, but most LLW is suitable for shallow land burial. To reduce its volume, it is often compacted or incinerated before disposal. LLW is divided into four classes: class A, class B, class C, and greater than class C.

3.25.1 Intermediate-Level Waste

Intermediate-level waste (ILW) contains higher amounts of radioactivity and in general require shielding, but not cooling (USNRC, April 3 2017). ILWs includes resins, chemical sludge, and metal nuclear fuel cladding, as well as contaminated materials from reactor decommissioning. It may be solidified in concrete or bitumen for disposal. As a general rule, short-lived waste (mainly nonfuel materials from reactors) is buried in shallow repositories, while LLW (from fuel and fuel reprocessing) is deposited in geological repository. US regulations do not define this category of waste; the term is used in Europe and elsewhere.

3.25.2 High-Level Waste

High-level waste (HLW) is produced by nuclear reactors. The exact definition of HLW differs internationally. After a nuclear fuel rod serves one fuel cycle and is removed from the core, it is considered HLW (Janicki 2013). Fuel rods contain fission products and transuranic elements generated in the reactor core. Spent fuel is highly radioactive and often hot. HLW accounts for over 95% of the total radioactivity produced in the process of nuclear electricity generation. The amount of HLW worldwide is currently increasing by about 12 000 MT every year, which is the equivalent to about 100 double-decker buses or a two-story structure with a footprint the size of a basketball court (Rogner 2010). A 1000-MW nuclear power plant produces about 27 T of spent nuclear fuel (unreprocessed) every year (Myths and Realities of Radioactive Waste, Feb. 2017). In 2010, there was very roughly estimated to be stored some 250 000 T of nuclear HLW (World Nuclear Association, July 2015) that does not include amounts that have escaped into the environment from accidents or tests. Japan estimated to hold 17 000 T of HLW in storage in 2015 (Geere 2010). HLWs have been shipped to other countries to be stored or reprocessed, and in some cases, shipped back as active fuel.

The ongoing controversy over high-level radioactive waste disposal is a major constraint on the nuclear power's global expansion (Humber 2015). Most scientists agree (Findlay 2010) that the main proposed long-term solution is deep geological burial, either in a mine or a deep borehole. However, almost six decades after commercial nuclear energy began, no government has succeeded in opening such a repository for civilian high-level nuclear waste (Humber 2015), although Finland is in the advanced stage of the construction of such facility, the Onkalo spent nuclear fuel repository. Reprocessing or recycling spent nuclear fuel options already available or under active development still generate waste and so are not a total solution, but it can reduce the sheer quantity of waste, and there are many such active programs worldwide. Deep geological burial remains the only responsible way to deal with high-level nuclear waste (World Nuclear Association 2015). The Morris Operation is currently the only de facto high-level radioactive waste storage site in the United States.

3.25.3 Transuranic Waste

Transuranic waste (TRUW) as defined by US regulations is, without regard to form or origin, waste that is contaminated with alpha-emitting transuranic radionuclides with half-lives greater than 20 years and concentrations greater than 100 nCi/g (3.7 MBq/kg), excluding HLW. Elements that have an atomic number greater than uranium are called transuranic ("beyond uranium"). Because of their long half-lives, TRUW is disposed more cautiously than either low- or intermediate-level waste. In the United States, it arises mainly from weapons production and consists of clothing, tools, rags, residues, debris, and other items contaminated with small amounts of radioactive elements (mainly plutonium).

Under US law, transuranic waste is further categorized into "contact-handled" (CH) and "remote-handled" (RH) on the basis of the radiation dose rate measured at the surface of the waste container. CH TRUW has a surface dose rate not greater than 200 mrem/h (2 mSv/h), whereas RH TRUW has a surface dose rate of 200 mrem/h (2 mSv/h) or greater. CH TRUW does not have the very high radioactivity of HLW nor its high heat generation, but RH TRUW can be highly radioactive, with surface dose rates up to 1 000 000 mrem/h (10 000 mSv/h). The United States currently disposes of TRUW generated from military facilities at the Waste Isolation Pilot Plant in a deep salt formation in New Mexico (Biello 2011).

3.26 Nuclear Waste Management

Of particular concern in nuclear waste management are two long-lived fission products, Tc-99 (half-life 220 000 years) and I-129 (half-life 15.7 million years), which dominate spent fuel radioactivity after a few thousand years. The most troublesome transuranic elements in spent fuel are Np-237 (half-life two million years) and Pu-239 (half-life 24 000 years) (Nuclear Decommissioning Authority 2014). Nuclear waste requires sophisticated treatment and management to successfully isolate it from interacting with the biosphere. This usually necessitates treatment, followed by a long-term management strategy involving storage, disposal, or transformation of the waste into a nontoxic form (Vandenbosch and Vandenbosch 2007). Governments around the world are considering a range of waste management and disposal options, though there has been limited progress toward long-term waste management solutions (Ojovan and Lee 2014).

In the second half of twentieth century, several methods of disposal of radioactive waste were investigated by nuclear nations (Brown 2004), which are as follows:

1) "Long term above ground storage," not implemented
2) "Disposal in outer space" (for instance, inside the Sun), not implemented – as it would be currently too expensive
3) "Deep borehole disposal," not implemented
4) "Rock-melting," not implemented
5) "Disposal at subduction zones," not implemented
6) "Ocean disposal," done by the USSR, the United Kingdom (World Nuclear Association 2011), Switzerland, the United States, Belgium, France, The Netherlands, Japan, Sweden, Russia, Germany, Italy, and South Korea (1954–1993). This is no longer permitted by international agreements.
7) "Sub seabed disposal," not implemented, not permitted by international agreements
8) "Disposal in ice sheets," rejected in Antarctic Treaty
9) "Direct injection," done by USSR and United States.

In the United States, waste management policy completely broke down with the ending of work on the incomplete Yucca Mountain Repository (The Independent London 1997). At present, there are 70 nuclear power plant sites where spent fuel is stored. A Blue Ribbon Commission was appointed by President Obama to look into future options for this and future waste. A deep geological repository seems to be favored (The Independent London 1997).

3.26.1 Initial Treatment

3.26.1.1 Vitrification

Long-term storage of radioactive waste requires the stabilization of the waste into a form which will neither react nor degrade for extended periods. It is theorized that one way to do this might be through vitrification (Blue Ribbon Commission 2012). Currently, at Sellafield the HLW (PUREX first cycle raffinate) is mixed with sugar and then calcined. Calcination involves passing the waste through a heated, rotating tube. The purposes of calcination are to evaporate the water from the waste, and denitrate the fission products to assist the stability of the glass produced.

The "calcine" generated is fed continuously into an induction heated furnace with fragmented glass (National Research Council 1996). The resulting glass is a new substance in which the waste products are bonded into the glass matrix when it solidifies. As a melt, this product is poured into stainless steel cylindrical containers ("cylinders") in a batch process. When cooled, the fluid solidifies ("vitrifies") into the glass. After being formed, the glass is highly resistant to water (Laboratory-scale vitrification and leaching of Hanford high-level waste for the...model verification 2009).

After filling a cylinder, a seal is welded onto the cylinder head. The cylinder is then washed. After being inspected for external contamination, the steel cylinder is stored, usually in an underground repository. In this form, the waste products are expected to be immobilized for thousands of years (Ojovan et al. 2006).

The glass inside a cylinder is usually a black glossy substance. All this work (in the United Kingdom) is done using hot cell systems. Sugar is added to control the ruthenium chemistry and to stop the formation of the volatile RuO_4 containing radioactive ruthenium isotopes. In the West, the glass is normally a borosilicate glass (similar to Pyrex), while in the former Soviet Bloc it is normal to use a phosphate glass (Nuclear Energy Agency, Paris 1994). The amount of fission products in the glass must be limited because some (palladium, the other Pt group metals, and tellurium) tend to form metallic phases which separate from the glass. Bulk vitrification uses electrodes to melt soil and wastes, which are then buried underground

(Ojovan and Lee 2010). In Germany a vitrification plant is in use; this is treating the waste from a small demonstration reprocessing plant which has since been closed down (Pacific Northwest National Laboratory, PNNL-15198, July 2005).

3.26.1.2 Ion Exchange

It is common for medium active wastes in the nuclear industry to be treated with ion exchange or other means to concentrate the radioactivity into a small volume. The much less radioactive bulk (after treatment) is often then discharged. For instance, it is possible to use a ferric hydroxide floc to remove radioactive metals from aqueous mixtures (Hensing and Schultz 1995). After the radioisotopes are absorbed onto the ferric hydroxide, the resulting sludge can be placed in a metal drum before being mixed with cement to form a solid waste form (Brünglinghaus, Waste Processing 2013). In order to get better long-term performance (mechanical stability) from such forms, they may be made from a mixture of fly ash, or blast furnace slag, and Portland cement, instead of normal concrete (made with Portland cement, gravel, and sand).

Problems

3.1 The suspended particulate concentration in an aqueous industrial stream has been determined to be 27.6 mg/l. Convert this value to units of μg/l, g/l, lb/ft^3, and lb/gal.

3.2 A biologically contaminated dairy liquid waste requires disinfection with chlorine prior to discharge into a nearby lake. Given a contact time of 30 minutes, a required chlorine dose for a 99.99% kill of the pathogenic organisms. To carry out the two-steps oxidation–reduction and complete disinfection, the chlorine doses were kept at 3.0 and 0.5 mg/l, respectively. Calculate how many pounds of chlorine would be required daily to disinfect 100 000 gal/day.

3.3 Polymer is supplied to an industrial wastewater pretreatment plant at a concentration of 0.5 lb/gal. The polymer feed pump delivers a flow of 0.1 gpm and the water flow rate to the filter is 3000 gpm. Calculate the dose (in mg/l) of polymer in the water applied to the filter.

3.4 Given the waste constituent and concentration shown in the following table, determine the quantity (in kilogram per day) of hydrated lime [$Ca(OH)_2$] required to neutralize the waste. Estimate the TDS after neutralization. Report your answer in milligram per liter (mg/l).

Constituent	Concentration (mg/l)	Flow (l/min)
HCl	100	5

3.5 Given the waste constituent and concentration shown in the following table, determine the quantity (in kilogram per day) of sulfuric acid [$H_2(SO)_4$] required to neutralize the waste. Estimate the TDS after neutralization. Report your answer in milligram per liter (mg/l).

Constituent	Concentration (mg/l)	Flow (l/min)
NaOH	250	20

3.6 The concentration of TCE in an air sample near a hazardous waste landfill is 34 ppm, the ambient air temperature is 31 °C, and the atmospheric pressure is 0.94 atm. What is the TCE concentration in mg/m^3?

3.7 A sample of air has been collected above an air stripping tower to evaluate its performance and to determine if toxic air emission standards are being violated. The temperature at the time of collection was 18 °C and the atmospheric pressure was 740 mm Hg. The concentrations in the sample were 4.8 mg/m^3 of benzene and 1.7 mg/m^3 per chloroethylene, and 2.0 mg/m^3 of methylene chloride. Calculate the concentrations in ppm using the equation as follows: $C_{ppm} = C(RT/PM) \times 10^6$, where C is the concentration of the sample in mg/m^3, $R = 8.21 - 10^{-5}$ (m^3·atm/mol·K), M is the molecular weight of the gas, and T is the absolute temperature in K.

3.8 Liquid alum is usually available at a concentration of 5.4 lb alum/gal. In an industrial wastewater pretreatment application, the alum feed pump delivers 88 ml/min and the flow to the filter is 3000 gpm. Calculate the dose of alum in the water applied to the filter.

3.9 An advanced treatment is implemented in industrial wastewater (addition of ferric chloride, $FeCl_3$, and polymer), the BOD is removed in primary treatment by 30–45%. Calculate the inlet BOD load in pounds per day, BOD load, and percent reduction in aeration basin, and the annual chemical cost using the data provided below:

Flow, $Q = 50\,\text{MGD}$
Influent BOD $= 175\,\text{mg/l}$
$FeCl_3 = 20\,\text{mg/l}$ at 0.012/lb
Polymer $= 0.2\,\text{mg/l}$ at \$1.70/lb

3.10 Determine the time, in hours, required for the concentration of toluene in a wastewater treatment process to be reduced to one-half of its initial value. Assume the first-order reaction rate constant for toluene is 0.07/h.

Calculate the time for the toluene to be reduced to 99% of its initial value. Also calculate the time for 99.99% of the toluene to react.

3.11 In a pretreatment, industrial wastewater is coagulated with a dosage of 30 mmg/l of aluminum sulfate as alum and an equivalent dosage of lime to neutralize the hydrogen ions produced in the hydrolysis of aluminum. Calculate (a) how many pounds of alum are needed per MG of processed water treated? (b) How many pounds of $Al(OH)_3$ sludge are produced per MG of water treated?

3.12 A tannery uses a filter press to dewater its raw sludge. The dewatered sludge contains 120 mg of total chromium/kg of sludge. The filter press has a surface area $150\,\text{ft}^2$ with a filtration rate of $10\,\text{gal/h·ft}^2$. Assume that it is 100% efficient in separating solids. Estimate the amount of total chromium disposed per year from this plant if the plant operates 16 h/day, and 7 day/week, 50 week/year. The following information is available.

Before dewatering:

Water content of raw sludge = 95%;
solid content of *raw* sludge = 5%

After dewatering:

Water content = 52%; solid content = 48%.

References

Allen, D. and Rosselot, K. (1997). Pollution Prevention for Chemical Processes. Hoboken, NJ: Wiley.

Arthur, C. (1997). Ministers admit nuclear waste was dumped in sea. *The Independent London* (1 July).

Bateson, T. and Schwartz, J. (2008). Children's response to air pollutants. *Journal of Toxicology and Environmental Health* 71 (3): 238–243.

Berglund, R., Wood, D., and Covin, T. (1989). Fugitive Emissions from the Acrolein Production Industry. Paper presented at the Air and Waste Management Association International Specialty Conference: SARA Title III, Section 313, Industry Experience in Estimating Chemical Releases.

Biello, D. (2011). Presidential Commission Seeks Volunteers to Store U.S. Nuclear Waste. *Scientific American* (29 July). https://www.scientificamerican.com/article (accessed 21 September 2019).

Blue Ribbon Commission (2012). America's Nuclear Future: Executive Summary.

Bond, R.G. and Straub, C.P. (1974). Wastewater, Treatment and Disposal. Cleveland, OH: CRC Press.

Brown, P. (2004) Shoot it at the Sun. Send it to Earth's Core. What to do with Nuclear Waste? *The Guardian* (14 April).

Brünglinghaus, M. (2013). Waste Processing. *Euronuclear.org.*

Chambers, A., Wootton, T., Moncrieff, J., and McCready, P. (2008). Direct measurement of fugitive emissions of hydrocarbons from a refinery. *Journal of the Air and Waste Management Association* 58 (8): 1047–1056.

Chaudhary, R. and Rachana, M. (2006). Factors affecting hazardous waste solidification/stabilization: A review. *Journal of Hazardous Materials* B137: 267–276.

Clearstone Engineering (1994). A National Inventory of Greenhouse Gas (GHG), Criteria Air Contaminant (CAC) and Hydrogen Sulphide (H2S) Emissions by the Upstream Oil and Gas. Industry, Volume 1, Overview of the GHG Emissions Inventory. Canadian Association of Petroleum Producers.

Cochran, R. (1999). The Nuclear Fuel Cycle: Analysis and Management, 52–57. La Grange Park, IL: American Nuclear Society.

Cosmic Origins of Uranium (2006). Continuous plutonium recycling in India: improvements in reprocessing technology. uic.com.au.

Crites, R.W. and Tchobanoglous, G. (1998). Small and Decentralized Wastewater Management Systems. New York, NY: WCB/McGraw-Hill.

Eckenfelder, W.W. (2000). Industrial Water Pollution Control, 3e. New York, NY: McGraw-Hill.

Findlay, T. (2010). Nuclear Energy to 2030 and Its Implications for Safety, Security and Nonproliferation: Overview. Nuclear Energy Futures Project.

Gabbard, A. (1993). Coal combustion. *ORNL Review* 26: 3–4.

Geere, D. (2010) Where Do You Put 250,000 tonnes of Nuclear Waste? (Wired UK). Wired.co.uk. *Global Defence News and Defence Headlines – IHS Jane's 360.*

Hallam, B. (2010). Techniques for efficient hazardous chemicals handling and disposal. *Pollution Equipment News* (April/May 2010), p. 13.

Haugen, O.A., Watson, K.M., and Ragatz, R.A. (1954). Chemical Process Principles – Part I: Materials and Energy Balances, 2e. New York: Wiley.

Hauser, C. (2018). Algal Bloom in Lake Superior Raises Worries on Climate Change and Tourisms. *The New York Times* (29 August).

Hensing, I. and Schultz, W. (1995). Economic Comparison of Nuclear Fuel Cycle Options. Cologne: Energiewirtschaftlichen Instituts.

Himmelblau, D.M. (1998). Basic Principles and Calculations in Chemical Engineering, 2e. Upper Saddle River, NJ: Prentice Hall.

Humber, Y. (2015). Japan's 17,000 Tons of Nuclear Waste in Search of a Home, Bloomberg. Idaho State University. Radioactivity in Nature.

Industrial Waste Treatment (IWT) (1999). V1&V2. California State University, USA. http://www.owp.csus.edu/courses/wastewater/industrial-waste-treatment-vol-i.php (accessed 21 September 2019).

International Atomic Energy Agency (IAEA). (2007). The Joint Convention.

Janicki, M. (2013). Iron boxes for ILW transport and storage. *Nuclear Engineering International* 58 (711): 47–48.

Kampa, M. and Castanas, E. (2008). Human health effects of air pollution. *Environmental Pollution* 151 (2): 362–367.

Karell, M. (2017). Estimating air emission rates. *Chemical Engineering Progress* 113 (7): 59–65.

King, J. (1995). The Environmental Dictionary, 745. New York: Wiley.

Krenkel, P.A. and Novotny, V. (1980). Water Quality Management. Cambridge, MA: Academic Press.

Chapman, C.C., Buell, J.L., Slate, S.C. et al. (2009). Laboratory-scale vitrification and leaching of Hanford high-level waste for the purpose of simulant and glass property models validation.

Latza, U., Gerdes, S., and Baur, X. (2008). Effects of nitrogen dioxide on human health: systematic review of experimental and epidemiological studies conducted between 2002 and 2006. *International Journal of Hygiene and Environmental Health* 212 (3): 271–287.

Metcalf & Eddy (2003). Wastewater Engineering: Treatment and Reuse, 4e. New York: McGraw Hill.

Moss, B. (2008). Water pollution by agriculture. *Philosophical Transactions of the Royal Society B: Biological Sciences* 363: 659–666.

Myths and Realities of Radioactive Waste. (2017).

National Research Council (1996). Nuclear Wastes: Technologies for Separation and Transmutation. Washington DC: National Academy Press.

Nuclear Decommissioning Authority (2014). Progress on Approaches to Management of Separated Plutonium. Cumbria: Nuclear Decommissioning Authority.

Nuclear Logging (2009). https://petrowiki.org/Nuclear_logging (accessed 21 September 2019).

OECD Nuclear Energy Agency (1994). The Economics of the Nuclear Fuel Cycle. Paris: OECD Nuclear Energy Agency.

Ojovan, M. and Lee, W. (2010). Glassy wasteforms for nuclear waste immobilization. *Metallurgical and Materials Transactions, A* 42 (4): 837.

Ojovan, M. and Lee, W. (2014). An Introduction to Nuclear Waste Immobilization. Amsterdam: Elsevier.

Ojovan, M., Lee, W., Hand, R., and Ojovan, N. (2006). Corrosion of nuclear waste glasses in non-saturated conditions: time-temperature behavior.

Orloff, K. and Henry, F. (2003). An international perspective on hazardous waste practices. *International Journal of Hygiene and Environmental Health* 206 (4–5): 291–302.

Pacific Northwest National Laboratory (2005). Waste Form Release Calculations for the 2005 Integrated Disposal Facility Performance Assessment, PNNL-15198. Richland: Pacific Northwest National Laboratory.

Pink, D.H. (2006). Investing in Tomorrow's Liquid Gold, Yahoo, Archived from the Original on 23 April 2006.

Regulation of TENORM (2012). Recycling spent nuclear fuel: the Ultimate Solution for the US. Archived from the original on 28 November 2012, "Regulation of TENORM". Tenorm.com.

Rogner, H. (2010). Nuclear power and stable development. *Journal of International Affairs* 64: 149.

Rouessac, F. and Rouessac, A. (2007). Chemical Analysis: Modern Instrumentation Methods and Techniques, 2e. Hoboken, NJ: Wiley.

Salvi, S. (2007). Health effects of ambient air pollution in children. *Pediatric Respiratory Reviews* 8 (4): 275–280.

Shibamoto, T., Yasuhara, A., and Katami, T. (2007). Dioxin formation from waste incineration. *Reviews of Environmental Contamination and Toxicology* 190: 1–41.

Stephenson, J. (2008). Electronic Waste: EPA Needs to Better Control Harmful U. S. Exports Through Stronger Enforcement and More Comprehensive Regulations. DIANE Publishing. Survey & Identification of NORM Contaminated Equipment. enprotec-inc.com.

Tammemagi, H.Y. (1999). The Waste Crisis. Oxford University Press.

Title 40 of the Code of Federal Regulations (CFR) (2018).

USEPA (1975). Technical Support Document: EPA Regulations for Preventing the Significant Deterioration of Air Quality. Research Triangle Park, NC: Office of Air Quality Planning and Standards, Publication EPA-450/2-75-001.

US Geological Survey (1997). Radioactive Elements in Coal and Fly Ash: Abundance, Forms, and Environmental

Significance, Fact Sheet FS-163-1997. Reston, VA: U.S. Department of the Interior, U.S. Geological Survey.

United Nations Scientific Committee on the Effects of Atomic Radiation (2008). Sources and Effects of Ionizing Radiation. UNSCEAR.

United States Code, Clean Water Act, Section 402(p) 33 U.S.C. §1342(p) (1999).

United States Code, Clean Water Act, Section 502(14) 33 U.S.C. §1362(14) (1999).

University of Pennsylvania (2016). Laboratory Chemical Waste Management Guidelines. Environmental Health and Radiation Safety, University of Pennsylvania.

USEPA (1985a). AP-42: Compilation of Air Pollutant Emission Factors, Volume I: Stationary Point and Area Sources, 4e, with Supplements A-D. Research Triangle Park, NC: Office of Air Quality Planning and Standards, Publication AP-42, 1985.

USEPA (1985b). A Model for Evaluation of Refinery and Synfuels VOC Emission Data, Volume I. Research Triangle Park, NC: Office of Air Quality Planning and Standards Publication EPA-600/7-85-022a, 1985.

USEPA (1988). Protocols For Generating Unit-Specific Emission Estimates For Equipment Leaks of VOC and VHAP. Research Triangle Park, NC: Office of Air Quality Planning and Standards, Publication EPA-450/3-88-010, 1988.

USEPA (1990). The Clean Air Act Amendments of 1990 Summary Materials. Washington, DC: USEPA, Office of Air and Radiation.

USEPA (1993). Protocols for Equipment Leak Emission Estimates. Research Triangle Park, NC: Office of Air Quality Planning and Standards, Publication, available through NTIS as PB 229219, June 1993.

USEPA (1995). Protocols for Equipment Leak Emission Estimates. Research Triangle Park, NC: Office of Air Quality Planning and Standards, Publication EPA-450/R-95-017 (1995).

USEPA (1998). Study of Hazardous Air Pollutant Emissions from Electric Utility Steam Generating Units: Final Report to Congress. Washington DC: Office of Air Quality Planning and Standards 453/R-8-004.

USEPA (2006a). Air Quality Criteria for Ozone and Related Photochemical Oxidants (Final Report). Washington, DC: USEPA, National Center for Environmental Assessment EPA. EPA/600/R-05/004aF-cF. http://cfpub.epa.gov/ncea/isa/recordisplay.cfm?deid=149923 (accessed 28 July 2018).

USEPA (2006b). Air Quality Criteria for Lead (Final Report). Washington, DC: USEPA, National Center for Environmental Assessment EPA/600/R-05/144aF-bF. http://cfpub.epa.gov/ncea/isa/recordisplay.cfm?deid=158823 (accessed 28 July 2018).

USEPA (2006c). TENORM Sources Radiation Protection. epa.gov.

USEPA (2008a). Integrated Science Assessment for Oxides of Nitrogen: Health Criteria (Final Report). Washington, DC: National Center for Environmental Assessment EPA/600/R-08/071.

USEPA (2008b). Integrated Science Assessment for Sulfur Oxides: Health Criteria (Final Report). Washington, DC: USEPA, National Center for Environmental Assessment EPA/600/R-08/047F. http://cfpub.epa.gov/ncea/isa/recordisplay.cfm?deid=198843 (accessed 29 July 2018).

USEPA (2009a). Integrated Science Assessment for Particulate Matter (Final Report). Washington, DC: USEPA, National Center for Environmental Assessment EPA/600/R-08/139F. http://cfpub.epa.gov/ncea/CFM/recordisplay.cfm?deid=216546 (accessed 29 July 2018).

USEPA. 2009b. National Ambient Air Quality Standards (NAAQS). http://www.epa.gov/ttn/naaqs (accessed 29 July 2018).

USEPA (2010a). Clean Air Act. USEPA, Office of Air and Radiation. http://www.epa.gov/air/caa (accessed 30 July 2018).

USEPA (2010b). Integrated Science Assessment for Carbon Monoxide (Final Report). Washington, DC: USEPA, National Center for Environmental Assessment EPA/600/R-09/019F. http://cfpub.epa.gov/ncea/cfm/recordisplay.cfm?deid=218686 (accessed 30 July 2018).

USEPA (2012). Municipal Solid Waste Generation, Recycling, and Disposal in the United States: Facts and Figures for 2012, EPA/530/R-012/001. Washington, DC: Office of Solid Waste and Emergency Response.

USEPA, (2018). "Harmful Algal Blooms: Nutrient Pollutant," USEPA. https://www.epa.gov/nutrientpollution/harmful-algal-blooms (accessed 29 September 2018).

USNRC (2017). Backgrounder on radioactive waste. www.nrc.gov (accessed 21 September 2019).

Vandenbosch, R. and Vandenbosch, S. (2007). Nuclear Waste Stalemate. Salt Lake City: University of Utah Press.

Weidenhamer, J. and Clement, M. (2007). Leaded electronic waste is a possible source material for lead-contaminated jewelry. *Chemosphere* 69 (7): 1111–1115.

Wigle, D., Arbuckle, T., Walker, M. et al. (2007). Environmental hazards: evidence for effects on child health. *Toxicology and Environmental Health Part B: Critical Reviews* 10 (1–2): 3–39.

World Nuclear Association (2011). Storage and disposal options. https://www.world-nuclear.org/information-library/nuclear-fuel-cycle/nuclear-wastes/storage-and-disposal-of-radioactive-waste.aspx (accessed 21 September 2019).

World Nuclear Association (2015). Radioactive Waste Management. Resources Conservation and Recovery Act. USEPA.

Yohe, T.M. and Rich, J.E. (1995). Textile Mill Builds for Today, Plans for Tomorrow, *Industrial Wastewater*, Water Environment Federation, 30–35.

4

Industrial Wastewater, Air Pollution, and Solid and Hazardous Wastes

Monitoring, Permitting, Sample Collections and Analyses, QA/QC, Compliance with State Regulations and Federal Standards

4.1 Introduction

This chapter describes and deals with important aspects of selecting the best remedial control technologies for pollutants, managing wastes, monitoring, sampling industrial water, air, and solid and hazardous materials, modes of sample collections, samples analyses by various analytical, physical-chemical methods approved by governmental agency, quality control (QC) and quality assurance (QA), laboratory audits, testing, monitoring, permitting, report keeping, reporting, and compliance with local, state, and federal governments for discharging wastewater, emitting air, pollutants, and safely disposing of solid and hazardous materials.

4.2 Industrial Process Water

In industrial process, the control of the quality of process water can be critical to the quality of the end product. Water is often used as a carrier of reagents and the loss of reagents to product must be continuously monitored to ensure that correct replacement rate is made. Parameters measured relate specifically to the process in use and to any of the expected contaminants that may arise as by-products. This may include unwanted organic chemicals appearing in an inorganic chemical process through contamination with oils and greases from machinery. Monitoring the quality of the wastewater discharged from industrial premises is a key factor in controlling and minimizing pollution of the environment. In this application monitoring, systems analyze for all possible contaminants arising within the process and especially contaminants that may have particularly adverse impacts on the environment such as cyanide and many organic species such as pesticides (USEPA 1973). In the nuclear industry, analysis focuses on specific isotopes or elements of interest. Where the nuclear industry makes wastewater discharges to rivers which have drinking water abstraction on them, radioisotopes which could potentially be harmful or those with long half-lives such as tritium will form part of the routine monitoring suite.

Depending on the type of industry, characteristics of process waters including physical-chemical-biological properties are likely to vary during operations, owing to breakdowns, spillovers, floor washing, change of a shift, and other causes. Because of potential variability of wastewaters, the sampling frequency, types of samples (e.g. instant grab or continuous), and location of collection, the sampling and the analysis of process waters usually require careful attention to details compared to sampling and analyzing municipal influents and effluents.

The purpose of sampling and analysis is to record the peak load concentration, the duration of peak loads or the occurrence of variation throughout the day.

4.3 Common Elements, Radicals, and Chemicals in Water Analysis

Names of elements common to water and wastewater technology along with their symbols, atomic weights, common valence, and equivalent weights are given in Table 4.1. Certain groupings of atoms act together as a unit in a large number of different molecules; these refer to radicals, are given special names, such as the hydroxyl group (OH^-). The most common radicals in ionized form are listed in Table 4.2. Radicals themselves are not compounds, but they join with other elements to form compounds. Data on inorganic compounds common to water and wastewater chemistry are given in Table 4.3. The proper name, formula, molecular weight, and equivalent weight are included for all of the

Industrial Environmental Management: Engineering, Science, and Policy, First Edition. Tapas K. Das.

Companion website: www.wiley.com/go/Das/IEM_1e

Table 4.1 Basic information on common elements.

Name	Symbol	Atomic valence	Common weight	Equivalent weight[a]
Aluminum	Al	27.0	3+	9.0
Arsenic	As	74.9	3+	25.0
Barium	Ba	137.3	2+	68.7
Boron	B	10.8	3+	3.6
Bromine	Br	79.9	1−	79.9
Cadmium	Cd	112.4	2+	56.2
Calcium	Ca	40.1	2+	20.0
Carbon	C	12.0	4−	3.0
Chlorine	Cl	35.5	1−	35.5
Chromium	Cr	52.0	3+; 6+	17.3
Copper	Cu	63.5	2+	31.8
Fluorine	F	19.0	1−	19.0
Hydrogen	H	1.0	1+	1.0
Iodine	I	126.9	1−	126.9
Iron	Fe	55.8	2+; 3+	27.5
Lead	Pb	207.2	2+	103.6
Magnesium	Mg	24.3	2+	12.2
Manganese	Mn	54.9	2+; 4+; 7+	27.5
Mercury	Hg	200.6	1+; 2+	100.3
Nickel	Ni	58.7	2+	29.4
Nitrogen	N	14.0	3−; 5−	
Oxygen	O	16.0	2−	8.0
Phosphorus	P	31.0	5+	6.0
Potassium	K	39.1	1+	39.1
Selenium	Se	79.0	6+	13.1
Silicon	Si	28.1	4+	6.5
Silver	Ag	107.9	1+	107.9
Sodium	Na	23.0	1+	23.0
Sulfur	S	32.1	2−	16.0

[a] Equivalent weight (combining weight) equals atomic weight divided by valence.

chemicals listed. Popular names, for example, alum for aluminum sulfate, are included in brackets. For chemicals used in water treatment, one common use is given; many have other applications which are not included. Equivalent weights for compounds and some combinations (e.g. $Ca(HCO_3)_2$) involved in treatment are provided.

Examples 4.1 Calculate the molecular and equivalent weights of ferric sulfate.

Table 4.2 Common radicals found in process water.

Name	Formula	Molecular weight	Electric charge	Equivalent weight
Ammonium	NH_4^+	18.0	1+	18.0
Hydroxyl	OH^-	17.0	1−	17.0
Bicarbonate	HCO_3^-	61.0	1−	61.0
Carbonate	CO_3^-	60.0	2−	30.0
Orthophosphate	PO_4^-	95.0	3−	31.7
Hydrogen phosphate	HPO_4^-	96.0	2−	48.0
Dihydrogen phosphate	$H_2PO_4^-$	97.0	1−	97.0
Bisulphate	HSO_4^-	97.0	1−	97.0
Sulfate	SO_4^-	96.0	2−	48.0
Bisulfite	HSO_3^-	81.0	1−	81.0
Sulfite	SO_3^-	80.0	2−	40.0
Nitrite	NO_2^-	46.0	1−	46.0.0
Nitrate	NO_3^-	62.0	1−	62.0
Hypochlorite	OCl^-	51.5	1−	51.5

Solution

The formula from Table 4.3 is $Fe_2(SO_4)_3$. Using atomic weight from Table 4.1

$$\begin{array}{ll} \text{Fe} & 2\times 55.8 = 111.6 \\ \text{S} & 3\times 32.1 = 96.3 \\ \text{O} & 3\times(4\times 16.0) = 192.0 \end{array}$$

$$\text{Molecular weight} = 399.9 \text{ or } 400\,\text{g}.$$

The ferric (oxide of iron) atom has a valence of 3^+, thus a compound with two ferric atoms has a total electrical charge of 6^+. (Three sulfate radicals have a total of 6^- charges.)

$$\begin{aligned} \text{Equivalent weight} &= (\text{molecular weight})/(\text{electrical charge}) \\ &= 400/6 = 66.7\,\text{g per equivalent weight (Table 4.3)} \end{aligned}$$

4.4 Purposes and Objectives for Inspecting and Sampling

Having *National Pollutant Discharge Elimination System* (NPDES) permit holder, each industrial facility with an approved pretreatment program with/without working with a publicly owned treatment works (POTW), must implement its compliance monitoring program that (i) meets the requirements and intent of the General Pretreatment Regulations, (ii) is effective and timely in determining

Table 4.3 Basic information on common use of inorganic chemicals in process water.

Name	Formula	Common usage	Molecular weight	Equivalent weight
Activated carbon	C	Taste and odor control	12.0	N.A.[a]
Aluminum sulfate (alum)	$Al_2(SO_4)_3\cdot14.3\ H_2O$	Coagulation	600	100
Aluminum hydroxide	$Al_2(OH)_3$	(Some combination)	78.0	26.0
Ammonia	NH_3	Chloroamine disinfection	17.0	N.A.
Ammonium fluorosilicate	$(NH_4)_2SiF_6$	Fluoridation	178	N.A.
Ammonium sulfate	$(NH_4)_2SO_4$	Coagulation	132	66.1
Calcium bicarbonate	$Ca(HCO_3)_2$	(Some combination)	162	81.0
Calcium carbonate	$CaCO_3$	Corrosion control	100	50.0
Calcium fluoride	CaF_2	Fluoridation	78.1	N.A.
Calcium hydroxide	$Ca(OH)_2$	Softening	74.1	37.0
Calcium hypochlorite	$Ca(OCl)_2\cdot2H_2O$	Disinfection	179	N.A.
Calcium oxide (lime)	CaO	Softening	56.1	28.0
Carbon dioxide	CO_2	Recarbonation	44.0	22.0
Chlorine	Cl_2	Disinfection	71.0	N.A.
Chlorine dioxide	ClO_2	Taste and odor control	67.0	N.A.
Copper sulfate	$CuSO_4$	Algae control	160	79.8
Ferric chloride	$FeCl_3$	Coagulation	162	54.1
Ferric hydroxide	$Fe(OH)_3$	(Some combination)	107	35.6
Ferric sulfate	$Fe_2(SO_4)_3$	Coagulation	400	66.7
Ferrous sulfate	$Fe_2(SO_4)\cdot7H_2O$	Coagulation	278	139
Fluorosilicic acid	H_2SiF_6	Fluoridation	144	N.A.
Hydrochloric acid	HCl	N.A.	36.5	36.5
Magnesium hydroxide	$Mg(OH)_2$	Defluoridation	58.3	29.2
Oxygen	O_2	Aeration	32.0	16.0
Ozone	O_3	Disinfection	48.0	N.A.
Potassium permanganate	$KMnO_4$	Oxidation	158	N.A.
Sodium aluminate	$NaAlO_2$	Coagulation	82.0	N.A.
Sodium bicarbonate	$NaHCO_3$	pH adjustment	84.0	84.0
Sodium chloride	$NaCl$	Ion-exchange regeneration	106	53.0
Sodium chloride	NaF	Fluoridation	58.4	58.4
Sodium hexametaphosphate	$(NaPO_3)_n$	Corrosion control	42.0	N.A.
Sodium hydroxide	$NaOH$	pH adjustment	N.A.	N.A.
Sodium hypochlorite	$NaOCl$	Disinfection	40.0	40.0
Sodium silicate	Na_4SiO_4	Coagulation aid	74.4	N.A.
Sodium fluorosilicate	Na_2SiF_6	Fluoridation	184	N.A.
Sodium thiosulfate	NaS_2O_3	Dechlorination	188	N.A.
Sulfur dioxide	SO_2	Dechlorination	64.1	N.A.
Sulfuric acid	H_2SO_4	pH adjustment	98.1	49.0
Water	H_2O	N.A.	18.1	N.A.

Source: Adapted from Hammer and Hammer (2012).
[a] N.A., not applicable.

compliance with categorical standards, local limits, and discharge prohibitions, (iii) provides representative data required to meet reporting requirements to POTW, the state and/or the EPA, and (iv) provides sampling data that would be admissible in court, if an enforcement action were to be undertaken by the POTW, state, or federal government.

Specifically, the objectives of the industrial facility's compliance monitoring program are as follows:

- Evaluate industrial user compliance
  - Verify the completeness, accuracy, and representativeness of self-monitoring data from the industrial user as well as whether required sampling and analysis techniques were used.
  - Verify production/flows at applicable production-based categorical industrial users to validate accurate equivalent concentration permit limits.
  - Determine compliance with all applicable requirements, such as industrial user permit conditions and sewer use ordinance provisions.
  - Evaluate *best management practices* (BMPs) and pollution prevention measures.
  - Evaluate the industrial user's *standard operating procedures* (SOPs), chemical waste handling procedures, preventative maintenance programs, process/pretreatment equipment and general housekeeping.
  - Determine if the industrial user has corrected problems identified in previous inspections.
  - Support past and future enforcement actions against noncompliant industrial users, including ensuring the industrial user complies with obligations imposed through prior enforcement actions.
  - Assess the potential for spills and slug discharges and evaluate the effectiveness of slug discharge control measures.
  - Evaluate the adequacy of the industrial user's hazardous waste management and disposal practices.
  - Consider potential compliance issues with other statutes or regulations [e.g. *Occupational Safety and Health Administration*, *Resource Conservation and Recovery Act* (RCRA), etc.].
- Develop and revise industrial user permit requirements
  - Confirm applicability of appropriate categorical standards for the facility.
  - Confirm the accuracy of information in the most current permit application (or survey form) with respect to the industrial user's operations. Specifically, assure raw materials and chemicals used, production, treatment, disposal practices, and all wastestreams are accurately captured in the application (or survey form).
  - Evaluate the impacts of the industrial user on the POTW's influent, treatment processes, sludge, and effluent (including water quality impacts from the effluent).
  - Assess the adequacy of the industrial user's self-monitoring program and permit, and support permit revisions as appropriate.
  - Confirm comprehensive process narrative and wastewater flow schematics coincide with current operations at the industrial user (i.e. have there been any changes in either that have not been reported?).
  - Evaluate and document pollution prevention (also known as P2) activities and opportunities (source reduction, *environmental management system* (EMS), BMPs, water/energy re-use/conservation, in-situ recycling, etc.).
  - Provide a basis for establishing the sampling requirements of the industrial user (above the federal minimums).
  - Evaluate the adequacy of the industrial user's operation and maintenance (O&M) activities on its pretreatment system.
- Maintain a strong local pretreatment program
  - Evaluate the need for revised local limits.
  - Inform the regulated community of pretreatment requirements.
  - Maintain current data on each regulated industrial user.
  - Generate data that can be used by the POTW in its annual report to the Approval Authority.
  - Establish and maintain good working relationships with the industrial user.
  - Provide a presence in the field so that industrial users understand that the POTW is serious about its regulatory oversight duties.

Industrial user sampling and inspections are the cornerstone of the POTW's pretreatment compliance and enforcement program. The inspector's presence helps to establish the POTW's role as a responsible public agency observing the actions of the regulated industry. Without the inspector on the scene, industrial processes/practices that violate federal, state, or local laws could jeopardize the POTW, the environment, and public health. The inspector may be the only person from the POTW who regularly visits the industrial user's facility. Furthermore, the knowledge that an inspection could occur unannounced encourages industrial plant managers to keep their operations in compliance.

4.4.1 Analytical Methods

POTWs with approved pretreatment programs may, at their discretion (as authorized by their local ordinances and regulations), require their indirect dischargers to achieve specific minimum levels when performing analyses or may require the use of specific analytical methods to enable the POTW to better characterize contributions into their system. Where a state or the EPA is the pretreatment Control Authority, the specific requirements for analytical methods can be specified in the control mechanism issued to the indirect discharger. The General Pretreatment Program requirements at 40 CFR part 136 and 40 CFR 403.12(g)(5) require that samples be analyzed using the approved methods listed in 40 CFR part 136 (CFR, Part 60, Appendix A Method 9 2017) (CFR, Part 60, Appendix A Method 22 2019). The approved analytical methods for analyzing wastewater samples can be found at 40 CFR part 136 (http://www.ecfr.gov/cgi-bin).

The EPA Office of Water analytical methods website is (https://www.epa.gov/cwa-methods).

The industry and POTW and laboratory analyst should consult the current regulations at 40 CFR part 136 as analytical methods may be updated and/or deleted (USEPA 1983). Table 4.4 shows selected analytical methods for corresponding pollutants in water samples.

4.4.2 State Waste Discharge Permit

In the United States, a *State Waste Discharge Permit* is required for a discharge of treated wastewater to waters of the state which includes surface water and groundwater. A state waste discharge permit is also required for any industrial or commercial operators discharging solid or liquid waste material into sewerage systems operated by municipalities or public entities which in turn discharge to waters of the state (State of Washington in the United States as an example). This permit is issued under authority of Chapter 90.48 RCW (revised code of Washington State) for water pollution control. A state waste discharge permit is valid for a maximum of five years. An applicant for a state waste discharge permit may receive a *Temporary State Waste Discharge Permit* if the state fails to act upon a complete application for a period of 60 days. A temporary state waste discharge permit is valid for five years or until state issues waste discharge permit. A temporary state waste discharge permit authorizes discharge as requested in the permit application. A facility covered under a temporary state waste discharge permit must reapply at least 60 days prior to expiration of the temporary permit.

4.4.3 NPDES Wastewater Discharge Permit

An *NPDES Permit* is required for a discharge of wastewater to waters of the United States (surface waters). This permit is issued by the state by delegated authority of the Clean Water Act (CWA 1972). Since waters of the United States are also waters of the state, NPDES permits are actually combined NPDES/state waste discharge permits issued

Table 4.4 Laboratory analytical methods.

Parameters	Methods
General Chemistry	
Turbidity	EPA Revised 1983: 180.1
Conductivity	EPA Revised 1983: 120.1
Alkalinity	EPA Revised 1983: 310.1
Hardness	EPA Revised 1983: 130.1
Fluoride	EPA Revised 1983: 340.3
Fluoride (soil)	EPA Revised 1983: 340.3
Solids	
TS	EPA Revised 1983: 160.3
TNVS	EPA Revised 1983: 160.3
TSS	EPA Revised 1983: 160.2
TNVSS	EPA Revised 1983: 160.2
% Solids	APHA, 1989: 2540G
% Volatile solids	EPA Revised 1983: 160.4
Oxygen Demand Parameters	
BOD_5	EPA Revised 1983: 405.1
COD	EPA Revised 1983: 410.1
TOC (water)	EPA Revised 1983: 415.1
TOC (soil/sediment)	EPA Revised 1983: 405.1
Nutrients	
NH_3-N	EPA Revised 1983: 350.1
$NO_2 + NO_3$-N	EPA Revised 1983: 353.2
P – total	EPA Revised 1983: 365.3
Organics	
VOC (water)	EPA Revised 1986: 8260
VOC (soil/sediment)	EPA Revised 1986: 8240
BNAs (water)	EPA Revised 1986: 8270
BNAs (soil/sediment)	EPA Revised 1986: 8270
Dioxin/Furans (water)	EPA Revised 1986: 8290
Pest/PCB (water)	EPA Revised 1986: 8080
Pest (soil/sediment)	EPA Revised 1986: 8080
PAH (water)	EPA Revised 1986: 8310
Metals	
PP Metals (water)	EPA Revised 1986: 200 - 299
PP Metals (soil/sediment)	EPA Revised 1986: 200 - 299
Aluminum – ICP	EPA Revised 1986: 220.7
Miscellaneous	
Oil and grease (water)	EPA Revised 1983: 413.1
Oil and grease (soil/sediment)	EPA Revised 1983: 413.1
Cyanide (total)	EPA Revised 1983: 335.2
Cyanide (wk & dis)	APHA, 1989: 4500 -CNI
Cyanide (wk & dis soil/sed)	APHA, 1989: 4500 -CNI
Fecal Coliform MF	APHA, 1989: 9222D

(*Continued*)

Table 4.4 (Continued)

Parameters	Methods
Bioassays	
Rainbow Trout, Salmonid (acute 100%)	Ecology, 1981
Daphnia magna (acute)	EPA, 1985
Ceriodaphnia (chronic)	EPA, 1989: 1002
Fathead Minnow (chronic)	EPA, 1989: 1000
Hyallela (acute)	Nebekar, 1984
Microtox (acute)	Beckman, 1982
Field Analysis	
pH	APHA, 1985: 423
Temperature	APHA, 1985: 212
Chlorine Residue	APHA, 1985: 408E

Method Bibliography

APHA -AWWA-WPCF, 1989, Standard Methods for the Examination of Water and Wastewater, 17th Edition.

Beckman Instruments, Inc., 1982. Microtox System Operating Manual.

Ecology, 1981. Static Acute Fish Toxicity Test, WDOE 80-12, revised July 1981.

EPA, Revised 1983, Methods for Chemical Analysis of Water and Wastes, EPA-600/4-79-020 (Rev. March, 1983)

EPA, 1985, Methods for Measuring the Acute Toxicity of Effluents to Freshwater and Marine Organisms, EPA/600/4-85/013.

EPA, 1986: SW846. Test Methods for Evaluating Solid Waste Physical/Chemical Methods, SW-846, 3rd. Ed., November, 1986.

EPA, 1989. Short-term Methods for Estimating the Chronic Toxicity of Effluents and Receiving Waters to Freshwater Organisms, second ed. EPA/600/4-89/100.

Nebeker, et al. 1984. "Biological Methods for Determining Toxicity of Contaminated Freshwater Sediments to Invertebrates." *Environmental Toxicology and Chemistry*, Vol. 3, 617–630.

under dual (State and Federal) authorities. NPDES permits are typically issued for a maximum of five years. The most important amendment to the CWA was the NPDES permit program. Technology-based effluent limits backed by plants discharging pollutants through a pipe or a conveyance (also see Section 4.9.3; Table 4.6).

4.4.4 General Wastewater Discharge Permit

General permits are wastewater discharge permits that are developed for a category of discharger instead of an individual facility. General permits may be issued under federal (actually combined NPDES/state) or state-only authority. The Waste Discharge General Permit Program rule, Chapter 173-226 WAC, clarifies the intent in its purpose statement: "Permits issued under this chapter are designed to satisfy the requirements for discharge permits under 307 and 402(b) of the federal Water Pollution Control Act (33 U.S.C. § 1251) and the state law governing water pollution control (chapter 90.48 RCW)." General permits are typically issued for five years. Facilities which are eligible for coverage under the permit apply for coverage and then are covered for the remaining length of the permit. Permittees must reapply for permit coverage at least 180 days before expiration.

4.4.4.1 Wastewater Flow Rate

Because the hydraulic design of both collection and treatment facilities is affected by variations in wastewater flow rates, the flow rate characteristics have to be analyzed carefully from existing records. In cases where only flow rate data in the collection system is available, it must be recognized that the flow rates may differ somewhat from the flow rate entering the treatment plant because the flow.

4.4.4.2 Quality Assurance

Very low detection and quantification limits must be achieved to reduce analytical variation and the number of samples below the analytical detection limits. To help reduce sample variation, QA procedures during sample collection and analysis should include careful attention to the major elements identified by U.S. Environmental Protection Agency (USEPA 1995a), including special cleaning procedures for sample containers and filter units, grab sample collection directly into pre-cleaned sample containers, field filtration with laboratory-cleaned filter units, field and laboratory blanks, standard reference materials, and field and laboratory duplicate samples. QA procedures and results should be documented in individual project plans and reports for any study.

4.5 Sampling and QA/QC Plan

In order to ensure the quality of POTW sampling, the POTW should develop a general plan that lays out all of its procedures for conducting sampling to meet the requirements under its pretreatment program. Here, this plan is referred to as the Sampling and Quality Assurance Plan, or simply *the plan*. Many POTWs have a plan(s) that encompasses the information suggested here, but they are called something different, such as SOPs. It does not matter what it is named; however, it is important for POTWs to have documented procedures for QA/QC, sample equipment maintenance, sample collection, sample documentation, sample shipping, chain-of-custody (COC) procedure, and safety. This plan should be used by all inspectors and/or sampling personnel so that samples are collected consistently and the data are of high quality and will be admissible in court, if necessary. The plan should ensure that representative samples are collected; 40 CFR part 136 sample collection and preservation procedures are followed; sample integrity is maintained; safe practices are followed during sampling events; and QA/QC procedures are followed. The inspector should contact the laboratory in advance of any sampling to discuss the sampling plan and QA/QC procedures, to allocate laboratory time, and obtain sample identification numbers and field and trip blanks.

The Sampling and Quality Assurance Plan should be a guide, or reference, for inspectors and sampling personnel. Based on review of the Sampling and Quality Assurance Plan, the inspector or sampling personnel should customize a sampling action plan specific for each industrial user. The Sampling and Quality Assurance Plan, in general, should lay out all preparation and sampling actions for the sampling team. Not all items in the POTW's plan will apply to all industrial user sampling events. In certain situations, the inspector and sampling personnel may need to deviate from procedures outlined in the plan due to uncontrollable circumstances at the industrial user facility. The inspector has discretion to deviate from the plan if, in the opinion of the inspector, circumstances at the facility warrant such a change. Whenever possible; however, the elements in the plan should be followed for consistency. If circumstances arise in the field that make it unrealistic or physically impossible for the procedures to be followed, the inspector should document any deviation from the plan and the reason for the deviation.

4.5.1 QA/QC Procedures

QA/QC procedures are tools that are necessary to maintain a specified level of quality in the measurement, documentation, and interpretation of sampling data. The QA/QC procedures the POTW will use to assure data of good quality for both industrial users and POTW sampling and for industrial user self-monitoring should be specified in the Sampling and Quality Assurance Plan (Keith 1991). QA is the program function used to prevent mistakes and assure the quality of measurement data, while QC is the process of carrying out those procedures stated in the QA Plan. The QA activities of the plan should be general while QC activities are specific. A QA program should *control* the quality of the data through use of set procedures to meet the program requirements. A QC program is designed to ensure the routine application of procedures necessary for the measurement process to meet prescribed standards of performance (e.g. through instrument calibration and analysis of reference unknowns). A program describing the schedule for calibration is QA, while the actual calibration procedures are QC.

All data generated or used by the POTW must be of known, defensible, and verifiable quality. This includes data that are generated through self-monitoring at the industrial facility. Therefore, the industrial user should also have QA/QC procedures in place to ensure the adequacy of the data submitted as part of its periodic compliance report. All inspections, and the data obtained as a result of the inspection, have the potential to be used in an enforcement proceeding and should be treated as potential evidence to be admitted in court.

To produce evidence that is admissible in an enforcement action, QA and QC procedures are necessary both in the field (during sampling) and in the laboratory (Keith 1991). The QA/QC procedures used in the field are separate from those used in the laboratory, but both are crucial for obtaining reliable data. Both laboratory and field QA/QC are discussed in this section. QA/QC procedures are used to obtain data that are both precise (degree of closeness between two or more samples) and accurate (degree of closeness between the results obtained from the sample analysis and the true value that should have been obtained). By following QA/QC procedures, the POTW's confidence in the validity of the reported analytical data is increased.

4.5.2 QA Procedures for Sampling

A QA program for sampling equipment and for field measurement procedures is necessary to ensure data of the highest quality. The inspector should recognize the importance of implementing QA in sample collection to minimize such common errors as improper sampling methodology, poor sample preservation, and lack of adequate mixing during compositing and testing. Again, each of these activities should be a part of the POTWs Sampling and Quality Assurance Plan, so that all POTW sampling personnel are familiar with the proper sampling procedures. QA checks will help the inspector determine when sample collection techniques are inadequate for the intended use of the data. In addition to the general plan components discussed in Section 4.5.1, the POTW's Sampling and Quality Assurance Plan should also contain these field QA elements which sampling technicians should understand and apply:

- The required analytical methodology for each regulated pollutant; special sample handling procedures; and the precision, accuracy, and detection limits of all required analytical methods.

- The basis for selecting the sampling and analytical methods. For example, each analytical method should consist of approved procedures. Where the method does not exist, the Sampling and Quality Assurance Plan should state how the new method will be documented, justified, and approved for use.
- The number of analyses for QC (e.g. the percentage of spikes, blanks, or duplicates), expressed as a percentage of the overall analyses [e.g. 10% (one duplicate sample per 10 samples)] to assess data validity. Generally, the QA program should be approximately 15% of the overall program, with 10% and 5% assigned to laboratory QC and field QC respectively. The Sampling and Quality Assurance Plan should include shifting these allocations or decreasing these allocations depending on the degree of confidence established for collected data.
- Frequency and procedures to calibrate and maintain field instruments and automatic samplers.
- A performance evaluation system that addresses the following areas:
  - Qualifications of personnel for a particular sampling situation
  - Determining the best representative sampling site
  - Sampling techniques (including the location of sampling points within the wastestream, the choice of grab or composite samples, the type of automatic sampler, special handling procedures, sample preservation procedures, and sample identification)
  - Flow measurement, where applicable
  - Completeness of data, data records, processing, and reporting
  - Calibration and maintenance of field instruments and equipment
  - Use of QC samples, such as field duplicates, or splits to assess the validity of the data
  - Training of all personnel involved in any function affecting data quality

By following these QA procedures, the inspector or sampling personnel can ensure that proper quality data is obtained during compliance sampling.

4.5.3 QC Procedures for Sampling

Sampling QC begins with calibration and preventative maintenance procedures for sampling equipment. The POTW's Sampling and Quality Assurance Plan should include a calibration plan and documentation record for all field sampling and analysis equipment used. A complete document record should be kept in a QC logbook, including equipment specifications, calibration date, and calibration expiration data, and maintenance due date. The person sampling should keep in mind that field analytical equipment should be recalibrated in the field prior to taking the sample. Personnel conducting sampling should be well-trained in the use, cleaning, calibration, and maintenance of all instruments or samplers used. Automatic sampler tubing, bottles, and the sampler itself should be cleaned prior to each sampling event. Automatic samplers should be calibrated for sample quantity, line purge, and the timing factor, if applicable. This calibration should be checked in the field to verify draw. The manufacturer's directions should be reviewed and followed for cleaning and calibrating all equipment.

In addition to calibration procedures, the person conducting field sampling should conduct control checks during the actual sample collection to assess the performance of sample collection techniques. In general, the most common monitoring errors usually are improper sampling methodology, improper preservation, inadequate mixing during compositing and splitting, and excessive sample holding time. In addition, the inspector or sampling personnel should collect and analyze the following QC samples to check sample collection techniques.

4.5.3.1 Blanks

- Trip blank – This is a sample vial(s) filled at the laboratory with analyte-free deionized water. The blank(s) follows the same handling and transport procedures as the samples collected during the event. The blank(s) functions as a check on sample contamination originating from sample transport, shipping, and from site conditions. Note: Expose the trip blank vial(s), to the same environmental conditions (e.g. light or temperature) as the sample vial(s) but do not open until it is time for analysis.
- Field blank/field reagent blank – These are similar to the trip blanks except they are prepared in the field with analyte-free deionized water exactly as the sample(s) that is collected. Field blanks are used to check for analytical artifacts and/or background introduced by sampling and analytical procedures.
- Equipment/rinsate blank. Collect a blank when using an automatic sampler or other nondedicated equipment during the sampling process. The blank is a check of the equipment cleanliness. For automatic samplers, prepare blanks prior to collecting samples, by pumping deionized water that is free of organics and other analytes, through the sampler and collecting the discharge purge water in a sample container for analysis for the constituents of concern.

4.5.3.2 Field Duplicate

This is a precision check on sampling equipment and sampling technique. At selected stations on a random time frame, duplicate samples are collected from two sets of field equipment installed at the site, or duplicate grab samples are collected from a single piece of equipment at the site. Duplicate samples check analytical precision as well as evaluate the "representativeness" of the sample aliquot.

4.5.3.3 Split Samples

These are samples that have been divided into two containers for analysis by separate laboratories. These samples provide an excellent means of identifying discrepancies in the industrial users' (or their contracted laboratory) analytical techniques and procedures. When filling split samples from a single composite jug, shake the composited sample well and half fill the sample container, then shake the composite again and fill half of the permittee's container. Repeat the procedure for each sample collected for a specific parameter or location of interest. Sometimes a sample is split between the industrial user and the POTW, both of whom send their split of the sample to a separate laboratory for analysis. Please note that oil and grease samples cannot be split due to the nature of the pollutant.

4.5.4 Laboratory QA/QC

Laboratory QA/QC procedures ensure correct and representative analyses through instrument calibration and the processing of control samples. The precision of laboratory findings refers to the reproducibility of results. In a laboratory QC program, a sample is analyzed independently (more than once) using the same methods and set of conditions. Precision is estimated by comparing the measurements. Accuracy refers to the degree of difference between observed values and known or actual values. The accuracy of a method may be determined by analyses of samples to which known amounts of a reference standard have been added.

Laboratory QA/QC procedures can be quite complex. Often, the analytical methods specify QA/QC requirements for calibration, interference checks (e.g. inductively coupled plasma mass spectrometry analyses), control samples, spiking (including the method of standard additions), blank contaminant level, and instrument tuning. Accuracy is normally determined through the analysis of blanks, standards, blank spikes, laboratory control samples, and spiked samples. Precision is determined through the comparison of duplicate results or duplicate spiked results for organic analysis. For more information on laboratory QA/QC, the POTW should contact their QA manager or approval authority.

The methods used by in-house or contract laboratories to analyze industrial user samples must be methods that are EPA-approved under 40 CFR part 136 and thus are acceptable to a court of law as the most reliable and accurate methods of analyzing water and wastewater. Although some field test kits are useful as indicators of current conditions (and, thus, may be used for process control considerations), they are not appropriate for sampling that is conducted to verify or determine compliance. If non EPA-approved methods are used for analysis, the data will not be considered admissible as evidence of noncompliance. When choosing a contract laboratory, POTWs should consult their state's accredited laboratory standards, when applicable, or obtain and review a copy of the laboratory's QA/QC plan. For more information on laboratory considerations, the POTW should see the EPA's *Guidance Manual for POTW Pretreatment Program Development* (USEPA 1983, 1985a).

The laboratories performing the sample analyses should use the following control measures as part of their QA/QC plan:

- *Prep/reagent blank* – A sample consisting of reagent(s), without the target analyte or sample matrix, introduced into the analytical procedure at the appropriate point and carried through all subsequent steps to determine the contribution of the reagents and/or the contribution of the analytical steps to any error in the observed value.
- *Quality control sample* – This is an uncontaminated sample matrix spiked with known amounts of analytes from a source independent from the calibration standards. Use this sample to establish intra-laboratory or analyst-specific precision and bias or to assess the performance of all or a portion of the measurements' system.
- *Matrix spike (MS) and matrix spike duplicate (MSD)* – Two aliquots of an environmental sample to which known quantities of the analytes of interest and surrogates (if applicable) are added in the laboratory. The MS/MSD are prepared and analyzed exactly like a field sample. Their purpose is to quantify any additional bias and imprecision caused by the sample matrix. The background concentrations of the analytes in the sample matrix must be determined in a separate aliquot and the measured values in the MS/MSD corrected for background concentrations.
  - Use an MS to verify accuracy of the analytical procedures.
  - An MSD is a duplicate of an MS sample. It measures the precision of the analysis in terms of relative percent difference.
- *Laboratory duplicate* – This is a sample that is received by the laboratory and then subsequently divided into two or more portions. Each portion is separately and identically prepared and analyzed. The results from laboratory duplicate samples check the laboratory precision.

4.5.5 Sampling Location

The first step in preparing to sample is to verify that the sample location is appropriate. The industrial user permit must specify the sampling location for compliance sampling [40 CFR 403.8(f)(1)(iii)(B)(4)] (CFR, §52.21(aa)(2)(v) 1980). Sampling points should be located where there is no discharge other than the discharge from the industrial user process effluent being monitored. This sampling location must be representative of the actual discharge from the facility. A representative sampling location is where a sample can be obtained that is as close as possible to the same composition as that in the larger volume of wastewater being discharged and is typical of the discharge from the facility on a normal operating day. Convenience, accessibility, and safety are important considerations, but they are secondary to the representativeness of the sample. The most representative samples will be drawn from a wastewater depth where the flow is turbulent and well mixed and the chance of solids settling is minimal. The depth of sample collection should be 40–60% of the wastestreams depth. To avoid contamination, the inspector should take care to collect samples from the center of the flow with the opening of the sampling device or container facing upstream. Wide channels or paths of flow may require dye testing to determine the most representative sampling site. If dye testing is inconclusive, multiple samples may need to be collected by cross-sectional sampling.

The inspector should use the sampling location specified in the industrial user's permit. If the sample location specified in the permit is not adequate to collect a representative sample, the inspector should determine an alternative location. This determination should be based on the inspector's knowledge of the plant itself, the production processes, and the outfalls. If there is a conflict between the sample location described in the permit and the location the inspector believes is most representative, samples should be collected at both sites. The reason for the conflict should be thoroughly documented for later resolution by the POTW. If necessary, the permit or other control mechanism must be amended to reflect the more representative sampling location(s). The CWA federal categorical standards apply at the end-of-process (or at the end of treatment, if treatment exists) unless the standard specifies a different location to collect the sample (e.g. in 40 CFR part 433, metal finishing, the sample location for cyanide is after the cyanide destruct system prior to dilution with other streams). If process effluent is mixed prior to treatment with unregulated wastestreams or dilution water, the CWA (USEPA 1985b) must be used (40 CFR 403.6). In this case, the POTW will designate, in the industrial user's permit or other control mechanism, where the sampling point(s) will be (end-of-pipe, end-of-process, or both). The POTW must ensure that any change in the industrial user's sampling point(s) will not allow the user to substitute dilution for adequate treatment to achieve compliance with applicable standards. Where a treated, regulated, process, wastestream is combined prior to treatment with wastewaters other than those generated by the regulated process, the industrial user may monitor either the segregated process wastestream or the combined wastestream for the purpose of determining compliance with applicable pretreatment standards.

4.5.6 Type of Sample

There are two basic types of samples: grab samples and composite samples. The type of sampling that will be undertaken at the industrial user must be understood clearly and should be outlined in the facility's Sampling and Quality Assurance Plan. It is very important that the facility establish specific procedures for collecting grab and/or composite samples. According to 40 CFR 403.12(g)(3), grab samples must be used for pH, cyanide, total phenols, oil and grease, sulfides, and volatile organic compounds. For all other pollutants, 24-hour composite samples must be obtained through flow-proportional composite sampling techniques unless time-proportional composite sampling or grab sampling is specified in the protocol. Where time-proportional composite sampling or grab sampling is recommended in the protocol, the samples must be representative of the discharge and the decision to allow the alternative sampling must be documented in the file for that facility or in any control mechanism.

The correct sample type for each regulated parameter must be included in the industrial user's permit or control mechanism and for compliance determination purposes the inspector must collect the specified type of sample. Any problems noted with the type of sample specified in the industrial user's permit or the control mechanism should be documented. If necessary, the permit or other control mechanism may need to be revised or amended.

4.5.6.1 Grab Samples

A grab sample is an individual sample collected over a period of time not to exceed 15 minutes. Grab samples are usually taken manually, and the sample volume depends on the number of analyses to be performed. The sampler must make sure that sufficient volume of sample is taken to conduct all necessary analytical procedures, including QA/QC. Grab samples represent the conditions that exist at the moment the sample is taken and do not necessarily represent conditions at any other time. Grab sampling is the appropriate method of sampling under the following conditions:

- 40 CFR 403.12(g)(3) requires grab samples for pH, temperature, cyanide, total phenols, oil and grease, sulfides, volatile organic compounds, microbiological tests, and residual chlorine.
- When the effluent is not discharged on a continuous basis (i.e. batch discharges of an hour or less), and only when the batch is continuously stirred or well-mixed and the pollutant can be safely assumed to be uniformly dispersed.
- When sampling a facility where a statistical relationship can be established between previous grab samples and composite data.
- When the effluent is being screened to see if a parameter is present (NOTE: This is only true when the sample is well-mixed and representative of the discharge).
- When the waste conditions are relatively constant (i.e. are well-mixed and homogeneous) over the period of the discharge. In lieu of complex sampling activities, a grab sample provides a simple and accurate method of establishing waste characteristics.
- When an industrial facility or POTW or state has adopted an instantaneous local limit which is based on grab samples.
- Grab samples are useful in characterizing an industrial user's fluctuations or extremes in wastewater flow and quality (i.e. changes in pollutant concentrations or loadings) and, therefore, are useful in identifying slug loads.

For additional guidance on when it is appropriate to use grab samples for determining compliance, see the EPA memorandum "The Use of Grab Samples to Detect Violations of Pretreatment Standards" at http://www.epa.gov/npdes/pubs/pretreatment_memo_grabsamplesdetectviolations.pdf.

4.5.6.2 Composite Samples

Composite samples are samples collected over time, either continuous sampling or by mixing discrete samples. Composite samples represent the average characteristics of the wastestream during the compositing period. Composite samples are collected when

- average pollutant concentration during the compositing period is desired
- mass per unit time loadings are calculated
- wastewater characteristics are highly variable

The primary methods of composite sample collection are time compositing and flow proportion compositing. The permit may specify which type of composite sample to use.

Composite samples are collected either manually by combining multiple grab samples of by using automatic sampling equipment. Inspectors should consider variability in wastestream flow rate, parameter concentrations, and approved EPA methods when choosing compositing methods, sampling equipment, and QA procedures (Figure 4.1). The compositing methods, all of which depend on either continuous or periodic sampling, are described in the following.

Time-proportional composite sampling: It is composed of constant volume discrete sample aliquots collected at constant time intervals. This method provides representative samples when the flow of the sampled stream is relatively constant (i.e. when the flow does not vary by more than 10% of the average flow rate over time).

Flow-proportional composite sampling: There are two methods used to collect a flow-proportional composite sample.

In the first method, the time between samples is constant, and the volume of each sample is proportional to the flow at that given moment in time (i.e. the volume of the sample varies over time as the flow changes). This is the preferred method of sampling when taking a manually composited sample. This method requires that discrete samples be collected over the operating day and then be manually composited. It is crucial, when using this method, to have accurate flow data continuously recorded during the sampling period (ISCO 2001).

Figure 4.1 ISCO automatic programmable composite sampler.

Table 4.5 Composite sampling methods: advantages and disadvantages.

Method	Advantages	Disadvantages	Comments
Time-proportional composite			
Constant sample volume, constant time interval between samples.	Minimal instrumentation and manual effort. Requires no flow measurement.	May lack representativeness especially for highly variable flows.	Widely used in both automatic samplers and manual handling.
Flow-proportional composite			
• Constant sample volume, time interval between samples proportional to stream flow.	Minimal manual effort.	Requires accurate flow measurement reading equipment.	Widely used in automatic as well as manual sampling.
• Constant time interval between samples, sample volume proportional to total stream flow at time of sampling.	Minimal instrumentation.	Samples must be manually composited. Discrete samples must be taken. Chance of collecting samples that are too small or too large for a given composite volume.	Used in automatic samplers and widely used as manual method.

The second flow-proportional sampling method involves collecting a constant sample volume for each volume of wastestream flow (e.g. 200 ml sample collected for every 5000 gal of flow) at time intervals inversely proportional to the stream flow. This is the preferred method when taking composite samples using an automatic sampler. This method is based on taking a sample after a set amount of wastewater has been discharged. This method provides representative samples of all wastestreams when the flow is measured accurately. A composite sample should be collected over a workday. If a facility operates and discharges 24 h/day, then the composite sample should be taken as a 24-hour composite (either flow proportional or time composite). If a facility operates 24 h/day but only discharges wastewater for 6 hours, a 6-hour composite sample should be collected. In general, composite samples should be collected to assess compliance with categorical standards and local limits, as long as the limits are daily, weekly, or monthly averages, except for those parameters listed above that must be taken by grab sample.

There are numerous cases where composites are inappropriate. Samples for some parameters such as pH, residual chlorine, temperature, cyanides, volatile organics, microbiological tests, oil and grease, and total phenols should not be composited. They are also not recommended for sampling batch or intermittent processes. Grab samples are needed in these cases to determine fluctuations in effluent quality. For bioassays, composite samples are used unless it is known that the effluent is most toxic at a particular time. Some toxic chemicals are short-lived, degrading rapidly, and will not be present in the most toxic form after lengthy compositing even with refrigeration or other forms of preservation. Require grab samples for bioassays to be taken under those circumstances. In the absence of an expressed sampling protocol in regulation, the duration of the compositing time period and frequency of aliquot collection is established by the state permit writer. Whether collected by hand or by an automatic device, the time frame within which the sample is collected should be specified in the NPDES permit. The number of individual aliquots which compose the composite should also be specified.

Table 4.5 lists the advantages and disadvantages of each composite sampling method. Either manual or automatic sampling techniques can be used. If a sample is composited manually, sample manipulation should be minimized to reduce the possibility of contamination. The inspector must use the sampling method specified in the permit or other control mechanism, which in turn must reflect the requirements at 40 CFR 403.12(g)(3) (CWA 1972).

4.5.7 Continuous Monitoring

Continuous monitoring is another option for a limited number of parameters such as pH, temperature, dissolved oxygen (DO), conductivity, and TOC. Reliability, accuracy, and cost vary with the parameter. Continuous monitoring can be expensive, so continuous monitoring will usually only be an appropriate requirement for the most significant dischargers with variable effluent. The environmental significance of the variation of any of these parameters in the effluent should be compared to the cost of continuous monitoring equipment available. The regulations concerning pH limits allow for a period of excursion when the effluent is being continuously monitored (40 CFR 401.17). Continuous monitoring or labor-intensive periodic monitoring by grab sampling is necessary where pH excursions are allowed.

4.5.8 Sample Preservation and Holding Times

Preservation techniques ensure that the sample remains representative of the wastestream at the time of collection. Since most pollutants in the samples collected are unstable (at least to some extent), this instability requires that the sample be analyzed immediately or that it be preserved or fixed to minimize changes in the pollutant concentration or characteristics between the time of collection and analysis. Because immediate analysis is not usually possible, most samples are preserved regardless of the time of analysis. This preservation must take place as soon as possible after collecting the sample. This means that preservation must take place in the field (CWA 1972 40 CFR 136.3). The most common procedures used for preserving samples include icing, refrigeration, pH adjustment, and chemical fixation. When chemical fixation is used, the chemical preservative must be added before the samples are transferred to the laboratory. Likewise, refrigeration should be supplied immediately upon taking the sample. For many samples, if preservation is not appropriately used, bacteria can quickly degrade certain pollutant constituents (e.g. phenols and phosphorous). Other constituents such as cyanide and sulfides may volatilize or in the case of hexavalent chromium may react to form different chemical species. Proper preservation and holding time for each parameter is essential for the integrity of the monitoring program. Problems may be encountered, however, when 24-hour composite samples are collected. Since sample deterioration can take place during the compositing process, it is necessary to preserve or stabilize the samples during the compositing in addition to preserving the aggregate sample before shipment to the laboratory. Preservation techniques vary depending on the pollutant parameter that is to be measured; therefore, the inspector must be familiar with the 40 CFR part 136 preservation techniques to ensure proper sample handling and shipment. It is important to verify that the preservation techniques for one parameter do not affect the analytical results of another parameter in the same sample. If this is a possibility, then two discrete samples should be collected and preserved independently. Sample preservation should be provided during compositing, generally by refrigeration to less than or equal to 6 °C (or icing) as required (CWA 1972 40 CFR 136.3).

Refrigeration is the most widely used preservation technique because it has no detrimental effect on the sample composition and does not interfere with any analytical methods. Refrigeration suppresses biological activity and volatilization of dissolving gases and organic substances. This technique is used at the start of sample collection in the field and during sample shipment, and continued until the sample is analyzed by the laboratory. Sample temperature should be verified and recorded by the inspector. This is particularly important if the analytical results are to be used in an enforcement action.

In addition to preservation techniques, 40 CFR part 136 indicates maximum holding times. Times listed are the maximum times between sample collection and analysis that are allowed for the sample to be considered valid. Some parameters have separate holding times from the time of sample collection to extraction/preparation, and from extraction/preparation to analysis. A composite wastewater sample becomes a sample when the last aliquot is collected. A grab sample holding time begins when the sample is collected. At that point, holding time limitations begin. A list of preservation methods and holding times are provided in 40 CFR 136.3 (CWA 1972). Inspectors should consult 40 CFR part 136 for the most current and detailed specifications. These sample preservation procedures and holding times were selected by the EPA because they retard sample degradation and minimize monitoring costs by extending holding times as long as possible.

4.5.9 Sample Documentation

Documentation is an integral part of any pretreatment program. The validity of the samples collected and the data obtained both in the field (e.g. pH and flow) and in the laboratory (i.e. chemical analyses) is ensured through documentation and record keeping. All the information documented must be complete and accurate. Failure to maintain records and documentation according to set procedures could result in these documents being deemed inadmissible as evidence in court. All field records containing data must be signed by the inspector at the time of collection, including all COC forms.

4.5.10 General Documentation Procedures

Since many sampling reports may be used in enforcement proceedings, the inspector must keep a precise record of all sample collection and data handling. The industrial facility or POTW's Sampling and Quality Assurance Plan should specify that the following information must be documented in the field record:

- *Unique sample or log number*: All samples should be assigned a unique identification number. If there is a serial number on the transportation case, the inspector should add this number to their field records.
- *Date and time of sample collection*: The date and time of sample collection must be recorded. In the case of composite samples, the sequence of times and aliquot size should be noted.

- *Source of sample (facility name and address)*: The name and address of the facility being sampled shall be recorded as well as a narrative description and/or diagram referring to the particular site where the sample was taken.
- *Name of sampling personnel*: The name(s) and person(s) taking the sample should be indicated. For a composite sample, the name(s) of the person(s) installing the sampler and the name(s) of the person(s) retrieving the sample should be included.
- *Sample type*: Each sample should indicate whether it is a grab or composite sample. If the sample is a composite, volume and frequency of individual samples should be noted.
- *Preservation method*: Any preservatives (including the amount) added to the sample should be recorded, including the preservative amount and the method of preservation (e.g. refrigeration).
- *Analysis required*: All parameters for which the sample should be analyzed shall be specified.
- *Field analysis*: Field measurements must be recorded at the time that the analysis is completed. Examples of analyses that should be conducted and recorded in the field include:

  pH, temperature, DO, residual chlorine, and sulfites. Field analyses should be treated the same as any other sample (i.e. the sampler must record the date of the sample, type of sample, name of the sampler, COC, etc.).
- *Flow*: If flow is measured at the time of the sampling, the flow measurement should be recorded. If the sample is to be used to determine compliance, flow must be measured.
- *Production rates*: Information on products manufactured and production rates should be included since many effluent limitations are based on production rates.
- *Date, time, and documentation of sample shipment*: The shipment method (e.g. air or rail) as well as the shipping papers or manifest number should be recorded.
- *Comments*: This refers to all relevant information pertaining to the sample or the sampling site. Such comments include the condition of the sample site, observed characteristics of the sample, environmental conditions that may affect the sample, and problems encountered in sampling.

Each of these items should be recorded by the facility/POTW inspector when conducting site sampling, but this information should also be kept by the industrial user when it conducts its required self-monitoring. This information should be available for review by the facility/POTW when conducting an on-site inspection or sampling visit.

4.5.11 COC Procedures

Once an appropriate sample has been obtained and the sample collection methods are properly documented, a written record of the possession of that sample should be made. "COC" refers to the documented account of changes in possession that occur for a particular sample or set of samples (see Section 4.5.11 for a definition of "custody"). To ensure that all necessary information is documented, a COC form should be developed by the POTW.

COC procedures are a critical aspect in monitoring industrial users, and these procedures should be incorporated into the POTW's Sampling and Quality Assurance Plan. The COC record allows an accurate step-by-step recreation of the sample path, from its origin through its final analysis in the laboratory. Every sample needs to be accompanied by a COC form that is properly signed and transferred to each person in the chain, from the original sampler to the final person involved in analyzing the sample. Since it is impossible to predict which violations will require legal action, it should be assumed that all data generated from sampling will be used in court. If a case ultimately goes to trial, the integrity of the data must be established. The sampling results will only be admissible in court if POTW personnel can prove that a sample has been properly collected, preserved, and analyzed, and has not been tampered with or mishandled.

The information that should be addressed in a COC form includes the following:

- Name of the person collecting the sample
- Sample identification number(s)
- Sample location
- Date and time of sample collection
- Parameters to be analyzed
- Container type(s) used
- Preservation method(s)
- Type of sample collected (i.e. grab, composite) for each pollutant to be analyzed
- Results of any analysis performed in the field (e.g. pH and temperature)
- Name(s), signature(s), dates, and times of all persons handling the samples in the field and in the laboratory

Note: Laboratories often have separate COC procedures, so the COC used in the field may end once the sample reaches the laboratory.

An example of a COC form used by POTWs is shown in Section D.1. All personnel handling the sample should receive a copy of the COC form showing the path of the sample from field to laboratory. All sample shipments must be accompanied by the COC record, while a copy of these forms should be retained by the originator. In addition, all receipts associated with the shipment should be retained. Carriers typically will not sign for samples; therefore, seals must be used to verify that tampering has not occurred during shipment.

When transferring possession of samples, the transferee must sign and record the date and time on the COC record. In general, custody transfers are made for each sample, although samples may be transferred as a group, if desired. Each person who takes custody must fill in the appropriate section of the COC record. Finally, the person or group at the POTW responsible for permitting, and/or compliance and enforcement, should receive a copy of the completed COC form.

COC records are crucial if the analytical data are to be used in an enforcement proceeding because they allow such data to be introduced as evidence without testimony of the persons who made the record. Therefore, it is important that all COC records be complete and accurate. To maintain the sample's integrity, COC records must show that the sample was properly collected, preserved, and analyzed, and was not tampered with during shipment. COC records also show that samples were received and analyzed in accordance with method requirements. Therefore, all compliance samples taken at an industrial user should follow COC procedures.

4.5.12 Sample Identification and Labeling

Each sample must be accurately and completely identified. At the time a sample is collected, a waterproof, gummed label or a waterproof tag that is able to withstand field conditions should be attached to the sample container. This label and/or tag is necessary to prevent any misidentification of samples, since it provides the laboratory with relevant information for analysis, such as the name of the sample collector, the sample identification number, the date and time of sample collection, the location of the sample collection, the preservatives used, the type of sample (grab or composite), and the identification of the parameters to be analyzed. Sample identification, therefore, is a crucial part of COC. Sample tags should be used as part of the COC process because they can be removed after the sample has been transported to the laboratory and placed in the evidence file for that sample, while labels are usually discarded with the sample. The tags can then be used as evidence in an enforcement proceeding. The sample identification procedures should be incorporated into the facility's/POTW's Sampling and Quality Assurance Plan.

Sample seals or cooler seals (i.e. seals placed around a cooler with similar samples inside) should be used to protect the sample's integrity from the time it is collected to the time it is opened in the laboratory. The seal should contain the collector's name, the date and time of sample collection, and the sample identification number. Information on the seal must be identical to the information on the label and tag. In addition, the seal must be attached so it must be broken to open the sample container or cooler. Caution should be observed to assure that glue on the sample seals and tag wires does not contaminate samples, particularly those containing volatile organics and metals.

4.5.13 Sample Packaging and Shipping

After the samples are properly labeled, they should be placed in a transportation case along with the COC form, pertinent field records, and analysis request forms. Glass bottles should be wrapped in foam rubber, plastic bubble wrap, or other material to prevent breakage during shipment. The wrapping can be secured around the bottle with tape. Samples should be placed in ice or a synthetic ice substitute that will maintain sample temperature at 6 °C or below throughout shipment. Ice should be placed in double-wrapped watertight bags to ensure the water will not drip out of the shipping case. Metal or heavy plastic chests make good sample transportation cases. Filament tape wrapped around each end of the ice chest ensures that it will not open during transport. Sampling records should be placed in a waterproof bag, envelope, or airtight sample bottle and taped to the inside lid or other appropriate place inside the transported container to prevent tampering or breach of custody. Shipping containers should also be sealed to prevent tampering. Sample packaging and shipping considerations should be included in the POTW's Sampling and Quality Assurance Plan (discussed in Section 4.5.13).

Most samples will not require any special transportation precautions except careful packaging to prevent breakage and/or spillage. If the sample is shipped by common carrier or sent through the US mail, it must comply with Department of Transportation (DOT) Hazardous Waste Materials Regulations (49 CFR parts 171–177) (CFR, Title 40, Part 51, 1977) (CWA 1972). Air shipment of hazardous waste materials may also be covered by requirements of the International Air Transport Association (IATA). Before shipping a sample, the inspector should be aware of, and follow, any special shipping requirements. Special packaging and shipping rules apply to substances considered hazardous as defined by IATA rules. Typical wastewater samples are not generally considered hazardous materials.

4.6 Whole Effluent Toxicity Testing

4.6.1 Introduction

For selected industrial wastewater treatment plants (WWTPs), whole effluent biological testing is required under their NPDES permit. Whole effluent toxicity (WET), also referred as biomonitoring, is the toxicity of an effluent sample measured directly with a toxicity test in order to assess the total toxic effect of all pollutants. WET testing is necessary because the Environmental Protection Agency (EPA) cannot develop water quality criteria for every toxic pollutant possible in wastewater discharges. The expense would be huge to try to measure every possible chemical. WET testing is also the only method for assessing the toxic interaction of all pollutants in a discharge. The approach is called "whole effluent toxicity" to contrast it with measuring the individual concentrations of the multiple toxic chemicals in an effluent for a one-by-one comparison to a water quality criterion if there happens to be one. Because WET testing often discovers effects from unknown toxicants, the toxic effects cannot be reduced until the toxicants involved have been identified.

4.6.2 The WET Testing

The bioassay is performed by exposing selected aquatic organisms, depending on freshwater or saltwater organisms, to process-water effluent in a controlled bioassay-laboratory environment. In the static short-term test, effluent and uncontaminated water (control) are placed in containers and test organisms are added to both containers and then monitored for toxic effects in a controlled laboratory environment. The objective is to estimate the safe, or no-observed-adverse-effect, level that would permit normal propagation of aquatic life in the receiving waters. The chambers are borosilicate glass or disposable poly-styrene ranging in volume from 250 to 1000 ml, depending on the size of the test organism. During the test period, the DO concentration should be near saturation, with the temperature held at 25 °C, and pH checked periodically. The effluent is a 24-hour composite filtered through a sieve to remove suspended solids (SS). During effluent preparation, aeration should be limited to prevent loss of volatile organic compounds. The common freshwater test organisms are rainbow trout (*Oncorhynchus mykiss*), fathead minnow (*Pimephales promelas*), daphnid (*Ceriodaphnia dubia*), daphnid shrimp, and green alga (*Selenastrum capricornutum*). Common marine saltwater species include *Champia parvula*, the red alga; *Mysidopsis bahia*, the mysid shrimp; *Menidia beryllina*, the inland silversides; and *Cyrinidon variegatus*, the sheepshead minnow. Some states in the United States have developed culturing and testing methods for indigenous species.

The test result is expressed in terms of mortality after a 24-hour exposure. A 25% reduction in survival is defined as the threshold of biological significance, indicating probable impairment of the receiving water. Impairment in the receiving water and effluent toxicity are related to dilution of the effluent within the receiving water. Therefore, a more definitive toxicity test is to set up a series of laboratory containers at decreasing effluent dilution, such as, 100, 50, 25, 12.5, and 6.25% (a geometric series). This range is likely to encompass the effluent flow corresponding to the 7-day average low flow with a recurrence interval of 10 years (noted 7Q10) is used. This 7Q10 flow in a river or stream is the basis for establishing the maximum allowable concentration of pollutants set by surface water standards. The dilution water is from the edge of the receiving stream above the outfall pipe or from the edge of the mixing zone (Section 4.8.1).

The bioassay tests may be conducted either acute 24-hour toxicity or chronic toxicity for monitoring up to 7 days. Based on the results, the no-observed-effect-concentration (NOEC) is determined for effluent in the diluted test samples. This is compared to the wastewater concentration that should be less than the NOEC, with a margin of safety (MOS).

Pretreatment of industrial wastewaters is essential to control toxic substances. Since many hazardous substances are biocides, reduction in their concentrations in raw wastewater are essential to effective biological processing. If the evidence of toxicity is less dramatic, their presence in the raw wastewater and the treatment plant effluent may pass unnoticed. A typical list of toxic pollutants include 13 priority pollutant metals, cyanide, 15 polynuclear aromatic hydrocarbons, 13 aromatic, 10 phenols, 21 aliphatic, 6 phthalates, 2 nitrogen compounds, 2 oxygenated compounds, and pesticides.

4.6.3 Toxicity Testing and Evaluation of Toxicity Test Results

Ecological toxicity tests are classified according to duration (short-term, intermediate, and/or long-term), method of adding test solutions (static, recirculation, renewal, or flow-through), and (to satisfy permit requirements for a proposed pollutant discharge elimination system, and to determine mixing zones, etc.). Detailed contemporary testing protocols are summarized in the regulatory literature (USEPA 1985b, c, 1988, 1989).

A number of terms are utilized in expressing toxicity test results. *Acute toxicity* is toxicity severe enough to produce a response rapidly (typically a response observed in 48 or 96 hours). "Acute" does not necessarily imply mortality. In studies of marine organisms, the LC_{50} is the concentration of effluent in dilution water that causes mortality to 50% of the test population; the EC_{50} is the effluent concentration that causes a measurable negative effect on 50% of the test population. The NOAEL (no observed acute effect level) is defined as the highest tested effluent concentration that causes 10% or less mortality.

Chronic toxicity is the toxicity impact that lingers or continues for a relatively long period of time, often 1/10 or more of the target organism's life span. Chronic effects could include mortality, reduced growth, or reduced reproduction. The NOEC is the highest measured continuous concentration of an effluent or toxicant that causes no observable effect based on the results of chronic testing. The LOEC (lowest observed effect concentration) is defined as the lowest observed concentration having any effect. The LOEC is determined by an analysis of variance techniques.

Toxicity data are analyzed using the procedures developed by Stephan (1977). The LC_{50} values are determined analytically using the Spearman–Kayber, moving average, binomial, and probit methods. Graphical methods, as illustrated in Example 4.1, can also be used to obtain estimated LC_{50} values. Typically, LC_{50} values are computed based on survival at both 48- and 96-hour exposures. Analysis of variance and Duncan's multiple comparison of means typically are utilized to compare chronic test results. Example 4.2 presents a simple comparison of results from acute toxicity testing.

Example 4.2 Analysis of toxicity data

Use the following hypothetical data to determine the 48- and 96-hour LC_{50} values in percent by volume.

Concentration of waste % by volume	No. of test animals	No. of test animals dead after[a]	
		48 hours	96 hours
40	20	17 (85)	20 (100)
20	20	12 (60)	20 (100)
10	20	6 (30)	14 (70)
5	20	0 (0)	7 (35)
3	20	0 (0)	4 (20)

[a] Percentage values are given in parentheses.

Solution

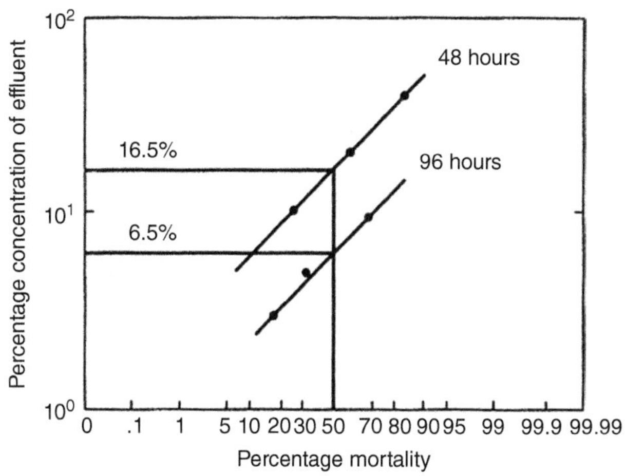

1) Plot the concentration of wastewater in percent by volume (log scale) against test animals surviving in percent (probability scale), as shown in the accompanying diagram.
2) Fit a line to the data points by eye, giving most consideration to the points lying between 16 and 84% mortality.
3) Find the wastewater concentration causing 50% mortality.

The estimated LC_{50} values are seen to be 16.5 vol% for 48 hours and 6.5 vol%.

Comment. The estimated values of the LC_{50} concentrations obtained graphically are usually quite close to the values obtained with formal prohibit analysis (Stephan 1977). It should be noted that confidence limits cannot be obtained in a graphical analysis (APHA 1989).

4.6.4 Toxic Units

The toxic units (TU) approach has become widely accepted for utilizing the toxicity test results. Both federal and state standards and/or criteria have been or are being formulated on the toxic unit basis. In the toxic units approach (USEPA 1985b, c), a TU concentration is established for the protection of aquatic life.

A toxic unit acute (TU_a) is the reciprocal of the effluent dilution that caused the acute effect by the end of the acute exposure period.

$$TU_a = \frac{100}{LC_{50}} \tag{4.1}$$

A toxic unit chronic (TU_c) is the reciprocal of the effluent dilution that caused no unacceptable effect on the test organisms by the end of the chronic exposure period:

$$TU_c = \frac{100}{NOEC} \quad (4.2)$$

Where, as before, NOEC is the no observable effect concentration.

Formerly, acute to chronic ratios (ACR) were determined by the equation ACR = LC_{50}/NOEC. The chronic data were determined using extrapolation from acute data. Acute to chronic ratios have been found to vary tremendously between species and between different toxicants. Use of the whole effluent approach prevents implementation of overly stringent as well as overly lenient requirements.

4.6.5 Application of Toxicity Test Results

Water quality criteria ensure protection of designated uses by including magnitude (quantity of toxicant allowable), duration (period of time over which instream concentration is averaged), and frequency (how often criteria can be exceeded without unacceptable receiving water ecological community impacts). Contemporary water quality criteria are designed to protect against short-term (acute) effects through use of the criterion maximum concentration (CMC), and against long-term (chronic) effects through use of the criterion continuous concentration (CCC). These criteria generally apply after mixing. Typical water quality criteria contain a concentration limit, an averaging period, and a return frequency. The CMC is typically the four-day average concentration not to be exceeded more than once every three years, on the average, and the CCC is typically the one-hour average concentration not to be exceeded more than once every three years, on the average.

4.6.6 Protection Against Acute Toxicity

For protection against acute toxicity, the CMC must not exceed $0.3\,TU_a$, as measured by the most sensitive result of the tests conducted.

$$CMC = \frac{TU_a}{CID} \le 0.3\,TU_a \quad (4.3)$$

where CID is the critical initial dilution. In ocean discharges, the CID is defined as the dilution achieved given "worst case" ambient conditions within the region close to the discharge, where mixing and dilution of the effluent plume are determined by the initial momentum and buoyancy of the discharge. In river discharges, the dilution achieved at the boundary of a mixing zone is usually taken as the CID. Based on the results of numerous 96-hour effluent toxicity tests, it was found that a factor of 0.3 accounted for 91% of the observed LC_{50} to LC_1 ratios (LC_1 is equal to the concentration of effluent in the dilution water that causes mortality to 1% of the test population). Consequently, for acute protection, the CMC should not exceed $0.3\,TU_a$ based on the most sensitive species tested. The acute criterion thus equals the CMC that approximates the LC_1.

4.6.7 Protection Against Chronic Toxicity

To prevent chronic effects from occurring outside the initial mixing zone, or zone of influence, resulting from the discharge, the CCC must not exceed $1.0\,TU_c$, based on the results obtained with the most sensitive of at least three species tested.

$$CCC = \frac{TU_c}{CID} \le 1.0\,Tu_c \quad (4.4)$$

Compliance is based on comparison of the toxicity criteria (both TU_a and TU_c), expressed in toxicity units, in the effluent with critical initial dilution to determine if the EPA's recommended criteria will be met. The application of these criteria is illustrated in Example 4.3.

Example 4.3 Application of Toxicity Test Results
A critical initial dilution of 225 : 1 is achieved for a treated effluent discharged to marine receiving waters. Toxicity tests were conducted with the WWTP effluent using three marine species. The following acute and chronic toxicity test results indicate that *C. parvula* exhibited the most sensitive species acute endpoint (2.59% effluent) in accordance with the LC_{50}, and also the most sensitive species chronic endpoint (1.0%) in accordance with the NOEC. Using the given toxicity data, determine the compliance with the CMC and CCC criteria.

Results of acute toxicity tests

Species	Exposure (hours)	Control survival (%)	Percent effluent	
			LC_{50}	NOAEL
Mysidopis bahia	96	100	18.66	10.0
Cyprinodon variegates	96	100	>100	50.0
Champia parvula	48/168	100	2.59	12.25

Results of chronic toxicity tests

Species	Exposure (days)	Control survival (%)	Percent effluent	
			NOEC	LOEC
Mysidopis bahia	7	82	6.0	10.0
Cyprinodon variegates	7	98.8	15.0	>15.0
Champia parvula	7	100	1.0	2.25

Solution

1) Check compliance with the CMC criterion.
   a) Based on data for the most sensitive species tested, we use Eq. (4.1) to find the number of acute toxic units:

$$TU_a = \frac{100}{LC_{50}} = \frac{100}{2.59} = 38.6\,\text{units}$$

   b) For acute protection, the criterion maximum concentration must not exceed $0.3\,TU_a$, as shown in Eq. (4.3). Following an initial dilution of 225, the CMC is

$$\begin{aligned} CMC = \frac{TU_a}{CID} &\le 0.3\,TU_a \\ \frac{38.6}{225} &\le 0.3(38.6) \\ 0.17 &\le 11.58 \end{aligned}$$

   c) The CMC (0.17) is considerably less than the value of $0.3\,TU_a$ (11.58) required for compliance with the CMC criterion.
2) Check compliance with the CCC criterion.
   a) Based on data for the most sensitive species tested, we use Eq. (4.2) to find the number of chronic toxic units:

$$TU_c = \frac{100}{NOEC} = \frac{100}{1.0} = 100\,\text{units}$$

   b) For chronic protection, the criterion continuous concentration must not exceed 1.0 toxic unit as shown in Eq. (4.4). Following an initial dilution of 225, the CCC is

$$\begin{aligned} CCC = \frac{TU_c}{CID} &\le 1.0\,TU_c \\ \frac{100}{225} &\le 1.0(100) \\ 0.44 &\le 100 \end{aligned}$$

The CCC (0.44) is considerably less than the value of $1.0\,TU_c$ (100) required for compliance with the CCC criterion.

The use of whole effluent toxicity testing affords a number of advantages. In this approach, the bioavailability of the toxics is measured and the effects of any synergistic interactions are also considered. Because the aggregate toxicity of all components of a wastewater effluent is determined, the toxic effect can be limited by limiting only one parameter, the effluent toxicity. Because contemporary receiving water management strategies are based on site-specific water quality criteria, toxicity testing facilitates comparison of effluent toxicity with site-specific water quality criteria designed to protect representative, sensitive species and allow for establishment of discharge limitations that will protect aquatic environments.

4.7 Flow Measurements

Pollutant limits in the industrial user's permit are often expressed in terms of mass loadings to an industrial facility or a POTW (e.g. categorical standards for conventional pollutants, organics, metals, microorganisms, etc.). To determine a mass loading, it is necessary to obtain and record accurate flow data. This section briefly describes the types of devices and procedures used to measure flow. In situations where flow-measuring devices such as those described in the following sections are not available, the facility may need to rely on the use of water consumption records at the facility. However, when a mass loading needs to be determined for assessing compliance, the facility should have the ability to assess directly the flow at the facility during the sampling event.

4.7.1 Open Channel Flow

Open channel flow, where the flow occurs in conduits that are not full of liquid, is the most prevalent type of flow at industrial user discharge points regulated by the pretreatment program. Partially full pipes that are not under pressure are classified as open channels as well. Open channel flow is measured using both primary and secondary devices (as described below).

4.7.1.1 Primary Devices

Primary devices are calibrated, hydraulic structures installed in the channel so flow measurements can be obtained by measuring the depth of liquid at a specific point in relationship to the primary device. Weirs and flumes are examples of primary devices.

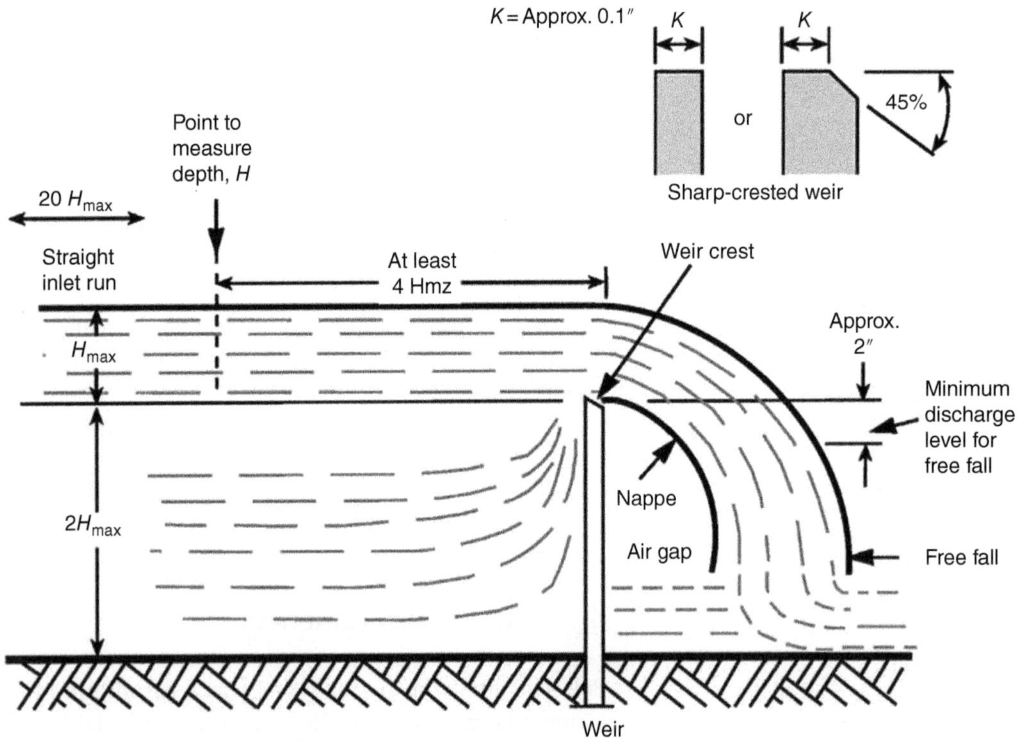

Figure 4.2 Profile and nomenclature of sharp-crested weirs.

The most common type of weir consists of a thin, vertical plate with a sharp crest that is placed in a stream, channel, or partially filled pipe. Figure 4.2 shows a profile of a sharp-crested weir and indicates the appropriate nomenclature. Four common types of weirs (rectangular, trapezoidal, sharp-crested, and v-notch or triangular) are shown in Figure 4.3. The crest is the upper edge of the weir to which water must rise before passing over the structure. The vertical distance from the crest of the weir to the top of the water surface is termed the "head." To determine flow rate, the inspector must measure the hydraulic head. The rate of flow over a weir is directly related to the height of water (hydraulic head) above the crest. To measure the hydraulic head, a measuring device is placed upstream of the weir at a distance of at least four times the head. To approximate the head, the inspector can measure at the weir plate. However, this value will provide only a rough estimate of flow.

Assuming negligible upstream velocity, the basic uncorrected weir equation is

$$Q = C_w \left(\frac{2}{3}\right) b\sqrt{g}\, H^{\frac{3}{2}} \tag{4.5}$$

where C_W is the weir coefficient, b is the width, and H is the weir height.

Example 4.4 Water flows over a broad-crested weir with a flow rate of 3 cfs when the weir head is 0.5 ft. Determine the flow rate expected when the head is increased to 0.8 ft. Assume that the weir coefficient remains the same for each case.

Solution

From Eq. (4.5), $Q = \dfrac{3\,\text{ft}^3}{\text{s}}$, $H = 0.5$ ft

$$\frac{3\,\text{ft}^3}{\text{s}} = C_w b \left(\frac{32.2\,\text{ft}}{\text{s}^2}\right)^{\frac{1}{2}} \left(\frac{2}{3}\right) (0.5)^{\frac{3}{2}}$$

Thus, $C_w b = 2.24$ ft; hence, with $H = 0.8$ ft,

$$Q = 2.24\,\text{ft}\,(32.2)^{\frac{1}{2}} \left(\frac{2}{3}\right) (0.8\,\text{ft})^{\frac{3}{2}} = \frac{6.07\,\text{ft}^3}{\text{s}}\ (\text{answer})$$

The flume is an artificial channel constructed so the wastestream flows through it. The wastestreams flow is proportional to the depth of water in the flume and is calculated by measuring the head. A flume is composed of three sections: (i) a converging upstream section; (ii) a throat or contracted section; and (iii) a diverging or dropping downstream section. The two principal types of flumes are the Parshall flume and the Palmer-Bowlus flume.

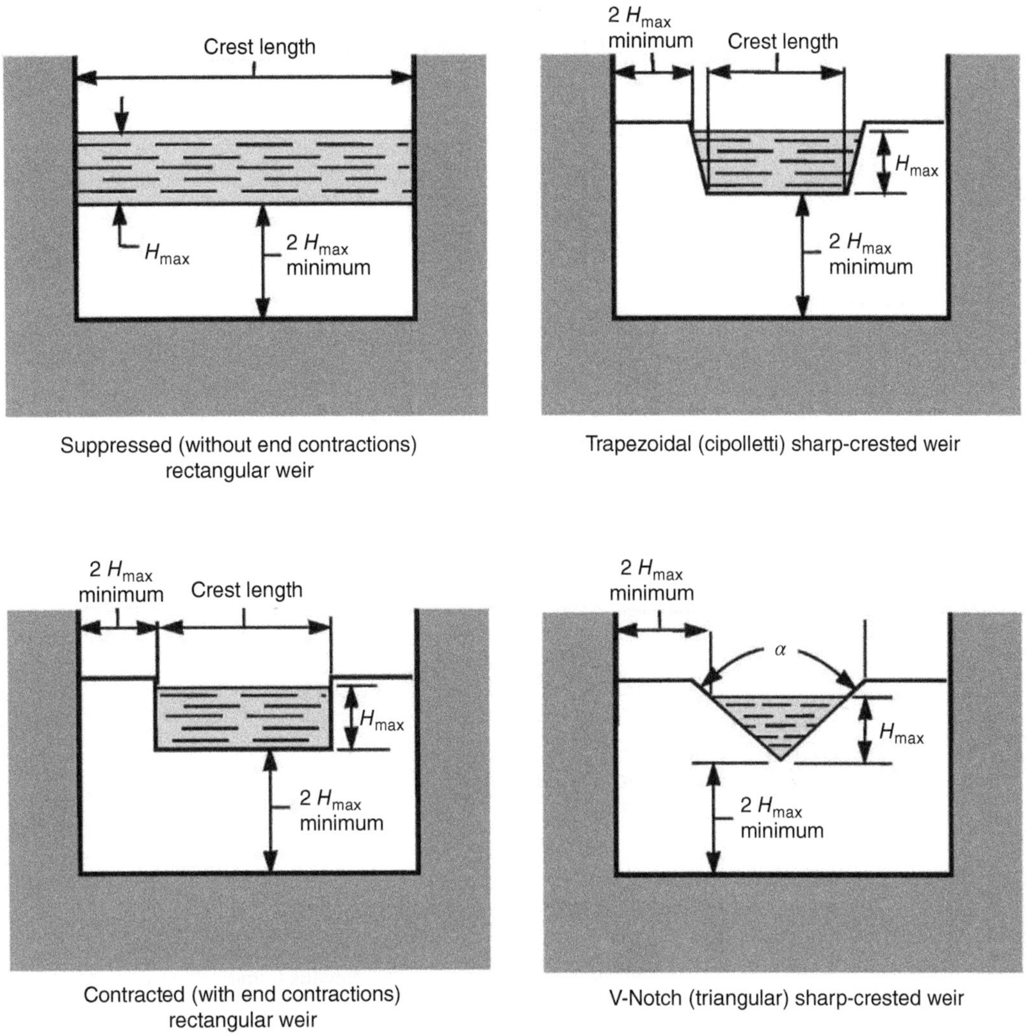

Figure 4.3 Four common types of sharp-crested weirs.

Figure 4.4 presents a sketch of the Parshall flume, identifying its different parts. In the Parshall flume, the floor level of the converging section is higher than the floor of the throat and diverging section. The flume operates on the principle that when water flows through a constriction in the channel, a hydraulic head is produced that is proportional to the flow. Flumes are good for measuring open channel waste flow because they are self-cleaning. Sand and SS are unlikely to affect the device's operation (Benefield and Judkins 1984).

Under free-flowing (unsubmerged) conditions, the Parshall flume is a critical-depth meter that establishes a mathematical relationship between a steady height, h, and discharge, Q. For a flume throat width (w) of at least 1 ft (0.305 m), but less than 8 ft (2.44 m), flow can be calculated using the following equation:

$$Q = 4WH^{(1.522)W^{0.026}} \tag{4.6}$$

where

Q = flow (cfs)
W = throat with (ft)
H = upper head (ft)

Example 4.5 What is the calculated wastewater flow through a Parshall flume with a throat width of 2.0 ft at the maximum free-flow head of 2.5 ft?

Solution

Using Eq. (4.6),

$$Q = 4 \times 2 \times 2.5^{(1.522)(2)^{0.026}}$$

$$Q = 4 \times 2.0 \times 2.5^{1.5496}$$

$$= 33.093\,\text{cfs} = 21.39\,\text{MGD} \quad (\text{answer})$$

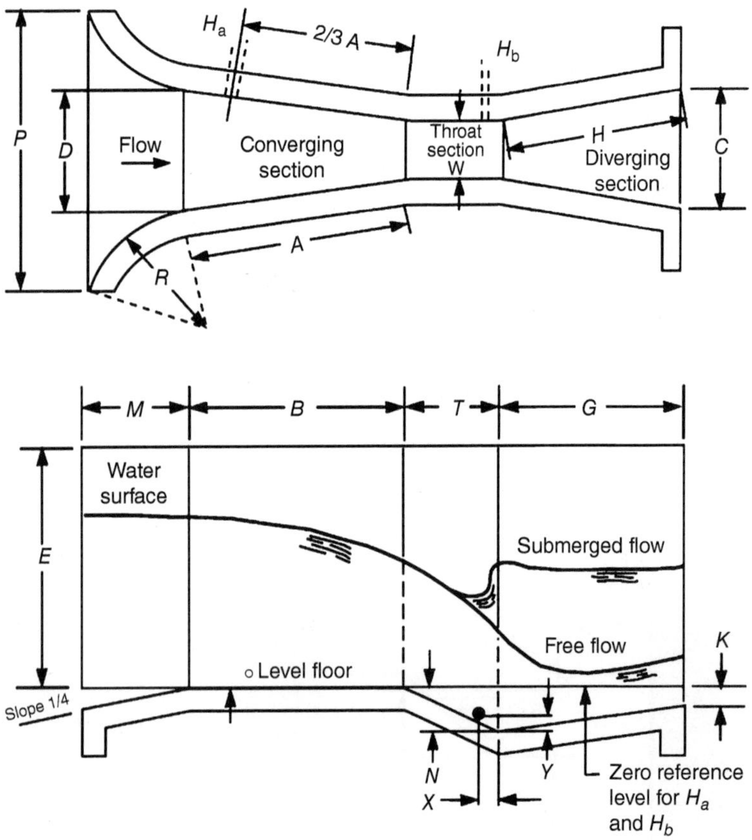

Figure 4.4 Plan view and cross section of a Parshall flume.

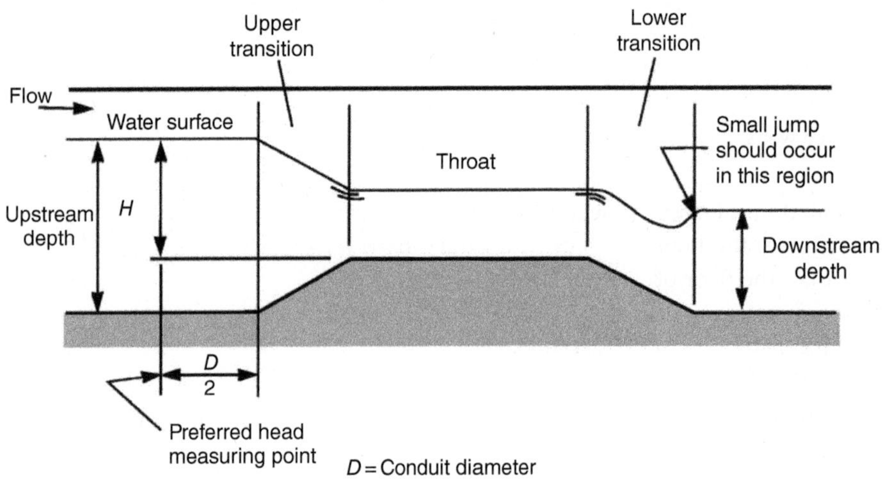

Figure 4.5 Free-flowing Palmer-Bowlus flume.

A Palmer-Bowlus flume, which may or may not have a constriction, has a level floor in the throat section and is placed in a pipe for approximately the length of the pipe's diameter. The depth of water above the raised step in the throat is related to the discharge rate. The head should be measured at a distance ($d/2$) upstream of the throat, where (d) is the width of the flume. The height of the step is usually unknown until the manufacturer's data are consulted, and it is difficult to measure manually the height of water above the step at an upstream point. The dimensions of each Palmer-Bowlus flume are different. Therefore, the manufacturer's data must be consulted to establish a relationship between the head and the discharge rate. Figure 4.5 contains a sketch of a free-flowing Palmer-Bowlus flume.

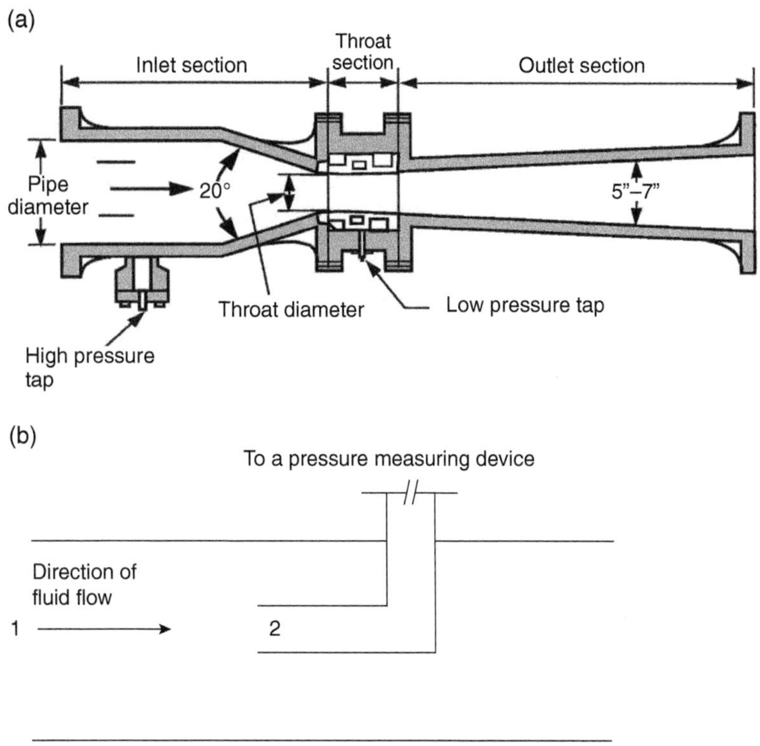

Figure 4.6 (a) Configuration and nomenclature of a Venturi meter. (b) A pitot tube.

4.7.1.2 Secondary Devices

Secondary devices are used in conjunction with primary devices to determine the actual flow passing the measuring point. Typically, secondary devices measure the depth of water in the primary device and convert the depth measurement to a corresponding flow, using established mathematical formulas. The output of the secondary device is generally transmitted to a recorder and/or totalizer to provide instantaneous and historical flow data to the operator. Outputs may also be transmitted to sampling systems to facilitate flow proportioning. Secondary devices can be organized into two broad categories:

1) A non-recording type with direct readout (e.g. a staff gauge) or indirect readout from fixed points (e.g. a chain, wire weight, or float)
2) A recording type with either digital or graphic recorders (e.g. float in well, float in flow, bubbler, electrical, or acoustic).

4.7.2 Closed Channel Flow

Closed channel flow is normally encountered between treatment units in a WWTP and after lift stations, where liquids and/or sludges are pumped under pressure. It is also encountered in submerged outfalls. Flow in closed channels is usually measured by a metering device inserted into the conduit. Examples of closed channel flow measuring devices are the Venturi meter and the electromagnetic flow meter.

The Venturi meter is one of the most accurate primary devices for measuring flow in closed channels. It is basically a pipe segment consisting of an inlet with a converging section, a throat, and a diverging outlet section, as illustrated in Figure 4.6a. The water velocity is increased in the constricted portion of the inlet section which results in a decrease in static pressure. The pressure difference between the inlet pipe and the throat is proportional to the flow

Flow in a Venturi meter can be calculated using the equation:

$$Q = CA_2 \left[2g \frac{P_1 - P_2}{p} \right]^{\frac{1}{2}} \tag{4.7}$$

where

Q = flow (cfs)
C = discharge coefficient, commonly in the range of 0.90–0.98
A_2 = cross-sectional area of throat (ft^2)
g = acceleration of gravity (ft/s^2)
$P_1 - P_2$ = differential pressure (lb/ft^2)
ρ = specific gravity of water (lb/ft^3)

Example 4.6 What is the calculated water flow through a Venturi meter with 5.063 in throat diameter when the measured pressure differential is 160 in H_2O? Assume $C = 0.95$ and water temperature of 50 °F.

Solution
Substituting in Eq. (4.7), we obtain

$$Q = 0.95 \times \frac{\pi}{4}\left(\frac{5.063\ \text{in.}}{12\frac{\text{in.}}{\text{ft}}}\right)^2 \left(2 \times \frac{32.2\ \text{ft}}{\text{s}^2} \times \frac{160\ \text{in.}}{\left(62.4\frac{\text{lb}}{\text{ft}^3}\right)12.0\frac{\text{in.}}{\text{ft}}}\right)^{0.5}$$

$$= 0.492\ \text{cfs} = 221\ \text{gpm}$$

4.7.3 Pitot Tube

Pitot tube or pitot probe (also known as an impact or stagnation tube) is simply a hollow tube that is placed longitudinally in the direction of fluid flow, allowing the flow to enter one end at the fluid's velocity of approach (Figure 4.6b). It is a pressure measurement instrument used to measure fluid flow velocity and finds uses in both subsonic and supersonic application. Pitot tube is widely used to determine the airspeed of an aircraft, water speed of a boat, and to measure liquid, air, and gas flow velocities at stack and in certain industrial applications. The pitot tube is used to measure the local flow velocity at a given point in the flow stream and not the average flow velocity in the pipe or conduit.

Bernoulli's equation can be used to predict the static pressure at the stagnation point. Since the velocity of the fluid within the pitot tube is zero, the upstream velocity can be calculated if the static and stagnation pressures are known.

$$\frac{P_1}{\rho} + \frac{v_1^2}{2} = \frac{p_2}{\rho} \tag{4.8}$$

$$V_1 = \sqrt{\frac{2(p_2 - p_1)}{\rho}}\ \ (\text{SI}) \tag{4.9}$$

$$V_1 = \sqrt{\frac{2g_c(p_i - p_s)}{\rho}}\ \ (\text{US}) \tag{4.10}$$

In reality, both friction and heating occur, and the fluid may be compressible. These errors are corrected by a correction factor known as the impact factor, C_i, which is applied to the derived velocity. C_i is usually very close to 1.00 (e.g. 0.99 or 0.995).

$$V_{\text{actual}} = C_i V_{\text{vindicated}}$$

Example 4.7 The static pressure of air (0.075 lbm/ft^3; 1.2 kg/m^3) flowing in a pipe is measured by a precision gauge, shown below, to be 10.00 psig (68.95 kPa). A pitot tube manometer indicates 20.6 in. (0.52 m) of mercury. Losses are insignificant. What is the velocity of the air in the pipe?

Solution in SI

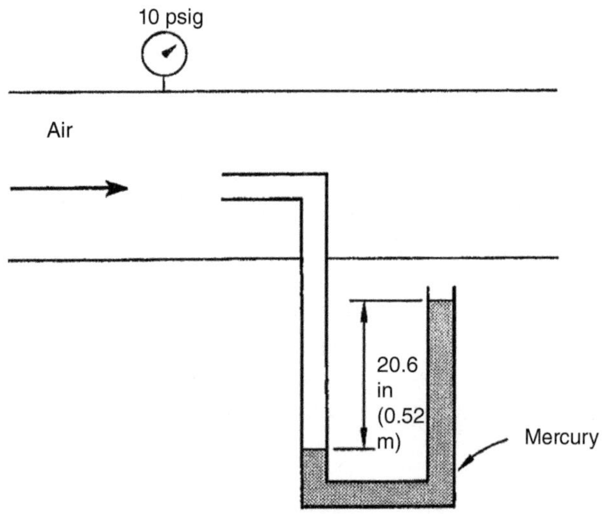

The density of mercury is 13 600 kg/m^3. The impact pressure is

$$P_i = \rho g h = \left(13600\frac{\text{kg}}{\text{m}^3}\right)\left(9.81\frac{\text{m}}{\text{s}^2}\right)\left(\frac{0.52\,\text{m}}{1000\frac{\text{Pa}}{\text{kPa}}}\right) = 69.38\,\text{kPa}$$

From Eq. (4.9), the velocity is

$$V = \sqrt{\frac{2(p_i - p_s)}{\rho_{\text{air}}}}$$

$$= \sqrt{\frac{(2)(69.38 - 68.95\,\text{kPa})\left(1000\frac{\text{Pa}}{\text{kPa}}\right)}{1.2\,\text{kg/m}^3}};\ \text{where}\ \rho_{\text{air}} = 1.2\,\text{kg/m}^3$$

$$= 26.8\,\text{m/s}$$

(customary US solution)

The pitot tube measures impact (stagnation) pressure. For mercury, pressure in equivalent psi is 0.491 psig/in. Therefore, $p_i = (20.6\,\text{in.})\left(0.491\frac{psi}{in}\right) = 10.11\,\text{psig}$.

Since impact pressure is the sum of the static and kinetic (velocity) pressure, the kinetic pressure is $p_v = p_i - p_s = 10.11 - 10.00 = 0.11\,\text{psig}$, using Eq. (4.10) we get,

$$V = \sqrt{\frac{2g_c\left(p_i - p_s\right)}{\rho_{air}}}$$

$$V = \sqrt{\frac{2\left(32.2\frac{ft}{s^2}\right)\left(0.11\frac{lb}{s^2}\right)\left(144\frac{in^2}{ft^2}\right)}{0.075\frac{lb}{ft^3}}};$$

where density of air $= 0.075\,lb/ft^3$

Therefore, $V = 116.6\,ft/s = 117\,ft/s$ (answer).

4.7.4 Electromagnetic Flow Meter

Electromagnetic flow meter operation is based on the fact that the voltage induced by a conductor moving at right angles through a magnetic field will be proportional to the velocity of that conductor as it moves through the field. In the case of the electromagnetic flow meter, the conductor is the stream of water to be measured, and the magnetic field is produced by a set of electromagnetic coils. A typical electromagnetic flow meter is shown in Figure 4.7.

4.8 The Point of Compliance with the Water Quality Standards

4.8.1 Mixing Zones

The disposal of wastewater from industrial WWTP in rivers and estuaries is considered in this section. The Water Quality Standards allow the use of mixing zones for discharges that would otherwise exceed the water quality criteria for aquatic life. Mixing zones are areas where the water quality standards may be exceeded, but the exceedances are small enough and short enough so as not to interfere with beneficial uses. Mixing zones are a regulatory recognition that the concentrations and effects of most pollutants diminishes rapidly after discharge due to dilution (Adams 1982). They are established in a manner which limits the duration of exposure for organisms passing through the effluent plume in order to minimize the risk from each discharge. The water quality standards for chronic protection must be met at the boundary of this zone and beyond. A smaller zone in which acute criteria may be exceeded can also be authorized. This zone must be small enough to limit exposure times and therefore not cause acute mortalities or interfere with passage of aquatic organisms in the water body. The mixing zone is not significant for DO analyses because BOD exertion is slow compared to the travel time within the mixing zone. Relative to other constituents, such as toxic chemicals, the characteristics of the mixing zone are important. Clearly, rapid mixing of the effluent with the ambient flow is desirable to decrease toxicity.

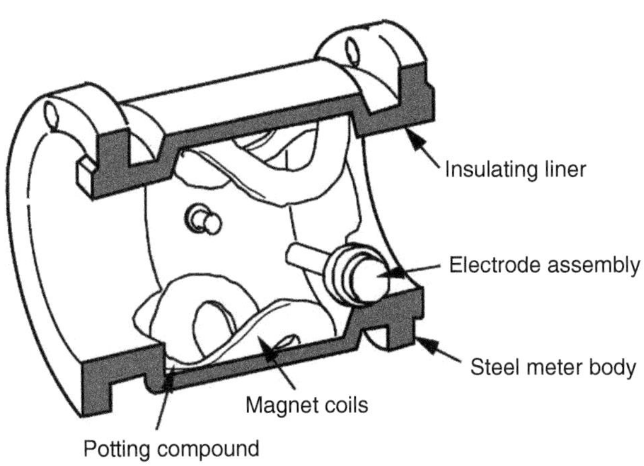

Figure 4.7 Electromagnetic flow meter.

The size of mixing zones depends not only on the effluent characteristics but also on the river flow. Typically, lower effluent dilutions and therefore larger mixing zones are achieved for low river flows. Analysis of flow records are therefore needed to establish the magnitude, duration, and frequency of low-flow period. Regulations in force in the United States often prescribe the duration and frequency of the low flow that must be used for mixing zone analyses, e.g. 7Q10.

For small river depths, vertically mixed conditions are reached within a short distance of the discharge point. Mixing zone analyses can be conducted by simulating the discharge as a plane source extending over the full river depth. For pipe discharges, the width of the source can be assumed to be equal to the river depth. The limited river width constraints the lateral spreading of the plume. The limited river width can be accounted for by introducing image sources with respect to the river banks, as illustrated in Figure 4.8.

Oceans and large lakes, such as Great Lakes, provide extensive assimilation capacity and are used for wastewater disposal by many industries and communities. The wastewater is typically carried to an offshore discharge point by a pipe laid on a buried in the ocean floor, or by a tunnel. The discharge in ocean or deep lakes can be through a single-port or multiport outfall structure. A configuration of a wastewater plume in the ocean is shown in Figure 4.9. In the first region, called the "initial mixing region" or "discharge nearfield," the effluent forms a buoyant plume, rapidly rising in the water column. This plume entrains large amounts of ambient water, thereby diluting the effluent (Muellenhoff et al. 1985).

An intermediate goal in point source pollution control which is consistent with the CWA goal of zero discharge is

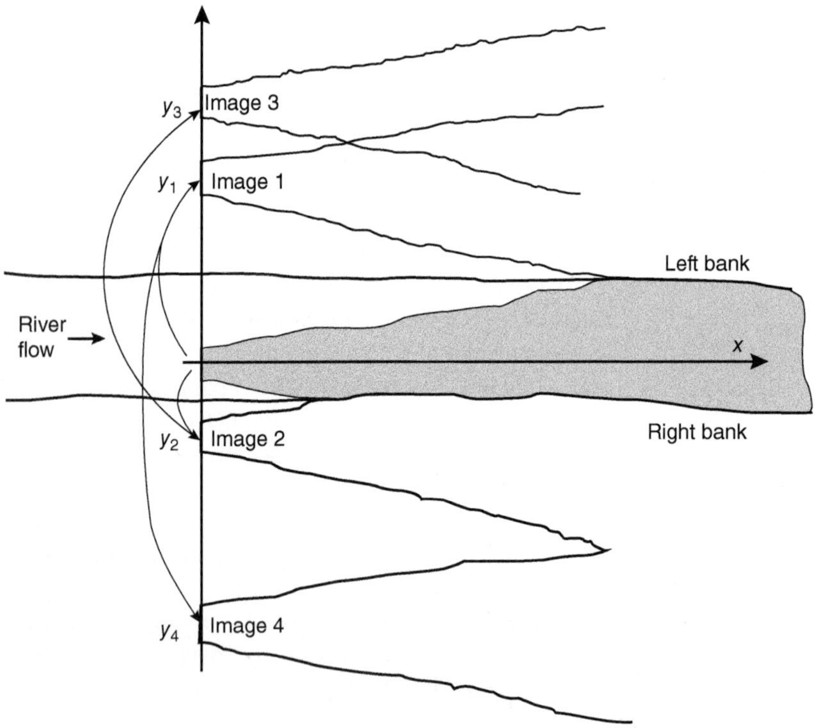

Figure 4.8 Mixing zone analysis with method of image in a shallow stream.

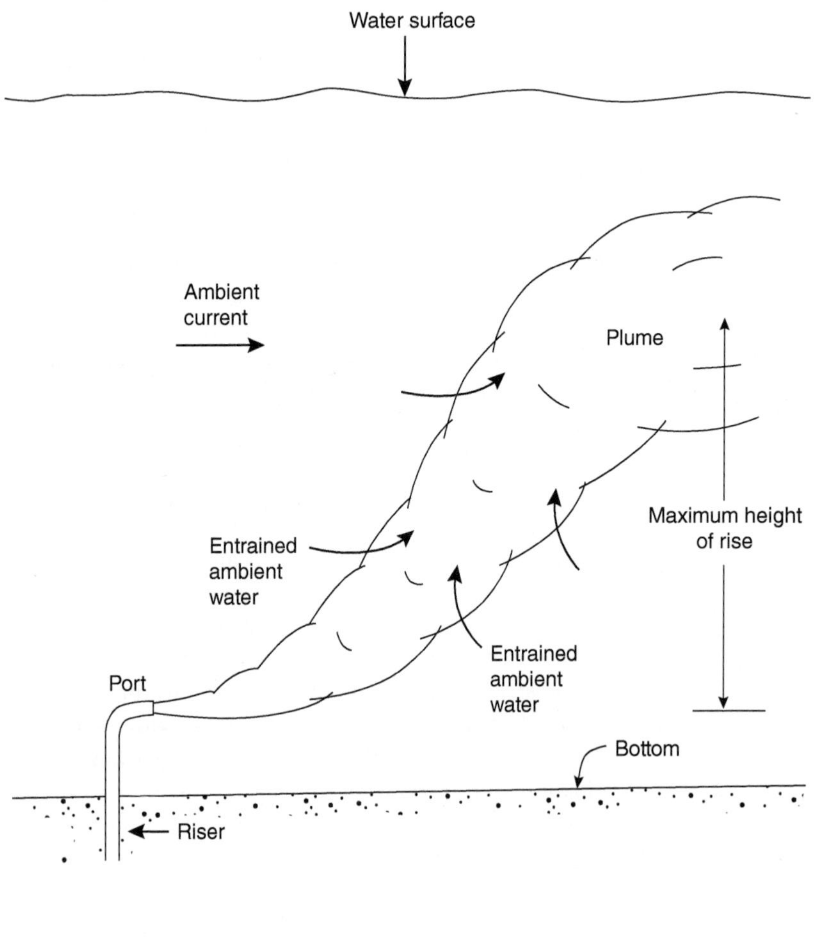

Figure 4.9 Wastewater discharge plume in deep receiving waterbody.

the elimination of the need for a mixing zone. The water quality standards do not prohibit a permit writer from granting a mixing zone for a shore discharge, but shore areas are important biological areas. A permit writer should be sure that granting of the mixing zone on a shoreline will not cause biological effects as specified in the regulation. This may require a demonstration on the part of the permittee in some cases.

4.8.2 Streeter–Phelps Equation and DO Sag Curve in a River

The Streeter–Phelps equation determined the relation between the dissolved oxygen concentration and the biological oxygen demand over time (Streeter and Phelps 1925). The concentration of DO in a river is an indicator of general health of the river. All rivers have some capacity for self-purification. As long as the discharge of oxygen demanding wastes is well within the self-purification capacity, the DO level will remain high, and a diverse population of plants and animals, including game fish, can be found. As the amount of the waste increases, the self-purification capacity can be exceeded, causing detrimental changes in plant and animal life. The stream loses its capacity to cleanse itself, and the DO level decreases. When the DO drops below about 4–5 mg/l, most game fish will have been driven out. If the DO is further reduced to or below 2 mg/l (critical point) (Figure 4.10), fish and other animals are killed or driven out, and extremely noxious conditions result. The water becomes foul-smelling as organic waste and dead animal life decompose under anaerobic conditions (i.e. without oxygen).

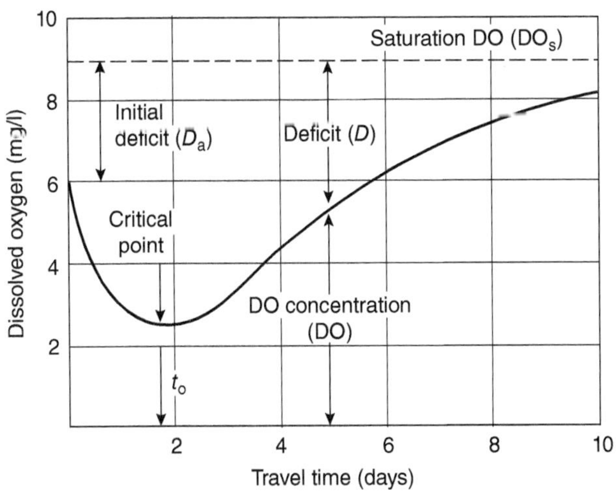

Figure 4.10 The DO sag curve derived from the Streeter–Phelps equation.

One of the major tools of water quality management in rivers is increasing the capacity of a stream to absorb a waste load (also discussed in Section 4.8.1 in river TMDL and WLA). This is done by determining the profile of DO concentrations downstream from a waste discharge. This profile is called the DO sag curve (Figure 4.10) because the DO concentration dips as oxygen-demanding materials are oxidized and then rises again further downstream as the oxygen is replenished from the atmosphere and photosynthesis processes.

4.8.3 Mixing of Wastewater in Rivers: Mass-Balance Approach

When two streams or rivers merge or wastewater is discharged to a stream, it is possible to determine the BOD and DO after mixing assuming steady-state conditions and instantaneous mixing (Jolánkai 1997). The two streams are considered as dilutions of each other; thus, the initial BOD and DO will be

$$L_a = \frac{\left(L_w Q_w + L_r Q_r\right)}{\left(Q_w + Q_r\right)} \tag{4.11}$$

and

$$\mathrm{DO} = \frac{\left(\mathrm{DO}_w Q_w + \mathrm{DO}_r Q_r\right)}{\left(Q_w + Q_r\right)} \tag{4.12}$$

where

L_a is the initial concentration of ultimate BOD after mixing, g/m^3 (mg/l)
L_w is the ultimate BOD of the wastewater g/m^3 (mg/l)
L_r is the ultimate BOD of the river, g/m^3 (mg/l)
DO_w is the dissolved oxygen content in the wastewater g/m^3 (mg/l)
DO_r is the dissolved oxygen content in the river g/m^3
Q_w is the flow rate of wastewater m^3/s
Q_r is the flow rate of the river m^3/s

Example 4.8 A food-processing plant in the town of Zella in Washington discharges 17 300 m^3/day of treated wastewater into a nearby Salmon Creek. The treated wastewater has a BOD_5 of 12 mg/l, and the Salmon Creek has a flow rate of 0.43 m^3/s. The DO of the wastewater is 1.0 mg/l and the DO in the Salmon Creek is 6.5 mg/l. Compute the DO after complete mixing.

Solution

The DO after mixing is given by Eq. (4.12). To use this equation we must convert the wastewater flow rate to compatible units, that is, to cubic meter per second.

$$Q_w = \frac{17300\ \text{m}^3/\text{day}}{86400\ \text{s}/\text{day}} = 0.20\frac{\text{m}^3}{\text{s}}$$

$$\text{The DO} = \left[\frac{\left(0.20\frac{\text{m}^3}{\text{s}}\right)\left(1.0\frac{\text{mg}}{\text{l}}\right)+\left(0.43\frac{\text{m}^3}{\text{s}}\right)\left(6.5\frac{\text{mg}}{\text{l}}\right)}{\left(0.20\frac{\text{m}^3}{\text{s}}+0.43\frac{\text{m}^3}{\text{s}}\right)}\right]$$

$$= 4.75\frac{\text{mg}}{\text{l}} \quad \text{or} \quad 4.75\,\text{ppm}$$

4.9 Water Quality Modeling

Water quality modeling involves the prediction of water pollution using mathematical simulation techniques. A typical water quality model consists of a collection of formulations representing physical mechanisms that determine position and momentum of pollutants in a water body. Models are available for individual components of the hydrological system such as surface runoff; there also exist basin-wide models addressing hydrologic transport and for ocean and estuarine applications. Often finite difference methods are used to analyze these phenomena, and, almost always, large complex computer models are required (USEPA 1985d; CORMIX 1996; Section D.2).

4.9.1 Formulations and Associated Constants

Water quality is modeled by one or more of the following formulations:

- Advective fate and transport formulation
- Dispersive fate and transport formulation
- Streeter–Phelps formulation
- Surface heat budget formulation
- DO saturation formulation
- Reaeration formulation
- Carbonaceous deoxygenation formulation
- Nitrogenous biochemical oxygen demand formulation
- Sediment oxygen demand formulation
- Photosynthesis and respiration formulation
- pH and alkalinity formulation
- Nutrients formulation (fertilizers)
- Algae formulation
- Zooplankton formulation
- Coliform bacteria formulation (e.g. *Escherichia coli*)

4.9.2 WWTP BOD, SS, and Fecal Coliform Removal Efficiencies: Meet Water Quality Standards

The major goals of the combined primary, secondary, and sometime tertiary treatment are to remove the conventional pollutants including organics and pathogens (e.g. BOD, COD, SS, fecal coliform), to meet the water quality standards and NPDES permit requirements (CWA 1972; Davis and Masten 2014; USEPA 1995b, 2010a). Primary treatment will typically remove 60% of the SS in raw wastewater and 35% of the BOD_5. The major goal of secondary treatment is to remove the soluble BOD_5 that escape the primary process and to provide added removal of SS. Secondary treatment is achieved by using biological processes shown in Figure 4.11, which is a schematic diagram of a pulp and paper mill's process WWTP (Das 1991, 1993). The WWTP includes effluent from both sides of the bleach plant that receives primary clarification. A surge basin, which is connected to the primary clarifier, is not commonly used. However, in the event of a spill (black liquor for example), overflows can be retained in the surge basin for about six hours, based on the mill's total wastewater flow.

Following primary clarification, nutrients (ammonia and phosphoric acid) are added to the primary effluent. A high-purity oxygen-activated sludge secondary treatment, under the trade name UNOX®, is followed by secondary clarification. The overflows from secondary five clarifiers are collected in a wet well, and the effluent is discharged to the Columbia River through a 310 ft long diffuser section on the submerged outfall pipe at a rate of 65 MGD (Figure 4.11).

This WWTP secondary treatment provides the same biological reactions that would occur in the receiving water if it had adequate capacity to assimilate the wastewater. The secondary treatment processes are designed to speed up these natural processes so that the breakdown of the degradable organic pollutants can be achieved in a relatively short time. Although secondary treatment may remove more than 85% of the BOD_5 and SS, it does not remove significant amounts of nitrogen, phosphorus, or heavy metals, nor does it completely remove pathogenic bacteria and viruses. The NPDES permit often requires effluent to meet at least 85% removal of BOD_5 and SS, fecal coliform counts 200 MPN in 100 ml effluent, and pass standard bio-assessment tests for selected freshwater (e.g. trout or salmon, fathead minnow) and marine saltwater species (e.g. mysid shrimp, *Champia parvula*, sand dollar, *Echinarachnius parma*).

In case where secondary levels of treatment are inadequate, additional treatment processes are applied to the secondary effluent. These tertiary treatment processes may involve chemical treatment and filtration of the

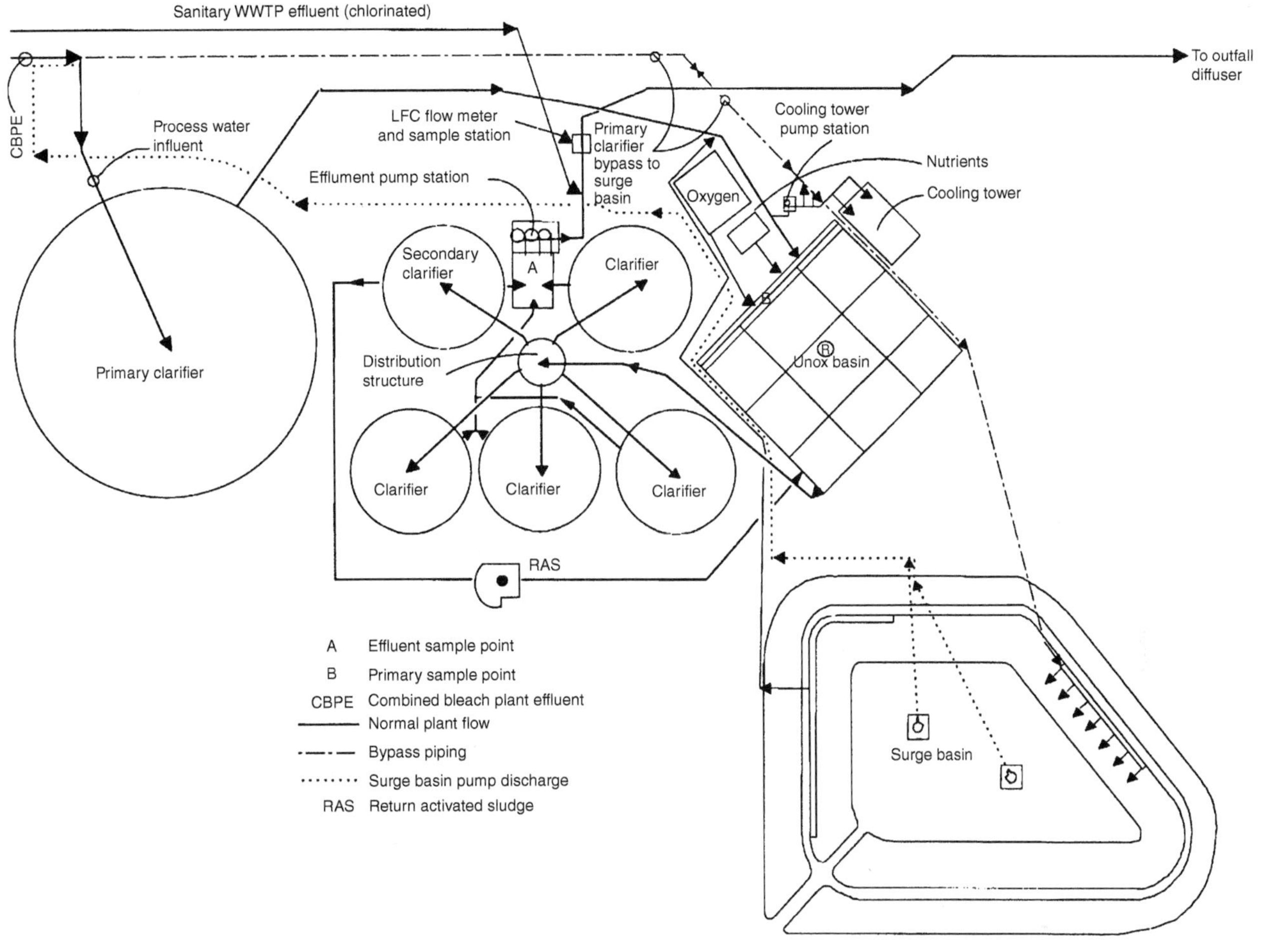

Figure 4.11 A Kraft pulp and paper mill's wastewater treatment plant general flow diagram (Das 1991).

wastewater. Some of these tertiary treatment processes can remove as much as 99% BOD_5, SS, total-phosphorus, bacteria, and 95% of the nitrogen. These processes can produce a sparkling clean, colorless, odorless effluent which could be applied on land and for other beneficial reuse.

Most of the impurities removed from the wastewater do not simply vanish. Some organic compounds are broken down into carbon dioxide and water. Most of the impurities are removed from wastewater as solid, that is sludge. Because most of the impurities removed from water are present in sludge, therefore, sludge handling and disposal must be carried out carefully to achieve the pollution control (also discussed this in Section 4.9.2).

4.9.3 NPDES Wastewater Discharge Permits for Point Sources

The NPDES permits program is authorized by the CWA (CWA Section 402, 33 U.S. C. 1342). The initial permits issued in the 1970s and early 1980s focused on POTWs and industrial wastewater – typically "process" wastewater and cooling water where applicable. An *NPDES Permit* is required for a discharge of wastewater to waters of the United States (surface waters). This permit is issued by a state by delegated authority of the USEPA (1972). Since waters of the United States are also waters of the state, NPDES permits are actually combined NPDES/State Waste Discharge permits issued under dual authorities. NPDES permits are issued for a maximum of five years. Conventional pollutants contained in industrial, commercial, or domestic wastewaters are BOD, SS, fecal coliform, pH, temperature, free and residual chlorine, oil and grease, and DO. Toxic pollutants include synthetic organic chemicals and heavy metals. Nonconventional pollutants are any additional substances not in conventional or toxic categories, such as ammonia as nitrogen, nitrate–nitride, and total phosphorus.

NPDES permits for discharging wastewaters contain the following standard conditions common to all permits, site-specific effluent limits, compliance monitoring and

Table 4.6 NPDES permit limits for a pulp and paper mill's wastewater treatment plant.

Effluent parameter	NPDES permit daily limit: Maximum	NPDES permit daily limit: Average	30-Day average concentration	7-Day average concentration
BOD_5	28 000 lb	20 000 lb		
	40 mg/l	35 mg/l	30 mg/l	45 mg/l
TSS	46 000 lb	32 000 lb		
	50 mg/l	45 mg/l	30 mg/l	45 mg/l
Flow			—	60 MGD
pH				6.0–9.0 all times
Rainbow trout	80% survival in 100% effluent			
Bioassay	65% effluent for a 96-hour period			
Fecal Coliform/	20 000 (#/100/ml)	5 000 (#/100/ml)		
% Klebsiella	N/A	98%		

Source: From Das (1991).
Note: Typical permit renewal in every five years. This table outlines only the quantitative limit. The entire permit is 26-pages long. BOD_5, five-day biochemical oxygen demand; MGD, million gallons per day; N/A, not applicable; pH, negative $\log_{10}$ of hydrogen ion concentration in gmol/l; TSS, total suspended solids.

reporting requirements (see Section 4.4.4), and site-specific conditions necessary for discharge control. Standard conditions describe the legal aspects of the permit and its revocability. They also describe the permittee's duties and obligations; for example, the permittee is required to report changed conditions, to allow compliance inspections, and to reapply for a permit renewal. Finally, the permittee is informed regarding penalties that may be assessed for permit violations. Effluent limits imposed are based on technology-based and water-quality based standards (USEPA 1995b, 2010a; Hammer and Hammer 2012; Section D.3). The management of wastewater sludge is also controlled by conditions in a NPDES permit. The EPA is responsible for implementing and enforcing the NPDES program; however, individual states have the authority to perform these functions with oversight by the USEPA. An example of an industrial NPDES permit limits for criteria pollutants is shown in Table 4.6.

The NPDES is a self-monitoring program whereby the owner of the treatment plant is required to be responsible for compliance monitoring and reporting the performance of wastewater and sludge processing, including any violation of permit conditions. Effluent limitations and monitoring provisions specify the procedures to be used for sampling, flow measurement, laboratory testing, and record keeping. Operation and maintenance require reporting of treatment system failure, by-passing of untreated wastewater, and process upset. Additional reporting requirements are any change in discharge, noncompliance notification, and compliance schedules. Enforcement for noncompliance is by administrative order to a monetary penalty. For serious violations, judicial action can be taken as either a civil or a criminal action (Hammer and Hammer 2012). Additional frequently asked questions and their answers pertinent to NPDES permitting are given in Section D.3.

Example 4.9 An activated sludge extended aeration WWTP with a flow of 6.0 MGD has BOD of 320 mg/l, SS of 280 mg/l, and a total phosphorus of 7.35 mg/l, keeping the effluent the same as 19 mg/l of BOD, 20 mg/l of SS, and 1.25 mg/l of total phosphorous.

Calculate the following:

1) The BOD loading
2) BOD and SS removals by the plant
3) Pounds of BOD and SS removed per day
4) Total settled SS and phosphorous sludges in lb/day
5) If NPDES permit condition requires 85% removal of BOD, is the plant in compliance?

Solution

1) $\text{BOD loading} = 320\,\frac{\text{mg}}{\text{l}} \times 6.0\,\text{MGD} \times 8.34 = 16\,012\,\frac{\text{lb}}{\text{day}}$

2) $\text{BOD removal} = \frac{(320-19)}{320} \times 100 = 94\%$

$$\text{SS removal} = \frac{(280-20)}{280} \times 100 = 93\%$$

3) $\frac{\text{lb}}{\text{day}}$ BOD removal $= (320-19)\ \frac{\text{mg}}{\text{l}} \times 6\,\text{MGD} \times 8.34$

$$\left(\frac{\text{lb}\cdot\text{l}}{\text{MG}\cdot\text{mg}}\right) = 15062\ \frac{\text{lb}}{\text{day}}$$

$$\frac{\text{lb}}{\text{day}}\ \text{SS removal} = (280-20)\ \frac{\text{mg}}{\text{l}} \times 6.0\,\text{MGD} \times 8.34$$

$$\left(\frac{\text{lb}\cdot\text{l}}{\text{MG}\cdot\text{mg}}\right) = 13010\,\text{lb}$$

4) $= [(280 - 20) + (7.35 - 1.25)] \times 6.0\ \text{MGD} \times 8.34$

$$\left(\frac{\text{lb}\cdot\text{l}}{\text{MG}\cdot\text{mg}}\right) = 13316\ \frac{\text{lb}}{\text{day}}$$

5) Yes, BOD removal is 94%, therefore the plant well exceeded the 85% removal requirement by the NPDES permit condition.

4.10 Example NPDES Permits (for Refinery and Aluminum Smelter are shown in Section D.1)

4.10.1 Total Maximum Daily Load (TMDL) Rule

Under the CWA, the states in the United States are required to identify surface waters that do not meet specified water-quality standards or do not meet the designated beneficial uses (USEPA 1991a). In the 1990s, a significant percentage of the nation's waters were found to be impaired based on quality standards or beneficial uses. For the most part, these waters were not suitable for fishing, swimming, or drinking. The major causes were excess sediments, SS, nutrients (phosphorus and nitrogen) from industries and municipalities, and human pathogens. Other causes were low DO, toxic metals, pesticides, pH, colors, algae, and habitat alterations. Sources of excess pollutants are from both point and nonpoint sources.

A total maximum daily load (TMDL) is the maximum of pollutant that a waterbody can receive and still maintain the water quality standards that the state has set. Therefore, a TMDL for each pollutant of concern must be established for all impaired waters. The TMDL will include a waste load allocation (WLA) for each point source of contamination, such as WWTP discharge, and a load allocation for each nonpoint sources as well as a MOS as shown in Eq. (4.13) and Figure 4.12.

$$\Sigma\text{WLA} + \Sigma\text{LA} + \text{MOS} = \text{TMDL} \tag{4.13}$$

TMDL can be specified in mass per units, such as pounds per day, toxicity units, or other measures, such as percent reduction. The TMDL concept is shown in Figure 4.12 where a water quality standard of 2 mg/l of ammonia as nitrogen and a stream flow of 30 MGD give a TMDL of 500 lb/day. With a given nonpoint source load allocation for agriculture runoff of 75 lb/day and a MOS of 50 lb/day, the WLA for the treatment plant is determined to be 375 lb/day, which would be written into the operating permit for the WWTP. The discharge permit for the treatment plant would be written so that this WLA would not be exceeded.

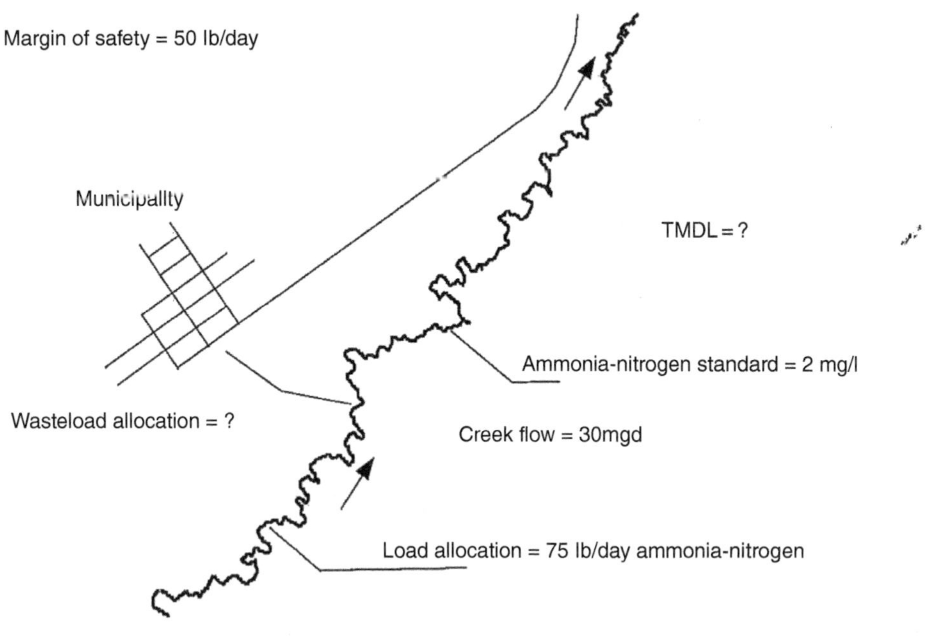

Figure 4.12 Total maximum daily load (TMDL) – a contributing component. (a) *Source:* Adopted from Viessman et al. 2009.

Example 4.10 TMDL Calculation

$$\text{TMDL} = \left(2\frac{\text{mg}}{1} \times 30\,\text{MGD} \times 8.34\frac{\text{lb}}{\text{mg}}\frac{1}{\text{MG}}\right) = 500\frac{\text{lb}}{\text{day}}$$

[where, mg = milligram; MG = million gallons]

Rearranging Eq. (4.13) gives

$$\begin{aligned}\text{WLA} &= \text{TMDL} - \text{LA} - \text{MOS} \\ &= 500 - 75 - 50 = \mathbf{375}\frac{\text{lb}}{\text{day}}\ (\text{answer})\end{aligned}$$

4.11 Air Pollution Perspective

Air pollution is woven throughout the fabric of our modern life. A by-product of the manner in which we build our cities, air pollution is waste remaining from ways we produce our goods, transport ourselves, and generate the energy to heat and light the cities where we live, work, and play. The major cause of all air pollutants is combustion which is an essential part of our civilization. When perfect or theoretical combustion occurs, the hydrogen and carbon in the fuel combine with oxygen from the air to produce heat, light, carbon dioxide, and water vapor. However, impurities in the fuel, poor fuel-to-air ratio, or too high or too low combustion temperatures cause the formation of such by-products as carbon monoxide, sulfur dioxide, nitrogen oxides, fly ash, and unburned hydrocarbons – all air pollutants.

Air pollution is of public health concern on the micro, meso, and macroscales (Section 4.11). Industrial and mobile sources contribute to mesoscale air pollution and contaminates the ambient air that surrounds us outdoors. Macroscale (or global) effects include transport of ambient air pollutants over large distances. Examples of macroscale effects include acid rain, ozone, and dust or particulate matter (PM) pollution. Global effects of air pollution results from sources that may potentially change the upper atmosphere, examples of which include depletion of good ozone layer and global warming. Although microscale and macroscale effects are of concern, our focus is predominantly on mesoscale air pollution in the troposphere where we live.

The 1970 Clean Air Act (CAA) required the USEPA to investigate and describe the environmental effects of any air pollutant emitted by stationary and/or mobile sources that may adversely affect human health and the environment. The EPA used these studies to establish the National Ambient Air Quality Standards (NAAQS). The NAAQS have been established for six criteria air pollutants – five primary (meaning emitted directly) and one secondary pollutant (because it is formed in the lower atmosphere by chemical reactions among primary pollutants).

These standards are for the ambient air, that is the outdoor air that normally surrounds us. The EPA calls the pollutants listed in Table 3.9 criteria pollutants because they were developed on health-based criteria. The primary standard was established to protect human health with an "adequate MOS." The secondary standards are intended to prevent environmental and property damage. In a conflict between ambient air quality and anticipated economic growth, regulations were proposed in August 1974 to prevent serious deterioration of air quality in areas where the air is already cleaner than required by federal standards. A threefold classification was put into effect with a goal that in no circumstances the ambient air quality of an area violate federal primary and secondary standards. The classifications are as follows:

Class I Area where almost no change from current air quality will be allowed (e.g. national parks and protected wilderness areas)

Class II Areas where moderate change will be allowed, but where stringent air quality constraints are desirable (e.g. residential areas)

Class III Areas where substantial industrial will be allowed and where the increase in concentration of pollutants up to the federal standards will be insignificant

In 1987 the EPA revised the NAAQs by replacing the standard for hydrocarbons (or VOCs) with total suspended PM with an aerodynamic diameter less than or equal to 10 μm. This standard is referred to as the PM_{10} standard. In 1997 a new $PM_{2.5}$ standard was added. Figure 4.13 should help you visualize the size of a micrometer or micron. Note that a hair has an average diameter of approximately 80 μm.

4.11.1 Causes, Sources, and Effects

In this section, we will briefly discuss some of the major pollutants with respect to their causes, sources, and effects. A cause is distinguished from a source in that a cause is fundamental, whereas a source is locational; that is, a cause explains why or how a pollutant is formed, whereas a source identifies what type of process, industry, or device discharges a particular pollutant.

Criteria air pollutants, or criteria pollutants, are a set of air pollutants that cause smog, acid rain, and other health hazards. These pollutants are typically emitted from many sources in industry, chemicals, mining, transportation, electricity generation, refineries, smelter, pharmaceuticals, agriculture, and others. In most cases, they are the products

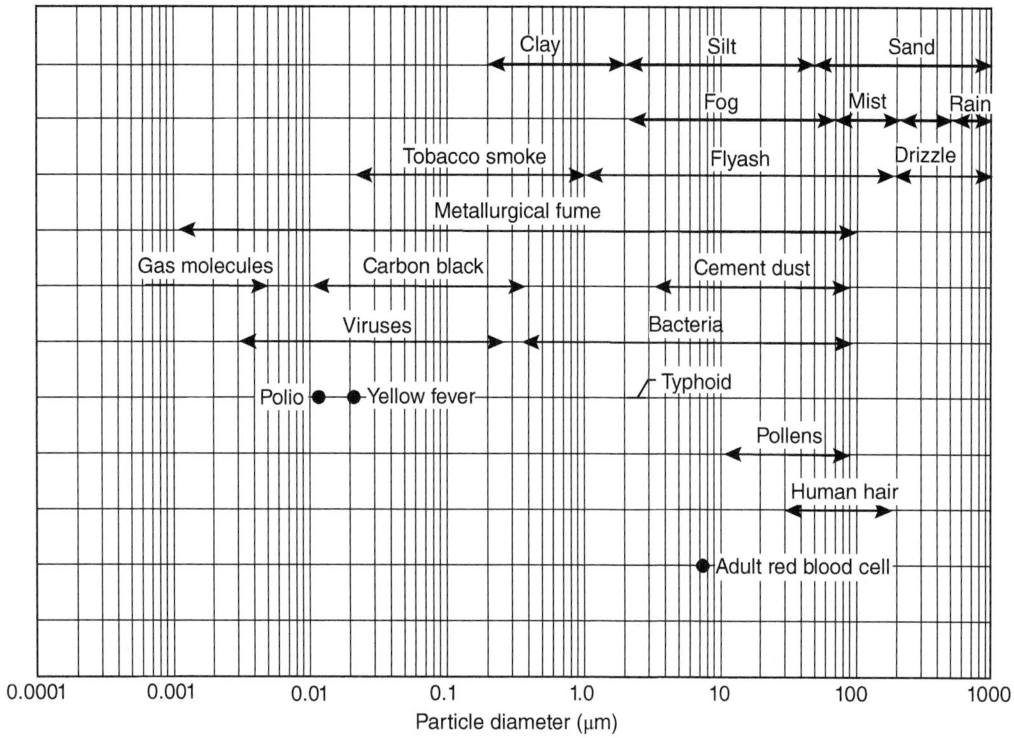

Figure 4.13 Relative sizes of small particles in micrometers.

of the combustion of fossil fuels or of industrial processes. The sources and the effects on public health and the environment were discussed and presented in Table 3.8.

The five primary criteria pollutants are *PM less than 10 μm* in diameter and *PM less than 2.5 μm* in diameter (PM −10 and PM −2.5), *sulfur dioxide* (SO_2), *nitrogen dioxide* (NO_2), *carbon monoxide* (CO), and *particulate lead* (Pb); the secondary criteria pollutant is *ozone* (O_3). Among these six pollutants, the first four are emitted in the United States (and other large industrialized countries) in quantities measured in millions of metric tons per year and are sometimes called *major* primary pollutants. Another class of pollutant – volatile organic compounds (VOCs) – though not a criteria pollutant, is recognized as a major primary pollutant because of its large emissions (both anthropogenic and biogenic sources) and its importance in the atmospheric reactions that form ground level ozone.

Table 4.7 presents data on emission rates from major sources, including industry, stationary, and mobile, for five major primary pollutants in the United States in 2006 and in 2016. We can see from the Table 4.7 that emissions from automobiles, trucks, buses, motorboats, lawns and garden equipment, construction vehicles, and other sources account for approximately 61 and 42% of all the NO_x and VOC emissions, respectively, and over 80% of all the CO emissions in the United States. Internationally, with continuing population growth particularly in Asia and South America, significance of mobile source pollution has been growing at an even faster rate than industrial pollution. Hence, the air pollution from the mobile sources is more severe each year. In various countries around the world, motor vehicles account for some 10–60% of total air pollution emissions. Here our focus is on air emissions from stationary sources and its industrial-scale best available control technologies (BACT) for all major primary pollutants (also see Section 6.7.1).

4.11.2 Air Toxics; Toxic Air Pollutants

Title III of the amendments deal with air toxics. Toxic air pollutants (TAPs) or air toxics, and also known as hazardous air pollutants (HAPs) are airborne pollutants that are known to have adverse human health effects, such as cancer. Currently, there are about 189 chemicals identified on the CAA list of HAPs or TAPs (Table C.2) (USEPA 1998). Examples of air toxics include the heavy metals mercury and chromium, and organic chemicals such as benzene, hexane, perchloroethylene (perc), 1,3-butadiene, dioxins, and polycyclic aromatic hydrocarbons. The HAPs are regulated under the national emissions standards for hazardous air pollutants (NESHAP) program were identified as causal agents for a variety of diseases. For example, arsenic, asbestos, benzene, coke oven emissions, and radionuclides may cause cancer. The Title III amendment requires the EPA to

Table 4.7 U.S. emissions of conventional pollutants, 2006 and 2016 (million tons per year).

	CO		NO_x		VOCs		PM_{10}		SO_2	
Source category	**2006**	**2016**	**2006**	**2016**	**2006**	**2016**	**2006**	**2016**	**2006**	**2016**
Industrial processes										
Chemicals and petroleum	0.52	1.09	0.44	0.87	1.14	3.30	0.06	0.06	0.41	0.22
Metals processing	0.83	0.61	0.07	0.07	0.05	0.03	0.08	0.06	0.18	0.11
Other	0.58	0.62	0.48	0.34	5.72	3.98	1.10	0.85	0.33	0.18
Subtotal	1.93	2.32	0.85	1.28	6.91	7.31	1.24	0.97	0.92	0.51
Waste disposal	1.50	2.16	0.13	0.11	0.32	0.20	0.27	0.31	0.03	0.04
Stationary sources										
Fuel combustion										
Electric utilities	0.63	0.72	3.59	1.05	0.05	0.04	0.56	0.23	9.55	1.19
Industrial furnaces	1.07	0.96	1.68	1.19	0.12	0.11	0.30	0.33	1.51	0.66
Comm., res., other	3.14	2.45	0.68	0.53	0.53	0.38	0.44	0.37	0.49	0.13
Subtotal	4.84	4.13	5.95	2.77	0.70	0.53	1.30	0.93	11.55	1.98
Subtotal stationary sources	8.27	8.61	6.93	4.16	7.93	8.04	2.81	2.211	2.50	2.53
Mobile sources										
On-road vehicles	40.05	17.87	7.99	3.61	3.46	1.77	0.38	0.29	0.11	0.02
Non-road vehicles	20.37	12.99	3.90	2.42	2.75	1.58	0.28	1.6	0.40	0.03[a]
Subtotal mobile sources	60.42	30.86	11.89	6.03	6.21	3.35	0.66	0.45	0.51	0.05
Area sources and miscellaneous	17.11	19.62	0.27	0.29	3.76	4.49	17.94	21.80	0.13	0.15
Total all sources	85.80	59.09	19.09	10.48	17.90	15.88	21.41	24.46	13.14	2.73

Source: Compile from data from USEPA, Clearinghouse for Inventories and Emissions Factors, U.S. EPA http://www.epa.gov/ttn/chief/trends/index.html. Misc. CO is almost all from fires; Misc. PM-10 is almost all fugitive dust from paved and unpaved roads, construction sites, and agriculture sources.

[a] Non-road includes boats, planes, trains, construction equipment, lawn, and garden equipment, and others.

list source categories that emit one or more of the air toxics and to publish a schedule for regulation of those categories.

The CAA defined a major source of HAPs as a stationary source that has the potential to emit 10 T/Y of any one HAP on the list or 25 T/Y of any combination of HAPs. Examples include chemical complexes and oil refineries. The CAA prescribes a very high level of pollution control technology for HAPs called MACT (maximum achievable control technology). Small area sources, such as dry cleaners, emit lower HAP tonnages, but taken together they are a significant source of HAPs. Emission reductions can be achieved by changes in work practices such as material substitution and other pollution prevention strategies.

HAPs affect human health via the typical inhalation or ingestion routes. HAPs can accumulate in the tissue of fish, and the concentration of the contaminant increases up the food chain to humans. Many of these persistent and bioaccumulative chemicals are known or suspected carcinogens. Mercury has been especially targeted for regulation because it is released during the combustion of coal. Thus, it is one of the few HAPs that is widespread in the environment. Of particular concern are children who are exposed to methylmercury prenatally (USEPA 1997, 2004a).

Title IV addresses acid deposition, calling for a nationwide SO_2 emissions reduction of 10 million T/Y from 1980 levels, primarily from electric utilities. It promotes energy conservation and clean coal technologies and utilizes an innovative market-based system for the SO_2 reductions, in which utility allowances for SO_2 emissions can be bought and sold. In addition, USEPA is required to develop a revised NO_x new source performance standards (NSPS) for utility boilers.

Title V establishes a new federal *operating permit* program to be administered by the states. States had three years to implement the permit program, subject to EPA approval, and by 1998 all states had complied. Permit fees will be assessed to pay for the program. Permit programs must include at least all CAA requirements applicable to each source, a schedule for compliance, and requirements for monitoring and reporting.

4.12 Prevention of Significant Deterioration (PSD) Permitting Process

4.12.1 Introduction

The Prevention of Significant Deterioration (PSD) program was established on 7 August 1977, after the US Congress substantially amended the federal CAA and outlined a detailed PSD program. The federal PSD program is set forth in Part C of the federal CAA (Sections 160 through 169B). The PSD program was developed to prevent significant environmental impacts on "attainment areas" from large industrial sources of air pollution. Attainment areas are regions of the US where air quality meets standards established by EPA (also called NAAQS). Before the PSD program was developed, construction projects in large industrial facilities were being permitted with the primary goal of not worsening air pollution in problematic areas, and improving it if possible. Congress realized that this needed to be extended to ensure that those areas with clean, healthy air would stay that way, and that industrial activity would not significantly impact protected (e.g. Class I, wilderness) areas.

4.12.2 The PSD Program Goals

- Economic growth in harmony with preserving clean air resources.
- To protect public health from adverse effects of air pollution.
- To preserve, protect, and enhance air quality in national parks and wilderness areas.

The PSD program does not completely stop sources from increasing emissions. In fact, for a project to become subject to PSD review, there must be a significant increase in emissions after the use of control equipment. Instead, the PSD program is designed to (USEPA 1990)

- prevent significant deterioration of ambient air quality
- ensure compliance with all NAAQS
- protect public health and welfare from any adverse effect which might occur even at air pollution levels lower than the NAAQS
- preserve, protect, and enhance the air quality in classified Class I areas (which include national parks, national wilderness areas, national monuments, national seashores, and other areas of special national or regional natural, recreational, scenic, or historic value)
- ensure that economic growth will occur in a manner consistent with the preservation of existing clean air resources and to prevent the development of any new non-attainment areas
- assure that any decision to permit increased air pollution in any area to which PSD permitting requirements apply is made only after careful evaluation of all the consequences of such a decision, and after adequate procedural opportunities for informed public participation in the decision-making process

The PSD program preserves local air quality and protecting areas of special value by reviewing PSD applications to ensure the requirement to apply BACT, and they comply with the NAAQS, the applicable PSD increment concentrations, as well as evaluating visibility impacts, energy and environmental impacts, soils and vegetation impacts, and growth impacts. The program includes an opportunity for public participation in the decision-making process.

The PSD regulations apply to new "major stationary sources" and "major modifications." A stationary source includes all pollutant-emitting activities which belong to the same industrial grouping, are located on contiguous or adjacent properties, and are under common control. A "major modification" means a physical change or a change in the method of operation of a major stationary source which would result in a contemporaneous significant net increase in the emissions of any regulated pollutant. In determining if a proposed increase would cause a significant net increase to occur, several detailed calculations must be performed.

PSD review is only triggered if the prospective or existing location of the major source or modification has been formally designated by EPA as "attainment" or "unclassifiable" for any pollutant for which a NAAQS exists. Conversely, non-attainment new source review (NNSR) applies to new major sources or major modifications at existing sources for pollutants where the source is located in an area not in attainment with the NAAQS. NNSR requirements are customized for the non-attainment area. All non-attainment NSR programs have to require

- the installation of the lowest achievable emission rate (LAER)
- emission offsets
- opportunity for public involvement

The entire preconstruction permitting program, including both the PSD and NNSR permitting programs, is referred to as the New Source Review (NSR) program. PSD review and NNSR are conducted on a pollutant-by-pollutant basis. For example, a facility may emit many air pollutants; however, depending on the magnitude of the emissions of each pollutant, only one or a few may be subject to the PSD permit requirements. Since EPA has not established a NAAQS for greenhouse gases (GHGs), the NNSR program does not apply to GHGs.

Table 4.8 Coal-fired power plant title V/PSD permit emission limits and related monitoring and reporting requirements for boiler.

Parameters	Limits and averaging period (shall not exceed)[a]	Monitoring and reporting
NO_x	0.21 lb/MMBtu (30-boiler operating day rolling average, both units combined)	Primary compliance is determined through use of continuous emissions monitoring systems (CEMS) operated in accordance with 40 CFR 75 (Acid Rain Program) for NO_x and SO_2, and 40 CFR 63 Subpart UUUUU for Hg. In case of Hg, it may also utilize sorbent trap monitoring system. *Reference test methods*: NO_x: EPA Method 7E SO_2: EPA Method 6 or 6C Hg: EPA Method 30B Hourly data must be reported to EPA and delegated air quality agencies quarterly.
SO_2	10 000 T/Y	
SO_2	250 ppmvd at 7% O_2 (1-hour average)	
SO_2	1350 lb/h (both units combined, 1-hour average)	
Hg	1.2 lb/TBtu (30-boiler operating day rolling average)	
CO	200 ppm (both units averaged together, annual average)	Primary compliance is determined through use of continuous emissions monitoring systems (CEMS) operated in accordance with 40 CFR 60, Performance Specification 4. *Reference test method*: EPA Method 10 Summary results must be reported quarterly.
PM (filterable)	0.010 gr/dscf corrected to 7% O_2 (1-hour average) 0.030 lb/MMBtu (or 0.30 lb/MWh) (30-boiler operating day average)	EPA Method 5 is the primary reference test method. Testing, consisting of three 1-hour test runs, is required each operating quarter. Additional compliance assurance monitoring consists of continuously monitoring the opacity in the ductwork upstream of the wet scrubber to assure the opacity of emissions remains below levels correlated with compliance.

Source: Southwest Clean Air Agency, Vancouver, Washington.
Typical permit renewal in every five years. This table outlines only the quantitative limits. The entire permit is 128-pages long.
dscf, dry standard cubic foot; gr, grain; ppm, parts per million; ppmvd, parts per million volume dry; TBtu, trillion British thermal unit.
[a] All deviations from permit requirements must be reported no later than 30 days after the end of the month during which the deviation is discovered. Deviations which represent a potential threat to human health or safety must be reported as soon as possible but no later than 12 hours after the deviation is discovered.

Table 4.8 presents a 1400 MW coal-fired power plant Title V/PSD permit emission limits and related monitoring and reporting requirements for boiler. This is only a brief summary of relevant emission limits. There are many additional details and requirements. Also, there are many different aspects of coal-fired power plants and many factors that result in different plants being subject to different emissions limits – the limits in the Table 4.8 are somewhat unique to this facility (Lamoreaux, personal communication).

4.13 An Overall Permitting Process

Federal laws governing air pollution are designed to protect ambient air quality and public health. The USEPA, acting with authority of such laws, establishes federal regulations that limit the emissions allowed from various sources. The specific regulations for various industrial categories are documented in the Code of Federal Regulations (e.g. 40 CFR Part 60) and include emission limits, performance standards, permitting procedures, testing and monitoring requirements, record keeping, and reporting. Most states in the United States adopt the federal emission limits and standards by reference within their state laws. The states may create standards that are more stringent than the federal standards if they so desire, but may not make standards less stringent.

States have established their own air pollution permitting and compliance programs. Once a state plan is approved by the USEPA, the state is delegated the authority to ensure clean and healthy air within its borders. The permitting process is the vehicle through which the states track and enforce their regulations. The state regulations set certain emission limits for various pollutants, above which a source may not emit without first obtaining a permit. Any source that fails to obtain a proper permit is subject to a significant fine. One purpose of the state permit program is to allow the state to keep track of major point sources of air pollution. Permits provide a means to endure that sources use appropriate technology to control pollution, and they demonstrate ongoing compliance with emissions standards through routine testing, monitoring, record keeping, and reporting.

4.13.1 Who Needs a PSD Permit?

There are 28 specific industries or industrial processes that must meet the requirements of the PSD program if they emit, or have the potential to emit, at least 100T/Y of any pollutant regulated by the federal CAA. These industries and processes are listed in Table 4.9. For all other industries and industrial processes, the PSD program applies if they emit, or have the potential to emit, at least 250T/Y of any regulated pollutant. Regulated pollutants that most commonly lead to source-wide PSD applicability include PM, carbon monoxide (CO), sulfur dioxide (SO_2), VOCs, nitrogen oxides (NO_x), and GHGs. Other relatively common regulated pollutants include fluorides, sulfuric acid mist (H_2SO_4), hydrogen sulfide (H_2S), and total reduced sulfur compounds.

In the case of a modification to an existing major stationary source, PSD review is triggered if a physical change or change in the method of operation at the major stationary source results in both a significant emissions increase and a significant net emissions increase at the source. The PSD program establishes specific applicability criteria for determining if the modification causes both a significant emissions increase and a significant net emissions increase at the source. The PSD applicability criteria are discussed in detail in Sections 4.12, 4.13, and D.4.

Special case when a facility is less than the above levels before the project, but the project itself is over 100 or 250T/Y level. Example 1: an 80T/Y SO_2 facility does a 50T/Y SO_2 expansion would not be required to get a PSD permit at this point. Example 2: the same facility does a 120T/Y expansion would need to get a PSD permit.

Table 4.9 List of 28 source categories with a 100T/Y major source threshold.

1. Fossil fuel–fired steam electric plants of more than 250 MBtu/h heat input
2. Coal cleaning plants (w/thermal dryers)
3. Kraft pulp mills
4. Portland cement plants
5. Primary zinc smelters
6. Iron and steel mill plants
7. Primary aluminum ore reduction plants (w/thermal dryers)
8. Primary copper smelters
9. Municipal incinerators capable of charging more than 250T of refuse per day
10. Hydrofluoric acid plants
11. Sulfuric acid plants
12. Nitric acid plants
13. Petroleum refineries
14. Lime plants
15. Phosphate rock processing plants
16. Coke oven batteries
17. Sulfur recovery plants
18. Carbon black plants (furnace process)
19. Primary lead smelters
20. Fuel conversion plants
21. Sintering plants
22. Secondary metal production plants
23. Chemical process plants (which does not include ethanol production facilities that produce ethanol by natural fermentation included in NAICS codes 325193 or 312140)
24. Fossil-fuel boilers (or combinations thereof) totaling more than 250 million British thermal units per hour heat input
25. Petroleum storage and transfer units w/a total storage capacity exceeding 300 000 barrels
26. Taconite ore processing plants
27. Glass fiber processing plants
28. Charcoal production plants

4.13.2 What Does the PSD Program Require of the Applicant?

To obtain a PSD permit, an applicant must (USEPA 1990) do the following:

1) Fully identify the scope of the project and what equipment is new, modified, or increased utilization of upstream and downstream equipment.
2) *Apply BACT for new or modified equipment.* A BACT considers energy, environmental, and economic impacts in determining the maximum degree of reduction achievable for the proposed source or modification. The BACT determination cannot result in an emission limitation which would not meet any applicable standard of performance under 40 CFR Parts 60, 61, and 63. Equipment with increased utilization is not subject to BACT review. PSD BACT only applies to the pollutants that have triggered PSD. Other pollutants may be evaluated by the minor permitting agency.
3) *Conduct an ambient air quality analysis.* Each PSD new source or modification of an existing source must

perform an air quality analysis to demonstrate that its new pollutant emissions would not violate either the applicable NAAQS, or the applicable PSD increment. The equipment that has an increase in yearly utilization impacts will be part of the long-term air quality analysis.

The Air Quality Impact Analysis (AQIA) starts with preliminary modeling for each pollutant to determine whether an applicant can forego detailed analysis and preconstruction monitoring. If the projected ambient concentration increase for a given pollutant is below the PSD *Significant Impact Levels* (SILs) and *Significant Monitoring Concentration* (SMCs) for each averaging period, no further analysis of the ambient impact is required for that pollutant. For those pollutants with averaging periods that have impacts greater than the SIL, a full impact analysis (taking into account other increment consuming sources) is used to demonstrate compliance with PSD increments and NAAQS. The baseline years for modeling should be the last two years' emissions.

The *Federal Land Managers'* (FLMs) guidance on evaluating impacts of major projects on Class I areas is the FLMs' *Air Quality Values Workgroup* (FLAG) Phase I Report – revised (National Park Service 2010). The FLAG effort focused on how air pollutants – primarily PM, nitrogen dioxide, sulfur dioxide, nitrates, sulfates, and ozone – affect health and status resources in areas managed by National Park Service, US Fish and Wildlife Service, and US Department of Agriculture Forest Service. FLAG subgroups concentrate on four key areas: (i) visibility; (ii) aquatic and terrestrial effects of wet and dry pollutant deposition; (iii) terrestrial effects of ozone; and (iv) process and policy issues.

1) *Analyze impacts to soils, vegetation, and visibility.* An applicant is required to analyze whether its proposed emissions increases would impair visibility, impact soils, or vegetation. Not only must the applicant look at the direct effect of source emissions on these resources but also must consider the impacts from general commercial, residential, industrial, and other growth associated with the proposed source or modification.
2) *Not adversely impact a Class I area.* If the state agency receives a PSD permit application that could impact a Class I area, the state environmental regulatory agency will notify the *Federal Land Manager* (FLM) and federal official charged with direct responsibility for managing these lands. These officials are responsible for protecting the air quality–related values in Class I areas and for consulting with the state agency to determine whether any proposed project will adversely affect a Class I area. If the FLM demonstrates that emissions from a proposed source or modification would impair air quality–related values, even though the emissions levels would not cause a violation of the NAAQS or allowable air quality increment, the FLM may recommend that the reviewing authority deny the permit. The state environmental regulatory agency can still issue the permit if it determines all the requirements have been met.
3) *Undergo adequate public participation.* Specific public notice requirements including State Environmental Policy Act and a public comment period are required before state takes final action on a PSD application.

4.14 Best Available Control Technology

4.14.1 Introduction

BACT is defined as

> An emission limitation ... based on the maximum degree of reduction for each pollutant ... taking into account energy, environmental, and economic impacts and other costs ... through application of production processes or available methods, systems, and techniques...

This section reviews the important elements in this definition and describes the approach that must be used to satisfy BACT requirements for a proposed new source or modification of an existing source.

There is a requirement for NSR and other air quality permits that new or modified sources be built with BACT. Because BACT is required for new or modified sources, it provides a substantial incentive for developers of control technologies to show that their product provides the maximum degree of reduction of pollutant emissions available better than competing products or at a lower cost. If a new technology can be shown to reduce emissions more than currently available equipment, taking into account energy, environmental, and economic impacts and other costs, BACT requires that it will be used. This kind of requirement, which persistently forces improved technologies on sources of air pollutant emissions, is termed *technology forcing*.

BACT is intended to include end-of-pipe controls, such as scrubbers or baghouses; process changes, such as changing from an organic solvent to a water-based coating; and operations, such as inspection and maintenance of valves and flanges to reduce fugitive emissions. The intent is that there is no technology feasible to use that would result in lower emissions.

However, the definition also provides that the evaluation of BACT for a source will be on a case-by-case basis in which the characteristics and constraints of the source will be considered. The definition calls for evaluation of energy, environmental, and economic impacts and other costs that are specific to the source being considered. If these impacts are too great for a pollution control technology applied to a specific source, that technology is not considered BACT *for that case*. In another case, different conditions can lead to a different conclusion.

In the following sections, several additional definitions will be offered for terms that are similar to BACT but apply to special situations. Next, strategies for use in BACT determination for a proposed project are discussed. Then the "top-down" approach to determine BACT is described. This is the approach required by most permitting agencies. Finally, several examples are given for determining BACT.

4.14.2 Control Technology Requirement Definitions

In addition to BACT, there are several other technology requirement terms that may be encountered in air quality permitting. All of them are conceptually similar to BACT – they define emission limits applicable to different source types, different pollutants, and/or different attainment status. They are determined in similar fashions and, in some cases, they may be more stringent. Several of them are discussed in the following:

There Are Several Different Control Technology Requirements

- RACT – Reasonably available control technology
- BACT – Best available control technology
- MACT – Maximum achievable control technology
- LAER – Lowest achievable emission rate
- GACT – Generally available control technology

Others include T-BACT, BARCT, RACM, and BACM.

4.14.2.1 Reasonably Available Control Technology

RACT is the term that is applied when a State Implementation Plan (SIP) calls for reductions in emissions of existing sources in non-attainment areas in order to progress toward attainment. Typically, RACT is implemented as a prohibitionary rule applicable to non-attainment areas of a state or region. An example would be a limitation on NO_x emissions from a boiler larger than 50 000 Btu/h design capacity. An RACT rule would require that all boilers above this size have NO_x emissions less than 0.5 lb/MMBtu of heat input, and that existing boilers would have three years after the date of promulgation of the rule to comply. RACT requirements are never more stringent than BACT and are usually less stringent because they must be retrofitted to existing equipment. Since RACT applies to existing sources, it is not a requirement which must be addressed for new sources. However, it may become an issue when a new source is seeking offsets from an existing source being taken out of service.

4.14.2.2 Best Available Retrofit Control Technology

Best available retrofit control technology (BARCT) is a Californian term that refers to retrofit controls for existing sources in non-attainment with California ambient air quality standards. It is defined in the California CAA. It is difficult to distinguish between RACT and BARCT. Both apply to existing sources. Both are only applicable in non-attainment areas. BARCT is sometimes more stringent than RACT for a given source type. The California Air Resources Board periodically publishes technical memoranda identifying BARCT for particular source types. As with RACT, BARCT is implemented through prohibitionary rules within each of the 34 California air pollution control districts.

4.14.2.3 Maximum Achievable Control Technology

MACT is the control level for hazardous air pollutant emission sources required under the Clean Air Act Amendments of 1990 (CAAA). The definition in the Act is

> ... the maximum degree of reduction in emissions of hazardous air pollutants subject to [Section 112 of the Act] (including a prohibition on emissions, where achievable) that the Administrator, taking into consideration the cost of achieving such emission reductions, and any non-air quality health and environmental impacts and energy requirements, determines is achievable for new or existing sources in the category or subcategory to which such emission standard applies.

MACT regulations are for major sources of HAPs, because HAPs are formed as by-products of incomplete combustion source. A major source is defined as any facility that emits more than 10 T/Y of any HAP or more than 25 T/Y of all HAPs combined. Thus, all large industrial boilers, process furnaces, and stationary internal combustion engines are potential major sources. In addition to major sources like steel plants or petrochemical facilities, small area sources like dry cleaners, gas stations, and printers are also regulated in the United States. A MACT can include process changes, materials substitutions, or air pollution equipment. MACT is distanced in its definition from LAER, as it is not intended to ignore economic considerations.

4.14.2.4 LAER

LAER is the control technology that is required for non-attainment pollutant sources. If an area is non-attainment for ozone, LAER will be required for new or modified sources that emit the ozone precursors, nitrogen oxides (NO_x) and VOCs. LAER is defined as follows:

> For any source, that rate of emissions which reflects
> a) the most stringent emission limitation which is contained in the implementation plan of any state for such class or category of source, unless the owner or operator of the proposed source demonstrates that such limitations are not achievable, or
> b) the most stringent emission limitation which is achieved in practice by such class or category of source, whichever is more stringent.

The stringency of LAER is its absence of extenuating circumstances such as those contained in the definition of BACT. Although there is still a provision "unless the owner or operator of the proposed source demonstrates that such limitations are not achievable," it only applies to controls identified in implementation plans, and any control "achieved in practice" must be adopted. The consequence of this language is that economic feasibility is not included, and a proposed source must install the most stringent control that has been "achieved in practice." As applied, there has been some debate about whether "achieved in practice" counts if the "practice" is outside the United States (USEPA 1995c).

The LAER is used by the EPA to determine if emissions from a new or modified major stationary source are acceptable under SIP guidelines. LAER standards are required when a new stationary source is located in a non-attainment air quality region. It is the most stringent air pollution standard above the best available control technology and reasonably available control technology standards, regardless of the cost of that control technology.

4.14.2.5 Generally Available Control Technology

GACT, like MACT, is a term that addresses hazardous air pollutant control. GACT is the control technology standard that is applied to nonmajor hazardous air pollutant sources. Since these sources emit smaller amounts of hazardous air pollutants, GACT may be less stringent than MACT, although for some hazardous air pollutant source categories for which MACT standards have been promulgated, the GACT standard has been the same.

4.14.2.6 Toxic Best Available Control Technology

Toxic best available control technology (T-BACT) was coined by South Coast Air Quality Management District (SCAQMD) in California to identify technologies that were suitable for control of toxic air pollutants. The term was adopted by several other districts in California and applies to new sources of toxic air pollutants in those jurisdictions that have estimated health risk above a threshold value of 1 in 100 000 cases of cancer due to those emissions. In practice, T-BACT control technologies are the same as technologies for control of PM, and VOC in almost every circumstance, so this term has not become widespread. For a more complete discussion of toxic air pollutants and health risk, refer to SCAQMD (1995) and CAPCOA (1993).

4.14.2.7 Reasonably Available Control Measures and Best Available Control Measures

Reasonably available control measures (RACM) and best available control measures (BACM) are control measure guidelines developed by EPA for inclusion in SIPs. Although these terms can be applied to any pollutant, they have been most commonly applied to PM_{10} sources. The terms RACM and BACM address moderate and serious PM_{10} non-attainment areas respectively. RACM and BACM technical guidance was mandated under Section 190 of the CAAA of 1990 for three major PM_{10} source categories: urban fugitive dust, residential wood combustion, and prescribed silvicultural and agricultural burning. These measures were developed to use as prohibitionary rules, similar to those implementing RACT, aimed at reducing PM_{10} emissions with the goal of reaching attainment for those PM_{10} non-attainment areas. Measures identified as RACM and BACM will almost certainly be on the list of alternative technologies to be considered as possible BACT for sources of PM.

4.14.3 BACT Selection Strategy

Most jurisdictions require that BACT be used in all new or modified sources. The regulations seldom specify how to determine BACT beyond defining the term. In Section 4.14.4, the "top-down" approach developed by EPA for formal determination of BACT is described. The "top-down" process is thorough, sequential, and complete. However, it can be expensive to conduct and may result in the applicant being required to use a technology not desired. This dilemma can sometimes be avoided.

In most cases, both the applicant and the permitting agency agree on what is the most stringent technology for the source being permitted. If the applicant decides to use that control technology, a "top-down" BACT process is not needed, and accepting that most stringent technology early in the permitting process can greatly speed up other approval steps.

In other cases, the number of technologies reviewed in the "top-down" process can be limited by prior agreement with the permitting agency. The number may be limited by citing BACT analyses that have been done for similar processes in similar settings and having the agency accept these analyses as applicable.

In a BACT protocol, the project description can be summarized, assumptions can be stated, and technologies that will be considered as part of the BACT analysis identified. This document can clearly set the boundaries of the BACT analysis and avoid surprises later in the process. Technical feasibility is usually used as the basis for including technologies, and infeasible technologies can be excluded prior to the "official" start of the analysis. After a BACT protocol has been submitted, a telephone call is often adequate to agree on the boundaries of the analysis.

There are many instances when only a few (often as few as two) technologies are considered in the BACT analysis, and in those situations the permitting agency may accept a much less formal analysis than the "top-down" approach. Brevity usually makes the analysis easier to understand and allows the opportunity to make important points clearly without unnecessary boilerplate.

4.14.4 Top-Down BACT Analysis

In 1987, EPA issued a memorandum that described a "top-down" approach to determining what constitutes BACT for a particular source. The memorandum was to be followed up with regulatory language in 1996. The fundamental guide of the "top-down" approach is that the examination of what control technology constitutes BACT will start with the most stringent technology – that is, the one with the greatest emission reduction. The applicant must demonstrate that this most stringent technology will not work in order to reject it. The basis for it not working can be technical infeasibility, or because of energy, environmental, economic, or other cost impacts that are unacceptable. The assumption starting the analysis is that the most stringent technology will work, and should be used, unless proven otherwise. The process of disproving feasibility of successively less stringent technologies continues until a control technology is found that will work, and that technology becomes BACT for that particular source. In the following subsections, the steps required in a "top-down" BACT analysis are described.

4.14.5 Identify Technologies

The first step in the "top-down" process is to identify the control technologies that will be considered in the analysis. Historically, intense arguments have arisen over this question, mostly related to whether a technology proposed by a regulatory agency for inclusion on the "top-down" list was available. When addressing this issue, the BACT provision to include "any other emission control technique found by the California Air Pollution Control Officers Organization to be technologically feasible ...." can give rise to suggestions to consider many different technologies that may not yet be commercially available (CAPCOA 1993).

The availability criteria used by SCAQMD contains most of the elements used elsewhere. The SCAQMD list includes the following:

1) The technology is commercially available.
2) The technology is reliable. There has been at least 12 months of demonstrated performance in comparable field conditions.
3) The technology is effective as demonstrated by a source test or other performance documentation.
4) The technology has been achieved in practice or is technologically feasible and cost effective.

Lists have been compiled of control technologies for different equipment or processes. Descriptions of control technologies are also available. These references, which contain very similar information, are adequate to provide a list of alternatives that must be considered in the "top-down" analysis for most jurisdictions, and they are updated periodically. Some of the lists include technologies considered LAER as the most stringent technology listed. The logic for including LAER is that this is the most stringent technology, and the applicant must disprove its use in the "top-down" process.

Process changes that result in lower emissions can also be considered BACT. In some instances, process changes such as changing from a solvent to a water-based coating can result in much lower emissions and fewer hazardous waste products.

There is opportunity for a great deal of judgment in the selection of technologies for the "top-down" list, and that judgment is exercised differently by agencies in different states or regions. Including technologies on the initial "top-down" list that have little or no likelihood of being technically feasible for a particular source increases the amount of work that must be done to exclude that technology from the process.

The term available with respect to control technologies refers to commercial availability. Can you purchase this technology and has it been demonstrated in a commercial setting? The commercial setting means that it must have been part of a commercial process, and not a research effort, pilot test, or demonstration.

There is sometimes disagreement about whether a technology that has been used commercially in a foreign country is considered "available" in the United States. Often there is not good information available about foreign "commercial technologies" to determine whether the use was a commercial setting and not subsidized by the technology vendor or an agreement that could reimburse the user if the technology were adopted as BACT in the United States. Usually, if there are several "apparently commercial" applications in other countries, the assumption of commercial availability is made for the purposes of including the technology in a BACT list. However, of the references cited above for lists of BACT technologies, foreign uses have not been included.

4.14.6 Determine Technical Feasibility

Each of the technologies identified must be technically feasible for the source being permitted. The technology must be used in the particular circumstances of the source being permitted. For example, a technology to reduce NO_x from combustion turbines is to use water injection into the combustion chambers. However, for a gas pipeline compressor station in a remote, arid area where water is not available, water injection is not considered technically feasible. Another example would be the use of floating roof organic liquid storage tanks. This technology is considered technically feasible in most areas, but in extremely cold, icy environments such as on the North Slope, the sliding seal on floating roof tanks tends to freeze in place, preventing the roof from moving. Because of this, floating roof tanks are considered technically infeasible in Arctic locations.

In some instances, prior agreement can be negotiated with the agency to exclude technologies that have been shown to be infeasible in recent similar analyses. Such agreements should always be confirmed in writing to assure that a notice of deficiency is not issued for the subsequent application.

There can be some uncertainty about whether a technology is technically infeasible for a particular application, or whether it would just be considered too costly. If the issue were cost, it would need to be addressed in the cost evaluation later in the "top-down" process.

For most of the arguments of technical infeasibility, an expensive approach can be imagined to make it feasible. For example, for the NO_x control by water injection case, drilling for water could be done, and at some depth water could probably be found. The cost of obtaining water would be extremely high but possible! The strategy to use in the BACT analysis should be to propose a limitation as to technical infeasibility. Then the additional effort needed to estimate costs is not necessary – the technology is off the list. If the technical infeasibility argument is rejected, additional analysis will be needed to demonstrate either that the technology is too expensive or that the technology could be defined as BACT. Note that in applications where LAER is required, the cost argument is not usable, so proposing technical infeasibility may be the only choice.

4.14.7 Rank Technically Feasible Alternatives

Once a "short list" of technically feasible control alternatives has been developed, the alternatives are ranked from the most to least effective in reducing emissions from the proposed source. Often it provides clarification to express control level in more than one way, such as percent control and concentration of pollutants in an exhaust stream. The most useful common measure of control effectiveness is the total emissions per year from the source, using the control technology identified. Technologies are listed "top-down" starting with the most effective technology that is technically feasible.

This short list should be presented to the permitting agency and agreed upon before continuing the BACT analysis. Once the list of technologies has been agreed to between the agency and the applicant, the remaining steps in the "top-down" process take place one technology at a time, starting with the "top" or most effective technology. The intent of the next steps is to determine whether there are reasons for rejecting the application of this "top" technology to the process or equipment being permitted.

4.14.8 Evaluate Impacts of Technology

The next step in the "top-down" process is to determine whether there are energy, environmental, or cost impacts from the use of this top technology that would render it infeasible.

4.14.8.1 Energy Impacts

Energy impacts include the energy consumed by the control technology or a reduction in energy produced, if an energy-production process is permitted. One example of an energy impact would be the need to add heat to a low-temperature flue gas to increase its temperature to the place where a selective catalytic reduction system that operates at a minimum temperature of about 600 °F (315 °C) could be used for NO_x control. Quantitative thresholds for energy impacts have not been published, so an energy impact rejection of a top technology is discretionary. Many years ago, an analysis was done of possible technologies for reducing CO_2 emissions from an electric utility power plant. At that time, the only process for reducing CO_2 was a high-pressure amine scrubber. The analysis of this alternative indicated that in order to scrub the CO_2 from the power plant exhaust, more energy would be required than was produced by the power plant. Such a technology could be eliminated based on energy impacts.

4.14.8.2 Environmental Impacts

If use of a proposed control technology results in other environmental impacts, the technology may be rejected because of other environmental impacts. For example, the use of the technology might result in the production of hazardous wastes, the creation of other air pollutants than those being controlled with the technology, excess water use in an area with inadequate water supply, generation of large quantities of solid wastes or solid wastes that are difficult to dispose of, or even destruction of critical habitat if the footprint of a control technology installation extended beyond currently disturbed areas.

4.14.8.3 Cost Effectiveness

The cost effectiveness of a control technology is measured in dollars per ton of pollutant removed. It is determined by first calculating the rate of pollutant removal (tons/year). This rate is the difference between the emissions before the control is installed and the rate after control. Then the annual cost of the technology (US$/year) is determined. Cost effectiveness is the ratio of the annual cost of the technology to the rate of pollutant removal (dollars/ton). The control efficiency and rate of pollution removal are discussed in Sections 3.7.6.7 and 7.11.

4.14.9 Plant-Wide Applicability Limitation

Plant-wide applicability limitation (PAL) is a voluntary limit based on plant-wide actual emissions, on a pollutant-specific basis (also called an actuals PAL). A PAL is an emission limitation expressed in T/Y on a 12-month rolling basis for a pollutant at a major stationary source. The PAL is enforceable as a practical matter and established source-wide in accordance with 40 CFR §52.21(aa)(1) through (15) (40 CFR).

A facility that has a PAL in place is allowed to make changes to the facility or individual emission units which result in increases in the source's or individual units' pollutant emissions, as long as its plant-wide actual emissions do not exceed its PAL limit. In return for this flexibility, the facility must monitor emissions from all emissions units under the PAL, and comply with record keeping, monitoring, and reporting requirements. The PAL must be included in a permit issued by the regulating state. Major NSR applicability provisions continue to apply to air pollutants which have no PAL. The PAL process has been rarely used in the United States of America.

4.15 Atmospheric Dispersion Modeling

Atmospheric dispersion modeling is the mathematical simulation of how air pollutants disperse in the ambient atmosphere. It is performed with computer programs that include algorithms to solve the mathematical equations that govern the pollutant dispersion. The dispersion models are used to estimate the downwind ambient concentration of air pollutants or toxins emitted from sources such as industrial plants, vehicular traffic, or accidental chemical releases. They can also be used to predict future concentrations under specific scenarios (i.e. changes in emission sources). Therefore, they are the dominant type of model used in air quality policy-making. They are most useful for pollutants that are dispersed over large distances and that may react in the atmosphere. For pollutants that have a very high spatiotemporal variability (i.e. have very steep distance to source decay such as black carbon) and for epidemiological studies, statistical land-use regression models are also used.

Dispersion models are important to governmental agencies tasked with protecting and managing the ambient air quality. The models are typically employed to determine whether existing or proposed new industrial facilities are or will be in compliance with the NAAQS in the United States and other nations. The models also serve to assist in the design of effective control strategies to reduce emissions of harmful air pollutants. During the late 1960s, the Air Pollution Control Office of the USEPA initiated research projects that would lead to the development of models for the use by urban and transportation planners (Fensterstock et al. 1971).

Air dispersion models are also used by public safety responders and emergency management personnel for emergency planning of accidental chemical releases. Models are used to determine the consequences of accidental releases of hazardous or toxic materials. Accidental releases may result from fires, spills, or explosions that involve hazardous materials, such as chemicals or radionuclides. The results of dispersion modeling, using worst-case accidental release source terms and meteorological conditions, can provide an estimate of location impacted areas, ambient concentrations, and be used to determine protective actions appropriate in the event a release occurs. Appropriate protective actions may include evacuation or shelter in place for persons in the downwind direction.

At industrial facilities, this type of consequence assessment or emergency planning is required under the CAA codified in Part 68 of Title 40 of the Code of Federal Regulations.

The dispersion models vary depending on the mathematics used to develop the model, but all require the input of data that may include the following:

- Meteorological conditions such as wind speed and direction, the amount of atmospheric turbulence (as characterized by what is called the "stability class"), the ambient air temperature, the height to the bottom of any inversion aloft that may be present, cloud cover, and solar radiation
- Source term (the concentration or quantity of toxins in emission or accidental release source terms) and temperature of the material
- Emissions or release parameters such as source location and height, type of source (i.e. fire, pool, or vent stack) and exit velocity, exit temperature, and mass flow rate or release rate
- Terrain elevations at the source location and at the receptor location(s), such as nearby homes, schools, businesses, and hospitals
- The location, height, and width of any obstructions (such as buildings or other structures) in the path of the emitted gaseous plume, surface roughness, or the use of a more generic parameter "rural" or "city" terrain

Many of the modern, advanced dispersion modeling programs include a preprocessor module for the input of meteorological and other data, and many also include a post-processor module for graphing the output data and/or plotting the area impacted by the air pollutants on maps. The plots of areas impacted may also include isopleths showing areas of minimal to high concentrations that define areas of the highest health risk. The isopleths plots are useful in determining protective actions for the public and responders.

The atmospheric dispersion models are also known as atmospheric diffusion models, air dispersion models, air quality models, and air pollution dispersion models.

4.15.1 Atmospheric Layers

Discussion of the layers in the Earth's atmosphere is needed to understand where airborne pollutants disperse in the atmosphere. The layer closest to the Earth's surface is known as the *troposphere*. It extends from sea-level to a height of about 18 km and contains about 80% of the mass of the overall atmosphere. The *stratosphere* is the next layer and extends from 18 km to about 50 km. The third layer is the *mesosphere* which extends from 50 km to about 80 km. There are other layers above 80 km, but they are insignificant with respect to atmospheric dispersion modeling. On a mesoscale, Figure 2.2 shows the four regions of the atmosphere, as temperature varies with altitude, so does density.

The lowest part of the troposphere is called the *atmospheric boundary layer (ABL)* or the *planetary boundary layer* and extends from the Earth's surface to about 1.5–2.0 km in height. The air temperature of the atmospheric boundary layer decreases with increasing altitude until it reaches what is called the *inversion layer* (where the temperature increases with increasing altitude) that caps the atmospheric boundary layer. The upper part of the troposphere (i.e. above the inversion layer) is called the *free troposphere* and it extends up to the 18 km height of the troposphere.

The ABL is of the most important with respect to the emission, transport, and dispersion of airborne pollutants. The part of the ABL between the Earth's surface and the bottom of the inversion layer is known as the mixing layer. Almost all of the airborne pollutants emitted into the ambient atmosphere are transported and dispersed within the mixing layer. Some of the emissions penetrate the inversion layer and enter the free troposphere above the ABL.

In summary, the layers of the Earth's atmosphere from the surface of the ground upwards are the ABL made up of the mixing layer capped by the inversion layer, the free troposphere, the stratosphere, the mesosphere, and others. Many atmospheric dispersion models are referred to as *boundary layer models* because they mainly model air pollutant dispersion within the ABL. To avoid confusion, models referred to as *mesoscale models* have dispersion modeling capabilities that extend horizontally up to a few hundred kilometers. It does not mean that they model dispersion in the mesosphere.

Example 4.11 A 500-MW coal-fired power plant is burning coal with a heating value of 12 000 Btu/lb. The coal contains 2.0% sulfur and 150 ppb of mercury. The plant has a thermal efficiency of 38% and uses an air/fuel ratio of 14 : 1 (mass basis). Calculate (a) the coal firing rate (lb/h), (b) the flue-gas flow rate (scfm), (c) the SO_2 removal efficiency needed to meet a standard of 1.4 lb/MWh gross electrical energy output, and (d) the overall mercury capture percentage that is required to meet a mercury standard of 0.015 lb/GWh of gross energy output.

Given: Heating value = 12 000 Btu/lb

Coal content: 2.0% S, 150 ppb Hg

$\eta_{thermal} = 38\%$

A/F = 14 : 1 (mass basis)

Solution

a) Coal-firing rate

$$\text{Energy input}: E_0 = \frac{500\,\text{MW}}{0.38} = 1316\,\text{MW}$$

$$E_0 = \left(\frac{3.412\times10^6\,\frac{\text{Btu}}{\text{h}}}{\text{MW}}\right)(1316\,\text{MW}) = 4.49\times10^9\,\frac{\text{Btu}}{\text{h}}$$

$$\text{Coal input} = \frac{4.49\times10^9\,\frac{\text{Btu}}{\text{h}}}{12000\,\frac{\text{Btu}}{\text{lb}}} = 374123\,\frac{\text{lb}}{\text{h}}$$

b) Flue-gas flow rate:

A/F = 14 : 1, so flue-gas mass flow is 15 times coal feed rate

$$Q_{FG} = \left(374\,123\,\frac{\text{lb}}{\text{h}}\right)\times(14+1) = 5\,611\,845\,\frac{\text{lb}}{\text{h}}$$

$$Q_{FG} = \left(5\,611\,845\,\frac{\text{lb}}{\text{h}}\right)\times\left(\frac{1\,\text{ft}^3}{0.075\,\text{lb}}\right)\times\left(\frac{1\,\text{h}}{60\,\text{min}}\right)$$

$$= 1\,247\,000\ \text{scfm}$$

c) SO_2 removal efficiency

$$\text{Sulfur in the fuel}: \left(374\,123\,\frac{\text{lb}}{\text{h}}\right)\times0.02 = 7482\,\text{lb}\,\frac{\text{S}}{\text{h}}$$

$$S + O_2 \rightarrow SO_2$$

$$\left(\frac{7\,482\,\text{lb S}}{\text{h}}\right)\times\left(\frac{\text{lbmol S}}{32\,\text{lb}}\right)\times\left(\frac{1\,\text{lbmol SO}_2}{1\,\text{lbmol S}}\right)\times\left(\frac{64\,\text{lb}}{1\,\text{lbmol SO}_2}\right)$$

$$= \frac{14\,965\ \text{lb SO}_2}{\text{h}}$$

$$\frac{\frac{149\,65\,\text{lb SO}_2}{\text{h}}}{500\,\text{MW}} = \frac{29.9\,\text{lb}\ \frac{\text{SO}_2}{}}{\text{MWh}}$$

$$\frac{\left[\frac{(29.9-1.4)\,\text{lb SO}_2}{\text{MWh}}\right]}{\left(\frac{29.9\,\text{lb SO}_2}{\text{MWh}}\right)}\times100$$

$= 95.3\%$ sulfur removal efficiency

d) Overall mercury capture percentage

$$m_{Hg} = \frac{\left(\frac{374\,123\,\text{lb coal}}{\text{h}}\right)\times(150\,\text{ppb Hg})}{10^9} = \frac{0.0561\ \text{lb Hg}}{\text{h}}$$

Gross energy output:

$$\left(\frac{500\,\text{MW}}{\text{h}}\right)\times\left(\frac{1\,\text{GWh}}{1000\,\text{MWh}}\right) = \frac{0.50\,\text{GWh}}{\text{h}}$$

Uncontrolled mercury emissions:

$$= \frac{\frac{0.0561\,\text{lb Hg}}{\text{h}}}{\frac{0.500\,\text{GWh}}{\text{h}}} = \frac{0.1122\,\text{lb}}{\text{GWh}}$$

Mercury capture percentage:

$$\frac{\left(\frac{0.1122\,\text{lb}}{\text{GWh}} - \frac{0.015\,\text{lb}}{\text{GWh}}\right)}{\left(\frac{0.1122\,\text{lb}}{\text{GWh}}\right)}\times100 = 86.66\%$$

4.16 Dispersion Models: Indoor Concentrations

Diffusion of contaminants in workplace air results in the net movement of the contaminants from regions of higher concentration to regions of lower concentration. The spread of the contaminant is aided by the advective mass transfer driven by the ventilation system. The combination of these influences results in movement of contaminants away from their source into the surrounding room.

The mass-balance model presumes a uniform concentration of the chemical contaminant in the work area. Dispersion models have a notable advantage; they describe the variation of contaminant concentration with distance from the source. The concentration gradient is described by the following equation when advection occurs in the x-direction only and dispersion occurs equally in all directions:

$$u\frac{\text{d}C}{\text{d}x} = \frac{D}{r^2}\frac{\text{d}}{\text{d}r}\left[\frac{r\text{d}c}{\text{d}r}\right] \tag{4.14}$$

where

u is the air velocity in the x-direction (length/time)
C is the concentration of airborne contaminant (mass/length^3)
D is the diffusion coefficient (length^2/time)
x is the distance downwind from the source (length)
r is the distance from the source to the sampling point (length)

This equation has been solved for concentrations resulting from emissions into an infinite space:

$$C = \frac{G}{4\pi Dr} \exp\left[\left(-u/2D\right)\left(r-x\right)\right] \quad (4.15)$$

where G is the contaminant emission rate from the source (mass/time).

The diffusion coefficient (D) can be derived from measurements at the sampling site or estimated from values available in the literature. Measurements of the diffusion coefficient in indoor industrial environments have ranged from 0.05 to 11.5 m^2/min, with 0.2 m^2/min being a typical value.

Example 4.12 Freon is emitted from an open-top vapor degreaser at a rate of 0.74 g/min. Estimate the concentration in the air inhaled by a worker 3 m downwind from the degreaser if the air velocity is 0.79 m/min.

Solution

Since the worker is downwind of the degreaser, $x = r$ in the diffusion convection equation (Eq. 4.15) and

$$C = \frac{0.74\ \text{g/min}}{(4)(3.1416)(0.2\ \text{m}^2/\text{min})(3\ \text{m})} = 0.1\ \text{g/m}^3$$

Molecular diffusion theory strictly applies to vapors and gases; however, PM with aerodynamic diameters less than 10 μm are distributed in workplace air in a similar manner. Using the dispersion models to describe the distribution of dusts and fumes is reasonable for small particles.

4.16.1 Gaussian Dispersion Model

The Gaussian dispersion model is most often used to characterize the dilution of toxic air pollutants with distance from the source. The model provides reasonable agreement with the experimental data and is, in its simplest form, calculations with. The mean concentration, C, resulting from emission at a continuous point source of strength Q at a height H above the totally reflecting Earth along the plume centerline is given by

$$C = Q\left(\pi\sigma_x\sigma_y U\right)^{-1} \times \exp\left(-0.5H^2\sigma_z^{-2}\right) \quad (4.16)$$

where

C is the concentration of toxic air pollutant (μg/m^3)
Q is the source release rate (μg/s)
U is the mean wind speed at the stack height (m/s)
H is the effective height of release above the Earth (m)
y is the distance in a direction transverse to the wind (m)
z is the height at which the observation is made (m)
σ_y and σ_z are the standard deviations of the concentrations of plume transverse to the wind and perpendicular to the Earth, respectively (m).

Published values of σ_y and σ_z are based on laboratory and field measurements of velocity fluctuations under a variety of atmospheric conditions. Atmospheric stability is used to represent the amount of mixing in the atmosphere and is generally classified as stable, neutral, or unstable. A stable atmosphere is characterized by temperatures that increase with distance from the surface of the Earth and reduced vertical mixing; nighttime atmospheric conditions are generally represented as stable. More vigorous atmospheric mixing is expected as the Sun warms the surface of the Earth and the warmer, less dense air accumulates near the Earth's surface; eventually, gravity will displace the warm air with cooler air from above. Daytime atmospheric conditions are typically represented as either neutral or unstable.

The product of the standard deviations has been represented by an equation of the form

$$\sigma_y\sigma_z = ax^b \quad (4.17)$$

where

a and b are constants (nondimensional)
x is the distance downwind from the source (length)

Kumar (1998, 1999) performed regression analysis to develop expressions for the constants, a and b, for urban and rural settings and neutral and stable atmospheric conditions. Urban settings are appropriate when there are many obstacles in the immediate area of the release; obstacles include buildings and trees. Rural settings are appropriate when there are no buildings in the immediate area of the release and the terrain is generally flat and unobstructed. Table 4.10 lists the results of the regression when the distance from the source is in meters.

Example 4.13 Hydrogen sulfide is released from a low-level vent in a rural area at a rate of 0.025 kg/s. Calculate the concentration at the plant boundary located 300 m downwind from the vent during daytime conditions when the wind speed is 4 m/s and at night when the wind speed is 2.5 m/s. (a) If the concentration of concern for hydrogen sulfide is 42 mg/m^3, will this concentration be exceeded at the plant boundary? (b) Estimate the exposure to hydrogen sulfide of a person living 300 m downwind from the facility described during (i) daytime and (ii) nighttime if the individual at rest breathes 0.9 m^3/h of the ambient air.

Solution

a) Equation (4.16) can be used to calculate the concentration at ground-level by setting the height above the Earth equal to zero. The appropriate correlations for the dispersion coefficients are given by items (a) and (c) in Table 4.10 for daytime and nighttime conditions, respectively. For daytime conditions,

$$C = \frac{0.025\ \text{kg/s}}{(\pi)(0.01082)(300\ \text{m})^{1.78}(4\ \text{m/s})}$$
$$= 7.17\times10^{-6}\ \text{kg/m}^3 = 7.17\ \text{mg/m}^3$$

For nighttime conditions,

$$C = \frac{0.025\ \text{kg/s}}{(\pi)(0.0049)(300\ \text{m})^{1.66}(2.5\ \text{m/s})}$$
$$= 5.02\times10^{-5}\ \text{kg/m}^3 = 50.2\ \text{mg/m}^3$$

The concentration of hydrogen sulfide is below the concentration of concern under daytime conditions but exceeds the concentration of concern at night when atmospheric mixing is less.

b) For daytime condition, $(7.17\ \text{mg/m}^3)(0.9\ \text{m}^3\text{/h}) = 6.45\ \text{mg/h}$

For nighttime conditions, $(50.2\ \text{mg/m}^3)(0.9\ \text{m}^3\text{/h}) = 45.2\ \text{mg/h}$

4.16.2 Modeling Protocol

The development of a modeling protocol is the first step in a dispersion modeling analysis performed as part of a regulatory permitting analysis. The modeling protocol describes how the modeling will be carried out. It is usually necessary to submit a modeling protocol to the regulatory agency in order to receive agency approval for the modeling prior to performance of the modeling. The modeling protocol contains the following elements:

- Dispersion model selection
- Modeling methodology and assumptions
- Source input selection and preparation
- Meteorological data selection and processing
- Basis for ambient pollutant concentrations
- Receptor grid development

Table 4.10 Regression equations for dispersion coefficients.

a. Rural release, neutral atmosphere, $x < 500$ m:	$\sigma_y\sigma_z = 0.01082\,x^{1.78}$
b. Rural release, neutral atmosphere, $x > 500$ m:	$\sigma_y\sigma_z = 0.04487\,x^{1.56}$
c. Rural release, stable atmosphere, $x < 2000$ m:	$\sigma_y\sigma_z = 0.0049\,x^{1.66}$
d. Rural release, stable atmosphere, $x > 2000$ m:	$\sigma_y\sigma_z = 0.01901\,x^{1.46}$
e. Urban release, neutral atmosphere, $x < 500$ m:	$\sigma_y\sigma_z = 0.0224\,x^{2}$
f. Urban release, neutral atmosphere, $x > 500$ m:	$\sigma_y\sigma_z = 0.394\,x^{1.54}$
g. Urban release, stable atmosphere, $x < 500$ m:	$\sigma_y\sigma_z = 0.008\,x^{2}$
h. Urban release, stable atmosphere, $x > 500$ m:	$\sigma_y\sigma_z = 0.34\,x^{1.37}$

4.16.3 Dispersion Model Selection

AQIAs are carried out using Gaussian dispersion models – models that use the basic assumption of Gaussian dispersion. There are several such models, and different models are applicable to different source types and terrain.

The primary reference for selection of models is EPA's *Guideline on Air Quality Models.* It identifies the models that have been developed and validated by EPA and are considered reliable for use in permitting. In particular, there are two models included in the Guideline that are most commonly used. They are

- AERSCREEN
- AERMOD

4.16.3.1 AERSCREEN

Rather than writing a separate screening model that would require updating whenever the refined model is updated, AERSCREEN is an interactive wrapper for AERMOD and associated auxiliary programs where the user supplies responses to program prompts. Those responses are used to write an input file for AERMOD and other required programs specific to the problem at hand.

AERSCREEN is the recommended screening model based on AERMOD. The model will produce estimates of "worst-case" one-hour concentrations for a single source, without the need for hourly meteorological data and also includes conversion factors to estimate "worst-case" 3-hour, 8-hour, 24-hour, and annual concentrations. AERSCREEN is intended to produce concentration estimates that are equal to or greater than the estimates produced by AERMOD with a fully developed set of meteorological and terrain data, but the degree of conservatism will vary depending on the application.

Before running AERSCREEN, users should consult and become familiar with the most recent versions of the following:

- AERMOD User's Guide (USEPA 2016a)
- AERMAP User's Guide (USEPA 2018b)

- AERMET User's Guide (USEPA 2019b)
- AERSCREEN User's Guide (USEPA 2016b)
- AERSURFACE User's Guide (USEPA 2008)
- Screening Procedures for Estimating the Air Quality Impact of Stationary Sources (USEPA 1992)

4.16.3.2 AERMOD

The AERMOD atmospheric dispersion modeling system is an integrated system that includes three modules (Prater and Midgley 2008; Turner 1994; USEPA 2004b):

1) A steady-state dispersion model designed for short-range (up to 50 km) dispersion of air pollutant emissions from stationary industrial sources.
2) A meteorological data preprocessor (AERMET) that accepts surface meteorological data, upper air soundings, and optionally data from on-site instrument towers. It then calculates atmospheric parameters needed by the dispersion model, such as atmospheric turbulence characteristics, mixing heights, friction velocity, Monin–Obukhov length, and surface heat flux.
3) A terrain preprocessor (AERMAP) whose main purpose is to provide a physical relationship between terrain features and the behavior of air pollution plumes. It generates location and height data for each receptor location. It also provides information that allows the dispersion model to simulate the effects of air flowing over hills or splitting to flow around hills.

AERMOD also includes PRIME (Plume Rise Model Enhancements) (Turner 1994; USEPA 2004b) which is an algorithm for modeling the effects of downwash created by the pollution plume flowing over nearby buildings.

4.16.4 CALPUFF

CALPUFF is an advanced, integrated Lagrangian puff modeling system for the simulation of atmospheric pollution dispersion (Code of Federal Register, 40 CFD). The model has been adopted by the USEPA in its *Guideline on Air Quality Models* as a preferred model for assessing long-range transport of pollutants and their impacts on Federal Class I areas and on a case-by-case basis for certain near-field applications involving complex meteorological conditions (Scire et al. 1990; Turner 1994).

The integrated modeling system consists of three main components and a set of preprocessing and post-processing programs. The main components of the modeling system are CALMET (a diagnostic three-dimensional meteorological model), CALPUFF (an air quality dispersion model), and CALPOST (a post-processing package). Each of these programs has a graphical user interface. In addition to these components, there are numerous other processors that may be used to prepare geophysical (land use and terrain) data in many standard formats, meteorological data (surface, upper air, precipitation, and buoy data), and interfaces to other models such as the Penn State/NCAR Mesoscale Model (MM5), the National Centers for Environmental Prediction Eta model, and the RAMS meteorological model.

The CALPUFF model is designed to simulate the dispersion of buoyant, puff, or continuous point and area pollution sources as well as the dispersion of buoyant, continuous line sources. The model also includes algorithms for handling the effect of downwash by nearby buildings in the path of the pollution plumes (Turner 1994).

More insights are given in Section D.5 on dispersion modeling and understanding how to predict the time-averaged concentration of contaminants emitted into the atmosphere from a single elevated source or complicated industrial complexes with multiple emission points.

4.16.5 Attainment and Non-Attainment Areas

States are divided into air quality control regions (AQRs). An AQR that has air quality equal to or better than the primary standard listed in NAAQS is called an attainment area. And those areas that do not meet the primary standard are called non-attainment areas. Figure 4.14 shows all non-attainment counties in the United States, including Alaska and Guam for six NAAQS pollutants.

4.17 State Implementation Plan

The CAA requires states to develop individualized SIP that outlines how they plan to achieve the NAAQS. The SIP is a collection of regulations and documents used by a state, territory, or local air district to reduce air pollution in areas that do not meet the NAAQS.

4.17.1 What National Standards must SIPs Meet?

EPA has established the NAAQS for six "criteria" air pollutants – which are widespread common pollutants known to be harmful to human health:

1) Carbon monoxide pollution
2) Lead air pollution
3) Nitrogen oxide pollution
4) Ozone pollution
5) PM pollution
6) Sulfur dioxide pollution

The CAA identifies specific requirements that states must submit to EPA as part of their SIP to attain and maintain the NAAQS. SIPs provide a plan for implementation, maintenance, and enforcement of the NAAQS in each state.

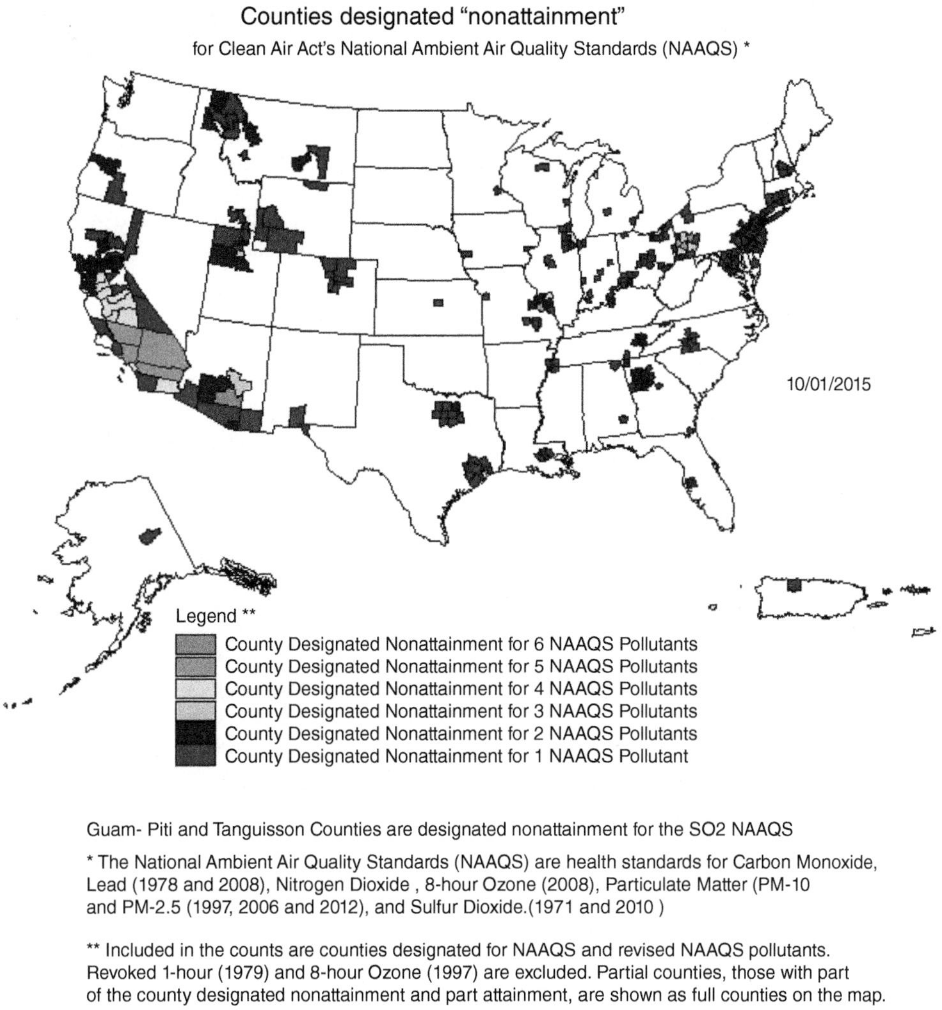

Figure 4.14 Non-attainment counties in the United States for NAAQS pollutants.

4.17.2 What Is Included in a SIP?

The contents of a typical SIP fall into three categories:

1) State-adopted control measures which consist of either regulations or source-specific requirements (such as orders and consent decrees)
2) State-submitted "non-regulatory" components (see list of examples below)
3) Additional requirements promulgated by EPA to satisfy a mandatory requirement in Section 110 or Part D of the CAA

Examples of EPA-approved documents and materials associated with the SIP include, but are not limited to, the following:

- SIP narratives
- PM plans
- Carbon monoxide plans
- Ozone plans
- Maintenance plans
- Vehicle inspection and maintenance (I/M) SIPs
- Emissions inventories
- Monitoring networks
- State statutes submitted for the purposes of demonstrating legal authority
- Part D non-attainment area plans
- Attainment demonstrations
- Transportation control measures (TCMs)
- Committal measures
- Contingency measures
- Non-regulatory and non-TCM control measures
- 15% Rate of progress plans
- Emergency episode plans
- Visibility plans

4.17.3 Who Is Responsible for Enforcing a SIP?

SIPs are generally enforced by the state. However, the EPA is authorized to take enforcement action against violators for federally approved SIPs.

If a SIP has been approved by a state but not yet approved by the EPA, then it is only state-enforceable and not federally enforceable until approved by the EPA.

Members of the public can also file citizen suits under the CAA to address violations of SIPs.

4.18 Compliance

4.18.1 Compliance Requirements

Once a source has received an air quality permit, it is important to stay in compliance with all of the permit conditions included. The federal PSD permitting process and Title V program require that each PSD permit, Title V permit specifies the regulations applicable to the permitted source. Furthermore, the PSD permit also specifies how compliance will be determined for each applicable regulation. The methods to determine compliance and the EPA Compliance Assurance Monitoring program are discussed in the following.

4.18.1.1 Methods to Determine Compliance

The PSD permit and Title V permit applications must identify both requirements applicable to the facility and any emission units and how compliance with those applicable requirements will be determined. The elements in demonstrating compliance include compliance monitoring, record keeping, and reporting.

4.18.1.2 Direct Measurement of Air Emission Rates

The basis of direct measurement is the group of procedures published by the EPA in 40 CFR Part 60, Appendix A (Title 40 – Protection of Environment; Part 60 – Standards of Performance for New Stationary Sources and all Test Methods revised as of 1 July 1991). Table 4.11 presents the EPA methods for direct measurement of air emission rates from stationary sources and their group of procedures presented in Appendix A including measurements of gaseous contaminants, specific metals, and VOCs. These methods contain precise standards for performing the procedure, with QA/QC steps, and specify which data to record. Not following the procedure to the latter seriously risk the integrity of the results and jeopardizes their approval. The general procedure for collecting a sample from an exhaust stack for analysis is found in Methods 1–5, as are guidelines on setting up a testing train. The integrity of any test depends on ensuring the sampling train is a closed system with minimal outside (clean) air leakage, which would dilute the concentration of the contaminant in the exhaust sample.

Table 4.11 USEPA methods for direct measurements of air emissions and the group of procedures published by the EPA in 40 CFR Part 60, Appendix A.

Methods, stationary sources	Procedures and pollutants of interests
1	Sample and velocity traverses for stationary sources
2	Determination of stack gas velocity and volumetric flow rate (type S pitot tube)
4	Determination of moisture content in stack gases
5	Determination of PM emissions from stationary sources
6	Sulfur dioxide
7	Nitrogen oxides
8	Sulfuric acid and sulfur dioxide
9	Opacity monitoring
10	Carbon monoxide
11	Hydrogen sulfide content in flue-gas streams in petroleum refineries
12	Lead
13	Fluoride
15	Hydrogen sulfide; carbonyl sulfide; carbon disulfide
16, 16A	Reduced sulfur compounds
18, 24, 25	VOCs
201, 202	PM_{10}[a]

[a] PM that is 10 µm or less in diameter.

Method 1 contains guidance for the selection of sampling ports and traverse points for acceptable sampling of a representative aliquot of exhaust. A key concern is collecting a uniform sample over time. It contains procedures for determining whether cyclonic or swirling flow exists. Method 2 contains the procedure for measuring exhaust velocity and volumetric flow rate.

Figure 4.15 is a schematic of a typical sampling train (this particular setup is used to measure PM emissions). Method 5 contains guidance for measuring PM from two sources: filterable PM and condensable PM (the latter is found condensed in impingers down-stream of the filter). Typically, the PM measured in both locations must be combined. Other Method 5 procedures exist for specific circumstances from specific sources types (i.e. wood heaters, asphalt industry, etc.). Methods 201 and 202 are tests that should be considered for condensable ultrafine PM (Cooper and Alley 2011; Karell 2017).

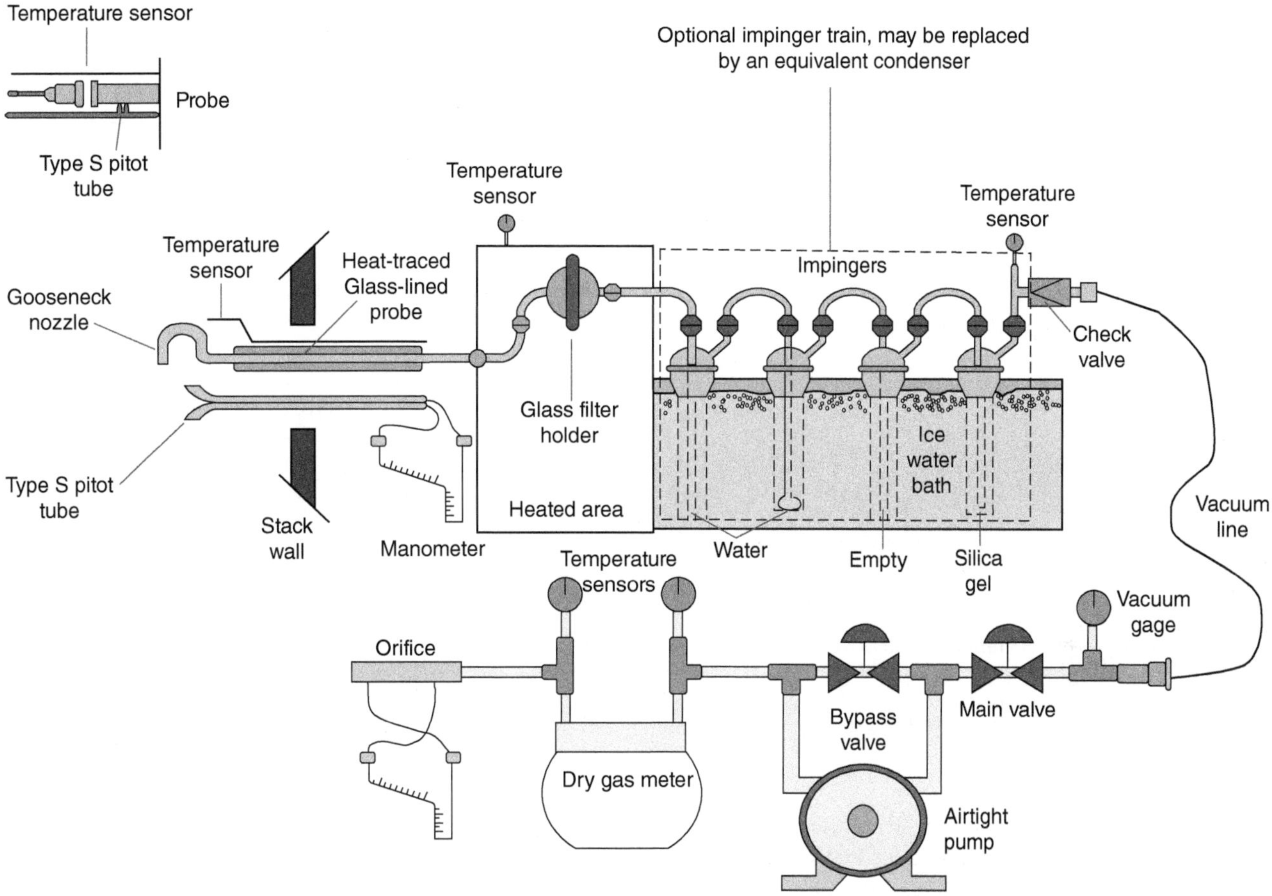

Figure 4.15 A process train to measure PM emissions. *Source:* Adopted from USEPA (1991b).

Example 4.14 Check for Emission Standards Compliance

As a consulting engineer, you have been contracted to modify an existing control device used in fly ash removal. The standards for emissions have been changed to a total numbers basis. Determine if the unit will meet an emission standard of $10^{5.7}$ particle/acf. Data for the unit are given below:

Average particle size, $d_p = 10\,\mu m$, assume constant
Particle specific gravity = 2.3
Inlet loading, IL = $3.0\,gr/ft^3$
Efficiency (mass basis), $E = 99.5\% = 0.995$

Solution

The outlet loading (OL) is given by

$$\begin{aligned} OL &= (1.0 - E)(IL) \\ &= (1.0 - 0.995)3.0\,gr/ft^3 \\ &= 0.015\,gr/ft^3 \end{aligned}$$

Assume a basis of $1.0\,ft^3$

Particle mass $- M = \rho_p V_p = \rho_p \left(\pi d_p^3\right)/6$

$$= \left(7000\frac{gr}{lb}\right)\left(\frac{\pi\left[(10\,\mu m)\left(0.328\times10^{-5}\frac{ft}{\mu m}\right)\right]^3 (2.3)62.4\,lb/ft^3}{6}\right)$$

$$= 1.881\times10^{-8}\,gr/particle$$

$$\text{Number of particles} = \frac{0.015\,gr/ft^3}{1881\times10^{-8}\,gr/particle}$$

$$= 8.0\times10^5\,\frac{particle}{ft^3}$$

Allowable number of particles/ft^3 = $10^{5.7}$ = 5.01×10^5

$$\frac{particle}{ft^3} < 8.0\times10^5\,\frac{particle}{ft^3}$$

Therefore, the unit will not meet the particle emission standard.

Example 4.15 A sample stream of dry gas is being withdrawn from a stack. The stack gases are 200 °C and 730 mm Hg. The stream flows through a heated filter, a set of cooled impingers, a small air pump, and then through a flow meter, as shown in Figure 4.15. The rate of flow is determined to be 30.0 l/min at 20 °C and 790 mm Hg.

1) Calculate the actual volumetric flow rate through the filter (at $T = 200\,°\mathrm{C}$ and $P = 730\,\mathrm{mm\,Hg}$).
2) If 1.32 mg of solid particles is collected on the filter in 30 minutes, calculate the concentration of particles in the stack gas (in μg/m³).

Solution

1) $Q_{\mathrm{filter}} = (30.0)\left(\frac{790}{730}\right)\left(\frac{473}{293}\right) = 52.41\,\mathrm{l/min}$
2) Total volume of gas sampled (at stack condition):

$$52.41\frac{1}{\mathrm{min}} \times \frac{1\mathrm{m}^3}{1000\mathrm{l}} \times 30\,\mathrm{minutes} = 1.572\mathrm{m}^3$$

$$C_{\mathrm{particulate}} = \frac{1.32\,\mathrm{mg}}{1.572\mathrm{m}^3} \times \frac{10^3\mu\mathrm{g}}{1\mathrm{mg}} = 841\ \mu\mathrm{g/m}^3$$

Example 4.16 A stack (T = 550 °F and P = 750 mm Hg) was sampled using EPA Method 5 (see Figure 4.15).

The total gas volume that flowed through the dry gas meter was 2.785 m³ (at $T = 60\,°\mathrm{F}$ and $P = 800\,\mathrm{mm\,Hg}$). The mass of PM collected was 1.50 g. Also, 72.0 g of H_2O was collected in impingers.

a) Calculate the concentration of PM in the stack (μg/m³) at stack conditions.
b) The stack gas exits at an average velocity of 20 m/s, and the stack has an inside diameter of 5 m. What is the emission rate of PM in kg/day?
c) What is the PM concentration in μg/dscm?

Given:

$$T_{\mathrm{stack}} = 550°\mathrm{F} = 560.85\mathrm{K} \quad P_{\mathrm{stock}} = 750\mathrm{mm\,Hg} = 0.987\mathrm{atm}$$

$$T_{\mathrm{DG}} = 60°\mathrm{F} = 288.71\mathrm{K} \quad P_{\mathrm{DG}} = 800\mathrm{mm\,Hg} = 1.053\mathrm{atm}$$

$$\mathrm{Vol}_{\mathrm{DG}} = 2.785\mathrm{m}^3 = 2785\mathrm{l} \quad m_{\mathrm{PM}} = 1.50\,\mathrm{gm}$$

$$m_{\mathrm{H_2O}} = 72.0\,\mathrm{gm} \quad V_{\mathrm{exit}} = 20\frac{\mathrm{m}}{\mathrm{s}}$$

$$D_{\mathrm{stock}} = 5.0\mathrm{m}$$

Solution

Part (a)

Calculate the moles of dry gas using the ideal gas law (Eq. 3.5)

$$R = 0.08206\frac{\mathrm{atm\,L}}{\mathrm{gmol\,K}}$$

$$n_{\mathrm{DG}} = \frac{P_{\mathrm{DG}}\mathrm{Vol}_{\mathrm{DG}}}{RT_{\mathrm{DG}}} = \frac{(1.053\mathrm{atm})(2785\mathrm{l})}{0.08206\frac{\mathrm{atm\,l}}{\mathrm{gmol\,K}}(288.71\mathrm{K})}$$

$$= 123.8\,\mathrm{gmol}$$

Calculate the moles of water

$$\mathrm{MW_{H_2O}} = 18\frac{\mathrm{g}}{\mathrm{gmol}}$$

$$n_{\mathrm{H_2O}} = \frac{m_{\mathrm{H_2O}}}{\mathrm{MW_{H_2O}}} = \frac{72.0\mathrm{g}}{18\frac{\mathrm{g}}{\mathrm{gmol}}} = 4.0\,\mathrm{gmol}$$

$$V_{\mathrm{stock}} = n \cdot R \cdot T\frac{T_{\mathrm{stock}}}{P_{\mathrm{stock}}}$$

$$= \frac{(123.8\,\mathrm{gmol})\left(0.08206\frac{\mathrm{atm}\cdot\mathrm{l}}{\mathrm{gmol\,K}}\right)(560.85\mathrm{K})}{0.987\mathrm{atm}}$$

$$= 5770\mathrm{l} = 5.77\mathrm{m}^3$$

Calculate the concentration of PM in the stack at stack conditions

$$C_{\mathrm{PM}} = \frac{M_{\mathrm{PM}}}{V_{\mathrm{stack}}} = \frac{1.50\mathrm{g}}{5.77\mathrm{m}^3}\left(\frac{10^6\mu\mathrm{g}}{1\mathrm{g}}\right)$$

$$C_{\mathrm{PM}} = \frac{2.6\times10^5\mu\mathrm{g}}{\mathrm{m}^3}$$

Part (b)

Calculate the flow and then the emission rate of PM.

$$A_{\mathrm{stack}} = \left[\frac{\pi\left(D_{\mathrm{stack}}\right)^2}{4}\right] = \frac{\pi(5.0\mathrm{m})^2}{4} = 19.63\mathrm{m}^2$$

$$Q_{\mathrm{stack}} = A_{\mathrm{stack}} \times V_{\mathrm{exit}} = \left(19.63\mathrm{m}^2\right)\left(20\frac{\mathrm{m}}{\mathrm{s}}\right) = 392.6\frac{\mathrm{m}^3}{\mathrm{s}}$$

$$\mathrm{Rate_{PM}} = Q_{\mathrm{PM}}C_{\mathrm{PM}}$$

$$= \left(392.6\frac{\mathrm{m}^3}{\mathrm{s}}\right)\left(\frac{2.6\times10^5\mu\mathrm{g}}{\mathrm{m}^3}\right)\left(\frac{1\mathrm{kg}}{10^9\mu\mathrm{g}}\right)\left(\frac{3600\ \mathrm{seconds}}{1\ \mathrm{hour}}\right)\left(\frac{24\,\mathrm{hours}}{1\mathrm{day}}\right)$$

$$\mathrm{Rate_{PM}} = 8820\frac{\mathrm{kg}}{\mathrm{day}}$$

Part (c)

Adjust the volume to STP conditions using Eq. (3.5).

$$P_{\text{standard}} = 1\,\text{atm}$$
$$T_{\text{standard}} = 298.15\,\text{K}$$
$$\text{Vol}_{\text{STP}} = \text{Vol}_{\text{DG}}\left(\frac{P_{\text{DG}}}{P_{\text{standard}}}\right)\left(\frac{T_{\text{standard}}}{T_{\text{DG}}}\right)$$
$$= 2.785\,\text{m}^3\left(\frac{1.053\,\text{atm}}{1\,\text{atm}}\right)\left(\frac{298.15\,\text{K}}{288.71\,\text{K}}\right) = 3.028\,\text{m}^3$$

Calculate the PM concentration.

$$C_{\text{PM}} = \frac{m_{\text{PM}}}{\text{Vol}_{\text{STP}}} = \frac{(1.50\,\text{gm})}{(3.028\,\text{m}^3)}\left(\frac{10^6\,\mu\text{g}}{1\,\text{gm}}\right)$$
$$C_{\text{PM}} = 4.95\times10^5\,\frac{\mu\text{g}}{\text{m}^3}$$

4.18.1.3 Monitoring

Monitoring can include everything from continuous emission monitoring (CEM) to keeping records of complaints received. Monitoring can be quite expensive. Even if the equipment used for monitoring is not expensive, there is always a frequency of monitoring that must be met, and doing the monitoring takes staff resources that, for a large facility, can be substantial.

One of the most promising emission monitoring techniques recently developed is called parametric emission monitoring system (PEMS). PEMS uses a multivarient computer code to correlate data from parametric monitors on an emission source with measurements from a portable leased CEM system. During initial installation, information from all available parametric monitors is provided to the computer. The PEMS code then evaluates which of the parameters are independent and identifies those parameters. Long-term monitoring is then limited to the independent parameters. The PEMS code then correlates parameter changes with emission changes – following testing in which the PEMS predicted emissions are compared to CEM data, the CEM system is removed, and PEMS provides emission estimates. Since parametric monitors such as fuel flow and temperature are usually much more reliable than CEM monitors, the availability of the PEMS is usually much higher than that of a CEM system. Operating costs also are lower, because daily calibrations and other CEM maintenance requirements are not necessary.

4.18.1.4 Opacity Monitoring

Opacity compliance must be addressed in virtually every Title V permit. Opacity limits are found both in agency prohibitionary rules, in NSPS, and in conditions for individual pieces of equipment. Several levels of opacity measurement can be considered, including the following:

- Continuous opacity monitoring systems (COMS)
- EPA Method 9 opacity monitoring
- EPA Method 22 opacity monitoring
- Complaint reporting and response

a) *Continuous opacity monitoring systems.* COMS are optical monitors at the top of a stack or vent that measure the attenuation of a light beam through stack gases. They provide a continuous reading of opacity that is recorded electronically. Although reliable, they require periodic cleaning of the optical elements to ensure that readings are of gas opacity and not extraneous contamination. Also, they are generally not able to distinguish between PM and water vapor. Opacity due to water vapor is not considered a violation, so this equipment limitation can be important. One technique for distinguishing between PM opacity and water vapor is to install the opacity monitor at a point in the stack where the temperature is above the condensation temperature of any moisture.

b) *EPA Method 9 opacity monitoring.* EPA Method 9 monitoring is visual monitoring of emissions from a point source by a qualified observer or "smoke reader": The certification is a course and examination offered by agencies or commercial schools, and to remain certified, a person must be recertified every six months. Monitoring consists of 24 observations of 15 seconds duration and comparison of the opacity of the emissions to a standard gray scale. It is considered a violation if average opacity of the 24 measurements exceeds the standard.

For *Method 9* to be used as a compliance demonstration, a facility must have a certified "smoke reader" available, and readings must be taken at intervals specified in permit conditions. Typically two employees are certified so that readings can be taken even when one of the certified persons is unavailable. One cement kiln that utilized Method 9 for demonstrating opacity compliance had 18 certified smoke readers on staff to ensure that readings were possible at any time during the 24-hour operation of the kiln. Notwithstanding, certified readings can only be taken during daylight hours when there is no precipitation (40 CFR Part 60).

Only COMS and Method 9 observations can be used to quantify opacity for comparison with a specified standard. The additional methods described below are qualitative.

c) *EPA Method 22 opacity monitoring*. EPA Method 22 is also a visual observation method, but it does not require a certified smoke reader. Method 22 suggests that observers be familiar with the techniques in Method 9, but asks only that an observer note the percentage of time visible opacity from emissions is observed during the observation period. The observation period must be at least six minutes in duration. The visible opacity must be sustained for the period specified in the applicable regulation to constitute a notable incident, and the requirement is that these incidents be documented, responded to, and reported (40 CFR Part 60, USEPA 1992). An alternate to *Method 9* is also available using laser radar (LIDAR 1980).
d) *Complaint reporting and response*. The simplest method of opacity monitoring is to make a commitment to keep records of any opacity complaints received by a facility and to respond to those complaints with corrective action to eliminate observed opacity. Documentation of the corrective action taken must also be maintained. The record of complaints and corrective action is then provided to the agency as part of periodic reporting.

One method commonly used to check opacity is a combination of Method 22 and the complaint recording and response. The plant manager or other staff regularly drives around the plant to see if there are any visible emissions. For a continuously operated plant, one plant manager conducts this inspection upon arrival at the plant each day.

4.18.1.5 Record Keeping

Record keeping must be used to provide documentation of monitoring that has been conducted. For material balance compliance demonstrations, record keeping of material used and the composition of the material is the essence of the demonstration. For methods that use a surrogate parameter, or activity data, along with an emission factor to demonstrate compliance, records must be kept of the activity, such as fuel use or operating hours.

A standard condition for both construction and Title V permits is that permitting agency staff or their representatives have authority to enter the premises of the facility at "any reasonable time," and without prior notification, to inspect both monitoring methods and the records that demonstrate compliance.

Records need not be hard copy. Magnetic or other electronic recording is acceptable provided it can be displayed and audited. In some instances, agencies have required that data be available electronically from agency offices, so that compliance can be verified at any time. Although this requirement is unusual, it is well within the state of the art of current monitoring and data transfer systems.

The discipline of good record keeping is essential to demonstrating compliance with a Title V permit. Use of ditto marks or photocopying of record-keeping pages in a log indicate poor record-keeping discipline, and often are indicators of record-keeping errors. One advantage of electronic record keeping is that repeated records can be copied, and then appropriate changes made to update the inputs.

Periodic checking of records by a person unfamiliar with the operation can quickly identify entries that are understandable only to the equipment operator or person making record-keeping entries. The records, including the signature or initials of the person making the entries, must be understandable to others.

4.18.1.6 Reporting

For Title V permit holders, periodic reports must be provided on the results of compliance monitoring. Reporting for Title V permits is required either every six months or annually. Reports are public records and can result in enforcement actions if noncompliance is reported.

4.18.1.7 Good Conduct Provisions

For any compliance monitoring that is included in a Title V permit, a good conduct provision should be considered. Such provisions have been included in NSPS. They specify that initially periodic monitoring be done frequently; if no noncompliance is monitored, the frequency of monitoring decreases. Several steps of decreases can be included, and if during any monitoring subsequent to the decrease in frequency noncompliance is observed, the monitoring steps back to a more frequent interval and begins the process again. This good-conduct provision provides an incentive for the facility to stay in compliance, thereby limiting the expense of periodic monitoring.

4.19 CAA Enforcement Provisions

There are several enforcement options that EPA can take under the CAA. These include the following:

- Issuing an administrative penalty order
- Issuing an order requiring compliance or prohibition
- Bringing civil action in court
- Requesting the Attorney General to bring criminal action

Each of these is discussed below. Individual states may have similar provisions for violation of state statutes and will have provisions at least as stringent as these for violation of state statutes that are delegation of federal statutes.

Civil Penalties Can Be Assessed Under the Clean Air Act

- Field citations: No more than $5 000/day up to a maximum of $25 000
- Civil penalties: Up to $25 000/day for each violation
- Whistleblower awards of $10 000 can be given by EPA
- Recovery of attorney and expert witness fees can be assessed

4.19.1 Administrative Penalty Orders

Administrative penalty assessments may be made for violations of any of the provisions of the CAA, including violation of permit conditions. Field citations can be issued by an inspector. These citations have a fine of $5 000/day, and they may accumulate up to a maximum of $25 000. Individuals receiving field citations may request a hearing if they do so within a reasonable time; they will have opportunity to present evidence in defense.

Civil penalties can be assessed for up to $25 000/day for each violation. Also under the category of administrative penalties are whistleblower awards of up to $10 000 that can be made by EPA to persons reporting a violation. A person receiving an administrative penalty may appeal the assessment within 30 days or may appeal a final ruling to the federal district court in an additional 30-day window. The administrator of EPA is authorized to compromise, modify, or place conditions on a penalty.

Persons receiving administrative penalties should promptly contact the regulating agency to clarify any uncertainty about the violation, identify extenuating circumstances, indicate that the cited noncompliance has been corrected, and negotiate reductions in the penalty. Extenuating circumstances are considered by the agency, and it is unusual for a final penalty to remain as high as the original amount.

4.19.2 Issuing an Order Requiring Compliance or Prohibition

Issuing an order is a more formal, and more serious, response to a violation. It is usually used when a specific problem must be addressed, and when there is the potential for impairing human health or of serious or irreversible damage to the environment. A hearing is required to address an order, and civil or criminal penalties may be assessed subsequently.

4.19.3 Bringing Civil Action in Court

Civil action enforcement is more serious, and potentially a great deal more expensive than the administrative actions discussed above. There is provision in the statute for recovery of attorney and expert witness fees from the defendant during civil action. Under Title VII of the CAAA of 1990, citizen suits can be brought as civil action. Such suits can be based on a Title V permit application that identifies current noncompliance, even if a compliance plan is included in the application. The exception to the ability to bring a citizen suit against current noncompliance is if the noncompliance has already been adjudicated and the compliance plan is part of a consent decree already approved by the court.

4.19.4 Requesting the Attorney General to Bring Criminal Action

Provisions under federal law as well as laws in many states provide for criminal sanctions against both organizations and individuals. Usually, criminal action is reserved for flagrant, intentional violations. It includes up to 2 years imprisonment for false statements or noncompliant monitoring or record keeping, 1 year for negligent endangerment, and up to 15 years and a fine of up to $1 000 000 for knowing endangerment of death or serious bodily injury.

The Clean Air Act Also Provides for Criminal Enforcement Including Imprisonment

- Two years for false statements, noncompliant monitoring and record keeping
- One year for negligent endangerment
- Fifteen years for knowing endangerment including knowing release of HAP or EHS causing imminent danger
- Five years for knowing violation of SIP, NSPS, NESHAP, Title V, or Section 114 requirements
- Fine of approximately $250 000 for individuals, approximately $500 000 for organizations

4.19.5 Emergency as a Defense

If there is a release of pollutants that would otherwise be considered a violation, it is possible to argue that it was an emergency. For this purpose, an emergency is defined as any sudden and reasonably unforeseeable event beyond the control of the source that requires immediate corrective action to restore normal operation and causes exceedence of technology-based emission limitations.

If the noncompliance is caused by improperly designed equipment, lack of preventative maintenance, careless or improper operation, or operator error, it is not considered an emergency. An emergency can be used as a defense if there are properly signed operating logs or other evidence that indicates an emergency occurred and the cause can be identified, the facility was being properly operated at the time, and the permittee took all reasonable steps to minimize levels of exceedences.

To use this defense, notice must be provided to the permitting agency within two working days containing a description of the emergency and steps taken to mitigate emissions and perform corrective actions. If the only way to reduce the emissions to allowable levels is to halt or reduce activity at the plant, and that is not done, the emergency defense cannot be used. That is, if you do not shut down, you are liable.

As with other compliance issues, good record keeping is vital to protection from prosecution when using the emergency defense. It is necessary to have records indicating that the plant had been operating normally and that efforts were made to halt or reduce the excess emissions.

4.19.6 Section 114: Fact-Finding

EPA has the authority to obtain information from a source in anticipation of prosecution. If a source receives a letter from EPA citing Section 114, it should be considered at least as serious as an Internal Revenue Service audit. EPA may ask questions or seek information about both emissions and the structure of environmental management in the company. EPA may request records, monitoring, sampling, or compliance certifications. EPA has the right of entry to company premises on which records are kept when credentials are presented.

Section 114 Is Fact-Finding in Anticipation of Prosecution

- EPA may request records, monitoring, sampling, or compliance certifications.
- Receipt of a Section 114 letter is equivalent to an IRS audit.
- Questions about both emissions and the way environmental management is organized are likely.
- Section 114 responses should be prompt, accurate, and limited to the questions asked.

Often a letter is stimulated by evidence obtained by EPA that excess emissions have occurred. This can include emissions reported as part of EPCRA 313 Form R reporting of releases to the environment, newspaper accounts, or notices of receipt of a variance from state emission limits.

Responses to Section 114 letters should be prompt, accurate, and limited to the questions asked. Counsel should participate in preparing or reviewing the response and in any subsequent meetings with the agency staff.

4.19.7 Inspection Protocol

Inspections are a standard activity for permitting agency staff, and a facility should expect to be inspected periodically. Some very large facilities, such as oil refineries, have a full-time inspector from the regulating agency assigned, and this person is at the facility every day. Other facilities have less frequent inspections. For any facility, an inspection protocol is worthwhile.

The inspection protocol should identify who is to respond when an inspector comes, what responses are appropriate, and what rights the inspector has as well as rights the company has.

The agency staff, or an authorized representative, has the right of entry to a facility. The receptionist or gate guard should be instructed about whom to call when an inspector presents his or her credentials and asks for entry. A member of the environmental staff should greet the inspector and, unless the inspection is anticipated and routine, take the inspector to a conference room for a pre-inspection conference. It is appropriate to ask the inspector what he or she wants to see. Although the company cannot bar access, safety provisions are often necessary for visiting certain areas of the plant, and planning such visits is usually required by plant safety rules. Inspectors should be required to use all safety equipment required of employees and other visitors, and in some instances a safety briefing is needed before visiting the plant.

Once the area to be visited or information sought has been identified, the inspector should be escorted promptly and directly to that area. It can be a safety hazard as well as present a risk of providing misleading information for an inspector to visit a facility without an escort. In the event that the inspector requests a stop at another area not covered by the original request, such a stop can be prohibited until another initial interview has been conducted, since the request was not covered in the original interview.

If an inspector uses a recorder or any test instruments, the company should also record the same information and test the same location. In some instances, it may be appropriate to videotape the inspection.

At the end of the inspection, an exit conference should be held. The inspector should be asked what information was obtained, and copies of any information should be requested.

Most companies develop good working relationships with air quality inspectors who regularly visit their plants and it is important that procedures used to respond to an inspection are not hostile or unnecessarily inhibiting. However, it is not unusual for an inspector, after working some time with a facility, to assume that he or she is as knowledgeable of facility operation as the plant operators. Since this is almost never the case, escorts and explanations of operations are needed, even for experienced inspectors.

4.19.8 Continuous Emission Monitoring

An alternative to performing one-time measurement is to perform continuous measurement of the contaminant using a CEM (also referred to as CEM system). A CEM has many of the same attributes as a stack test (Sections 4.19.8 and 4.19.9). A probe is inserted in the stack and the sample is conditioned appropriately before it enters an analyzer for determination of concentration. A data acquisition system (DAS) calculates, stores, and organizes data. Typically, the CEM takes samples at regular intervals and results can be averaged as needed (Karell 2017).

Continuous emission monitoring systems (CEMS) were historically used as a tool to monitor flue gas for oxygen, carbon monoxide and carbon dioxide to provide information for combustion control in industrial settings (Stultz and Kitto 2005).They are currently used as a means to comply with air emission standards such as the USEPA Acid Rain Program (CFR, Title 40, Part 72, Subpart A 2011), or other federal emission programs, or state permitted emission standards.

The standard CEM system consists of a sample probe, filter, sample line (umbilical), gas conditioning system, calibration gas system, and a series of gas analyzers which reflect the parameters being monitored. Typical monitored emissions include: sulfur dioxide, nitrogen oxides, carbon monoxide, carbon dioxide, hydrogen chloride, airborne PM, mercury, volatile organic compounds, and oxygen. CEM systems can also measure air flow, flue gas opacity and moisture.

In the United States, the EPA requires a data acquisition and handling system to collect and report the data. SO_2 emissions must be measured in pounds per hour using both an SO_2 pollutant concentration monitor and a volumetric flow monitor. For NO_x, both a NO_x pollutant concentration monitor and a diluent gas monitor are required to determine the emissions rate (lb/MBtu). Opacity must also be monitored. NO_x measuring is not a current requirement, however if monitored, a CO_2 or oxygen monitor plus a flow monitor should be used. In monitoring these emissions, the system must be in continuous operation and must be able to sample, analyze, and record data at least every 15 minutes and then averaged hourly (USEPA 2009a).

4.19.8.1 Operation

A small sample of flue gas is extracted, by means of a pump, into the CEM system via a sample probe. Facilities that combust fossil fuels often use a dilution-extractive probe to dilute the sample with clean, dry air to a ratio typically between 50 : 1 and 200 : 1, but usually 100 : 1. Dilution is used because pure flue gas can be hot, wet and, with some pollutants, sticky. Once diluted to the appropriate ratio, the sample is transported through a sample line (typically referred to as an umbilical) to a manifold from which individual analyzers may extract a sample. Gas analyzers employ various techniques to accurately measure concentrations. Some commonly used techniques include: infrared and ultraviolet adsorption, chemiluminescence, fluorescence, gas chromatography, flame ionization detector, mass spectrometer, and beta ray absorption. After analysis, the gas exits the analyzer to a common manifold to all analyzers where it is vented out of doors. A DAS receives the signal output from each analyzer in order to collect and record emissions data (PNM 2018).

Another sample extraction method used in industrial sources and utility sources with low emission rates is commonly referred to as the "hot dry" extractive method or "direct" CEMS. The sample is not diluted, but it is carried along a heated sample line at high temperature into a sample conditioning unit. The sample is filtered to remove PM and dried, usually with a chiller, to remove moisture. Once conditioned, the sample enters a sampling manifold and is measured using the same methods above. One advantage of this method is the ability to measure % oxygen in the sample, which is often required in the regulatory calculations for emission corrections. Since dilution mixes clean dry air with the sample, dilution systems cannot measure % oxygen (PNM 2018).

4.19.9 QA and QC in Air Emission Rates

The EPA specifies QA/QC procedures for ensuring a CEM's accuracy for compliance purposes. Calibration and evaluation of drift from zero must be performed at intervals designated by regulatory requirements, such as running a sample with a known concentration of the compound being measured and running a "relative accuracy" test of the CEM versus another, recently calibrated CEM (relative accuracy) at the same time.

Direct measurement involves collecting a representative sample of gas from the exhaust or stack during process operation, preheating the gas sample to remove any components that may interfere with the testing, and performing a chemical analysis to measure the concentration of the component of interest. Simultaneously measuring the exhaust flow rate enables conversion of concentration into an emission rate (typically in pounds per hour). QA and/or QC procedures accompany each step.

Accuracy of the system is demonstrated in several ways. An internal QA check is achieved by daily introduction of a certified concentration of gas to the sample probe. The EPA also allows for the use of Continuous Emissions Monitoring Calibration Systems which dilute gases to generate calibration standards (USEPA 1993). The analyzer reading must be accurate to a certain percentage. The percent accuracy can vary, but most fall between 2.5 and 5%. In power stations affected by the Acid Rain Program, annual (or bi-annual) certification of the system must be performed by an independent firm. The firm will have an independent CEM system temporarily in place to collect emissions data in parallel with the plant CEMS. This testing is referred to as a Relative Accuracy Test Audit (RATA).

In the United States, periodic evaluations of the equipment must be recorded and reported (USEPA 2009a). This includes daily calibration error tests, daily interference tests for flow monitors, and semiannual (or annual) RATA and bias tests (USEPA 2018a). CEMS equipment is expensive and not always affordable for a facility. In such cases, a facility might install non-EPA compliant analysis equipment at the emissions point. Once yearly, for the equipment evaluation, a mobile CEMS company measures emissions with compliant equipment. The results are then compared to the noncompliant analyzer system (Nova Gas 2018).

4.19.10 Performing Stack Tests

Performing stack tests can be very complicated. Many things could go wrong or introduce error in the results. For this reason, virtually all agencies require that emissions be reported as an average of at least three successful runs, each of which consists of sufficient sample collection (e.g. by isokinetic sampling) and a successful leak test. Each run generally spans one hour. However, for contaminants found a dilute concentrations (e.g. trace metals), a longer sampling time is necessary to collect a sufficient sample to measure.

Because of the complexity of the test methods, most regulatory agencies require a formal written protocol that must be approved before the test begins. Normally prepared by the testing firm, these protocols should list the test methods, describe the method and sampling equipment thoroughly, and provide necessary calculations, such as number and locations of traverse points.

Once the protocol is approved, the facility and testing firm should set a mutually agreeable data with the regulatory agency. It is critical to coordinate testing with plant operations to ensure that the equipment in question is operating as intended during the stack test, and any deviations from standard operating conditions are recorded.

After the field work is complete, a stack test report that summarizes the methodology presents the results and includes raw data must be submitted to the regulatory agency. Results are often provided as a concentration or mass emission rate based on the measured volumetric flow through the exhaust. A lower limit is often defined below which the contaminant cannot be measured (Karell 2017; USEPA 2009b).

Example 4.17 Calculations for Standard Volume

Most atmospheric sampling techniques use a sampling train whereby air containing the pollutant of interest enters the train and passes through a sample collection device. The weight of the pollutant collected is compared to the volume of air drawn through the exhaust and train to enable the calculation of the concentration of the pollutant. The concentration is typically expressed as a concentration corrected to EPA's standard condition to be 298 K (25 °C) and 760 mmHg (1 atm).

The equation used to correct sample volume (V_s) to EPA standard volume (V_{std}) is based on Charles' Law.

$$V_{STD} = V_s\left(\frac{P_{atm}}{P_{std}}\right)\left(\frac{T_{std}}{T_{atm}}\right) \tag{4.18}$$

where

V_{std} = corrected volume of the gas; V_s = sampled volume of the gas

T_{std} = EPA standard temperature; T_{atm} = sample standard temperature

P_{std} = EPA standard pressure; P_{atm} = sample standard pressure

Determine the corrected sample volume if 10 ft^3 of air was sampled from exhaust air at an average temperature of 80 °F and pressure 14.3 psia.

Solution

Convert ambient temperature and pressure to appropriate units.

$$T(\text{K}) = (F-32)/1.8+273 = (80-32)/1.8+273 = 300\,\text{K}$$
$$P(\text{atm}) = P_s(1\text{atm}/14.7\,\text{psi}) = 14.3\,\text{psi}(1\text{atm}/14.7) = 0.97\,\text{atm}$$

Calculate the corrected volume using the Eq. (4.15).

$$V_{std} = V_s\left(\frac{P_{atm}}{P_{std}}\right)\left(\frac{T_{std}}{T_{atm}}\right) = 10\,\text{ft}^3\left(\frac{0.97\,\text{atm}}{1\,\text{atm}}\right)\left(\frac{273\,\text{K}}{300\,\text{K}}\right)$$
$$= \mathbf{8.83\,ft^3}$$

Example 4.18 Stack Velocity
The exhaust gas flow rate from a facility is 1500 scfm. All of the gas is vented through a small stack which has an inlet area of 2.3 ft^3. The exhaust gas temperature is 350 °F. What is the velocity of the gas through the stack inlet in ft/s. Assume standard conditions to be 70 °F and 1 atm. Neglect the pressure drop across the stack.

Solution
First note that the temperature must be converted to an absolute scale. Thus,

$$T_a = 350 + 460 = 810°R$$
$$T_s = 70 + 460 = 530°R$$

and

$$Q_a = Q_s\left(\frac{T_a}{T_s}\right)$$
$$= 1500\left(\frac{810}{530}\right) = 2292 \text{ acfm}$$

The average velocity may now be calculated:

$$V = \frac{Q_a}{A}$$
$$= \frac{2292 \text{ acfm}}{\left(2.3 \text{ ft}^2\right)\left(60 \text{ seconds}\right)} = \mathbf{16.61 \text{ ft/s}}$$

4.20 Industrial Solid Wastes and Its Management

4.20.1 Solid Waste Treatment: Some Perspectives on Recycling

In 2010, Americans generated about 250 million T of municipal solid waste (MSW) (before recycling), an about 8% increase over 2000, an increase of 17% over 1990, and 36% over 1980 (USEPA 2002a, 2014). Thus, management of MSW continues to be an important challenge facing the United States and other highly industrialized nations in the twenty-first century. Solid waste management is critical in the developing world, as well, where age-old traditions of wasting nothing are often followed without regard to the significance and implications of the zero discharge concept.

4.20.2 Why Recycle?

Each ton of solid waste diverted from disposal, whether reused, recycled, converted in a waste-to-energy program, or composted, is one less ton of solid waste requiring disposal. To see the value of reusing, recycling, and composting solid waste, one need simply to consider the amount of disposal space required to accept that material. By implementing environmentally benign waste management strategies (as well as resource-management strategies), a population can reduce its dependence on incinerators and landfills. And when recycled materials are substituted for virgin plastics, metal ores, minerals, glass, and trees, there is less pressure to expand the chemical, mining, and forestry industries. Supplying industry with recycled materials is preferable to extracting virgin resources from mines and forests not only because it conserves scarce natural resources but because it reduces dangerous air and water pollutants, such as GHG emissions, and saves energy.

Saving energy is an important environmental benefit of recycling because generating energy usually requires fossil-fuel consumption and results in emissions that pollute the air and water. The energy required to manufacture paper, plastics, glass, and metal from recycled materials is generally less than the energy required to produce them from virgin materials. Additionally, the collection, processing, and transportation of recycled materials typically uses less energy than the extraction, refinement, transportation, and processing steps to which virgin materials must be subjected before industry can use them.

As is well known, a great amount of energy used in industrial processes and in transportation comes from the burning of fossil fuels. Recycling helps stem the dangers of global climate change by reducing the amount of energy used by industry, thus reducing GHG emissions, as well.

4.20.3 What Is Recycling

Recycling is the industrial process in which mechanical, chemical, and/or biological means are used to reprocess or convert materials in discarded products into recyclates, feedstock, or energy that can be used again. Recycling is more than a waste-management strategy; it is also an important strategy for reducing the environmental effects of industrial production.

The aim of recycling is to close the loop of a material's flow through the stages of manufacturing, marketing, consumption, and disposal. In closed-loop recycling, an assigned set of operations is repeated in each of successive cycles and the operations vary according to the type of waste material being recycled. Thus, recycling is different from reuse, which is the further use, without any change in properties, of discarded products that have been recovered at the end of their service life.

There are three major ways of recycling waste materials:

1) *Mechanical recycling*, which can be divided into *primary recycling*, the in-house processing of clean, single-graded production offcuts into products equivalent in quality to the original items, and *secondary (postconsumer) recycling*, the reprocessing of discarded wastes from various sources into recycled products of inferior quality (for plastics and paper) or equivalent quality (for aluminum, glass, and metal) to the original items.
2) *Tertiary or chemical or feedstock recycling processes*, such as composting biodegradable wastes, transforming mixed plastics into monomer or low molecular weight oligomers in liquid or gaseous form, and application of mixed plastic flakes as a reduction agent for steelmaking.
3) *Quaternary recycling processes*, such as waste incineration with energy recovery, in which part of the energy content of combustible municipal or agricultural wastes is recouped and used, for example to raise steam (Brophy et al. 1997; Frisch et al. 1999; Mader 1997; Van Beukering 1999; Walker 1995).

Since recycling is an industrial process, its level of success depends highly on economic, technical, social, and political factors, which in turn vary greatly country to country (Bell 1998; Haque 1998). However, mechanical recycling of wastes is long-established and tends to be the most popular practice in both the developed and developing parts of the world (Gandy 1994; Haque 1998; Shah 2000).

4.20.4 A Brief Overview of Recycling in the United States and United Kingdom

The history of primary recycling of discarded products, particularly of plastic scrap, goes back nearly as far as the industrial manufacture of products from virgin materials (Saba and Pearson 1995). In fact, waste management systems in New York City and London routinely collected for secondary recycling reusable items (mainly paper, cloth, glass bottles, and metals) from street-cleaning enterprises, commercial premises, garbage dumps, and refuse transfer stations. The first refuse sorting plant in the United States opened in New York in 1898, and in the early 1920s London boasted a model sorting plant equipped with a long traveling belt from which recyclable materials could be removed (Gandy 1994).

In the years that followed World War II, planners who thought beyond the economic value of waste materials realized that recycling could reduce the volume of the wastestreams and that separation of incombustible glass and metals could increase the operational efficiency of incineration plants.

In the United States today, most locations recycle about 20–30% of their trash, approaching, meeting, or exceeding the national recycling target set decades ago. In the United Kingdom, wastes generated by commerce and industry are more readily recycled than domestic waste, a discrepancy reflecting the higher level of purity of the wastes of the former sector. The country recycles 25% of industrial wastes and 5% of domestic wastes (Haque 1998). The national government has set a revised target of 45% recovery of municipal wastes, including 30% recycling and composting, by 2010 (DETR 1999).

Schemes for waste collection and separation in both countries included providing households with outdoor receptacles for recycling, use of strategically located bottle and can banks, and sorting in a central material reclamation facility. Cans, paper products, glass, lead-acid batteries, cars, and textile goods have well established recycling routes. In contrast, discarded plastic products and packaging – an increasing component of the wastestream – present many problems in recycling. Technical difficulties and high costs associated with identifying, separating, and recycling plastics of different types are compounded by the lack of markets for recyclate.

4.20.5 Recycling Today

Today in the United States and the United Kingdom, pressures for recycling arise largely from environmental concerns and from industrial policies minimizing consumption of energy and materials, aimed at lowering the raw materials and production costs. Further support for recycling activities is generated by concerns over the environmental pollution stemming from landfill and incineration, the conventional disposal treatments for solid waste.

Therefore, in many developed countries, the recycling of postconsumer waste is being actively encouraged by new legislation, as well as by economic incentives in place since the 1970s. A German law requiring retailers to take back their packaging has resulted in the availability of enormous quantities of material for recycling. Indeed, the success of this initiative created surpluses that depressed prices of recovered materials in Europe, thus hindering recycling efforts elsewhere. The German experience was perhaps anticipated by a former USEPA administrator, who called recycling "a good idea that has gone too far" (Bickerstaffe 1997).

In contrast to developed countries, the generation of waste per capita is substantially less than in less industrialized nations, which nevertheless post a recycling percentage for certain materials of close to 100%. Recycling was introduced to developing countries in the 1920s but did not take hold on a large scale until at least the late 1950s (Nahar 1990; Ratra 1994). From the start, these waste recycling systems have been driven by economic necessity, since the poor earn a living by scavenging discarded glass, plastics, paper, metals, and other materials which they then sell via an informal system for recycling in facilities that vary in their sophistication.

4.20.6 Recycling as a Route to Sustainable Productivity and Growth

Despite the massive recycling programs and pollution-prevention efforts now under way, the amount of waste to be disposed of, as noted at the outset, is on the rise. We are using natural resources much faster than nature can replenish them. After we have used resources extracted from the Earth, we discard them, often in more harmful forms, into the same environment we depend on to provide air to breathe, water to drink, and soil to grow food. Concerns about this practice are voiced repeatedly and are well known.

Another concern that is related to waste, although it has broader ramifications as well, is community safety. The public is becoming more aware of how many tons of hazardous and nonhazardous materials alike are hauled over road, rail, and sea. Given the risk of accidents, especially when hazardous wastes are involved, the volume of transported substances continues to grow, concerns mount.

Plans for sustainable production and growth focus on leading the way to a future in which few materials and toxic substances will be required to support the needs of society, a future in which materials are reused or recycled rather than being incinerated, buried, or dumped. This shift will take place as we redesign processes, consumer and corporate behaviors, reuse more materials, and improve technologies. Industry, business, and government at all levels will need to capitalize on successful programs and invest in infrastructure and activities to meet these goals.

Interestingly, the economics of recycling is inversely related to the level of economic development of the recyclers, and in more general terms, the propensity to recycle is also inversely related to their socio-economic status. In developed nations, facilities for reprocessing plastic wastes operate under known conditions and yield products of known quality. In developing countries, however, plastic wastes generally are recycled several times under undocumented conditions and exposed to the cumulative effects of extraneous contaminants. The deleterious effects on material quality of such uncontrolled and repeated processing are not hard to imagine. Thus, it is suggested that limits to the maximum number of times that plastics may feasibly be recycled vary with the socio-economic conditions in different countries.

4.20.7 Resource Conservation and Recovery Act

The RCRA, enacted in 1976, is the principal federal law in the United States governing the disposal of solid waste and hazardous waste (RCRA 1976).

4.20.7.1 History and Goals

Congress enacted RCRA to address the increasing problems the nation faced from its growing volume of municipal and industrial waste. RCRA amended the Solid Waste Disposal Act of 1965. It set national goals for

- protecting human health and the natural environment from the potential hazards of waste disposal
- energy conservation and natural resources
- reducing the amount of waste generated, through source reduction and recycling
- ensuring the management of waste in an environmentally sound manner (USEPA 2010b)

It is now most widely known for the regulations promulgated under RCRA that set standards for the treatment, storage, and disposal (TSD) of hazardous waste in the United States.

4.20.7.2 RCRA Implementation

The USEPA has published waste management regulations, which are codified in Title 40 of the Code of Federal Regulations at parts 239 through 282 (USEPA 2011a). Regulations regarding management of hazardous waste begin in part 260 (USEPA 2011b). As noted in the following, most states have enacted laws and created regulations that are at least as stringent as the federal regulations. Furthermore, the RCRA statute authorizes states to carry out many of the functions of the federal law through their own hazardous waste programs (as well as their state laws) if such programs have been approved by the EPA.

4.20.7.3 Identification and Characterization of Hazardous Waste

For a material to be classified as hazardous waste, it must first be considered a waste and also at least one of the following criteria:

1) It exhibits any characteristic of a hazardous waste.
2) It has been named a hazardous waste and listed as such in the EPA regulations (Table C.2).
3) It is a mixture containing a listed hazardous waste and a nonhazardous solid waste.
4) It is a mixture containing a listed or characteristic hazardous waste and special nuclear material.
5) It is a waste residue generated from the TSD of a listed hazardous waste (i.e. is a "derived-from" waste).

4.20.7.4 Characteristics

Characteristic wastes are those that exhibit measurable properties indicating that a waste poses enough of a threat to deserve regulation as a hazardous waste. The USEPA decided that the characteristics of hazardous waste should be detectable by suing a standardized test method or by applying general knowledge of the waste properties. Given these criteria, the EPA established four hazardous waste characteristics: ignitability, corrosivity, reactivity, and toxicity.

1) Ignitability – Ignitable wastes are easily combustible, or flammable, and they may cause fires during their transport, storage, or disposal. Examples include paint waste, certain degreasers, or other solvents. Regulations describing the characteristic of ignitibility are found in 40 CFR 261.21. A waste that exhibits any of the following properties is considered to be hazardous waste identified by the waste code D001:
   - A liquid, except aqueous solutions, containing less than 24% alcohol and that has a flash point of less than 60 °C (140 °F).
   - A nonliquid capable of spontaneous and sustained and sustained combustion under normal conditions.
   - An ignitable, compressed gas as defined by US DOT regulations.
   - An oxidizer as defined by US DOT regulations.
2) Corrosivity – Corrosive wastes can react dangerously with other wastes, dissolve metal or other materials, or burn the skin. Examples include waste rust removers, waste acid or alkaline cleaning fluids, and waste battery acid. The regulations describing the corrosivity characteristic are found in 40 CFR 261.22. A waste that exhibits either of the following properties is considered to be hazardous waste identified by the waste code D002.
   - An aqueous material with a pH of less than or equal to 2 or of greater than or equal to 12.5.
   - A liquid that corrodes steel at a rate greater than 0.25 in./year at a temperature of 55 °C (130 °F).
3) Reactivity – Reactive wastes are unstable and undergo a rapid or violent chemical reaction with water or other materials. Examples include cyanide plating wastes, waste bleaches, and other waste oxidizers. Reactivity characteristic is described in the regulations at 40 CFR 261.23. A waste that exhibits any of the following properties is considered to be a hazardous waste identified by the waste code D003.
   - Normally unstable and reacts violently without detonating.
   - Reacts violently with water (i.e. causes sudden flash of fire accompanied by spattering).
   - Forms an explosive mixture with water (i.e. causes an explosion) that may damage the container and anything or anyone near it.
   - Generate toxic gases, vapors, or fumes when mixed with water that are not necessarily accompanied by a violent reaction or explosion.
   - Contains cyanide or sulfide and generates toxic gases, vapors, or fumes at a pH of between 2 and 12.5.
4) Toxicity – Toxic wastes are considered to be hazardous because of the presence of toxic constituents in the wastes at greater than the established regulatory levels. To determine if a waste displays a toxicity characteristic, a test method (i.e. the Toxicity Characteristic Leaching Procedure or TCLP) is performed. The TCLP regulations describing the toxicity characteristic are described in 40 CFR 261.24.
5) Other characteristics – Medical waste that may be infectious include used bandages and hypodermic needle, human tissue and blood, and biological substances generated by hospitals and research institutions. Because of its infectious nature, this waste is handled and disposed of according to specific EPA guidelines.

4.20.7.5 Listed Hazardous Wastes

Listed hazardous wastes are generated by specific industries and processes and are automatically considered hazardous waste based solely on the process that generates them and irrespective of whether a test of the waste shows any of the "characteristics" of hazardous waste. Examples of listed wastes include

- many sludges left over from electroplating processes
- certain waste from iron and steel manufacturing
- wastes from certain cleaning and/or degreasing processes

Hazardous wastes are incorporated into lists published by the USEPA. These lists are organized into three categories:

1) *The F-list* (nonspecific source wastes). This list identifies wastes from common manufacturing and industrial processes, such as solvents that have been used in cleaning or degreasing operations. Because the processes producing these wastes can occur in different sectors of industry, the F-listed wastes are known as wastes from nonspecific sources (see 40 CFR 261.31).
2) *The K-list* (source-specific wastes). This list includes certain wastes from specific industries, such as petroleum refining or pesticide manufacturing. Certain sludges and wastewaters from treatment and production processes in these industries are examples of source-specific wastes (see 40 CFR 261.32).

3) *Discarded wastes*: *P-list* and *U-list* wastes are actually sublists of the same major list applying to discarded wastes. These wastes apply to commercial chemical products that are considered hazardous when discarded and are regulated under the following U.S. Federal Regulation: 40 C.F.R. 261.33 (e) and (f). P-list wastes are wastes that are considered "acutely hazardous" when discarded and are subject to more stringent regulation. Nitric oxide is an example of a P-list waste and carries the number P076. U-listed wastes are considered "hazardous" when discarded and are regulated in a somewhat less stringent manner than P-listed wastes. Acetone is an example of a U-listed waste and carries the number U002.

4.20.8 Few RCRA Provisions: Cradle-to-Grave Requirements

The EPA's cradle-to-grave hazardous waste management system is an attempt to track hazardous waste from its generation point (the "cradle") to its ultimate disposal point (the "grave") (Figure 4.16). Arguably, the most notable provisions of the RCRA statute are included in Subtitle C, which directs EPA to establish controls on the management of hazardous wastes from their point of generation, through their transportation and TSD. Because RCRA requires controls on hazardous waste generators (i.e. sites that generate hazardous waste), transporters, and treatment, storage and disposal facilities (TSDFs) (i.e. facilities that ultimately treat/dispose of or recycle the hazardous waste), the overall regulatory framework has become known as the "cradle-to-grave" system. The program imposes stringent record keeping and reporting requirements on generators, transporters, and operators of TSDFs handling hazardous waste.

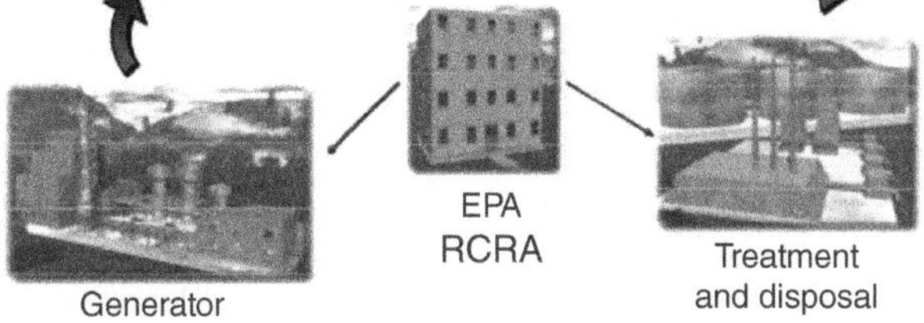

Figure 4.16 Hazardous wastes from cradle to grave.

The uniform hazardous waste manifest (the *manifest*) is the key to cradle-to-grave waste management (Section D.6). Through the use of a manifest, generators can track the movement of hazardous waste from point of generation to the point of ultimate TSD.

4.20.8.1 Nonhazardous Solid Wastes

Nonhazardous solid wastes include certain hazardous wastes which are exempted from the Subtitle C regulations, such as hazardous wastes from households and from conditionally exempt small quantity generators. Oil and gas exploration and production wastes, such as drill cuttings, produced water, and drilling fluids are categorized as "special wastes" and are also exempt from Subtitle C (USEPA 2002b), Subtitle D also includes garbage (e.g. food containers, coffee grounds), nonrecycled household appliances, residue from incinerated automobile tires, refuse such as metal scrap, construction materials, and sludge from industrial and municipal wastewater facilities and drinking water treatment plants.

4.20.8.2 Underground Storage Tanks

Background

The operation of underground storage tanks (USTs) became subject to the RCRA regulatory program with enactment of the Hazardous and Solid Waste Amendments of 1984. At that time there were about 2.1 million tanks subject to federal regulation, and the EPA program led to closure and removal of most substandard tanks. As of 31 March 2019 there were approximately 548 700 active USTs at 198 000 sites subject to federal regulation (USEPA 2019a).

The new regulations include a number of exclusions:

- Hazardous waste UST systems
- Regulated wastewater treatment facilities
- Any UST system of less than 41.5 l
- Any UST system containing a *de minimis* (negligible) concentration of regulated substances
- Any emergency spill or overflow containment system that is expeditiously emptied after use

4.20.8.3 Transportation of Hazardous Waste

This section covers important aspects of transporting hazardous wastes. This should familiarize readers with the complexity of the risks and challenges involved. In 1975, the US Congress passed the Hazardous Materials Transportation Act. The intent of this law was to improve regulatory and enforcement activities by giving the Secretary of Transportation broad authority to regulate all aspects of hazardous material transport. The mode of such transportation includes aircraft, water-craft, motor vehicle, and railway.

Table 4.12 Classification of hazardous materials for shipping and packaging.

Hazard	Class	Division
Forbidden materials	—	—
Forbidden explosives	—	—
Explosives (with a mass explosion hazard)	1	1.1
Explosives (with a mass projection hazard)	1	1.2
Explosives (with predominantly a fire hazard)	1	1.3
Explosives (with no significant blast hazard)	1	1.4
Very intensive explosives, blasting agents	1	1.5
Extremely insensitive detonating substances	1	1.6
Flammable gas	2	2.1
Nonflammable compressed gas	2	2.2
Poisonous gas	2	2.3
Flammable and combustible liquid	3	—
Flammable solid	4	4.1
Spontaneously combustible material	4	4.2
Dangerous when wet material	4	4.3
Oxidizer	5	5.1
Organic peroxide	5	5.2
Poisonous materials	6	6.1
Infectious substance (etiological agent)	6	6.2
Radioactive material	7	—
Corrosive material	8	—
Miscellaneous hazardous material	9	—
Other regulated materials	—	—

Source: 40 CFR 173.

The U.S. Federal Aviation Administration enforces regulations pertaining to air transportation. The U.S. Coast Guard enforces regulations pertaining to water transportation. Similarly, the U.S Federal Highway Administration enforced regulation associated with hazardous waste transportation by highways and railroads, respectively.

Table 4.12 lists the classifications of hazardous materials that are subject to federal transportation regulations. The purpose is to regulate materials likely to cause injury to public and the carrier personnel if they are accidently released during transport. The Hazardous Materials Table in 49 CFR 172.101 further defines the labeling, packing, and carrier requirements for specific hazardous chemicals (see Section 4.5.13).

4.20.9 TSDFs Permits

TSDFs manage hazardous waste under RCRA Subtitle C and generally must have a permit in order to operate. While most facilities have RCRA permits, some continue to operate under what is called "interim status." Interim status requirements appear in 40 CFR Part 265 (USEPA 2012).

The permitting requirements for TSDFs appear in 40 CFR Parts 264 and 270. TSDFs manage (treat, store, or dispose) hazardous waste in units that may include container storage areas, tanks, surface impoundments, waste piles, land treatment units, landfills, incinerators, containment buildings, and/or drip pads. The unit-specific permitting and operational requirements are described in further detail in 40 CFR Part 264, Subparts J through DD (USEPA 1999).

4.20.9.1 Recycling

Some hazardous wastes can be recycled into new products. Examples may include lead-acid batteries or electronic circuit boards. When heavy metals in these types of ashes go through the proper treatment, they could bind to other pollutants and convert them into easier-to-dispose solids, or they could be used as pavement filling. Such treatments reduce the level of threat of harmful chemicals, like fly and bottom ash while also recycling the safe product.

4.20.9.2 Portland Cement

Another commonly used treatment is cement-based solidification and stabilization (Chaudhary and Rachana 2006). Cement is used because it can treat a range of hazardous wastes by improving physical characteristics and decreasing the toxicity and transmission of contaminants. The cement produced is categorized into five different divisions, depending on its strength and components. This process of converting sludge into cement might include the addition of pH adjustment agents, phosphates, or sulfur reagents to reduce the settling or curing time, increase the compressive strength, or reduce the leach ability of contaminants.

4.20.9.3 Incineration, Destruction, and Waste-to-Energy

The importance of the three T's of incineration – temperature, time, and turbulence – has been recognized for many years. Recent general guidelines for three T's include higher temperature (1200–2000 °F) and longer residence times (0.2–2.0 seconds) in order to promote more complete destruction of HAPs. Hazardous waste may be "destroyed." For example, by incinerating it at a high temperature, flammable wastes can sometimes be burned as energy sources. For example, many cement kilns burn hazardous wastes like used oils or solvents. Today, incineration treatments not only reduce the amount of hazardous waste but also generate energy from the gases released in the process. It is known that this particular waste treatment releases toxic gases produced by the combustion of by-product or other materials which can affect the environment. However, current technology has developed more efficient incinerator units that control these emissions to a point where this treatment is considered a more beneficial option (e.g. waste-to-energy – see Section 7.13.5 and Appendix G).

Example 4.19 Mercury Waste
An incinerator burns mercury-contaminated medical waste. The waste material has an ash content of 1%. The solid waste feed rate is 1000 lb/h and the gas flow rate is 20000 dscf. It is reported that the average mercury content of the particulates was 2.42 μg/g, when the vapor concentration was 0.3 mg/dscm (dry standard cubic meter). For the case where the incinerator emissions meet the particulate standard of 0.08 gr/dscf (0.1832 g/dscm) with a 99.5% efficient electrostatic precipitator (ESP), calculate

1) the amount of mercury bound to the fly ash, which is captured in the ESP in grams/day
2) the amount of mercury leaving the stack as a vapor and with the fly ash in grams/day.

Solution
The amount of ash leaving the stack is:

$$\left(0.08\frac{\text{gr}}{\text{dscf}}\right)\left(\frac{1}{7000}\frac{\text{lb}}{\text{gr}}\right)\left(20000\frac{\text{dscf}}{\text{min}}\right)\left(60\frac{\text{min}}{\text{h}}\right)\left(24\frac{\text{h}}{\text{day}}\right)$$

$$=\mathbf{329\,lb\frac{ash}{d}}$$

The amount of ash collected in the ESP with a 99.5% efficient ESP is

$$=\left(\frac{329\frac{\text{lb}}{\text{day}}}{(1.0-0.995\,\text{collected})}\right)=\mathbf{65800\frac{lb}{day}}$$

The amount of mercury leaving the stack with the fly ash is

$$=\left(329\frac{\text{lb ash}}{\text{day}}\right)\left(2.42\times10^{-6}\frac{\text{g Hg}}{\text{g ash}}\right)$$

$$=7.96\times10^{-4}\,\text{lb}\frac{\text{Hg}}{\text{day}}\times454\frac{\text{g}}{\text{lb}}=\mathbf{0.361\,g\frac{Hg}{day}}$$

The amount of mercury leaving the stack as vapor is

$$=\left(3\times10^{-4}\frac{\text{g Hg}}{\text{dscm}}\right)\left(20000\frac{\text{dscf}}{\text{min}}\right)\left(\frac{1}{35.31}\frac{\text{m}^3}{\text{ft}^3}\right)$$

$$\left(60\frac{\text{min}}{\text{h}}\right)\left(24\frac{\text{h}}{\text{day}}\right)$$

$$=244.689\frac{\text{g}}{\text{day}}$$

Therefore, total mercury leaving the stack = 244.689 + 0.361 = **245 g/day.**

4.20.9.4 Pyrolysis

Some hazardous waste types may be eliminated using pyrolysis in an ultrahigh temperature electrical arc, in inert conditions to avoid combustion. This treatment method may be preferable to high-temperature incineration in some circumstances such as in the destruction of concentrated organic waste types, including PCBs, POHCs, pesticides, and other persistent organic pollutants.

USEPA regulations for hazardous waste incineration (HWI) facility requires a minimum of 99.99% destruction and removal efficiency (DRE) of the principal organic hazardous constituents (e.g. PCBs, POHCs) being burned. In these cases, the final operating permits issued by the EPA almost always require a high temperature (1800 °F or higher) and minimum of 2.0 seconds residence time in the afterburner. Hospital (biomedical) waste incinerators often utilize a HWI type of facility.

The DRE is determined for each designated POHC from a mass balance of the waste introduced into the incinerator:

$$\text{DRE}=\frac{(W_{\text{in}}-W_{\text{out}})}{W_{\text{in}}}\times100\% \tag{4.19}$$

where

DRE = destruction and removal efficiency (percent)
W_{in} = mass feed rate of POHC fed to an incinerator (lb/h or kg/h)
W_{out} = mass emission rate of POHC fed to the atmosphere (lb/h or kg/h)

Example 4.20 During a trial burn, an incinerator was operated at a waste feed rate of 5000 lb/h and 7% O_2 dry volume in the stack gases. The gas flow rate measured in the stack was 19200 dscfm. Under these conditions, the measured concentrations of the principal organic hazardous component were

Trichlorethylene (1) = 4.9 μg/dscf
1,1,1-Trichloroethane (2) = 1.0 μg/dscf
Methylene chloride (3) = 49 μg/dscf
Perchloroethylene (4) = 490 μg/dscf

Each hazardous component listed here constitutes approximately 5% of the total waste feed rate. Calculate the DRE of the POHCs. Is the unit in compliance with regulations?

Solution
Each hazardous component constitutes 5% of the total waste feed, therefore,

$$W_{\text{in}}=(0.05)(\text{total feed rate})=(0.05)(5000\,\text{lb/h})$$
$$=250\,\text{lb/h}$$

The mass flow rate of each hazardous component in the stack is given by

$$\left(W_{\text{out}}\right)_i = \frac{q_s C_i}{7.57 \times 10^6}$$

where

$$q_s = \text{volumetric gas flow rate}\left(\text{dscfm}\right)$$
$$C_i = \text{concentration of component } i\left(\mu\text{g/dscf}\right)$$
$$7.57 \times 10^6 = \text{concentration factor that converts } \mu\text{g/min to lb/h}$$

(Note that (454 g/h)(10^6 μg/g)(h/60 min) = 7.57×10^6)
Therefore,

$$\left(W_{\text{out}}\right)_1 = \left(\frac{(19200)(4.9)}{7.57 \times 10^6}\right) = 0.01243\,\text{lb/h}$$

$$\left(W_{\text{out}}\right)_2 = \left(\frac{(19200)(1.0)}{7.57 \times 10^6}\right) = 0.00254\,\text{lb/h}$$

$$\left(W_{\text{out}}\right)_3 = \left(\frac{(19200)(49)}{7.57 \times 10^6}\right) = 0.1243\,\text{lb/h}$$

$$\left(W_{\text{out}}\right)_4 = \left(\frac{(19200)(490)}{7.57 \times 10^6}\right) = 1.243\text{lb/h}$$

The DRE for component i, DRE_i, is given by Eq. (4.16):

$$\text{DRE} = \left[\left(\frac{\left(W_{\text{in}}\right)_i - \left(W_{\text{out}}\right)_i}{W_{\text{in}}}\right)\right]100\%$$

$$\text{DRE}_1 = \left(\frac{250 - 0.01243}{250}\right)100 = 99.995\%$$

$$\text{DRE}_2 = \left(\frac{250 - 0.00254}{250}\right)100 = 99.999\%$$

$$\text{DRE}_3 = \left(\frac{250 - 0.1243}{250}\right)100 = 99.95\%$$

$$\text{DRE}_4 = \left(\frac{250 - 1.1243}{250}\right)100 = 99.55\%$$

Regulations require that minimum 99.99% DRE be achieved. In this case, only two of the principal organic hazardous components (1 and 2) achieved this status. Assuming that either component 3 or 4 (or both) was a designated POHC, the unit would be out of compliance.

4.21 Hazardous Waste Landfill (Sequestering, Isolation, etc.)

Hazardous waste may be sequestered in a hazardous waste landfill or permanent disposal facility. "In terms of hazardous waste, a landfill is defined as a disposal facility or part of a facility where hazardous waste is placed or on land and which is not a pile, a land treatment facility, a surface impoundment, an underground injection well, a salt dome formation, a salt bed formation, an underground mine, a cave, or a corrective action management unit" (40 CFR 260.10).

4.21.1 Final Disposal of Hazardous Waste

Historically, some hazardous wastes were disposed of in regular landfills. This resulted in unfavorable amounts of hazardous materials seeping into the ground. These chemicals eventually entered to natural hydrologic systems. Many landfills now require countermeasures against groundwater contamination. For example, a barrier has to be installed along the foundation of the landfill to contain the hazardous substances that may remain in the disposed waste. Currently, hazardous wastes must often be stabilized and solidified in order to enter a landfill and must undergo different treatments in order to stabilize and dispose them. Most flammable materials can be recycled into industrial fuel. Some materials with hazardous constituents can be recycled, such as lead-acid batteries.

Example 4.21 Land Disposal Daily Cover
Landfill-DNR asks: How much material do we need for daily cover for our proposed hazardous waste landfill? How much of our total volume (air space) will be used as the daily cover during the life of the landfill? How much longer could the landfill be used if daily cover were not employed? The landfill is 500 m × 250 m in average plan and 15 m high. Daily filling rates are 15 m × 10 m × 2 m and the daily cover is 0.3 m.

Solution

The total landfill air space is

$$V_1 = 500 \times 250 \times 15 = 1.88 \times 10^6\,\text{m}^3.$$

Each daily layer (lift) has a volume of

$$V_2 = 15 \times 10 \times 2 = 300\,\text{m}^3$$

The daily cover has a volume of

$$V_3 = 1.5 \times 10 \times 0.3 = 45\,\text{m}^3$$

The volume of air space (in percent) used by daily cover is

$$V_{\text{pc}} = \left(V_3/V_1\right) \times 100 = 45/300 \times 100 = 15\%$$

The life of the landfill with daily cover is

$$t_1 = V_1/\left(V_2 + V_3\right) = 5449\text{ days}$$
$$= 21\text{ years at 5 days/week}$$

The life of the landfill without daily cover is

$$t_2 = V_1/V_2 = 6267\,\text{days}$$
$$= 24 \text{ years at 5 days/week}$$

Therefore, the landfill life will be extended by

$$t_3 = t_2 - t_1 = 818 \text{ days or about 2.24 years.}$$

4.22 Industrial Waste Generation Rates

Industrial waste is the waste produced by industrial activity which includes any material that is rendered useless during a manufacturing process such as that of factories, industries, mills, and mining operations. It has existed since the start of the Industrial Revolution. Some examples of industrial wastes are chemical solvents, pigments, liquid waste and sludges, metals, plastic, ash, paints, sandpaper, paper products, industrial by-products, electronic waste, metals, glass, and radioactive wastes.

Given the shortcomings in the available data, it is clear that there are significant uncertainties in the United States estimates of industrial waste generation. Because there is no single source of data on waste generation that is truly comprehensive, trying to develop a national picture of industrial waste generation is much like trying to develop a picture of an elephant by having a dozen blindfolded people touch different parts of the animal. Some waste databases touch the tail, other the trunk, only a few touch the main body of the elephant.

The best estimates available from national waste generation data indicate that more than 12 billion T of combined industrial waste are generated in the United States annually. Only 800 million T, or less than 10% of this waste mass, are defined as hazardous under RCRA. Based on EPA data from sources such as the Biennial Report System, the Toxic Chemical Release Inventory, and the National Survey of Hazardous Waste Generators, more than 25 000 generators of hazardous waste can be identified. The precise number of generators of hazardous waste is subject to considerable uncertainty, due to definitions of hazardous waste and reporting limits that vary from inventory to inventory and from state to state. Despite the uncertainties, it is clear that there are tens of thousands of industrial hazardous generators in the United States producing 300–800 million T of hazardous waste annually. The large uncertainty in the amount of waste generated is again due to differing definitions.

4.22.1 Generator Requirements and Responsibilities

Generators of hazardous waste are the first link in the cradle-to-grave chain of hazardous waste management established under RCRA. The regulatory requirements for hazardous waste generators include (USEPA 1986) the following:

1) Obtaining an EPA identification (ID) number
2) Handling of hazardous waste before transport
3) Manifesting of hazardous waste
4) Record keeping and reporting

Generators are responsible for determining whether the wastes they generate are hazardous and, if so, for managing them in with 40 CFR 261.20-261.24. Determine if materials are hazardous wastes involves classifying them as waste, ensuring they are not excluded from RCRA regulations, and evaluating them for the hazardous properties outlined in the regulations.

The USEPA recognized that hazardous waste regulations would impose a substantial burden on small generators if they were brought entirely within the Subtitle C regulatory system. Thus, the EPA has exempted small-quantity generators from most hazardous waste requirements. The EPA defines generators as follows:

- Large-quantity generators (LQGs) are those facilities producing more than 1000 kg of hazardous waste per calendar month (about 2200 lb). LQGs have permit to store hazardous waste in 55-gal drums on-site and keep them on-site for 90 days or less (may be extended by regulator).
- Small-quantity generators (SQGs) are those facilities producing more than 100 kg but less than 1000 kg of hazardous waste per calendar month. SQGs may store hazardous waste in 55-gal drums and keep them on-site for 180 days or less; if a disposal site is over 200 miles away, up to 270 days.
- *Conditionally exempt small-quantity generators* are those facilities producing less than 100 kg of hazardous waste or less than 1 kg of acutely hazardous waste per calendar month. The requirements for this category additionally limit the facility's waste accumulation to less than 1000 kg of hazardous waste, 1 kg of acute hazardous waste, or 100 kg of any residue from the cleanup of spill of acutely hazardous waste at any time.

4.22.2 Environmental Audits

The USEPA began proposing environmental audits as an element of enforcement case settlements during the mid-1970s (Federal Register 1994). Such settlements are frequent The EPA defines environmental audits as a systematic, documented, periodic, and objective review by regulated entities of facility operations and practices related to meeting environmental requirements. Audits can be designed to verify compliance with environmental requirements, evaluate the effectiveness of environmental management systems already in place, and assess risks from both regulated and unregulated materials and practices.

Environmental auditing is a management tool that is considered to generally improve the overall environmental performance of manufacturing facilities. These audits usually focus on operating facilities that may involve processing, manufacturing, material procurement, storage, maintenance, distribution, and waste management. Environmental auditing should detect problems such as noncompliance and improper or inadequate waste management operations and, hopefully, permit timely correction. Managers of hazardous waste facilities, however, are doubtful and are concerned about the confidentiality of trade secret and possible discoveries by outsiders who may compete for similar manufacturing or processing business.

4.23 Comprehensive Environmental Response, Compensation, and Liability Act and Superfund

Superfund is a United States federal government program designed to fund the cleanup of sites contaminated with hazardous substances and pollutants. Sites managed under this program are referred to as "Superfund" sites. It was established as the Comprehensive Environmental Response, Compensation, and Liability Act of 1980 (CERCLA) (USEPA 1980). It authorizes federal natural resource agencies, primarily the EPA, states, and Native American tribes to recover natural resource damages caused by hazardous substances, though most states have and most often use their own versions of CERCLA. CERCLA created the Agency for Toxic Substances and Disease Registry. The EPA may identify parties responsible for hazardous substances releases to the environment (polluters) and either compel them to clean up the sites, or it may undertake the cleanup on its own using the Superfund (a trust fund) and costs recovered from polluters by referring to the U.S. Department of Justice.

Approximately 70% of Superfund cleanup activities historically have been paid for by parties responsible (PRPs) for the cleanup of contamination. The exceptions occur when the responsible party either cannot be found or is unable to pay for the cleanup. Until the mid-1990s, most of the funding came from a tax on the petroleum and chemical industries, reflecting the polluter pays principle, but since 2001 most of the funding for cleanups of hazardous waste sites has come from taxpayers. Despite the name, the program has suffered from under-funding, and Superfund cleanups have decreased to a mere 8 in 2014, out of over 1200. As a result, the EPA typically negotiates consent orders with the potentially responsible parties (PRPs) to study sites and develop cleanup alternatives, subject to EPA oversight and approval of all such activities.

The EPA and state agencies use the Hazard Ranking System (HRS) to calculate a site score (ranging from 0 to 100) based on the actual or potential release of hazardous substances from a site. A score of 28.5 places a site on the National Priorities List, eligible for long-term remedial action (i.e. cleanup) under the Superfund program. As of 12 June 2019, there were 1344 sites listed, an additional 255 had been delisted, and 41 new sites have been proposed.

4.23.1 History

CERCLA was enacted by Congress in 1980 in response to the threat of hazardous waste sites, typified by the Love Canal disaster in New York, and the Valley of the Drums in Kentucky. The initial trust fund to clean up a site where a polluter could not be identified, could not or would not pay (bankruptcy or refusal) consisted of about $1.6 billion (Beins and Lester 2015).

The EPA published the first HRS in 1981, and the first National Priorities List (NPL) in 1983. Implementation during early years, the two terms of the Reagan administration was ineffective, as only 16 of the 799 Superfund sites were cleaned up, and only $40 million of $700 million in recoverable funds from responsible parties were collected. Reagan's policies were described as laissez-faire.

The Superfund Amendments and Reauthorization Act of 1986 added minimum cleanup requirements in Section 121 and required that most cleanup agreements with polluters be entered in federal court as a consent decree subject to public comment (section 122). This was to address sweetheart deals between industry and the Reagan-era EPA, that Congress had discovered.

According to U.S. Government Accountability Office (2015) report, since 2001, most of the funding for cleanups of hazardous waste sites has come from taxpayers; a state pays 10% of cleanup costs in general and at least 50% of it is operated by the facility responsible for contamination. By 2013 funding had decreased from $2 billion in 1999 to less than $1.1 billion (Ryan 2004).

4.23.2 Provisions

CERCLA authorizes two kinds of response actions:

1) *Removal actions*. These are typically short-term response actions, where actions may be taken to address releases or threatened releases requiring prompt response.

Removal actions are classified as (i) emergency; (ii) time-critical; and (iii) non-time critical. Removal responses are generally used to address localized risks such as abandoned drums containing hazardous substances, and contaminated surface soils posing acute risks to human health or the environment.

2) *Remedial actions.* These are usually long-term response actions. Remedial actions seek to permanently and significantly reduce the risks associated with releases or threats of releases of hazardous substances and are generally larger more expensive actions. They can include measures such as using containment to prevent pollutants from migrating, and combinations of removing, treating, or neutralizing toxic substances. These actions can be conducted with federal funding only at sites listed on the EPA NPL in the United States and the territories. Remedial action by responsible parties under consent decrees or unilateral administrative orders with EPA oversight may be performed at both NPL and non-NPL sites, commonly called Superfund Alternative Sites in published EPA guidance and policy documents.

A PRP is a possible polluter who may eventually be held liable under CERCLA for the contamination or misuse of a particular property or resource. Four classes of PRPs may be liable for contamination at a Superfund site:

1) the current owner or operator of the site
2) the owner or operator of a site at the time that disposal of a hazardous substance, pollutant, or contaminant occurred
3) a person who arranged for the disposal of a hazardous substance, pollutant, or contaminant at a site
4) a person who transported a hazardous substance, pollutant, or contaminant to a site, who also has selected that site for the disposal of the hazardous substances, pollutants, or contaminants.

4.23.3 Procedures

Map of Superfund sites. Red indicates currently on final National Priority List, yellow is proposed, green is deleted (usually meaning having been cleaned up). This map is as of 13 September 2018. Upon notification of a potentially hazardous waste site, the EPA conducts a Preliminary Assessment/Site Inspection (PA/SI), which involves records reviews, interviews, visual inspections, and limited field sampling. Information from the PA/SI is used by the EPA to develop a HRS score to determine the CERCLA status of the site. Sites that score high enough to be listed typically proceed to a Remedial Investigation/Feasibility Study (RI/FS). The RI includes an extensive sampling program and risk assessment that defines the nature and extent of the site contamination and risks. The FS is used to develop and evaluate various remediation alternatives. The preferred alternative is presented in a Proposed Plan for public review and comment, followed by a selected alternative in a record of decision (ROD). The site then enters into a Remedial Design phase and then the Remedial Action phase. Many sites include long-term monitoring. Five-year reviews once the Remedial Action has been completed are required whenever hazardous substances are left on-site above levels safe for unrestricted use.

- The CERCLA information system (CERCLIS) is a database maintained by the EPA and the states that lists sites where releases may have occurred, must be addressed, or have been addressed. CERCLIS consists of three inventories: the CERCLIS Removal Inventory, the CERCLIS Remedial Inventory, and the CERCLIS Enforcement Inventory.
- The *Superfund Innovative Technology Evaluation* (SITE) program supports development of technologies for assessing and treating waste at Superfund sites. The EPA evaluates the technology and provides an assessment of its potential for future use in Superfund remediation actions. The SITE program consists of four related components: the Demonstration Program, the Emerging Technologies Program, the Monitoring and Measurement Technologies Program, and Technology Transfer Activities.
- A *reportable quantity* is the minimum quantity of a hazardous substance which, if released, must be reported.
- A *source control action* represents the construction or installation and start-up of those actions necessary to prevent the continued release of hazardous substances (primarily from a source on top of or within the ground, or in buildings or other structures) into the environment (40 C.F.R. 300.5).
- A *section 104(e) letter* is a request by the government for information about a site. It may include general notice to a potentially responsible party that CERCLA-related action may be undertaken at a site for which the recipient may be responsible (42 U.S.C. 9604 e). This section also authorizes the EPA to enter facilities and obtain information relating to PRPs, hazardous substances releases, and liability, and to order access for CERCLA activities. The 104(e) letter information-gathering resembles written interrogatories in civil litigation.
- A *section 106 order* is a unilateral administrative order issued by EPA to PRP(s) to perform remedial actions at a Superfund site when the EPA determines there may be an imminent and substantial endangerment to the

public health or welfare or the environment because of an actual or threatened release of a hazardous substance from a facility, subject to treble damages and daily fines if the order is not obeyed.

- A *remedial response* is a long-term action that stops or substantially reduces a release of a hazardous substance that could affect public health or the environment. The term remediation, or cleanup, is sometimes used interchangeably with the terms remedial action, removal action, response action, remedy, or corrective action.
  - A *nonbinding allocation of responsibility* (NBAR) is a device, established in the Superfund Amendments and Reauthorization Act, that allows the EPA to make a nonbinding estimate of the proportional share that each of the various responsible parties at a Superfund site should pay toward the costs of cleanup.
- *Relevant and appropriate requirements* are those United States federal or state cleanup requirements that, while not "applicable," address problems sufficiently similar to those encountered at the CERCLA site that their use is appropriate. Requirements may be relevant and appropriate if they would be "applicable" except for jurisdictional restrictions associated with the requirement (40 C.F.R. 300.5).

4.23.4 Implementation

As of 12 June 2019, there were 1344 sites listed on the National Priority List, an additional 255 had been delisted, and 41 new sites were proposed.

Historically, about 70% of Superfund cleanup activities have been paid for by PRPs. When the party either cannot be found or is unable to pay for the cleanup, the Superfund law originally paid for toxic waste cleanups through a tax on petroleum and chemical industries. The chemical and petroleum fees were intended to provide incentives to use less toxic substances. Over five years, $4.2 billion was collected, and the tax went to a trust fund for cleaning up abandoned or uncontrolled hazardous waste sites.

4.23.5 Hazard Ranking System

The HRS is a scoring system used to evaluate potential relative risks to public health and the environment from releases or threatened releases of hazardous wastes at uncontrolled waste sites. Under the Superfund program, the EPA and state agencies use the HRS to calculate a site score (ranging from 0 to 100) based on the actual or potential release of hazardous substances from a site through air, surface water, or groundwater. A score of 28.5 places the site on the National Priorities List, making the site eligible for long-term remedial action (i.e. cleanup) under the Superfund program.

4.23.6 Environmental Discrimination

Federal Actions to address the disproportionate health and environmental disparities that minority and low-income populations face through Executive Order (E.O) 12898 required federal agencies to make environmental justice central to their programs and policies. Superfund sites have been shown to impact minority communities the most. Despite legislation specifically designed to ensure equity in Superfund listing, marginalized populations still experience a lesser chance of successful listing and cleanup than areas with higher income levels. After the executive order had been put in place, there persisted a discrepancy between the demographics of the communities living near toxic waste sites and their listing as Superfund sites, which would otherwise grant them federally funded cleanup projects. Communities with both increased minority and low-income populations were found to have lowered their chances of site listing after the executive order, while on the other hand, increases in income led to greater chances of site listing (O'Neil 2007). Of the populations living within 1 mile radius of a Superfund site, 44% of those are minorities despite only being around 37% of the nation's population. It has also been shown that the government responds slower to community demands from minority communities than from white communities. Superfund sites near white communities have seen better clean up and harsher penalties for the polluters than minority communities (Bullard 2012).

4.23.7 Case Studies in African American Communities

In 1978, residents of the rural black community of Triana, Alabama, were found to be contaminated with DDT (Dichlorodiphenyltrichloroethane) and PCB (Polychlorinated byphenyl), some of whom had the highest levels of DDT ever recorded in human history (Bullard 2012). The DDT was found in high levels in Indian Creek, which many residents relied on for sustenance fishing. Although this major health threat to residents of Triana was discovered in 1978, the federal government did not act until five years later after the mayor of Triana filed a class-action lawsuit in 1980.

In West Dallas, Texas, a mostly African American and Latino community, a lead smelter poisoned the surrounding neighborhood, elementary school, and day cares for more than five decades. Dallas city officials were informed in 1972 that children in the proximity of the smelter were being exposed to lead contamination. The city sued the lead smelters in 1974, then reduced its lead regulations in 1976. It wasn't until 1981 that the EPA commissioned a study on the lead contamination in this neighborhood, and found the same results that had been found a decade earlier. In 1983, the surrounding day cares had to close due to the lead exposure while the lead smelter remained operating. It was later revealed that EPA Deputy Administrator John Hernandez had deliberately stalled the clean-up of the lead-contaminated hot spots. It wasn't until 1993 that the site was declared a Superfund site, and at the time it was one of the largest ones. However, it was not until 2004 when the EPA completed the clean-up efforts and eliminated the lead pollutant sources from the site.

The Afton community of Warren County, North Carolina is one of the most prominent environmental injustice cases and is often pointed to as the roots of the environmental justice movement. PCBs were illegally dumped into the community and then it eventually became a PCB landfill. Community leaders pressed the state for the site to be cleaned up for an entire decade until it was finally detoxified (Braithwaite et al. 2009). However, this decontamination did not return the site to its pre-1982 conditions. There has been a call for reparations to the community which has not yet been met.

Bayview-Hunters Point, San Francisco, a historically African American community, has faced persistent environmental discrimination due to the poor remediation efforts of the San Francisco Naval Shipyard, a federally declared Superfund site. The negligence of multiple agencies to adequately clean this site has led Bayview residents to be subject to high rates of pollution, gentrification, and has been tied to high rates of cancer, asthma, and overall higher health hazards than other regions of San Francisco.

4.23.8 Case Studies in Native American Communities

One example is the Church Rock uranium mill spill on Navajo Nation. It was the largest radioactive spill in the United States, but received a long delay in government response and cleanup after being placed as a lower priority site. Two sets of five-year clean-up plans have been put in place by US Congress, but contamination from the Church Rock incident has still not been completely cleaned up. Today, uranium contamination from mining during the Cold War era remains throughout the Navajo Nation, posing health risks to the Navajo community.

Cuts to the EPA's funding and resources would hinder the regulation and remediation of Superfund sites. This would perpetuate the exposure to health risks that adjacent communities face from proximity to the Superfund site. Delays in government response to Superfund conditions increase the exposure of health risks to proximate communities (Konisky 2015; U.S. Government Accountability Office 2015).

4.24 Industrial Waste Management in India: Shifting Gears

With India's growing population, coupled with rapid industrialization and urbanization, waste generation is projected to increase from 62 million T/Y to about 165 million T in 2030 (ASSOCHAM India 2017; TERI 2014). India has been witnessing a paradigm shift in policy, perspective, and industrialization (make in India initiative), along with the *Swachh Bharat Abhiyan* (Clean India Campaign) initiative. Industries and corporations are committed to play a catalytic role in promoting the concept of a "green economy" by urging all stake holders including city, state, and federal waste management to work together as part of the "Clean India Initiative." Figure 4.17 shows a general classification of waste including industrial hazardous and non-hazardous, urban, e-waste, and biomedical. These are generated during the extraction of raw materials into intermediate and final products, the consumption of final products, and other human activities.

4.24.1 Integrated Solid Waste Management

Integrated solid waste management (ISWM) refers to the strategic approach to sustainable management of solid wastes covering all sources and all aspects such as generation, segregation, transfer, sorting, treatment, recovery, and disposal in an integrated manner, with an emphasis on maximizing resource use efficiency. An effective ISWM system considers how to prevent, recycle, and manage solid waste in ways that most effectively protect human health and the environment. ISWM involves evaluating local needs and conditions, and then selecting and combining the most appropriate waste management activities for those conditions. The major ISWM activities are waste prevention, recycling and composting, and combustion (waste-to-energy) and disposal in properly designed, constructed, and managed landfills.

The selection of the most appropriate waste management systems and sustainable technologies is also needed to deliver an optimum and sustainable ISWM system. In combination with economic and social considerations, this approach would help waste managers to design more sustainable and waste management systems.

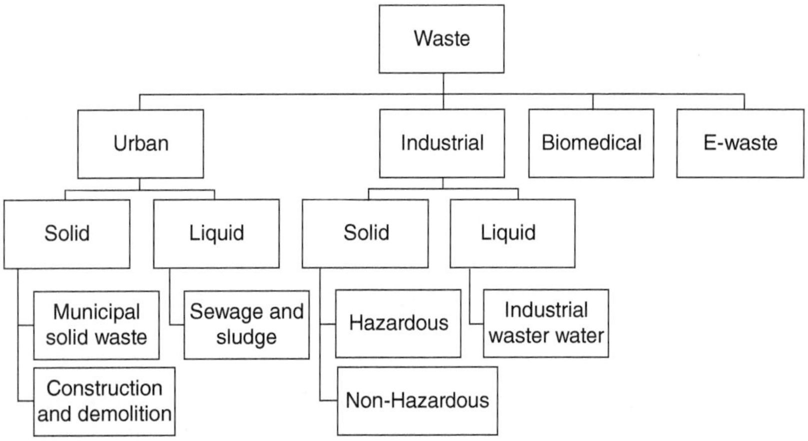

Figure 4.17 Classification of wastes in India. *Source:* From TERI (2014) and ASSOCHAM India (2017).

Thus, for the management of solid waste, the following is the preferred hierarchy of approaches:

- Reduction at *source:* meaning incorporation of the tenets of waste management at every stage of consumption from design, manufacture, purchase, or use of materials to reduce the amount or toxicity of waste generated.
- Environmentally suitable reuse and recycling: to conserve natural resources and energy through systematic segregation, collection, and reprocessing.

ISWM concept has to be adapted with a view that effective management schemes need the flexibility of design, adaptation, and systems which can best meet current social, economic, and environmental conditions. These are likely to change over time and vary by location. ISWM with respect to three perspectives are life-cycle waste generation and waste management (TERI 2014).

4.24.2 Hazardous Waste Handling and Management Rule

To address industrial waste issues, the Ministry notified the Hazardous Wastes (Management and Handling) Rules in 1989 (amended in 2000 and 2003) and Hazardous Wastes (Management, Handling and Transboundary Movement) Rules in 2008. The 1989 Rules were drafted to enable regulatory authorities to control the handling, movement, and disposal of hazardous wastes generated within the country. The amendment in 2000 and 2003 were largely aimed at harmonizing the definition of hazardous waste with provisions of the Basel Convention. The 2008 Rules bring issues of e-waste management into the ambit of hazardous waste management (Gadgil et al. 2011; Khwaja et al. 2010).

4.24.3 Biomedical Waste Rule

The Biomedical Waste (Management and Handling) Rules were issued in 1998 (amended in 2000 and 2003). The Rules are based on the principle of segregation of general waste from biomedical waste (BMW). They lay out color codes for containers, and treatment and disposal options for 10 categories of waste. The state governments have taken initiatives for setting up of Common Biomedical Waste Treatment Facilities (CBWTFs) for processing and disposal of waste. For treatment and disposal of BMW generated in the country, there are 177 common treatment and disposal facilities in operation which were developed by the private entrepreneurs (Gadgil et al. 2011; Khwaja et al. 2010).

4.24.4 E-Waste Rule

The E-waste (management and handling) Rules were notified in May 2011, which became operational from May 2012. Concept of Extended Producer Responsibility was introduced in this rule, making producers responsible for environmentally sound management of their end of life products, including collection and their channelization to registered dismantler or recycler. These rules will apply to every producer, consumer, or bulk consumer involved in the manufacture, sale, and purchase and processing of electrical and electronic equipment or components as specified in Schedule I, collection center, dismantler, and recycler of E-waste (Gadgil et al. 2011).

4.24.5 Plastic Nonhazardous Waste Rule

The Ministry of Environment and Forests notified the draft “Plastics Manufacture, Usage and Waste Management Rules, 2009” to replace the Recycled Plastics Manufacture and Usage Rules, 1999 (amended in 2003) to regulate the manufacture and usage of plastic carry bags. The draft rules were widely published for public comments. An expert committee was constituted by the Ministry to examine these comments and to suggest economic instruments. These Rules were finalized as Plastic Waste (Management and Handling) Rules 2011 and notified on 4 February 2011 (Khwaja et al. 2010).

Problems

4.1 Water flows through a Parshall flume at a rate of $47.8\,ft^3/s$. The throat width is 4.0 ft. Calculate the flume upstream head in feet.

4.2 Calculate the flume upper head in a Parshall flume with a throat width of 1.5 ft for a flow of $5.56\,ft^3/s$.

4.3 A pitot-static tube is mounted on an airplane to measure airspeed. At an altitude of 10 000 ft, where the temperature is 23 °F and the pressure is 10 psia, a pressure difference corresponding to 10 in. of water is measured. What is the airspeed? Water and air properties at 23 °F are given: $\rho_{H_2O} = 62.4\,lb/ft^3$; $R_{air} = 1716\,J/kg\,K$.

4.4 A differential pressure gage is connected to a Pitot-static tube in a wind tunnel. In the wind tunnel test, the gauge indicates a ΔP of 712 Pa. What is the air velocity in the tunnel in m/s? The pressure and temperature in the tunnel are 98 kPa absolute and 23 °C, respectively. Air properties at 20 °C is given: $R_{air} = 287\,J/kg\,K$.

4.5 Water flows through a Parshall flume with a throat width of 4.0 ft at a depth 7.5 ft. Determine the flow rate.

4.6 Water flows through a Parshall flume with a throat width of 5.0 ft at a depth 3.4 ft. Determine the flow rate.

4.7 A rectangular, sharp-crested weir with end contraction is 1.6 m long. How high should it be placed in a channel to maintain an upstream depth of 2.5 m for $0.5\,m^2/s$ flow rate?

4.8 For a sharp-crested suppressed weir ($C_w = 3.33$) of length $B = 8.0$ ft, $P = 2.0$ ft, and $H = 1.0$ ft, determine the discharge over the weir. Neglect the velocity of approach head.

4.9 The head on a 90° V-notch weir is 1.5 ft. Determine the discharge.

4.10 Determine the weir coefficient of a 90° V-notch weir for a head of 18 cm for a flow rate of 20 l/s.

4.11 Determine whether or not a pulverized coal, dry bottom, wall-fired boilers using bituminous coal at a power plant rate at 60 MW meet the NSPS for SO_2, PM, and NO_x. The power plant burns bituminous coal with a sulfur content of 1.8% and ash content 6.2%. The coal has a heating value of 14 000 Btu/lb. The boiler efficiency is 35%. Use the EPA emission factors to estimate the emissions. Assume the efficiency of SO_2 control is 85% and the efficiency of PM control equipment is 99%.

4.12 What does the acronym NPDES refer to? List several ideas behind this permit program. How is the NPDES program monitored in the United States?

4.13 If a river with a seasonal low flow of 200 cfs has a water quality standard for phosphorus of 0.02 mg/l at total-P, find out the TMDL for P. Dispersed agricultural runoff contributes about 10 lbs of P/day. If the MOS is set at 4 lbs of P/day, what would be the maximum WLA of P from a pulp and paper mill's WWTP discharging to the river? Ignore all other point and nonpoint sources.

4.14 List the types of load applications and waste load applications that may be used in developing a TMDL. What other information is needed?

4.15 Define RACT, BACT, MACT, and LAER. What are the differences between these regulatory terminologies in the CAA?

4.16 What is SIP? How SIP works in area of noncompliance with air quality standards.

4.17 Calculate the SO_2 emission rate in units of pounds per hour from a power plant using the following information provided by a continuous emission rate monitoring system:

A Exhaust stack flow is 1 300 000 dry cubic feet per minute at EPA standard conditions (68 °F, 1 atm).

B SO_2 concentration is 25 ppm, dry volume basis.

4.18 A test burn waste mixture consisting of three principal organic hazardous constituent (POHCs) (chlorobenzene, toluene, and xylene) is incinerated at 1000 °C. The waste feed rate and the stack discharge are shown in the following table. The stack gas flow rate is 375.24 dscm/min. Is the unit in compliance?
(Note: HCl emission limit is 1.8 kg/h).

Compound	Inlet (kg/h)	Outlet (kg/h)	Compound	Inlet (kg/h)	Outlet (kg/h)
Chlorobenzene (C_6H_5Cl)	153	0.010	HCl	n/a	1.2
Toluene (C_6H_8)	432	0.037	Particulates	n/a	3.615
Xylene (C_8H_{10})				435	0.070

Note: Outlet concentrations were measured in the stack after APC (air pollution control) equipment.

4.19 The POHCs from a trial burn are shown in the table below. The incinerator was operated at a temperature of at 1100 °C. The stack gas flow rate was 5.90 dscm/s. Assume that all the chlorine in the feed is converted to HCl. The allowable particulate emission is 180 mg/m^3.

Is the unit in compliance if the emissions are measured downstream of the APC equipment?

Compound	Inlet (kg/h)	Outlet (kg/h)
Benzene	913.98	0.2436
Chlorobenzene	521.63	0.0494
Xylenes	1378.91	0.5670
HCl	n/a	4.85
Particulates	n/a	10.61

4.20 A waste mixture consisting of toluene and cyclobenzene is being incinerated at 2050 °F with a 2.2 seconds residence time. The waste inlet feed rate and the stack discharge rate (including particulate catch) are given below. The stack gas flow rate has been determined to be 13 250 dscfm (60 °F, 1 atm) with an O_2 concentration of 4% dry O_2 volume. Is the unit in compliance with existing regulations? Component (1) and (2) are the designated POHCs.

Compound	Inlet (lb/h)	Outlet (lb/h)
1) C_7H_8 (toluene)	953	0.082
2) C_6H_5Cl (chlorobenzene)	337.5	0.022
3) HCl	—	—
4) Particulates	—	—

4.21 A hazardous waste incinerator operates at a chemical plant to treat a liquid slurry production wastestream. Your manager requests that in preparation for a trial burn, you follow a procedure for obtaining compliance with emission regulations. The wastestream feed rate 800 lb/h and the stack gas flow rate is 22 760 scfm (60 °F, 1 atm). The incinerator is to operate at a DRE for hexachlorobenzene.

The analytical laboratory informs you that the detection limit of hexachlorobenzene is 10 μg/l. If the sample is concentrated in 25 ml of solvent, what is minimum volume of the flue gas that must be collected to detect 99.995% DRE and the time that would take to collect the flue gas?

4.22 Describe the RCRA permit process.

4.23 Provide details on the MACT standards for hazardous waste incinerators, cement kilns, and lightweight aggregate kilns.

4.24 Define small quantity generators.

4.25 The following data are from total solids and total volatile solids tests on a pulp and paper mill's wastewater sample. Calculate the total and volatile solids, SS, and dissolved solids.

Total solids data
Weight of empty dish = 68.942 g
Weight of dish plus dry solids = 69.049 g
Weight of dish plus volatile (ignited) solids = 69.003 g
Volume of wastewater sample = 100 ml

4.26 Listed below are total solids and SS data on an industrial wastewater sample. Calculate the total and volatile solids, SS, and dissolved solids.

Total solids data
Weight of empty dish = 85.337 g
Weight of dish plus dry solids = 69.049 g
Weight of dish plus volatile (ignited) solids = 85.375 g
Volume of wastewater sample = 85 ml

SS data
Weight of glass-fiber filter disk = 0.1400 g
Weight of disk plus dry solids = 0.1530 g
Weight of disk plus (volatile) ignited solids = 0.1426 g
Volume of wastewater sample = 200 ml

4.27 An unseeded BOD test is conducted on a polluted surface water sample by adding 100 ml to a 300-ml BOD bottle and filling with dilution water. The initial DO measured 8.2 mg/l, and the final concentration after 5 days of incubation at 20 °C measured 2.9 mg/l. Calculate the BOD.

4.28 A BOD test was conducted on the unchlorinated effluent of an industrial treatment plant. The wastewater portion added to a 300-ml BOD bottle was 30 ml, and the DO values listed below were measured using a DO probe. Plot a BOD vs. time curve and determine the five-day BOD value.

Time (days)	DO (mg/l)	Time (days)	DO (mg/l)
0	8.7	6.0	4.9
2.0	6.7	10.0	3.9
4.0	5.7	14.0	0.7

4.29 What are the major steps in performing the WET test?

References

Adams, E.E. (1982). Dilution analysis for unidirectional diffusers. *Journal of the Hydraulic Division* 108 (HY3): 327–342.

American Public Health Association (APHA) (1989). *Standard Methods for the Examination of Water and Wastewater*, 17e. New York: American Public Health Association.

ASSOCHAM, India (2017). Waste management in India: shifting gears. www.pwc.in (accessed 16 August 2018).

Beins, K. and Lester, S. (2015). Superfund: Polluters Pay so Children Can Play.

Bell, J.N.B. (1998). Recycling, *Microsoft® Encarta® 98 Encyclopedia*. © 1993–1997. Microsoft Corporation.

Benefield, L.D. and Judkins, J.F. (1984). *Treatment Plant Hydraulics for Environmental Engineers*. Upper Saddle River, NJ: Prentice Hall.

Bickerstaffe, J. (1997). The effect of environmental consideration and legislative developments on packaging. In: *Chemical Aspects of Plastics Recycling* (eds. W. Hoyle and D.R. Karsa), 3–12. Cambridge: The Royal Society of Chemistry.

Braithwaite, R.L., Taylor, S.E., and Treadwell, H.M. (eds.) (2009). *Health Issues in the Black Community*, 3e. Hoboken: Wiley.

Brophy, J.H., Hardman, S., and Wilson, D.C. (1997). Polymer cracking for feedstock recycling of mixed plastics waste. In: *Chemical Aspects of Plastics Recycling* (eds. W. Hoyle and D.R. Karsa), 214–219. Cambridge: The Royal Society of Chemistry.

Bullard, R. (2012). *The Wrong Complexion for Protection: How the Government Response to Disaster Endangers African American Communities*, 100–125. New York: NYU Press.

California Air Pollution Control Officers Association (CAPCOA) (1993). *A Compilation of California BACT Determinations Received by the CAPCOA BACT Clearinghouse*, 2e. Sacramento: CAPCOA.

CFR (Code of Federal Regulations) (1977). Appendix W, Part 51, 40 CFR, Title 40.

CFR (Code of Federal Regulations) (1980). 40 CFR §52.21(aa)(2)(v).

CFR (Code of Federal Regulations) (1991). Title 40, § Part 60. Appendix A to Part 60 – Test Methods.

CFR (Code of Federal Regulations) (2011). Title 40, Part 72, Subpart A.

CFR (Code of Federal Regulations) (2017). Part 60, Appendix A Method 9 Visual Determination of the Opacity of Emissions from Stationary Sources.

CFR (Code of Federal Regulations) (2019). Part 60, Appendix A Method 22 Visual Determination of Fugitive Emissions from Material Sources and Smoke Emissions from Flares.

Chaudhary, R. and Rachana, M. (2006). Factors affecting hazardous waste solidification/stabilization: a review. *Journal of Hazardous Materials* B137: 267–276.

Clean Water Act (CWA) (1972). The Federal Water Pollution Control Act of 1948; As Amended in 1972, the Law Became Commonly Known as the Clean Water Act (CWA) (1972).

Cooper, C.D. and Alley, F.C. (2011). *Air Pollution Control: A Design Approach*, 4e. Long Grove, IL: Waveland Press.

CORMIX (1996). Mixing processes and mixing zone regulations: pp. 3–15; General features of the CORMIX system: pp. 17–23; CORMIX data input: pp. 25–45. In: *User's Manual for CORMIX: A Hydrodynamic Mixing Zone Model and Decision Support System for Pollutant Discharges into Surface Waters* (eds. G.H. Jirka, R.L. Donekar and S.W. Hinton). Washington, DC: USEPA, Office of Science and Technology.

Das, T.K. (1991). Longview Fiber Company NPDES Permit Inspection Report. Washington Department of Ecology, Environmental Investigations and Laboratory Services Program, Olympia, WA.

Das, T.K. (1993). Pulp and paper wastewater characterization to determine compliance with water quality objectives. *American Institute of Chemical Engineers Summer National Meeting, Paper No. 79a*, Seattle, WA (7–10 June 1993).

Davis, M.L. and Masten, S.J. (2014). *Principles of Environmental Engineering and Science*, 3e. New York, NY: McGraw-Hill.

Department of the Environment, Transport and the Regions (DETR) (1999). *A Way with Waste: A Draft Waste Strategy for England and Wales, Part One*, 24–25. DETR.

Federal Register (1994). 59 Federal Register, Vol 59, Issue 65.

Fensterstock, J.C., Kurtzweg, J.A., and Ozolins, G. (1971). Reduction of air pollution potential through environmental planning. *Journal of the Air Pollution Control Association* 21 (7): 395–399.

Frisch, K.C., Klempner, D., and Prentice, G. (1999). *Recycling of Polyurethanes: Advances in Plastics Recycling*, vol. 1. Lancaster, PA: Technomic Publishing Company Inc.

Gadgil, M., Pisupati, B., Sankar, U. et al. (2011). Report to the People on Environment and Forests, Ministry of Environment and Forests, Government of India.

Gandy, M. (1994). *Recycling and the Politics of Urban Waste*, 38–76. London: Earthscan.

Hammer, M.J. and Hammer, M.J. (2012). *Water and Wastewater Technology*, 7e. Upper Saddle River, NJ: Prentice Hall.

Haque, A. (1998). Limits of plastics recycling in the developed and developing economy. PhD thesis, University of London.

ISCO (2001). *Open Channel Flow Measurement Handbook*. Lincoln, NE: ISCO.

Jolánkai, G. (1997). *Basic River Water Quality Models, Computer Aided Learning (CAL) Program on Water Quality Modelling (WQMCAL version 1.1)*, International Hydrological Program, Technical Documents in Hydrology, vol. 13. Paris: UNESCO.

Karell, M. (2017). Estimating air emission rates. *Chemical Engineering Progress* 113 (7): 59–65.

Keith, L.H. (1991). *Environmental Sampling and Analysis: A Practical Guide*. Chelsea, MI: CRC Press, Lewis.

Khwaja, R.H., Ghosh, N., Biswas, M. et al. (2010). Report to the People on Environment and Forests, Ministry of Environment and Forests, Government of India.

Konisky, D.M. (2015). *Failed Promises: Evaluating the Federal Government's Response to Environmental Justice*. MIT Press.

Kumar, A. (1998). Estimating hazard distances from accidental release. *Chemical Engineering*: 121–128. August 1998.

Kumar, A. (1999). Estimate dispersion for accidental releases in rural areas. *Chemical Engineering*: 91–94. July 1999.

LIDAR (1980). An Alternate to Method 9 is also available using laser radar (LIDAR).

Mader, F. (1997). Plastics waste management in Europe: the need for an integrated approach. In: *Chemical Aspects of Plastics Recycling* (eds. W. Hoyle and D.R. Karsa), 13–27. Cambridge: The Royal Society of Chemistry.

Muellenhoff, W.P., Soldate, A.M., Baumgartner, D.J. et al. (1985). *Initial Mixing Characteristics of Municipal Ocean Discharges*, Report EPA/600/3-85/073. Washington, DC: USEPA.

Nahar, N. (1990). Solid waste management in Dhaka City. MSc thesis, University of Dhaka, Dhaka, pp. 35–38.

National Park Service (2010). Federal Land Manager's Air Quality Value Workgroup (FLAG) Phase I Report – Revised. NPS/NRPC/NRR – 2010/232.

Nova Gas (2018). Flue gas and emissions analyzers. https://www.nova-gas.com/analyzers/flue-emissions (accessed 19 February 2018).

O'Neil, S.G. (2007). Superfund: evaluating the impact of executive order 12898. *Environmental Health Perspectives* 115 (7): 1087–1093.

PNM (2018). Tier 2 Generation Capital Projects. www.pnm.com (accessed 18 February 2018).

Prater, E.T. and Midgley, C. (2008). A new air dispersion modeling system is helping create more accurate industrial source models. *Environmental Protection* 17 (3), Stevens Publishing.

Ratra, O.P. (1994). Plastic waste and the environment. *Conference Proceedings on Plastics and the Environment*, Federation of Indian Chambers of Commerce and Industry and Indian Plastics Institute, New Delhi (7–8 September 1994), pp. 18–28.

Resource Conservation and Recovery Act (RCRA) (1976). P.L. 94-580, 90 Stat. 2795, 42 U.S.C. § 6901.

Ryan, H. (2004). Neoliberalism and environmental justice in the united states environmental protection agency: translating policy into managerial practice in hazardous waste remediation. *Geoforum* 35 (3): 285–297.

Saba, R.G. and Pearson, W.E. (1995). Curbside recycling infrastructure: a pragmatic approach. In: *Plastics, Rubber, and Paper Recycling: A Pragmatic Approach* (eds. C.P. Rader, S.D. Baldwin, D.D. Cornell, et al.), 2–10. Washington, DC: American Chemical Society.

Scire, J.S., Strimatis, D, and Yamartino, R.J. (1990). Model Formulation and User's Guide for the CALPUFF Dispersion Model. Report No. A025-2, California Air Resources Board, Sacramento, CA.

Shah, K.L. (2000). *Basics of Solid and Hazardous Waste Management Technology*. Upper Saddle, NJ: Prentice Hall.

South Coast Air Quality Management District (SCAQMD) (1995). Draft Best Available Control Technology Methodology Report.

Stephan, C.E. (1977). Methods for calculating an LC_{50}. In: *Aquatic Toxicology and Hazard Evaluation*, ASTM STP, vol. 634 (eds. F.L. Mayer and J.L. Hamelink), 65–84. Philadelphia: American Society for Testing and Materials.

Streeter, H.W. and Phelps, E.B. (1925). *A Study of the Pollution and Natural Purification of the Ohio River. III. Factors Concerned in the Phenomena of Oxidation and Reaeration*, Public Health Bulletin, vol. 146. Washington, DC: U.S. Public Health Service.

Stultz, S.C. and Kitto, J.B. (2005). *Steam: Its Generation and Use*, 41e. The Babcock & Wilcox Company.

The Energy and Resources Institute (TERI) (2014). *Waste to Resources: A Waste Management Handbook*. New Delhi, India: TERI.

Turner, D.B. (1994). *Workbook of Atmospheric Dispersion Estimates: An Introduction to Dispersion Modeling*, 2e. Boca Raton, FL: CRC Press.

U.S. Government Accountability Office (2015). Superfund: Trends in Federal Funding and Cleanup of EPA's Nonfederal National Priorities List Sites. GAO Report Number GAO-15-812.

USEPA (1972). Laws and regulations. History of the Clean Water Act. https://www.epa.gov/laws-regulations/history-clean-water-act (accessed 2018).

USEPA (1973). *Handbook for Monitoring Industrial Wastewater*, EPA/625/6-73/002. Washington, DC: USEPA, Technology Transfer.

USEPA (1980). Summary of the Comprehensive Environmental Response, Compensation, and Liability Act (Superfund), 42 U.S.C. §9601.

USEPA (1983). *Methods for the Chemical Analysis of Water and Wastes*, EPA600/4-79-020, Cincinnati, OH: Environmental Monitoring and Support Laboratory, Office of Research and Development, USEPA. https://www.epa.gov/cwa-methods (accessed 28 August 2018).

USEPA (1985a). *Guidance Manual for the Use of Production-Based Pretreatment Standards and the Combined Wastestream Formula*. Washington, DC: Industrial Technology Division, Office of Water Regulations and Standards and Permits, Office of Water Enforcement and Permits.

USEPA (1985b). *Methods for Measuring the Acute Toxicity of 4 Freshwater and Marine Organisms*, LEFIA-600/4-85/013. Cincinnati, OH: EPA Environmental Monitoring and Support.

USEPA (1985c). *Short Term Methods for Estimating Chronic Effluent and Receiving Waters to Freshwater Organisms*, EPA-660/4-85-014. Cincinnati, OH: Environmental Monitoring and Support Laboratory, Office of Research and Development, USEPA.

USEPA (1985d). *Rates, Constants and Kinetics Formulations in Surface Water Quality Modeling*, 2e. Athens, GA: Environmental Research Laboratory Document no. EPA/600/3-85/040.

USEPA (1986). *RCRA Orientation Manual*, Solid Waste Program, EPA/530-SW-86-001. Washington, DC: USEPA.

USEPA (1988). *Short Term Methods for Estimating the Chronic Effluents and Receiving Waters to Marine and Estuarine Organisms*, EPA-600/4-88/028. Cincinnati, OH: USEPA, Office of Research and Development.

USEPA (1989). *Short Term Methods for Estimating Chronic Effluents and Receiving Waters to Freshwater Organisms*, EPA-660/2nd. Washington, DC: USEPA, Office of Water.

USEPA (1990). *New Source Review Workshop Manual: Prevention of Significant Deterioration and Non-attainment Permitting, Office of Air, Noise, and Radiation*. Research Triangle Park, NC: Office of Air Quality Planning and Standards.

USEPA (1991a). *Guidance for Water Quality-Based Decisions: The TMDL Process*, EPA 440/4-91-001. Washington, DC: USEPA.

USEPA (1991b). *Method 5: Determination of Particulate Matter Emissions from Stationary Sources*. (40 CFR 60, Appendix A to Part 60). US Government Publishing Office.

USEPA (1992). *Fugitive Dust Background Document and Technical Information Document for Best Available Control Measures*, EPA-450/2-92-004. Research Triangle Park, NC: Office of Air Quality Planning and Standards.

USEPA (1993). *Traceability Protocol for Assay and Certification of Gaseous Calibration Standards*, EPA-600/R93/224. Research Triangle Park, NC: USEPA, Office of Research and Development, Revised September 1993.

USEPA (1995a). *Method 1669: Sampling Ambient Water for Trace Metals at EPA Water Quality Criteria Levels*, EPA 821-R-95-034. Washington, DC: USEPA, Office of Water.

USEPA (1995b). National pollutant discharge elimination system permit application requirements for publicly owned treatment works and other treatment works treating domestic sewage, proposed rules. *Federal Register* 60 (234), USEPA, Office of Water. Washington, DC.

USEPA (1995c). BACT/RACT/LAER Clearinghouse Information System (BLIS), EPA Office of Air Quality Planning and Standards (OAQPS) Technology Transfer Network (TTN), Research Triangle Park, NC.

USEPA (1997). *1997 Mercury Study Report to Congress*. USEPA www.ep.gov.

USEPA (1998). *Study of Hazardous Air Pollutant Emissions from Electric Utility Steam Generating Units: Final Report to Congress*, EPA-453/R-98-004a, vol. 1. Research Triangle Park, NC: USEPA.

USEPA (1999). "Part 264 – Standards For Owners And Operators Of Hazardous Waste Treatment, Storage, And Disposal Facilities." 40 C.F.R. 264. "Part 270 – EPA Administered Permit Programs: The Hazardous Waste Permit Program." 40 C.F.R. 270.

USEPA (2002a). *Municipal Solid Waste in the United States: Facts and Figures*, EPA/530/R-002/001. Washington, DC: Office of Solid Waste and Emergency Response.

USEPA (2002b). *Office of Solid Waste, Exemption of Oil and Gas Exploration and Production Wastes from Federal Hazardous Waste Regulations*. Washington, DC: USEPA, Solid Waste and Emergency Response.

USEPA (2004a). *EPA Fact Sheet*. USEPA. www.epa.gov (accessed 1 September 2018).

USEPA (2004b). *Description of Model Formulation*, EPA-454/R-03-004. Research Triangle Park, NC: Office of Air

Quality Planning and Standards, Emission Monitoring and Analysis Division.
USEPA (2008). *AERSURFACE User's Guide*, EPA-454/B-08-001. Research Triangle Park, NC: USEPA.
USEPA (2009a). *Continuous Emissions Monitoring Fact Sheet*. USEPA. Archived from the original on 11 February 2009.
USEPA (2009b). *Clean Air Act National Stack Testing Guidance*, Office of Enforcement and Compliance Assurance, Office of Quality Planning & Standards, Washington, DC, Issued on 27 April 2009. www.epa.gov/compliance/resources/polices/civil/caa (accessed 1 September 2018).
USEPA (2010a). *NPDES Permit Writers' Manual*. "Chapter 6: *Water Quality-Based Effluent Limitations*." September, EPA-833-K-10-001. Washington, DC: USEPA.
USEPA (2010b). History of RCRA.
USEPA (2011a). Non-hazardous Waste Regulations.
USEPA (2011b). Hazardous Waste Regulations.
USEPA (2012). Part 265: Interim Status Standards for Owners and Operators of Hazardous Waste Treatment, Storage, and Disposal Facilities. 40 C.F.R. 265.
USEPA (2014). *Municipal Solid Waste Generation, Recycling, and Disposal in the United States: Facts and Figures for 2012, Office of Solid Waste & Emergency Response*, EPA-530F-14-001. Washington, DC: USEPA.
USEPA (2016a). *User's Guide for the AMS/EPA Regulatory Model (AERMOD)*, EPA-454/B-16-001. Research Triangle Park, NC: USEPA.
USEPA (2016b). *AERSCREEN User's Guide*, EPA-454-/B-16-004. Research Triangle Park, NC: USEPA.
USEPA (2018a). Specifications and test procedures for total hydrocarbon continuous monitoring systems in stationary sources. www3.epa.gov (accessed 23 February 2018).
USEPA (2018b). *User's Guide for the AERMOD Terrain Preprocessor (AERMAP)*, EPA-454/B-18-004. Research Triangle Park, NC: USEPA.
USEPA (2019a). Overview of the Federal UST Program.
USEPA (2019b). *User's Guide for the AERMOD Meteorological Preprocessor (AERMET)*, EPA-454/B-19-028. Research Triangle Park, NC: USEPA.
Van Beukering, P. (1999). Plastics Recycling in China: An International Life Cycle Approach. CREED Final Report, Vrije Universiteit Amsterdam and International Institute for Environment and Development (IIED), Amsterdam, pp. 9–22.
Viessman, W., Hammer, M.J., Perz, E., and Chadik, P.A. (2009). *Water Supply & Pollution Control*, 8e. Upper Saddle River, NJ: Pearson-Prentice Hall.
Walker, P.M.B. (1995). *Dictionary of Science and Technology*, 859. Edinburgh: Larousse Plc.

5

Assessment and Management of Health and Environmental Risks

Industrial and Manufacturing Process Safety

The concept of risk, the possibility of loss or injury, is familiar in every human endeavor. Both chemical engineers and environmental engineers routinely consider the risks of adverse health and environmental impacts from exposures to industrial by-products released to air, water, and land. Risk assessment, thus, is an organized process for describing and estimating the likelihood of such events. Risk management is the decision-making process by which planners, who are not necessarily chemical or environmental engineers, attempt to minimize risks without undue harm to societal values.

This chapter explains the paradigms used in evaluating the impacts of chemical releases on human health and on the environment. Tools and procedures are described for monitoring exposure to chemicals and for communicating to the public the results of analysis of risk data. The coverage of risk management decisions should be made independently of risk analysis. However, the important concepts in this field are explained and illustrated (Omenn and Faustman 2002; Slovic et al. 2004).

5.1 Health Risk Assessment

Risks to human health may be assessed on the basis of information collected from environmental monitoring; alternatively, the data may be incorporated into models of human activity and exposure, in the workplace and elsewhere, to permit analysts to draw conclusions on the likelihood of adverse effects. In both cases, risk assessment is an essential tool used by those responsible for making decisions with environmental consequences (APHA 1989; Allen and Shonnard 2002; National Academy of Sciences USA 2006).

The results of environmental risk assessment are generally intended to be incorporated into decisions that affect the public, along with the economic, social, technological, and political consequences of a proposed action. The need for such analytical data is indicated in Table 5.1, which lists selected US laws that require or suggest human health risk assessment before regulations are promulgated. The list is enormous and probably will grow with time.

In contrast to the specific laws addressing concerns about environmental pollutants presented in Table 5.1, Figure 5.1 offers a generalized scheme of the routes by which these pollutants enter the environment, ending with human exposure.

The paradigm for assessing human health risk consists of formulation of the problem, assessment of the exposure, assessment of any toxic effects, and risk characterization (USEPA 2003).

5.1.1 Air Pollution

Air pollution is an important public health issue in the United States and worldwide. The World Health Organization (WHO) estimated that in 2012, over seven million people died (about one-eighth of all mortalities) from illness related to exposure to air pollutants (WHO 2014). In addition to the human costs, there also financial costs, including higher public health care costs associated with increased hospitalizations, as well as lost productivity. The high frequency of respiratory ailments, as well as the necessity of wearing an antipollution mask to reduce inhalation of pollutants in certain countries, is well documented in news media.

In the United States, concern over air pollution is significant and is going to be one of the most challenging environmental health issues of our time. A 2015 Gallup poll showed that 38% of Americans were concerned about air pollution "a great deal" (Jones 2015). A 2013 study by researchers at the Massachusetts Institute of Technology estimated about 200 000 Americans die prematurely each year because of air pollution, although this toll is brought on by all sources of air pollution – including nonindustrial ones – such as indoor combustion (e.g. automobiles, trucks, etc.) (Chu 2013).

Industrial Environmental Management: Engineering, Science, and Policy, First Edition. Tapas K. Das.

Companion website: www.wiley.com/go/Das/IEM_1e

Table 5.1 United States safety, health, and environmental statutes that imply risk assessment.

Environmental Protection Agency	
Atomic Energy Act (also NRC)	42.U.S.C.2011
Comprehensive Environmental Response, Compensation and Liability Act (CERCLA, or Superfund)	42.U.S.C.9601
Clean Air Act	42.U.S.C.7401
Clean Water Act	33.U.S.C.1251
Emergency Planning and Community Right to Know Act	42.U.S.C.11001
Federal Food and Drug, and Cosmetics Act (also HHS)	21.U.S.C.301
Federal Insecticide, Fungicide, and Rodenticide Act	7.U.S.C.136
Lead Contamination Control Act of 1988	42.U.S.C.300j-21
Marine Protection, Research, and Sanctuaries Act (also DA)	16.U.S.C.1431
Nuclear Waste Policy Act	42.U.S.C.10101
Resource Conservation and Recovery Act	42.U.S.C.6901
Safe Drinking Water Act	42.U.S.C.300f
Toxic Substances Control Act	7.U.S.C.136
Food Quality Protection Act of 1996	7.U.S.C.6
Consumer Product Safety Commission	
Consumer Product Safety Act	15.U.S.C.2051
Federal Hazardous Substance Act	15.U.S.C.1261
Lead-Based Paint Poisoning Act (also HHS and HUD)	42.U.S.C.4801
Lead Contamination Control Act of 1988	42.U.S.C.300j-21
Poison Prevention Packaging Act	15.U.S.C.1471
Department of Agriculture	
Eggs Products Inspection Act	21.U.S.C.1031
Federal Meat Inspection Act	21.U.S.C.601
Poultry Products Inspection Act	21.U.S.C.451
Department of Labor	
Federal Mine Safety and Health Act	30.U.S.C.801
Occupational Safety and Health Act	29.U.S.C.651
Department of Transportation	
Hazardous Liquid Pipeline Safety Act	49.U.S.C.1671
Hazardous Materials Transportation Act	49.U.S.C.1801
Motor Carrier Safety Act	49.U.S.C.2501
National Traffic and Motor Vehicle Safety Act	15.U.S.C.1381
National Gas Pipeline Safety Act	49.U.S.C.2001

Source: From National Academy of Sciences (2006), Hoppins et al. (2006), Asante-Duah (2002), Kammen and Hassenzahl (1999), Roberts and Abernathy (1997) and Federal Focus (1991).

5.1.2 Problem Formulation

Problem formulation comprises such activities as definition of the goals and spatial and temporal scale of the ecological risk assessment (ERA), development of a site conceptual model (SCM), and selection of endpoint and nonhuman receptor species. The final step is to identify contaminants of potential concern and to determine whether, for example, threshold limit values (TLVs) have been exceeded. Examples 5.1 and 5.2 illustrate how the results of environmental monitoring of three common industrial pollutants can be used in calculations of the quantities present.

Example 5.1 Monitoring

An air sampler collects 16 mg of methyl chloride during 6-hour work shift. The sampler has the following characteristics:

Average flow rate = 1.2 l/min

Collection efficiency = 0.9

What is the eight-hour total weighted average (TWA) of methyl chloride?

Solution

The total air collected was

$$1.2\,\text{l/min} \bullet 60\,\text{min/h} - 432\,\text{l}$$

$$\frac{432\ \text{l}}{1000\ \text{l/m}^3} = 0.432\,\text{m}^3$$

Sampler content was 16 mg or $\frac{16}{0.9} = 17.8$ mg in the air sampled

Methylene chloride concentration in air sampled = $17.8\,\text{mg}/0.432\,\text{m}^3 = 41.2\,\text{mg/m}^3$ (ppm)

Average air concentration during the six-hour collection time

$$= 41.2\frac{\text{mg}}{\text{m}^3} \bullet \frac{6\,\text{hours}}{8\,\text{hours}} = 31\frac{\text{mg}}{\text{m}^3}(\text{ppm})$$

Was the TLV exceeded?

TWA for methyl chloride is 50 ppm (mg/m^3) and was not exceeded.

Example 5.2 Continuous Monitoring

Lead fumes are produced in a lead-burning operation. A continuous monitor in the work room is operated for five consecutive days. The following are known:

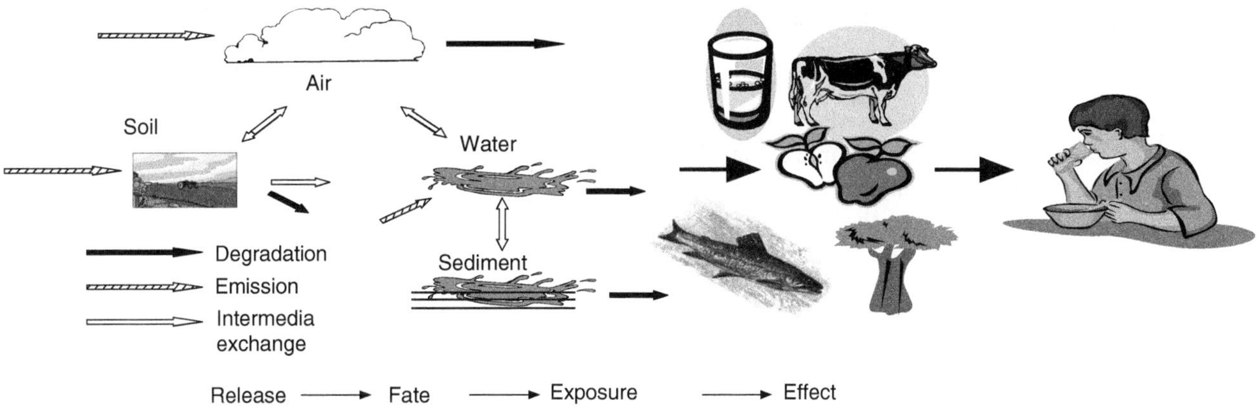

Figure 5.1 Generalized scenario for exposure by human to environmental pollutants.

Initial sample rate	1 cfm (cubic feet per minute)
Final sample rate	0.7 cfm
Filter efficiency	0.999 for particulates ≥20 μm average mean diameter (AMD)
	0.83 for particulates <20 μm AMD
Mass of dust collected	1.7 g
% of lead in dust	23

Solution

Determine the TWA lead concentration in the work air.

$$\text{Quality of air sampled} = \text{average sample rate} \times \text{sampling time}$$

Assume approximately linear reduction in sampling rate as filter loads. Hence, average sampling rate is

$$\frac{1+0.7}{2} = 0.85\,\text{cfm}$$

and total air sampled is

$$0.85\,\text{cfm} \bullet 5\,\text{days} \bullet 1440\,\text{min/day} - 6120\,\text{ft}^3$$

Total dust collected = 1.7 g or 1.7 • 0.23 = 0.39 g of lead, since 23% of dust is lead. Since lead was present as a fume, AMD < 20 μm and total lead collected is

$$\frac{0.39g}{0.83} = 0.47\,\text{g}$$

Average lead concentration is the total amount divided by sample volume or

$$\frac{0.47\,\text{g}}{6120\,\text{ft}^3} = \frac{470\,\text{mg}}{6120\,\text{ft}^3} \bullet \frac{1}{35.31\ \text{m}^3/\text{ft}^3} = 0.0022\,\text{mg/m}^3$$

Assuming lead generation occurs only during eight-hour workday, the TWA is

$$\text{TWA} = \frac{0.0022\ \text{mg}}{\text{m}^3} \bullet \frac{24}{8} = 0.007\ \text{mg/m}^3$$

Dusts pose a particular problem with respect to determining acceptable air concentrations. The particle size distribution needs to be taken into account, and the specific respirable fractions, as published in the literature, must be used to determine exposure. Dust particle sizing can be accomplished by size-selective samplers, typically multistage impactors.

These sample examples only scratch the surface of techniques used in problem formulation. They are, however, illustrative. Sometimes, as in Example 5.1, monitoring reveals that existing pollution prevention measures are working properly and that no problem exists. Example 5.2 shows that determining the TWA of a pollutant is not always sufficient to formulate a problem.

5.1.3 Exposure Assessment

Measures of ongoing exposure to health risks call for determination of concentration. In the Section 5.1.3.1, Example 5.3 shows how to determine bioconcentration, and Example 5.4 explains the use of slope factors, often used in quantifying risks of cancer.

5.1.3.1 Partition Coefficient and Bioconcentration Factor

While extensive research continues to refine existing methods and to develop new markers of exposure and effect, experience has identified several biomarkers of value to hazardous waste management decisions. Exposure to chemicals that bioaccumulate in tissue (see, e.g., Table 5.2) is easily accomplished by analyzing tissue residue. In the absence of tissue residues, empirical constants often derived and used by ecotoxicologists to predict bioaccumulation include the octanol–water partition coefficient (K_{ow}), which approximates a chemical's behavior between the lipid water phase and the bioconcentration factor (BCF), which describes the relationship between the chemical concentration in an organism and the concentration in the exposure medium (typically water). The BCF is empirically expressed as

$$C_{org} = (\mathrm{BCF})(C_w) \quad \text{or} \quad \mathrm{BCF} = \frac{C_{org}}{C_w}$$

Example 5.3 illustrates the use of this relationship.

Example 5.3 Predicting Bioconcentration
Routine monitoring of sediments and surface water downstream from a hazardous waste site indicates that reportable concentrations of the pesticide DDT average, respectively, 33 mg/kg and 0.25 μg/l. Considering that the US Food and Drug Administration has identified levels of 5 ppm or higher of DDT as requiring action, is there a reason to be concerned about potential accumulation of DDT in fish tissue near this site?

Solution
Since the routine monitoring does not provide information regarding DDT residues in fish tissue, we refer to Table 5.2 and BCF. Transforming the log BCF of 4.78, we estimate a BCF of 60.256 (mg/kg DDT in tissue per mg/l DDT in water or l/kg).

Table 5.2 Selected physicochemical parameters and bioconcentration factors for typical compounds which bioaccumulate.

Compound	Log K_{ow}	Log BCF (fish)	Log K_{oc}	S (mg/l)
Chlordane	5.15	4.05	4.32	0.0056
DDT	5.98	4.78	4.38	0.0017
Dieldrin	5.48	3.76	4.55	0.022
Lindane	4.82	2.51	2.96	0.150
Chlorpyrifos	4.99	2.65	4.13	0.3
Cyhexatin	5.38	2.79	<3.64	<1.0
2,4D	1.57	1.30	1.30	900
2,4,5T	0.60	1.63	1.72	2.38
TCDD	6.15	4.73	5.67	0.0002

BCF, bioconcentration factor (l/kg); K_{oc}, organic carbon partition coefficient; K_{ow}, octanol-water partition coefficient (unitless); S, water solubility.

Substituting our monitoring data and the empirical BCF for DDT in the following equation yields

$$C_{org} = \left(\frac{60\ 256 \frac{\mathrm{mg}}{\mathrm{kg}}}{\frac{\mathrm{mg}}{\mathrm{l}}} \right) \times \left(0.000\ 25 \frac{\mathrm{mg}}{\mathrm{l}} \right) = 0.015 \frac{\mathrm{mg}}{\mathrm{kg}}$$

We estimate a DDT tissue residue of 0.015 mg/kg. Based on this estimate, it is probable that DDT residues in tissue are an issue that should be evaluated.

The calculation of carcinogenic risk involves the use of a carcinogen potency factor (CPF). A CPF is basically the slope or steepness of the dose–response curve at very low exposures, and is now referred to as slope factor. The dimensions of a slope factor are expressed as the reciprocal of the daily dose $(\mathrm{mg/kg/day})^{-1}$. Having derived a slope factor, the calculation of carcinogenic risk is straightforward. Quantification of carcinogenic risk of an exposure simply requires converting the dose to the appropriate terms (mg/kg/day) and multiplying it by the slope factor as demonstrated in Example 5.4.

Example 5.4 Use of Slope Factors
A petroleum refinery is performing a quantitative risk assessment on the atmospheric release of volatile organic compounds (VOCs) from the facility, some of which are toxic. Assess the risk of benzene released to the air. What is the maximal number of excess lifetime cancer cases expected for a population of 100 000 adults with a daily inhalation dose of 0.14 mg of benzene per adult body weight per day?

Solution

1) Inhalation carcinogenic risk:

   $$\text{Inhalation dose} \times \text{dose response carcinogenic slope factor} = I_{\mathrm{INH}} \times \mathrm{SF}$$

2) The slope factor for benzene is 0.029 $(\mathrm{mg/kg\ BW\text{-}day})^{-1}$ taken.
3) Assuming an adult weight of 70 kg, the lifetime cancer risk is calculated as

$$\begin{aligned} \text{Individual cancer risk} &= \left(0.14 \frac{\mathrm{mg\,benzene}}{70\,\mathrm{kg\,BW\text{-}day}}\right) \left(0.029 \frac{\mathrm{mg\,benzene}}{\mathrm{kg\,BW\text{-}day}}\right)^{-1} \\ &= 0.000058 \text{ or } 0.0058\% \\ &= \text{risk} \times \text{exposed population} \\ &= 0.000\ 058 \times 100\ 000 \\ &= 6 \text{ lifetime cancer cases} \end{aligned}$$

5.1.3.2 Exposure Point Concentrations

The risk analyst must next estimate the concentration of contaminants at the exposure points, which define the locations of the receptors, that is the human or animal population whose exposure to a pollutant is of concern. Receptors are identified for each exposure scenario simply by overlaying the demographic information with the exposure pathway. An exposure point may be as close as the sources of waste at the site itself (e.g. the trespasser scenario) or at a considerable distance, particularly for pathways involving the food chain. It may take any pathway: air, ground and surface water, soils, sediments, and food (e.g. plants and fish). For present exposures, actual monitoring data at the exposure point should be used wherever possible. For example, contaminant concentrations should be obtained for drinking water wells in the vicinity of a site. For on-site exposure points, representative concentrations in soil or groundwater may be calculated as the arithmetic or geometric mean (depending on the statistical distribution of the site analytical concentration data).

Future conditions can differ starkly. A plume of gases may not have yet migrated to a potential exposure point. Remediation will, of course, drastically reduce migration. In a comprehensive risk assessment, the exposure concentrations for each for remedial alternative would be estimated including the present situation (baseline conditions of no remedial action). Determining the concentration of a contaminant at the exposure point for future conditions often requires the use of fate and transport modeling methods and standard references. The major effort with these models is calibration; once calibrated, successive runs can be made relatively easily to estimate concentrations for a range of conditions and assumptions.

For groundwater contaminants, hydrogeologic models can be used to estimate the future concentration at a downstream well. For VOCs released to the atmosphere, a Gaussian diffusion model can be employed to estimate future downwind concentrations. In general, the level of effort employed in data collection and modeling will depend on the estimated severity of the risk. Nominal risks do not warrant the same level of analysis as the clearly significant risks.

All mathematical models require the making of assumptions. This is true, as well, of the mechanistic models on which many analyses are based.

5.1.3.3 Mechanistic Models

In mechanistic models, it is assumed that a certain number of reactions, events, or "hits" (concept derived from radiation biology), or transition stages, related to a critical target in the cell (DNA), are necessary to transform a normal cell to a cancer cell. These models are important because there seems to be general consensus that tolerance distribution models (e.g. Mantel–Bryan) should not be used unless the available data are accurate enough to exclude models involving a linear component in the low-dose region, or a genotoxic mechanism might be involved.

The multistage model, first proposed by Armitage and Doll (1954), is an extension of the one-hit model developed for ionizing radiations. One generalized version of this model where at least one of the stages is assumed to be dose-related takes the form:

$$P(D) = 1 - \exp\left[-\left(q_0 + q_1 D + q_2 D^2 + \cdots + q_k D^k\right)\right],\ k > 1 \quad (5.1)$$

where

$P(D)$ = the probability of cancer at dose D
q_k = coefficients to best fit to the data
D^k = applied does raised to the kth power
k = number of stages (usually set arbitrarily at the number of dose levels minus one.

The most likely estimate at very low does becomes increasingly unstable with a small change in the response at experimental does. Therefore, a further development in modeling was the replacement of the linear term in the polynomial function by its 95% confidence limit to achieve more stable estimates of risk above background than are obtained for the most likely estimates. This so-called linearized multistage model is used routinely by the Environmental Protection Agency (EPA) and has, therefore, become the most widely used model for estimation of cancer risk. At low doses, the function becomes essentially linear. The multistage model is very flexible in fitting data sets because it is a polynomial function of dose. The so-called carcinogen potency factors provided by the EPA have been based on this model and are calculated from animal data using commercially available computer programs.

The multistage model is favored by the EPA because it generally gives conservative risk estimates for low exposures, but the model has several weaknesses.

As an example, pica is a real phenomenon and should be considered, but the assumption that toddlers who eat 200 mg/day of soil will continue this behavior for a 70-year lifetime is not realistic. Thus, the impact of assumptions on the exposure point concentration and, ultimately, the risk values should be examined to identify which of the myriad of input variables have the most significant impact on the resulting risk value. Such a refinement is called a sensitivity analysis.

5.1.3.4 Receptor Doses

The final step in the exposure assessment stage is to estimate the doses of the different classes of chemicals to which receptors are potentially exposed at the exposure points. There are three types of dose: the administered dose (the amount ingested, inhaled, or in contact with the skin), the intake dose (the amount absorbed by the body), and the target dose (the amount reaching the target organ).

For purposes of calculating risks, the dose should be in the same form as that of the dose–response relationship reported for the specific chemical and the exposure route under study. This will almost always be either administered dose or absorbed dose. Given the concentration of the contaminant at the exposure point, the calculation of administered dose is straightforward. In contrast, the calculation of absorbed dose based on administered dose requires consideration of some complex factors. The key factors influencing the uptake of contaminants by the body are simplified as follows:

Ingestion	Contaminant concentration in the ingested medial
	Amount of ingested material
	Bioavailability to the gastrointestinal system
Inhalation	Concentration in air and dust
	Particle size distribution
	Bioavailability to the pulmonary system
	Rate of respiration
Dermal contact	Concentration in soil and dust
	Rate of deposition of dust from air
	Direct contact with soil
	Bioavailability
	Amount of skin exposed

Other factors to be considered in determining the intake of contaminants include considerations of life style, frequency and duration of exposure (e.g. chronic, subchronic, or acute), and the body weight of the receptor. In the majority of hazardous waste sites, long-term (i.e. chronic) exposures are frequently of greatest concern.

The calculation of an administered dose is summarized in the following generic equation:

$$I = \frac{C \times CR \times \mathrm{EF} \times \mathrm{ED}}{\mathrm{BW} \times \mathrm{AT}} \tag{5.2}$$

where

I = intake (mg/kg of body weight-day)
C = concentration at exposure point (e.g. mg/l in water or mg/m^3 in air)
CR = contact rate (e.g. l/day or mg/m^3/day)
EF = frequency (day/year)
ED = exposure duration (year)
BW = body weight (kg)
AT = averaging time (days)

Equation (5.2) is typical modified for specific exposure pathways. For example, the intake dose from the inhalation of fugitive dust may be calculated as

$$I = \frac{C \times \mathrm{CR} \times \mathrm{EF} \times \mathrm{ED} \times \mathrm{RR} \times \mathrm{ABS}}{\mathrm{BW} \times \mathrm{AT}} \tag{5.3}$$

where

RR = retention rate (decimal traction)
ABS = absorption into bloodstream (decimal fraction)

For fugitive dust, the concentration in the air is determined by

$$C - C_s \times P_c$$

where

C_s = concentration of chemical in fugitive dust (mg/mg)
P_c = concentration of fugitive dust in air (mg/m^3)

Often, appropriate parameters may be found in the literature. Considerable research has been done in recent years to define many basic parameters, such as skin surface areas, soil ingestion rates, and inhalation rates. Other parameters, such as exposure frequency and duration, are often based on site-specific information (if available) or professional judgment. For example, in the evaluation of a trespasser scenario, observations of trespassers during site investigation activities may dictate the values used for exposure frequency (e.g. number of days per year or number of events per year) and exposure duration (e.g. number of years that the activity occurred). Common sense also plays an important role in the selection of exposure parameters. For example, if one were evaluating residential use of two sites, one in North Dakota and one in Florida, it would not be reasonable to assume the exposure frequency (e.g. days per year that an activity occurred) would be the same in the two risk assessments for a child potentially exposed to surface soils as a result of playing in the backyard.

Averaging time (AT) is another important parameter which must be defined in the intake equation. The averaging time selected will depend on the type of constituent being evaluated. For example, to assess long-term or chronic effects associated with exposure to noncarcinogens, the intake is averaged over the exposure duration (expressed in days). Exposure to carcinogens, however, is averaged over a lifetime (assumed to be 70 years or 25 550 days), to be consistent with the approach used to develop slope factors.

Some parameters used for this type of calculation are shown in Table 5.3. It should be noted that these values can

Table 5.3 Standard parameters for calculation of dosage and intake determined for a landfill.

Parameter	Adults (males)	Child age 6–12	Child age 2–6
Average body weight (kg)	70	29	16
Skin surface area (cm^2)	18150	10470	6980
Water ingested (l/day)	2	2	1
Air breathed (m^3/h)	0.83	0.46	0.25
Retention rate (inhaled air) (%)	100	100	100
Absorption rate (inhaled air) (%)	100	100	100
Soil ingested (mg/day)	100	100	200
Bathing duration (minutes)	30	30	30
Exposure frequency (days)	365	365	365
Exposure duration (years)	30	6	4

vary greatly depending on the assumed exposure conditions (i.e. the selected exposure scenario). As an example, the air-breathing rate for adult males is 0.83 m^3/h in Table 5.3. However, this rate can vary by an order of magnitude from 0.6 m^3/h at rest to 7.1 m^3/h for vigorous physical exercise.

Examples 5.5 and 5.6 illustrate the calculation of contaminant intake via the inhalation and dermal route, respectively.

Example 5.5 Calculation of Contaminant Intake
Determine the chronic daily inhalation intake, by adults, of a noncarcinogenic chemical as a function of concentration in fugitive dust at the ABC Landfill (Table 5.3).

Solution
For an adult exposed to a noncarcinogenic constituent, the intake I_N may be calculated from Eq. (5.2):

$$I_N = \frac{C \times \mathrm{CR} \times \mathrm{EF} \times \mathrm{ED} \times \mathrm{RR} \times \mathrm{ABS}}{\mathrm{BW} \times \mathrm{AT}}$$

From Table 5.3, we take the air-breathing rate for adults as 0.83 m^3/h and write

CR = $0.83 \times 24 = 19.92\ \mathrm{m^3/day}$
EF = 365 days
ED = 30 years for chronic exposure to noncarcinogens

In the absence of better information, a conservative approach would assume the RR and the absorption into bloodstream would both equal 1.0.
BW = 70 kg (Table 5.3)
AT = 365 days × 30 years

$$I_N = \frac{C \times 19.92 \times (365 \times 30) \times 1.0 \times 1.0}{70(365 \times 30)}$$

$$I_N = 0.285\ \mathrm{m^3/kg/day} \times C$$

$$C = \text{exposure point concentration } (\mathrm{mg/m^3})$$

Example 5.6 Average Daily Intake from Dermal Contact with Soil

Determine the average daily intake of chlorobenzene over one year of exposure for on-site workers from dermal contact of the soils:
Assume the following additional parameters:
A = skin exposed = 20% – 0.2 × 18150 cm^2 - 3630 cm^2
DA = dust adherence – 0.51 mg/cm^2
ABS = skin absorption rate – 6%
SM = effect of soil matrix = 15% (i.e. due to soil matrix only 15% of contamination is actually available for contact)
EF = two exposure events per day; 156 exposure days per year
ED = 1 year
BW = 70 kg
AT = 365 days

Solution

$$I_N = \frac{C\left(\frac{\mathrm{mg}}{\mathrm{kg}}\right) \times A\left(\mathrm{cm^2}\right) \times \frac{\mathrm{DA}}{\mathrm{exp.event}}\frac{\mathrm{mg}}{\mathrm{cm^2}} \times \mathrm{ABS} \times \mathrm{SM} \times \frac{2\,\mathrm{exp.events}}{\mathrm{day}} \times \frac{156\ \mathrm{days}}{\mathrm{year}} \times \mathrm{ED}\frac{\mathrm{kg \cdot soil}}{10^6\ \mathrm{mg \cdot soil}}}{\mathrm{BW} \times \mathrm{AT}}$$

$$I_N = \frac{C \times 3630 \times 0.51 \times 0.06 \times 0.15 \times 2 \times 156 \times 10^{-6}}{70 \times 365}$$

$$I_N = C \times 2.03 \times 10^{-7}/\mathrm{day}$$

The given average concentration of chlorobenzene in soil is 1.39 mg/kg. The daily intake of chlorobenzene is

$$1.39\ \mathrm{mg/kg} \times 2.03 \times 10^{-7}/\mathrm{day}$$

Thus,

$$I_N = 2.82 \times 10^{-7}\ (\mathrm{mg/kg \cdot day})$$

Again, the foregoing examples represent only two approaches to calculating contaminant intake in the course of an exposure assessment. The next factor to be analyzed is toxicity.

5.1.4 Toxicity Assessment

Broadly speaking, toxicity is the degree to which a substance is poisonous. Most chemicals, toxic or otherwise, enter the body through eyes, respiratory tract, digestive tract, and skin. Two levels of toxicity are defined: acute "short-term" exposure that initiates poisoning and chronic "long-period" exposure that causes anemia, leukemia, and death.

In the context of the third stage of the risk assessment process, toxicity is defined by the dose–response relationship for each surrogate chemical. The output takes the form of mathematical constants for insertion into risk calculation equations. In addition to providing a set of mathematical constants for calculating risk, the toxicological assessment should also analyze the uncertainty inherent in these numbers and describe how this uncertainty may affect the estimates of risk.

In a prudent approach to public health protection, the EPA has built several safety factors into its methods for establishing reference doses and carcinogenic slope factors (USEPA 2002a, 2002b, 2005). Virtually, all errors of uncertainty are made in the direction of public health protection to insure that risks are overestimated rather than underestimated. Four examples of this protective approach follow:

- For noncarcinogens, extrapolation of animal reference doses to humans utilizes two safety factors of ten, one for animal-to-human extrapolation, a second for variation for toxic sensitivities within the human population.
- For carcinogens, the linearized multistage model assumes the upper-bound 95% confidence level of extrapolated data (i.e. it is likely to be conservative 19 out of 20 times).
- For carcinogens, the linearized multistage model extrapolates data from the 10–90% carcinogenesis range observed in experimental animals to the regulatory target of 0.0001% carcinogenesis, a step which could overstate risk by several orders of magnitude.
- Although evidence indicate that, like noncarcinogens, nongenotoxic carcinogens have thresholds below which they fail to influence cellular differentiation or division, they are treated mathematically like genotoxic carcinogens according to the linearized multistage dose–response model.

5.1.5 Risk Characterization

The final stage of a four-stage human health risk assessment is to estimate risks. Cancer is not the only undesired health consequence of pollution: exposure to ionizing radiation and coal dust, for example, can lead to radiation sickness and birth defects, as well as cancer. To illustrate risk characterization; however, we shall focus on the calculations that yield quantitative estimates of both the carcinogenic and noncarcinogenic risks to receptors for all exposure scenarios considered. Estimates are typically calculated for all three exposure routes and for the maximally exposed individual as well as the most probable exposed population. Such calculations are straightforward and yield quantitative estimates of risk. The challenge lies not in making the calculations but in interpreting the results such that they are applied properly in decision-making. The overall effort is referred to as risk characterization. Risks from toxic chemicals, depending on the context, may be defined, described, and calculated in different ways.

Risk is normally defined as the probability for an individual to suffer an adverse effect from an event. What is the probability that certain types of cancer will develop in people exposed to aflatoxin in peanut products or benzene from gasoline? What is the likelihood that workers exposed to lead will develop nervous system disorders? In the context of this text, a chemical release is an example of an event. As with any relationship expressing or using probability, there is no defined way of expressing (mathematically or with scientific rigor) a single deterministic value of a phenomenon that is probabilistic. A conceptual way of expressing chemical risk is to write it as a function of hazard and exposure.

Hazard is the potential for a substance or situation to cause harm or to create adverse impacts on persons or the environment. The magnitude of the hazard reflects the potential adverse consequences, including mortality, shortened life-span, impairment of bodily function, sensitization to chemicals in the environment, or diminished ability to reproduce. Exposure denotes the magnitude and the length of time the organism is in contact with an environmental contaminant, including chemical, radiation, or biological contaminants.

When risk is in terms of probability, it is expressed as a fraction, without units. It has values from 0.0 (absolute certainty that there is no risk) to 1.0 (absolute certainty that an adverse outcome will occur).

For chemicals the term hazard is typically associated with the toxic properties of a chemical specific to the type of exposure. Similar chemicals would have similar innate hazards. However, one must examine the exposure to that hazard to determine the risk. For example, let us say you have three pumps that are all transporting the same chemical (same hazard), but one pump has a seal leak. Which pump poses the greatest risk to the worker? The pump with the seal leak has the greatest potential for exposure, while the hazards are equal (same chemical), so the seal leak pump poses the greatest risk. To expand, let us say we have three pumps that are transporting different chemicals; which one poses the greatest risk to the worker? In this case, the engineer would need to examine the hazard, or innate inherent toxicity, of each of the chemicals, as well as the operation of the pumps to determine which poses the greatest risk.

5.1.5.1 Risk for Average and Maximum Exposures

In the exercises required for Superfund sites, both average and maximum exposure point concentrations are used to estimate risk. Performing a specific risk calculation using both values permits the estimation of a range of potential risks, which can frequently be useful in providing perspective regarding the potential hazards associated with a particular set of exposure conditions. The significance of either measurement depends on the amount of data and the associated confidence in that data. However, in general, calculation of potential risk using an average concentration permits a better estimate of risk associated with chronic exposures, since the average value represents a more likely estimate of the exposure point concentration to which a receptor would be exposed over time. Use of a maximum value is best in the estimation of shorter-term, subchronic risks, although its use can provide a useful upper bound estimate of potential risk.

It should be noted that current risk assessment guidance emphasizes the use of a single, upper bound estimate of exposure point concentration in the calculation of potential risks. This value is often taken to be 95% CI, the 95th percentile upper confidence limit of the arithmetic mean. Use of this number generally provides a worst-case estimate of risk, and can result in a significant overestimate of potential risk, especially when this value is used in combination with other worst-case assumptions to define a reasonable maximum exposure.

5.1.5.2 Carcinogenic Risk

Carcinogenic risk may be defined as the chronic daily intake dose (developed in the exposure assessment) multiplied by the carcinogenic slope (selected by the toxicity assessment). The product is a real term: the probability of excess lifetime cancer from exposure to this chemical. The computation is as follows:

$$\text{Risk} - \text{CDI} \times \text{SF} \tag{5.4}$$

where

CDI = chronic daily intake (mg/kg/day)
SF = carcinogen slope factor $(\text{mg/day/mg})^{-1}$

Characterization of carcinogenic risk first involves determining the intake for each chemical appropriate to the exposure route and pathway under study then multiplying each one by the proper exposure point concentration and slope factor. The classical methodology for risk assessment assumes additivity of risks from individual toxicants. For carcinogens this means that the total carcinogenic risk equals the sum by exposure route of carcinogenic risks from all individual substances. These calculations are repeated for each exposure scenario and exposed population. It must be emphasized that slope factors are specific to the exposure route (e.g. oral) and may be used only when the exposure data apply to the same route. With this caveat in mind, we can proceed to a consideration of assessing the risks associated with some familiar environmental pollutants.

5.2 Assessing the Risks of Some Common Pollutants

Risk assessment is built on the principle that small exposures carry with them some risk of an untoward health effect such as development of a malignant tumor or leukemia at some time in the future. Such risks are generally considered to be stochastic or probabilistic in nature and are expressed in terms of a risk coefficient, an expression of the probability or chance of the specific health effect occurring per unit of exposure. Although response to ionizing radiation is generally considered to be linear with dose, the risk from chemical carcinogens may not be; thus, risk evaluation needs to consider such effects as the existence of a threshold dose below which the effect does not occur, as well as the shape of the dose–response curve.

Also, the latency period, the time between the exposure and the onset of disease, needs to be taken into account. For example, if the latency period is 25 years and exposure occurs at age 50, the individual may die from other causes prior to developing the specific health effect under consideration. Even if the individual lives long enough for the health effect to be manifested, the number of years of life lost, and productive years of life lost, will be less than if the exposure had occurred at an earlier age. Health and safety risks are sometimes expressed in terms of the number of years of productive (or health) life lost per thousand worker-years, as shown in Example 5.7.

Example 5.7 Cancer Risk Assessment

Estimate the total number of fatal leukemias in 50 years to a population of 100 000 exposed to an average annual dose of 500 μGy if the risk coefficient is 0.03 Gy^{-1}.

Solution

$$\text{Fatal leukemias} = \frac{0.03}{\text{Gy}} \times \frac{500\,\mu\text{Gy}}{\text{person year}} \times 50\,\text{years} \times 100000\,\text{persons}$$
$$= 75$$

5.2.1 NO_x, Hydrocarbons, and VOCs: Ground-Level Ozone

Ground-level ozone is one of the most pervasive and intractable air pollution problems in the United States. In more than 20 urban areas, its concentration exceeds one or more of the ambient air quality standards. In the language of regulators, these areas are said to "non-attainment." (We emphasize difference between the "bad" ozone created at or near ground level from the "good" or stratospheric ozone that protects us from UV radiation.)

Ground-level ozone, a component of photochemical smog, is actually a secondary pollutant in that certain precursor contaminants are required to create it. The precursor contaminants are nitrogen oxides (NO_x, primarily NO and NO_2) and hydrocarbons. The oxides of nitrogen along with sunlight cause ozone formation, but the role of hydrocarbons is to accelerate and enhance the accumulation of ozone.

5.2.1.1 Smog Formation

The most important process for ozone formation in the lower atmosphere is photodissociation of NO_2:

$$NO_2 + h\upsilon \rightarrow O(^3P) + NO$$

$$O(^3P) + O_2 + M \rightarrow O_3 + M$$

$$O_3 + NO \rightarrow NO_2 + O_3$$

Where M is molecular oxygen or nitrogen oxide. This cycle results in O_3 concentration being in a photostationary state dictated by the NO_2 photolysis rate and ratio of $[NO_2]/[NO]$. The role of VOCs in smog formation is to form radicals which convert NO to NO_2 without causing O_3 destruction, thereby increasing the ration $[NO_2]/[NO]$, and increasing O_3.

$$VOC + \cdot OH \rightarrow \cdot RO_2 + \text{other oxidation products}$$

$$\cdot RO_2 + NO \rightarrow NO_2 + \text{radicals}$$

$$\text{Radicals} \rightarrow \cdot OH + \text{other oxidation products}$$

The tendency of individual VOCs to influence O_3 levels depends upon its hydroxyl radical (■OH) rate constant and elements of its reaction mechanism, including radical initiation, radical termination, and reaction which removes NO_x. Simplified smog formation potential indices have been proposed based only on VOC hydroxyl radical rate constants.

Oxides of nitrogen (NO_x) are formed in high-temperature industrial and transportation combustion processes. In 2016, US transportation sources accounted for 56.6% and non-transportation fuel combustion contributed 34.4% of total NO_x emissions. Health effects associated with short-term exposure to NO_2 (<3 hours at high concentrations) are increased respiratory illness in children and impaired respiratory function in individuals with preexisting respiratory problems.

Major sources of hydrocarbon emissions are the chemical and oil-refining industries, and motor vehicles. In 2016, industrial processes accounted for 80.5%, while the transportation sector contributed 19.5% of the total of manmade (non-biogenic) hydrocarbon sources, while transportation sources accounted for 18.1% and non-transportation sources for 81.9% of total VOCs (both anthropogenic and biogenic sources), respectively. Solvents comprise 66% of the industrial emissions and 34% of total VOC emissions. It should be noted that there are natural (biogenic) sources of HCs/VOCs, such as isoprene and monoterpenes that can contribute significantly to regional hydrocarbon emissions and low-level ozone levels.

Also, ground-level ozone concentrations are exacerbated by certain physical and atmospheric factors. High-intensity solar radiation, low-prevailing wind speed (dilution), atmospheric inversions, and proximity to mountain ranges or coastlines (stagnant air masses) all contribute to photochemical smog formation. Figure 5.2 shows a picture of downtown Los Angeles sky covered with smog cause poor visibility (New York Times, 17 February 2018b).

Human exposure to ozone can result in both acute (short-term) and chronic (long-term) health effects. The high reactivity of ozone makes it a strong lung irritant, even at low concentrations. Formaldehyde, peroxyacetylnitrate (PAN), and other smog-related oxygenated organics are eye irritants. Ground-level ozone also affects crops and vegetation adversely when it enters the stomata of leaves and destroys chlorophyll, thus disrupting photosynthesis. Since ozone is an oxidant, materials with which, such as rubber and latex painted surfaces, undergo deteriorations.

The perfumes, deodorants, soap that keep us smelling good are fouling the air with a harmful type of pollution – at levels as high emissions from today's cars and trucks. Researchers found that petroleum-based chemicals used in perfumes, paints, and other consumer products can, taken together, emit as much air pollution in the form of VOCs, as motor vehicles do. The VOCs interact with other particles in the air to create the building blocks of smog, namely ozone, which can trigger asthma and permanently scar the lungs, and another type of pollution known as $PM_{2.5}$, fine particles that are linked to heart attacks, strokes, and lung cancer. Smog is generally associated with cars, but since the 1970s regulators pushed automakers to invest in technologies that have substantially reduced VOC

Figure 5.2 Smog in Los Angeles, California. *Source:* The New York Times, 17 February (2018b). Photograph by Frederic J. Brown/ Agence France-Presse-Getty Images.

emissions from automobiles. California Air Resources Board's consumer products and air quality's preliminary results also suggested that emissions, both point and fugitive sources, from those products and processes were higher than previously estimated (New York Times, 17 February 2018b).

There are tens of thousands of chemicals in consumer products and researchers have not yet pin-pointed which chemicals are most likely to form ozone or $PM_{2.5}$ particles. We need to remind ourselves that our energy sources and consumer products we use every day are continuously changing the composition of our atmosphere.

5.2.2 Carbon Monoxide

Carbon monoxide (CO) is a colorless, odorless gas formed primarily as a by-product of incomplete combustion. The major health hazard posed by CO is its capacity to bind with hemoglobin in the blood stream and thereby reduce the oxygen-carrying ability of the blood. Transportation sources account for the bulk (76.6%) of total national CO emissions. Ambient CO concentrations have decreased significantly in the past two decades, primarily due to improved control technologies for vehicles. Areas with high-traffic congestion generally will have high ambient CO concentrations: 80 urban areas in the United States are non-attainment for CO. High localized and indoor CO levels can come from cigarettes (second-hand smoke), wood-burning fireplaces, and kerosene space heaters. Example 5.8 considers exposure to CO in the workplace.

Example 5.8 Total Weighted Average Concentration

a) In a plant, employees were exposed to ambient CO throughout the day at different levels as follows:

Hour	Conc. (ppm)
8–9 a.m.	0
9–11 : 30 a.m.	225
11 : 30–12 : 30 p.m.	100
12 : 30–2 : 00 p.m.	165
2 : 00–4 : 00 p.m.	45

What is the TWA in ppm? If national ambient air quality standard is 50 ppm for 8-hour, did employees' exposure levels exceed the standard?

b) CO concentration in a working environment is 150 ppm. Working area is 1 150 000 ft^3. How much air is needed to ventilate in the room to reduce the CO concentration to 50 ppm?

c) For safety reasons, ventilation dilution is not recommended. Why not? (Note that the molecular weight of CO is similar to that of air.)

Solutions

a) Eight-hour weight average

$$\text{TWA} = \frac{1}{8}\int_0^{t_w} C(t)\,dt$$

$$\text{TWA} = \frac{C_1 t_1 + C_2 t_2 + -- + C_n T_n}{\sum t_u}$$

$$= \frac{(0\times1)+(225\times2.5)+(100\times1)+(165\times1.5)+(45\times2)}{1+2.5+1+1.5+2}$$

$$= 125\,\text{ppm}$$

Therefore, 125 ppm 8-hour TWA concentration exceeded the national ambient air quality standard 50 ppm.

b) Assure steady state with perfect mixing and exponential decay.

$$C = C_o \times \exp\left(\frac{-t}{t_d}\right)$$

$$50 = 150\exp\left(\frac{-1}{t_d}\right)$$

$\therefore t_a = 0.91\,\text{hours} = 54.6\,\text{minutes}$

$\therefore$ volume of air need to ventilate $= \frac{\text{area}}{\text{time}} = \frac{1\,500\,000}{54.6}$ $= 27\,472\,\text{cfm}$

c) Incomplete combustion in the heater unit may increase the concentration of CO and add other noxious contaminants to the air in the warehouse. It is true, as well, that fixed dilution ventilation rate may not be adequate for CO removal during peak generation period. In an isolated warehouse environment, moreover, dilution ventilation may not adequately remove pockets of air that are high in CO concentration. In any event, CO has an acute toxicity, and dilute ventilation is not recommended for acutely toxic gases.

5.2.3 Lead and Mercury

Lead in the atmosphere is primarily found in fine particulates, up to 10 μm in diameter, which can remain suspended in the atmosphere for significant periods of time. Tetraethyl lead was used as an octane booster and antiknock compound for many years before its full toxicological effects were understood. The Clean Air Act of 1970 banned all lead additives and the dramatic decline in lead concentrations and emissions has been one of the most important yet unheralded environmental improvements of the past 25 years. In 1997, industrial processes accounted for 74.2% of remaining lead emissions, with 13.3% resulting from transportation, and 12.6% from non-transportation fuel combustion (USEPA 1997).

Lead also enters waterways in urban runoff and industrial effluents and adheres to sediment particles in the receiving water body. Uptake by aquatic species can result in malformations, death, and aquatic ecosystem instability. There is a further concern that increased levels of lead can occur locally due to acid precipitation that increases lead's solubility in water and thus its bioavailability. Lead persists in the environment and is accumulated by aquatic organisms.

Lead enters the body by inhalation and ingestion of food (contaminated fish), drinking water, soil, paint, and airborne dust. It is a corrosion by-product from high-lead solder joints in copper piping, old lead-pipe goosenecks connecting to service lines to the water main, and old brass fixtures. The primary human health effect of lead in the environment is its effect on organs, tissues, and the brain development, especially in children. There is a direct correlation between elevated levels of lead in the blood and decreased IQ, especially in the urban areas of developing countries that have yet to ban lead as a gasoline additive.

According to EPA, coal-fired power generation companies currently emit over 50 T of mercury every year in the United States. There was no quibbling that these levels were high and a potential health concern to humans and wildlife. Eating mercury-tainted fish can trigger a variety of problems, ranging from hair loss and chronic fatigue in adults to nervous system impairment of fetuses and children.

Mercury taints the atmosphere worldwide, but there are large variations in how much of it drops onto land or water at any location. Recent experiments have begun identifying oxidizing gases, such as ozone and molecules containing the halogens bromine and chlorine, as triggers for that mercury fallout. Which oxidants dominate that process appears to depend on the environment, the season, the altitude of the airborne mercury, and even the amount of daylight.

Mercury enters the air easily. It is released when coal is burned, gold is mined, some chlorine is manufactured, and even when a fluorescent lightbulb breaks. Some 99% of the airborne metal is elemental. Fairly insoluble and unreactive in this form, it can circumnavigate the globe for up to two years. What is contaminating the Everglades, therefore may have originated in Miami, India, or Siberia.

However, atmospheric chemists have discovered that when elemental mercury encounters certain oxidants, it changes into so-called reactive gaseous mercury. Unlike the element, this form is both highly reactive and water soluble, so it remains airborne only hours to days and falls – in rain or snow or attached to dust – near where it is formed. In a lake or ocean, bacteria transform it into methylmercury, the harmful form of the metal that fish and, in turn, people and other predators accumulate in their tissues. Symptoms of methylmercury poisoning include mental disturbance and impairment of speech, hearing, vision, and movement.

Example 5.9 Mercury Removal

Medical sludge containing mercury is burned in an incinerator. The mercury feed rate is 0.92 lb/h. The resulting 500 °F product (40 000 lb/h of gas; MW = 32) is quenched with water to a temperature of 150 °F. The resulting stream is filtered to remove all particulates. What happens to the mercury? Assume the process pressure is 14.7 psi and that the vapor pressure of Hg at 150 °F is 0.005 psi.

Solution

For the mercury to be removed by the filter, it must condense and form particulates. Therefore, the question to be answered relates to the partial pressure of mercury during removal compared to its vapor pressure at 150 °F.

Molar flow rate of Hg = (9.3 lb/h)/(200.6 lb Hg/lbmol) = 0.046 lbmol/h

Molar flow rate of gas = (40 000 lb gas)/(32 lb gas/lbmol) = 1250 lbmol/h

$$y_i = \left(\frac{\text{lbmol Hg}}{\text{lbmol Hg} + \text{lbmol gas}}\right) = \left(\frac{0.046\dfrac{\text{lbmol}}{\text{h}}}{0.046\dfrac{\text{lbmol}}{\text{h}} + 1250\dfrac{\text{lbmol}}{\text{h}}}\right)$$

$$= 3.68 \times 10^{-5}$$

Partial pressure $= y_i P = 3.68 \times 10^{-5}(14.7\,\text{psi}) = 5.41 \times 10^{-4}$ psi.

Since the partial pressure is much less than the vapor pressure of 5×10^{-3} psi, mercury will NOT condense and thus will NOT be removed by the filter.

Example 5.10 Inhalable PCBs and PAHs

Several industrial facilities emit inhalable pollutants including ethylene oxide (EO), poly-chlorobiphenyls (PCBs) and polycyclic aromatic hydrocarbons (PAHs). The worker's annual average exposure concentration of EO, PCBs, and PAHs are 10, 2, and 5 μg/m^3, respectively.

Calculate the cancer risk caused by each pollutant and total cancer risk. Express results in additional cancer per million people. Use the following data given in a tabular form to solve this problem, along with the Eq. (5.5).

$$\text{Cancer risk of pollutant} = (\text{annual average concentration})(\text{unit risk}) \tag{5.5}$$

Unit risk data

Pollutant	Unit risk (m^3/μg)
EO	8.8×10^{-5}
PCBs	1.4×10^{-3}
PAHs	1.7×10^{-3}

Solution

The cancer risk caused by ethylene oxide is established using the Eq. (5.5), and data presented in the problem statement:

$$\begin{aligned}\text{Cancer risk of EO} &= 10\,\mu\text{g/m}^3\left(8.8 \times 10^{-5}\,\text{m}^3/\mu\text{g}\right)\\ &= 88 \times 10^{-5} = 880 \times 10^{-6}\\ &= 880\,\text{excess cancer per million people.}\end{aligned}$$

The cancer risk by PCBs is estimated to be

$$\begin{aligned}\text{Cancer risk of PCBs} &= 2\,\mu\text{g/m}^3\left(1.4 \times 10^{-3}\,\text{m}^3/\mu\text{g}\right)\\ &= 2.8 \times 10^{-3} = 2800 \times 10^{-6}\\ &= 2800\,\text{excess cancer per million people}\end{aligned}$$

The cancer risk by PAHs is estimated to be

$$\begin{aligned}\text{Cancer risk of PAHs} &= 5\,\mu\text{g/m}^3\left(1.7 \times 10^{-3}\,\text{m}^3/\mu\text{g}\right)\\ &= 8.5 \times 10^{-3} = 8500 \times 10^{-6}\\ &= 8500\,\text{excess cancer per million people}\end{aligned}$$

The total cancer risk by inhalation is calculated from the arithmetic sum of individual cancer risks calculated above, assuming no interaction of pollutants in terms of carcinogenic effects.

Total cancer risk = 880 + 2800 + 8500 = 12 180 excess cancer cases per million people.

This rate is extremely high, 12 180 times higher than one in million cancer risk normally used as a basis for management of air toxics. This situation should be rectified as soon as possible by a significant reduction in air toxics releases. Also note that PAHs contributes the majority of risk to the workers at this facility.

5.2.4 Particulate Matter

Particulate matter (PM) is the general term for microscopic solid or liquid phase (aerosol) particles suspended in air. PM exists in a variety of sizes, with diameters ranging from a few angstrom units to several hundred micrometers. Particles are either emitted directly from primary sources or are formed in the atmosphere by gas-phase reactions (secondary aerosols).

Since particle size determines how deep into the lung a particle is inhaled, there are two NAAQS for PM, $PM_{2.5}$, and PM_{10}, where the subscripts indicate particle diameter in micrometers. Particles smaller than 2.5 μm, called "fine," are composed largely of inorganic salts (primarily ammonium sulfate and nitrate), organic species, and trace metals. Fine PM can deposit deep in the lung where removal is difficult. Particles larger than 2.5 μm are called "coarse" particles and are composed largely of suspended dust. Coarse PM tends to deposit in the upper respiratory tract, where removal is more easily accomplished. In 1997, industrial processes accounted for 42.0% of the emission rate for traditionally inventoried PM_{10}. Non-transportation fuel combustion and transportation sources accounted for

34.9 and 23.0%, respectively. As with the other criteria pollutants, PM_{10} concentrations and emission rates have decreased modestly due to pollution control efforts.

Coarse particle inhalation frequently causes or exacerbates upper respiratory difficulties, including asthma. Fine particle inhalation can decrease lung functions and cause chronic bronchitis. Inhalation of specific toxic substances such as asbestos, coal mine dust, or textile fibers are now known to cause specific associated cancers (asbestosis, black lung cancer, and brown lung cancer, respectively).

An environmental effect of PM is limited visibility in many parts of the United States including some national parks. In addition, nitrogen- and sulfur-containing particles deposited on land increase soil acidity and alter nutrient balances. When deposited in water bodies, the acidic particles alter the pH of the water and lead to death of aquatic organisms. PM deposition also causes soiling and corrosion of cultural monuments and buildings, especially those that are made of limestone. More than 100 areas in the United States are non-attainment for PM_{10}.

5.2.4.1 Visibility Impairment

The Clean Air Act established a national visibility protection goal of preventing any further and remedying of any existing impairment of visibility in Class I federal areas in which impairment results from anthropogenic pollution. Both NO_x and SO_x contribute to regional haze and disturbance of the biochemical cycling of other nutrients and metals in eco-systems (*The New York Times*, 17 February 2018b). Nitrogen deposited on the land contributes to the land becoming nitrogen-saturated causing more available nitrogen to run off into nearby waters leading to increased acidification of both the soils and waters. Increased nitrate nitrogen removes calcium and magnesium from soil.

Sulfates and nitrates are secondary aerosols formed from conversion of SO_2 and NO_x to particulate form. The rates of conversion in the atmosphere depends upon a number factors, including weather conditions, relative humidity, and availability of chemicals like ammonia in the atmosphere. Both sulfate (SO_4) and nitrate (NO_3) aerosols are considered to be $PM_{2.5}$. In presence or in absence of humidity, these fine aerosols in atmosphere absorb and scatter lights that can cause poor visibility in the region (Figure 5.2).

5.2.5 SO_2, NO_x, and Acid Deposition

Sulfur dioxide (SO_2) is the most commonly encountered of the sulfur oxide (SO_x) gases and is formed upon combustion of sulfur-containing solid and liquid fuels (primarily coal and oil). SO_x are generated by electric utilities, metal smelting, and other industrial processes. Nitrogen oxides (NO_x) are also produced in combustion reactions; however, the origin of most NO_x is the oxidation of nitrogen in the combustion air. After being emitted, SO_x and NO_x can be transported over long distances and are transformed in the atmosphere by gas phase and aqueous phase reactions to acid components (H_2SO_4 and HNO_3). The gas phase reactions produce microscopic aerosols of acid-containing components, while aqueous phase reactions occur inside existing particles. The acid is deposited to the Earth's surface as either *dry deposition* of aerosols during periods of no precipitation or *wet deposition* of acid-containing rain or other precipitation. There are also natural emission sources for both sulfur and nitrogen-containing compounds that contribute to acid deposition. When natural sources of sulfur and nitrogen acid rain precursors are considered, the "natural" background pH of rain is expected to be about 5.0. As a result of these considerations, "acid rain" is defined as having a pH less than 5.0. Figure 5.3 shows the major environmental cause and effect steps for acidification of surface water by acid rain.

In the United States, major sources of SO_2 emissions were non-transportation fuel combustion (72.6%), industrial processes (18.7%), transportation (5.8%), and miscellaneous (2.9%) (see Table 4.6). Fifteen US urban areas are non-attainment for SO_2. Emissions are expected to continue to decrease as a result of implementing the Acid Rain Program established in 1997 by EPA under Title IV of the Clean Air Act. The goal of this program is to decrease acid deposition significantly by controlling SO_2 and other emissions from utilities, smelters, and sulfuric acid manufacturing plants, and by reducing the average sulfur content of fuels for industrial, commercial, and residential boilers.

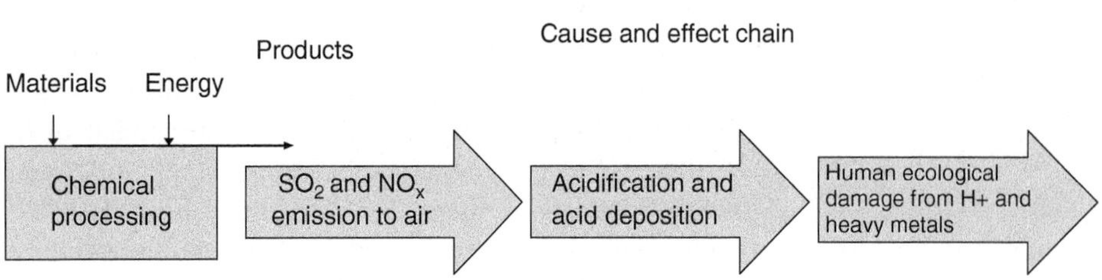

Figure 5.3 Environmental cause and effect for acid rain.

There are a number of health and environmental effects of SO_2, NO_x, and acid deposition (Figure 5.4a, b). SO_2 is absorbed readily into the moist tissue lining the upper respiratory system, leading to irritation and swelling of this tissue and airway constriction. Long-term exposure to high concentrations can lead to lung disease and aggravate cardiovascular disease. Acid deposition on soil, plants, lakes, and rivers causes destruction of forest and plants, acidification of surface water, damage vegetation, monuments, etc., especially in regions of high SO_2 concentrations and low buffering and ion exchange capacity of soil and surface water. Acidification of water can harm fish populations, by exposure to heavy metals, such as aluminum which is leached from soil. Excessive exposure of plants to SO_2 decreases plant growth and yield and has been shown to decrease the number and variety of plant species in a region (McLaughlin and Taylor 1985; USEPA 1998a).

(a)

(b)

Figure 5.4 (a) Trees killed by acid rain at Mt. Mitchell in North Carolina. (b) Corn tree killed by SO_2.

5.2.6 Air Toxics

Hazardous air pollutants (HAPs), or air toxics, are airborne pollutants that are known to have adverse human health effects, such as cancer. There are over 180 chemicals identified on the Clean Air Act list of HAPs given in Section C.2 (USEPA 1998c). Examples of air toxics include the heavy metals mercury and chromium, and organic chemicals such as benzene, hexane, perchloroethylene (perc), 1,3-butadiene, dioxins, and PAHs.

The Clean Air Act defined a major source of HAPs as a stationary source that has the potential to emit 10 T/Y of any one HAP on the list or 25 T/Y of any combination of HAPs. Examples of major sources include chemical complexes and oil refineries. The Clean Air Act prescribes a very high level of pollution control technology for HAPs called maximum achievable control technology. Small area sources, such as dry cleaners, emit lower HAP tonnages but taken together are a significant source of HAPs. Emission reductions can be achieved by changes in work practices such as material substitution and other pollution prevention strategies.

HAPs affect human health via the typical inhalation or ingestion routes. HAPs can accumulate in the tissue of fish, and the concentration of the contaminant increases up the food chain to humans. Many of these persistent and bioaccumulative chemicals are known or suspected carcinogens.

5.3 Ecological Risk Assessment

The four major components of the ERA paradigm are problem formulation, exposure assessment, effects assessment, and risk characterization (Anderson and Albert 1998; Suter et al. 2000; USEPA 1992, 1997, 1998a). An ERA begins with problem formulation. Activities occurring during this phase include definition of the goals and spatial and temporal scale of the ERA, development of a SCM, selection of endpoint and nonhuman receptor species selection, and preliminary identification of contaminants of potential concern. Exposure assessment and effects assessment follow and can be performed simultaneously. Exposure assessment evaluates the fate, transport, and transformation of chemicals in the environment, and quantitative uptake and intake of these substances in receptor organisms. Effects assessment establishes the relationship between exposure levels and toxic effects in receptors. Risk characterization is the last step in the ERA and is where exposure and toxic effect information are combined to describe the likelihood of adverse effects in receptors. Many of the evaluation criteria needed to evaluate an ERA are identical to those already presented for human health risk assessment (HHRA). This section focuses primarily on the unique aspects of ERAs and does not repeat material covered under HHRA that applies to both subjects.

5.3.1 Technical Aspects of Ecological Problem Formulation

Determining how many data are needed to address the ERA goals is part of the process of meeting a project's Data Quality Objective (DQO). All risk assessment stakeholders (e.g. the US EPA, the State, the Fish and Wildlife Service, etc.) should be involved in this process. The DQO is ascertained at the beginning of an assessment, to define both the amount and quality of data required to complete the assessment. Scheduling time to complete DQOs at the beginning of the ERA may save the project time and money in the end. Once the goals and DQOs have been determined, the remainder of the problem formulation may be conducted. The ultimate goal of problem formulation is the SCM.

A wide range of ecosystem characteristics may be considered during problem formulation. These include abiotic factors (e.g. climate, geology, soil/sediment properties) and ecosystem structure (e.g. abundance of species at different trophic levels, habitat size, and fragmentation). The environmental description may be documented using recent photographs and maps. Plant and animal species lists should be compiled.

The scale of the assessment is especially important if a large, complex site has been subdivided into several smaller sites. It also is not uncommon for Superfund sites to be located adjacent to each other. Hence, the areal extent of the assessment must be defined. For example, if an off-site area is included in the assessment, the distance off-site must be specified. The development of the SCM and the selection of assessment endpoints will be directly related to the spatial scale. For example, due to their large home ranges, effects of soil contamination on deer would not be assessed if the site encompasses only two acres; assessment of endpoint species with smaller home ranges, such as small mammals, would be more appropriate.

It is necessary to decide if the assessment must consider temporal changes. All historical information should be evaluated. Then, it may be determined how much new information is needed to adequately evaluate impacts and risks. Certain parts of the year may need to be included in the sampling season for the assessment. For example, environmental exposures may change over the course of a year, or over several years, due to various seasonal influences in either chemical form or organism behavior (e.g. salmon returning to a contaminated river to spawn; migrating birds making temporary use of a site).

5.3.1.1 Site Conceptual Models

The SCM describes a series of working hypotheses regarding how contaminants or other stressors may affect ecological receptors (ASTM 2003). An SCM clearly illustrates the contaminated media, exposure routes, and receptors for the risk assessment. In addition to a written description, a diagrammatic SCM is easy to understand and is useful for ensuring that no relevant component is omitted from the assessment. The model ensure that all exposure scenarios are considered and allows for full documentation of the rationale behind selection and omission of pathways and receptors. The idea behind the SCM is that although many hypotheses may be developed during problem formulation, only those that are expected to contribute significantly to risks at the site are carried through the remainder of the ERA process.

During SCM development, all contaminant sources are identified (e.g. landfills, burial grounds, lagoons, air stacks, effluent pipes), and all contaminated media are represented (e.g. soil, water, sediment, air, biota). Groundwater usually is not considered an exposure medium until it becomes surface water, but contaminants in groundwater can migrate from soil to surface water and biota. An exception is shallow groundwater or seeps where plants may be exposed via their roots. All exposure pathways are represented, unless adequate rationale can be provided to exclude a pathway from the assessment. For example, an effluent pipe releasing metals into a stream would not need an air exposure pathway, and the only soils that would need to be considered are those of the floodplain. Thus, terrestrial receptors would be exposed by direct contact with or drinking from the stream, living in floodplain soils, or obtaining contaminated food from the stream and floodplain. An appropriate food web must be presented. A food web going from contaminated soil to earthworm to shrew may be appropriate for a one acre site, but a significantly larger site may require the food web to continue up to larger predators which have larger home ranges (see Figure 5.5).

For nonchemical stressors such as water level or temperature changes, or habitat disturbances, the SCM describes which ecological receptors are exposed to the physical disturbance, and the temporal and spatial scales of the alterations.

ERAs may have more than one SCM. In predictive ERAs, impacts on different components of the ecosystem from various activities may require several SCMS. In retrospective ERAs, a hypothetical future scenario often requires assessment. For example, an industrial area that now provides little habitat for wildlife (hence little exposure and

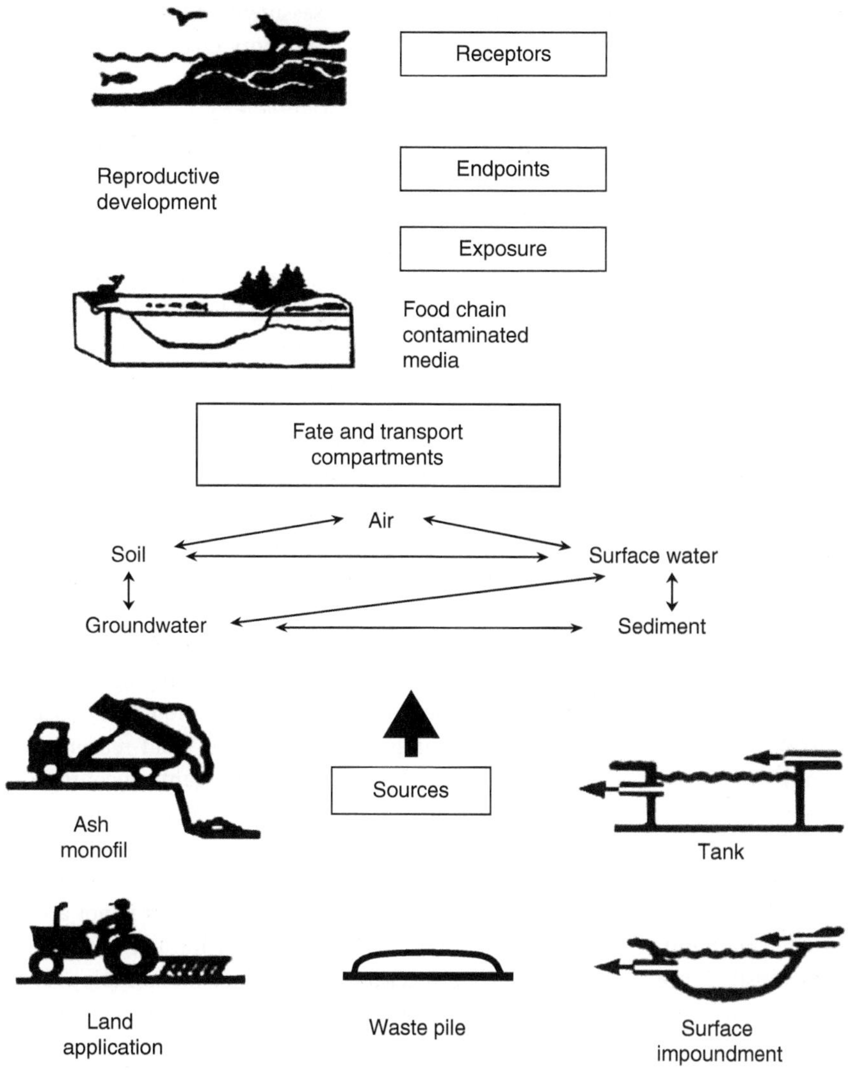

Figure 5.5 Environmental risk assessment multipathways analysis. *Source:* Adopted from USEPA (1995).

little risk) may in future become covered in vegetation. It then would be more attractive as wildlife habitat, and hence the risk of exposure to contaminants would become greater. Similarly, a plume of contaminated groundwater that does not now pollute a pond may become a water source in several years. The future risk must be evaluated.

5.3.1.2 Identifying Endpoints

Before the SCM can be completed, the assessment endpoints of the ERA must be defined and rationale given for their selection. An assessment endpoint is the actual environmental value that is to be protected (Suter 1993; Suter et al. 2000). An example of an assessment endpoint would be "no less than a 20% decrease in the survival, growth, or reproduction in the largemouth bass population in the creek." Desirable characteristics for assessment endpoint species include the following (Suter 1993; Suter et al. 2000):

- An assessment endpoint must be relevant to decision making.
- The structure and function of components of the ecosystem must be understood in order to determine the ecological relevance or importance of the endpoint. Species that control the abundance and distribution of other species and those that are involved in nutrient cycling and energy flow are generally considered to be ecologically relevant.
- Selection of endpoints may be influenced by societal involvement and concern.
- Only species that are present, or likely to be present at the site, should be used to evaluate risks, regardless of the value or importance of the species.
- Since only some species at a site can be evaluated, endpoint species must be selected which are sensitive to the contaminants at the site and are likely to receive high exposures. In this way, other species that may be less

sensitive or receive lower exposures will also be protected. Other information necessary for each receptor species includes diet composition; habitat preference/needs; home range size; intake rates of food, water, sediment, air, and soil; and body weight.

- Finally, an assessment endpoint must be measured or modeled. If there is no method available to measure or model effects on an endpoint, evaluation of risk cannot be completed.

5.3.1.3 Selecting Measurement Endpoints

Because there are so many species and other ecosystem characteristics from which to choose assessment endpoints, stakeholders must agree on the appropriate assessment endpoints early in the ERA process. The remainder of the assessment cannot be completed until these have been chosen. After assessment endpoints have been selected, ecological risk assessors can select appropriate measurement endpoints for each assessment endpoint. Measures of exposure and effect are measurable environmental characteristics related to the valued characteristic chosen as an assessment endpoint (Suter 1993; Suter et al. 2000). There are three categories of measures (USEPA 1989). "Measures of effect" are measurable changes in an attribute of an assessment endpoint in response to a stressor to which it has been exposed (formerly referred to as "measurement endpoints"). "Measures of exposure" are measures of stressor existence and movement in the environment and this contact or co-occurrence with the assessment endpoint. "Measures of ecosystem and receptor characteristics" are measures of ecosystem characteristics that influence the behavior and location of assessment endpoints, the distribution of a stressor, and life history characteristics of the assessment endpoint that may affect exposure or response to the stressor. These three difference measures are especially important when completing a complex ERA.

ERAs that involve Superfund remedial actions must meet federal and state standards or criteria called ARARs, for "applicable or relevant and appropriate requirements" (USEPA 1989). ARARs which may need to be considered at a site include those specified in the Clean Water Act, Clean Air Act, Endangered Species Act, Fish and Wildlife Conservation Act, Wild and Scenic Rivers Act, Migratory Bird Treaty Act, and many others. If numerical ARARs exist, modeled or measured chemical concentrations in site media cannot exceed these values.

During problem formulation, historical data and/or site investigation data are used to prepare a preliminary list of contaminants of potential ecological concern (COPEC). In order to obtain a meaningful ERA, selection of COPECs must ensure that all contaminants that may contribute significantly to risk are included. Reasoning must be provided for exclusion of chemicals from the COPEC list. In this initial screening of contaminants, valid reasons may include (but not be limited to) contaminant concentrations at or below background levels; concentrations below ARARS, other regulatory concentrations, or toxicity benchmarks; or chemicals infrequently detected. Exclusion of COPECs because the HHRA excluded them is not a valid reason. This is because protection of human health does not guarantee protection of nonhuman biota. Table 5.4 presents several aspects of this apparent paradox.

Table 5.4 Differences between human health and ecological risk assessments.

Component	Human health risk assessment (HHRA)	Ecological risk assessment (ERA)
Institutional controls	Institutional controls may be considered when selecting exposure parameters	Nonhuman organisms are not excluded from waste sites by controls, such as fences or signs
Standard exposure factors	The USEPA provides standard exposure parameters and toxicological benchmarks for humans	Risk assessors must generate their own exposure parameters and toxicity data
Receptor species	Humans only	Nonhuman organisms (flora and fauna) and ecosystem properties (e.g. nutrient flow)
Exposure routes	Ingestion of food and water, incidental ingestion of soil, inhalation of contaminants from air, dermal contact, ingestion of fish fillets	As well as the exposure routes common to HHRA, other routes exist, such as fish respiring water, benthic organisms consuming sediments, small mammals burrowing in soil leading to enhanced exposure, fish-eating wildlife consume the entire fish and chemicals accumulate to a different degree in different organs
Chemical form	Total metals in water are assumed to be available to humans	Dissolved metals are available to aquatic biota for gill uptake
Spatial scale	Often assumes a residential scenario at the site, regardless of appropriateness	Scale is important, since a small site (e.g. a few acres) cannot support a population of larger organisms (e.g. deer, hawks), but it could support small animal populations (e.g. shrews)
Temporal scale	Often only considered when seasonality may change chemical concentrations	Seasonality is more important in ERA, often because of habitat changes or changes in organism behavior

5.3.2 Ecological Exposure Assessment

ERA has several considerations that HHRA lacks. One of the most important factors affecting the exposure assessment is the spatial and temporal scale of the assessment. Spatially, exposure estimates must take into account the home range of, and the availability of, suitable habitat for the receptor species, relative to the areal extent of contamination. Temporal considerations include whether the receptor species is a resident or migrant species, and whether contaminant concentrations vary over the course of the year due to seasonal changes.

Another concept that is not often addressed in HHRA is the different level of protection afforded to different species. HHRAs are designed to protect individuals. In ERA, only threatened and endangered species, or other species of special legal (e.g. migratory birds) or public concern are evaluated for impacts at the individual level. For other species, protection is primarily afforded at the population level. For example, it is important to protect a population of deer at a site; individual deer will not be protected. Practically, this means that impacts on measures relevant to the population as a whole, such as survival and reproduction, are evaluated. Individual quality of life is not considered.

As in HHRA, for an exposure pathway to be complete, there must be a contaminated medium, a transport medium, receptor species, and an exposure route that enables the contaminant to enter the organism. However, ERA has unique exposure routes, such as respiration of water by fish. In the exposure assessment, contaminant concentrations at an exposure point are determined, or intake rates calculated. In the risk characterization, these concentrations are related to toxicological benchmarks; which are contaminant concentrations that are assumed not to be hazardous to the receptor species.

The exposure scenario in an ERA may not be the same scenario as the HHRA. ERA does not have a default "residential scenario," or "industrial scenario." However, hazardous waste sites often are industrial in nature. Scenarios are developed which are appropriate to the current land use. Like the human health assessment, the ERA may make assumptions regarding future land use. This future scenario may assume the site is abandoned and undergoes natural succession. Therefore, it is unreasonable to assume that the same wildlife species will be present in the current and future scenarios, especially if the habitat changes. All assumptions regarding exposure scenarios must be documented early in the ERA process.

During characterization of the exposure environment, the relationship between the receptor species and the environment is detailed. Ecosystem characteristics can modify the nature and extent of contaminants. Chemicals may be transformed by microbial communities or through physical processes such as hydrolysis and photolysis. The environment also may affect bioavailability of contaminants. Physical stressors such as stream siltation and water temperature fluctuations may have considerable impact on ecological risks, and, therefore, must be described.

As part of the characterization of the exposure environment, it is also important to consider both the habitat requirements of receptor species and the amount of suitable habitat available at the site. Availability of habitat will determine the amount of use that a site receives. Because exposure cannot occur if receptor species are not present and receptor species will not be present if suitable habitat is not available, it is important to identify habitat requirements and availability early in the exposure assessment. Selecting exposure routes depends on the endpoints to be evaluated. Several endpoint categories and exposure routes are discussed in Sections 5.3.2.1–5.3.2.5.

5.3.2.1 Fish Community

Fish are exposed to contaminants in surface water through respiration and dermal absorption. They also may be exposed through the consumption of contaminated sediment or food. There are two important considerations for the fish community. The first is that for inorganic contaminants, it is the dissolved fraction of the contaminant in the surface water that the fish are exposed to by inhalation (i.e. gill uptake). Practically speaking, this involves filtering the water sample through a nanometer filter prior to analysis. HHRA calculates exposures using the total inorganic concentration in water. However, the particulate-bound fraction is not available to fish at the gill. Secondly, dermal absorption as a separate exposure route is not evaluated, because existing toxicity data for fish were generated either by feeding contaminated food to fish or exposing fish to contaminants in the water, without attempting separate evaluations of the various uptake routes.

5.3.2.2 Benthic Macroinvertebrate Community

Benthic macroinvertebrates live in or on contaminated sediments. They may be exposed through ingestion of the sediment or contaminated food. Also, benthic organisms may respire overlaying water or the sediment pore-water. Special considerations for this endpoint include the need for bulk sediment contaminant concentrations and pore-water analyses, in order to compare these concentrations to benchmark concentrations. For nonionic/nonpolar organic contaminants, bulk sediment concentrations are used. The organic carbon content of the sediment is also required.

For ionic/polar organic contaminants, the sediment pore water must be analyzed. For inorganic contaminants, either analysis is adequate.

5.3.2.3 Soil Invertebrate Species

Soil invertebrates, such as earthworms, are in direct contact with contaminated soil. Also, the earthworm ingests large amounts of soil during feeding. Contaminants are in contact with and may be absorbed by the gut of the worm.

5.3.2.4 Terrestrial Plants

Plants may take up contaminants from the soil at the root. Also, contaminants in shallow groundwater may be taken up by the plant roots. Airborne contaminants (e.g. ozone, acid gases) also may enter the plant through the leaf stomata and cause damage.

5.3.2.5 Terrestrial Wildlife

As terrestrial wildlife move through the environment, they may be exposed to contamination via three pathways: oral, dermal, or inhalation. Oral exposure occurs through the ingestion of contaminated food, water, or soil. Dermal exposure occurs when contaminants are absorbed directly through the skin. Inhalation exposure occurs when volatile compounds or fine particulates are respired into the lungs. While methods are available to assess dermal and inhalation exposure to humans, data necessary to estimate dermal and inhalation exposure are generally not available for wildlife. However, these routes are generally considered to be negligible relative to other routes.

Ideally, seasonal data would provide the most complete evaluation of contaminants present in the environment. Wherever possible, site-specific data should be used, rather than modeled data. Where EPCs must be modeled, the same methods and considerations are applicable to ERA as in HHRA. EPCs are developed differently according to endpoint. For the fish community, the concentration of contaminant in water or sediment is used as the EPC. No exposure models are required. The upper 95% confidence limit on the mean water concentration may be used instead of the mean or maximum detected concentration. This is because chronic exposures of the maximally exposed aquatic organisms would be to spatially and temporally varying contaminant concentrations. For the benthic, soil invertebrate and plant communities, the concentration in the sediment or soil at each sample location is used as the EPC. Again, no exposure models are required. However, in each of these cases, the maximum concentration in the sediment or soil should be used as the EPC because these organisms are not particularly mobile. The entire community could be exposed to the maximum concentration present in the medium.

For wildlife species, contaminant concentrations in food, water, and soil are used in exposure models to estimate dose. Because wildlife are mobile, use various portions of a site, and are exposed through multiple media, the upper 95% confidence limit on the mean best represents the spatial and temporal integration of contaminant exposure wildlife will experience. Exposure estimates for wildlife are usually expressed in terms of a body weight-normalized daily dose or mg contaminant per kilogram body weight per day (mg/kg/day). Exposure estimates expressed in this manner may then be compared to toxicological benchmarks for wildlife, or to doses reported in the toxicological literature. A very few wildlife consume diets that consist exclusively of one food type.

To account for varying contaminant concentrations in different food types, exposure estimates should be weighted by the relative proportion of daily food consumption attributable to each food type, and the contaminant concentration in each food type. Each parameter in a wildlife contaminant intake equation must be obtained from the literature because few site-specific values are likely to be available. The EPA *Wildlife Exposure Factors Handbook* (USEPA 1993) contains a compilation of values for parameters such as diet composition, food intake rate, body weight, and home range for 15 birds, 11 mammals, and 8 reptiles and amphibians. The primary and secondary literature must be consulted for any parameter values not contained in this document or if the values provided are not appropriate for the site or become outdated.

One advantage that ERA has over HHRA is the ability to sample the receptor species itself. Rather than introducing modeling uncertainties, fish, benthic macro-invertebrates, soil invertebrates, plants, and some wildlife species (e.g. small mammals) can be sampled directly to give an indication of the bioavailability of environmental contaminants. Of course, it is not acceptable to destructively sample many species, such as rare, threatened, and endangered species, or those with high societal value or low abundance. However, when possible the additional sampling and analytical costs will be worth the added certainty in the exposure assessment and risk characterization.

Ideally, contaminant analysis of whole fish are used when conducting an exposure assessment on piscivorous species. However, fish body burdens may be estimated using bioaccumulation factors. Professional judgment is required when selecting a parameter value for the exposure model. Full rationale for the selection of any parameter value must be provided in the exposure assessment. Exposure assessments will use a variety of data with varying degrees of uncertainty associated with them. Each assumption made will be a result of professional judgment, but it will still have some uncertainty. It is important that the exposure assessment document and characterize each source of uncertainty, including those associated with analytical data, exposure model variables, contaminant distribution and bioavailability, receptor species presence and sensitivity, and other incomplete exposure information.

5.3.3 Ecological Effects Assessment

An ecological effects assessment includes a description of ecotoxicological benchmarks used in the assessment, toxicity profiles for contaminants of concern, and results of the field sampling efforts. The field data may include field survey information and toxicity test results.

Ecotoxicological benchmarks represent concentrations of chemicals in environmental media (i.e. water, soil, sediment, biota) that are presumed not to be hazardous to biota. There may be several benchmarks for each medium and each endpoint species, which allows for estimation of the magnitude of effects that may be expected based on the contaminant concentrations at the site. For example, there may be a benchmark for a "no-effect level," a "low-effect level," "chronic-effect level," a "population-effect level" and an "acute-effect level." Using all of these benchmarks will provide more information for decision makers than any one of the above.

There are few federal or state benchmarks currently available in the United States or elsewhere. Criteria that are used as benchmarks are the National Ambient Water Quality Criteria for the Protection of Aquatic Life (NAWQC) (USEPA 1986). These are ARAR, and are used as benchmarks for the fish community and other water-column species (e.g. invertebrates such as daphnids). However, not all contaminants have these criteria. Therefore, other benchmarks are needed. Benchmarks for the fish, benthic, soil invertebrate, and plant communities, and wildlife are described briefly in Sections 5.3.3.1–5.3.3.4. The primary source of toxicity information used in the development of these benchmarks is the open literature (Risk Assessment Forum 1996, 1999).

5.3.3.1 Fish Community

The acute and chronic NAWQC or state water quality criteria are ARARs and must be used as benchmarks. However, these were developed as broadly applicable values, and thus it may be more appropriate to determine benchmarks for the geographical location and species present at the site. The literature should be reviewed for chronic values in systems similar to that at the site, whether it be a freshwater, estuarine, marine, hard-water, or soft-water system. Laboratory toxicity tests have been conducted on many different aquatic species for many contaminants. In fact, the aquatic system currently has the largest readily available database of contaminant concentration/effects data.

5.3.3.2 Benthic Community

There are several methods that may be used for calculating sediment benchmarks for the benthic community. For nonionic/nonpolar organic contaminants, the equilibrium partitioning approach is often employed. For inorganic contaminants, existing bulk sediment toxicity values from the literature may be used, or pore water concentrations of contaminant may be compared to existing NAWQC. Unfortunately, the database of single-contaminant exposure/effects data for sediments is limited. The majority of the data come from contaminated sites and, therefore, multiple contaminants were present. However, sediment contamination is receiving more attention, and risk assessors and managers must stay current with respect to advances in the areas of sediment toxicology and policy.

5.3.3.3 Soil Invertebrate and Plant Communities

The plant community plays a dominant role in energy flow and nutrient cycling in ecosystems. Soil invertebrates and plants form the bases of many food webs. There is an extensive database for soil contaminants. However, the majority of endpoints used by researchers have been food crop species. While this information is crucial to human health risk assessors, it is not directly applicable to ecological risk issues. The primary literature will be the major source of toxicity information that must be used in the development of toxicity benchmarks. Soil contamination impacts on plant, invertebrate, and even microbial communities are recent important issues. Again, this is an area within ERA in which it is imperative to remain current.

5.3.3.4 Wildlife

Wildlife benchmarks are particularly complicated because wildlife may be exposed to contaminants in their drinking water, the soil around them, and in their diet (from both plant and animal sources). Therefore, wildlife benchmarks must account for these multiple exposure routes. Benchmarks may be derived for each exposure route separately (for cases where exposure is through only one route) and also for total exposure. In the case of exposures from multiple routes, a benchmark such as NOAEL or LOAEL (no/lowest observed adverse effects level) is selected and expressed as a dosage, such as milligrams of contaminant per kilogram of body weight (per day). Benchmarks for wildlife are species specific, in order to account for different species sensitivities, body weights, foraging habits, and diets. In the selection of appropriate benchmark values, the toxicological literature must be consulted, with emphasis on reproduction endpoints. Contaminant toxicity profiles assist risk assessment readers to clearly understand the toxic effects of contaminants in the environment. Toxicity profiles in a risk assessment can provide a concise summary of relevant toxicity information. It is worth repeating that the information must be relevant to the waste site and endpoints of concern. That is, the profile should not simply be a list of median lethal doses (LD_{50}s) for rats and mice. Dose–response information should be compiled for the contaminants that are found at the site, and for the receptor species of interest there.

Toxicity profiles and biological effects data for the species of interest are also useful for helping risk assessors and risk managers evaluate the extent and magnitude of risk. Contaminant concentrations at which lethal and sublethal effects (including behavioral modifications) are observed should be presented (i.e. dose–response information). Information such as the mobility of the chemical (e.g. water solubility, soil sorption, octanol–water partition coefficient), persistence in the environment, and interactions with other contaminants will help risk managers make an informed decision and educate the public.

5.3.4 Additional Components of Ecological Risk Assessments

5.3.4.1 Sampling and Surveys

Although general sampling issues will have necessarily been addressed before the ERA reached the effects assessment stage, it is worthwhile to note a few of them here. This will ensure that the risk assessor has mentioned and considered the potential impacts of these issues. Field surveys and ambient media chemical analyses are also addressed.

5.3.4.2 Field Sampling

Before sample locations are determined, sampling "reaches" must be defined. These are areas that may be impacted by specific contaminant sources. For example, one stream may have several contaminant sources along its length; a reach may be defined as that area between two sources. Sampling in reaches allows for the determination of the relative contribution of various sources to observed toxicity. It is important not to forget to sample an appropriate background (or reference) site. In fact, it is better to have a few reference sites, to account for natural variability in the environment. In the past, there was a distinction between background (meaning pristine) and reference (meaning not impacted by this particular site). Since; however, the distinction is losing currency, it is necessary to know which definition is being used.

One facet of field sampling that is often forgotten when schedules are set is the problem of seasonality in field parameters. For a large portion of the country, winter hinders sampling efforts. For example, it is difficult to sample worms or fish when the ground and creeks are frozen. Also, bats hibernate during the winter, birds migrate, and rare plants are more difficult to identify when they are not in bloom. It is better to delay completion of a risk assessment than to collect data at an inappropriate time.

A waste site investigation will necessarily involve the coordination of a variety of investigators covering the various sampling tasks. The coordination is important in order to obtain results useful for the ERA. For example, whereas water concentrations may change dramatically over a short period of time, water samples being tested for toxicity should be taken at the same time and from the same location as those taken for chemical analysis. It is less critical to coordinate other activities, such as collection of sediment samples, since sediment concentrations take longer period to show contamination.

In ERA, samples of ambient media do not consist exclusively of groundwater, surface water, sediment, soil, and air; the biota must be included as well, to permit evaluation of contaminant exposure and effects. This important source of information, available to ecological risk assessors, may allow greater certainty in the ERA results.

5.3.4.3 Field Surveys

Field surveys have the advantage of giving a real-world indication of effects. However, the cause of any observed effects is likely to be unknown. For example, a decrease in young of the year fish may be due to contaminants that impact fish eggs or larvae or may be due to natural causes, such as a storm event which caused increased water flow that eroded the spawning beds. Another disadvantage is that small changes are unlikely to be detected. Usually a greater than 20% decrease in a field parameter (e.g. population size, number of species) is necessary for deleterious change to be detected. Field surveys may be further complicated because without appropriate and comparable reference sites, interpretation of effects observed at the site is extremely difficult.

In the case of predictive ERAS, field surveys provide information on the environment that may receive contaminants in the future. It is important to have this information in order to document any future adverse impacts. Surveys may include threatened and endangered species surveys, aquatic and terrestrial community surveys, and wetland survey. We mention the latter in passing before turning to a consideration of speciation.

In the United States, a wetland survey must be done for the site to identify and, if necessary, delineate wetlands. Note, it is easier (and less expensive) to identify than to delineate wetlands. It would only be necessary to delineate a wetland if remediation or other activities necessitated the destruction of all or part of the wetlands.

5.3.4.4 Speciation

Information on the speciation of the chemical in various media may be useful for contaminants, such as arsenic or chromium that have species with very different relative toxicities. Before the samples are sent for analysis, it must be ascertained whether the analytical method used will have detection limits below the regulatory concentrations of interest (e.g. ARAR) and the concentration that would produce an unacceptable risk, unless this is not technically or economically feasible. If these detection limits cannot be met, there will be added uncertainty in the risk assessment, because it will not be known whether these contaminants are present or not, and hence whether they constitute a risk.

Chemical concentrations in media at a site, along with the abundant single chemical toxicity data available in the literature, may be used to determine the specific causes of the impacts observed in the field surveys or toxicity tests, and define the sources of the contamination. These data are used in predictive ERAs to model effects of contaminant exposures. However, the measured concentrations may not be indicative of the bioavailable fraction (e.g. chemicals may be bound to soil particles and hence not be available for uptake by organisms).

5.3.4.5 Sources of Other Effects Information

Supplementary information that may be useful in the interpretation of ecological data includes an analysis of biomarkers. Biomarkers serve as sensitive indicators in individual organisms of exposure to contaminants or other sublethal stressors. They are typically physiological or biochemical responses, such as enzyme concentrations, genetic abnormalities, histopathological abnormalities or body burdens of contaminants. While biomarkers give an indication of exposure to stressors, they cannot yield information on the impacts of this exposure on individuals. This is because ERA is concerned primarily with the viability of organism populations, not physiological effects in a single fish or bird. However, some biomarkers are chemical-specific, and hence may provide valuable information on the potential cause of observed toxic effects. For example, increased blood levels of the enzyme δ-amino-levulinic acid dehydratase indicates exposure to lead.

5.3.4.6 Additional Effects that Figure in Many ERAs

Global Warming The atmosphere allows solar radiation from the sun to pass through without significant absorption of energy. Some of the solar radiation reaching the surface of the Earth is absorbed, heating the land and water. Infrared radiation is emitted from the Earth's surface, but certain gases in the atmosphere absorb this infrared radiation, and redirect a portion back to the surface, thus warming the planet and making life, as we know it, possible. This process is often referred to as the *greenhouse effect*. The surface temperature of the Earth will rise until a radiative equilibrium is achieved between the rate of solar radiation absorption and the rate of infrared radiation emission. Human activities, such as fossil-fuel combustion, deforestation, agriculture, and large-scale chemical production, have measurably altered the composition of gases in the atmosphere. Some believe that these alterations will lead to a warming of the Earth–atmosphere system by enhancement of the greenhouse effect.

The primary greenhouse gases (GHGs) are water vapor, carbon dioxide, methane, nitrous oxide, chlorofluorocarbons, and tropospheric (ground-level) ozone. Water vapor is the most abundant GHG, but it is omitted because it is generally not from anthropogenic sources. Carbon dioxide contributes significantly to global warming due to its high emission rate and concentration. The major factors contributing to global warming potential of a chemical are infrared absorptive capacity and residence time in the atmosphere.

5.3.5 Tropospheric Ozone Pollution and Its Effects on Plants

Ozone is formed in the troposphere when sunlight causes complex photochemical reactions involving oxides of nitrogen (NO_x), and VOCs that originate chiefly from gasoline engines and burning of other fossil fuels. Woody vegetation is another major source of VOCs (e.g. isoprene or terpene, formaldehyde, acetaldehyde, methyl-ethyl-ketone, acetone, etc.). NO_x and VOCs can be transported long distances by regional weather patterns before they react to create ozone in the atmosphere, where it can persist for several weeks. Major photochemical reactions that cause ozone formation are given in the following.

$$2NO + O_2 \Rightarrow 2NO_2$$

$$RH + OH\bullet \Rightarrow RH\bullet \Rightarrow\Rightarrow RO_2^\bullet + NO \Rightarrow NO_2$$

$$NO_2 + h\nu \Rightarrow NO + O$$

$$O + O_2 + M \Rightarrow O_3$$

$$\text{Net reaction}: NO_x + RH + h\nu \Rightarrow O_3$$

5.3.5.1 Visibility and Regional Haze

The oxidizing agents, ozone (O_3), PAN, peroxybenzoyl nitrate, and other trace substances react with nitrogen oxides and VOCs and produce smog. These components of photochemical smog which are most damaging to plants and detrimental to human health are these photochemical oxidants. The aerosols formed during the chemical reactions that create smog caused a marked reduction in visibility and regional haze, and give the atmosphere a brownish cast. Regional haze occurs at distances where the plume has become evenly dispersed into the atmosphere, such that there is no definable plume. The primary cause of regional haze are sulfates and nitrates (primarily as ammonium salts), which are formed from SO_2 and NO_x through chemical reactions in the atmosphere (Figure 5.2).

5.3.5.2 Eutrophication: A Widespread Ecological Effect

Eutrophication is a natural process of aging of a body of water. It is a result of a very slow process of natural sedimentation of microscopic organisms which takes geologic times to complete. The completion of the process results in the extinction of the water body.

The process of eutrophication is propelled by increasing concentrations of nutrients necessary for biological activity. First, envision clean and clear water. In this condition, since the nutrients available are minimal, there is no significant biological activity in the water column that can support sedimentation. The water body is healthy and the condition is called *oligotrophic*. Over time, however, nutrients can build up. A water body with nutrient concentration supporting biological activity that is not objectionable but above that of oligotrophic conditions is considered mesotrophic. In the next stage of the life cycle, the water becomes eutrophic – this is characterized by murky water with an accelerated rate of sedimentation of microorganisms. The final life stage before extinction is a pond, marsh, or swamp.

Although under natural conditions eutrophication occurs very slowly, over geologic times, human activity encourages the production of nutrients and shortens the cycle. People who have a green lawn are prolific producers of nutrients. Farmlands are an excellent source of nutrients. The Chesapeake Bay in Maryland, for example, is loaded with nutrients coming from farmlands as far away as New York.

To survive, microorganisms must be supplied with nutrients containing their components: carbon, hydrogen, oxygen, nitrogen, phosphorus, and trace elements. Since carbon, hydrogen, and oxygen are already in abundance in the environment, eutrophication control focuses on limiting the input of nitrogen and phosphorus. Nitrogen is utilized by organisms in the form of NH_4^+, NO_2^-, and NO_3^-; phosphorus is utilized in the form of orthophosphorus. Algae, the prime cause of eutrophication, can be controlled by limiting their access to nitrogen or phosphorus. Which should be chosen? It has been found that blue-greens algae, which can survive almost anywhere and can easily outgrow other algae under adverse conditions, can fix nitrogen from the air. For this and other reasons, the default choice, phosphorus control is recommended.

Eutrophication is our final example of the effects included in ecological risk analyses. There are, of course, many others; the literature is ample. Now, however, we are ready to consider another step in an ERA: toxicity testing.

5.3.6 Toxicity Testing

Ecological toxicity tests are classified according to duration (short-term, intermediate, and/or long-term), method of adding test solutions (static, recirculation, renewal, or flow-through), and (to satisfy permit requirements for a proposed pollutant discharge elimination system, and to determine mixing zones, etc.). Detailed contemporary testing protocols are summarized in the regulatory literature (USEPA 1985a, b, 1988, 1989).

Toxicity testing has been widely validated in recent years. Even though organisms vary in sensitivity to effluent toxicity, the EPA has documented that effluent toxicity correlates well with toxicity found in the receiving waters when effluent dilution was measured and that predictions of impacts from both effluent and receiving water toxicity tests compare favorably with ecological community responses in the receiving waters. The EPA has conducted nationwide tests with freshwater, estuarine, and marine ecosystems. Methods include both acute as well as chronic exposures.

Current bioassay methods can assess several different phylogenic groups in four to seven days (USEPA 1988). The tests are based on species of nearly national distribution for which a large body of life history and toxicity sensitivity data are available. Proper testing protocol involves assessment of a range of sensitivities of test species to a particular effluent. Typically, two or three species are considered to eliminate uncertainty for this factor.

Common marine species include *Champia parvula*, the red alga; *Mysidopsis bahia*, the mysid shrimp; *Menidia beryllina*, the inland silversides; and *Cyrinidon variegatus*, the sheepshead minnow. Common freshwater species include *Pimephales promelas*, fathead minnow, and *Ceriodaphnia dubia*, the daphnid shrimp.

At the termination of the test, survival is determined and growth is measured (as an increase in dry weight) compared with control. The acute endpoint is the death of the fish.

5.3.6.1 Evaluation of Toxicity Test Results

A number of terms are utilized in expressing toxicity test results. *Acute toxicity* is toxicity severe enough to produce a response rapidly (typically a response observed in 48 or 96 hours). "Acute" does not necessarily imply mortality. In studies of marine organisms, the LC_{50} is the concentration of effluent in dilution water that causes mortality to 50% of the test population; the EC_{50} is the effluent concentration that causes a measurable negative effect on 50% of the test population. The NOAEL (no observed acute effect level) is defined as the highest tested effluent concentration that causes 10% or less mortality.

Chronic toxicity is the toxicity impact that lingers or continues for a relatively long period of time, often 1/10 or more of the target organism's life span. Chronic effects could include mortality, reduced growth, or reduced reproduction. The NOEC (no observable effect concentration) is the highest measured continuous concentration of an effluent or toxicant that causes no observable effect based on the results of chronic testing. The LOEC (lowest observed effect concentration) is defined as the lowest observed concentration having any effect. The LOEC is determined by an analysis of variance techniques.

The use of whole effluent toxicity testing affords a number of advantages. In this approach, the bioavailability of the toxics is measured and the effects of any synergistic interactions are also considered. Because the aggregate toxicity of all components of a wastewater effluent is determined, the toxic effect can be limited by limiting only one parameter, the effluent toxicity. Because contemporary receiving water management strategies are based on site-specific water quality criteria, toxicity testing facilitates comparison of effluent toxicity with site-specific water quality criteria designed to protect representative, sensitive species and allow for establishment of discharge limitations that will protect aquatic environments.

5.3.6.2 Ecological Risk Characterization

Historically, the most common approach to risk characterization was the calculation of hazard quotients. This was adopted from the HHRA field, where this approach is still used. Simply, it compares chemical concentrations in ambient media to some toxicity benchmark. If the quotient exceeds 1, there is a potentially unacceptable risk. This approach has found use in predictive assessments and in screening-level (otherwise known as preliminary or tier I) retrospective ERAS, where it is used to refine the list of contaminants of concern and to focus a subsequent, more detailed assessment. However, for a baseline ERA, the quotient method should be used with caution. It is especially important to realize that the magnitude of the excess in the hazard quotient has no quantitative relation to the magnitude of potential toxic effects. Calculating several hazard quotients using different benchmarks (e.g. derived from different toxicity data, such as acute, chronic, or population level effects) has more direct applicability than using a single benchmark.

Because ecological effects can be measured in a retrospective ERA, an epidemiological, weight-of-evidence approach can be used. This approach depends upon weighing multiple lines of evidence, such as those provided by the field surveys, toxicity tests, and ambient media chemical analyses and literature toxicity data. Risk assessors, risk managers, and the public will have more confidence in a risk assessment that uses the weight-of-evidence approach, because it integrates all sources of information, attempts to reconcile conflicting data, and can account for the bioavailable fraction of chemicals in the environment, and the effects of multiple contaminants.

The primary line of evidence in the weight-of-evidence approach is the field survey data. Field surveys monitor actual ecological impacts, and therefore are the most credible line of evidence. However, as discussed in Section 5.3.4.1, field surveys have their limitations. Also, many ERAs will not have the budget necessary to conduct field surveys, and nocturnal, migratory, secretive, or wide-ranging species are not easily surveyed. Also, small impacts are not readily apparent in field surveys. Therefore, other lines of evidence are used as support.

Toxicity tests give an indication of whether ambient media are toxic. When several contaminants exceed benchmarks and there is an impact in the toxicity tests or field surveys, it is important and necessary to evaluate the magnitude of the effect caused by the contaminants which exceeded benchmarks. Using media contaminant analysis and the information provided in the toxicity profile, compiled in accordance with Section 5.3.5, an evaluation is conducted of which contaminants could be responsible for the observed toxicity. Combining all of these lines of evidence will present a picture of actual or potential impacts at the site and contaminants responsible for the impacts. In some cases, benchmarks may indicate unacceptable risk while field observations show no measurable impacts. Therefore, the weight of evidence suggests no unacceptable risks to a community, even though contaminant concentrations exceeded benchmarks. Reconciling multiple lines of evidence is difficult and requires experience and understanding of the ecosystem being evaluated.

5.4 Risk Management

Risk management, the process of decision-making that attempts to minimize risks without undue harm to other societal values, should be performed independently of risk analysis (NRC 1983). Since, however, the effectiveness of a risk analysis depends on the successful interaction of the risk analyst and the risk manager, we discuss risk management briefly. The risk analysts do not perform the decision-making process for the risk manager, and the risk manager does not dictate what data, models, and assumptions the analysts use, but they must communicate for the risk analysis to be relevant and the decision to be well-founded.

As detailed earlier, this relationship should have begun with the definition of the source, environment, and endpoints of concern. Because these activities of the scientist/risk analyst require that value judgments be made that are not based on science, the only way to avoid improper injection of the analyst's values is by interjecting the values of the risk manager through either policy statements or ad hoc judgments. However, risk managers often are not environmental scientists, so the risk analysts are obliged to assist the risk managers in acquiring enough scientific background to make informed judgments. In addition, the risk analyst must understand the needs and interests of the risk manager well enough to perform the risk characterization.

Figure 5.6 is a simple diagram of the factors involved in setting criteria for the allowable concentration of chemical in air, water, or other ambient medium. The ideal input from the risk analysis is a function relating the probability of unacceptable effects on the endpoint of concern to the criterion level or, more likely, functions for each of a number of endpoints. One former director of the EPA has said that the probability distributions are clearer to decision-makers than numeric values (Ruckelshaus 1984). From this function, the risk manager may choose a criterion that is unlikely to be overprotective from the upper end of the curve, one that is unlikely to be under protective from the lower end of the curve, or an intermediate criterion. Although not all risk managers are equally comfortable with probabilities, pressure on regulatory agencies to acknowledge and quantify uncertainty is increasing (Finkel 1990).

In addition to the results of the human health and ERAs, the decision will be based on a variety of social, political, legal, engineering, and economic considerations. Legal considerations include the phrasing of the law being implemented and any judicial decisions that indicate the appropriate degree of protection. The principal engineering consideration is the availability of practical technologies to meet the criterion. The cost of meeting the criterion and the value of the resource at risk may be the subject of a formal risk–benefit or cost–benefit analysis. Other, more nebulous considerations include the degree of public concern and the political philosophy of the currently governing administration. The manager may set a criterion or may defer a decision and require that more testing, measurement, or research be performed to reduce the contribution of uncertainty to the slope of the function.

5.4.1 Valuation of Ecological Resources

In order to perform cost-benefit analyses or to set fines or compensation levels for environmental damage, it is

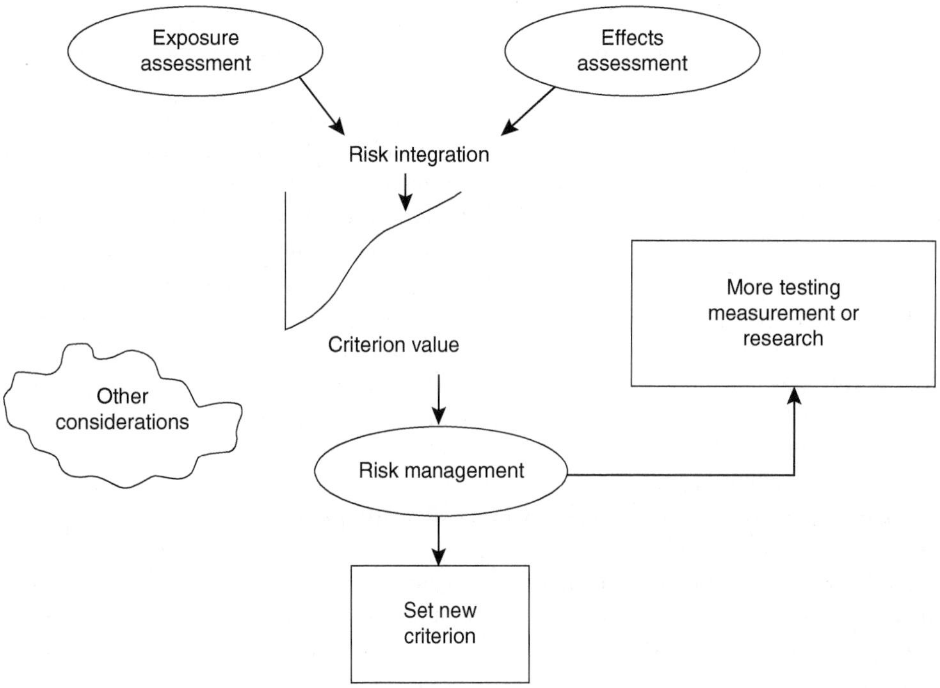

Figure 5.6 Diagrams of risk management for setting an environmental quality criteria.

necessary to predict, measure, or estimate the potential or actual loss of value of the system, and then convert that loss into monetary or equivalent units. Many ecologists and environmentalists have objected to the practice of placing monetary values on components of the natural environment, because they feel that economic analyses inevitably undervalue the environment relative to the value of a new chemical or development, or relative to the costs of treatment or cleanup. A counter argument is that decision-makers inevitably consider economics in their judgments and, if environmental scientists do not develop methods for monetarizing the values of the environment, decision-makers are likely to underestimate the economic value of the environment. In the United States, for example, the EPA is required to justify new environmental regulations by performing cost-benefit analyses.

A fundamental argument against valuation of environmental components is that they have intrinsic values that are independent of human values (Rolston 1988). However, even if this premise is accepted, it is impossible for a human to determine what a forest's values are (i.e. would it rather be old growth or young and vigorously growing?), so a poetic description of the intrinsic worth of a forest is not utilitarian, but it is no less anthropocentric than a dollar value. All value systems devised by humans are inevitably based on human values.

In valuing ecological resources, it is helpful to consider that there are various classes of utilitarian and nonutilitarian values that can be assigned. Each class of environmental values must have its own methods of estimation and valuation.

5.4.1.1 Utilitarian

Utilitarian values are often classified in terms of commodities, potential commodities, recreation, and services of nature.

Commodity Values Trees, rangeland, commercial fish, and other species and communities are sold or leased, so they have market values that are easily determined by conventional market surveys.

Potential Commodity Values Plants and animals are sources of chemicals that may be marketed directly or be synthesized and marketed as drugs, pesticides, etc. In addition, all species constitute genetic resources that may be useful for genetic engineering of crops, livestock, industrial microbe, etc. Hence, all species have potential commodity values that are very difficult to estimate. However, this value is lost only in cases of extinction.

Recreation Fish and wildlife have recreational value for fishermen, hunters, bird watchers, and others who enjoy seeing animals in their natural habitats. Natural areas have recreational value for campers, hikers, and students of the environment.

Services of Nature Ecosystems moderate floods, control the local and global climate, degrade and sequester pollutants, reduce soil and nutrient loss, and perform other services that have palpable economic benefits. These can be estimated as costs of replacing the service (e.g. additional waste treatment) or costs resulting from the lack of the service (e.g. flood damage).

5.4.1.2 Nonutilitarian

Nonutilitarian values range from the existential and aesthetic to the scientific and cultured.

Existence Species and ecosystems have value to many people simply because they exist. Those people may not expect to see a golden lion marmoset or the Arctic National Wildlife Refuge, but they would experience a loss if the marmoset became extinct or the refuge was converted to an oil field.

Aesthetic The aesthetic experience provided by organisms and ecosystems has a value equivalent to that of paintings or dance performances, but it is not so easily quantified.

Scientific Both species and ecosystems may have scientific value because they may serve as a particularly apt illustration of an unrecognized property of nature or because they are the subjects of long-term studies which may be disrupted. Of these, only the commodity values and recreational values are reasonably well characterized by existing methods. Research is needed to determine whether defensible methods can be developed for estimating losses of the other values, and monetarizing those losses. It can be argued that attempts to monetarize nonmarket values are inherently artificial and the balancing of apples and oranges should be performed by decision-makers rather than economists (NRC 1975). Other issues relating to category errors, as well as to the appropriateness of discounting any values and to obligations to future generations are discussed by Costanza (1991), Desvousges and Skahen (1987), and many contributors to the journal *Ecological Economics*.

Cultural and Ethical Certain ecosystems and species have cultural significance. This value is illustrated by the greater emotional impact in the United States of the decline of the bald eagle than the concurrent and more serious declines of the peregrine falcon and brown pelican.

Finally, it should be emphasized that cost-benefit analysis should not be considered to be the sole reliable basis for decision-making. It is one input that describes one particular criterion. In particular, it ignores moral and ethical considerations. Many, if not most, members of modern Western cultures would argue that we have moral duties toward the environment that supersede considerations of utility. However, these duties must be balanced against duties toward the poor, the hungry, practitioners of traditional cultures, etc. (Randall 1991). Therefore, environmental values are not absolute, even in moral terms, but must be balanced against competing values. Complexities like these put risk management beyond the scope of formal analysis.

5.4.2 Modeling Risk Management

Determining risk severity is one of a risk manager's first tasks. On this basis, he or she must decide whether a given risk is unacceptable, hence must be reduced or prevented with monitoring, or can be accepted, at least for the present. The risks associated with possible alternatives must be considered, as well. For example, the alternative to a pesticide that kills *x* birds per year may be clean farming which destroys refugia for pests but also eliminates habitat for *y* birds. Even if *y* is much larger than *x*, clean farming may be more desirable because it is more readily embraced by environmental and animal rights groups. The integration of such considerations is nearly always an informal process.

Our single brief example is presented as if there is only one exposure model, one effects model, and one set of assumptions about how to parameterize them and combine them into a risk estimate. This approach is generally applicable to human health risk assessments where endpoints, models, and treatment of parameters are standardized. However, it is not appropriate for ERAs. One might estimate effects on fisheries with an ecosystem model, a population model, and a statistical model of an organism-level toxicity test. Ecosystem-level effects might be represented by results from a microcosm test, a lake ecosystem model, and a stream ecosystem model. Each risk estimate will have its own assumptions and associated uncertainties and those uncertainties may not be expressed equivalently. The separate lines of evidence must be evaluated, organized in some coherent fashion, and explained to the risk manager so that a weight-of-evidence evaluation can be made.

One approach to presenting alternate lines of evidence is to present them graphically of common axes. The axes should be chosen from the four dimensions of risk estimation so as to clarify the differences in the risk estimates that are most important to the decision. That is, does the decision depend primarily on the concentration at which an effect occurs, the time to recovery, the number of organisms dying, or some combination of axes? The output of the alternate models, assumptions, or data should be plotted on those axes.

5.4.3 Other Considerations for Risk Characterization

Those charged with risk management for human health are fully aware that some risks have consequences so dire that they must be avoided no matter what the cost. Risks so obviously unacceptable are called de manifestis. The companion concept, de minimis, which we shall explore shortly, points to the existence of risks with consequences so harmless that they are not worth treating, no matter how little the cost. Whipple (1987) has written about de minimis risks, and Travis et al. (1987) and Kocher and Hoffman (1991) have discussed both concepts in relation to cancer prevention.

Risk levels between the de manifestis and de minimis bounds are balanced against costs, technical feasibility of remediation, and other considerations to determine their acceptability. Because there have been no standard assessment endpoints for ecological effects, there have been no de minimis or de manifestis ecological risk levels. However, the protection provided to endangered species in the United States provides one basis for de manifestis risk levels because it ostensibly precludes economic and sociopolitical considerations. Establishment of de manifestis and de minimis risks would allow assessors to quickly dispense with clearly unacceptable or trivial risks after minimal assessments that do not establish the exact expected risk level, or balance costs and benefits. Consideration of what levels and types of ecological effects are clearly and invariably acceptable, and which are clearly and invariably unacceptable would provide risk assessors and risk managers with a good basis for working out the relation between societal values and ecological principles.

A final consideration in risk characterization and management is the need for audiences other than the risk manager to understand the risks and the bases for the decision. These include the public, environmental advocacy groups, and regulated parties. Risk communication to the nonprofessional has been a major topic of research and discussion with respect to health risks, but not ecological risks (NRC 1989). This is understandable because the average citizen is more concerned with personal and family health than with ecological effects. However, when outside parties become concerned about potential ecological effects, the diversity of ecological endpoints and the uncertainty of their valuation make risk communication highly complex. Earlier we cited a hypothetical situation in which a nonpesticide approach to eliminating crop-destroying insects that would harm some wildlife was likely to be acceptable to groups opposed on principle to pesticides. Under different circumstances, there might be scientific support for replacing a pesticide with another substance that posed a higher risk to avian populations but a lower risk to zooplankton. Yet arguments for accepting more dead birds in exchange for fewer dead zooplankton would not be easily conveyed to a diverse audience.

5.4.4 Conceptual Bases for De Minimis Risks

The concept of de manifestis risk is not controversial because some effects are clearly unacceptable. However, the idea that some exposures to and effects of pollutants are acceptable (de minimis) is controversial. The use of the de minimis concept is based on the following considerations:

1) Some exposures and effects are manifestly trivial. For example, it is absurd to assess effects of exposure to one molecule per organism or to consider restricting a pollutant exposure because it causes a transient physiological response or causes one copepod species to replace another.
2) There are many genuinely significant risks, so it is inappropriate to set them aside while assessing the trivial ones. As Weinberg (1987) remarked, one does not swat gnats while being charged by elephants.
3) Regulatory scientists and decision-makers do not have infinite time and resources. The benefits from their time and resources should be maximized by quickly setting aside trivial risks.
4) Very low level exposures and effects are very difficult to accurately quantify. Therefore, any attempt to balance their costs against the costs of treatment or prevention, or the benefits of a product are likely to be confusing or misleading rather than enlightening.
5) Small increases in exposure to naturally occurring chemicals or other hazardous agents are likely to have trivial effects, because of adaptation to the background levels. This is particularly true if the increase is also small relative to variation in the naturally occurring background levels.
6) Effects that are small relative to natural temporal or spatial variability of the biological parameter are likely to have trivial implications either at other levels of biological organization or for other components of the same level of organization. For example, an apparently large (e.g. 50%) temporary depression of the decomposition rate is unlikely to significantly affect primary production or abundance of birds because they are adapted to much larger variance in the rate due to wetting and drying cycles and other natural variables.
7) Treating a trivial risk as if it were potentially significant unnecessarily raises public concerns that are not entirely extinguished by a subsequent finding that the risks are small.

Since we cannot prevent all human effects on the environment, and we cannot even carefully assess the benefits and costs of all human actions, it is desirable to develop criteria for eliminating clearly trivial risks from further consideration.

5.4.5 Ecological Risk Assessment of Chemicals

5.4.5.1 One-Dimensional Models

Most ecological assessments depend on one-dimensional models of toxicant/organism interaction, and in most cases that dimension is concentration. Because toxicology is historically the science of poisons, the fundamental paradigm of toxic effects is the single lethal dose. To the poisoner, duration of exposure, severities other than mortality and even the exact proportion responding are unimportant, so the only dimension of interest is concentration. This paradigm is appropriate in weed and pest control and, to a somewhat lesser extent, to unintentional acute poisoning, as when wildlife are sprayed with pesticides.

Use of only the concentration dimension simplifies assessment because the outcome is determined by the relative magnitudes of the exposure concentration and the effective concentration. However, most assessments in environmental toxicology involve some element of time and some concern for the severity and extent of effects, so use of only the concentration dimension requires that these other dimensions be collapsed. This collapsing of the other dimensions is most commonly done by using a standard test endpoint as the effective concentration. Standard test endpoints are used because (i) the assessor does not have the skill, time, latitude, or inclination to develop alternate models; (ii) the assessor has a vaguely defined assessment endpoint and is willing to accept the toxicologists judgments as to what constitutes an appropriate duration of exposure, severity of effect, and frequency of effect; or (iii) by luck the assessment endpoint corresponds to a standard test endpoint. The test endpoints most often used median lethal concentrations and doses, LC_{50} and LD_{50}, were introduced earlier. With these endpoints, proportion responding, severity, and time are collapsed by considering only 50% response, only mortality, and only the end of the test.

Occasionally, time is the basis of a one-dimensional assessment. If the concentration of a release is relatively constant, then the risk manager needs to know how long it can be released without unacceptable risk. For example, a waste treatment plant might be taken off-line for repairs, and the operator would like to know how long he or she can release untreated waste. The most common temporally defined test endpoint is the median lethal time (LT_{50}). In fact, many releases have relatively constant concentrations and time would be a more useful single dimension than concentration for assessment. However, because time is generally felt to be less important than concentration, it is more often considered in higher-dimensional models.

The most common model for risk estimation, the quotient method (Barnhouse et al. 1982; Urban and Cook 1986), relies on a one-dimensional scale. The method consists of dividing the expected environmental concentration by a test endpoint concentration (i.e. determining their relative position on the concentration scale); the risk is assumed to increase with the magnitude of the quotient.

5.4.5.2 Two-Dimensional Models

Concentration/Response The most common two-dimensional model in ecological toxicology is the concentration–response function. This preeminence is explained by the desire to show that response increases with concentration of a chemical, thereby establishing a causal relationship between the chemical and the response. In this model, time is collapsed by only considering the end of the test, and either severity or proportion responding is eliminated so as to have a single response variable. Most commonly, proportion responding is preserved and severity is collapsed by considering only one type of response, usually mortality. Severity is most often used when a functional response of a population or community of organisms such as primary production is of interest, rather than the distribution of response among individuals.

A probit, logit, or other function is fit to the concentration–response data obtained in the test. Typically, that function is used to generate the standard LC_{50} or LD_{50} or median effective dose (ED_{50}) by inverse regression, thereby reverting to a one-dimensional model and throwing out much information. Because assessments are more often concerned with preventing mortality or at least preventing mortality to some significant proportion of the population, it would be more useful to calculate an LC_1, LC_{10}, or other threshold level, but information would still be thrown away. If both dimensions are preserved, the concentration–response function can be used in each new assessment to estimate the proportion responding at a predicted exposure concentration or to pick a site-specific threshold for significant effects. When only an LC_{50} is available and the assessment requires estimation of the proportion responding, it is possible to approximate the concentration–response curve by assuming a standard slope for the function and using the LC_{50} to position the line on the concentration scale. Concentration–response functions are an improvement over being stuck with an LC_{50}, but in many cases the proportion responding is less important than the time required for response to begin.

One improvement in concentration–response functions would be to match the test durations to the exposure durations, rather than using standard test durations and assuming that the exposure durations match reasonably well. Ideally, the temporal pattern of exposure would be anticipated and reproduced in the toxicity tests. For example, if a power plant will periodically "blowdown" chlorinated cooling water, on a regular schedule for a constant time period, that intermittent exposure can be reproduced in a laboratory toxicity test (Brooks and Seegert 1977; Heath 1977).

Time–Response Functions Time–response functions, like concentration–response functions, are primarily generated in order to calculate a one-dimensional endpoint, in this case the LT_{50}. Time–response functions are useful when concentration is relatively constant, but the duration of exposure is variable. With the time–response function defined, it is possible for the assessor to consider the influence of changes in exposure time on the severity of effects or the proportion responding.

Time–Concentration The most useful two-dimensional model of toxicity that includes time is the time–concentration function or toxicity curve (Sprague 1969). This is created from experimental data by recording data at multiple times during the test and, for each time, calculating the LC_{50}, EC_{50}, or the concentration causing some other response proportion or severity, as described above. These concentrations are then plotted against time and a function is fit (Figure 5.7). Such functions have been advocated by eminent scientists in prominent publications (Lloyd 1979; Sprague 1969), but they are seldom used even though they require no more data than is already required by standard methods for flow-through LC_{50}s for fish (ASTM 1991; USEPA 1982). If the response used corresponds to a threshold for significant effects on the assessment endpoint, then this function can be used to determine whether any combination of exposure concentration and duration will exceed that threshold (i.e. the area above a line, as in Figure 5.8).

When time is a concern but temporal test data are not available, it is necessary to approximate temporal dynamics.

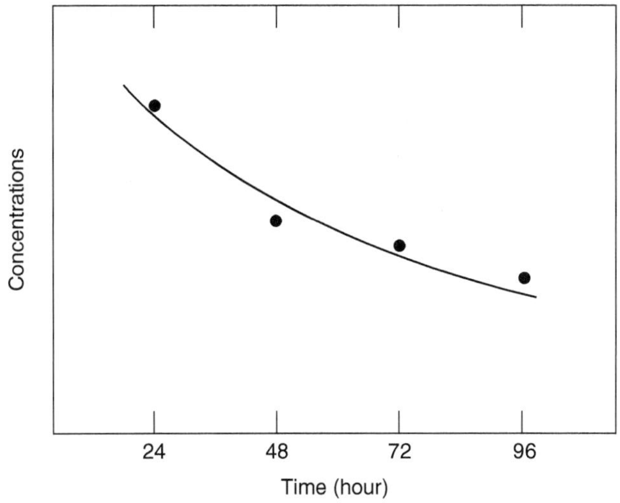

Figure 5.7 Toxic effects as a function of concentration and time. The function is derived by plotting LC_{50}s or other test end points against the times at which they were determined (i.e. 24, 38, 72, and 96 hours).

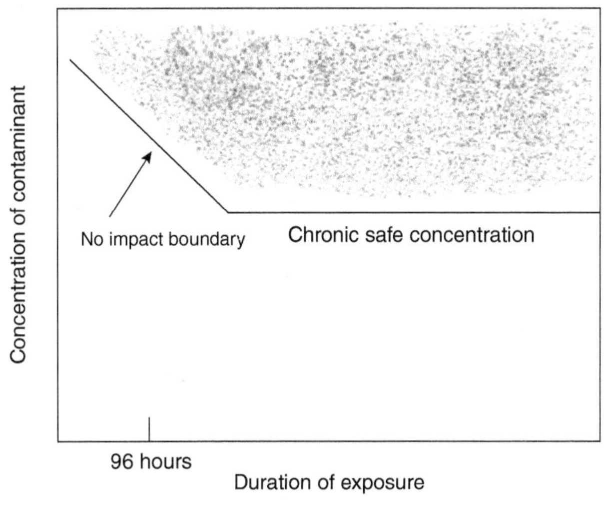

Figure 5.8 Toxic effects as a function of concentration and time, with time shown a dichotomous variable, acute and chronic. Acute time ends at 96 hours, the end of a standard acute lethality test for fish.

The simplest approach is to treat the level of effects and the exposure concentration as constants within set categories of time. This approach is used in most environmental assessments. The assessor, faced with toxic effects expressed as standard test endpoints, tries to match the test durations to the temporal dynamics of the pollution in some reasonable manner. Time is divided into acute and chronic categories, and standard acute and chronic test endpoints are used as benchmarks separating acceptable and unacceptable concentrations. There are serious conceptual problems with this categorization, and however we address these now.

"Acute" and "Chronic" as Temporal Categories Ambiguities of terminology complicate the matching of test durations to exposure durations. Because "acute" and "chronic" refer to short and long periods of time, respectively, it is tempting to relate endpoints from acute tests to short exposures, and those from chronic tests to long exposures. However, the terms have acquired additional connotations and using them to describe severity as well as duration leads to complications. Acute exposures and responses are assumed to be both of shorter duration and more severe than chronic exposures and toxicities. The implicit model behind this assumption is that chronic effects are sublethal responses that occur because of the accumulation of the toxicant or of toxicant-induced injuries over long exposures. Conversely, because of the cost of chronic toxicity tests, toxicologists have attempted to reduce testing costs by identifying responses that occur quickly and that are severe enough to be easily observed, but that occur at concentrations that are as low as those that cause effects in chronic tests (McKim 1995). As a result, the relationship between the acute-chronic dichotomy and gradients of time and severity has become confused.

This confusion is illustrated by the standard test endpoints for fish. The standard acute endpoint is the 96-hour median lethal concentration (LC_{50}) for adult or juvenile fish (ASTM 1991; OECD 1981; USEPA 1982). The standard chronic test endpoint has been the maximum acceptable toxicant concentration (MATC, also termed the "chronic value"), which is the threshold for statistically significant effects on survival, growth, or reproduction (ASTM 1991; USEPA 1982). Because this chronic endpoint is based on only the most sensitive response, life stages that appeared to be generally less sensitive have been dropped from chronic tests, so that those tests have been reduced from life cycle (12–30 months) to early life stages (28–60 days) (McKim 1995). Tests that expose larval fish for only 11 or 5–7 days have now been proposed as equivalent to the longer chronic tests. As a result, the chronic test endpoint for fish is now tied to events of short duration (the presence and response of larvae), whereas the acute endpoint is applicable to exposures or similar duration and to life stages that are continuously present.

Even the severity distinction between acute and chronic tests is not clear. Although the LC_{50} clearly indicates a severe effect on a high proportion of the population, the fact that the MATC is tied to a statistical threshold rather than a specified magnitude of effect means that it too can correspond to severe effects on much of the population. For example, more than half of female brook trout exposed to chlordane failed to spawn at the MATC.

It would be advantageous to clarify the distinction between acute and chronic toxicity by restoring the original temporal distinction and expressing effects in common terms. Without that clear distinction, concentration–time functions like Figure 5.8 are uninformative.

Concentration and Duration as Replacement for Concentration and Time In the absence of good time–concentration information from the test data, a concentration–duration function may be assumed. For example, the first version of the model of marine spill effects for type A damage assessments simply assumed a linear function (USDOI 1986, 1987). Lee and Jones (1982) combined this linearity assumption for temporal effects in acute exposures with an assumption of time independence in chronic exposures to generate the concentration–duration function shown in Figure 5.8. This linearity assumption is chosen for its simplicity rather than any theoretical or empirical evidence. Parkhurst et al. (1981), assuming that LC_{50} values are available for 24, 48, and 96 hours, linearly interpolated between these values and assumed time independence for exposures beyond 96 hours to generate the function shown in Figure 5.9.

These approaches depend on the assumption that temporal dynamics can be treated in terms of various durations of exposure to prescribed concentrations. In reality, organisms are subjected to a continuous spectrum of fluctuations in exposure concentrations due to variation in aqueous dilution volume or atmospheric dispersion, variation in effluent quality and quantity, intermittent release of effluents, and accidental spills or upset effluents. These can be treated conventionally if (i) the time between episodes is sufficient for recovery so they can be treated as independent, (ii) the fluctuations are of sufficient frequency and of sufficiently low amplitude that the organisms effectively average them, or (iii) certain frequencies predominate, so that temporal categories of exposure can be identified as discussed above. An example of the third possibility is Tebo's (1986) categorization of fluctuations in aqueous, point-source effluents as ponded and well mixed-wastes that are fairly uniform in character, wastes subject to short-term, daily fluctuations, and batch process wastes subject to severe fluctuations.

Worst Case If it is not possible to characterize fluctuations in one of these ways, one can define a worst-case concentration and duration and assume that if that event does not cause unacceptable effects when it occurs in isolation, then it will also be acceptable when it occurs as part of a history of fluctuating exposures (Figure 5.10). This assumption is commonly adopted in effluent regulation. For example, the worst-case dilution condition for aqueous effluents has traditionally been the minimum flow which occurs for 7 days with an average recurrence frequency of 10 years, referred to as the 7Q10. The EPA recommends use of the lowest one-hour average and four-day average dilution flows that recur with an average frequency of three years (USEPA 1985c). These correspond to highest one-hour and four-day exposures. The three-year recurrence frequency is assumed to allow for recovery of the system so that the peak exposures can be treated as independent events.

Finding a Middle Ground The best solution would be to avoid the acute–chronic dichotomy and worst-case assumptions by identifying characteristic temporal patterns of exposures or biological responses, and either conducting toxicity tests to simulate those patterns or classifying test results into the environmentally based temporal category in which their durations fall. That is, one can scale time to the characteristic temporal scales of the processes determining

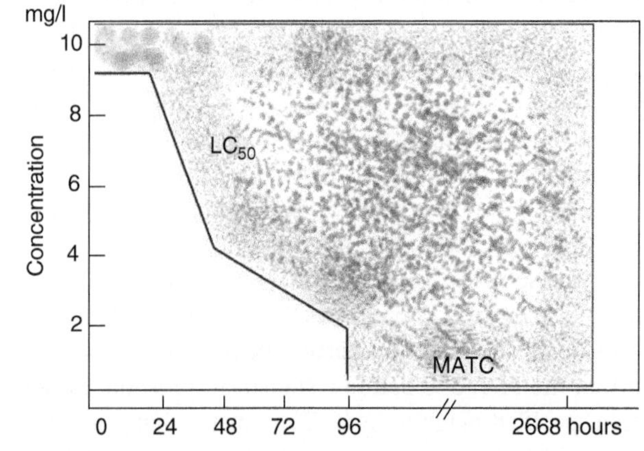

Figure 5.9 Toxic effects as a function of concentration and time, with acute time interpolated between measured values and extrapolated to zero and to the value for chronic time, which is expressed a constant discrete variable. *Source:* From Parkhurst et al. (1981).

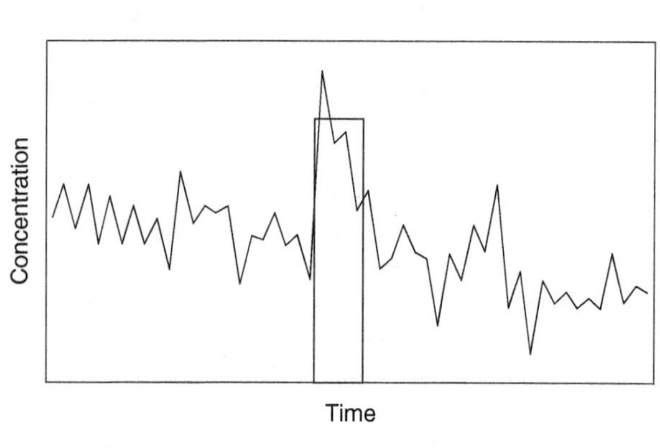

Figure 5.10 A Fluctuating ambient exposure concentration can be represented in an assessment by a continuous exposure corresponding to the worst-case peak exposure, as highlighted by the vertical bar.

the risk. For example, exposures to gaseous pollutants from point sources might be classified as (i) plume strikes (an hour or less), (ii) stagnation events (hours to a week), and (iii) the growing season average exposure. Existing data on concentrations of an air pollutant causing phytotoxic effects might be plotted against time in terms of these categories (Figure 5.11), and then compared with estimated ground-level concentrations for each of the three categories of events. In any case, the matching of exposure durations with toxicological endpoints should be based on an analysis of the situation being assessed rather than on preconceptions about acute and chronic toxicity.

Dose–Response Functions Although dose is defined in a variety of ways, all definitions concern the amount of material taken up by an organism. Traditionally, dose is simply the amount of material that is ingested, injected, or otherwise administered to the organism at one time. Exposure duration is not an issue for this definition (although time-to-response may be), and results are represented as dose–response functions. These functions, like the LD_{50} values that are calculated from them, are useful to environmental assessors in cases like the application of pesticides in which very short duration exposures occur due to ingestion, inhalation, or surface exposure.

An alternate definition of dose is the product of concentration and time, as discussed earlier in connection with Figure 5.7, or, if concentration is not constant, the integral over time of concentration. This concept is applied to exposure concentrations as well as body burdens, allowing the calculation of dose–response functions for exposures to polluted media. It has been commonly used in the assessment of air pollution effects on plants and is referred to as "exposure dose." For example, McLaughlin and Taylor (1985) compiled data on field fumigations with SO_2, of soy beans and snap beans, and plotted percent reduction in yield against dose in ppm-hours. Newcombe and MacDonald (1991) found that the severity of effects of suspended sediment on aquatic biota was a log linear function of the product of concentration and time.

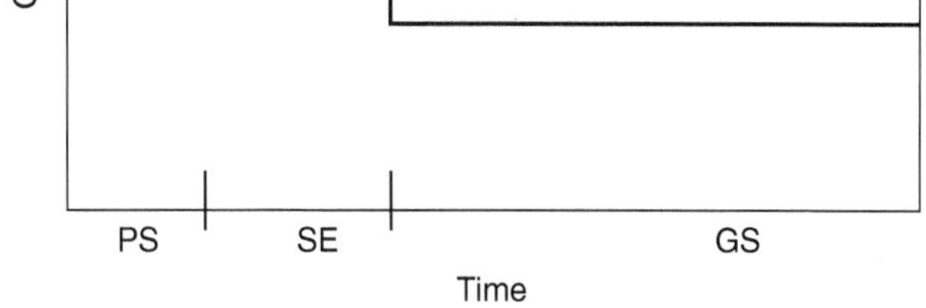

Figure 5.11 Toxic effects as a function of concentration and time with time expressed as a quartered variable defined in terms of categories of exposure (PS, plume strike; SE, stagnation event; and GS, growing season).

The most strictly defined form of dose is delivered dose, the concentration or time integral of concentration of a chemical at its site of toxic action. This concept is applied empirically by measuring the concentration of the administered chemical in the target organ or tissue, in some easily sampled surrogate tissue such as blood, or in the whole body in the case of small organisms. By relating effects to internal concentrations, rather than (or in addition to) ambient exposure concentrations, this approach can provide a better understanding of the action of the chemical observed during tests, and it is useful as an adjunct to environmental monitoring. It allows body burdens of pollutants in dead, moribund, or apparently healthy organisms collected in the field to be compared to controlled test results to explain effects in the field. Toxicokinetic models provide a means of predicting dose to target organs from external exposure data.

Rather than simply being a function of peak body burden (i.e. mg/kg), effects may be a function of the product of body burden and time or, more generally, the time integral of body burden. This variable, termed "dose commitment," is used in estimating effects of exposures to radionuclides and may be appropriate to some heavy metals and other environmental pollutants.

5.4.5.3 Three-Dimensional Models

Three-dimensional representations of toxic effects are rare. Only concentration–time–response models can be readily derived from conventional test data. Functions of this sort can be derived from the data that are required from flow-through 96-hour LC_{50} tests of fish by ASTM and EPA protocols (ASTM 1991; USEPA 1982). Such models have the obvious advantage of allowing the assessor to estimate the level of effects from various combinations of exposure concentration and duration. The response surfaces developed by Richardson and Burton (1981) for two estuarine species exposed to ozonated water and for trees exposed to SO_2 are good examples.

Concentration–time–response relationships can also be a useful tool for dealing with diverse data. If test data are diverse with respect to both duration and severity of effects, then both of these dimensions must be related to concentration if regulatory criterion concentrations are to be calculated. Dourson

(1986) explained how this is done to estimate allowable daily intake for humans. The various exposures are categorized as having no observed effect, no observed adverse effect, adverse effects, or frank effects. This severity scale is then translated into a series of symbols that are plotted on a dose and time surface. Lines are then fitted by eye separating the categories of severity and establishing the no adverse effect level, adverse effect level, and frank effect level. The same approach has been used by the EPA to set air quality criteria for phytotoxic effects of air pollutants. The U.S. Department of Agriculture reported that ground-level ozone causes more damage (growth and yield) to plants than other air pollutants combined (Figures 5.12 and 5.13a, b) (USDA 2011).

Having characterized the risk of exposure to chemical release to air, water, and land, and the degree of uncertainty associated with the risk, the next step is to use the information to improve the basis for making decisions. This often involves the public, or at least embrace public concerns and attitude. It requires an examination of the question, "what is acceptable risk?" An essential part of this is risk communication – communication in terms of openness and transparency, understanding and engaging stakeholders, as well as providing balanced information to allow public make decisions on how to deal with risk.

5.4.5.4 Plastic Waste Choking Oceans, Rivers, Landfills

About 60% of the 9.1 billion T of plastic produced throughout history has ended up as waste, with more than three-fourths of that going into landfills or the natural environment. Only 14% of plastic packing is currently collected for recycling. Figure 5.14 shows some of the plastic waste that washed up at the site of the old Queenhithe dock on the *River Thames* in London (Seattle Times, 22 February 2018). It is estimated that in 2010 alone, between 4.4 million and 13.2 million T of plastic entered the marine environment. The report also highlighted that the weight of plastic in the ocean would equal that of fish by 2050 if current trend continue.

In an article published in *The New York Times* (23 March 2018a issue), it is reported that about 87 000 T of plastic are present in the Pacific Ocean region. In the Pacific Ocean between California and Hawaii, hundreds of miles from any major city, plastic bottles, children's toys, broken electronics, abandoned fishing nets, and millions more fragments of debris are floating in the water. This notorious mess has become known as the Great Pacific Garbage Patch, a swirling oceanic graveyard where everyday objects get deposited by the currents. The plastic eventually disintegrate into tiny particles that are often eaten by fish and may ultimately enter our food chain (New York Times, 23 March 2018a).

According to these reports, Amcor, Ecover, Evian, L'Oreal, Mars, M&S, Pepsi Co, Coca-Cola, Unilever, Walmart, and Warner & Mertz – which together use more than 6 million MT of plastic packing a year – has committed to use only reusable, recyclable, or compostable packing by 2025. Amid growing evidence of the risks posed by plastic waste worldwide, some companies are embracing conservation and more recycling (see RCRA in Section 4.20.7 on

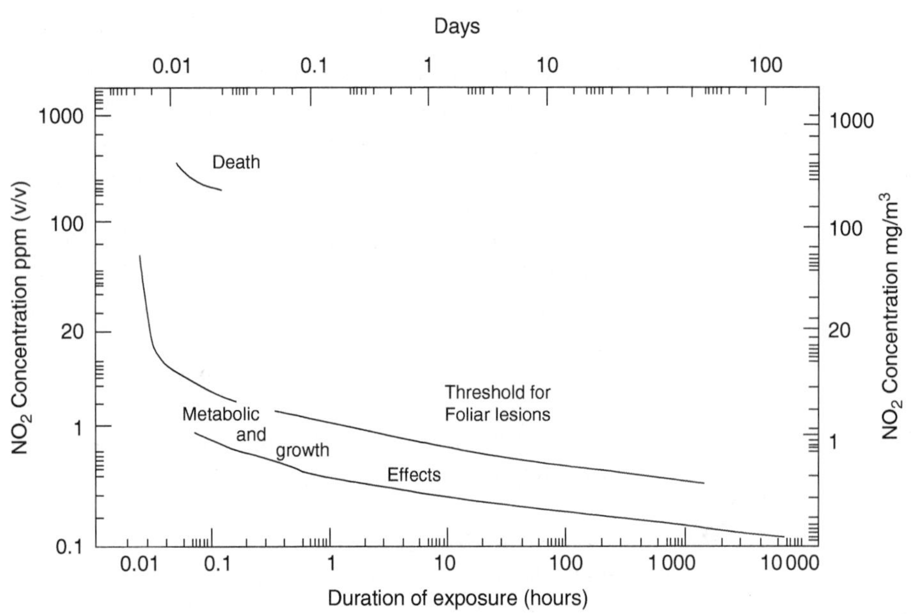

Figure 5.12 Categorization of effects of NO_2 on plants arrayed with respect to concentrations and duration of exposure. *Source:* From USEPA (1982).

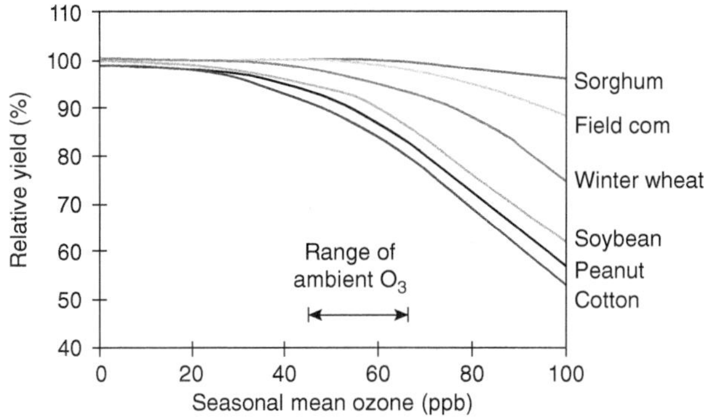

Figure 5.13 (a) Ozone damage to radish plant resulting in smaller radish. *Source:* From Haggestad (2001). (b) Effects of ozone on plants' growth and yield. Heck et al. (1983).

Figure 5.14 Plastic waste in River Thames in London.

resource conservation and recovery and for more on plastic recycling in Section 4.20.6).

More than 48 lb of plastic, including disposable dishes, corrugated tubes, shopping bags, and a detergent package with its bard code still visible, were found inside a dead sperm whale (Figure 5.15). This was the latest in a grim international collection of whale carcasses burdened by dozens of pounds of plastics trash. Europe is the second largest plastics manufacturer in the world, dumping 150–500 thousand T of macroplastics and 70–130 thousand T of microplastics in the sea every year. "Plastic pollution is one of the worst enemies of marine species in the world today."

5.5 Communicating Information on Environmental and Health Risks

A key component of successful environmental and health risk management for any industry is an effective community outreach, or social responsibility program. This is even more important today than in the past, given the current explosion in public access and exposure to, and interest in, information regarding local health and environmental issues. The internet

Figure 5.15 Plastics junk found in a dead whale. *Source:* Photograph by Seame Sardinia Onlus, Associated Press. The New York Times, 3 April 2019.

provides information on industrial facilities that is available to the public 24-hours a day; Web sites such as those provided in the United States by the EPA and the Environmental Defense Fund offer up-to-date information on the status of local industry to all users at the click of a mouse.

Facilities that have not taken a proactive approach to communication may find a public besieged with a wealth of negative information about a plant's operations and no positive data or reasons to trust the industry. In fact, in the absence of open communications or a public perception of information accessibility, a company may acquire a negative image even if there are no environmental or health risks associated with facility operations.

The chemical industry offers a good example of the potential divergence between real and perceived risk in the absence of communication. While major reductions in releases of both toxic chemicals and suspected carcinogens were documented between 1991 and 1997, and the accident rate for the chemical industry declined 32%, the public perception of the chemical industry did not improve. In fact, only 25% of people living near chemical plants have a favorable view of the industry. In general, the public does not hear about positive innovation or advances in risk reduction and is skeptical of what they hear. These industries have to work at establishing trust. They actively must improve their public image if they want to mitigate and/or avoid future conflict with neighbors.

A commitment by industry to build and maintain an ongoing dialogue with local communities is essential. This dialogue needs to become an integral part of almost every industrial facility's day-to-day operations, with management providing access to information, in addition to learning and responding to the public's questions and concerns.

As Section 5.5.1 demonstrates, facilities that have not opened the lines of communication with their neighbors may find themselves unwillingly drawn into the process – under circumstances that will leave them on the defensive and subject to public concern, or even outrage. The only effective solution is to build relationships over time with local communities and other key audiences.

5.5.1 From Concern to Outrage: Determinants of Public Response

What is likely to make people concerned or angry about a real or perceived risk? The answer depends on a number of risk perception and environmental vulnerability factors. A number of these environmental vulnerability factors are listed in Table 5.5. Without regular communication between industry and the public, people tend to use their general impressions to form an opinion regarding the acceptability of a facility in their neighborhood. Impressions may be based on the "look" of the plant, for instance, whether the buildings and grounds are well cared for and maintained. If smoke or vapors are emitted, or if neighbors can smell or hear plant operations, a facility's environmental vulnerability is perceived to be greater. Environmental vulnerability will also be high if the plant has a history of community or employee health complaints or if high-profile substances such as radioactive materials are known to be on site. And also, the appropriateness or inappropriateness of the facility in the community where it is located. People will be particularly sensitive to the risks imposed by the operations of facilities located near schools, hospitals, or important scenic or cultural sites.

The concept of environmental vulnerability helps to identify the factors that heighten public concern. However, environmental vulnerability does not clarify why people become outraged over certain kinds of risks and not others. In his classic discussion of "outrage factors," Sandman (1993) outlines the characteristics of a risk that help to determine the level of anger or fear in people's reaction to a risk. According to Sandman, the public's perception of risk goes beyond the narrow definition favored by scientists and engineers. The latter group uses a technical definition of risk, which can be established by multiplying the magnitude of the hazard by the probability of exposure, as follows:

Table 5.5 List of environmental vulnerability factors.

Environmental vulnerability factors
• Things people can see, smell, hear (e.g. odors, clouds of steam, noise)
• High volumes of emissions or hazardous wastes
• History of community or employee health complaints
• History of unexplained odors or releases
• Presence of "dreaded" substances known to cause cancer
• Poor facility, housekeeping, or appearance
• Proximity to sensitive locations, including schools, nursing homes, and hospitals
• Reports of health problems among school children or staff
• Proximity to important scenic or cultural sites
• Active presence of organized environmental groups
• History of poor community outreach
• Lack of risk and crisis communication program

$$\text{Risk} = \text{hazard} \times \text{exposure} \tag{5.6}$$

In this definition, risk increases with the hazardousness of the process or material (e.g. a chemical's toxicity) and with the amount and length of exposure by the person affected.

The public's understanding of risk, however, requires the incorporation into the equation of what Sandman has dubbed "outrage." Outrage accounts for all factors associated with a risk, not including the technical hazard. These factors, which are extremely important to public perception, include whether the risk is assumed voluntarily, as in smoking or not wearing a seatbelt, or involuntarily, as in dying in a plant explosion, whether the risk is controlled by the individual or by the system, and whether it is industrial or natural. The perceived risk associated with an industry or incident will be high if its effects are dreaded, concentrated in time and space, involuntary, memorable, and have no visible benefits, among other factors. With the incorporation of outrage, the layperson's view of risk can then be summarized and can be expressed as follows:

$$\text{Risk} = \text{hazard} + \text{outrage} \tag{5.7}$$

where risk and hazard are the same as in Eq. (4.9) and outrage represents the incorporation of outrage factors. While the technical level of risk will also concern people, and they will expect something to be done, the outrage factors will help to determine their level of anger and will influence whether a risk is perceived as more or less dangerous. A qualitative breakdown of Sandman's outrage factors is given in Table 5.6.

Clearly, it is beneficial to industry to communicate to the public trustworthy information about potential or perceived problems before citizen response has reached the outrage state.

5.5.2 Sustainable Strategies for Environmental and Health Risk Communication

While many industrial facilities may recognize the benefits of communication, many are reluctant to devote the necessary time and money to initiating community

Table 5.6 Outrage factors.

More risky	Less risky
Involuntary	Voluntary
Industrial/artificial	Natural
Exotic/unusual	Familiar
Memorable	Not memorable
Dreaded	Not dreaded
Concentrated in time and space	Not concentrated in time and space
Unfair	Fair
Morally relevant	Morally neutral
Untrustworthy	Trustworthy
Closed process	Open process
No visible benefits	Visible benefits

Source: From Sandman (1993).

outreach programs. Some managers believe that if people are already getting information in ways over which industry has no control, there is no point in fighting the process. In fact, they are wise not to challenge the technical basis of environmental groups' statements: industry does not have the credibility that these groups have, and such challenges are likely to be seen as obfuscatory, further damaging the industry's image. However, the absence of communication espouses a negative perception. Communication should not be approached as a reaction to negative publicity or as combative. It needs to become an essential component of day-to-day operations in order to be effective.

5.5.2.1 Breaking Down the Barriers to Communication

> *The single biggest problem in communication is the illusion that it has taken place*
>
> – George Bernard Shaw.

Industry commonly uses the following rationalizations to defend a position of noncommunication.

We're Too Small to Do Outreach A low-profile or "no-profile" facility can actually find itself subject to increased public scrutiny because it lacks a history of outreach. Companies without a budget, activities, or personnel allotted to community outreach may feel this is appropriate to their size and risk-potential. However, even a small chemical facility can be the object of the public's wrath, especially if people feel the risk has been hidden from them.

Our Facilities and Operations Are Environmentally Sound A solid environmental performance is certainly essential to a facility's efforts to be a good neighbor. However, a successful relationship can be achieved only when environmental performance is joined with regular communication and trust. Especially, if there is no record of communication between industry and the public, people will base an opinion about whether a given operation is a good neighbor on the basis of general impressions of its facilities. Every facility is vulnerable in some way to public questions, concerns, and even outrage (Sheader 2014).

If We Talk to People, They Will Just Get Upset Sometimes after years or even decades of silence, an industrial facility will suddenly find itself engaged in an active and heated dialogue with its neighbors. The event precipitating the communication could be a minor chemical release, something that poses a small risk but requires a large response from the fire department and other emergency responders. While the technical level of risk will influence public response, the level of outrage will be in a large part determined by prior experiences with the company. Communication outrage per se does not create. Anger results when the appearance of unfairness leads to distrust and the perception that risks cannot or will not be controlled.

Our People Are Already Overworked Risk communication and community outreach need to be part of each facility's day-to-day business plan and its employees' job descriptions. Section 5.5.2.2 discusses how these goals can be achieved. In the long run, time and money can be saved if public outrage is kept low and confrontation is avoided. A risk communication program should be thought of as a form of insurance against future problems with neighbors. By keeping in touch with community concerns, industry can largely avoid being blindsided by reactions to circumstances that should not be causes for alarm.

5.5.2.2 Building a Communication Program

Having characterized the risk of exposure to constituents of criteria and hazardous pollutants and the degree of uncertainty associated with the risks, next move is to use the information to improve the basis for making decisions. This often involves the public, or at least embraces public concerns and attitudes.

The rationale for an effective risk communication program has been discussed, as have the theoretical underpinnings for determining public response or outrage to risk. On a practical level, industry needs to operationalize these concepts. While each facility will develop its own approach, three key steps to building sustainable strategies for risk communication are applicable to all operations.

- *Step one: Begin with facility employees*

While the commitment to communication comes from top management, the success of an outreach program must involve employees at all levels. Employees are the best communication resource. They affect the way the facility is perceived, whether or not they have a specific responsibility for communication procedures. In fact, frontline employees, who themselves are exposed to whatever risk a facility might pose, possess greater credibility with the public than does management. If these employees do not believe the workplace is safe and environmentally sound, or that top management is responsive to their concerns, the community will have a similar outlook.

Employees also need to be the first audience for communications. If the goal is consistent communication with the public at large, the employees need to be informed about risks. Frontline employees must be provided with risk and crisis communication training. They need the tools that will allow them to feel comfortable in fulfilling their communication responsibilities. Communication is a skill that can be learned, and if employees are expected to take an active role in the process, they need to be given a solid foundation for addressing public concern.

- *Step two: Make communication a daily and ongoing commitment*

Building trust and credibility with the public takes time. People reluctant to change long-held opinions of industry will do so only when they are satisfied that a new viewpoint is merited. To build trust, a company must make itself accountable to the public. This means establishing a track record of reaching out, following through, and hearing and responding to people's concerns.

It is better to begin outreach with a series of small events rather than trying to address all audiences at one time. It is beneficial to seek out informal, as well as formal ways to interact, such as through meetings of civic organizations or homeowner's associations, and presentations at local schools or conferences. These interactions demonstrate an interest in the community and a willingness to participate on the public's terms and to relinquish control over communication infrastructure.

- *Step three: Focus on listening and understanding*

The key to communication is hearing what the public has to say. People want to receive information, but they also want to know that a company will listen to them and respond to answers. Never should conversation be limited to technical risk issues. A facility must be prepared to address a wide range of other issues and to take them seriously. Only then can industry construct a record of accountability.

5.5.3 Case Study: Environmental and Health Risk Communication Neglected Until After an Accident

To demonstrate the principles of risk communication, we present the simple case of a small oil-and-water separation plant that briefly released black smoke after a heavy rain caused oil to seep into a natural gas burner. Although the smoke was under control in less than 10 minutes, in the aftermath of this seemingly small incident, the facility was nearly forced to shut down, and the company had to invest thousands of dollars and countless hours retain its operating permit.

Management at the plant had been sure that the facility did not require a formal community-outreach program to explain its safe, unobtrusive oil/water separation process. A mixture of oil and water was piped in from offshore, heated, and transferred into settling tanks. The oil was then piped out through a pipeline, while the water was treated and discharged into the sewer system. The whole operation was so simple that for much of the time the plant was unstaffed.

Operations were quiet and odorless and produced no emissions. Likely due to the facility's unobtrusiveness, no one had ever asked about it, despite its location across the street from an elementary school. There was no signage on the perimeter fences, and most local residents could not identify the facility from the street.

5.5.3.1 The Accident and Aftermath

Black smoke associated with burning oil brought the plant to the attention of its neighbors, who were concerned and had many questions regarding plant safety. One resident was particularly alarmed. He went to speak to the fire department and the facility manager, only to find that his questions were not taken seriously. Facility representatives refused to acknowledge that citizen's concerns had any validity, which served to make him more angry and suspicious. He began to go door-to-door throughout the neighborhood, compiling a list of anecdotal health concerns, including rashes, coughs, and problematic pregnancies. A number of residents teamed together with the objective of getting the facility shut down. The anger and level of activism among residents increased. Neighbors circulated a petition to the city council, requesting the plant's closure. News media picked up the story, which received considerable play in the papers and on television. The local air-pollution control district became involved, and the city came very close to withdrawing the plant's operating permit.

Why were the repercussions of a single and short-lived incident so extreme? The primary cause for the extended crisis was the fact that, in its 30 years of operation, the facility had not made an effort to talk to its neighbors or to learn about their potential concerns. In fact, most residents in the neighborhood were unaware that the plant existed. When people suddenly discovered the facility, via the black smoke emissions, they were angry. Residents felt that they had a right to know what kind of industrial operations existed in their neighborhood and what kinds of risks the facility posed to their environment and health.

5.5.3.2 The Costs of Noncommunication

The oil–water separator facility was able to continue operations only after the company had funded two health risk assessments, established health-protective levels for chemicals used during cleaning and turnaround, promised to provide extensive environmental monitoring during nonroutine operations, and hosted four community meetings. Despite these efforts to gain the community's confidence, when the company attempted to reopen a similar facility 20 miles away, residents from the first neighborhood protested and succeeded in keeping the second facility from obtaining an operating permit.

Company representatives had discovered too late that fallout from the concerns of an ignored public could be very costly. Issues of concern to the local community include basic facts about the industries in the locality and the potential risks posed by their operations. In the United States, as a result of the Freedom of Information Act and many provisions of environmental statutes, people have a right to such information. If the oil–water separator facility had initiated a dialogue with the community early on, the meeting with the public would not have been associated with a developing crisis.

This case study illustrates how quickly a situation can escalate to crisis when there is no record of consistent, two-way communication between industry and the public. Several of the environmental vulnerability factors outlined in Table 5.5 came into play: people initially became concerned about an unexplained emission from an unexpected source; they wondered belatedly about the advisability of having the facility located so near to an elementary school and residential properties; and those who had experienced unusual health problems began to suspect a connection to the oil–water separation facility.

In cases such as this one, where the public has little or no reason to trust the industry or facility that is creating the risk, the effect of the outrage factors is also likely to be intense (see Table 5.6). The unusual and memorable release of smoke led to concern that people had no control over the kinds of industry allowed in their own community. Management of the separation plant was not seen as trustworthy. There was resentment, as well, over the company's failure to communicate with its neighbors prior to the incident (Boholm 2008).

5.5.4 Lessons Learned

The key to communicating with the public lies not in simply explaining the technical hazards but also in understanding and responding to the sources of people's outrage. Public communication does not end with the distribution of right-to-know information. To the contrary – as more and more community members acquire information – the tendency is to want more input into the entire process of risk management. Industry must be ready not only to communicate but also to share decision-making power with local communities. Such efforts will succeed only if community involvement becomes an integral part of the facility's day-to-day routine and of the company's culture. Building good relationships with the public takes time. However, it is these relationships that will bring trust and credibility to the company. This does not mean that there will never be conflict, but it does mean that the facility will not be starting from ground zero when attempting to explain complex, technical information about health, environmental, and safety risks to the public.

5.6 Environmental Information Access on the Internet

The Internet offers instant access to a wealth of resources for environmental (as well as chemical and other) information. The Internet makes available many materials that can be useful for environmental risk assessment and risk management. Not all sources are equally valuable, however. Section 5.6.2 gives some important caveats to those who use data from the following sources:

- Searchable databases of federal, state and local government regulations
- Technical information from universities and research organizations
- Data on chemical and environmental trends and specific events, as reported by hundreds of corporations and analyzed by trade associations, government agencies, and nongovernmental organizations
- Enforcement and inspection guidance, and penalty alternatives offered by the U.S. Occupational Safety and Health Administration (OSHA) and U.S. Environmental Protection Agency (USEPA)
- Product listings and capability statements from providers of environmental and chemical engineering support services, and directories of such firms

- Links to material exchanges, as well as pollution prevention techniques and other resources for reducing the costs and liabilities associated with the handling, release, and disposal of hazardous materials
- Online news services
- Publications, meeting announcements, and other opportunities for professional exchange

Following are some of the best environmental resources available on the Web, as well as lesser known but incredibly valuable sites for data to support environmental and chemical decision-making, models and resources available from regulatory agencies, assistance with outreach and public communication, and environmental publications (Bolstridge 2003).

5.6.1 Internet Sources

5.6.1.1 EPA's Integrated Risk Information System

http://www.epa.gov/iriswebp/iris/index.html. Integrated Risk Information System (IRIS) is EPA's database of dose–response data on human health effects from chemical exposure. The searchable database contains information on more than 500 substances and is intended for use in cancer risk assessments, decision-making, and regulatory activities (USEPA 2005).

5.6.1.2 EPA Chemical Accident Histories and Investigations

http://yosemite.epa.gov/oswer/ceppoweb.nsf/content/preventingaccident.htm. This page permits accessing documents and reports on recent chemical accidents, investigations by the EPA Chemical Accident Investigation Team, and Incident Summaries by the National Response Center, as well as related advisories and notices. EPA's Accidental Release Information Program database contains information on accidental releases of hazardous chemicals that have occurred in the United States at fixed facilities between 1986 and 2000. Facilities are asked by EPA Regional Offices to complete a questionnaire of 23 questions about the facility, the incident's circumstances and causes, any control and prevention practices and technologies that were in place prior to the event, and changes made as a result of the release. The questionnaire focuses on accident prevention, including hazards assessment, training, emergency response, public notification procedures, mitigation techniques, and prevention equipment and controls. The entire database (in compressed format) can be downloaded and searched using any database program that can handle dbase format files. A copy of the questionnaire and a data dictionary to explain the database fields are also included in the compressed file.

5.6.1.3 International Toxicity Estimates for Risk database

http://www.tera.org/iter. Integrated Risk Information System database is a compilation of human health risk values for more than 500 chemicals that have been identified as environmental concerns by health organizations around the world. Values for actual risks (e.g. noncancer oral; cancer, oral; noncancer inhalation; cancer, inhalation) can be obtained for any listed chemical, as well as detailed information on the study or research that generated the risk value reported, including information on peer review and detailed bibliographic references.

5.6.1.4 AIRNow's Links to Real-Time Air Pollution Data

http://www.epa.gov/airnow. This site provides access to national air quality information, including air quality forecasts and real-time air quality for over hundred US cities. Local air quality forecasts are available, as well as links to state and local websites that provide real-time air quality monitoring and forecasting data. Each state/local Internet site provides different types of information, but much is focused on the urban areas' ozone and air quality indices that are commonly reported in the news media.

5.6.1.5 Air and Waste Management Association's State Agency Listing

http://www.awma.org/resources/gad/state.htm. The Air and Waste Management Association site provides links to state agencies involved in environmental regulations and compliance enforcement.

5.6.1.6 Acid Rain, Atmospheric Deposition and Precipitation Chemistry

http://btdqs.usgs.gov/acidrain. The U.S. Geological Survey's National Atmospheric Deposition Program and National Trends Network collects data from 220 sites throughout the United States to determine if regulatory actions to reduce air pollution are improving the quality of precipitation. There is a five- to six-month time lag between data collection and availability of results, as would be expected when laboratory analysis is required. Online data access requires completing a form and providing the requestor's name and address and the intended use of the data. Data are collected for daily precipitation, field and laboratory pH, and weekly, monthly, and annual average concentrations and dry deposition data for sulfate (SO_4), nitrate (NO_3), ammonia (NH_4), calcium (Ca), magnesium (Mg), potassium (K), sodium (Na), chloride (Cl), and nitrogen (N) from NO_3 and NH_4.

5.6.1.7 Predict Mixed Chemicals' Reactions

http://response.restoration.noaa.gov/chemaids/react.html. The National Oceanic and Atmospheric Administration's

(NOAA) Chemical Reactivity worksheet is a free software package that contains reactivity data on 4000 chemicals and determines reactivity effects of accidental mixing of chemicals. Based on the classification of the chemical into one or more reactive groups, the software predicts the kind of reactivity likely to occur when members of the groups are mixed together. It does not limit the number of chemicals that can be selected to be mixed, but it becomes less effective for more than 10 chemicals.

5.6.1.8 Model an Oil Spill

http://response.restoration.noaa.gov/software/gnome/gnome.html. NOAA provides software that allows a scenario for an oil spill to be specified. The General NOAA Oil Modeling Environment model produces a predicted trajectory of the spill, using data on tides and currents.

5.6.1.9 Hazards Analysis for Toxic Substances, Version 3

http://yosemite.epa.gov/oswer/ceppoweb/content/ds-epds.htm#hats. Hazards analysis for toxic substances is a menu-driven screen-show presentation for emergency planners and responders. It introduces EPA's Emergency Planning and Community Right-to-Know Act program and the use of EPA's approach to analyzing the hazards of toxic substances to the community surrounding manufacturing plants where such materials are manufactured or used.

5.6.1.10 EPA Center for Exposure Assessment Modeling

http://www.epa.gov/ceampubl/index.htm. This site provides access to tabulated listings of exposure-assessment software packages to address modeling of organic chemicals and metals in groundwater, surface water, food chains, and multimedia systems. The models are available for downloading along with descriptions of the packages' capabilities.

5.6.1.11 Regulatory Air Models

http://www.epa.gov/ttn/scram. EPA's Support Center for Regulatory Air Models provides access to the source code, data sets, and user's guides for various air models, including regulatory models (ISCST3, UAM, BLP, CTDMP), receptor models (CMB7, CMB8), screening models (CAL3QHC, CTSCREEN, LONGZ, SHORTZ, VALLEY), other air models (DEGADIS, PLUVUE2, TOXST, MESOPUFF, RPM-IV, FDMI), related programs (CALMPRO, RAMMET, WRPLOT, MPRM, MIXHTS), and non-EPA models (SLAB, ADAM, AFTOX, SCREEN).

5.6.1.12 EPA's Risk Management Consequence Analysis (RMP*Comp)

http://www.epa.gov/ceppo/tools/rmp-comp/rmp-comp.html. EPA provides Version 1.07 of this free software to perform the risk management consequence analysis that is required under Section 112(r) of the Clean Air Act.

5.6.1.13 Agency for Toxic Substances and Disease Registry's Primer on Health Risk Communication

http://atsdr1.atsdr.cdc.gov/HEC/primer.html. The Agency for Toxic Substances and Disease Registry provides a framework of principles and approaches for communication of health risk information to diverse audiences. It is intended for spokespersons who must respond to public concerns about exposure to hazardous substances in the environment. The document contains a "do's and don'ts" list that describes approaches to use and pitfalls to avoid when communicating risk information.

5.6.1.14 Disaster Communication from Indiana Law University

http://www.law.indiana.edu/webinit/disaster/index.html. This site summarizes national laws and international conventions that affect global information and communication regarding natural and manmade disasters, including improving relief efforts and reducing overall impacts. The site provides access to documents that address the role of the news media in disaster reduction and links to other organizations involved in disaster communication.

5.6.1.15 EPA's National Service Center for Environmental Publications

http://www.epa.gov/ncepihom/ordering.htm. This site taps into a central repository for all EPA documents that can either be downloaded or obtained by mail for free. Over 5500 titles are available in paper and/or electronic format.

5.6.1.16 The National Academies' Reports Online

http://nap.edu/info/browse.htm. The National Academies Press publishes the reports issued by the National Academy of Sciences, National Academy of Engineering, the National Research Council, and the Institute of Medicine. Many texts on a variety of environmental, health, natural resources, and engineering topics can be reviewed or ordered online.

5.6.2 Implications and Limitations of Using the Internet

The Internet has become the ultimate data source for science and technology. However, caution must be used when applying Internet data to scientific efforts or regulatory compliance. Often, information is published on the Internet without any description of the study or research that was used to develop the data. When such documentation is lacking, consider sending an e-mail message to the organization or individual that maintains the site to request the reference information. Some online data are severely out-of-date or represent student research efforts that have not been reviewed for accuracy or quality. Finally, data provided on the Internet may have been developed for purposes other than environmental or chemical analysis, and as a result it may contain hidden biases or inaccuracies that make them inappropriate for your intended use. Realize that you may not be getting exactly what you think you have when downloading free data from the Internet.

Reliance on official government sites can help to avoid being misled by the content of some Internet sites. However, university and nonprofit organizations can offer more timely or detailed information on recent events and issues. Always review the organization behind any Internet location that is to be used as a resource. Focus on locating the types of funding sources, peer review and data quality evaluations, as well as the date of the last update, for additional insight into the true value represented by a particular site. Much of this information will be available online or can be obtained by sending an e-mail to the site's developer. Chemical and environmental engineering professionals require many types of information. Much of the required data, and even entire software packages, are now available to be downloaded from the Internet, free of charge. In fact, as the Internet matures, vast amounts of data and software are becoming available to everyone. However, converting data to information, and more importantly converting information into knowledge, has become the major challenge facing scientific professionals.

5.7 Health and Occupational Safety

5.7.1 Occupational Safety and Health Administration

The OSHA Act was enacted to ensure so far as possible every working man and woman in the nation safe and healthful working conditions and to preserve our human resources (OSHA 2018). Under the Act, the OSHA was created within the U.S. Department of Labor to

1) encourage employers and employees to reduce workplace hazards and to implement new or improve existing safety and health programs
2) provide for research to develop innovative ways to dealing with occupational safety and health problems
3) establish "separate but dependent responsibilities and rights" for employers and employees for achievement of better safety and health conditions
4) maintain a reporting and record-keeping system to monitor job-related injuries and illnesses
5) establish training programs to increase the number and competence of occupational safety and health personnel
6) develop mandatory job safety and health standards and enforce them effectively
7) provide for the development analysis, evaluation, and approval of state occupational safety and health programs.

5.7.1.1 "Right-To-Know" Laws

Probably one of the most important safety and health standards ever adopted is the OSHA hazard communication standard, emore popularly known as the "right-to-know" law. The hazard communication standard requires employers to communicate information to the employees on hazardous chemicals that exist within the workplace. The program requires employers to craft a within-hazardous communication program, to keep material safety data sheets (MSDS) for all hazardous chemicals, and to ensure that proper warning labels are in place.

The Hazardous Waste Operations and Emergency Response Regulation enacted by OSHA addressed the safety and health of employees involved in operations at mandate, and in certain hazardous waste treatment, storage, and disposal operations conducted under RCRA. The standard provided for employee protection during initial site characterization and analysis, monitoring activities, training, and emergency response.

Four major areas are under the steps of the regulation:

1) Cleanup operations at uncontrolled hazardous waste sites that have been identified for cleanup by a governmental health or environmental agency.
2) Routine operations at hazardous waste TSD (transportation, storage, and disposal) facilities or those portions of any facility regulated by 40 CFR parts 264 and 265.
3) Emergency response operations at sites where hazardous substances have or may be released.
4) Corrective actions at RCRA sites.

The regulation address three specific populations of workers at the above operations. First, it regulates hazardous substance response operations under CERCLA, including initial investigations at CERCLA sites before the presence or absence of a hazardous substance has been ascertained; corrective actions taken in cleanup operations under RCRA; and those hazardous waste operations at sites that have been designated for cleanup by state or local government authorities. The second worker population to be covered are those employees engaged in operations involving hazardous waste TSD facilities. The third employee population to be covered are those employees engaged in emergency response operations for release or substantial threat of release of hazardous substances, and post emergency response operations to such facilities.

5.7.1.2 National Institute of Occupational Safety and Health

National Institute of Occupational Safety and Health (NIOSH) recommends standards for industrial exposure that OSHA uses in its regulations. OSHA has the power to enforce all safety and health regulations and standards recommended by NIOSH. Although OSHA is in the U.S Department of Labor, NIOSH is in the U.S. Department of Health and Human Services and is an agency established to help viding research, information, education, and training in the field of occupational injury, disability, and death by gathering information, conducting scientific research, and translating the knowledge gained into products and services. NIOSH's mission is critical to the health and safety of every American workers. NIOSH objectives include the following:

1) Conduct research to reduce work related illness and injuries
2) Promote safe and healthy workplaces through interventions, recommendations, and capacity building.
3) Enhance global workplace safety and health through international collaborations.

The NIOSH Pocket Guide may be found on-line at: http://www.cdc.gov/niosh/npg/pgdstari.html.

It includes safety information, some chemical properties, and OSHA Permissible Exposure Limit concentrations, or PELs. The lower the permissible concentration, the greater the hazard to human health. By law, these concentration values cannot be exceeded in the workplace. Concentration measurements of these chemicals must be taken so that exposure levels are known and any documented overexposures can be addressed.

5.7.1.3 Material Safety Data Sheets

The MSDS is a document developed by chemical manufacturers. The MSDS contains safety and hazardous information, physical and chemical characteristics, and precautions on safe handling and use. It may also include hazards to animals, especially aquatic species. The manufacturer is required to keep it up to date. Any employer that purchases a chemical is required by law to make the MSDS available to employees. Development of an MSDS is required under OSHA's Hazardous Communication Standard.

5.7.1.4 Integrated Risk Information System

IRIS is a database maintained by the USEPA. IRIS stands for Integrated Risk Information System. It is available through http://www.epa.gov/ngispgm3/iris/index.html.

IRIS is a database of human health effects that may result from exposure to various substances found in the environment.

5.7.1.5 Nanotechnology Legislation

Comment on how and to what degree new legislation and rulemaking would be necessary for environmental control/concern of nanotechnology (operators) from an OSHA perspective.

Workplace exposure to chemical substances (including nanoparticles) and the potential for pulmonary toxicity is subject to regulation by the OSHA under the OSHA Act, including the requirement that potential hazards be disclosed on MSDS. (An interesting question arises as to whether carbon nanotubes, chemically carbon but with different properties because of their small size and structure, are considered the same as or different from carbon black for MSDS purpose). Both governmental and private agencies can be expected to develop the requisite TLVs for workplace exposure (Theodore 2006).

5.8 Industrial Process Safety System Guidelines

An *industrial safety system* is a countermeasure crucial in any hazardous plants such as oil and gas plants and nuclear plants. They are used to protect human, industrial plants, and the environment in case of the process going beyond the allowed control margins.

As the name suggests, these systems are not intended for controlling the process itself but rather protection. Process control is performed by means of process control systems (PCS) and is interlocked by the safety systems so that immediate actions are taken should the PCS fail.

Process control and safety systems are usually merged under one system, called *Integrated Control and Safety System*. Industrial safety systems typically use dedicated systems that are safety integrity level (SIL) certified. SIL is defined as a relative level of risk-reduction provided by a safety function, or to specify a target level of risk reduction. In simple terms, SIL is a measurement of performance required for a safety instrumented function. SIL applies to both hardware and software requirements such as cards, processors redundancy, and voting functions. Under any circumstances, however, safety must not be compromised with cost (CCPS 2012; Crowl and Louvar 2011; Das and Prabhudesai 1996).

5.8.1 Types of Safety Systems

There are two main types of *safety systems* in chemical and allied products industries:

1) Process safety system or process shutdown system (PSS)
2) Safety shutdown system (SSS): This includes emergency shutdown (ESD) and emergency depressurization (EDP) systems

A third system also exist which acts as a barrier and contains the spray out of hot oil and gases from flanges, valves, and pipe joints. These systems are popularly known as safety spray shields and flange guards. The use of spray guards is mandated by OSHA.

5.8.1.1 Emergency Shutdown System

These systems may also be redefined in terms of ESD/EDP levels as follows:

- ESD level 1: In charge of general plant area shutdown, can activate ESD level 2 if necessary. This level can only be activated from main control room in the process industrial plants.
- ESD level 2: This level shuts down and isolates individual ESD zones and activates if necessary EDP.
- ESD level 3: This level provides "liquid inventory containment."

5.8.1.2 Safety Shutdown System

The SSS shall shut down the facilities to a safe state in case of an emergency situation, thus protecting personnel, the environment, and the asset. The SSS shall manage all inputs and outputs relative to ESD functions (environment and personnel protection). This system might also be fed by signals from the main fire and gas system (FGS).

5.8.1.3 Fire and Gas System

The main objectives of the fire and gas system (FGS) control are to protect personnel, environment, and plant (including equipment and structures). The FGS shall achieve these objectives by

- detecting at an early stage, the presence of flammable gas
- detecting at an early stage, the liquid spill (LPG and LNG)
- detecting incipient fire and the presence of fire
- providing automatic and/or facilities for manual activation of the fire protection system as required
- initiating environmental changes to keep liquids below their flash point
- initiating signals, both audible and visible as required, to warn of the detected hazards
- initiating automatic shutdown of equipment and ventilation if 2 out of 2 or 2 out of 3 detectors are triggered
- initiating the exhausting system

5.8.1.4 Emergency Depressurization Systems

Due to closing ESD valves in a process, there may be some trapped flammable fluids, and these must be released in order to avoid any undesired consequences (such as pressure increase in vessels and piping). For this, EDP systems are used in conjunction with the ESD systems to release (to a safe location and in a safe manner) such trapped fluids.

5.8.1.5 Pressure Safety Valves

Pressure safety valves or PSVs are usually used as a final safety solution when all previous systems fail to prevent any further pressure accumulation and protect vessels from rupture due to overpressure by their designed action.

5.9 Industrial Hygiene

Industrial hygiene or *occupational hygiene* is the anticipation, recognition, evaluation, control, and confirmation of protection from hazards at work that may result in injury, illness, or affect the well-being of workers. These hazards or stressors are typically divided into the categories biological, chemical, physical, ergonomic, and psychosocial. The risk of a health effect from a given stressor is a function of the hazard multiplied by the exposure to the individual or group. For chemicals, the hazard can be understood by the dose response profile most often based on toxicological studies or models. *Industrial hygienists* work closely with toxicologists for understanding chemical hazards, physical therapist for physical hazards, and physicians and microbiologists for biological hazards. Environmental and industrial hygienists are considered experts in exposure science and exposure risk management. Depending on an individual's type of job, a hygienist will apply their exposure science expertise for the protection of workers, consumers, and/or communities (Johnson 2017).

5.9.1 Toxicology

The degree to which a substance is poisonous is termed its toxicity. A toxicant is any chemical substance which when inhaled, ingested, absorbed through the skin or when applied to, injected into, or developed within the body, in relatively small amounts may cause, by its chemical reaction, damage to the body structure or disturbance to human function or even death. Most chemicals enter the body through respiratory tract, digestive tract, eyes, and skin.

Two levels of toxicity are defined: (i) acute "short-term" exposure that causes initiation poisoning, (ii) chronic "long-period" exposure that causes anemia, leukemia, and death.

5.9.2 TLVs and Exposure Limits

Toxic TLVs are the greatest concentrations of a toxicant in air that can be tolerated for a given length of time without any toxic effects. Definitions of commonly used limit values are given in the following.

The different types of TLVs and exposure limits are as follows:

TLV-TWA	The TWA for a normal 8-hour workday or 40-hour work week to which a worker may be exposed without adverse effect
TLV-STEL	The short-term exposure limit is the maximum concentration to which a worker may be exposed continuously up to 15 minutes without adverse effect
TLV-C	The ceiling limit is the concentration which should not be exceeded at any time
PEL	Permissible exposure limit is the maximum exposure allowed by OSHA 29CFR1910.1000
IDLH	Immediately dangerous to life or health. It is the maximum concentration from which one could escape within 30 minutes without escape-impairing symptoms or any irreversible health effects

TLVs are expressed either in parts per million by volume (ppm) or milligram per cubic meter (mg/m^3).

$$1\,\text{ppm} = \frac{P \cdot M}{0.08205T}\,\text{mg/m}^3 0.04\ \text{M at } 25^\circ\text{C and } 1\ \text{atm}$$

where

P = pressure in atm;
T = temperature in K;
M = molecular mass

5.10 Atmospheric Hazards

Atmospheric hazards are comprised of three primary types: oxygen deficient atmospheres, explosive/flammable atmospheres, and toxic atmospheres. These types of hazards require air monitoring and ventilation before entering a manhole or other confined space. Air monitoring equipment is discussed later in this chapter. It should be noted that atmospheric hazards caused or contributed to by an industrial user discharge that result in the presence of toxic gases, vapors, or fumes within factories and workplaces in a quantity that may cause acute worker health and safety problems is violation of the specific prohibition at 40 CFR 403.5(b)(7).

5.10.1 Oxygen Deficient Atmosphere

The minimum OSHA requirement for oxygen concentration in the atmosphere is 19.5%. A Self- Contained Breathing Apparatus is necessary to enter an atmosphere with less than 19.5% oxygen. An oxygen enriched atmosphere, which exists when the oxygen concentration is greater than 25%, is also considered hazardous because of its ability to support combustion. Normal air consists of the gases listed in Table 5.7. Other gases, such as nitrogen and carbon dioxide, which are harmless under normal conditions, may build up in confined spaces in quantities large enough to displace the oxygen necessary to support life. When the concentration of oxygen in the atmosphere falls to 10–16%, a person will experience shortness of breath. Loss of consciousness will occur at a 6–10% oxygen concentration and death will occur rapidly when the concentration of oxygen falls below 6%.

5.10.2 Toxic Atmosphere

There are various guidelines for assessing chemical hazards in the atmosphere. The TLVs are guidelines developed and published by the American Conference of Governmental Industrial Hygienists to be used for identifying and controlling potential hazards. One form of the TLV, the TWA, refers to the vapor phase concentration a worker may be exposed to for an 8-hour workday or 40-hour work week without chronic or acute health effects. TLV-TWA numbers are sometimes used to calculate industrial user discharge screening levels for VOCs. See TLVs for various chemical substances at http://www.acgih.org/tlv-bei-guidelines/tlv-chemicalsubstances-introduction.

The EPA has issued a guidance document entitled *Guidance to Protect POTW Workers from Toxic and Reactive Gases and Vapors* (EPA 812-B-92-001) June 1992. This document should be used to evaluate the potential for exposure to toxic atmospheres and necessary steps for avoiding contact

Table 5.7 Atmospheric constituents.

Gas	Volume (%)
Nitrogen	78.09
Oxygen	20.95
Argon	0.93
Carbon dioxide	0.03

with such atmospheres. Hydrogen sulfide (H_2S) is a common gas found in the collection system. The gas, which is formed by anaerobic decomposition of organic matter, is heavier than air and tends to collect at the bottom of an enclosed space. At low concentrations, hydrogen sulfide has an odor of rotten eggs, at higher concentrations, however, the olfactory system becomes impaired and the gas cannot be detected by smell. The TLV-TWA for hydrogen sulfide is 100 ppm. At higher concentrations of H_2S, damage can occur to the eyes, nervous system, and respiratory system. A caustic solution (sodium sulfide) is formed when the gas comes into contact with moist tissue, such as in the eyes and respiratory tract, which causes the irritation and danger from the chemical. At concentrations of 500–1000 ppm, the respiratory system is paralyzed and death will occur.

Hydrogen cyanide gas may be generated when cyanide salts react with an acidic wastestream. Cyanide salts are often found in plating baths and metal-finishing facilities. The gas causes death by preventing the transfer of oxygen within the bloodstream.

Chlorine (Cl_2) gas may be encountered at plating facilities where it is used for cyanide destruction. Aside from the irritating odor, chlorine gas is corrosive in the presence of moisture. It combines with moisture in the lungs and the respiratory system to form hydrochloric acid. Pulmonary edema (fluid in the lungs) may occur at 50 ppm and at 1000 ppm death occurs rapidly.

Carbon monoxide (CO) is another gas that is generated in a collection system from anaerobic decomposition of organic materials. Asphyxiation occurs from exposure to this gas because the hemoglobin of the blood has 300 times more affinity for carbon monoxide than for oxygen. Carbon monoxide combines with hemoglobin to form carboxy hemoglobin. As a result, blood cells with CO cannot transport oxygen to body tissues, and death occurs.

Toxic vapors also present a hazard to inspectors and sampling personnel. Vapors are the volatile form of substances that are normally in a solid or liquid state. Chlorinated solvents used in degreasing or photoresist developing operations generate vapors that may accumulate in the collection system. Some of the vapors have an anesthetic effect when inhaled. In addition, the vapors are generally heavier than air, causing oxygen in a confined space to be displaced which may create an oxygen deficiency in that space.

5.10.3 Chronic Industrial Exposure

Chronic toxicity in hazardous waste management is most often caused by long-term, low-level exposure to hazardous chemicals. Chronic toxicity is difficult to quantify because less is known about the long-term effects of chemicals compared to acute toxicity evaluations. A typical dose–response curve for chronic toxicity is shown in Figure 5.16.

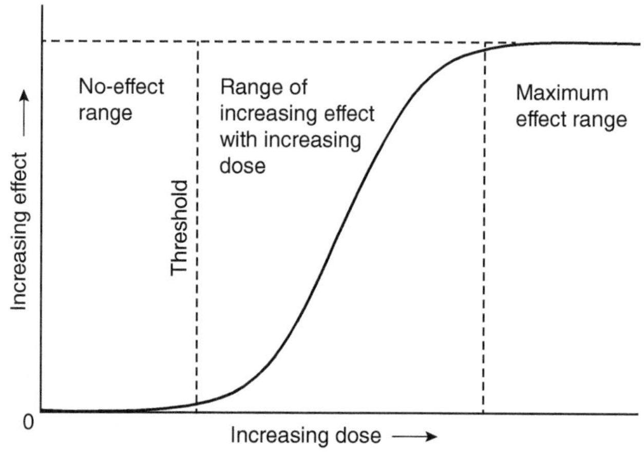

Figure 5.16 Dose–response function with no-effect region.

At low contaminant concentrations, essentially no evidence of toxicity is exhibited, and this is called the *no-effect region*. The basis for the lack of response from zero dose to the threshold dose is a biochemical or physiological defense, such as detoxification or excretion, that prevents the effect from occurring. The middle segment rises from a point where some toxicity is shown (the threshold), and then increases in a sigmoidal fashion until a maximum effect is found. As the exposure to the chemical is increased, the detoxification mechanisms are overwhelmed and the effects are exhibited.

5.10.4 Accidental Chlorine Gas Release: Case Study

The Pierce County Chambers Creek Wastewater Treatment facility in Tacoma, Washington uses chlorine for disinfection; the gas is stored on site in a 1-T container. The maximum anticipated storage inventory is ten containers (20 000 lb), which exceeds the threshold for applicability of the Risk Management Plan (RMP) requirements specified in the Accidental Release Prevention Program (USEPA 1996). The threshold for chlorine is 2500 lb. The potential off-site consequences of an accidental release of this poisonous gas are serious, and an evaluation of the situation is called for. The methodology used for this analysis follows the approach outlined in EPA's *risk management program guidance for wastewater treatment plants* (USEPA 1998b) and in the guidance provided in the White's *Handbook of chlorination and alternative disinfectants* (2010).

According to the Accidental Release Prevention Program (USEPA 1996), the toxic gas worst case scenario must assume the release of the single largest vessel or container over a period of 10 minutes. The largest vessel at Chambers Creek is the 1-T container noted earlier.

For worst-case scenarios, releases are assumed to take place without consideration for physical cause or likelihood of occurrence. Only passive mitigation measures (e.g. dikes, enclosures) can be considered, while active mitigation measures cannot be included. Furthermore, the rule states that toxic gases should be evaluated at ambient temperatures unless the gas is liquefied by refrigeration at ambient pressure. The chlorine containers at the treatment facility can be stored either inside or outside the building. For a worst-case scenario, it was assumed that a container was ruptured outside the building.

The release rate, calculated for the characteristics of a vapor release, was 200 lb/min for 10 minutes. An alternative release scenario, assuming a slower rate of release owing to the use of active mitigation measures, was calculated as 31.3 lb/min over 20 minutes. For the remainder of this case study, we consider both the worst-case scenario and the alternative release possibly just described.

5.10.5 Determination of Toxic Endpoint Distance

The distance to the toxic endpoint was calculated for both the worst and alternate case release scenarios. The toxic endpoint for chlorine is 0.0087 mg/l (3 ppm). This value is based on the Emergency Response Planning Guidelines Level 2 (ERPG-2) value. ERPG-2 is defined as the maximum airborne concentration below which it is believed that nearly all individuals could be exposed for up to one hour without experiencing or developing irreversible or other serious health effects, or symptoms that could impair an individual's ability to take protective action.

Highly specialized computers software was used to estimate the distance to the toxic endpoint for each scenario. The programs, RMP*Comp, version 1.06, is a mathematical tool developed by EPA and NOAA specifically to determine the endpoint distances from various releases under the RMP rule (for details, visit the U.S. government website http://response.restoration.noaa.gov/chemaids/rmp/rmp.html, http://www.epa.gov/rmp/rmpcomp; USEPA 1996).

Table 5.8 shows the input used in RMP*Comp. RMP*Comp assumes a meteorological condition of class F stability with a 1.5 m/s wind speed for worst-case releases and, for alternate releases, a meteorological condition of class D stability and a 3 m/s wind speed. The model requires choosing between urban and rural dispersion patterns. Urban dispersion was chosen for this analysis because the site is surrounded by residential and industrial development.

5.10.6 Determination of Exposed Population to this Scenario

Population information was obtained using LandView III, an electronic mapping software system that includes database information from the EPA, the Bureau of Census, the U.S. Geological Survey, and other federal agencies. LandView III was developed at the EPA's Chemical Emergency Preparedness and Prevention Office in Washington, DC. The software includes two different calculation methods for estimating the population within a radius of a given point using an updated Census data. The methodology used for this analysis was the default method, called the Block Group Proration method, which sums the data for each census block group that has any portion falling within the radius of the circle and then prorates the result based on the ratio of the area of the circle to the total land area of the block groups. The results of the calculation are summarized for this scenario in Table 5.9.

Table 5.8 RMP*Comp inputs.

	Worst-case inputs	Alternative case inputs
Release type	**Vapor**	**Vapor**
Amount released (lb)	2000	900
Release rate (lb/min)	200	31.3
Release duration (min)	10	20
Wind speed (m/s)	1.5	3.0
Atmospheric stability class	F	D
Roughness type	URBAN	URBAN
Passive mitigation measures	Release outside building	Release inside building
Active mitigation measures	None	None
Tank parameters		
Hole area	Not considered	1.25×10^{-4} in^2 (0.15 in. diameter)
Tank temperature (°C)	25	25
Tank pressure (psi)	120	120

5.10.7 Chronic Industrial Exposure: TWA and TLV

The use of the TWA, which has been adopted by OSHA, enables in assessing exposure during a eight-hour periods of time as follows:

$$\text{TWA} = \frac{C_a T_a + C_b T_b + \cdots + C_n T_n}{8} \tag{5.8}$$

where

C_a, C_b,...,C_n = concentrations of first, second, ..., and final durations of exposure periods, respectively.

T_a, T_b,...,T_n = time of first, second, ..., and final time exposure periods (hours), respectively.

Table 5.9 Population estimates within endpoint circles.

	Worst case	Alternate case
Distance to endpoint (miles)	1.3	0.1
Estimated residential population within distance to endpoint	13322	5
Public receptors within distance to endpoint	Residences, commercial and industrial facilities, recreation areas, hospital	Residences, recreation areas
Environmental receptors within distance to endpoint	Parks	None

Chronic toxicity in hazardous waste management is most often caused by long-term, low-level exposure to hazardous chemicals. Chronic toxicity is difficult to quantify, because it is less known about the long-term effects of chemicals compared to acute toxicity evaluations. A parameter used by industrial hygienists, the *TLV*, is the concentration of a chemical in air to which workers may be exposed safely over a working over their occupational lifetime (i.e. for 8 h/day, 5 days/week, over a working lifetime). TLVs have been defined for hundreds of airborne contaminants. Some contaminants (e.g. carcinogens) have been given a zero TLV because no threshold exists. The TLV is a copyrighted term of the American Conference of Governmental Industrial Hygienists (ACGIH), a group that develops policy and continually updates TLV data. Table 5.10 presents TLVs of some common hazardous compounds. The TLV ceiling value is the concentration that should not be exceeded at any time (ACGIH 1986).

Example 5.11 Estimate the time weighted concentration for 8 hours if a worker is exposed to concentrations of 100 ppm for 1 hour, 90 ppm for 2 hours and 80 ppm for 5 hours.

Solution
Using Eq. (5.8) we get

$$\text{TWA} = \frac{100 \times 1 + 90 \times 2 + 80 \times 5}{1 + 2 + 5} = 85\,\text{ppm}$$

If more than one chemical is present in the workplace, one procedure is to assume that the effects of the toxicants are additive. The combined exposures from multiple toxicants with different TLV–TWAs are determined from the equation:

Table 5.10 TLVs of some common hazardous compounds.

Compound	TLV
Acetone	750 ppm
Aldrin	0.25 mg/m^3
Atrazine	5 mg/m^3
Benzene	10 ppm
Carbaryl	5 mg/m^3
Carbon tetrachloride	5 ppm
Chlorobenzene	10 ppm
Chloroform	10 ppm
Chrysene	Suspected human carcinogen
Cyclohexane	300 ppm
Cyclohexene	300 ppm
DDT	1 mg/m^3
2,4-D (Dichlorophenoxyacetic acid)	0.1 mg/m^3
Dieldrin	0.25 mg/m^3
Dimethyl phthalate	5 mg/m^3
Ethylbenzene	100 ppm
n-Heptane	400 ppm
Hexachlorobenzene	0.025 mg/m^3
Hexachlorocyclopentadiene	0.01 ppm
Malathion	10 mg/m^3
Methyl butyl ketone	50 ppm
Methyl ethyl ketone (MEK)	200 ppm
Methyl isobutyl ketone (MIBK)	50 ppm
Methylene chloride	50 ppm
Naphthalene	10 ppm
n-Octane	300 ppm
Parathion	0.1 mg/m^3
Pentachlorophenol	0.5 mg/m^3
Perchloroethylene (PCE)	50 ppm
Phenol	5 ppm
Picric acid	0.1 mg/m^3
Polychlorinated biphenyls (PCBs)	0.5 mg/m^3
Toluene	100 ppm
1,1,2-Trichloroethane (TCA)	10 ppm
Trichloroethylene (TCE)	50 ppm
2,4,6-Trinitrotoluene (TNT)	0.5 mg/m^3
o-Xylene	100 ppm
m-Xylene	100 ppm
p-Xylene	100 ppm

Source: From ACGIH (1986).

$$\sum_{i=1}^{n} \frac{C_1}{\left(\text{TLV} - \text{TWA}\right)_i} \quad (5.9)$$

where

n is the total number of toxicants

C_i is the concentration of chemical i with respect to the other toxicants

$(\text{TLV–TWA})_i$ is the TLV–TWA for chemical species i.

If this sum in Eq. (5.9) exceeds 1, then the workers are overexposed.

The mixture TLV–TWA can be computed from

$$\text{TLV-TWA}_{\text{mix}} = \frac{\sum_{i=1}^{n} C_i}{\sum_{i=1}^{n} \frac{C_1}{\left(\text{TLV} - \text{TWA}\right)_i}} \quad (5.10)$$

If the sum of the concentration of the toxicants in the mixture exceeds this amount, then the workers are overexposed.

For mixture of toxicants with different effects (such as an acid vapor mixed with lead fume) the TLVs cannot be assumed to be additive (Crowl and Louvar 2011).

Example 5.12 Determine the mixture of TLV–TWA for a worker exposed to 42 ppm PCE, 38 ppm TCE, 86 ppm MEK, and 35 ppm MIBK in the air of a RCRA solvent recycling operation. Are the concentrations of chemicals exceeding the $\text{TLV–TWA}_{\text{mixture}}$?

From Table 5.10, the TLVs for the chemicals are

Compound	TLV (ppm)
PCE	50
TCE	50
MEK	200
MIBK	50

Using Eq. (5.9), these values are applied to determine the $\text{TLV–TWA}_{\text{mixture}}$ over an 8-hour workday.

$$\left(\text{TLV} - \text{TWA}\right)_{\text{mix}} = \frac{42 + 38 + 86 + 35}{\frac{42}{50} + \frac{38}{50} + \frac{86}{200} + \frac{35}{50}}$$

$$\left(\text{TLV} - \text{TWA}\right)_{\text{mix}} = 73.6\,\text{ppm}$$

The total exposure, $\Sigma C_i = 42 + 38 + 86 + 35 = 201\,\text{ppm}$; because the ratio of $\Sigma C_i/(\text{TLV} - \text{TWA})_{\text{mix}}$ is greater than 1, the concentrations of these RCRA chemicals exceeded the limit. The air quality in the solvent recycling facility is unhealthy based on the TLVs.

5.11 Safety Equipment

Recognizing the physical and atmospheric hazards associated with working in various industries and manufacturing plants is important. Simply acknowledging the hazards, however, does not guarantee safe working conditions. An employer is obligated by OSHA to provide safety equipment and establish a training program for employees. The important thing to remember is that the responsibility for using the equipment and following safety procedures rests with the individual employee. The following discussion will acquaint the employees and plant workers with proper safety techniques.

5.11.1 Personal Protective Equipment

If engineering controls cannot eliminate an exposure, the first step in selecting PPE is to minimize the portions of the body requiring protection (eyes, face, ears, toes, full body, etc.). Personal protective equipment (PPE) selection depends on the area of coverage and the exposure severity and type. Higher-risk exposures require more careful selection of PPE and rigorous training and certification (Ibrahim et al. 2018).

Companies should select PPE certified to comply with widely accepted standards to ensure that the PPE is sufficient for the intended use. Table 5.11 lists standards and best practices for various hazards and their applicability. It is important to understand the limits of each PPE as well as the employee's potential exposure. Exposure route(s) need to consider both the source of the hazard and the target receptor. Common hazards as well as environmental hazards include cold air and surfaces, heat stress, and smoke. Target receptors can include skin and eyes (dermal route) and lungs (inhalation route) (Ibrahim et al. 2018; ISO 2015; NFPA 2006, 2015, 2017, 2018a, b, c, d).

5.11.1.1 Heat Exposures

Most industry-specific guidance focuses on PPE guarding against flame and heat. NFPA 2113 (NFPA 2015) recommends a workplace hazard analysis to determine where and when flame-resistant garments, gloves are required. Section A.5.2 of the standard lists specific materials and tasks for which flame-resistant garments are recommended (NFPA 2015).

5.11.1.2 Electric Exposures

PPE requirements for electrical hazards are well-specified in NFPA 70E, electrical safety in workplace, which covers a broad spectrum of electrical activities (NFPA 2018a). The 2018 edition of NFPA 70E has added a comprehensive list of PPE for electrical hazards and nonelectric hazards, such

Table 5.11 Safety standards can help in selecting appropriate PPE.

Regulation	Target area	Hazard	Intended users
NFPA 70E – Standard for Electrical Safety in the Workplace	Full body	Electrical	All
NFPA 1852 – Standard on Selection, Care, and Maintenance of Open-Circuit Self-Contained Breathing Apparatus (SCBA)	Inhalation	Biological, chemical, radiological, and/or nuclear (CBRN)	All
NFPA 1891 – Standard on Selection, Care, and Maintenance of Hazardous Materials Clothing and Equipment	Dermal	Physical and health	All
NFPA 1986 – Standard on Respiratory Protection Equipment for Technical and Tactical Operations	Inhalation	Health	First responders
NFPA 1992 – Standard on Liquid Splash-Protective Ensembles and Clothing for Hazardous Materials Emergencies	Dermal	Liquid chemicals	First responders and trained personnel
NFPA 2112 – Standard on Selection, Care, Use, and Maintenance of Flame-Resistant Clothing for Protection of Industrial Personnel against Short-Duration Thermal Exposures from Fire	Dermal	Fire	Industrial personnel
NFPA 2113 – Standard on Selection, Care, Use, and Maintenance of Flame-Resistant Garments for Protection of Industrial Personnel against Short-Duration Thermal Exposures from Fire	Dermal	Fire	Industrial personnel
ISO 11612 – Protective Clothing – Clothing to Protect against Heat and Flame – Minimum Performance Requirements	Dermal	Fire	All

as fall protection, head protection, foot protection, and more (Ibrahim et al. 2018).

5.11.1.3 Chemical Exposures

Safety data sheets (or formerly *material safety data sheets*) (SDS/MSDS) generally recommend PPE for use with chemical under standard temperature and pressure (STP) and typically do not consider the chemical's full potential. The hazard potential of a chemical process would depend on the severity of the process operating conditions. Certain process operating parameters (e.g. high pressures) may elevate the hazard of the chemical, while others (low temperatures) may reduce it. Certain protective measures, such as hermetic sealing of containers, can also reduce the hazard potential of a chemical.

The PPE assessment should also consider any potential by-products. It is possible for the reaction of two relatively benign chemicals to form a particularly hazardous chemical. For instance, a dilute nitric acid solution containing elemental copper will form nitric oxide, a highly toxic gas.

5.11.1.4 Inhalation Risks

Protection against inhalation hazards can be one of the more arduous tasks of PPE selection. The NIOSH offers an online tool, the *certified equipment* list that provides information on how respirators and breathing apparatus protect against chemical, biological, radiological, and/or nuclear (CBRN) hazards (NIOSH 2016). The reference and attached URL allows users to search for NIOSH-certified equipment to meet their needs. When identifying PPE for toxic inhalants, it is important to understand the type of toxicity, the effect of chronic vs. acute exposure, and other characteristics of toxicity.

5.11.1.5 Eye Risks

Eye protection is extremely important and should be worn to protect workers' eyes. Safety glasses provide protection from external debris, dusts, and should provide side protection via a wrap-around design or side shields. Also, googles provide better protection and are effective in preventing eye injury from chemical splashes, impact, and dusty environments. Welding requires special goggles or shielding (ANSI 2015). The ANSI Z87.1 standard for safety glasses at work outlines eye protection requirements for a variety of hazards such as impact, splash, dust, and fine dust (including flammable dust and/or outside factory building "wind-blown dusts," from piles of coal and saw dust, for example). When selecting eye protection, consider the comfort level of personnel as well as some employees' need for prescription eyewear. Also account for interactions between certain chemicals and contact lenses.

5.11.2 Personal Protective Clothing

Protective clothing is an important aspect of safety, and the guidelines presented here should be followed when working at industrial facilities.

- *Hard hat* – All persons entering a confined space are required to wear a hard hat. A full-strength hard hat with a brim and chin strap provides protection from head injuries. Hard hats may also prevent injury during process area walk-throughs.

- *Coveralls* – A person's skin should be covered as much as possible to prevent scrapes and cuts and to avoid skin contact with hazardous substances.
- *Gloves* – Hand protection is necessary when working with and/or operating machines and handling liquid, gas, and solid materials. In addition to preventing absorption of hazardous chemicals (e.g. solvents, biological hazards, etc.) and/or heat through the skin, gloves will protect the hands from cuts, scratches, or heat stress.
- *Shoes* – Rubber-soled, non-skid, protective-toed shoes (e.g. steel, composite) and boots must always be worn in or around a confined space, in process areas, chemical storage areas, etc. Safety shoes are designed to protect against impact and/or hazardous chemicals.
- *Ear plugs* – Ear plugs should be worn when working in areas with high noise levels (e.g. general manufacturing areas). This equipment will protect employees and personnel from the cumulative effects of loud noises in the work place.
- *Safety goggles* – Safety goggles are necessary during working in manufacturing industries to prevent eye contact with hazardous substances. Contact lenses are often prohibited around some industrial processes such as plating operations. If a hazardous substance comes into contact with a person's eye through splashing or exposure to mists or vapors, that substance may become trapped behind the lens where it would be difficult to flush out and could cause severe eye damage.
- *Safety vests* – Safety vests are necessary warning devices in industrial workplace areas with vehicle traffic.

5.12 Communication Devices

Two-way radios and cell phones are effective ways to maintain contact with a main safety office. For hazardous sites, a sampling team should make contact upon arrival and departure from each sampling location. Industrial off-sites are often located in unpopulated areas; therefore, it is important that the radio or cell phones be kept in working order in the event it is necessary to call for help. If an accident should occur, the rescuer must call for help before any assistance is given to the victim.

5.12.1 Air Monitoring Devices

Before entering in a confined space, tests should be done for (i) explosive gases, (ii) the presence of toxic gases, and (iii) oxygen deficiency. The most effective method for detecting these conditions is with an atmospheric monitor. The gas detectors discussed in the following are the most commonly used for atmospheric monitoring.

- *Single purpose detector* – Designed to detect specific gases, such as carbon monoxide, methane, or hydrogen sulfide. These gases are commonly present in various systems and confined spaces. Single detection units or tubes are available for measuring gases that are less common.
- *Dual purpose detector* – Capable of detecting lack of oxygen and explosive conditions in an area.
- *Combination detector* – Capable of detecting a lack of oxygen, explosivity, and the presence of toxic gases. This type of meter provides maximum protection by detecting the presence of all three hazards. Hydrogen sulfide and carbon monoxide are the gases usually measured because they occur most frequently in a confined area or in an anoxic system.

The atmosphere in a confined space can change suddenly; therefore, a detector that continuously monitors the atmosphere is recommended. In addition, detectors should be equipped with an audible and visual alarm that is activated in response to specific hazardous conditions or a low battery, thus eliminating the need for taking the time to read a dial or gauge.

It is important to remember that using an atmospheric tester does not ensure safe conditions. Gas detectors are only one source of information pertaining to a potentially hazardous situation. Most gas detectors are designed to test for common gases such as hydrogen sulfide and are not effective for detecting less common substances such as trichloroethylene vapors. There are test kits available for detecting the less common gases. The kits consist of a bellows-type pump and glass tubes containing an indicator chemical which are sealed at the ends until they are used. The indicator chemical in the detector tube is specific to a particular contaminant or group of gases. A predetermined volume of air is drawn through the tube and the contaminant reacts with the indicator chemical, producing a color change that can be compared to a color calibrated chart to determine an approximate concentration.

When measuring explosivity, gas meters measure the percentage of the lower-explosive limit (LEL) of a calibration gas, which is usually methane. Gases are combustible throughout a range of air mixtures.

The meters do not differentiate between gases, but they only indicate explosivity relative to the calibration gas. The range begins with the LEL, which is the lowest concentration of a combustible gas or vapor in air that is necessary to support combustion. The explosive range extends upward to the upper explosive limit (UEL) which is the maximum concentration that will support combustion. If the concentration of gas is below the LEL, there is insufficient fuel to support ignition. Alternatively, if the concentration is above the UEL, there is insufficient oxygen to support combustion. These limitations in the atmospheric monitoring equipment emphasize the need for constant ventilation and awareness of potential hazards.

5.12.1.1 Estimate of Dilution Air

When volatile substances are emitted in working spaces such as a laboratory, fresh air is ventilated into the working space to maintain average concentration of the volatile matter below TLV–TWA. The quantity of such dilution air in CFM (cubic feet/minute) may be estimated by

$$\text{CFM} = \frac{(3.87 \times 108)W}{M \times \text{TLV} \times K} \tag{5.11}$$

where

W = mass of volatile matter evaporated/time (lb/min)
M = molecular mass of the volatile matter
TLV = threshold limit value of the volatile substance (ppm)
K = nonideal mixing factor = 0.1 to 0.5 for most situations

5.12.2 Ventilation Devices

Few confined spaces have adequate natural or mechanically induced air movement, and in most spaces, it is necessary to remove harmful gases or vapors by ventilation with a blower, fan, and capturing hoods. The most common method of ventilation uses a large flexible hose attached at one end to a blower with the other end lowered into the space. The blower will push fresh air into the space to purge the area of hazardous substances. The blower allows the fresh air to enter the space at the lowest point possible. Because the atmosphere in a confined space can change quickly, ventilation should be continuous. The OSHA requires the establishment of numerous performance and design standards for ventilation systems. There are OSHA standards for specific substances, for operations, and for specific hoods. However, the recommended minimum design capture velocity could be from 500 to 1000 ft/min if contaminants are released with high velocity into turbulent air (Danielson 1973; McDermott 1976).

Example 5.13 A cleaning bath for electronic parts emits 0.5 g/s of CFC-12 into a small work room of dimensions 3 m × 2.45 m high. Calculate the concentration in the room under average and poor ventilation conditions if the air velocity in the room is 0.3 m/s and compare the results to the OSHA PEL for CFC-12 = 4950 mg/m^3.

Solution
When the air speed is 0.3 m/s, the volume of air flowing through the room will be

$$Q = 0.3\,\text{m/s} \times 3\,\text{m} \times 2.45\,\text{m} = 2.21\,\text{m}^3/\text{s}$$

and the concentration of CFC-12 in the air will be

$$\text{Average ventilation } C = 0.5\,\text{g/s} / \left(0.5 \times 2.21\,\text{m}^3/\text{s}\right) = 0.45\,\text{g/m}^3$$

$$\text{Poor ventilation } C = 0.5\,\text{g/s}/\left(0.1 \times 2.21\,\text{m}^3/\text{s}\right) = 2.27\,\text{g/m}^3$$

A comparison with the PEL of 4950 mg/m^3 indicates that even under poor ventilation conditions, the OSHA PEL will not be exceeded and respiratory protection will not be needed to safeguard the health of a person working in the room.

Example 5.14 Calculate the concentration of Freon in the cube, 1 m on a side, surrounding the top of the cleaning bath in Example 5.13 if the air exchange rate with the remainder of the room is 1 m^3/s. Repeat the calculation for an air exchange rate of 0.5 m^3/s.

Solution
From Example 5.13, the rate of release of Freon from the cleaning unit is 0.5 g/s and the ventilation rate in the room is 2.21 m^3/s. Thus, in the area closest to the cleaning bath, the concentration of Freon can be calculated as follows:

$$C = (0.5\,\text{g/s}) \times \left[1\,\text{m}^3/\text{s} + 2.21\,\text{m}^3/\text{s}\right] / \left[1\,\text{m}^3/\text{s} \times 2.21\,\text{m}^3/\text{s}\right] = 0.73\,\text{g/m}^3 = 730\,\text{mg/m}^3.$$

or

$$C = (0.5\,\text{g/s}) \times \left[0.5\,\text{m}^3/\text{s} + 2.21\,\text{m}^3/\text{s}\right] / [(0.5\,\text{m}^3/\text{s} \times 2.21\,\text{m}^3/\text{s}] = 1.23\,\text{g/m}^3 = 1230\,\text{mg/m}^3.$$

As would be expected, the local concentration of the chemical increases when the ventilation is less. Localized ventilation is an effective method of dispensing airborne chemicals and reducing exposures of workers to chemicals.

5.12.3 Safety Harness and Retrieval System

Any entry into a confined space must always be performed by a team consisting of at least two people. A standby person must be stationed outside of the confined space. OSHA says that the duty of the attendant is to "maintain communication with and keep an accurate account of those workers entering the permit space." All personnel required to enter the confined space must wear a safety harness. A full-body parachute-type harness with a lifeline attached at the shoulders is recommended. This type of harness will keep a body vertical and prevent a limp body from falling out when being pulled out of an area.

The lifeline should be attached to a retrieval system which includes a fall arrest mechanism. This type of safety system works in a similar manner as an automobile safety belt where a centrifugal locking mechanism is activated when a fall occurs. If a rescue attempt is necessary, this type of retrieval system eliminates the need to enter the space. Approximately one half of all fatalities that occur in confined spaces are unplanned rescue attempts where a worker instinctively rushes into the confined space to assist an injured coworker.

The retrieval system should be purchased from a reputable manufacturer or authorized distributor as complete systems, including repair and training for proper use. Most components of a retrieval system must meet certain manufacturer specifications and substitution of these components may result in liability for personal injury.

Safety equipment must be maintained and inspected on a regular basis. A safety harness and rescue rope should be examined for frayed strands of fibers, cuts or tears, chemical damage, decay, and kinks or extreme stiffness. Visual inspection of this equipment should be made prior to each use, and formal procedures should be implemented for periodic inspection and maintenance.

5.12.4 Respirators

The primary function of a respirator is to prevent exposure to hazardous atmospheres. It is important to choose a respirator based on the job to be performed and the potential hazards to which an employee may be exposed. The basic types of respirators are as follows:

- *Air purifying* – Masks that filter dangerous substances from the air.
- *Air supplying* – Devices that provide a supply of safe breathing air from a tank.

An air-purifying respirator will remove particles of dust and light concentrations of gas or vapors, but it will not protect against heavy gas concentrations. In addition, this type of respirator provides no oxygen other than what is filtered through the mask. Air-purifying respirators include the following types:

- *Gas and vapor respirators* – Contaminated air is passed through charcoal that traps gases and vapors.
- *Particulate respirator* – Contaminated air is passed through a filter for removal of particles.
- *Powered air-purifying respirator* – A blower passes contaminated air through a device that removes contaminants and provides purified air to a facepiece.
- *Combination gas, vapor, and particulate respirator (referred to as a gas mask)* – Gas, vapors, and particles are filtered out by passing contaminated air through a canister containing a filter and a device that will absorb gases and vapors.

Air-purifying respirators must *never* be used in oxygen-deficient atmospheres. Due to this limitation, the air-supplying respirator is more appropriate for entering confined spaces that are subject to oxygen deficiency.

Air supplying respirators differ from air-purifying respirators in that the air the wearer breathes is provided from an outside source. Types of air-supplying respirators include the following:

- *Self-contained respirators* – Offer protection from toxic gases and oxygen deficiency by providing a transportable supply of breathable air. An example is the self-contained breathing apparatus (SCBA).
- *Supplied-air respirators* – Breathable air is supplied to the wearer from a compressor through an air hose.
- *Combination self-contained and air-supplying respirators* – An auxiliary cylinder provides assistance if the supply from the air hose is cut off.

Personnel must be aware of the atmospheric hazards they may encounter and should use respirators that are appropriate for those hazards. Only use respirators that have been approved by the National Institute for Occupational Safety and Health (see 29 CFR 1910.134) found at https://www.osha.gov/pls/oshaweb/owadisp.show_document?p_id=12716&p_table=standards. In an emergency, respirator devices become an essential piece of protective equipment. Failure of this equipment could result in injury or death. Instructions for use, cleaning, and maintenance must be strictly followed for safe use of any respirator. Employees must receive thorough training in the use and limitations of their respirators.

5.12.5 Confined Space Entry

Confined space entry must always be considered a hazardous task due to the potential exposure to physical and respiratory hazards. Upon arrival to a sampling site subject to vehicular and pedestrian traffic, the sampling team must immediately position the work vehicle and properly place traffic control cones and warning flags, as necessary. Once this has been completed, the atmospheric conditions of the confined space must be tested for oxygen deficiency, explosivity, and toxic gases. The following air-monitoring procedures should be followed:

- Calibrate air-monitoring equipment. Each gas detector has its own calibration method, and personnel must be trained in the calibration and operating procedures for each piece of equipment available for their use. In addition, personnel must know how to interpret the readings of the equipment. Calibration data should be documented and included in field records.
- Place the tip of the air-monitoring probe into the vent hole of a manhole to test the atmosphere. Nonmetallic probes should be used so that a spark is not accidently produced.
- If the reading is acceptable, remove the manhole cover and test the manhole. Because gases and vapors may be heavier than air, the atmosphere in the manhole should be tested at several depths, from the top to the bottom.
- If readings inside the confined space are acceptable, proceed with the confined space entry procedures. If the readings are not acceptable due to an explosive atmosphere, the manhole cover may cause a spark and consequently should not be opened. When an explosive atmosphere is encountered, the proper authorities (local fire department) should be contacted immediately. If the reading is unacceptable due to a lack of oxygen or accumulation of gases, the space must be ventilated until the reading is acceptable. The length of ventilation time depends on the size of the space.

All personnel entering a confined space must wear a full-body harness in addition to required safety equipment (i.e. hard hat, safety gloves, etc.). Confined spaces should never be entered alone or unassisted. There must always be a second person available for emergency situations. In addition, when entering a confined space, a SCBA should always be worn in addition to the harness, unless you have proper ventilation and air monitoring equipment. The harness must be attached to a safety line that in turn is attached to a retrieval system located above the ground. In an emergency, the standby person should use the retrieval system to quickly remove the injured person.

5.12.6 Safety Training

The importance of proper safety training cannot be emphasized enough. If an emergency should arise, sampling personnel must be able to determine the cause of the accident and be prepared to act accordingly. Far too often, the rescuer, because of inadequate preparation, becomes another victim. Safety procedures should become routine, and the best way to accomplish this is through continuous training. Training in the following areas should be provided to all employees required to work in confined space or with a team that performs work in a confined space:

- Cardiopulmonary resuscitation and basic first aid
- Use of applicable safety equipment
- Confined space entry procedures

Entry and exit procedures include the following steps:

- *Identification* – Sampling personnel must be able to identify the potential hazards associated with inspections and sampling. Training should emphasize the possible results of improper safety procedures.
- *Testing* – All confined spaces must be tested prior to entry.
- *Evaluation* – Tests must be evaluated for oxygen level, explosivity, and potentially toxic substances. Sampling and working personnel should also consider necessary safety equipment.
- *Monitoring* – The atmosphere in a confined space is subject to change. Therefore, the area should be continuously monitored during the sampling and other activities.

Rescue procedures must be developed for each type of confined space that may be encountered by the sampling personnel. A written record of training and safety drills should be kept. Rescue procedures should be practiced frequently enough to ensure proficiency in any necessary rescue situations. In developing a successful training program, the industry is encouraged to call on other agencies (e.g. local fire and rescue department) with expertise in any of the areas described above.

Also, a summary of chemical and allied industrial plant hygiene method are given in Table E.1.

5.13 Noise

Noise, commonly defined as unwanted sound, is an environmental phenomenon to which we are exposed before birth and throughout life. Noise is considered as an environmental pollutant, a waste product generated in conjunction with various anthropogenic activities. Noise is any sound – independent of loudness – that can produce an undesired physiological or psychological effect in an individual and that may interfere with the social ends of an individual or group. These social ends include all of our activities – communication, work, rest, recreation, and sleep.

5.13.1 Occupational Noise Exposure

Noise, or unwanted sound, is one of the most pervasive occupational health problems. It is also a by-product of many industrial processes. Sound consists of pressure changes in a medium (usually air), caused by vibration or turbulence. These pressure changes produce waves emanating away from turbulence or vibrating source. Exposure to high levels of noise causes hearing loss and may cause other harmful health effects as well. The extent of damage depends primarily on the intensity of the noise and the duration of the exposure. Noise-induced hearing loss can be temporary or permanent. Temporary hearing loss results from short-term exposure to noise, with normal hearing returning after a period of rest. Generally, prolonged exposure to high-noise levels over a period of time gradually causes permanent damage. OSHA's hearing conservation program is designed to protect workers with significant occupational noise exposures from hearing impairment even if they are subject to such noise exposures over their entire working lifetimes.

5.13.2 Basics of Occupational Noise and Hearing Protection

An estimated 25 million Americans – many of whom are workers – are exposed to noise that poses a threat to their hearing. Everyone at some time or another has experienced the effects of noise pollution. Many people are unaware that the sounds that cause them so much serious health threats are a result of noise pollution. Noise pollution is traditionally not placed among the top environmental problems facing the nation; however, it is one of the more frequently encountered sources of pollution in everyday life. Noise is typically thought of as a nuisance rather than a source of pollution. This is due in part because noise does not leave a visible impact on the environment as do other sources of pollution.

The damage done by the traditional pollution of air and water is widely recognized. The evidence is right before one's eyes – in contaminated water, algae bloom, oil spills, and dying fish, as well as in smog that burns the eyes and sears the lungs. However, noise is more subtle pollutant. It is estimated over 20 million Americans are exposed to dangerous noise levels without knowing it from trucks, airplanes, motorcycles, stereos, lawnmowers, and kitchen appliances. Sound travels in waves through the air like waves through water. The higher the wave, the greater its power. The greater the number of waves a sound has, the greater is its frequency and pitch.

The strength of sound or sound level is measured in decibels (dB). The decibel scale ranges from 0, which is regarded as the threshold of hearing for normal, healthy ears, to 94, which is regarded as the theoretical maximum for pure tones. Since the decibel scale, like the pH scale, is logarithmic, 20 dB is 100 times louder than 0; 30 dB is 1 000 times louder; 40 dB is 10 000 time louder, etc. Thus at high levels, even a small reduction in level values can make a significant difference in noise intensity.

5.13.3 Noise: Physical Principles

The frequency of sound refers to the rate at which a complete cycle of high- and low-pressure regions is produced by the source. This is measured in hertz (Hz): 1 Hz is one complete cycle per second. The range of audible frequencies for a young person is from 20 Hz to 20 KHz. As a person age or is exposed to excessive noise, the ability to hear the higher frequencies is reduced. Older adults may have an effective cutoff of 10 KHz or less.

The sound intensity (I) is the average rate at which sound energy is transmitted through a unit area normal to the direction of sound propagation. It is expressed as an energy fluency rate, normally in terms of decibels. The decibel has been defined as 10^{-12} W/m^2. Sound intensity is a function of the sound power (P) of the source. The relationship of sound intensity to sound power is given by

$$I = P / 4\pi d^2 \tag{5.12}$$

Sound power should not be confused with sound pressure (p). Sound pressure refers to the root mean square (rms) value of the pressure changes above ambient atmospheric level produced by a sound source. It is typically measured in units of Pascal, Newton per square meter (N/m^2) or dB. The dB, as used for sound pressure, is referred to 20 μPa or 2×10^{-5} N/m^2, which is considered the threshold of hearing for normal person. The range of sound pressures encountered in everyday life is quite broad. Representative levels are shown in Table 5.12.

It is measured by a sound meter level instrument, sound pressure level (SPL), L_p, is characterized in units of dB and is expressed in dB.

$$L_p = 20\log\left(\frac{p}{p_0}\right) \tag{5.13}$$

where p is the measured rms sound pressure and p_0 is the reference value, 0.000 02 N/m^2.

The relationship between sound intensity I and sound reference intensity is

$$L_I = 10\log\left(\frac{I}{I_0}\right) \tag{5.14}$$

The reference intensity, I_0, is 10^{-12} W/m^2.

Table 5.12 Representative sound pressure levels.

Activity	dB	N/m²	Watts
Maximum perception level	1	0.000 02	10^{-12}
Night, quiet rural location	20	0.000 2	10^{-10}
Quiet room	40	0.002	10^{-8}
Normal conversation (3 ft)	60	0.02	10^{-6}
Heavy vehicular freeway traffic (automobiles, 50 ft)	80	0.20	10^{-4}
Inside helicopter	90	0.70	10^{-3}
Power lawn mower	95	1.0	0.04
Jet aircraft	110	7.0	0.10
Pain threshold	120	20	1.0

Example 5.15 Sound Pressure/Intensity Relationship
A new noise source with a sound level of 83 dB is installed in a plant with background noise level of 75 dB. Will the combined level exceed the 8-hour day duration TLV for noise?

Solution
The ACGIH guide gives the 8-hour noise TLV as 85 dB. The calculation can be done by using Eq. (5.14).

$L = \log(I/I_0)$ dB; the reference value $I_0 = 10^{-12}$ W/m² (Table 5.12) and calculating pressure from the new source

$$L = 83 = 10\log\left(\frac{I}{10^{-12}}\right) \quad \text{or} \quad \frac{83}{10} = \log\left(\frac{I}{10^{-12}}\right)$$

Taking antilog and rearranging gives $I = 2\times10^{8}\times10^{-12} = 2\times10^{-4}$.

Performing the same operation for the background SPL gives

$75 = 10\log(I/10^{-12})$
$3.2\times10^{7}\times10^{-12} = I$; therefore, $I = 3.2\times10^{-5}$.

Adding the two sound intensities gives $2\times10^{-4} + 0.32\times10^{-4} = 2.32\times10^{-4}$ which can then be used to determine the combined SLP.$L = 10\log(2.32\times10^{-4}/10^{-12}) = 10\log(2.32\times10^{8}) = 10\times8.37 = 83.7$ dB (average) which is less than 85 dB (average) TLV limit.

5.13.4 Noise Exposure and Noise Protection

For continuous exposure to noise over an 8-hour day, OSHA has a set limit of 90 dB average as an 8-hour TWA. Employers are required to provide a hearing conservation program, noise monitoring, annual audiograms, and hearing protection for employees whose average daily 8-hour exposure exceeds 85 dBA. Higher levels are permitted for shorter intervals; a 5 dBA increase is allowed up to a limit of 115 dBA for each half of exposure time (0.4 h/15 min). Thus, 95 dBA would be permitted for 4 h/day, 100 dBA for 2 hours, and so on. In general, noise levels above 85 dBA make telephone use impossible and voice communication very difficult and limited, requiring shouting at a distance of only 2 to 3 ft.

Employers must provide hearing protectors to all workers exposed to 8-hour TWA noise level of 85 dB or above. This requirement ensures that employees have access to protectors before they experience any hearing loss:

1) For any period exceeding 6 months from the time they are first exposed to 8-hour TWA noise levels of 85 dB or above until they receive their baseline audiograms.
2) If they have incurred standard threshold shifts that demonstrate they are susceptible to noise.
3) If they are exposed to noise over the permissible exposure limit of 90 dB over an 8-hour TWA.

Example 5.16 Noise Exposure
A worker is exposed to a noise level of 94 dBA for 30 minutes and to 88 dBA for 3 hours each day. Does this exposure exceed the TLV for noise?

Solution
The TLV for 94 dBA noise is 1 hour.
The TLV for 85 dBA is 4 hours.
The combined noise TLV can be expressed by

$$C_1/T_1 + C_2/T_2 + \cdots + C_n/T_n = \text{fraction of TLV}$$

For exposure of two noise levels,

$$0.5/1 + ¾ = 1.25,$$

which is greater than 1; therefore, the TLV is exceeded.

5.13.5 Noise Control

Noise control is affected in a number of ways. Machinery can be redesigned to reduce noise. The basic principle is similar to controlling indoor air pollution: eliminating the source of the noise. This can be accomplished through improved machinery design and maintenance or by damping vibrations by increasing stiffness or using rubber or plastic bumpers or shock absorbers. Larger, slower machines typically produce less noise than smaller and faster devices. Belt drives can be used to replace gears and cams, and hydraulic or electric devices can be substituted for mechanical ones.

Noise screens of various types can be designed and constructed. Noise baffles are effective, particularly for higher frequencies. Cosmetic correction, such as tree plantings or visual screens, may provide better acceptance of noise levels, but do little to reduce actual SPLs. Distance is a very effective noise control strategy. It is noted that higher sound frequencies are more easily attenuated than lower frequencies.

There are other means and ways to reduce noise from moving parts and machineries as follows:

- Reduce frictional resistance
- Reduce rotating areas
- Reduce noise leakage
- Isolate and dampen vibrating elements
- Provide mufflers or other silencers
- Noise control in the transmission path
- Apply noise absorbing materials
- Enclosures

Example 5.17 Noise Control

A turbine has an acoustic power output of 10 W. A workstation is to be located 10 ft from the turbine in an acoustically shielded booth providing a 29 dB reduction. What will the SPL be in the booth? Will it be low enough to permit normal conversation?

Solution

For a free-field, omnidirectional source, use Eq. (5.12) to calculate the intensity at $d = 10\,\text{ft}$.

$$\begin{aligned} I &= 10/4\pi(3.05)^2 \left[\begin{array}{l}\text{converting } d \text{ to meter:} \\ 10\,\text{ft} \times 0.305\,\text{m/ft} = 3.05\,\text{m}\end{array}\right] \\ &= 0.086\,\text{W/m}^2 \end{aligned}$$

Converting to sound pressure using Eq. (5.14):

$$L_\text{t} = 10\log(I/I_0) = 10\log\left(0.086/10^{-12}\right) = 109.3\,\text{dB}$$

With no attenuation

$$\text{SPL inside booth} = 109.3 \quad 29 = 80.3\,\text{dB}$$

80.3 dB (unweighted), which is higher than 60 dB for normal conversation (Table 5.12); therefore, it will interfere with normal conversation.

5.14 Radiation

5.14.1 Definition

Radiation is defined as the propagation of energy through a vacuum or matter. Any electromagnetic or particulate radiation capable of producing ions, directly or indirectly, by interaction with matter is referred to as ionization radiation. Medical uses generally involve X-rays and gamma rays as well as high-energy electrons and low-energy beta particles. The corpuscular emission from radioactive substances or other sources, commonly alpha particles, beta particles, heavy ions, and neutrons, are also forms of ionizing radiation. Radiation in wave form is classified according to wavelength. The relationship of radiation wavelength, velocity, and frequency is shown by

$$\text{Wavelength of radiation} = \frac{\text{velocity of radiation}}{\text{frequency of radiation}} \tag{5.15}$$

5.14.2 Different Sources of Radiation

The sun is a great nuclear furnace. It makes most of the cosmic radiation that strikes the outside of the Earth's atmosphere. The atmosphere stops most of the cosmic rays. People who live at high elevations or crews of jet airplanes have less atmosphere above them and they get more cosmic-ray radiation than other people. People are also exposed to man-made radiation. Most comes from medical and dental use of X-rays and gamma rays. These are equal on average to about one-fourth of the natural background.

We can make uranium atoms split apart (fission) to generate large amounts of energy. Fission also produces large amounts of radioactive isotopes ("fission products") and neutrons. Stable atoms and neutrons can combine to produce radioactive isotopes. Most fission products give off beta particles and gamma rays. Beta particles are high-energy electrons. Gamma rays are waves of high energy. Some radioactive atoms give off alpha particles. Alpha particles are high-energy particles made up of two protons and two neutrons. They are thousands of times heavier than beta particles.

5.14.3 External Exposure and Internal Exposure

External exposure: Gamma rays can penetrate into or through the body. Beta particles can penetrate through the skin. Alpha particles cannot penetrate even the skin's outer layer. Except for the eye, alpha particles are harmless from outside the body. Beta particles are stopped by thin layers of common light materials like paper or aluminum. Thick layers of common materials like concrete or layers of very dense material like lead are needed to stop gamma rays.

Internal exposure: Any radioactive materials can enter the body with food or drink (ingestion), or by breathing (inhalation), or through an open wound (dermal). If that happens, any kind of radiation can directly harm living cells. There are differences in the damage a particular amount of radioactive material can do. That depends on how quickly it decays and the kind of radiation it produces. It also depends on the kind of element it is. Some elements decay quickly, while some elements are eliminated by normal body functions. But some are long-lived and attach to tissues. They can stay in the body for a long time. Some stay in the body indefinitely.

5.14.4 Radionuclide Decay

All radionuclides, i.e., alpha, beta, etc. are characterized by a specific physical or radiological half-life, the time required for half of the radioactive nuclei to decay. Radioactive decay is an exponential process and is described by the fundamental decay law, which is expressed mathematically as

$$A = A_0 e^{-\lambda t} \tag{5.16}$$

where A refers to the activity at any time t, and A_0 refers to the activity at zero time. The decay constant λ is expressed in units of reciprocal time and is an indication of the rate or probability of decay. The larger the value of λ, the shorter the half-life, $T_{½}$, which is expressed by

$$T \;= 0.693/\lambda \tag{5.17}$$

A rule of thumb is that after 7 half-lives, the activity is reduced to less than 1% of the original amount, and after 10 half-lives, to less than 0.1%. Activity refers to the number of radioactive transformation per unit of time. In the old system of units, activity was expressed in curie (Ci) units. One Ci was defined as 3.7×10^{10} disintegration per second. The comparable unit in the SI unit is Becquerel (Bq), which is simply one disintegration per second.

Example 5.18 Radioactive decay
How long will it take for 750 Bq of a radionuclide to decay to 200 Bq if the decay constant is 0.086/day? What is the half-life?

Solution
$A_0 = 750\,\text{Bq}$
$A = 200\,\text{Bq}$
$\Lambda = 0.086/\text{day}$
Time to decay, using Eq. (5.16):

$$A = A_0 5^{-\lambda t};\quad 200 = 750^{-0.086 \times t};\quad \ln(200/750) = 0.086t$$
$$t = 15.4\,\text{days}$$

Half-life, using Eq. (5.17)

$$T \;= \frac{0.693}{0.086} = 8.06\,\text{days}$$
$$= \frac{0.693}{0.086} - 8.06\,\text{days}$$

5.14.5 Radiation Dose

Fundamentally, the harmful consequences of ionizing radiations to a living organism are due to the energy absorbed by the cells and tissues of the organism. The absorbed energy or dose produces chemical decomposition of the molecules present in the living cells. The mechanism of the decomposition appears to be related to ionization and excitation interactions between the radiation and atoms within the tissue. The amount of ionization or number of ion pairs produced by ionizing radiations in the cells or tissues provides some measure of the amount of decomposition or physiological damage that might be expected from a given quantity or dose. The ideal basis for radiation dose measurement could be, therefore, the number of ion pairs (or ionizations) taking place within the medium of interest. For certain practical reasons, the medium chosen for defining exposure does is air.

5.14.5.1 Radiation Dose Units

Exposure Dose: The Roentgen The exposure dose of X-rays or gamma radiation within a unit used for expressing the exposure to X-rays or gamma radiation is the roentgen (R). Its merit lies in the fact that the magnitude of the exposure does in roentgens can usually be related to the absorbed dose, which is of importance in predicting or quantifying the expected biological effect (or injury) resulting from the radiation.

The *roentgen* is an exposure dose of X-rays or gamma radiation such that the associated corpuscular emission per 0.001 293 g of air produces, in air, ions carrying one electrostatic unit of quantity of electricity of either sign. Note that 1 cm^3 of air at STP has a mass of 0.001 293 g.

Absorbed Dose: The Gray The absorbed dose of any ionizing radiation is the energy imparted to matter by ionizing radiation per unit mass of irradiated material at the place of interest. The SI unit of absorbed dose is the *gray* (Gy). One gray is equivalent to the absorption of 1 J/kg (Joule per kilogram). The former unit of absorbed dose was the rad. One *rad* is equivalent to the absorption of 100 ergs/g. 1 Gy – 100 rads. The Seivert (Sv) is obtained by multiplying the absorbed dose in GY by a radiation weighting factor.

5.14.6 Biological Effects of Ionizing Radiation

Radioactive materials is taken into the body are removed by two mechanisms: physical decay and biological processes, each with its own characteristic half-life. The combined removal by these two mechanisms is known as the effective half-life, T_{eff}, which is determined by

$$T_{eff} = \left(\frac{T_{rad} \times T_{bio}}{T_{rad} + T_{bio}}\right) \tag{5.18}$$

The effective half-life is always smaller than either the biological or radiological half-life. Or the biological half-life is very much smaller than the other, it will approximate the effective half-life.

Example 5.19 Biological Half-Life
A worker inhales a radionuclide in a single acute exposure. Body fluid sample shows a half-life of 6.1 days. The nuclide has a decay constant of 0.08/day. What is the biological half-life in the worker?

Solution
The radiological half-life is determined by the Eq. (5.17):

$$T = 0.693/\lambda = 0.693/0.08 = 8.7\,\text{days}$$

The biological half-life is obtained from Eq. (5.18):

$$\text{T}_{eff} = \left(\frac{T_{rad} \times T_{bio}}{T_{rad} + T_{bio}}\right)$$

$$6.1 = \left(\frac{8.7 \times T_{bio}}{8.7 + T_{bio}}\right)$$

$$6.1(8.7) + 6.1T_{biol} = 8.7T_{biol}$$

$$53.1 = 8.7T_{biol};\ -6.1T_{bio} = 2.6T_{bio};\ \text{Therefore, } T_{biol} = 20.41\,\text{days}$$

5.14.7 Radiation Protection Principles

The basic radiation exposure limit is 5 rem (50 mSv) annually, referred to as the *maximum permissible dose* (MPD). Doses from all sources, whether from internal or external radiations, are included. There are specific *annual limit on intake* (ALI) established for various radionuclides. Intake of one ALI will produce an MPD equivalent of 5 rem (50 mSv). Permissible air concentration for specific radionuclides are obtained by dividing the ALI by the annual breathing rate and converted to activity per volume of air. This is known as the *derived air concentration* which is commonly expressed in units of μCi/ml in the old system or Bq/m^3 in the SI unit.

The fundamental decay law described above has the same form as the shielding equation for X-rays and gamma rays. The basic shielding or attenuation may be expressed as follows:

$$-\frac{dI}{dx} = \mu I_0 \tag{5.19}$$

where
- dI = reduction of radiation
- I_0 = incident radiation
- μ = proportionality constant
- dx = thickness of absorber or shielding material

Integrating yields

$$I = I_0 5^{-\mu x} \tag{5.20}$$

The ratio I/I_0 is sometimes called the transmission. The linear attenuation coefficient, μ, is a specific function of the energy of the photons and the shielding material, and it is usually given in units of reciprocal centimeter. This is the linear attenuation coefficient divided by the density and symbolized by μ/x. If the mass attenuation coefficient is used, the thickness of the shielding material has to be expressed in comparable reciprocal units, which are obtained by multiplying the linear thickness by the density. Tables and graphs are available that give values of μ determined experimentally or that give transmission values for varying thickness or different shielding materials (Figures 5.17 and 5.18).

5.14.7.1 Reduction of External Radiation Hazards

Three fundamental methods are employed to reduce external radiation hazards: shielding, distance, and reduction of exposure time.

Shielding Shielding is one of the most important methods for radiation protection. It is accomplished by placing some absorbing material between the source and the person to be protected. Radiation is attenuated in the absorbing medium. Typical shielding materials used are lead, concrete, and iron, depending on the sources of radiation.

Distance Distance is not only very effective, but also in many instances the most easily applied principle of radiation protection. Beta particles of a single energy have a finite range in air. Sometimes the distance afforded by the use of remote control handling devices will supply complete protection.

The inverse square law for reduction of radiation intensity applies for point sources of X-rays, gamma, and neutron radiation. The inverse square law states that radiation intensity from a point varies inversely as the square of the distance from the source.

$$\frac{I_1}{I_2} = \frac{(R_2)^2}{(R_1)^2} \tag{5.21}$$

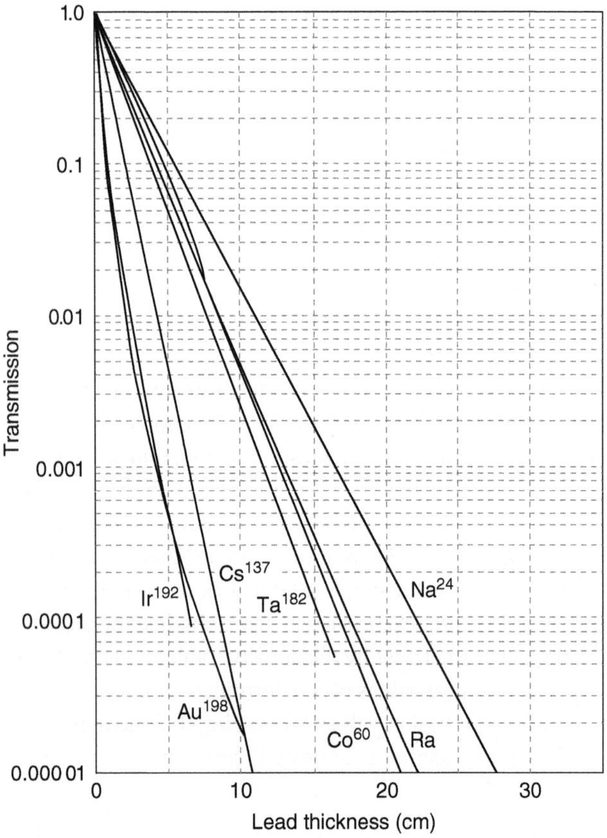

Figure 5.17 Lead shielding material thickness vs. radiation transmission.

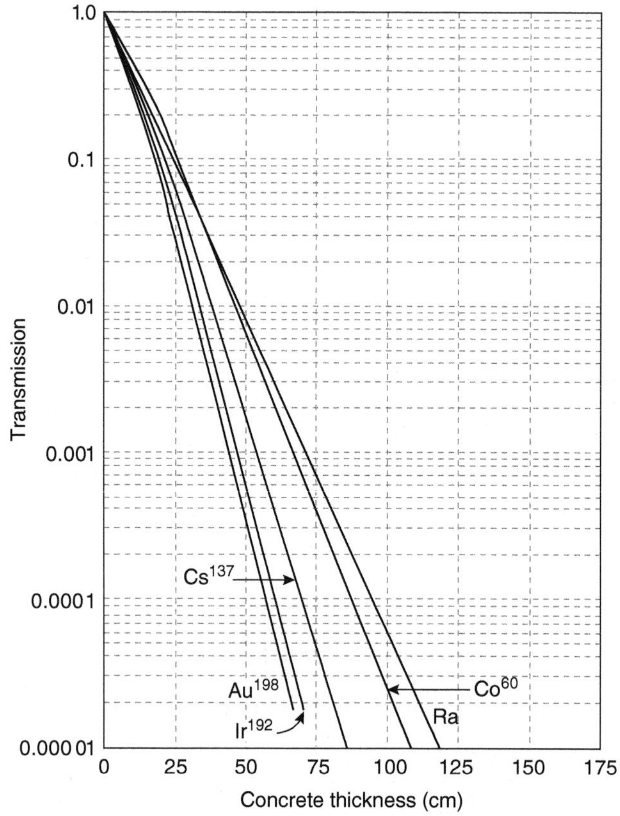

Figure 5.18 Concrete shielding material thickness vs. radiation transmission.

where I_1 is the radiation intensity at distance R_1 from the source, and I_2 is the radiation intensity at distance R_2 from the source. The inverse law does not apply to extended sources or to radiation fields from multiple sources.

Reduction of Exposure Time By limiting the duration of exposure to all radiation sources and by providing ample recuperative time between exposures, the untoward effects of radiation can be minimized. In emergency situations, it may occasionally be necessary to work in areas of very high dose rates. This can be done with safety by limiting the total exposure time so that the average permissible value for a day based on the radiation protection guide dose of 1×10^{-3} Gy (0.1 rad) per week is not exceeded.

Example 5.20 Lead Thickness
What thickness (in cm) of lead is required to shield a ^{60}Co source so that the transmission is reduced to 99.5%?

Solution
Given:
^{60}Co source, transmission reduction of 99.5% required using lead shield
Compute transmission allowed

a) $1.000 - 0.995 = 0.005$
b) From Figure 5.17 using ^{60}Co line
Lead thickness required = 8.5 cm
(Allow some tolerance for graph reading)

Example 5.21 Concrete Thickness
What is the equivalent thickness (in cm) of concrete to accomplish the same shielding as the lead in Problem 5.20?

Solution

a) Using a transmission of 0.005 from Problem 5.20 and ^{60}Co line in Figure 5.18
Concrete thickness = 53 cm
(Allow some tolerance for graph reading).

Example 5.22 Radioactive Decay
When an X-ray unit is operated at 70 kV and 5 mA, it produces an intensity of I_1 per minute at 1.0 m from the source. What intensity will it produce at 2.0 m from the source?

Solution
Given:
X-ray unit operated at 70 kVp and 5 mA produces intensity of I_1 per minute at 1.0 m from source

Apply Eq. (5.21)

$$\frac{I_1}{I_2} = \frac{(R_2)^2}{(R_1)^2}$$

$$\frac{I_1}{I_2} = \frac{(2.0)^2}{(1.0)^2}$$

$$I_2 = \frac{1.0}{4.0} I_1 = 0.25 I_1$$

Example 5.23 Radioactive Decay

As described earlier, exponential first-order decay can be described using either a reaction rate coefficient (k) or a half-life (τ). The equation that relates these two parameters is as follows:

$$\tau = 0.693 / k$$

Exponential decay can be expressed by the following equation:

$$N = N_0 5^{-kt}$$

where

k = reaction rate coefficient (per time)
N_0 = initial amount
N = amount at time t

Determine how much of a 100-g sample of P_0-210 is left after 5.52 days using:

1) The reaction rate coefficient
2) The half-life
3) Calculate the percent error between the two methods

The half-life for P_o-210 is 1.38 days.

Solution

1) Determine the reaction rate coefficient:
   $k = 0.693/\tau - 0.693/1.38\,\text{day} = 0.502/\text{day}$
   The amount of substance left after 5.52 days is as follows:
   $N = N_0 e^{-kt} = 100 e^{-0.502 \times 5.52} = 6.26\,\text{g}$
2) The first step is to determine how many half-lives the 100 g sample has undergone in the given time period:
   Number of half-lives = (5.52 days)/(1.38 days) = 4.0
   Therefore, in a 5.52 period, the 100 g sample has undergone 4 half-lives.
   The amount of the substance left after one half-life is calculated as follows:
   $(0.5)^1(100\,\text{g}) = 50\,\text{g}$
   Therefore, the amount of substance left after four half-lives is
   $(0.5)^4(100\,\text{g}) = 6.25\,\text{g}$
3) Since the two equations are, in principle, identical to each other, there is no difference between the two values. The small difference between the two results arises because of round-off errors.

5.15 Effects of Global Warming: Climate Change – The World's Health

Roughly 40 years ago, a small group of scientists and policy makers began to realize that humanity was on a dramatic collision course, as the rapidly growing world economy and population threatened to collide with the planet's finite resources and fragile ecosystems. The danger was first highlighted globally at the 1972 UN Conference on the Human Environment in Stockholm. A famous and influential book published that same year, *Limits to Growth*, warned that business as usual could lead to a collapse in the twenty-first century.

The globe is heating up. Both land and oceans are warmer now than record-keeping began in 1880, and temperatures are still ticking upward. This temperature rise, in a nutshell, is global warming. Altogether, climate change is therefore one very tough issue, and time is running out! The emissions of the main GHGs that lead to human-induced climate change are increasing each year, and the threats to the planet are growing as well. We are losing time even though the stakes for the planet are incredibly high.

Here are the bare numbers, according to the NOAA. Average surface temperatures rose a total of 0.95 °C (1.71 °F) between 1880 and 2016. The pace of change has been an additional 0.07 °C (0.13 °F) per decade, with the land surface warming faster than the ocean surface – 0.10 °C (0.18 °F) vs. 0.06 °C (0.11 °F) per decade, respectively. The average global temperature anomaly for the past 140 years is plotted in Figure 3.2. As can be seen in Figure 3.2, the bulk of the warming has occurred in two periods – from 1910 to 1940 and from 1980 to the present.

The Paris Agreement, ratified by 174 nations of 197 nations as of the summer 2017, aims to halt that warming at 1.5 °C (2.7 °F) above Earth's average temperature during preindustrial times – a goal most scientists and policy makers agree will be a challenge to meet. Here's how humanity has managed to heat up the planet.

5.15.1 The Greenhouse Effect

The basic physics and human-induced climate change were already worked out by scientists in the nineteenth century. One great scientific genius, Svante Arrhenius, a Swedish Noble laureate in chemistry, calculated accurately by hand, without a computer, the effects of doubling the atmospheric concentration of carbon dioxide (CO_2) (Arrhenius 1896). And he did so back in 1896! He correctly calculated that a doubling of the CO_2 in the atmosphere would cause a rise in the mean temperature of the planet at around 5 °C, an estimate that is within the likely range today based on advanced computer models and vastly more extensive data than Arrhenius had at his disposal.

The main driver of today's warming is the combustion of fossil fuels. The main contributor to global warming and climate change over the past century has been carbon dioxide; however, three other gases are now significant – methane, nitrous oxide, and CFCs. These hydrocarbons heat up the planet via the greenhouse effect, which is caused by the interaction between Earth's atmosphere and incoming radiation from the sun. According to Weiner (1990) and Pilson (2006), Arrehenius summed all up as follows:

We are evaporating our coal mines into the air.....[which must eventually cause] a change in the transparency of the atmosphere.

5.15.2 Greenhouse Gases

Since the beginning of the industrial revolution, humans have been rapidly changing the balance of gases in the atmosphere. Burning fossil fuels like coal and oil releases water vapor, carbon dioxide (CO_2), methane (CH_4), ozone (O_3), and nitrous oxide (N_2O) – the primary GHGs. Carbon dioxide is the most common GHG. Between about 800 000 years ago and the beginning of the Industrial Revolution, its presence in the atmosphere amounted to about 280 ppm. Today, it's about 405 ppm (Figure 5.19). (This number means there are 405 molecules of carbon dioxide in the air per every million air molecules.)

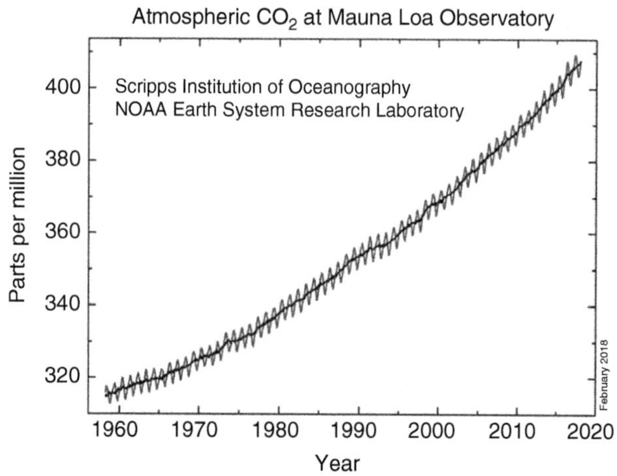

Figure 5.19 Global atmospheric CO_2 rise during last 60 years. *Source:* From National Oceanic a Atmospheric Administration, Earth System Research Laboratory (2019).

Example 5.24 Calculating GHG Mass-Based and CO_2e (Equivalent)-Based Emissions

For any source, since GHG emissions may be a mixture of up to six compounds, the amount of GHG emissions calculated for the PSD permit applicability analysis is a sum of the compounds emitted at the emissions unit. The following example illustrates the method to calculate GHG emissions on both a mass basis and CO_2e basis.

A proposed emissions unit emits five of the six GHG compounds in the following amounts:

50 000 TPY of CO_2
60 TPY of methane
1 TPY of nitrous oxide
5 TPY of HFC-32 (a hydrofluorocarbon)
3 TPY of PFC-14 (a perfluorocarbon)

The GWP for each of the GHGs used in this example are as follows:

GHG	GWP[a]
Carbon dioxide	1
Nitrous oxide	310
Methane	21
HFC-32	650
PFC-14	6500

[a] see 40 CFR Part 98, Subpart A, Table A.1.

The *GHGs mass-based emissions* of the unit are calculated as follows:

$$50\,000\,\text{TPY} + 60\,\text{TPY} + 1\,\text{TPY} + 5\,\text{TPY} + 3\,\text{TPY} = 50\,069\,\text{TPY of GHGs}$$

The *CO_2e-based emissions* of the unit are calculated as follows:

$$\left(50\,000\,\text{TPY} \times 1\right) + \left(60\,\text{TPY} \times 21\right) + \left(1\,\text{TPY} \times 310\right) + \left(5\,\text{TPY} \times 650\right) + \left(3\,\text{TPY} \times 6500\right)$$

$$= 50\,000 + 1260 + 310 + 3250 + 19\,500 = 74\,320\,\text{TPY}\,CO_2\text{e}$$

In 2015, CO_2 accounted for about 82% of all US GHG emissions, according to an EPA inventory. "We know through high-accuracy instrumental measurements that there is an unprecedented increase in CO_2 in the atmosphere." We know that CO_2 absorbs infrared radiation [heat] and because of that the global mean temperature is increasing. CO_2 makes its way into the atmosphere through a variety of routes. Burning fossil fuels releases CO_2 and is by far the primary way that US emissions warm the globe. According to the USEPA's 2015 report, US fossil-fuel combustion, including electricity generation, releases just over 5.5 billion T (5 billion MT) of CO_2 into the atmosphere annually. Other processes – such as nonenergy use of fuels, iron and steel production, cement production, and waste incineration – boost the total annual CO_2 release in the United States to almost 6 billion T (5.5 billion MT).

Example 5.25 CO_2 emission from cars

Assume an average car in the United States gets 20 miles/gal of gasoline, is driven 12 000 miles/year, and weighs

3 500 lb. Further assume that gasoline weighs 5.9 lb/gal and contains 85% carbon by weight. Is there any truth to the statement that each car emits its weight in carbon dioxide each year? Next, given that there are about 800 million vehicles worldwide, estimate the annual global carbon emissions from motor vehicles. Give your answer in Teragrams (1 Tg = 1 trillion grams = 1 MT) per year.

Solution

The carbon contained in the gasoline burned annually is

$$\left(1200\,\text{mi/year}\right)\times\left(1\,\text{gal}/20\,\text{mi}\right)\times\left(5.9\,\text{lb/gal}\right)\times 0.85$$
$$= 3010\,\text{lb C/year}$$

The carbon dioxide emitted is

$$\left(3010\,\text{lb C/year}\right)\times\left(44\,\text{lb CO}_2/12\,\text{lb C}\right)=11040\,\text{lb CO}_2/\text{year}$$

So the average US car emits much more than its own weight in carbon dioxide each year!

To estimate worldwide emissions from vehicles, we must make a number of gross assumptions as to the average vehicle in the world (including cars, trucks, buses, moto cycles, and so on). Let us assume that the average vehicle in the world travels 24 000 km/year, gets 9 km/l, and burns fuel with a density of 0.75 kg/l and with a carbon content of 87%. With these assumptions, annual carbon emissions

$$\left(24\,000\,\text{km/year}\right)\times\left(1\,\text{l}/9\,\text{km}\right)\times\left(0.75\,\text{kg/l}\right)$$
$$\times\, 0.87\times 800\times 10^6\ \text{vehicles}$$
$$\times\left(1\,\text{Tg}/10^9\,\text{kg}\right)=1392\,\text{Tg/year}$$
$$\left(\text{about } 1.4\ \text{billion MT/Y or about } 1.4\,\text{Gt/year}\right).$$

Deforestation is also a large contributor to excessive CO_2 in the atmosphere. In fact, deforestation is the second largest anthropogenic (human-made) source of carbon dioxide. When trees are killed, they release the carbon they have stored during photosynthesis. According to the 2010 Global Forest Resources Assessment, deforestation releases nearly a billion tons of carbon into the atmosphere per year.

Methane is the second most common GHG, but it is much more efficient at trapping heat. In 2012, the gas accounted for about 9% of all US GHG emissions, according to the EPA. The EPA reports that methane has 20 times more impact than carbon dioxide on climate change over a 100-year period.

5.15.3 Are the Effects of Global Warming Really Concerns for Our Future?

Short answer: Yes. Even a seemingly slight average temperature rise is enough to cause a dramatic transformation of our planet.

Eight degrees Fahrenheit – it may not sound like much, perhaps the difference between wearing a sweater and not wearing one on an early-spring day. But for the world in which we live, which climate experts project will be at least 8 °F warmer by 2100 should global emissions continue on their current path, this small rise will have grave consequences, ones that are already becoming apparent, for every ecosystem and living thing – including us.

According to the National Climate Assessment, human influences are the number one cause of global warming, especially the carbon pollution we cause by burning fossil fuels and the pollution-capturing we prevent by destroying forests. The carbon dioxide, methane, soot, and other pollutants we release into the atmosphere act like a blanket, trapping the sun's heat and causing the planet to warm. Evidence shows that the years 2000–2009 was hotter than any other decade in at least the past 1300 years. This warming is altering the Earth's climate system, including its land, atmosphere, oceans, and ice, in far-reaching ways.

5.15.4 More Frequent and Severe Weather

Higher temperatures are worsening many types of disasters, including storms, heat waves, floods, droughts, and wildfire. A warmer climate creates an atmosphere that can collect, retain, and drop more water, changing weather patterns in such a way that wet areas become wetter and dry areas drier. "Extreme weather events are costing more and more," according to the NRDC's Clean Power Plan initiative (2015). "The number of billion-dollar weather disasters is expected to rise."

According to the National Oceanic and Atmospheric Administration, in 2015 there were 10 weather and climate disaster events in the United States – including severe storms, floods, drought, and wildfires – that caused at least \$1 billion in losses. For context, each year from 1980 to 2015 averaged \$5.2 billion in disasters (adjusted for inflation). If you zero in on the years between 2011 and 2015, you see an annual average cost of \$10.8 billion.

The increasing number of droughts, intense storms, and floods we're seeing as our warming atmosphere holds – and then dumps – more moisture poses risks to public health and safety, too. Prolonged dry spells mean more than just scorched lawns. Drought conditions jeopardize access to clean drinking water, fuel out-of-control wildfires, and result in windblown dust storms, extreme heat events, and flash flooding in the States. Elsewhere around the world, lack of water is a leading cause of death and serious disease. At the opposite end of the spectrum, heavier rains cause streams, rivers, and lakes to overflow, which damages life and property, contaminates drinking water, creates hazardous-material spills, and promotes mold infestation and unhealthy air. A warmer, wetter world is also a boon for food-borne and waterborne illnesses and disease-carrying insects such as mosquitoes, fleas, and ticks.

5.15.5 Higher Death Rates

Today's scientists point to climate change as "the biggest global health threat of the twenty-first century." It's a threat that impacts all of us – especially children, the elderly, low-income communities, and minorities – and in a variety of direct and indirect ways. As temperatures spike, so does the incidence of illness, emergency room visits, and death.

5.15.6 Dirtier Air

Rising temperatures also worsen air pollution by increasing ground level ozone, which is created when pollution from cars, factories, and other sources react to sunlight and heat. Ground-level *ozone* is the main component of smog, and the hotter things get, the more of it we have. Dirtier air is linked to higher hospital admission rates and higher death rates for asthmatics. It worsens the health of people suffering from cardiac or pulmonary disease. And warmer temperatures also significantly increase airborne pollen, which is bad news for those who suffer from hay fever and other allergies.

5.15.7 Higher Wildlife Extinction Rates

As humans, we face a host of challenges, but we're certainly not the only ones catching heat. As land and sea undergo rapid changes, the animals that inhabit them are doomed to disappear if they don't adapt quickly enough. Some will make it, and some won't. According to the Intergovernmental Panel on Climate Change's (IPCC) 2014 assessment, many land, freshwater, and ocean species are shifting their geographic ranges to cooler climes or higher altitudes, in an attempt to escape warming. They're changing seasonal behaviors and traditional migration patterns, too. And yet many still face "increased extinction risk due to climate change." Indeed, a 2015 study showed that vertebrate species – animals with backbones, like fish, birds, mammals, amphibians, and reptiles – are disappearing 114 times faster than they should be, a phenomenon that has been linked to climate change, pollution, and deforestation.

5.15.8 More Acidic Oceans

The Earth's marine ecosystems are under pressure as a result of climate change. Oceans are becoming more acidic, due in large part to their absorption of some of our excess emissions. As this acidification accelerates, it poses a serious threat to underwater life, particularly creatures with calcium carbonate shells or skeletons, including mollusks, crabs, and corals. This can have a huge impact on shellfisheries. Indeed, as of 2015, acidification is believed to have cost the Pacific Northwest oyster industry nearly $110 million. Coastal communities in 15 states that depend on the $1 billion nationwide annual harvest of oysters, clams, and other shelled mollusks face similar long-term economic risks.

5.15.9 Higher Sea Levels

The polar regions are particularly vulnerable to a warming atmosphere. Average temperatures in the Arctic are rising twice as fast as they are elsewhere on Earth, and the world's ice sheets are melting fast. This not only has grave consequences for the region's people, wildlife, and plants; its most serious impact may be on rising sea levels. By 2100, it's estimated our oceans will be one to four feet higher, threatening coastal systems and low-lying areas, including entire island nations and the world's largest cities, including New York, Los Angeles, and Miami as well as Mumbai, Sydney, and Rio de Janeiro.

5.15.10 Effects of Global Warming on Humans

Climate change has brought about possibly permanent alterations to Earth's geological, biological, and ecological systems (NAP 2011).These changes have led to the emergence of a not so large-scale environmental hazards to human health, such as extreme weather (MPR 2013), ozone depletion, increased danger of wildland fires, (Tang et al. 2014) loss of biodiversity (Sahney et al. 2010a), stresses to food-producing systems, and the global spread of infectious diseases (McMichael et al. 2003). The WHO estimates that 160 000 deaths, since 1950, are directly attributable to climate change.

To date, a neglected aspect of the climate change debate, much less research has been conducted on the impacts of climate change on health, food supply, economic growth, migration, security, societal change, and public goods, such as drinking water, than on the geophysical changes related to global warming. Human impacts can be both negative and positive. Climatic changes in Siberia, for instance, are expected to improve food production and local economic activity, at least in the short to medium term. Numerous studies suggest, however, that the current and future impacts of climate change on human society are and will continue to be overwhelmingly negative.

The majority of the adverse effects of climate change are experienced by poor and low-income communities around the world, who have much higher levels of vulnerability to environmental determinants of health, wealth, and other factors, and much lower levels of capacity available for coping with environmental change. A report on the global human impact of climate change published by the Global Humanitarian Forum in 2009 estimated more than 300 000 deaths and about $125 billion in economic losses each year and indicated that most climate change induced mortality is due to worsening floods and droughts in developing countries.

5.16 Key Vulnerabilities

5.16.1 Health

Climate change poses a wide range of risks to population health – risks that will increase in future decades, often to critical levels, if global climate change continues on its current trajectory (McMichael et al. 2003). The three main categories of health risks include (i) direct-acting effects (e.g. due to heat waves, amplified air pollution, and physical weather disasters), (ii) impacts mediated via climate-related changes in ecological systems and relationships (e.g. crop yields, mosquito ecology, marine productivity), and (iii) the more diffuse (indirect) consequences relating to impoverishment, displacement, resource conflicts (e.g. water), and post-disaster mental health problems.

5.16.2 Extreme Weather Events

This trend toward more variability and fluctuation is perhaps more important, in terms of its impact on human health, than that of a gradual and long-term trend toward higher average temperature (Epstein 2002). Infectious disease often accompanies extreme weather events, such as floods, earthquakes, and drought. These local epidemics occur due to loss of infrastructure, such as hospitals and sanitation services, but also because of changes in local ecology and environment.

5.16.3 Environment

Climate change may dramatically impact habitat loss, for example, arid conditions may cause the deforestation of rainforests, as has occurred in the past (Sahney et al. 2010b).

5.16.4 Temperature

A sustained wet-bulb temperature exceeding 35 °C is a threshold at which the resilience of human systems is no longer able to adequately cool the skin. A study by NOAA from 2013 concluded that heat stress will reduce labor capacity considerably under current emissions scenarios (Dunne et al. 2013). There is evidence to show that high temperatures can increase mortality rates among fetuses and children. Although the main focus is often on the health impacts and risks of higher temperatures, it should be remembered that they also reduce learning and worker productivity, which can impact a country's economy and development.

5.16.5 Water

The freshwater resources that humans rely on are highly sensitive to variations in weather and climate. In 2007, the IPCC reported with high confidence that climate change has a net negative impact on water resources and freshwater ecosystems in all regions. The IPCC also found with very high confidence that arid and semiarid areas are particularly exposed to freshwater impacts (Kundzewicz et al. 2007).

As the climate warms, it changes the nature of global rainfall, evaporation, snow, stream flow, and other factors that affect water supply and quality. Specific impacts include the following:

- Warmer water temperatures affect water quality and accelerate water pollution (Miller 2007).
- Sea-level rise is projected to increase saltwater intrusion into groundwater in some regions. This reduces the amount of freshwater available for drinking and farming (Miller 2007; USEPA 2015).
- In some areas, shrinking glaciers and snow deposits threaten the water supply. Areas that depend on melted water runoff will likely see that runoff depleted, with less flow in the late summer and spring peaks occurring earlier (Miller 2007). This can affect the ability to irrigate crops. (This situation is particularly acute for irrigation in South America, for irrigation and drinking supplies in Central Asia, and for hydropower in Norway, the Alps, and the Pacific Northwest of North America.)
- Increased extreme weather means more water falls on hardened ground unable to absorb it, leading to flash floods instead of a replenishment of soil moisture or groundwater levels.
- Increased evaporation will reduce the effectiveness of reservoirs.
- At the same time, human demand for water will grow for the purposes of cooling and hydration.

5.16.6 Social Effects of Extreme Weather

As the World Meteorological Organization explains, "recent increase in societal impact from tropical cyclones has largely been caused by rising concentrations of population and infrastructure in coastal regions." Pielke et al. (2008) normalized mainland US hurricane damage from 1900 to 2005 values and found no remaining trend of increasing absolute damage. The 1970s and 1980s were notable because of the extremely low amounts of damage compared to other decades. The decade 1996–2005 has the second most damage among the past 11 decades, with only the decade 1926–1935 surpassing its costs. The most damaging single storm is the 1926 Miami hurricane, with $157 billion of normalized damage (Pielke et al. 2008).

The American *Insurance Journal* predicted that "catastrophe losses should be expected to double roughly every 10 years because of increases in construction costs, increases in the number of structures and changes in their characteristics." The Association of British Insurers (ABI) has stated that limiting carbon emissions would avoid 80% of the projected additional annual cost of tropical cyclones by the 2080s. The cost is also increasing partly because of building in exposed areas such as coasts and floodplains. The ABI claims that reduction of the vulnerability to some inevitable effects of climate change, for example through more resilient buildings and improved flood defenses, could also result in considerable cost-savings in the long term (ABI 2005).

5.17 Energy Sector

5.17.1 Oil, Coal, and Natural Gas

Oil and natural gas infrastructure is vulnerable to the effects of climate change and the increased risk of disasters such as storm, cyclones, flooding, and long-term increases in sea level. Minimizing these risks by building in less disaster prone areas can be expensive and impossible in countries with coastal locations or island states. All thermal power stations depend on water to cool them. Not only is there increased demand for freshwater, but climate change can increase the likelihood of drought and freshwater shortages. Another impact for thermal power plants is that increasing the temperatures in which they operate reduces their efficiency and hence their output. The source of oil often comes from areas prone to high natural disaster risks such as tropical storms, hurricanes, cyclones, and floods. An example is Hurricane Katrina's impact on oil extraction in the Gulf of Mexico, as it destroyed 126 oil and gas platforms and damaged 183 more (Urban and Mitchell 2011). However, previously pristine arctic areas will now be available for resource extraction.

5.17.2 Nuclear

Climate change, along with extreme weather and natural disasters, can affect nuclear power plants in a similar way to those using oil, coal, and natural gas. However, the impact of water shortages on nuclear power plants cooled by rivers will be greater than on other thermal power plants. This is because old reactor designs with water-cooled cores must run at lower internal temperatures and thus, paradoxically, must dump more heat to the environment to produce a given amount of electricity. This situation has forced some nuclear reactors to be shut down and will do so again unless the cooling systems of these plants are enhanced to provide more capacity. Nuclear power supply was diminished by low river flow rates and droughts, which meant rivers had reached the maximum temperatures for cooling. Such shutdowns happened in France during the 2003 and 2006 heat waves. During the heat waves, 17 reactors had to limit output or shut down. In France 77% of electricity is produced by nuclear power; and in 2009 a similar situation created a 8 GW shortage, and forced the French government to import electricity. Other cases have been reported from Germany, where extreme temperatures have reduced nuclear power production nine times due to high temperatures between 1979 and 2007.

5.17.3 Hydroelectricity

Changes in the amount of river flow will correlate with the amount of energy produced by a dam. Lower river flows because of drought, climate change, or upstream dams and diversions will reduce the amount of live storage in a reservoir and therefore reduce the amount of water that can be used for hydroelectricity. The result of diminished river flow can be a power shortage in areas that depend heavily on hydroelectric power. The risk of flow shortage may increase as a result of climate change. Studies from the Colorado River in the United States suggest that modest climate changes (such as a 2 °C change that could result in a 10% decline in precipitation) might reduce river run-off by up to 40%. Brazil, in particular, is vulnerable due to its having reliance on hydroelectricity as increasing temperatures, lower water flow, and alterations in the rainfall regime, could reduce total energy production by 7% annually by the end of the century (Urban and Mitchell 2011).

5.17.4 Transport

Roads, airport runways, railway lines, and pipelines (including oil pipelines, sewers, water mains etc.) may require increased maintenance and renewal as they become subject to greater temperature variation. Regions already adversely affected include areas of permafrost, which are subject to high levels of subsidence, resulting in buckling roads, sunken foundations, and severely cracked runways.

Problems

5.1 A RCRA landfill is characterized by diurnal variations in the volatilization of contaminants as a result of changes in temperatures, wind speed, and activity at the site. The most prevalent contaminant, benzene, follows the pattern:

Duration (hour)	Benzene concentration (ppm)
8–9 a.m.	5
9–10 a.m.	8
10–11 a.m.	8
11 a.m.–12 p.m.	9
12–1 p.m.	11
1–2 p.m.	14
2–3 p.m.	11
3–4 p.m.	10

Determine the TWA for benzene concentrations at the landfill.

5.2 An air sample collected at a RCRA 90-day storage facility contains 8 ppm toluene, 12 ppm PCE, and 25 ppm MEK.
A) What is the TLV–TWA for the mixture?
B) Has the TLV been exceeded?

5.3 Determine the mixture TLV–TWA for a worker exposed to 41 ppm PCE, 38 ppm TCE, 86 ppm MEK, and 35 ppm MIBK in air of RCRA solvent recycling operation. Are the concentrations of chemicals exceeding the $TLV–TWA_{mixture}$?

5.4 A groundwater plume has developed from a leaking underground storage tank and fumes have migrated to the basement of a nearby factory. A sample of the factory indoor air indicates toluene and benzene concentrations of 50 and 70 $\mu g/m^3$, respectively.

Determine the resulting health risk for an adult male working in the plant who is exposed to these vapors for 15 years, assuming a breathing rate of 15 m^3/day of both toluene and benzene. The carcinogenic slope factor-risk specific doses via the inhalation route are 0.021 $(mg/kg\,day)^{-1}$ and 0.029 $(mg/kg\,day)^{-1}$ for toluene and benzene. Assume the environmental risks for toluene and benzene are additive.

5.5 The OSHA identifies basic duties that must be carried out by employers. Discuss these basic duties. Also, state the major roles of NIOSH and OSHA.

5.6 Provide a short (in couple of sentences) explanation of the need for each piece of information on a typical MSDS sheet.

5.7 When responding to a chemical emergency, ambient oxygen measurement is frequently used to assist in specifying the level of PPE required for worker safety. An oxygen deficient atmosphere, on a vol% basis, is defined by NIOSH/OSHA as (select one)

1) 25 or less
2) 19.5 or less
3) 15 or less
4) 12.5 or less
5) 10 or less

5.8 Which factors will most affect worker mortality rates due to particulate air pollution (PM-2.5)?

5.9 A workplace plans to install five machines. The individual noise levels are listed below. Predict what the SPL or total noise level will be.

Machine	SPL (dBA)
1	86.0
2	88.1
3	88.1
4	82.2
5	84.5

5.10 Five machines, each measured at 85 dB are to be moved into the same work area. Predict the appropriate total SPL that would result.

The following data is available for an employee's exposure during a shift:

dBA	90	88	90	82	88	92
C_i	1	2	1	1	2	1
T_i	8	10.6	8	24.3	10.6	6.1

Determine the TLV for a uniform mixture of dusts containing the following particles:

Type of dust	Concentration (wt, %)	TLV (mppct)
Dust A	80	20
Dust B	20	2

Note: mppcf (millions of particles per cubic foot).

5.11 The following data is available for an employee's exposure during a shift:

dBA	90	88	90	82	88	92
C_i	1	2	1	1	2	1
T_i	8	10.6	8	24.3	10.6	6.1

5.12 Determine the TLV for a uniform mixture of dusts containing the following particles:

Type of dust	Concentration (wt, %)	TLV (mppct)
Dust A	80	20
Dust B	20	2

Note: mppcf (millions of particles per cubic foot).

5.13 Determine whether the following noise level is permissible with no additional control measure:

Noise level (dBA)	Duration (wt %)	Maximum allowed (hour)
85	3.6	16
95	3.0	4
110	0.5	0.5

5.14 How much fresh outside air is needed to maintain the workplace air concentration of methyl ethyl ketone below 100 ppm, the TLV–TWA. During degrease operations, an estimated 3 gal of MEK evaporated in an 8-hour shift, while ventilation quality is rated as average. Also, compute the air required if the operation is carried out in an enclosed hood with an opening of 50 ft^2 and a face velocity of 100 ft/min. The temperature is 81 °F and the pressure is 1 atm. The specific gravity of MEK is 0.81 and its molecular weight is 72.1. And the k-value is 1/8 (0.125). Use the following equation, calculate required ventilation dilution air.

$$Q_v = \left(\frac{Q_m R_g T}{k C_{ppm} PM} \times 10^6 \right)$$

where Q_v is the ventilation dilution air flow in ft^3/min, Q_m is the evaporation rate, M is the molecular weight of a volatile chemical, R_g is the ideal gas constant, T is the absolute temperature, k is the ventilation safety factor, and C is the vapor concentration in ppm.

5.15 Highlight and briefly explain the types and techniques for reducing worker's exposures at workplace.

5.16 Is biodegradability a solution to plastic waste pollution in ocean and land?

5.17 Define "cradle to grave."

References

Allen, D.T. and Shonnard, D.R. (eds.) (2002). *Green Engineering: Environmentally Conscious Design of Chemical Processes*. Upper Saddle River, NJ: Prentice Hall PTR.

American Conference of Governmental Industrial Hygienists (ACGIH) (1986). *Documentation of the Threshold Limit Values and Biological Exposure Indices*, 5e. Cincinnati, OH: ACGIH.

American National Standards Institute (ANSI) (2015). *American National Standard for Occupational and Educational Personal Eye and Face Protection Devices*, Publication Z87.1. Washington, DC: ANSI.

American Public Health Association (APHA) (1989). *Standard Methods for the Examination of Water and Wastewater*, 17e. Washington, DC: American Public Health Association.

American Society for Testing and Materials (ASTM) (1991). *Annual Book of ASTM Standards, Sec. 11, Water and Environmental Technology*. Philadelphia, PA: ASTM.

American Society for Testing and Materials (ASTM) (2003). *Standard Guide for Developing Conceptual Site Models for Contaminated Sites*, E1689-95. Philadelphia: ASTM.

Anderson, E.L. and Albert, R.E. (1998). *Risk Assessment and Indoor Air Quality*. Boca Raton, FL: Lewis Publishers.

Armitage, P. and Doll, R. (1954). The age distribution of cancer and a multistage theory of carcinogenesis. *British Journal of Cancer* 8: 1–12.

Arrhenius, S. (1896). On the influence of carbonic acid in the air upon the temperature of the ground. *The London, Edinburgh, and Dublin Philosophical Magazine and Journal of Science* 4 (251): 273–276.

Asante-Duah, K. (2002). *Public Health Risk Assessment for Human Exposure to Chemicals*. London: Kluwer Academic.

Association of British Insurers (ABI) (2005). Financial risks of climate change. http://insurance.lbl.gov/documents/abi-climate.pdf (accessed 12 November 2018).

Barnhouse, L.W., Deangelis, D.L., Gardner, R.H. et al. (1982). *Methodology for Environmental Risk Analysis*, ORNL/TM-8167. Oak Ridge, TN: Oak Ridge National Laboratory.

Boholm, A. (2008). New perspectives on risk communication uncertainty in a complex society. *Journal of Risk Research* 11 (1–2): 1–3.

Bolstridge, J. (2003). Access environmental information on the internet. *Chemical Engineering Progress*: 50–54. (January).

Brooks, A.S. and Seegert, S.R. (1977). The effect of intermittent chlorination on rainbow trout and yellow perch. *Transactions of the American Fisheries Society* 106: 278–286.

Center for Chemical Process Safety (CCPS) (2012). *Guidelines for Engineering Design for Process Safety*, 2e. Hoboken, NJ: Wiley.

Chu, J. (2013). Air pollution causes 200,000 early deaths each year in the U.S., MIT Laboratory for Aviation and the Environment. *MIT News* (29 August).

Costanza, R. (1991). *Ecological Economics: The Science and Management of Sustainability*. New York: Columbia University Press.

Crowl, D.A. and Louvar, J.F. (2011). *Chemical Process Safety: Fundamentals with Applications*, 3e. Englewood Cliffs, NJ: Prentice Hall.

Danielson, J.A. (1973). *Air Pollution Engineering Manual*, AP-40, 2e. Washington, DC: USEPA.

Das, D.K. and Prabhudesai, R.K. (1996). *Chemical Engineering License Review Course*, 2e. San Jose, CA: Engineering Press.

Desvousges, W.H. and Skahen, V.A. (1987). *Techniques to Measure Damages to Natural Resources*, Final Report, PB88-100136. Springfield, VA: National Technical Information Service.

Dourson, M.L. (1986). New approaches in the derivation of acceptable daily intake (ADI). *Comments Toxicology* 1: 35–48.

Dunne, J.P., Stouffer, R.J., and John, J.G. (2013). *Heat stress reduces labor capacity under climate warming*. Geophysical Fluid Dynamics Laboratory. Princeton University Forrestal Campus, Princeton, NJ.

Epstein, P. (2002). Climate change and infectious disease: stormy weather ahead? *Epidemiology* 13 (4): 373–375.

Federal Focus, Inc. (1991). *Towards Common Measures: Recommendations for a Presidential Executive Order on Environmental Risk Assessment and Risk Management Policy*. Washington, DC: Federal Focus Inc., and The Institute for Regulatory Policy.

Finkel, A. (1990). *Confronting Uncertainty in Risk Management*. Washington, DC: Resources for the Future.

Haggestad, H.E. (2001). Ambient ozone and plant health. *Plant Disease*, 85, 1–11. https://www.coursehero.com/file/p621p44/SULFUR-DIOXIDE-DAMAGE-TO-CORN-OZONE-DAMAGE-TO-RADISH-PLANT-RESULTING-IN-SMALLER (accessed 4 December 2018).

Heath, A.G. (1977). Toxicity of intermittent chlorination to freshwater fish: influence of temperature and chemical form. *Hydrobilogia* 56: 39–47.

Heck, W.W., Adams, R.M., Cure, W.W. et al. (1983). A reassessment of crop loss from ozone. *Environmental Science and Technology* 17 (12): 572A–581A.

Hoppins, J.A., Adgate, J.L., Eberhart, M. et al. (2006). Environmental exposure assessment of pesticides in farmworker homes. *Environmental Health Perspective* (6): 929–935. https://doi.org/10.1289/ehp.8530.

Ibrahim, F., Lebowitz, J., and Hughes, J. (2018). Essentials of job hazard analysis. *Chemical Engineering Progress* 114 (11): 52–56.

International Organization for Standardization (ISO) (2015). *Protective Clothing: Clothing to Protect Against Heat and Flame – Minimum Performance Requirements*, Publication 11612. Geneva, Switzerland: ISO.

Johnson, D.L. (2017). *Statitistical Tools for the Comprehensive Practice of Industrial Hygiene and Environmental Health Sciences*, 1e. Hoboken, NJ: Wiley.

Jones, J.M. (2015). In U.S., Concern About Environmental Threats Bases. *Politics* (15 March 2015).

Kammen, D. and Hassenzahl, D. (1999). *Should We Risk It? Exploring Environmental, Health, and Technological Problem Solving*. Princeton, NJ: Princeton University Press.

Kocher, D.C. and Hoffman, F.O. (1991). Regulating environmental carcinogens: where do we draw the line? *Environmental Science and Technology* 25: 1986–1989.

Kundzewicz, Z.W., Mata, L.J., Arnell, N.W. et al. (2007). Freshwater resources and their management. In: *Climate Change 2007: Impacts, Adaptation and Vulnerability. Contribution of Working Group II to the Fourth Assessment Report of the Intergovernmental Panel on Climate Change* (eds. M.L. Parry et al.). Cambridge and New York: Cambridge University Press.

Lee, G.F. and Jones, R.A. (1982). An approach to evaluating the potential significance of chemical contaminants in aquatic habitat assessment. In: *Acquisition and Utilization of Aquatic Habitat Inventory Information* (eds. M.B. Bain and N.J. Stevenson), 294–302. Washington, DC: American Fisheries Society.

Lloyd, R. (1979). Toxicity tests with aquatic organisms. Lectures presented at the Sixth FAO/SIDA Workshop on Aquatic Pollution in Relation to Protecting Living Resources, U.N. Food and Agricultural Organization, Rome.

Max Plank Research (MPR) (2013). *Reichstein, M. Extreme Meteorological Events and Global Warming: A Vicious Cycle?* Jena, Germany: Max Planck Institutes.

McDermott, H.J. (1976). *Handbook of Ventilation for Contaminant Control*. Ann Arbor, MI: Arbor Science.

McKim, J.K. (1995). Early life toxicity tests. In: *Fundamentals of Aquatic Toxicology* (eds. G.M. Rand and S.R. Petrocelli). Washington, DC: Hemisphere Publishing.

McLaughlin, S.B. and Taylor, G.E. (1985). SO_2 effects on dicot crops: some issues, mechanisms, and indicators. In: *Sulfur Dioxide and Vegetation* (eds. W.E. Winner, H.A. Mooney and R.A. Goldstein), 227–249. Stanford, CA: Stanford University Press.

McMichael, A., Campbell-Lendrum, D., Corvalan, C. et al. (2003). *Global Climate Change and Health: An Old Story Writ Large*. Geneva, Switzerland: World Health Organization.

Miller, K. (2007). *Climate Change Impacts on Water*, National Center for Atmospheric Research, Boulder, CO. https://

www.isse.ucar.edu/water_climate/impacts.html (accessed 27 November 2018).

National Academies Press (NAP) (2011). *America's Climate Choices. The Average Temperature of the Earth's Surface Increased by About 1.4°F (0.8°C) over the Past 100 Years with About 1.0°F (0.6°C) of This Warming Occurring over Just the Past Three Decades*. Washington, DC: National Academy Press.

National Academy of Sciences of the United States of America (2006). *Fluoride in Drinking Water: A Scientific Review of EPA's Standard*. Washington, DC: National Academies Press.

National Fire Protection Association (NFPA) (2006). *Standard on Selection, Care, and Maintenance of Hazardous Materials Clothing and Equipment*, NFPA 2113. Quincy, MA: NFPA.

National Fire Protection Association (NFPA) (2015). *Standard on Selection, Care, Use, and Maintenance of Flame-Resistant Garments for Protection of Industrial Personnel against Short-Duration Thermal Exposure from Fire*, NFPA 2113. Quincy, MA: NFPA.

National Fire Protection Association (NFPA) (2017). *Standard on Respiratory Protection Equipment for Tactical and Technical Operations*, NFPA 1986. Quincy, MA: NFPA.

National Fire Protection Association (NFPA) (2018a). *Standard for Electrical Safety in the Workplace*, NFPA 70E. Quincy, MA: NFPA.

National Fire Protection Association (NFPA) (2018b). *Standard on Selection, Care, and Maintenance of Open-Circuit Self-Contained Breathing Apparatus (SCBA)*, NFPA 1852. Quincy, MA: NFPA.

National Fire Protection Association (NFPA) (2018c). *Standard on Liquid Splash-Protective Ensembles and Clothing for Hazardous Materials Emergencies*, NFPA 1992. Quincy, MA: NFPA.

National Fire Protection Association (NFPA) (2018d). *Standard on Flame-Resistant Clothing for Protection of Industrial Personnel against Short-Duration Thermal Exposure from Fire*, NFPA 2112. Quincy, MA: NFPA.

National Institute of Occupational Safety and Health (NIOSH) (2016). *Certified Equipment List*, Atlanta, GA, NIOSH. https://www.cdc.gov/niosh/npptl/topics/respirators/cel/default.html (accessed 12 December 2018).

National Oceanica Atmospheric Administration, Earth System Research Laboratory (n.d.) Trends in atmospheric carbon dioxide. http://www.esr.noaa.gov/gmd/ccgg/trends/full.html (accessed 23 September 2019).

National Research Council (NRC) (1975). *Decision Making for Regulating Chemicals in the Environment*. Washington, DC: National Academy Press.

National Research Council (NRC) (1983). *Risk Assessment in the Federal Government: Managing the Process*. Washington, DC: National Academy Press.

National Research Council (NRC) (1989). *Improving Risk Communication*. Washington, DC: National Academy Press.

New York Times (2018a). 87,000 tons of plastic and counting in Pacific. *New York Times* (23 March 2018).

New York Times (2018b). Want to save the planet? Try using less deodorant. *New York Times* (17 February 2018).

New York Times (2019). 48 pounds of plastic junk are found in Dead Whale. *New York Times* (3 April 2019).

Newcombe, C.P. and MacDonald, D.D. (1991). Effects of suspended sediments on aquatic ecosystems. *North America Journal of Fisheries Management* 11: 72–82.

NRDC, National Resources Defense Council (2015). Clean Power: The Case for Carbon Pollution Limits. Report No. R-15-06-B.

Omenn, G.S. and Faustman, E.M. (2002). Risk assessment. In: *Risk Assessment and Risk Management* (eds. R. Detels et al.). Oxford University Press.

Organization for Economic Cooperation and Development (OECD) (1981). *OECD Guidelines for Testing of Chemicals*. Paris: OECD.

Parkhurst, M.A., Onishi, Y., and Olsen, A.R. (1981). A risk assessment of toxicants to aquatic life using environmental exposure estimates and laboratory toxicity data. In: *Aquatic Toxicology and Hazard Assessment*, ASTM/STP 737 (eds. D.R. Branson and K.L. Dickson), 59–71. Philadelphia, PA: American Society for Testing and Materials.

Pielke, R.A., Gratz, J., Landsea, C.W. et al. (2008). Normalized hurricane damage in the United States: 1900–2005. *Natural Hazards Review* 9 (1): 29–42.

Pilson, M.E. (2006). We are evaporating our coal mines into the air. *AMBIO: A Journal of the Human Environment* 35 (3): 130–133.

Randall, A. (1991). The value of diversity. *AMBIO* 20: 64–68.

Richardson, L.B. and Burton, D.T. (1981). Toxicity of ozonated estuarine water to juvenile blue crabs (*Calinects sapidus*) and juvenile Atlantic menhaden (*Brevoortia tyrannus*). *Bulletin of Environmental Contamination and Toxicology* 26: 171–178.

Risk Assessment Forum (1996). *Summary Report for the Workshop on Monte Carlo Analysis*. Washington, DC: USEPA.

Risk Assessment Forum (1999). Guiding Principles for Monte Carlo Analysis, Washington, 1997. Risk Assessment Forum, Report of the Workshop on Selecting Input Distributions for Probabilistic Assessments, Washington.

Roberts, W.C. and Abernathy, C.O. (1997). *Risk assessment: principles and methodologies*. In: *Toxicology and Risk Assessment* (ed. A.M. Fan). Academic Press.

Rolston, H. (1988). *Environmental Ethics*. Philadelphia, PA: Temple University Press.

Ruckelshaus, W.D. (1984). Risk in a free society. *Risk Anal* 4: 157–162.

Sahney, S., Benton, M.J., and Falcon-Lang, H.J. (2010a). Rainforest collapse triggered Pennsylvanian tetrapod diversification in Euramerica. *Geology* 38 (12): 1079–1082.

Sahney, S., Benton, M.J., and Ferry, P.A. (2010b). Links between global taxonomic diversity, ecological diversity and the expansion of vertebrates on land. *Biology Letters* 6 (4): 544–547.

Sandman, P. (1993). *Responding to Community Outrage: Strategies for Effective Risk Communication*. Fairfax, VA: American Industrial Hygiene Association.

Seattle Times (2018). As plastic piles up, companies vow to clean up: waste is chocking oceans, rivers, landfills. *Seattle Times* (22 February 2018).

Sheader, G. (2014). 10 ideas that will breakdown the barriers to effective workplace communication. The Manufacturer Alliance, Champion for UK Manufacturing Group.

Slovic, P., Finucane, M., Peters, E., and MacGregor, D.G. (2004). Risk as analysis and risk as feelings: some thoughts about affects, reason, risk, and rationality. *Risk Analysis* 24 (2): 1–12.

Sprage, J.B. (1969). Measurement of pollutant toxicity to fish I. Bioassay methods for acute toxicity. *Water Research* 3: 793–821.

Suter, G.W. (ed.) (1993). *Ecological Risk Assessment*. Boca Raton, FL: Lewis.

Suter, G.W., Efroymson, R.A., Sample, B.E., and Jones, D.S. (2000). *Ecological Risk Assessment for Contaminated Sites*. Boca Raton, FL: *Lewis*.

Tang, Y., Zhong, S., Luo, L. et al. (2014). The potential impact of regional climate change on fire weather in the United States. *Annals of the Association of American Geographers* 105 (1): 1–21.

Tebo, L.B. (1986). Effluent monitoring: historical perspective. In: *Environmental Hazard Assessment of Effluents* (eds. H.L. Bergman, R.A. Kimerle and A.W. Maki), 13–31. New York: Pergamon Press.

Theodore, L. (2006). *Nanotechnology: Basic Calculations for Engineers and Scientists*. Hoboken, NJ: Wiley.

Travis, C.C., Richter, S.A., Crouch, E.A.C. et al. (1987). Cancer risk management. *Environmental Science and Technology* 21: 415–420.

U.S. Department of Agriculture (USDA) (2011). *Effects of Ozone Air Pollution on Plants*. Raleigh, NC: Agricultural Research Service, Plant Science Research.

U.S. Department of the Interior (USDOI) (1986). Natural resource damage assessment: final rule. *Federal Register* 51: 27674–27753.

U.S. Department of the Interior (USDOI) (1987). Natural resource damage assessment: final rule. *Federal Register* 52: 9042–9100.

U.S. Occupational Safety and Health Administration (OSHA) (2018). Title 29, Code of Federal Regulation (CFR), – Part 1910 – Occupational Safety and Health Standards. Washington, DC. https://www.osha.gov/laws-regs (accessed December 2018).

Urban, D.J. and Cook, N.J. (1986). *Hazard Evaluation, Standard Evaluation Procedure, Ecological Risk Assessment*, EPA-540/9-85-001. Washington, DC: USEPA.

Urban, F. and Mitchell, T. (2011). *Climate Change, Disasters and Electricity Generation*, Overseas Development Institute and Institute of Development Studies, London, UK. https://www.odi.org/sites/odi.org.uk/files/odi-assets/publications-opinion-files/7151.pdf (accessed 16 January 2019).

USEPA (1982). *Guidelines and Support Documents for Environmental Effect Testing*. Washington, DC: USEPA.

USEPA (1985a). *Methods for Measuring the Acute Toxicity of 4 Freshwater and Marine Organisms*, USEPA Environmental Monitoring and Support L EFIA-600/4-85/013. Cincinnati, OH: USEPA.

USEPA (1985b). *User's Guide to the Conduct and Interpretation of Effluent Toxicity Tests at Estuarine/Marine Sites*, EPA-600/X-86/224. Washington, DC: USEPA.

USEPA (1985c). *Technical Support Document for Water Quality-Based Toxics Control*, EPA-440/4-85/032. Washington, DC: USEPA Office of Water.

USEPA (1986). *Quality Criteria for Water*. Washington, DC: *Office of Water*.

USEPA (1988). *Short Term Methods for Estimating the Chronic Effluents and Receiving Waters to Marine and Estuarine Organisms*, EPA-600/4-88/028. Narragansett, RI: USEPA.

USEPA (1989). *Short Term Methods for Estimating Chronic Effluents and Receiving Waters to Freshwater Organisms*, EPA-660/2nd. Cincinnati, OH: USEPA.

USEPA (1992). *Framework for Ecological Risk Assessment*. Washington, DC: *Risk Assessment Forum*.

USEPA (1993). *Wildlife Exposure Factors Handbook*, vol. 1. Washington, DC: Office of Research and Development.

USEPA (1995). *Development of Human Health Based and Ecologically Based Exit Criteria for the Hazardous Waste Identification Project*. Washington, DC: USEPA.

USEPA (1996). *Chemical Accidental Prevention and Risk Management Programs*, 40 CFR 68 Subpart B (June). USEPA, Washington, DC. http://www.epa.gov/rmp/rmpcomp; http://response.restoration.noaa.gov/chemaids/rmp/rmp.html (accessed 12 January 2019).

USEPA (1997). *Ecological Risk Assessment Guidance for Superfund Process for Designing and Conducting Ecological Risk Assessments*. Washington, DC: Interim Final, Emergency Response Team.

USEPA (1998a). *Guidelines for Ecological Risk Assessment*. Washington, DC: Risk Assessment Forum.

USEPA (1998b). *EPA's Risk Management Program Guidance for Wastewater Treatment Plants, 40CFR68*, EPA 550-B-98-010. Washington, DC: USEPA, Office of Solid Waste and Emergency Response, Chemical Emergency Preparedness and Prevention Office.

USEPA (1998c). *Study of Hazardous Air Pollutant Emissions from Electric Utility Steam Generating Units: Final Report to Congress*, EPA-453/R-98-004a, vol. 1. Washington, DC: USEPA.

USEPA (2002a). *Short Term Methods for Estimating Chronic Effluents and Receiving Waters to Freshwater Organisms*, EPA 821-R-02-013, 4e. Washington, DC: USEPA.

USEPA (2002b). *A Review of the Reference Dose and Reference Concentration Processes*, EPA/630/P-02/002F. Washington, DC: Risk Assessment Forum.

USEPA (2003). *Framework for Cumulative Risk Assessment*, EPA/630/P-02/001F. Washington, DC: Risk Assessment Forum.

USEPA (2005). *Guidelines for Carcinogen Risk Assessment*, Washington, DC, Risk Assessment Forum, EPA/630/P-03/001B. http://www.epa.gov/iris/cancer032505.pdf (accessed 12 January 2019).

USEPA (2015). Climate change indicators: sea level rise. https://www.epa.gov/climate-indicators/climate-change-indicators-sea-level (accessed 12 January 2019).

Weinberg, A. (1987). *Science and its limits: the regular's dilemma*. In: *De Minimus Risk* (ed. C. Whipple), 27–38. New York, NY: Plenium Press.

Weiner, J. (1990). *The Next One Hundred Years: Shaping the Fate of Our Living Earth*. New York, NY: Bantam Books.

Whipple, C. (ed.) (1987). *De Minimus Risk*. New York, NY: Plenum Press.

White, G.C. (2010). *White's Handbook of Chlorination and Alternative Disinfectants*, 5e. Hoboken, NJ: Wiley.

World Health Organization (WHO) (2014). Million Premature Deaths Annually Linked to Air Pollution. WHO Media Center.

6

Industrial Process Pollution Prevention

Life-Cycle Assesvsment to Best Available Control Technology

6.1 Industrial Waste

Industrial wastes are the wastes produced by industrial activities which include materials that are rendered useless during manufacturing processes such as that of factories, industries, mills, and mining operations. This has existed since the start of the Industrial Revolution. Some examples of industrial wastes and sources are chemicals and allied products, solvents, pigments, sludge, metals, ash, paints, furniture and fixtures, paper and allied products, plastics, rubber, leather, textile mill products, petroleum refining and related industries, electronic equipment and components, industrial by-products, metals, radioactive wastes, miscellaneous manufacturing industries, and the list goes on. Hazardous or toxic wastes, chemical waste, industrial solid waste, and municipal solid waste are also designations of industrial wastes.

More than 12 billion T of industrial wastes are generated annually in the United States alone. This is roughly equivalent to more than 40 T of waste for every man, woman, and a child in the United States. The sheer magnitude of these numbers is cause for big environmental concern and drives us to identify the characteristics of the wastes, the various industrial operations that are generating the waste, the manner in which the waste are being managed, and the industrial pollution prevention policy and strategies. The first portion of this chapter is devoted to pollution prevention hierarchy. Next there is an overview of how life cycle assessment (LCA) tools can be applied to choose best available technologies (BACT) to minimize the waste at various stages of manufacturing processes of products. Finally, a few case studies on industrial competitive processes and products applying LCA tools are reviewed; hence, selections of BACT to demonstrate hierarch pollution prevention (P^2) and environmental performance strategies.

6.1.1 Waste as Pollution

A waste is defined as an unwanted by-product or damaged, defective, or superfluous material of a manufacturing process. Most often, in its current state, it has or is perceived to have no value. It may or may not be harmful or toxic if released to the environment. Pollution is any release of waste to environment (i.e. any routine or accidental emission, effluent, spill, discharge, or disposal to the air, land or water) that contaminates or degrades the environment. Waste is a form of inefficiency, and an "economic system cannot be considered efficient, or ultimately competitive, if it generates waste" (Pauli 1996).

6.1.2 Pollution Prevention in Industries

Pollution prevention (P^2) reduces the amount of pollution generated by industries, agriculture, or consumers. In contrast to pollution control strategies, which seek to manage a pollutant after it is produced and to reduce its impact on the environment, the pollution prevention approach seeks to increase efficiency of a process, reducing the amount of pollution generated. Although there is wide agreement that source reduction is the preferred strategy, some professionals also use the term pollution prevention.

With increasing human population, pollution has become a great concern. The US Environmental Protection Agency (EPA) works to introduce pollution prevention programs to reduce and manage waste (USEPA 1992a). Reducing and managing pollution may decrease the number of deaths and illnesses from pollution-related diseases. As an environmental management strategy, pollution prevention shares many attributes with cleaner production, a term used more commonly outside the United States. Pollution prevention encompasses more specialized subdisciplines,

Industrial Environmental Management: Engineering, Science, and Policy, First Edition. Tapas K. Das.

Companion website: www.wiley.com/go/Das/IEM_1e

including green chemistry and green design (also known as environmentally conscious design).

We define industrial pollution prevention fairly broadly as any action that prevents the release of harmful materials to the environment. This definition manifests itself in the form of a pollution prevention hierarchy, with safe disposal forms at the base of the pyramid and minimizing the generation of waste at the source at the peak (Figure 6.1).

In contrast, the USEPA definition of pollution prevention recognizes only source reduction, which encompasses only the upper two tiers in the hierarchy – minimize generation and minimize introduction (USEPA 1992a). The EPA describes the seven-level hierarchy of Figure 6.1 as "environment management options." The European Community, on the other hand, includes the entire hierarchy in its definition of pollution prevention. The tiers in the pollution prevention hierarchy are broadly described as follows:

- *Sources reduction.* Reduce to a minimum the formation of nonsalable by-products in chemical reaction steps and waste constituents (such as tars, fines, etc.) in all chemical and physical separation steps and cut and down as much as possible on the amounts of process materials that pass through the system unreacted or are transformed to make waste. This implies minimizing the introduction of materials that are not essential ingredients in making the final product. For examples, plant designers can decide not to use water as a solvent when one of the reactants, intermediates, or products could serve the same function, or they can add air as an oxygen source, heat sink, diluent, or conveying gas instead of large volumes of nitrogen.
- *Reuse.* Avoid combining waste streams together with no consideration to the impact on toxicity or the cost of treatment. It may make sense to segregate a low-volume, high-toxicity wastewater stream from high-volume, low-toxicity wastewater streams. Examine each waste stream at the source and identify any that might be reused in the process or transformed or reclassified as valuable coproducts.
- *Recycling.* It is the process of converting waste materials into new materials and objects. It is an alternative to "conventional" waste disposal that can save material and help lower greenhouse gas (GHG) emissions. Recycling can prevent the waste of potentially useful materials and reduce the consumption of fresh raw materials, thereby reducing energy usage, air pollution (from incineration), and water pollution (from landfilling).

  Recycling is a key component of modern waste reduction and is the third component of the "Reduce, Reuse, and Recycle" waste hierarchy (Lienig and Bruemmer 2017). Thus, recycling aims at environmental sustainability by substituting raw material inputs into and redirecting waste outputs out of the economic system (Geissdoerfer et al. 2017).

  There are some ISO standards related to recycling such as ISO 15270:2008 for plastics waste and ISO 14001:2004 for environmental management control of recycling practice. Materials to be recycled are either brought to a collection center or picked up from the curbside, then sorted, cleaned, and reprocessed into new materials destined for manufacturing.

  In the strictest sense, recycling of a material would produce a fresh supply of the same material – for example, used office paper would be converted into new office paper or used polystyrene foam into new polystyrene. However, this is often difficult or too expensive (compared with producing the same product from raw materials or other sources), so "recycling" of many products or materials involves their *reuse* in producing different materials (e.g. paperboard) instead. Another form of recycling is the salvage of certain materials from complex products, either due to their intrinsic value (such as lead from car batteries, or gold from circuit boards) or due to their hazardous nature (e.g. removal and reuse of mercury from thermometers and thermostats).
- *Recover energy value in waste.* As a last resort, spent organic liquids, gaseous streams containing volatile organic compounds, and hydrogen gas can be burned for their fuel value. Often the value of energy and resources required to make the original compounds is much greater than that which can be recovered by burning the waste streams for their fuel value (also see in Figure G.1).
- *Treat for discharge.* Before any waste stream is discharged to the environment, measure should be taken to lower its toxicity, turbidity, global warming potential, pathogen

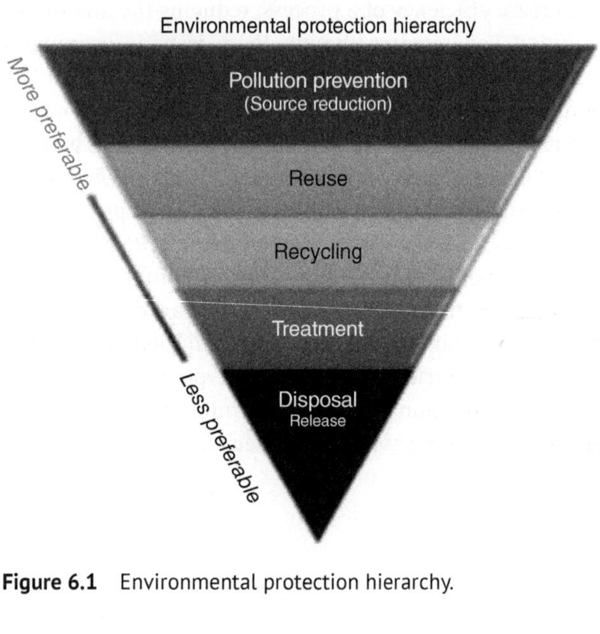

Figure 6.1 Environmental protection hierarchy.

content, and so on. Examples include biological wastewater treatment, carbon adsorption, filtration, and chemical oxidation.

- *Safe disposal.* Render waste streams completely harmless so that they do not adversely impact the environment. In this book, we define this as total conversion of waste constituents to carbon dioxide, water, and nontoxic minerals. An example would be post treatment of a wastewater treatment plant effluent in a private wetland. So-called secure landfills do not fall within this category unless the waste is totally encapsulated in granite.
- *Incineration.* It is a disposal method in which solid organic wastes are subjected to combustion so as to convert them into residue and gaseous products. This process reduces the volumes of solid waste by 80–95%. Incineration and other high temperature waste treatment systems are sometimes described as "thermal treatment." Incinerators convert waste materials into heat, gas, steam, and ash. Incineration is carried out both on a small scale by individuals and on a large scale by industry. It is used to dispose of solid, liquid, and gaseous waste. It is recognized as a practical method of disposing of certain hazardous waste materials (such as biological medical waste). Incineration is a controversial method of waste disposal, due to issues such as emission of gaseous pollutants.

  Incineration is common in countries such as Japan where land is more scarce, as the facilities generally do not require as much area as landfills. Waste-to-energy or energy-from-waste (see Appendix G) are broad terms for facilities that burn waste in a furnace or boiler to generate heat, steam, or electricity. Combustion in an incinerator is not always perfect and there have been concerns about pollutants in gaseous emissions from incinerator stacks. Particular concern has focused on some very persistent organic compounds, such as dioxins, furans, and PAHs, which may be created and which may have serious environmental consequences.

6.1.3 Defining Process Pollution Prevention (P^3)

The fundamentals and strategies of "process pollution prevention" (P^3) is the simultaneous realization of further waste reduction (environmental impact) and improvement of production (economic incentive).

$$P^3 = \text{Waste} \downarrow + \text{Production} \uparrow \tag{6.1}$$

Any time a pound is reduced in waste stream it is likely that it would end up in a product (Eq. 6.1). The goal is a closed loop in the economic subsystem, so that wastes inevitably created by human activities do not escape to contaminate the environment. There is a demand for technologies to manage and convert today's wastes into usable feedstocks to enhancing profitable pollution prevention. Chemical process design engineer and consulting firms will provide focal services to meet this demand through technology development, process intensification, system integration, and facility operation. Effective process and product stewardship requires designs that optimize performance throughout the entire life cycle. The next section would focus on process LCA, while similar concepts and tools are applied to desired product life cycles, highlighted with a few streamlined case studies.

6.2 What Is Life Cycle Assessment?

As environmental awareness increases, industries and businesses have started to assess how their activities affect the environment. Society has become concerned about the issues of natural resource depletion and environmental degradation. Many businesses and industries have responded to this awareness by providing "greener" products and using "greener" processes. The environmental performance of products and processes has become a key issue, which is why some companies are investigating ways to minimize their effects on the environment. Many companies have found it advantageous to explore ways of moving *beyond* compliance using pollution prevention strategies and environmental management systems to improve their environmental performance. One such tool is called life cycle assessment (LCA). LCA was first defined in the way we know it today at the Vermont Conference of the Society of Environmental Toxicology and Chemistry (SETAC 1991). The concept is holistic, cradle-to-grave environmental approach which provides a comprehensive view of the environmental aspects of a process or product throughout its life cycle by promoting analysis, quantification, and understanding of all the environmental impacts associated with an activity. But more importantly LCA identifies the potential transfer of environmental impacts from one media to another and/or from one life cycle stage to another. If an LCA were not performed, these trade-offs might not be recognized and properly included in the analysis because it is outside of the typical scope or focus of the decision-making process. The provision of such information aids in decision making and helps in the formulation of environmental strategy and policy as such, and LCA has been accepted into the mainstream of environmental thought and management (Azapagic 2000; Cooper 2003; Curran 2003, 2012, 2015; Das 2002, 2005; ENDS 1996; Pineda et al. 2002; SETAC 1991, 1993; UNEP-SETAC 2000, 2011; USEPA 2006).

The LCA process is a systematic, phased approach. Its components, discussed in details in Section 6.5.3.2, can be listed briefly as follows:

1) *Goal definition and scoping* – Define and describe the product, process, or activity. Establish the context in which the assessment is to be made and identify the boundaries and environmental effects to be reviewed for the assessment.
2) *Inventory analysis* – Identify and quantify energy, water and materials usage, and environmental releases (e.g. air emissions, solid waste disposal, wastewater discharge).
3) *Impact assessment* – Assess the human and ecological effects of energy, water, and material usage and the environmental releases identified in the inventory analysis.
4) *Interpretation* – Evaluate the results of the inventory analysis and impact assessment in terms of the goal established at the outset. This makes it possible to select the preferred product, process, or service with a clear understanding of the uncertainty and the assumptions used to generate the results.

By performing an LCA, researchers can

1) develop a systematic evaluation of the environmental consequences associated with a given product
2) analyze the environmental trade-offs associated with one or more specific products/processes to help gain stakeholder (state, community, etc.) acceptance for a planned action
3) quantify environmental releases to air, water, and land in relation to each life cycle stage and/or major contributing process
4) assist in identifying significant shifts in environmental impacts between life cycle stages and environmental media
5) assess the human and ecological effects of material consumption and environmental releases to the local community, region, and world
6) compare the health and ecological impacts between two or more rival products/processes or identify the impacts of a specific product or process
7) identify impacts to one or more specific environmental areas of concern

6.2.1 Benefits of Conducting an LCA

An LCA will help decision-makers select the product or process that results in the least impact to the environment. This information can be used with other factors, such as cost and performance data to select a product or process. LCA data identify the transfer of environmental impacts from one media to another (e.g. eliminating air emissions by creating a wastewater effluent instead) and/or from one life cycle stage to another (e.g. from use and reuse of the product to the raw material acquisition phase). Without an LCA, the transfer might be overlooked and excluded from the analysis because it is outside the typical scope or focus of product selection processes.

For example, when selecting between two rival products, it may appear that Option 1 is better for the environment because it generates less solid waste than Option 2. However, LCA might indicate that the first option actually creates larger cradle-to-grave environmental impacts when measured across air, water, and land. Perhaps, for example, it is seen to cause more chemical emissions during the manufacturing stage. Therefore, the second product although it produces solid waste may actually produce less cradle-to-grave environmental harm or impact than the first technology because its chemical emissions are lower.

This ability to track and document shifts in environmental impacts can help decision-makers and managers fully characterize the environmental trade-offs associated with product or process alternatives.

6.2.2 Limitations of LCAs as Tools

Performing an LCA can be resource and time intensive. Depending upon how thorough an LCA the users wish to conduct, gathering the data can be problematic and the unavailability of crucial data can greatly impact the accuracy of the final results. Therefore, it is important to weigh the availability of data, the time necessary to conduct the study, and the financial resources required against the projected benefits of the LCA.

LCA will not determine which product or process is the most cost effective or works the best. Nor does it take into account broader issues of acceptability. Therefore, the information developed in an LCA study should be used as one component of a more comprehensive process of assessing the trade-offs between performance and economic, geopolitical and social costs.

6.2.3 Conducting an LCA

To see the interrelatedness of the four components of an LCA, it is useful to refer to the flowchart shown in Figure 6.2.

6.2.3.1 Goal Definition and Scoping

The LCA process can be used to determine the potential environmental impacts from any product, process, or service. The goal definition and scoping phase will determine the time and resources needed. The defined goal and scope will guide the entire process to ensure that the most meaningful results

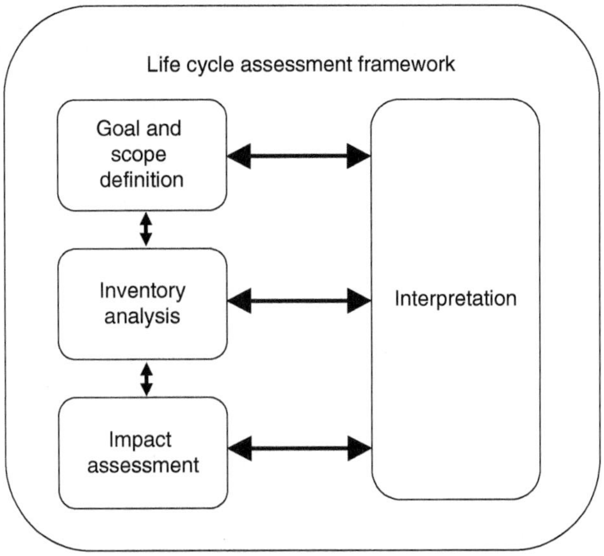

Figure 6.2 Life cycle stages (ISO 1997).

are obtained. Every decision made during goal definition and scoping impacts either how the study will be conducted or the relevance of the final results. Goal definition and scoping will result in the determination of the following:

1) The goal(s) of the project
2) The type of information is needed to inform the decision-makers
3) How the data should be organized and the results displayed
4) What will or will not be included in the LCA
5) The accuracy required of the data
6) Ground rules for performing the work

Each decision made in the fleshing out of these six areas has an impact on the LCA process, as explained in Sections 6.2.3.1 through 6.2.3.6.

6.2.3.2 Define the Goal(s) of the Project

The primary goal of the LCA is to choose the best product, process, or service with the least effect on human health and the environment. There may also be secondary goals for performing an LCA, which would vary depending on the type of project. Some typical secondary goals are as follows:

- To prove one product is environmentally superior to a competitive product
- To identify stages within the life cycle of a product or process where a reduction in resource use and emissions might be achieved
- To determine the impacts to particular stakeholders or affected parties
- To establish a baseline of information on a system's overall resource use, energy consumption, and environmental loadings
- To help guide the development of new products, process, or activities toward a net reduction of resource requirements and emissions

6.2.3.3 Determine the Type of Information Needed to Inform the Decision-Makers

LCA can help answer a number of important questions. Identifying the questions that the decision-makers care about will help define the study parameters. Some examples include the following:

- What is the impact to particular interested parties and stakeholders?
- Which product or process causes the least environmental impact (quantifiably) overall or in each stage of its life cycle?
- How will changes to the current product/process affect the environmental impacts across all life cycle stages?
- Which technology or process causes the least amount of acid rain, smog formation, or damage to local trees (or any other impact category of concern)?
- How can the process be changed to reduce a specific environmental impact of concern (e.g. global warming)?

Once the appropriate questions have been identified, the types of information needed to answer them will be apparent.

6.2.3.4 Determine How the Data Should Be Organized and the Results Displayed

LCA practitioners like to organize data in terms of a *functional unit* that appropriately describes the function of the product/process being studied. Comparisons between products and processes must be made on the basis of the same function, quantified by the same functional unit. This ensures that the activities being compared are true substitutes for each other. Careful selection of the functional unit to measure and display the LCA results will improve the accuracy of the study and the usefulness of the results.

An LCA study comparing two types of wall insulation to determine environmental preferability must be evaluated on the same function, the ability to decrease heat flow. Six square feet of 4-in. thick insulation Type A is not necessarily the same as 6 ft^2 of 4-in. thick insulation Type B. Insulation type A may have an R factor equal to 10, whereas insulation type B may have an R factor equal to 20. Therefore, types A and B do not provide the same amount of insulation and cannot be compared on an equal basis. If Type A decreases heat flow by 80%, you must determine how thick Type B must be to also decrease heat flow by 80%.

6.2.3.5 What Will and Will not Be Included

Ideally, an LCA includes all four stages of a product or process life cycle: raw material acquisition, manufacturing, use/reuse/maintenance, and recycle/waste management. These product stages are explained in more detail in the following. To determine whether one or all of the stages should be included in the scope of the LCA, the following must be assessed: the goal of the study, the required accuracy of the results, and the available time and resources. Figure 6.3 presents a set of life cycle stages that could be included in a manufacturing project related to treatment technologies. Note the "system boundary," which encompasses all aspects of the LCA. Additional examples of the four life cycle stages are explained in more detail in the following.

Raw Materials Acquisition The life cycle of a product begins with the removal of raw materials and energy sources from the Earth. For instance, the harvesting of trees or the mining of nonrenewable materials would be considered raw materials acquisition. Transportation of these materials from the point of acquisition to the point of processing is also included in this stage (USEPA 1993).

Manufacturing During the manufacturing stage, raw materials are transformed into a product or package, which is then delivered to the consumer. The manufacturing stage is broken into three parts: materials manufacture, product fabrication, and filling/packaging/distribution (USEPA 1993). While those activities are self-explanatory, it is noted that distribution in which finished products are transported to retail outlets or directly to the consumer entails environmental effects due to mode of transportation trucking, shipping, or others.

Use/Reuse/Maintenance Once the product is in the consumer's hand, all activities associated with the useful life of the item must be identified: energy demands and

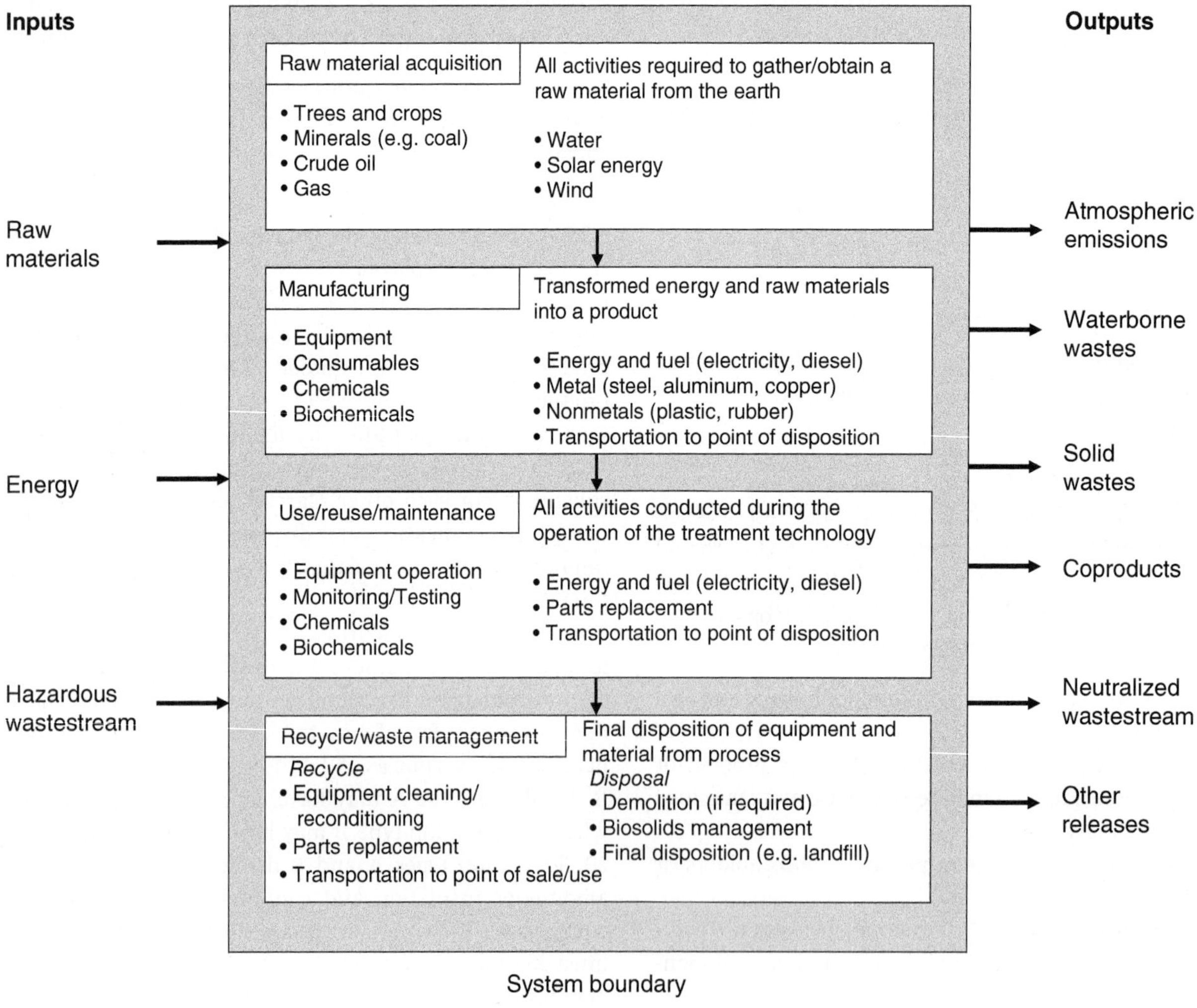

Figure 6.3 Sample life cycle stages for industrial manufacturing process.

environmental wastes from product storage and consumption, as well as any reconditioning, repaired or servicing may be required (USEPA 2012). When the consumer no longer needs the product, it will be recycled or disposed of.

Recycle/Waste Management Disposition of any product or material whether by recycling, incinerating, dumping, or other mode of waste management requires energy and results in other environmental wastes (USEPA 1993, 2012). These must be anticipated and listed.

6.2.3.6 Accuracy Required of the Data

The required level of data accuracy for the project depends on the use of the final results and the intended audience. (Will the results be used to support decision making in an internal process? In a public forum?) For example, if the intent is to use the results in a public forum to support product/process selection to a local community or regulator, then estimated data or best engineering judgment may not be accurate enough to justify basing policy decisions on them. In contrast, if the LCA is for internal decision-making purposes only, then estimates and best engineering judgment may be applied more frequently. This may reduce the overall cost and time required to perform the LCA, as well as enable completion of the study in the absence of precise, first-hand data. The criticality of the decision to be made and the amount of money at stake also come into play in determining the required level of data accuracy.

6.2.3.7 Ground Rules for Performing the Work

Prior to moving on to the inventory analysis phase it is important to define some of the logistical procedures for the project.

1) *Documenting assumptions* – All assumptions or decisions made throughout the entire project must be reported alongside the final results of the LCA project. If assumptions are omitted, the final results may be taken out of context or easily misinterpreted. As the LCA process advances from phase to phase, additional assumptions and limitations to the scope may be necessary to accomplish the project with the available resources.
2) *Quality assurance procedures* – Quality assurance procedures are important to ensure that the goal and purpose for performing the LCA will be met at the conclusion of the project. The level of quality assurance procedures employed for the project depends on the available time and resources and how the results will be used. If the results are to be used in a public forum, a formal review process is recommended. Evaluators might include internal and external LCA experts and interested parties whose support of the final results is sought. If the results are to be used for internal decision-making purposes only, then an internal reviewer who is familiar with LCA practices and is not associated with the LCA study may effectively meet the quality assurance goals. A formal statement from each reviewer documenting his or her assessment of each phase of the LCA process should be included with the final project report.
3) *Reporting requirements* – To ensure that the LCA meets appropriate expectations, participants should know from the outset how the final results are to be documented and exactly what is to be included in the final report. When reporting the final results, or results of a particular LCA phase, it is important to thoroughly describe the methodology used in the analysis. The report should explicitly define the systems analyzed and the boundaries that were set. The basis for comparison among systems and all assumptions made in performing the work should be clearly explained. The presentation of results should be consistent with the purpose of the study. The results should not be oversimplified solely for the purposes of presentation.

6.2.4 Life Cycle Inventory

A life cycle inventory (LCI) quantifies energy and raw material requirements, atmospheric emissions, waterborne emissions, solid wastes, and other releases for the entire life cycle of a product, process, or activity (USEPA 1993). Such an inventory is in the form of a list of the quantities of pollutants released to the environment and the amounts of energy and materials consumed. The results can be segregated by life cycle stage, by media (air, water, land), by specific processes, or any combination thereof.

Without an LCI, no basis exists to evaluate comparative environmental impacts or potential improvements. The level of accuracy and detail of the data collected is reflected throughout the remainder of the LCA process. LCI analyses can be used by industry for comparing products, processes, and materials. Government policy makers, too, can use LCI analyses in the development of regulations targeting resource use and environmental emissions.

The USEPA published two guidance documents, *Life-Cycle Assessment: Inventory Guidelines and Principles* (USEPA 1993) and *Guidelines for Assessing the Quality of Life-Cycle Inventory Analysis* (USEPA 1995). These federal guidelines provide the framework for performing an inventory analysis and assessing the quality of the data used and the results. The two documents define the following steps of an LCI:

- Develop a flow diagram of the processes being evaluated
- Develop a data collection plan
- Collect data
- Evaluate and report results

6.2.4.1 Step 1: Develop a Flow Diagram

A flow diagram is a tool to map the inputs and outputs to a process or system. The system boundary varies for the LCA, as established in the goal definition and scoping phase and expanded to include process inputs and outputs serves as the system boundary for the flow diagram. Unit processes inside the system boundary link together to form a complete life cycle picture of the required inputs and outputs (material and energy) to the system. Figure 6.4 illustrates the components of a generic unit process within a flow diagram for a given system boundary. The more complex the flow diagram, the greater the accuracy and utility of the results. Unfortunately, increased complexity also means more time and resources must be devoted to this step, as well as the data collecting and analyzing steps.

Flow diagrams are used to model all alternatives under consideration (e.g. both a baseline system and alternative systems). For a comparative study, it is important that both the baseline and alternatives use the same system boundary and are modeled to the same level of detail. If not, the accuracy of the results may be skewed.

6.2.4.2 Step 2: Develop an LCI Data Collection Plan

An LCI data collection plan ensures that the quality and accuracy of data, characterized as part of the goal definition and scoping phase, meet the expectations of the decision-makers.

Key elements of a data collection plan include the following:

- Defining data quality goals
- Identifying data sources and types
- Identifying data quality indicators
- Developing a data collection worksheet and checklist

Define data quality goals – Data quality goals provide a framework for balancing available time and resources against the quality of the data required to make a decision regarding overall environmental or human health impact (USEPA 1989a). Data quality goals, which are closely linked to overall study goals, both aid LCA practitioners in structuring an appropriate approach to data collection and serve as data quality performance criteria.

Although the number and nature of data quality goals necessarily depend on the level of accuracy required for a given LCA, the following list of hypothetical data quality goals is typical. Site-specific data are required for raw materials and energy inputs, water consumption, air emissions, water effluents, and solid waste generation. Approximate data values are adequate for the energy data category. Air emission data should be representative of similar sites in the United States. A minimum of 95% of the material and energy inputs should be accounted for in the LCI.

Identify data quality indicators – Data quality indicators are benchmarks against which the collected data can be measured to determine if data quality requirements have been met. Selection depends on which of the available indicators are most appropriate and applicable to the specific data sources being evaluated. Examples of indicators are precision, completeness, representativeness, consistency, and reproducibility.

Identify data sources and types – For each life cycle stage, unit process, or type of environmental release, the data source and/or type that will provide sufficient accuracy and quality to meet the study's goal is specified. Doing this prior to data collection helps to reduce costs and the time required to collect the data. Data sources include

- meter readings from equipment operating logs or journals; industry data reports, databases, or consultant's laboratory test results; government documents, reports, and databases; other publicly

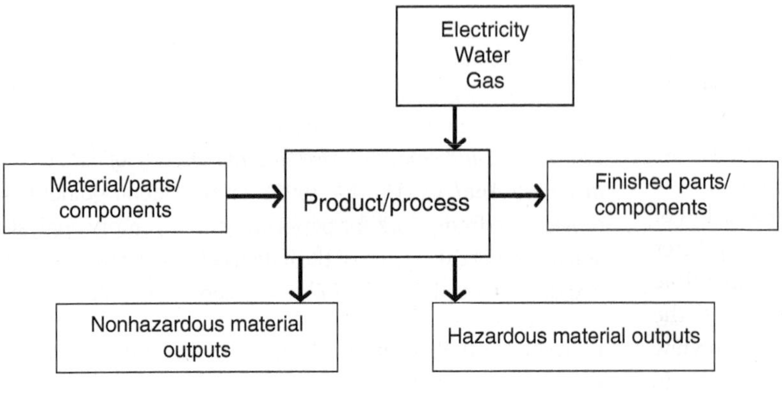

Figure 6.4 Unit process input/output template

available databases or clearinghouses; journals, and papers, books; reference books; patents; and trade associations. Related LCI studies are also useful, as are equipment and process specifications best engineering judgment.

Examples of data types include full measured, modeled, and sampled data; non site-specific (i.e. surrogate) data; non-LCI data (i.e. not intended for use in an LCI); and vendor data.

The required level of aggregated data should also be specified. For example, the reader should be able to ascertain quickly whether data are representative of one process or several processes.

Develop a data collection worksheet and checklist – The LCI checklist should cover most of the decision areas in the performance of an inventory. This document can be prepared to guide data collection and validation and can enable construction of a database to store collected data electronically. The following general decision areas should be addressed on the inventory checklist:

- purpose of the inventory
- system boundaries
- geographic scope
- types of data used
- data collection procedures
- data quality measures
- computational model construction
- presentation of results

All inputs and outputs for each process modeled in the flow diagram should be recorded on an accompanying data worksheet.

The checklist and worksheet are valuable tools for ensuring completeness, accuracy, and consistency. They are especially important for large projects when several people collect data from multiple sources. The checklist and worksheet should be tailored to meet the needs of a specific LCI.

6.2.4.3 Step 3: Collect Data

The flow diagram(s) developed in step 1 provides the road map for data to be collected. Step 2 specifies the required data sources, types, quality, accuracy, and collection methods. Step 3 consists of finding and filling in the flow diagram and worksheets with numerical data. This may not be a simple task. If some data are difficult or impossible to obtain, and available data are difficult to convert to the appropriate functional unit; therefore, the system boundaries or data quality goals of the study will have to be refined to describe the results that can reliably be obtained from the data available. This iterative process is common for most LCAs.

Data collection efforts involve a combination of research, site-visits, and direct contact with experts which generate large quantities of data. An electronic database or spreadsheet can be useful to hold and manipulate the data. Alternatively, it may be more cost effective to buy a commercially available LCA software package (see Section 6.3). Prior to purchasing an LCA software package the decision-makers or LCA practitioner should insure that it will provide the level of data analysis required.

A second method to reduce data collection time and resources is to obtain non-site specific inventory data. Several organizations have developed databases specifically for LCA that contain some of the basic data commonly needed in constructing an LCI. Some of the databases are sold in conjunction with LCI data collection software; others are stand-alone resources. Many companies with proprietary software also offer consulting services for LCA design.

6.2.4.4 Step 4: Evaluate and Document the LCI Results

When the data have been collected and organized, the accuracy of the results must be verified. In documenting the results of the LCI, it is important to thoroughly describe the methodology used in the analysis, define the systems analyzed and the boundaries that were set, and state all assumptions made in performing the inventory analysis. Use of the checklist and worksheet (see step 2) supports a clear process for documenting this information. The outcome of the inventory analysis is a list containing the quantities of pollutants released to the environment and the amount of energy and materials consumed. The information can be organized by life cycle stage, by media (air, water, land), by specific process, or any combination thereof that is consistent with the ground rules.

If the sensitivity of the LCI data collection efforts has not been properly determined before the next stage, life cycle impact assessment (LCIA), is begun, the LCA itself may have to be repeated because the data are found to be insufficient to permit the drawing of the desired conclusions.

6.2.5 Life Cycle Impact Assessment

The LCIA phase is the evaluation of potential human health and environmental impacts of the environmental resources and releases identified during the LCI. Impact assessment should address ecological and human health effects; it can also address resource depletion. An LCIA attempts to establish a linkage between the product or process and its potential environmental impacts. For example, an LCIA could determine whether one product or process causes more GHGs than other, or could potentially kill more fish.

The key concept in this component is that of stressors. A stressor is a set of conditions that may lead to an impact. For example, if a product or process is emitting GHGs, the increase of GHGs in the atmosphere *may* contribute to global warming. Processes that result in the discharge of excess nutrients into bodies of water *may* lead to eutrophication. An LCIA provides a systematic procedure for classifying and characterizing these types of environmental effects.

6.2.5.1 Why Conduct an LCIA?

Although much can be learned about a process by considering LCI data, an LCIA provides a more precise basis to make comparisons. Thus, we know that large releases of both carbon dioxide and methane are harmful; an LCIA can determine whether 9000 T of CO_2 or 5000 T of methane would have the greater potential impact. Using science-based characterization factors, an LCIA can calculate the impacts each environmental release has on problems such as smog or global warming. An impact assessment can also incorporate value judgments. In an air non-attainment zone, for example, air emissions could be of relatively higher concern than the same emission level in a region with better air quality (ISO 2000).

6.2.5.2 Key Steps of a LCIA

The following steps comprise an LCIA.

1) *Selection and definition of impact categories* – identifying relevant environmental impact categories (e.g. global warming, acidification, terrestrial toxicity).
2) *Classification* – assigning LCI results to the impact categories (e.g. classifying CO_2 emissions to global warming).
3) *Characterization* – modeling LCI impacts within impact categories using science-based conversion factors (e.g. modeling the potential impact of CO_2 and methane on global warming).
4) *Normalization* – expressing potential impacts in ways that can be compared (e.g. comparing the global warming impact of CO_2 and methane for the two options).
5) *Grouping* – sorting or ranking the indicators (e.g. sorting the indicators by location: local, regional, and global).
6) *Weighting* – emphasizing the most important potential impacts.
7) *Evaluating and reporting LCIA results* – gaining a better understanding of the reliability of the LCIA results.

The International Organization of Standardization standard for conducting an impact assessment states that impact category selection, classification, and characterization are mandatory steps for an LCIA and data evaluation (step 7) (ISO 1998a). Whether the other steps are used will depend on the goal and scope of the study.

Step 1: Select and Define Impact Categories The impact categories that will be considered as part of the overall LCA were selected as part of the initial goal and scope definition phase. To guide the LCI data collection process and required reconsideration for an LCIA, impacts are defined as the consequences due to the input and output streams of a system on human health, on plants and animals (i.e. ecological health), or on the future availability of natural resources (i.e. resource depletion). Table 6.1 shows some of the more commonly used impact categories.

Step 2: Classification For LCI items that contribute to only one impact category, the classification procedure is a straightforward assignment. For example, carbon dioxide emissions can be replaced in the global warming category.

For LCI items that contribute to two or more different impact categories, a rule must be established for classification. There are two ways of assigning LCI results to multiple impact categories (ISO 1998a):

- Allocate a representative portion of the LCI results to the impact categories to which they contribute. This is typically allowed in cases when the effects are dependent on each other.
- Assign all LCI results to all impact categories to which they contribute. This is typically allowed when the effects are independent of each other.

For example, since one molecule of sulfur dioxide (SO2) can stay at ground level or travel up into the atmosphere, it has the potential to affect either human health or acidification (but not both at the same time). Therefore, SO_2 emissions typically are divided between those two impact categories (e.g. 50% allocated to human health and 50% allocated to acidification). On the other hand, since nitrogen dioxide (NO_2) could potentially affect both ground level ozone formation and acidification simultaneously, the entire quantity of NO_2 would be allocated to both impact categories (e.g. 100% to ground level ozone and 100% to acidification). The allocation procedure must be clearly documented.

Step 3: Characterization Impact characterization uses science-based conversion factors, called characterization factors, to convert and combine the LCI results into representative indicators of impacts to human and ecological health. Characterization factors also are commonly referred to as equivalency factors. Characterization factors translate different LCI inputs – for example, the toxicity data for lead, chromium, and zinc – into directly comparable impact indicators. With such data in hand, estimates of the relative terrestrial toxicity of these metals could be made.

Table 6.1 Commonly used life cycle impact categories.

Impact category	Scale	Relevant LCI data (i.e. classification)	Common characterization factor	Description of characterization factor
Global warming	Global	Carbon dioxide (CO_2) Nitrogen dioxide (NO_2) Methane (CH_4) Chlorofluorocarbons (CFCs) Hydrochlorofluorocarbons (HCFCs) Methyl bromide (CH_3Br)	Global warming potential	Converts LCI data to carbon dioxide (CO_2) equivalents Note: global warming potentials can be 50, 100, or 500year potentials
Stratospheric ozone depletion	Global	Chlorofluorocarbons (CFCs) Hydrochlorofluorocarbons (HCFCs) Halons Methyl bromide (CH3Br)	Ozone depleting potential	Converts LCI data to trichlorofluoromethane (CFC-11) equivalents
Acidification	Regional local	Sulfur oxides (SO_x) Nitrogen oxides (NO) Hydrochloric acid (HCl) Hydrofluoric acid (HF) Ammonia (NH_4)	Acidification potential	Converts LCI data to hydrogen (H^+) ion equivalent
Eutrophication	Local	Phosphate (PO_4) Nitrogen oxide (NO) Nitrogen dioxide (NO_2) Nitrates Ammonia (NH_4)	Eutrophication potential	Converts LCI data to phosphate (PO_4) equivalents
Photochemical smog	Local	Non-methane hydrocarbon (NMHC)	Photochemical oxidant creation potential	Converts LCI data to ethane (C_2H_6) equivalents
Terrestrial toxicity	Local	Toxic chemicals with a reported lethal concentration to rodents	LC_{50}	Convert LC_{50} data to equivalents
Aquatic toxicity	Local	Toxic chemicals with a reported lethal concentration to fish	LC_{50}	Convert LC_{50} data to equivalents
Human health	Global, regional, local	Total releases to air, water, and soil	LC_{50}	Convert LC_{50} data to equivalents
Resource depletion	Global, regional, local	Quantity of minerals used Quantity of fossil fuels used	Resource depletion potential	Converts LCI data to a ratio of quantity resource used versus quantity of resource left in reserve
Land use	Global, regional, local	Quantity disposed of in a landfill	Solid waste	Converts mass of solid waste into volume using an estimated density

The impact categories listed in Table 6.1 have many possible endpoints, including the following:

Global impact	
Global warming	Polar melt, soil moisture loss, longer seasons, forest loss/change, and change in wind and ocean patterns
Ozone depletion	Increased ultraviolet radiation
Resource depletion	Decreased resources for future generations
Regional impacts	
Photochemical smog	"Smog," decreased visibility, eye irritation, respiratory tract and lung irritation, and vegetation damage
Acidification	Building corrosion, water body acidification, vegetation effects, and soil effects
Local impacts	
Human health	Increased morbidity and mortality
Terrestrial toxicity	Decreased production and biodiversity and decreased wildlife for hunting or viewing
Aquatic toxicity	Decreased aquatic plant and insect production and biodiversity and decreased commercial or recreational fishing
Land use	Loss of terrestrial habitat for wildlife and decreased landfill space

Impact indicators are typically characterized using the following equation:

$$\text{Inventory data} \times \text{characterization factor} = \text{impact indicators}$$

For example, all GHGs can be expressed in terms of carbon dioxide equivalents by multiplying the relevant LCI results by a CO_2 characterization factor and then combining the resulting impact indicators to provide an overall indicator of global warming potential.

The Intergovernmental Panel on Climate Change provides conversion factors for a number of industrial pollutants. In the following example, the global warming impacts of different amounts of chloroform ($CHCl_3$) and methane (CH_4) are characterized. The value of the conversion factor, called GWP, for global warming potential, is 9 for $CHCl_3$ and 21 for CH_4.

Thus, we write:

$$GWP_{CHCl_3} = 9 \times 20 = 180$$
$$GWP_{CH_4} = 21 \times 10 = 210$$

Use of the conversion factors shows that 10 lb of methane has a larger impact on global warming than 20 lb of chloroform.

The key to impact characterization is using the appropriate characterization factor. For some impact categories, such as global warming and ozone depletion, there is a consensus on acceptable characterization factors. For other impact categories, such as resource depletion, a consensus is still being developed. Table 6.1 includes descriptions of possible characterization factors for some of the commonly used life cycle impact categories. A properly referenced LCIA will document the source of each characterization factor to ensure that they are relevant to the goal and scope of the study. For example, many characterization factors based on studies conducted in Europe cannot be applied to American data unless it can be verified that they are appropriate to conditions in the United States.

Step 4: Normalization Normalization is an LCIA tool used to express impact indicator data in a way that can be compared among impact categories. In this procedure, the indicator results are divided by a reference value selected for the purpose. Reference value may be chosen from among numerous methods.

The following are representatives:

The total emissions or resource use for a given area that may be global, regional, or local.
The total emissions or resource use for a given area on a per capita basis.
The ratio of one alternative to another (i.e. the baseline).
The highest value among all options.

The goal and scope of the LCA may influence the choice of an appropriate reference value. Note that normalized data can only be compared within an impact category. For example, the effects of acidification cannot be directly compared with those of aquatic toxicity because the characterization factors were calculated using different scientific methods.

Step 5: Grouping Grouping assigns impact categories into one or more sets to facilitate the interpretation of the results into specific areas of concern. Typically grouping involves sorting or ranking indicators. The ISO (1998a) lists two possible ways to group LCIA data:

1) Sorting indicators by characteristics such as emissions (e.g. air and water emissions) or location (e.g. local, regional, or global)
2) Sorting indicators by a system of ranking based on value choices, such as high, low, or medium priority.

Step 6: Weighting Weighting (also referred to as valuation) assigns relative values to the different impact categories based on their perceived importance or relevance.

Weighting is important because the impact categories should also reflect study goals and stakeholder values. But since weighting is not a scientific process, its methodology must be clearly explained and documented. The weighting stage is the least developed of the impact assessment steps and also is the one most likely to be challenged for integrity. In general, weighting includes the following activities:

Identifying the underlying values of stakeholders
Determining weights to place on impacts
Applying weights to impact indicators

Weighted data should not be combined across impact categories unless the weighting procedure is explicitly documented. The unweighted data should be shown together with the weighted results to ensure a clear understanding of the assigned weights.

In some cases, the impact assessment results are so straightforward that a decision can be made without the weighting step. For example, when the best-performing alternative is significantly and meaningfully better than the others in at least one impact category and equal to the alternatives in the remaining impact categories, then *one alternative is clearly better.*

Several issues make weighting a challenge. The first issue is subjectivity. According to ISO 14042, any judgment of preferability is a subjective statement of the relative importance of one impact category over another (ISO 1998a). Additionally, these value judgments may change with location or time of year. For example, a resident of Los Angeles may place more importance on the values for photochemical smog than someone in Cheyenne, Wyoming. The second issue is derived from the first: how should users fairly and consistently make decisions based on environmental preferability, given the subjective nature of weighting? Trying to develop a truly objective (or universally agreeable) set of weights or weighting methods is not feasible. However, several approaches to weighting do exist and are used successfully for decision making.

Step 7: Evaluate and Document the LCIA Results Now that the impact potential for each selected category has been calculated, the accuracy of the results must be verified. Documentation of the results of the LCIA entails thoroughly describing the methodology used in the analysis, defining the systems analyzed and the boundaries that were set, and setting forth all assumptions on which the inventory analysis was passed.

The LCIA, like all other assessment tools, has inherent limitations, including the following:

- Lack of spatial resolution (e.g. a 4000 gal ammonia release is worse in a small stream than in a large river)
- Lack of temporal resolution (e.g. a 5 T release of particulate matter during a one month period is worse than the same release spread through the whole year)
- Inventory speciation (e.g. broad inventory listing such as "VOC" or "metals" do not provide enough information to accurately assess environmental impacts)
- Threshold and nonthreshold impact (e.g. 10 T of contamination is not necessarily 10 times worse than 1 T of contamination)

The selection of more complex or site-specific impact models can help reduce the limitations of the impact assessment's accuracy. It is important to document these limitations and to include a comprehensive description of the LCIA methodology, as well as, a discussion of the underlying assumptions, value choices, and known uncertainties in the impact models with the numerical results of the LCIA to be used in interpreting the results of the LCA.

6.2.6 Life Cycle Interpretation

Life cycle interpretation is a systematic technique to identify, quantify, check, and evaluate information from the results of the LCI and the LCIA, and communicate them effectively. Life cycle interpretation is the last phase of the LCA process. The ISO has defined the following objectives of life cycle interpretation:

1) Analyze results, reach conclusions, explain limitations, and provide recommendations based on the findings of the preceding phases of the LCA and to report the results of the life cycle interpretation in a transparent manner.
2) Provide a readily understandable, complete, and consistent presentation of the results of an LCA study, in accordance with the goal and scope of the study (ISO 1998b).

It is not always possible to use an LCA as the basis for starting, as one alternative is better than the others. This does not imply that efforts have been wasted. Uncertainty in the final results notwithstanding the LCA process still provides decision-makers with a better understanding of the environmental and health impacts associated with each alternative, where they occur (locally, regionally, or globally), and the relative magnitude of each type of impact potentially attributed to each of the proposed alternatives investigated. This information more fully reveals the pros and cons of the alternatives.

The purpose of conducting an LCA is to better inform decision-makers by providing a particular type of information (often unconsidered), with a life cycle perspective of environmental and human health impacts associated with each product or process. However, LCA does not take into account technical performance, cost, or political and social acceptance. Therefore, it is recommended that LCA be used in conjunction with these other parameters.

6.2.6.1 Key Steps to Interpreting the Results of the LCA

The guidance provided thus far summarizes the information on life cycle interpretation in the ISO draft standard *Environmental Management – Life Cycle Assessment – Life Cycle Interpretation* (ISO 1998b). The ISO draft standard covers the following steps in conducting a life cycle interpretation:

1) Identify significant issues
2) Evaluate the completeness, sensitivity, and consistency of the data
3) Draw conclusions and recommendations

Figure 6.5 illustrates the steps of the life cycle interpretation process in relation to the other phases of the LCA process.

Step 1: Identify Significant Issues Significant "issues" are the data elements that contribute most to the results of both the LCI and LCIA for each product, process, or service. Examples include

- inventory parameters (e.g. energy use, emissions, waste)
- impact category indicators (e.g. resource use, emissions, waste)
- essential contributions for life cycle stages to LCI or LCIA results such as individual unit processes or groups of processes (e.g. transportation, energy production)

When these issues have been identified, the results are used to evaluate the completeness, sensitivity, and consistency of the LCA study (step 2). The identification of significant issues guides the evaluation step.

Before determining which parts of the LCI and LCIA have the greatest influence on the results for each alternative, the previous phases of the LCA (e.g. study goals, ground rules, impact category weights, results, and external involvement, etc.) should be reviewed in a comprehensive manner.

A review of the information collected and the presentations of results developed indicates that if the goal and scope of the LCA study have been met. If they have, the significance of the results can be determined. Several analytical approaches are possible.

- *Contribution analysis* – The contribution of the life cycle stages or groups of processes are compared to the total result and examined for relevance.
- *Dominance analysis* – Statistical tools or other techniques, such as quantitative or qualitative ranking (e.g. ABC Analysis), are used to identify significant contributions to be examined for relevance.
- *Anomaly assessment* – Based on previous experience, unusual or surprising deviations from expected or normal results are observed and examined for relevance.

Step 2: Evaluate the Completeness, Sensitivity, and Consistency of the Data The evaluation step of the interpretation phase establishes the confidence in and reliability of the results of the LCA. This is accomplished by performing completeness, sensitivity, and consistency checks to ensure that products/processes are fairly compared.

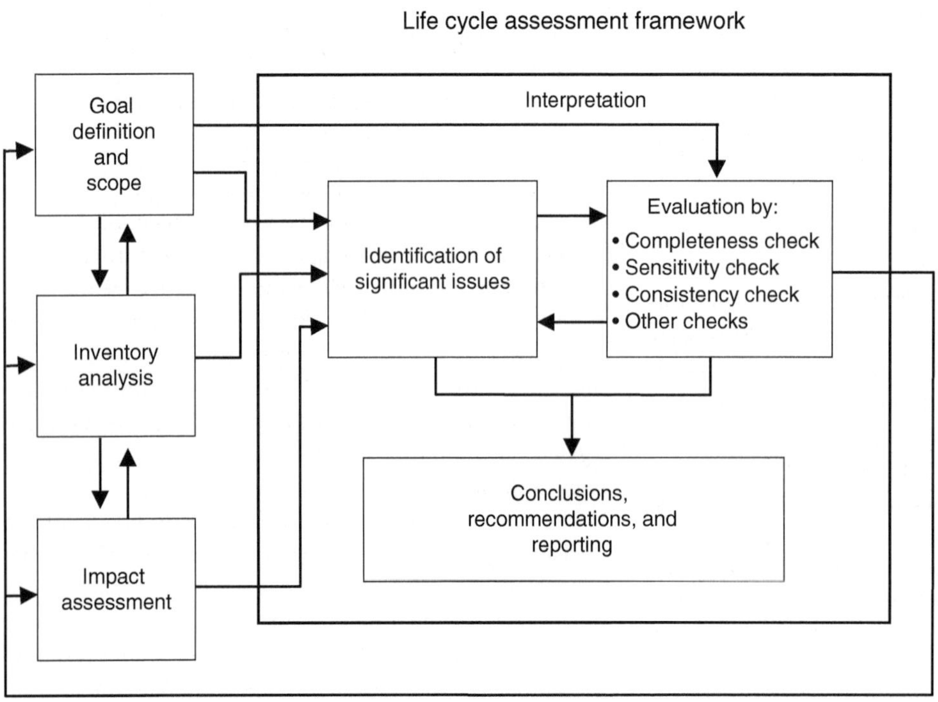

Figure 6.5 Relationship of interpretation steps with other phases of LCA (ISO 1998b).

Completeness check – The completeness check ensures that all relevant information and data needed for the interpretation are available and complete. A checklist should be developed to indicate each significant area represented in the results. Using the established checklist, it is possible to verify that the data comprising each area of the results are consistent with the system boundaries (e.g. all life cycle stages are included) and that the data is representative of the specified area (e.g. accounting for 90% of all raw materials and environmental releases).

The result of this effort will be a checklist indicating that the results for each product/process are complete and reflective of the stated goals and scope of the LCA study. If deficiencies are noted, an attempt must be made to remedy them. If this is not possible because data are not available, areas inadequately characterized because of insufficient data must be highlighted in the final results and their impact on the comparison estimated either quantitatively (percent uncertainty) or qualitatively (alternative A's reported result may be higher because "X" is not included in its assessment).

Sensitivity check – The objective of the sensitivity check is to evaluate the reliability of the results by determining whether the uncertainty in the significant issues identified in step 1 affect the decision-maker's ability to confidently draw comparative conclusions. Three common techniques for data quality analysis can be used in performing sensitivity checks.

1) Gravity analysis – Identifies the data that has the greatest contribution on the impact indicator results.
2) Uncertainty analysis – Describes the variability of the LCIA data to determine the significance of the impact indicator results.
3) Sensitivity analysis – Measures the extent that changes in the LCI results and characterization models affect the impact indicator results.

Additional guidance on how to conduct a gravity, uncertainty, or sensitivity analysis can be found in the EPA document entitled "Guidelines for Assessing the Quality of Life Cycle Inventory Analysis" (USEPA 1995). If one of these analyses has been conducted as part of the LCI and LCIA phases, these results can be used. Then the sensitivity check will serve to verify that the goals for data quality and accuracy defined early in have been met. If deficiencies exist, additional efforts are required to improve the accuracy of the LCI data collected and/or impact models used in the LCIA. If better data or impact models cannot be obtained, the deficiencies for each relevant significant issue must be reported and its impact on the comparison estimated either quantitatively or qualitatively, as with the completeness check.

Consistency check – The consistency check determines whether the assumptions, methods and data used throughout the LCA process are consistent with the goal and scope of the study, and for each product/process evaluated. Verifying and documenting that the study was completed as intended at the conclusion increases confidence in the final results. A formal checklist should be developed to communicate the results of the consistency check. Table 6.2 lists seven categories and provides examples of inconsistencies that can creep into the data. The goal and scope of the LCA determines which categories should be used.

If, after completion of steps 1 and 2, it is determined that the results of the impact assessment and the underlying inventory data are complete, comparable, and acceptable as bases for drawing conclusions and making recommendations then stop! If any inconsistency is detected, document the role it played in the overall consistency evaluation. Although some inconsistency may be acceptable, depending upon the goal and scope of the LCA, the presence of inconsistencies usually means that it is necessary to repeat steps 1 and 2 until the results are able to support the original goals for performing the LCA.v

Table 6.2 Examples of checklist categories and potential inconsistencies.

Category	Example of inconsistency
Data source	Alternative A is based on literature and Alternative B is based on measured data.
Data accuracy	For Alternative A, a detailed process flow diagram is used to develop the LCI data. For Alternative B, limited process information was available and the LCI data developed was for a process that was not described or analyzed in detail.
Data age	Alternative A uses 1980s era raw materials manufacturing data. Alternative B used a one-year-old study.
Technological representation	Alternative A is bench scale laboratory model. Alternative B is a full-scale production plant operation.
Temporal representation	Data for Alternative A describe a recently developed technology. Alternate B describes a technology mix, including recently built and old plants.
Geographical representation	Data for Alternative A were data from technology employed under European environmental standards. Alternative B uses the data from technology employed under US environmental standards.
System boundaries, assumptions, and models	Alternative A uses a Global Warming Potential model based on 500-year potential. Alternative B uses a Global Warming Potential model based on 100-year potential.

Step 3: Draw Conclusions and Recommendations The objective of this step is to interpret the results of the LCIA (not the LCI) to determine which product/process has the overall least impact on human health and the environment, and/or on one or more specific areas of concern as defined by the goal and scope of the study. Depending upon the scope of the LCA, the results of the impact assessment will return either a list of unnormalized and unweighted impact indicators for each impact category for the alternatives or a single grouped, normalized, and weighted score for each alternative. In the latter case, the recommendation may simply be to accept the product/process with the lowest score. The assumptions underlying the analysis should be borne in minds, however. If an LCIA stops at the characterization stage, the LCIA interpretation is less clear-cut. The conclusions and recommendations rest on balancing the potential human health and environmental impacts in light of study goals and stakeholder concerns.

It is essential to understand and communicate the uncertainties and limitations in the procedures that have produced the final recommendations. Perhaps no one product or process is better than another because of underlying uncertainties and limitations in the methods used to conduct the LCA. Perhaps insufficient good data were available, or restrictions on time or resources prevented analysts from thoroughly exploring certain aspects of the problem. Even so, the results of the LCA can be used to help inform decision-makers about the human health and environmental pros and cons and understand the significant impacts of each. Such LCA results will reveal whether effects are occurring locally, regionally, or globally, and will provide at least a rough estimate of the magnitude of each type of impact in comparison to the proposed alternatives being investigated.

6.2.6.2 Reporting the Results

When the LCA has been completed, the materials must be assembled into a comprehensive report documenting the study in a clear and organized manner. This will help communicate the results of the assessment fairly, completely, and accurately to others interested in the results. The report presents the results, data, methods, assumptions and limitations in sufficient detail to allow the reader to comprehend the complexities and trade-offs inherent in the LCA study.

If the results will be communicated to parties who were not involved in the LCA study (e.g. stakeholders), the report will serve as a reference document, and it can help prevent any misrepresentation of the results.

6.2.6.3 Conclusion

Adding LCA to the decision-making process provides a level of understanding of human health and environmental impacts that traditionally has not been available to those responsible for selecting a product or process. This valuable information provides a way to account for the full impacts of decisions, especially those made off-site, that are directly influenced by the selection of a product or process. As emphasized earlier, LCA is a tool to better inform decision-makers and other decision criteria such as cost and performance must be weighed to reach a well-balanced decision.

As we have seen, LCA and LCI can be valuable tools in environmental analysis. To make sense of the many large data sets collected in any such study is clearly a task too complex to be undertaken without the assistance of computers. Section 6.3 describes some of the software tools that have been developed for this purpose.

6.3 LCA and LCI Software Tools

Table 6.3 lists the LCA and LCI tools we shall discuss in this section, along with the vendor or developer of each and that organization's Internet address.

6.3.1 ECO-it 1.0

ECO-it is a database tool used to assist an LCI and LCIA. ECO-it comes with over 100 indicator values for commonly used materials such as metals, plastics, paper, board, and glass, as well as production, transport, energy, and waste treatment processes.

6.3.2 EcoManager

EcoManager is LCI tool designed to be used by persons who have or who gain a working knowledge of the LCI methodology for internal planning, screening, and evaluation. EcoManager uses a software program developed by Pira International of the United Kingdom, in combination with US LCI data from the Franklin Associates, Ltd. (FAL) database. Developed for use with Microsoft Excel, EcoManager utilizes spreadsheets and program codes called "macros" to guide the user through the construction of a LCI, access data from the databases, perform file management functions, edit existing and produce graphics and reports. The system provides online help at every stage of operation. The system also provides graphic and report features. And because the program has been created in Excel, all of the power of Excel for graphics, exporting, and file management are available for customized outputs.

6.3.3 Eco Bat 2.1

Eco Bat is an LCIA and design for environment (DfE) tool that models product life cycles with flow chart diagrams.

Table 6.3 LCA and LCI software tools.

Name	Developer/vendor	URL
1. ECO-it	PRé Consulting	http://www.pre.nl.eco-it.html
2. EcoManager	Franklin Associates, Ltd	http://www.fal.com/software/ecoman.html
3. EcoPro 1.5	EcoPerformance Systems, Switzerland	http://www.sinum.com/
4. Gabi 4	Institute for Polymer Testing and Polymer Science University of Stuttgart	http://www.pre-product.de/english/main/software.htm
5. IDEMAT	Delft University of Technology	http://www.io.tudelft.nl/research/mpo/idemat/idemat.htm
6. LCAID	Battelle Memorial Institute/US Department of Energy	http://www.estd.battelle.org/sehsm/lca/LCAdvantage.html
7. LCAiT	Chalmers Industriteknik, Sweden	http://www.ekologik.cit.chalmers.se/lcait.htm
8. REPAQ	Franklin Associates, Ltd.	http://www.fal.com/software/repaq.html
9. SimaPro 7	PRé Consulting	http://www.pre.nl/simapro.html
10. TEAM	Ecobalance	http://www.ecobalance.com/software/team/team_ovr.htm
11. TRACI	U.S. Environmental Protection Agency	http://www.epa.gov/ORD/NRMRL/std/sab/iam_traci.htm
12. Umberto NXT CO_2	Institute for Energy and Environmental Research Heidelberg, Hamburg GmbH	http://www.ifu.com/software/umberto-e/

This Windows-based application offers an online help, a toolbar, and icons, including a large database.

6.3.4 GaBi 4

GaBi is a life cycle engineering or life cycle management (LCM) and life-cycle engineering (LCE) tools that use many predefined data objects from industry and literature. Users can link data sets supplied with the GaBi database to own data in order to calculate both life cycle inventories and impact assessments. It allows clear weak point analyses of inventories and valuated balances. The structure is open to alterations and extensions.

6.3.5 IDEMAT

Idemat is a computer database of over 365 materials from the Department of Environmental Product Development of the faculty of Industrial Design Engineering at the Delft University of Technology. It provides technical information about materials and processes in words, numbers, and graphics, and puts emphasis on environmental information. The program was developed to be used by students of technically oriented academic disciplines like industrial design engineering, civil engineering, material science, and aerospace engineering. Users must be quite familiar with the principles of LCA and the methods for characterization and evaluation of the environmental impacts as published by SETAC and the Center for Environmental Studies at the University of Leiden (CML), and used in SimaPro.

6.3.6 EIOLCA

EIOLCA: The economic input–output life cycle assessment (EIOLCA) model, developed by Carnegie Mellon University, estimates the materials and energy resources required for, and the environmental emissions resulting from, activities in our economy. It is a free, fast, and easy LCA model available online (www.eiolca.net).

6.3.7 LCAD

Life-cycle advantage, or LCAD5, is a life cycle modeling tool that has a graphical user interface and database structure. LCAD can model process flow diagrams with material and energy balances, and labor and revenue inputs. LCAD can also assess the data reliability. The LCAD system includes a basic commodity database for the United States, covering fuels production and distribution, power generation, and cradle-to-gate operations for selected forest products, paper, metals, cement, and basic chemicals and plastics.

6.3.8 LCAiT

LCAiT is a LCI tool that aids in generating an energy and materials balance. LCAiT also contains a cradle-to-gate information regarding certain materials.

6.3.9 REPAQ

REPAQ is a LCI software program that permits users to examine energy and environmental emissions for the entire life cycle of a product, beginning with raw material extraction and continuing through refining and processing, material manufacture, product fabrication, and disposal. Products, processes, and packaging can be evaluated with REPAQ. Users may access the REPAQ database and enter their own data through the Custom Materials feature, which allows entry of data for any process for which LCI data can be gathered.

6.3.10 SimaPro 7

SimaPro is a full-featured LCA software tool that facilitates the comparison and analysis of complex products capable of dealing with complex tools and software: LCM, LCIA, LCI, LCE, substance/material flow analysis (SFA/MFA), DfE, product stewardship, supply chain management, life cycle sustainability (LCS), and life cycle cost (LCC). The process databases and the impact assessment databases can be edited and expanded without limitation. SimaPro can trace the origin of any result that has been implemented. Special features include multiple impact assessment methods, multiple process databases, and automatic unit conversion. SimaPro comes with several well-known impact assessment methods, including CML, Eco-points, and the Eco-indicator method developed by PRé. All impact assessment data can be edited and expanded.

6.3.11 TEAM (Tool for Environmental Analysis and Management)

TEAM is an LCA software program that allows the user to build and use a database and model any system representing the operations associated with products, processes, and activities (waste management options, means of transportation, etc.). It is designed to describe and model complex industrial systems and to calculate the associated life cycle inventories (LCM, SFA/MFA, DfE, DfR), life cycle potential environmental impacts, legal compliance checks, product stewardship, and process-oriented life cycle costing.

6.3.12 TRACI: A Model Developed by the USEPA

The tool for the reduction and assessment of chemical and other environmental impacts (TRACI) is described along with its history, the research and methodologies it incorporates, and the insights it provides within individual impact categories.

TRACI, a stand-alone computer program developed by the U.S. Environmental Protection Agency (USEPA), facilitates the characterization of environmental stressors that have potential effects, including ozone depletion, global warming, acidification, eutrophication, tropospheric ozone (smog) formation, ecotoxicity, human health criteria–related effects, human health cancer effects, human health noncancer effects, fossil fuel depletion, and land-use effects. TRACI was originally designed for use with LCA, but it is expected to find wider application in the future (Bare 2003; USEPA 2003).

6.3.13 Umberto NXT CO_2

Umberto is an LCA tool that uses the following instruments: LCM, LCIA, LCI, LCE, a graphical interface to model material flow networks to enter and track material and energy flows, SFA/MFA, supply chain management, LCS, and LCC.

6.3.14 International Organizations and Resources for Conducting Life Cycle Assessment

Appendix F provides a list of organizations in various countries, a brief description of each organization, and LCA centers and societies across all continents including Africa, Asia, Australia and New Zealand, Europe, North America, and South America.

6.4 Evaluating the Life Cycle Environmental Performance of Chemical-, Mechanical-, and Bio-Pulping Processes

6.4.1 Introduction

Pulp and paper manufacturing constitutes one of the largest industry segments in the United States in terms of water and energy usage and total discharges to the environment. More than many other industries, however, this industry plays an important role in sustainable development because its chief raw material – wood fiber – is renewable. This industry provides an example of how a resource can be managed to provide a sustained supply to meet society's current and future needs. The objective of this work is to present streamlined environmental LCA between chemical (kraft–sulfate), mechanical (or thermomechanical), and biopulping processes. This LCA would help us to evaluate the industry's current experience and practices in terms of environmental stewardship, regulatory and nonregulatory forces, life cycles of its processes and products, and future developments.

The pulping industry has been traditionally using mechanical or chemical pulping methods, or a combination of the two, to produce pulps of desired characteristics. Mechanical pulping accounts for about 25% of the wood pulp production in the world today. Mechanical pulping, with its high yield, is viewed as a way to extend the forest resources. However, mechanical pulping is electrical energy–intensive and yields paper with less strength compared to that produced by the chemical pulping process. These disadvantages limit the use of mechanical pulps in many grades of paper. Chemical pulping accounts for about 75% of the wood pulp production in the world. This process produces paper with very high strength. However, the process has the disadvantages of being capital- and energy-intensive, giving relatively low yields, producing troublesome waste products, and producing by-products that are of relatively low values.

A new technology that offers a biopulping process with the potential to ameliorate some of these problems is being tested on a commercial scale. Biopulping treats wood chips with a natural wood-decaying fungus before mechanical pulping and can save substantial amounts of electricity, significantly reduce the amount of air pollutants (including CO_2 and some odor-causing total reduced sulfur compounds) and water pollutants (BOD, COD, TSS) compared with conventional pulping, improve paper quality, and enhance economic competitiveness.

The USEPA's TRACI (USEPA 2003) model was used to assess the chemical, environmental, and human health impacts attributed to three pulp and papermaking processes. The results obtained from the biopulping process indicate a significant reduction in environmental and human health impacts. The biopulping process proves to be more sustainable in terms of economic advantage, and environmental and human health benefits.

6.4.2 Application of LCA

LCA is a tool, which has been described in Sections 6.4 and 6.5, for evaluating the environmental performance of a process, product, or activity, starting from raw material extraction, through manufacturing, to use, and final disposal. This is known as a "cradle-to-grave" approach (Figure 6.3). Because of its holistic approach to system analysis, LCA is becoming an increasingly important decision-making tool in environmental system management. Its main advantage over other, site-specific, methods for environmental analysis, such as environmental impact assessment or environmental audit, lies in broadening the system boundaries to include all burdens and impacts in the life cycle of a product or a process, and not focusing solely on the emissions and wastes generated by the plant or manufacturing site.

As an environmental management tool, LCA has two main objectives. Its first objective is to quantify and evaluate the environmental performance of a process or a product and so help decision-makers choose between alternative products or processes. Another objective of LCA is to provide a basis for assessing potential improvements of the environmental performance of an existing or newly designed system. This can be of particular importance to engineers and environmental managers because it can advise them on how to modify or design a system to decrease its overall environmental impact. LCA can thus be used both internally by a company or externally by industry, policy makers, planners, educators, and other stakeholders. If the results of LCA are to be used internally by a company, then possible areas where LCA can be useful include, but are not limited to, the following:

- strategic planning or environmental strategy development
- problem solving in the system
- identification of opportunities for, and tracking of, environmental improvements
- process and product design, innovation, improvement, and optimization

External applications of LCAs include uses of LCA as a marketing tool, to support environmental labeling or claims, or informational purposes, or to support policy decisions. LCA, green chemistry, green engineering, design-for-environment, industrial ecology, and other emerging areas of chemical engineering and environmental engineering provide greater opportunities for developing sustainable and innovative processes and products (Das and Scott 2003).

In this case study, we focus only on LCA related to three processes: chemical-, mechanical-, and biopulping using wood chips. This streamlined LCA approach will help to identify which process is more sustainable by considering environmental impacts, human health and safety, and process economics.

6.4.3 The Pulping Processes

6.4.3.1 Mechanical Pulping

Mechanical pulping, as the name implies, relies on mechanical energy to convert wood to pulp. Current mechanical pulp manufacturing processes constitute several high-energy grinding and refining systems, including refiner mechanical pulping (RMP) process, the thermomechanical pulping (TMP) process, the chemimechanical pulping process, and the chemithermomechanical pulping process (Someshwar and Pinkerton 1992; Smook 1992). Mechanical pulping accounts for about 25% of the wood pulp production in the world today. Mechanical pulping, with its high yield, is viewed as a way to extend these resources. However, mechanical pulping is electrical energy–intensive and yields paper with lower strength compared with that produced by chemical pulping which are used in the pulping process. Figure 6.6 shows the basic flow diagram for TMP (Smook 1992). TMP is a modification of RMP; it involves steaming the raw material for a short period of time before and during several refining stages. The steaming serves to soften the chips, with the result that the pulp produced has a greater percentage of long fibers and less shives than RPM. Most often, heating and refining are both done under pressure (TMP), but some systems refine under atmospheric pressure. Chemicals are sometime added at the various stages of refining. In the TMP process, the chips are usually steamed under a pressure of 20–40 psi for 2–4 minutes before refining.

Table 6.4 presents an average quantity (in %) of make-up chemicals that are typically needed for mechanical/thermomechanical pulping (USDA, personal communication). Table 6.5 presents the energy consumption in various operations of TMP and papermaking process (US Pulp and Paper Industry 1988). Table 6.6 shows a range of effluent loads and other environmental impacts of the TMP process (Hynninen 1998; Sundholm 1999). The data from Tables 6.4–6.6 are used to quantify the environmental and human health impacts that will be discussed later in Section 6.4.3.7.

A study by the National Council of the Paper Industry for Air and Stream Improvement (NCASI 1983) on estimating volatile organic compounds (VOCs) are likely to be released along with the steam during the cooking and refining process. The range of VOC emissions can vary between 1.0 and 7.0 lb of carbon per ton of pulp, depending on the wood species. No emissions other than VOC are expected from the mechanical pulping process.

6.4.3.2 Chemical Pulping Processes

The following discussion centers on chemical pulping processes, which supply more than two-thirds of world's wood pulp. The most widely used chemical pulping process in papermaking is the kraft, or sulfate, process. Other chemical-pulping processes (mainly using acid sulfite and soda) are sometimes combined with various chemical-recovery subprocesses. First used over a century ago, these processes now result in recovery of 90% of the inorganics, which are used in the pulping process. Nearly 100% of the

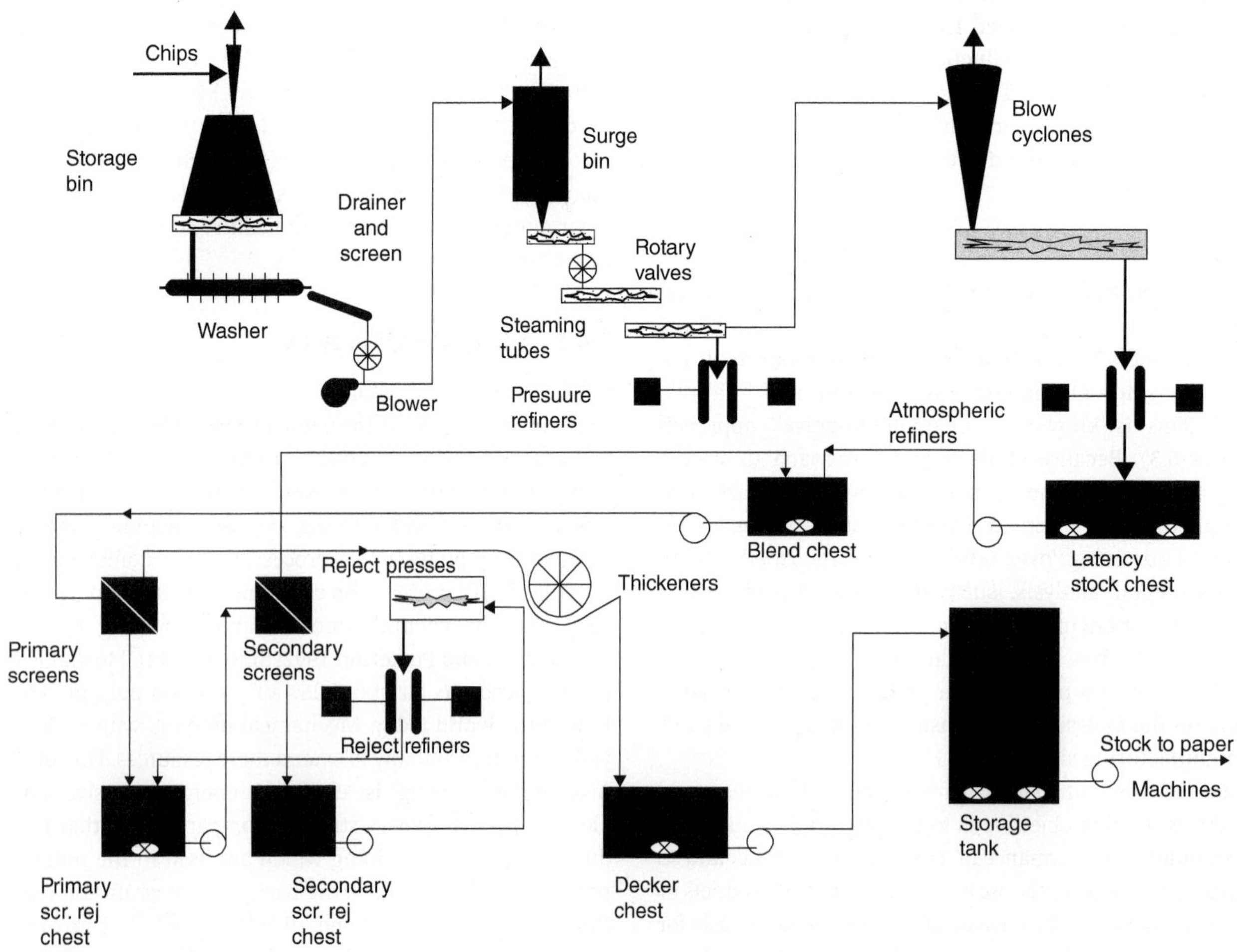

Figure 6.6 Basic flow diagram for TMP process. *Source:* From Das and Houtman (2004).

Table 6.4 Make-up chemical, and energy, water, and land use per 1000T oven dry (OD) pulp and paper production, using chemical-, mechanical-, and bio-pulping.

Component	Chemical (kraft/sulfite)	Mechanical	Bio-TMP
H_2O_2	0.5%	2.0%	2.5%
Na_2O–SiO_2	—	2.0%	2.5%
$MgSO_4$	0.5%	0.5%	0.5%
NaOH	3.0%	1.0%	1.5%
$NaSO_4$	2.5%	—	—
O_2	4.8%	—	—
CIO_2	3.0%	—	—
H_2SO_4	1.0%	—	—
Corn steep liquor	—	—	0.5%
Energy use[a] (MJ/1000T/day OD pulp)	3.15×10^7	8.24×10^7	5.25×10^7
Water use (gal/1000T/day OD pulp)	1.6×10^7	6.0×10^6	6.0×10^6
Land use (acre/1000T/day OD pulp)	20	20	25

[a] Derived from Table 6.5.

dissolved organics are converted to energy. The typical kraft process involves turning logs into wood chips, which are then pulped (Figure 6.7) (Someshwar and Pinkerton 1992). The major steps in chemical pulp and papermaking processes include wood yard operations, pulping, deknotting, washing and screening, chemical recovery, pulp bleaching, pulp drying, and paper making (Smook 1992; Someshwar and Pinkerton 1992). Various gaseous, particulate, liquid, and solid waste streams are produced within a kraft mill (BOD, COD, TSS, NO_x, SO_2, CO, CO_2, PM, TRS, odor, and ash). A wealth of information in the existing literature addresses various treatment methods and pollution prevention and process optimization techniques for these waste streams. Some of the pollution-prevention advancements and a summary of reduction in total environmental discharges (wastewater, air emissions, and solid and hazardous wastes) from kraft and sulfite mills across the United States over the last 20–25 years are discussed in the literature cited in the reference section (Das 2005; Das and Jain 2001).

Table 6.4 presents an average quantity (in percentage) of make-up chemicals that are typically needed for chemical pulping. Chemical pulping uses 63% more chemical than mechanical pulping, and 53% more than biopulping. Table 6.5 presents the energy consumption in various operations of chemical pulping and papermaking processes. Table 6.6 shows a range of effluent loads and other environmental impacts of chemical pulping and papermaking processes. The data from Tables 6.4–6.6 are used to quantify the environmental and human health impacts that will be discussed later in Section 6.4.3.7.

6.4.3.3 Biopulping: A Review of a Pilot Project

Biopulping is the treatment of wood chips and other lignocellulosic materials with lignin-degrading fungi before pulping. Ten years of industry-sponsored research has demonstrated the technical feasibility of the technology for mechanical pulping at a laboratory scale. Two 50-T outdoor chip pile trials conducted at the USDA Forest Service, Forest Products Laboratory (FPL) in Madison, WI, have established the engineering and economical feasibility of the technology. After refining the control and the fungus-treated chips through a thermomechanical pulp (TMP) mill, the resulting pulps were made into papers on the pilot-scale paper machine at FPL. In addition to the 30% savings in electrical energy consumption during refining, improvements in the strength of the resulting paper were seen as a result of fungal pretreatment. Because of the stronger paper, we were able to substitute at least 5% kraft pulp in a blend of mechanical and kraft pulps. This recent work has clearly demonstrated that economic benefits can be achieved with biopulping technology through both the energy savings and substitution of the stronger biopulped TMP for more expensive kraft, while maintaining the paper quality.

6.4.3.4 Introduction to Biopulping

Biopulping, which uses natural wood decay organisms, has the potential to overcome these problems. Fungi alter the lignin in the wood cell walls, which has the effect of "softening" the chips. This substantially reduces the electrical energy needed for mechanical pulping and leads to improvements in the paper strength properties. The fungal

Table 6.5 Summary of specific energy consumption in current pulp and papermaking processes (million Btu/T pulp unless specified otherwise).

	Range of energy variation		
Process	**Minimum**	**Maximum**	**Typical average**
Wood preparation component[a]	0.16	0.25	0.19
Debarking[a]			
Debarking			
Chipping	0.27	0.38	0.32
Pulping component: Chemical	1.5	8.00	4.7
Pulping			
Digesting			
Washing	0.32	1.10	0.57
Refining/screening	2.0	4.20	3.0
Drying	4.0	5.00	4.2
Mechanical pulping	13.2	15.20	14.5
Stone grinding/refining			
Mechanical refining			18.6
Thermomechanical refining	19.6	23.10	19.7
Wastepaper pulping	4.0	5.82	4.3
Bio-TMP pulping	13.72	16.17	13.8
Bleaching component	1.2	14.20	7.5
Chemical recovery component	2.05	5.35	4.4
Evaporation/concentration			
Recovery furnace auxiliary	2.00	4.00	2.64
Recausticizing	1.00	1.05	1.02
Lime kiln	0.94	2.50	2.0
Kraft recovered energy[b]	−4.00	−20.0	−15.0
Papermaking component[a]	2.7	5.01	3.38
Stock preparation			
Sheet formation	0.13	0.40	0.26
Pressing			0.32
Drying	5.0	20.00	7.40
Auxiliary component[a]	0.42	2.00	1.68
Lighting and space heating			
Power plant	1.00	1.50	1.25

[a] Million Btu/T of paper.
[b] Energy recovered as steam from kraft recovery boiler.

pretreatment is a natural process; therefore, no adverse environmental consequences are foreseen (Kirk et al. 1993, 1994; Sykes 1994). Based on the results of previous work and discussions with mill personnel, we envision a fungal treatment system that fits into existing mill operations with minimal disturbance. Figure 6.8 is a conceptual overview of the biotreatment process in relation to existing wood yard operations. Wood is harvested and transported to the mill site for debarking, chipping, and screening. Chips are decontaminated by steaming, maintaining a high temperature for a sufficient time to decontaminate the wood chip surfaces, and then cooled so that the fungus can be applied. The chips are then placed in piles that can be ventilated to maintain the proper temperature, humidity, and moisture content for fungal growth and subsequent biopulping. The retention time in the pile is 1–4 weeks. Recent efforts have focused on bringing the successful laboratory-scale procedures up to the industrial level. Our laboratory process

Table 6.6 Effluent loads from the manufacture of forest products.

Product	Effluent (m^3/MT)	Suspended solids (kg/MT)	BOD (kg/MT)	COD (kg/MT)	*N* (g/MT)	*P* (g/MT)
Pulp	20–60	12–15	5–10	20–30	200–400	80
Sulfate pulp, unbleached						
Sulfate pulp, conventional bleaching[a]	60–100	12–18	18–25	60–120	300–500	120
Sulfate pulp, ECF or TCF[b]	30–50	10–15	14–18	25–40	400–600	100
Sulfite pulp, conventional bleaching	150–200	20–40	30–40	60–100	100–200	60
Groundwood, unbleached	6–10	10–30	10–15	30–50	100–200	50
TMP, unbleached	6–15	10–30	15–25	40–80	100–200	70
TMP, peroxide bleached	6–5	10–30	20–40	60–100	200–300	100
Recycled fiber, deinked	10–20	5–10	20–40	40–80	100–200	40
Paper and paperboard	30–50	10–20	3–8	10–20	50–100	5
Fine paper, coated						
Newsprint	10–25	5–10	1–3	2–4	10–20	5
Folding boxboard	10–25	5–10	2–4	3–6	50–100	8
Sack paper	15–30	5–10	2–4	4–8	100–200	15
Tissue	20–40	5–10	1–3	3–6	50–80	8

The figures in the table were compiled from measurements made at Finish mills. ECF, elemental chlorine free; TCF, total chlorine free.
[a] Kappa number 20–30, depending on wood species.
[b] Extended cooking, oxygen delignification, kappa number 8–14, depending on wood species.

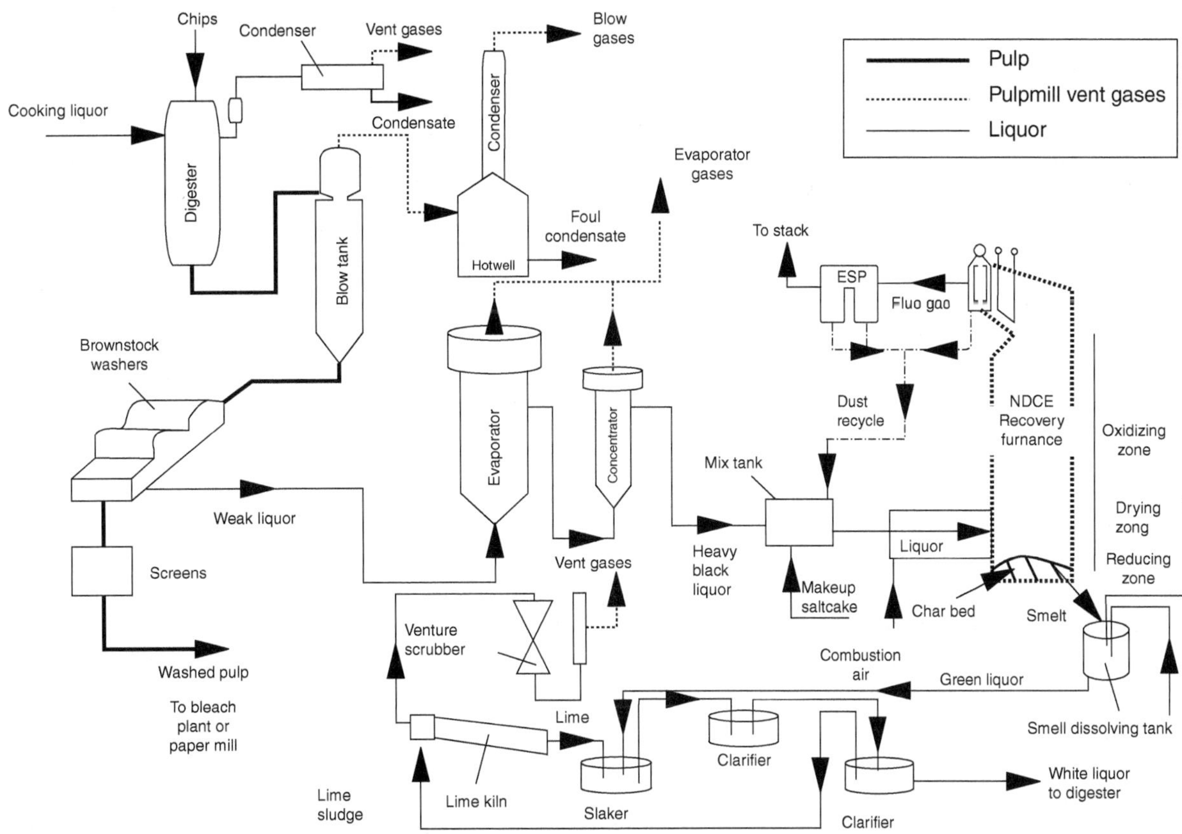

Figure 6.7 Schematic diagram of a typical kraft-sulfate pulping and recovery process. *Source:* From Someshwar and Pinkerton (1992).

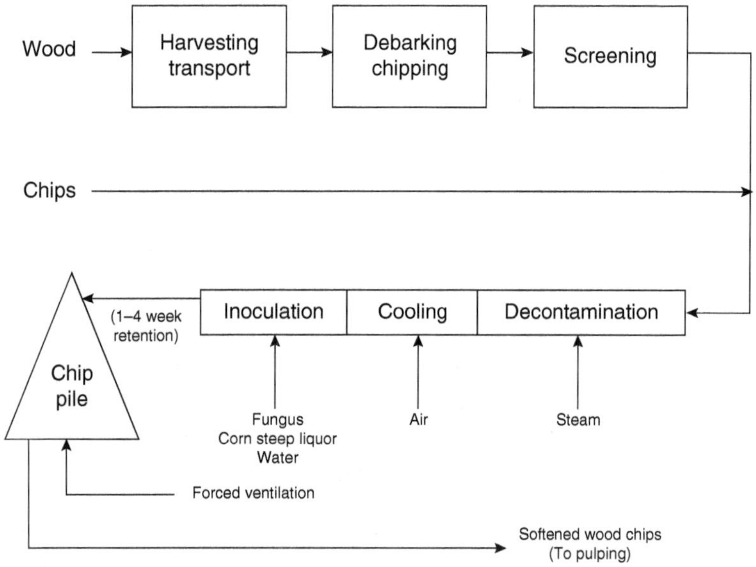

Figure 6.8 Overview of the biopulping process showing how the biotreatment process fits into an existing mill's wood handling systems. *Source:* From Das (2005).

treats approximately 1.5 kg of chips (dry weight basis) at one time. In our scale-up experiments that took biopulping from this lab scale to larger scales, certain implementations of each step were chosen. The chosen implementation allowed us to reach our two goals for this phase of the project: (i) to demonstrate that chips can be decontaminated and inoculated in a continuous process rather than as a batch process and (ii) to demonstrate that the process scaled as expected from an engineering standpoint. The following discusses the results obtained during the scale-up trials in relation to industrial processes.

In reactor scale-up studies, Kirk and others (1993) investigated two types of reactor systems: tubular reactors and chip piles. The tubular reactors have an advantage in obtaining the necessary engineering and kinetic data for scaling-up the process. The one-dimensional nature of the system is easy to analyze and model. The reactor also allows for well-controlled air flow in the system with air flow patterns that are well known. Heat loss from the system is easily controlled with exterior insulation, thus achieving conditions that would be experienced in the center of large chip piles. Details on the configurations of these reactors and the chip piles have been published.

6.4.3.5 Large-Scale Implementation of Biopulping

On a large scale, decontamination and inoculation must be done on a continuous basis and not batchwise as in a laboratory trial (Figure 6.9). To achieve this, the FPL investigators built a treatment system based on two screw conveyers that transport the chips and act as treatment chambers. Figure 6.9 is an overview of the continuous process equipment used in 5- and 50-T trials. Steam is injected into the first screw conveyer, which heats and decontaminates the chip surface. A surge bin, located between the two conveyers, acts as a buffer. From the bottom of the surge bin, a second screw conveyer removes the chips, which are subsequently cooled with filtered air. In the second half of the second screw conveyer, the inoculum suspension containing fungus, unsterilized corn steep liquor, and water is applied and mixed with the chips through the tumbling action. From the screw conveyer, the chips fall into the pile or reactor for a two-week incubation. In the first scale-up trial, 5 T of spruce wood chips were inoculated and incubated at a throughput of approximately 0.5 T/h.

In successful outdoor trials with the biopulping fungus *Ceriporiopsis subvermispora*, nearly 50 T of spruce was treated at a throughput of about 2 T/h (dry weight basis) continuously for over 20 hours. During the two weeks, the chip pile was maintained within the temperature growth range for the fungus, despite the outdoor exposure to ambient conditions.

6.4.3.6 Economics and Environmental Benefits of the Biopulping Process

The economic benefits of biopulping, evaluated based on the process studies and engineering data, result from several effects. Energy reduction at the refiner was used as the primary criterion for the effectiveness of biopulping. For a two-week process, the savings should be a minimum of 25% under the worst-case conditions of wood species and minimal process control, whereas up to nearly 40% can be achieved under some circumstances. Additionally, mills that are currently throughput-limited as a result of refiner capacity may achieve total capacity increases as a result of

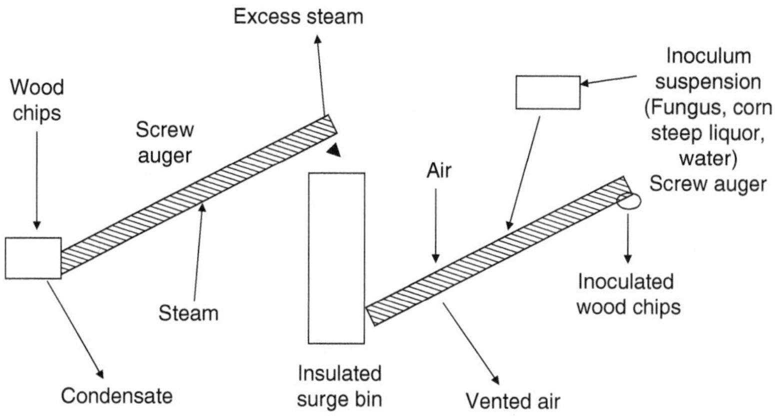

Figure 6.9 Continuous treatment system to decontaminate, cool, and inoculate wood chips. Wood chips are steamed in the first screw conveyor before being placed into a surge bin. The second screw conveyor then picks up the chips, cools them, and applies the inoculum. Pile Storage and Ventilation. *Source:* From Das (2005).

biopulping. The improved strength of the biomechanical pulps would allow the required strength of the blend to be achieved with a lower percentage of the kraft pulp in those cases where the pulps are blended. Finally, only benign materials are used, and no additional waste streams are generated. An economic analysis of the process yielded very favorable results. The analysis of a 200 T/day TMP mill are summarized here. Under different scenarios and assumptions for utility costs, equipment needs, and operating costs, the net savings can range from \$10 to more than \$26/T of pulp produced, with an estimated capital investment of \$2.5 million. Mills that are refiner-limited can experience throughput increases of over 30% from the reduction in refining energy by running the refiners to a constant total power load. Even a modest throughput increase of 10%, coupled with the energy savings of 30%, results in a payback in less than 1 year. This is equivalent to a savings of \$34/T at a 15% rate of return on capital. Furthermore, many mills blend mechanical pulps and kraft pulps to achieve the optical and strength properties desired. Additional benefits of over \$10/T can be realized when the anticipated stronger biopulped TMP is partially substituted for kraft at a 5% rate. The results of a 600 T/day analysis have also been published (Scott et al. 1998a, b). This preliminary analysis is subject to appropriate qualifications. The capital costs are subject to some variability, in particular the costs associated with integrating the new facility into an existing site. The additional advantages of biopulping, including the environmental benefits and pitch reduction, have not been quantified in this report. Finally, much of this analysis is site-specific, depending on the operating conditions at the particular mill considering incorporating biopulping into its operations.

Although the environmental benefits of biopulping for mechanical pulping have not been quantitatively determined in an economic sense, we can discuss the benefits qualitatively. Sykes (1994) determined that there were no environmental problems with the effluent from a biopulping process. In terms of GHGs, there is a net benefit to the biopulping process. The process itself is oxidative, producing carbon dioxide as the fungus metabolizes the various wood components. However, based on the typical generation efficiency of electrical energy, the reduced energy requirements (~30%) significantly lower the overall generation of carbon dioxide when the entire system is analyzed. Finally, in terms of biological issues with using fungi in the wood yard, there are benefits to using the biopulping fungus. The biopulping fungus outcompetes the normal cohort of fungi and other organisms that will grow in an uncontrolled wood chip pile. Some of these other organisms, such as *Aspergillis*, can be detrimental to human health. The biopulping fungi are naturally occurring in the forests of the world and have no known health effects on humans. More information on benefit–cost analysis and economic input–output model (EIO-LCA) for chemical process and manufacturing industries are available in Appendix F.

6.4.3.7 Quantitative Analyses Using the USEPA Model: TRACI

The tool for the reduction and assessment of chemical and other environmental impacts (TRACI) is described along with its history, the research and methodologies it incorporates, and the insights it provides within individual impact categories.

TRACI, a stand-alone computer program developed by the US Environmental Protection Agency (USEPA), facilitates the characterization of environmental stressors that have potential effects, including ozone depletion, global warming, acidification, eutrophication, tropospheric ozone (smog) formation, ecotoxicity, human health criteria–related effects, human health cancer effects, human health noncancer effects, fossil fuel depletion, and land-use effects. TRACI was originally designed for use with LCA, but it is expected to find wider application in the future (Bare 2003; UNEP-SETAC 2000).

The categories of odor, noise, radiation, waste heat, and accidents are outside of the USEPA's purview and are usually not included within case studies in the United States for various reasons, including, perhaps, because the perceived threat from these categories is often considered minimal, local, or difficult to predict. The resource depletion categories are recognized as being of significance in the United States, especially for fossil fuel, land, and water use.

For the development of TRACI, each of the above impact categories was considered and its current state of development and perceived societal value were assessed. The traditional pollution categories of ozone depletion, global warming, human toxicology, ecotoxicology, smog formation, acidification, and eutrophication were included within TRACI because various programs and regulations within the USEPA recognize the value of minimizing effects from these categories. The category of human health was further subdivided into cancer, noncancer, and criteria pollutants (with an initial focus on particulates) to better reflect the focus of renewable fuels, and purchased electricity used. Emissions from wastewater treatment plant and other biological processes are not included. In Washington State, 60% energy used in the pulp and paper mills are generated from renewable sources (almost all from hydro) and 40% from purchased electricity generated from natural gas, coal, and oil. The CO_2 emissions from renewable energy sources assumed zero. Typical values of 70% combustion efficiency and 60% electricity transmission efficiency were assumed. For an overall efficiency factor of 42% in calculating CO_2 emissions from nonrenewable sources, power-generating plants to pulp mills were used.

USEPA regulations and to allow methodology development consistent with US regulations, handbooks, and guidelines (USEPA 1989b, c). Smog-formation effects were kept independent and not further aggregated with other human health impacts because environmental effects related to smog formation would have become masked and/or lost in the process of aggregation. Criteria pollutants were maintained as a separate human health impact category, allowing a modeling approach that can take advantage of the extensive epidemiological data associated with these well-studied impacts.

6.4.3.8 Model Input and Output

It is important to note that data quality is fundamental to LCA and its interpretation. Some of the data quality issues such as reliability and consistency can be overcome by using standardized database, which are starting to emerge after years of data compilation and their incorporation into publicly and commercially available databases. Table 6.7 presents some chemical release data that are used in TRACI. Table 6.8 presents the overall environmental and human health impacts, including global warming, smog formation, human health criteria pollutants, acidification, eutrophication, and ecotoxicity. The results indicate that biopulping has a lower environmental impact than that of the TMP process because of its reduced energy consumption. Biopulping also has a lower impact than that of chemical pulping in all the measures except global warming. One should note that the total CO_2 production per ton was the highest for chemical pulping, but because kraft recovery furnaces are fired with lignin derived from the wood, a renewable resource, this contribution to GHGs was assumed to be net zero. Figure 6.10 shows global warming characterization results predicted by TRACI.

These results likely indicate that biopulping is the preferred process when the electrical supply is dominated by renewable energy sources. In cases where the fuel mix for producing electricity is largely fossil fuels, chemical pulping provides a process powered by renewable energy but at a cost of higher hazardous waste emissions.

In conclusion, this streamlined LCA clearly indicates that biopulping in the papermaking process has

Table 6.7 Emissions from chemical-, mechanical-, and bio-pulping and papermaking processes, 1000 T OD per day.

Pollutant	Chemical	Mechanical	Bio (TMP)
Acetone (lb)	1.5×10^3	—	—
BOD (lb)	4.4×10^3	1.76×10^3	1.76×10^3
CO_2,[a] (lb)	3.5×10^6	8.0×10^6	5.5×10^6
CO (lb)	5.5×10^4	5.5×10^3	5.5×10^3
Chloroform (lb)	8.5×10^2	7.5×10^2	7.5×10^2
COD (lb)	6.0×10^3	1.0×10^3	5.0×10^2
Ethanol (lb)	5.0×10^2	5.0×10^1	—
H_2S (lb)	2.35×10^3	—	—
Isoprene (lb)	1.0×10^3	1.0×10^3	1.0×10^3
Methane (lb)	8.2×10^3	2.7×10^3	2.7×10^3
Methanol (lb)	2.3×10^3	6.0×10^2	6.0×10^2
Nitrogen (lb)	7.5×10^2	—	—
NO_2 (lb)	3.5×10^4	1.0×10^3	5.0×10^2
NO (lb)	3.5×10^4	1.0×10^3	5.0×10^2
SO_2 (lb)	1.75×10^4	—	—

[a] CO_2 emissions would depend on actual nonrenewable energy sources.

Table 6.8 Environmental and human health impacts characterization results in percentage for chemical-, mechanical-, and bio-pulping processes using TRACI.

Impacts %	Chemical	Mechanical (TMP)	Bio-pulping (TMP)
Global warming	21	47	32
Photochemical smog	44	28	28
Acidification	97	2	1
Ecotoxicity	36	32	32
Human health cancer	36	32	32
Human health noncancer	36	32	32

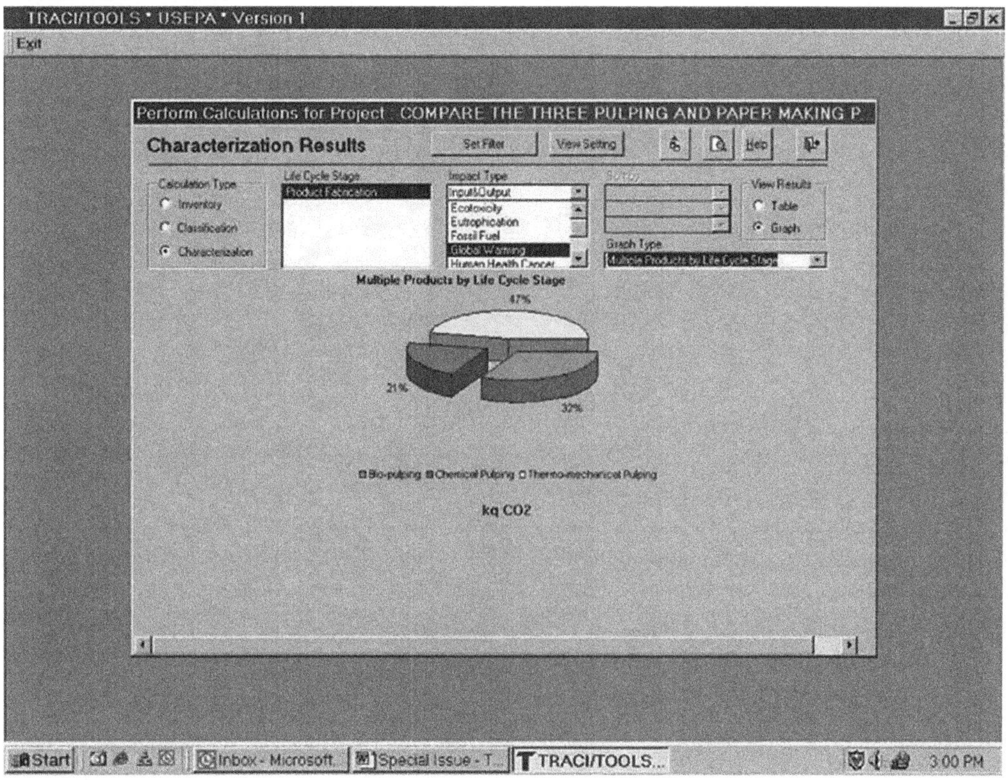

Figure 6.10 TRACI's global warming characterization for chemical-, mechanical-, and bio-pulping. *Source:* From Das and Houtman (2004).

several advantages over chemical and mechanical processes as follows:

- Reduced electrical energy consumption (at least 30%) over mechanical pulping
- Potential 30% increase in mill throughput for mechanical pulping
- Improve paper strength properties
- The TRACI model indicates a significant reduction in environmental and human health impacts
- The biopulping process proves to be more sustainable in terms of not only economic advantage but also environmental and human health benefits

6.5 Evaluating the Life Cycle Environmental Performance of Two Disinfection Technologies

With increasing emphasis on promoting a sustainable ecological future and concern over introducing toxic chemicals in water, disinfection process design is leaning toward technologies that destroy pathogens while balancing the effects of the disinfected wastewater aquatic biota or on a drinking water supply. Since ultraviolet (UV) irradiation is not a chemical additive, it does not leave or produce toxic by-product in the wastewater, unlike traditional chlorination and dechlorination processes; therefore, the use of UV does not affect a drinking water supply or the aquatic biota in receiving waters.

The objective of this work is to present streamlined environmental (LCAs) between two competitive disinfection processes: chlorine vs. UV. This section discusses the use of LCA to quantify the environmental and public health benefits of UV disinfection technology as opposed to chlorination and dechlorination methods. LCA tools are used to evaluate the short-term and long-term environmental effects of both processes and to select the best sustainable process. Our approach applies environmental LCA to these disinfection processes, incorporating economic criteria and all aspects of the environment: chemical use, electricity use, and releases to water, air, and land.

6.5.1 The Challenge

With increasing emphasis on promoting sound ecological practices and concern over introducing toxic chemicals into water, designs for disinfection processes are increasingly leaning toward technologies that destroy pathogens while balancing the effects of this disinfected wastewater on aquatic biota or a drinking water supply.

In the United States and Canada, the use of ultraviolet light irradiation for the disinfection of wastewater has become the accepted alternative to chlorination or chlorination/dechlorination. There are several reasons for this move away from a proven technology. For example, because of current Uniform Fire Codes, containment requirements for volatile gases like chlorine, and public health concerns, municipalities are limited in the amounts of chlorine that can be stored in a water treatment plant. Moreover, the dechlorination process uses yet another chemical pollutant, sulfur dioxide, to remove chlorine from the effluent before discharged into the receiving water. Thus, and because chlorination/dechlorination of wastewater produces possible carcinogens in addition to destroying aquatic biota in the receiving waters, USEPAs started to look for an alternative wastewater disinfection system.

Various governments, municipalities (Das 2002, 2004; Das and Ekstrom 1999; Ecology 1998; LOTT 1994; USEPA 1986, 1992b), and corporations have sponsored research (Loge et al. 1996; Scheible 1987; Scheible and Forndran 1986; White 2010) that shows that the UV disinfection of wastewater was effective and economical. Another most important development was the parallel flow open channel modular UV system. This new design of the UV system for wastewater in the early 1980s opened up UV disinfection for both the retrofit market and new wastewater treatment plants. To promote a friendlier discharge to the marine environment, designers have begun to prefer alternative disinfection technologies, which emphasize sustainable and clean ecological disinfectants – such a clean technology is UV disinfection.

6.5.2 The Chlorination (Disinfection) Process

Despite the acknowledged advantages of disinfection by means of UV irradiation, however, chlorine continues to be the most widely used chemical for the disinfection of wastewater in the United States and elsewhere. The major advantages of chlorine over alternative disinfectants are cost-effectiveness, reliability, and efficacy against a host of pathogenic organisms. We turn now to a detailed examination of the chlorination process.

When chlorine (Cl_2) is dissolved in freshwater, a mixture of hypochlorous acid (HOCl) and hydrochlorite ion (OCl^-) is formed (Eq. 6.2). Chlorine exists predominantly as HOCl below pH 7.6 and as OCl^- above pH 7.6. HOCl and/or OCl^- is defined as free available chlorine, with the hypochlorous acid being the primary disinfectant.

$$Cl_2 + H_2O \leftrightarrow HOCl + OCl^- + H^- \tag{6.2}$$

Mono-, di-, and tri-chloramines (NH_2Cl, $NHCl_2$, and $NHCl_3$) are formed when chlorine reacts with nitrogen present in secondary effluent in the form of ammonia. Municipal effluents usually contain all these forms of chlorine in some proportion and taken together they are known as "total residual chlorine" (TRC). Because saltwater contains bromide, the addition of chlorine to saltwater will also form hypobromous acid (HOBr), hypobromous ion (OBr^-), and bromamines.

Chlorine is typically supplied as liquefied gas in cylinders. Chlorinators apply gaseous chlorine to a feed stream which is then injected into a mixing zone in the chlorine contact chambers. Initial mixing and effective contact times are essential for good process performance. Generally, contact periods of 15–30 minutes are required at peak flow (USEPA, 1985).

6.5.2.1 Limitations

Although chlorine disinfection is a largely reliable and effective process, it has certain limitations. For examples, chlorine reacts with certain chemicals in the wastewater, leaving only the residual for disinfection. Wastewater components that readily combine with chlorine include reduced iron and sulfur compounds, ammoniated-nitrogen, organic nitrogen, tannins, uric and humic acid, cyanides, phenols, and unsaturated organics. Cysts of *Entamoeba histolytica*, *Giardia lamblia*, and *Mycobacterium tuberculosis*, some viruses, and eggs of parasitic worms show resistance to chlorine. Consistent disinfection in effluents containing organic nitrogen may pose problems, even when a measured free chlorine residual is present.

6.5.2.2 Human Health and Environmental Impact

Chlorine is toxic to aquatic, estuarine, and marine organisms. An additional hazard is the carcinogenic potential of chloro-organic compounds. Chlorine gas is potentially toxic when inhaled, and chlorine transport poses a risk (*see Section 5.10.5 and Tables 5.8 and 5.9*). Special handling is required and emergency response plans are required under right-to-know regulations for on-site storage of gaseous chlorine. Chlorine gas and the hypochlorites are also highly corrosive. Chlorine gas concentrations of 15–20 ppm for 30–60 minutes are dangerous; higher concentrations for very brief periods can be fatal. Chlorination can result in the formation of carcinogenic chloro-organics.

The EPA has established toxicity criteria for TRC in receiving waters. In freshwaters the acute level is 19 mg/l (1-hour average) and the chronic level is 11 mg/l (4-day average). The saltwater acute and chronic criteria are 13 mg/l (1-hour average) and 7.5 mg/l (4-day average), respectively. Due to the toxicity of chlorine residuals at such low concentrations and the high limit of analytical detection (50–100 mg/l), chlorine induced toxicity in the receiving stream is difficult to control.

6.5.3 Dechlorination with Sulfur Dioxide

In the past decade, concerns over chlorine toxicity and protection of fish and wildlife have led to a dramatic growth in the practice of dechlorination to remove all or part of the chlorine residual and halogenated organics remaining after chlorination. Dechlorination also reduces or eliminates toxicity harmful to aquatic life in receiving waters.

Sulfur dioxide gas successively removes free chlorine, monochloramine, dichloramine, nitrogen trichloride, and polychlorinated compounds. When sulfur dioxide is added to wastewater, the following reactions occur:

$$SO_2 + H_2O \rightarrow HSO_3^- + H^+ \tag{6.3}$$

$$HOCl + HSO_3^- \rightarrow Cl^- + SO_4^{-2} + 2H^+ \tag{6.4}$$

$$SO_2 + HOCl + H_2O \rightarrow Cl^- + SO_4^{-2} + 3H^+ \left(\text{free chlorine}\right) \tag{6.5}$$

$$SO_2 + H_2O \rightarrow HSO_3^- + H^+ \tag{6.6}$$

$$NH_2Cl + HSO_3^- + H_2O \rightarrow Cl^- + SO_4^{-2} + NH_4^+ + H^+ \tag{6.7}$$

$$SO_2 + NH_2Cl + 2H_2O \rightarrow Cl^- + SO_4^{-2} + NH_4^+ + 2H^+ \quad (\text{combined chlorine}) \tag{6.8}$$

For the overall reaction between sulfur dioxide and chlorine (Eq. 6.5), the stoichiometric weight ratio of sulfur dioxide to chlorine is 0.9 : 1. In practice, it has been found that about 1.0 mg/l of sulfur dioxide will accomplish for the dechlorination of 1.0 mg/l of chlorine residue (expressed as Cl_2). Because the reactions of sulfur dioxide with chlorine and chloramines are nearly instantaneous, contact time is not usually a factor and contact chambers are not used; however, rapid and positive mixing at the point of application is an absolute requirement.

The ratio of free chlorine to the total combined chlorine residual before dechlorination determines whether the dechlorination process is partial or proceeds to completion. If the ratio is less than 85%, it can be assumed that significant organic nitrogen is present and that it will interfere with the free residual chlorine process (Metcalf and Eddy 2003).

Since dechlorination removes most of the TRC from disinfected wastewaters, it reduces the toxicity of disinfected wastewater effluent to aquatic wildlife. In most situations, sulfur dioxide dechlorination is a very reliable unit process in wastewater treatment, provided the precision of the combined chlorine residual monitoring service is adequate. Excess sulfur dioxide dosages should be avoided not only because of the chemical wastage but also because of the oxygen demand (BOD and COD) exerted by the excess sulfur dioxide.

6.5.3.1 Limitations

Chlorination/dechlorination is more complex to operate and maintain than chlorination alone. Major difficulties are the inability to measure residual SO_2 and problems in the continuous measurement of a zero or low chlorine residual. Many halogenated organics are also rapidly formed upon chlorine addition and are unaffected by application of SO_2. Heltz and Nweke who examined many plants, reported that the amount of residual chlorine in dechlorinated effluents considerably exceeded EPA criteria for receiving waters (Heltz and Nweke 1995).

6.5.3.2 Environmental Impact

Sulfuric acid and hydrochloric acid are products of SO_2 dechlorination in small amounts but are generally neutralized in the wastewater. Based on laboratory experiments, residuals of sulfite dechlorination are at least three orders of magnitude less toxic than chlorine or ozone.

No cases of sulfur compounds affecting dissolved oxygen consumption or pH change in receiving waters or in dechlorinated effluents have been reported. In pilot studies, no significant oxygen depletion occurred until sulfur dioxide overdoses exceeded 50 mg/l. It is not uncommon, however, to find plants with post aeration after dechlorination to assure that dissolved oxygen requirements are met. Dechlorination with SO_2 would significantly reduce toxicity due to chlorination disinfection process.

Sulfur dioxide, which is used for dechlorination purposes, is stored on site in small quantities. The maximum anticipated storage inventory is quite small, and there need be no concerns about threats to the health of workers or of residents of nearby populated areas.

6.5.3.3 Effects of Chlorine on Aquatic Life

The potential environmental and chemical effects of chlorine toxicity on 14 aquatic species are summarized in Table 6.9. Rainbow trout was the most sensitive of the species tested, followed by the golden shiner and three-spined stickleback. A calculated chlorine residual of 0.03 mg/l, based on dilution of a measured concentration of 2.0 mg/l, reduced plankton photosynthesis by more than 20% of the value obtained with a dilution of effluent having no chlorine residual. Dechlorination with sodium bisulfite also eliminated chlorine-related toxicity. Dechlorination with sulfur dioxide also greatly reduced the acute and chronic toxicity to fish and invertebrate species.

In considering these data, it should be borne in mind that the toxicity of chlorine wastes in rivers depends not on the amount of chlorine added but on the concentration of chlorine remaining in solution (Merkens 1958). The toxicity of this residual chlorine will depend on its composition (i.e. the relative proportions of free chlorine and chloramines). Arriving at this ratio, which in turn depends on at

Table 6.9 Select summary of acute and chronic toxic effects of residual chlorine on aquatic life.

Species	Effect endpoint	Measured residual chlorine concentration (mg/l)
Coho salmon	7-day TL_{50}[a]	0.083
Pink salmon	100% kill (1–2)	0.08–0.10
Coho salmon	100% kill (1–2)	0.13–0.20
Pink salmon	Maximum nonlethal	0.05
Coho salmon	Maximum nonlethal	0.05
Brook trout	7-day TL_{50}	0.083
Brook trout	Absent in streams	0.015
Brown trout	Absent in streams	0.015
Brook trout	67% lethality (4)	0.01
Brook trout	Depressed activity	0.005
Rainbow trout	96-h TL_{50}	0.14–0.29
Rainbow trout	7-day TL_{50}	0.08
Rainbow trout	Lethal (12)	0.01
Trout fry	Lethal (2)	0.06
Yellow perch	7-day TL_{50}	0.205
Largemouth bass	7-day TL_{50}	0.261
Smallmouth bass	Absent in streams	0.1
White sucker	7-day TL_{50}	0.132
Walleye	7-day TL_{50}	0.15
Black bullhead	96-h TL_{50}	0.099
Fathead minnow	96-h TL_{50}	0.05–0.16
Fathead minnow	7-day TL_{50}	0.082–0.115
Fathead minnow	Safe concentration	0.0165
Golden shiner	96-h TL_{50}	0.19
Fish species diversity	50% reduction	0.01
Send	Safe concentration	0.00.34
Send	Safe concentration	0.012
Daphnia magna	Safe concentration	0.003
Protozoa	Lethal	0.1

[a] $TL_{50,}$ median tolerance limit (50% survival) (Brungs 1973).

least five other variables, is a complex undertaking. Although free chlorine is more toxic than chloramines, research that has stood since the 1950s suggests that the difference between the toxicity to fish of free chlorine and chloramine is not very great (Dondoroff and Katz 1950; Merkens 1958; Hart 1973; Kozloff 1974).

Results of Laboratory Bioassays Several studies (Dondoroff and Katz 1950; Merkens 1958; Thatcher 1977) indicated that salmonids were the most sensitive fish species. Laboratory bioassays support this generalization. A residual chlorine concentration of 0.006 mg/l was lethal to trout fry in 2 days, and the 7-day median tolerance limits, or TL_{50}, for rainbow trout was 0.08 mg/l with an estimated median period of survival of 1 year at 0.004 mg/l (Merkens 1958). The maximum nonlethal (in 23 days) concentration of residual chlorine for pink and coho salmon in seawater was 0.05 mg/l (Holland 1960). Rainbow trout were killed at 0.01 mg/l in 12 days, and they avoided a concentration of 0.001 mg/l (Sprague and Drury 1969). Brook trout had a mean survival time of 9 hours at 0.35 mg/l, 18 hours at 0.08 mg/l, and 48 hours at 0.04 mg/l. Mortality was 67% after 4 days at 0.01 mg/l (Dandy 1967, 1972). Fifty percent of brown trout were killed at 0.02 mg/l within 10.5 hours and at 0.01 mg/l within 43.5 hours (Pike 1971). Trout, salmon, and some fish-food organisms are more sensitive than warm-water fish, snails, and crayfish (Tables 6.9 and 6.10).

6.5.4 UV Disinfection Process

A UV disinfection system transfers electromagnetic energy from a mercury arc lamp to an organism's genetic material, the chromosomes which contain DNA and RNA. When UV radiation penetrates the cell wall of an organism, it destroys the cell's ability to reproduce. In the disinfection process, UV radiation, generated by an electrical discharge through mercury vapor, penetrates the genetic material of microorganisms and causes molecular rearrangements that retard their ability to reproduce. Das gave a detailed review of the mechanisms of germicidal action, how does UV light works, how UV-damaged DNA is repaired, and the effects of wastewater quality parameters on disinfection efficiency (Das 2004).

Table 6.10 Summary of results of brief exposures of fish to residual chlorine.

Species	Effect endpoint	Time	Measured residual chlorine concentration (mg/l)
Chinook salmon	First death	2.2 h	0.25
Brook trout	Median mortality	90 min	0.5
Brook trout	Mean survival time	9 h	0.35
Brook trout	Mean survival time	18 h	0.08
Brook trout	Mean survival time	48 h	0.04
Brown trout	Depressed activity	24 h	0.005
Rainbow trout	Slight avoidance	10 min	0.001
Rainbow trout	Lethal	2 h	0.3
Fingerling Rainbow trout	Lethal	4–5 h	0.25
Trout fry	Lethal	Instantly	0.3
Yellow perch	TL_{50}[a]	1 h	<0.88
Yellow perch	TL_{50}	12 h	0.494
Smallmouth bass	Median mortality	15 h	0.5
White sucker	Lethal	30–60 min	1.0
Largemouth bass	TL_{50}	1 h	<0.74
Largemouth bass	TL_{50}	12 h	0.365
Fathead minnow	TL_{50}	1 h	<0.79
Fathead minnow	TL_{50}	12 h	0.26
Miscellaneous	Initial kill	15 min	0.28
Miscellaneous	Erratic swimming	6 min	0.09

[a] TL_{50}, median tolerance limit (50% survival) (Brungs 1973).

UV disinfection systems are not mass-produced. Because their efficiency strongly depends on effluent characteristics that act to decrease the UV intensity in wastewater, each application must have a custom-designed system. Table 6.11 shows the major parameters that must be taken into consideration when a UV disinfection system is being designed for wastewater. In comparison to chlorination/dechlorination, UV disinfection offers a reduction of potential chlorinated hydrocarbons (including potential carcinogens) in the receiving waters, as well as considerably greater safety to wastewater treatment plant operators and to nearby populated areas.

6.5.4.1 UV Transmission or Absorbance

UV light's ability to penetrate wastewater is measured in a spectrophotometer at the same wavelength (254 nm) that is produced by germicidal lamps. This measurement is called the *percent transmission* or *absorbance* and it is a function of all the factors that absorb or reflect UV light. As the percent transmission gets lower (higher absorbance) the ability of the UV light to penetrate the wastewater and reach the target organisms decreases.

It cannot be estimated by simply looking at a sample of wastewater with the naked eye. The range of effective transmittances (*T*) will vary depending on the secondary treatment systems. In general, suspended growth-treatment processes produce effluent with *T* varying from 60 to 65%. Fixed film processes range from 50 to 55% *T* and lagoons 35–40% *T*. Industries that influence UV transmittance include textile, printing, pulp and paper, food processing, meat and poultry processing, photo developing, and chemical manufacturing. A discussion on other major parameters affecting the UV disinfection efficiency of wastewater is given by Das (2004).

6.5.4.2 Disinfection Standards

The level of disinfection required to obtain an EPA National Pollutant Discharge Elimination System (NPDES) permit is commonly less than 200 fecal coliform unit per 100 ml as a 30-day geometric mean. In general, a UV dose of 20–30 mW·s/cm^2 is required to achieve this level of disinfection in secondary-treated wastewater with a 65% transmittance and total suspended solids (TSS) below 20 mg/l. The UV dose requirement to meet specific limits depends on the nature of the particle with respect to numbers, size, and composition. Therefore, UV dose requirements will vary (Table 6.11). A more stringent limit (<2.2 total coliform units per 100 ml) is required for water reuse in California and Hawaii. In such cases, filtered effluents with TSS of 2 mg/l or less and 65% transmittance may require UV dose as high as 120 mW·s/cm^2 to achieve this level of disinfection. The concentrations of solids, bacteria in the particles, and the particle size distribution are the main limiting factors in the design of systems that must meet stringent disinfection limits. It appears that the UV dose required to achieve the traditional coliform limits will achieve better virus inactivation than the comparable chlorine dose. Figure 6.11, which illustrates the relative doses of UV and chlorine required to inactivate selected organisms compared to fecal coliform indicator, also shows that the chlorine doses required to inactivate most organisms are much higher than comparable UV doses that would achieve the same level of disinfection (Das 2002, 2004; Trojan Technologies Inc. 2010).

Table 6.11 Major parameters affecting the UV disinfection of wastewater.

Parameters	Acceptable values/conditions
Percent transmittance (T) or absorbance	35–65
Total suspended solids (TSS) mg/L	5–10
Particle size distribution (PSD) μm	10–40
Flow rate or hydraulics design	Ideal plug flow
Iron (mg/l)	<0.3
Hardness (mg/l)	<300

6.5.4.3 Operation, Maintenance, and Worker Safety

The proper operation and maintenance (O&M) of a UV disinfection system ensures that sufficient UV radiation is transmitted to the organisms to render them sterile. All surfaces between the UV radiation and the target organisms must be cleaned, and the ballasts, lamps, and reactor must be functioning at peak efficiency. Inadequate cleaning is one of the most common causes of a UV system's ineffectiveness. In all cases, the quartz sleeves or Teflon tubes need to be cleaned regularly by mechanical wipers, ultrasound equipment, or chemicals. The cleaning frequency is very site-specific: some systems needing to be cleaned more often than others.

Chemical cleaning is most commonly done with citric acid. Other cleaning agents include mild vinegar solutions and sodium hydrosulfite. A combination of cleaning agents should be tested to find the agent most suitable for the wastewater characteristics without producing harmful or toxic by-products.

UV is generated on-site and poses no significant safety concerns to surrounding communities. Worker safety requirements are directed to protection from exposure (primarily of the eyes to skin) from UV light, as well as to

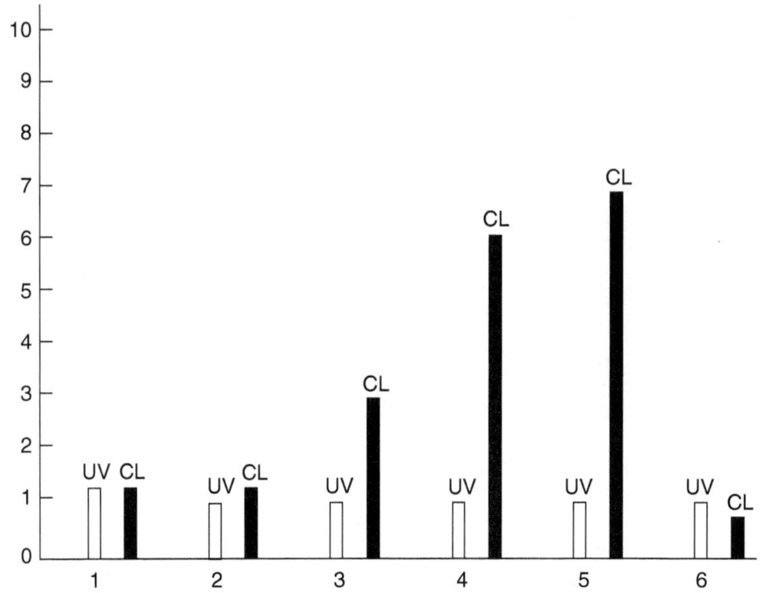

Figure 6.11 Comparison of the relative effectiveness of chlorine and UV on bacteria and viruses. (i) *Escherichia coli* (ii) *Salmonella typhosa* (iii) *Staphylococcus aureus* (iv) polio virus type 1 (v) Coxsackie AZ virus (vi) adenovirus type (Das 2002, 2004; Trojan Technologies Inc. 2010).

strict monitoring of electrical hazards and safe handling and disposal of expended lamps, quartz, and ballasts (US EPA 1996; US EPA 1998).

6.5.4.4 Costs

Specifically, consideration must be given to how environmental matters affect our economic thinking and, conversely, how economic decisions affect the environment. Cost considerations are an integral part of the decision-making process at the stage of identifying potential improvements to a process, a product, or an activity.

Estimated capital costs for 1, 10, and 50 million gal per day (mgd) (1 mgd = 3785 m^3/day) plants are $150 000, $700 000, and $3 000 000 respectively. O&M costs are estimated to be 2.6, 1.4, and 0.8 US cents/1000 gal (or 3.785 m^3), respectively, at an average dose of 6 mg/l. These increase to 4.0, 3.0, and 1.5 US cents/1000 gal at a dose of 10 mg/l.

Chemical costs of chlorine gas and hypochlorite vary considerably depending upon the locality, demand, and availability. Current prices for Cl_2 are varying in the range: $0.0425–0.06/lb (or 454 g) for 90 T (or ~90 000 kg) tank cars; $0.10/lb for 55 T (or ~55 000 kg) rail cars; $0.25–0.275/lb for 1 T (or ~1 000 kg) cylinders; and $0.50–0.55/lb for 150 lb (or ~68 kg) cylinders. Liquid sodium hypochlorite (12.5%) prices quoted were $0.50–1.75/gal.

A brief cost analysis for both chlorination/dechlorination and UV disinfection processes treating a wastewater treatment facility is presented in Figure 6.12, capable of processing 18 mgd. These costs can vary considerably depending on the locality, demands, and supply. The annualized cost values for alternatives were based on the design average plant flow of 18 mgd over a 20-year period and 5.5% interest rate (F. Soroushian, personal communication; Temmer et al. 2000).

Figure 6.12 indicates that, although the construction cost for chlorination system is considerably less than the UV, annualized costs are about the same for both systems. There is no more real economic benefit for pursuing chlorination disinfection system.

6.5.4.5 Environmental Impacts of Energy Sources and Implications of Renewables

Like chlorination, UV disinfection systems demand energy supplies. One of the most important factors that can contribute to achieving sustainable development is the requirement for a supply of energy resources that is itself fully sustainable. Effective and efficient utilization of energy resources calls for such resources to be readily available at reasonable cost utilized for all required tasks without causing negative societal impact. Clearly, there is an intimate connection between renewable energy sources and sustainable development (Das 2002).

Table 6.12 gives a brief summary of cost of generating powers, air quality, and environmental impacts of renewable and other forms of energy sources. It illustrates that renewable energy sources can make a significant contribution to reducing greenhouse and acid gas emissions. Renewable sources have their own environmental impacts but these are often small, site-specific, and local in nature.

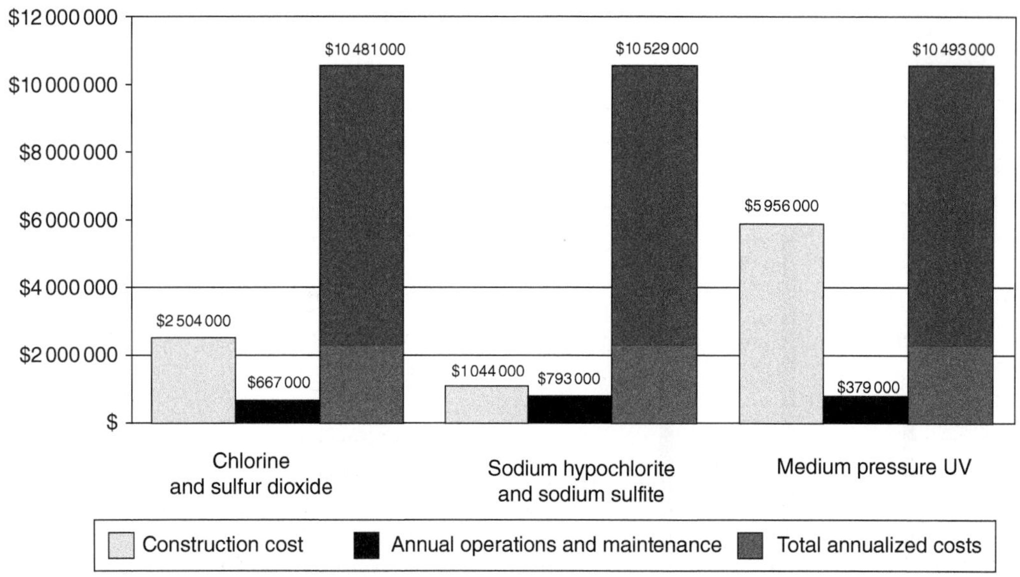

Figure 6.12 Disinfection system cost comparison of an 18-mgd facility.

Table 6.12 Cost of generating power, and air quality and environmental impacts.

Source	Cost	Potential environmental impact and pollutants of concern
Solar photovoltaic (large scale)	~18–20 US cents/kWh	Corrosive acid, potentially toxic and hazardous substance, land use and loss of habitat, visual intrusion, uncontrolled dumping in landfills
Solar photovoltaic (small scale)	~22–24 US cents/kWh	Potentially toxic and hazardous substance
Wind	~4.5–10 US cents/kWh	Land use and habitat damage, noise, bird strike
Biomass	~6–8 US cents/kWh excluding ethanol-based fuels	CO_2, NO_x, PM_{10}, CO, VOC, toxic air pollutants (TAPs)
Geothermal	~5–8 US cents/kWh	Emission of H_2S during operation, groundwater and soil contamination, surface water, land use, visual impact
Solar thermal	~5–25 US cents/kWh	Thermal or chemical pollution of surface water
Large-scale hydro	Ranges from 2.0 to 7.8 US cents/kWh	Fish and other aquatic lives, terrestrial ecosystems, local climate, public health, water flow and supply, population displacement, loss of agricultural land
Existing hydro	ranges from to 2.0 US cents/kWh	Fish and other aquatic lives, water supply, irrigation, SO_2
Nuclear	5.5–6.0 US cents/kWh	Radioactive substance, public health
Coal-fired generators	4.8–5.5 US cents/kWh	CO_2, NO_x, SO_2, PM_{10}, CO, VOC, TAPs
Diesel generators	10–15 US cents/kWh	CO_2, NO_x, SO_2, PM_{10}, CO, VOC, TAPs
Natural gas combined Cycle combustion turbine	(at natural gas price of $5/MMBtu) = 5 US cents/kWh	CO_2, NO_x, PM_{10}, CO, VOC, TAPs
Simple cycle combustion turbine	6 US cents/kWh at $5/MMBtu for natural gas	CO_2, NO_x, PM_{10}, CO, VOC, TAPs

Source: Energy Ideas Clearinghouse (2018).

6.6 Case Study: LCA Comparisons of Electricity from Biorenewables and Fossil Fuels

A series of LCAs was conducted on biomass, coal, and natural gas systems to quantify the environmental benefits and drawbacks of each. All those evaluations were conducted in a cradle-to-grave manner to cover all processes necessary for the operation of the power plant including raw material extraction, feed preparation, transportation, waste disposal, and recycling. We summarize data on energy balance, GWP, air emissions, and resource consumption for each system (NREL 2011).

The generation of electricity and the consumption of energy in general result in consequences to the environment. Using renewable resources and incorporating advanced technologies such as integrated gasification combined cycle (IGCC) may result in less environmental damage, but to what degree, and with what trade-offs? LCA studies have been conducted on various power generating options in order to better understand the environmental benefits and drawbacks of each technology.

Material and energy balances were used to quantify the emissions, energy use, and resource consumption of each process required for the power plant to operate. These include feedstock procurement (mining coal, extracting natural gas, growing dedicated biomass, collecting residue biomass), transportation, manufacture of equipment and intermediate materials (e.g. fertilizers, limestone), construction of the power plant, decommissioning, and any necessary waste disposal.

The following systems were studied:

- A biomass-fired IGCC system using a biomass energy crop (hybrid poplar)
- A pulverized coal (pc) boiler with steam cycle, representing the average for coal-fired power plants in the United States today
- A system cofiring biomass residue with coal (15% by heat input will be presented here)
- A direct-fired biomass power plant using biomass residue (urban, primarily)
- A natural gas combined cycle (NGCC) power plant.

Each study was conducted independently and can therefore stand alone, giving a complete picture of each power generation technology. However, the resulting emissions, resource consumption, and energy requirements of each system can ultimately be compared, revealing the environmental benefits and drawbacks of the renewable and fossil-based systems.

6.6.1 Results

6.6.1.1 System Energy Balance

The total energy consumed by each system includes the fuel energy consumed plus the energy contained in raw and intermediate materials that are consumed by the systems. Examples of the first type of energy use are the fuel spent in transportation and fossil fuels consumed by the fossil-based power plants. The second type of energy is the sum of the energy that would be released during combustion of the material (if it is a fuel) and the total energy that is consumed in delivering the material to its point of use. Examples of this type of energy consumption are the use of natural gas in the manufacture of fertilizers and the use of limestone in flue-gas desulfurization (FGD). The combustion energy calculation is applied where nonrenewable fuels are used, reflecting the fact that the fuel has a potential energy that is being consumed by the system. The combustion energy of renewable resources, those replenished at a rate equal to or greater than the rate of consumption, is not subtracted from the net energy of the system. This is because, on a life cycle basis, the resource is not being consumed. To determine the net energy balance of each system, the energy used in each process block is subtracted from the energy produced by the power plant. The total system energy consumption by each system is shown in Table 6.13.

To examine the process operations that consume the largest quantities of energy within each system, two energy measurement parameters were defined. First, the energy delivered to the grid divided by the total fossil-derived energy consumed by each system was calculated. This measure, known as the net energy ratio, is useful for assessing how much energy is generated for each unit of fossil fuel consumed. The other measure, the external energy ratio, is defined to be the energy delivered to the grid divided by the total non-feedstock energy to the power

Table 6.13 Total system energy consumption.

System	Total energy consumed (kJ/kWh)
Average coal	12 575
Natural gas IGCC	8 377
Biomass/coal cofiring (15% by heat input)	10 118
Biomass-fired IGCC using hybrid poplar	231
Direct-fired biomass power plant using biomass residue	125

Source: From NREL (2011).

plant. That is, the energy contained in the coal and natural gas used at the fossil-based power plants is excluded. The external energy ratio assesses how much energy is generated for each unit of upstream energy consumed.

Because the energy in the biomass is considered to be both generated and consumed within the boundaries of the system, the net energy ratio and external energy ratio will be the same for the biomass-only cases (biomass-fired IGCC and direct-fired biomass). In calculating the external energy ratio, we are essentially treating the coal and natural gas fed to the fossil power plants as renewable fuels, so that upstream energy consumption can be compared. The energy results for each case studied are shown in Figure 6.13.

As expected, the biomass-only plants consume less energy overall, since the consumption of nonrenewable coal and natural gas at the fossil plants results in net energy balances of less than one. The direct-fired biomass residue case delivers the most amount of electricity per unit of energy consumed. This is because the energy used to provide a usable residue biomass to the plant is fairly low. Despite its higher plant efficiency, the biomass IGCC plant has a lower net energy balance than the direct-fired plant because of the energy required to grow the biomass as a dedicated crop. Residue resource limitations, however, may necessitate the use of energy crops in the future. Cofiring biomass with coal slightly increases the energy ratios over those for the coal-only case, even though the plant efficiency was derated by 0.9 percentage points.

In calculating the external energy ratios, the feedstocks to the power plants were excluded, essentially treating all feedstocks as renewable. Because of the perception that biomass fuels are of lower quality than fossil fuels, the external energy ratios for the fossil-based systems were expected to be substantially higher than those of the biomass-based systems. The opposite is true, however, due to the large amount of energy that is consumed in upstream operations in the fossil-based systems. The total non-feedstock energy consumed by the systems is shown in Table 6.14. In the case of coal, 35% of this energy is consumed in operations relating to flue-gas cleanup, including limestone procurement.

Table 6.14 Non-feedstock energy consumption.

System	Total energy consumed (kJ/kWh)
Average coal	702
Natural gas IGCC	1718
Biomass/coal cofiring (15% by heat input)	614
Biomass-fired IGCC using hybrid poplar	231
Direct-fired biomass power plant using biomass residue	125

Source: From NREL (2011).

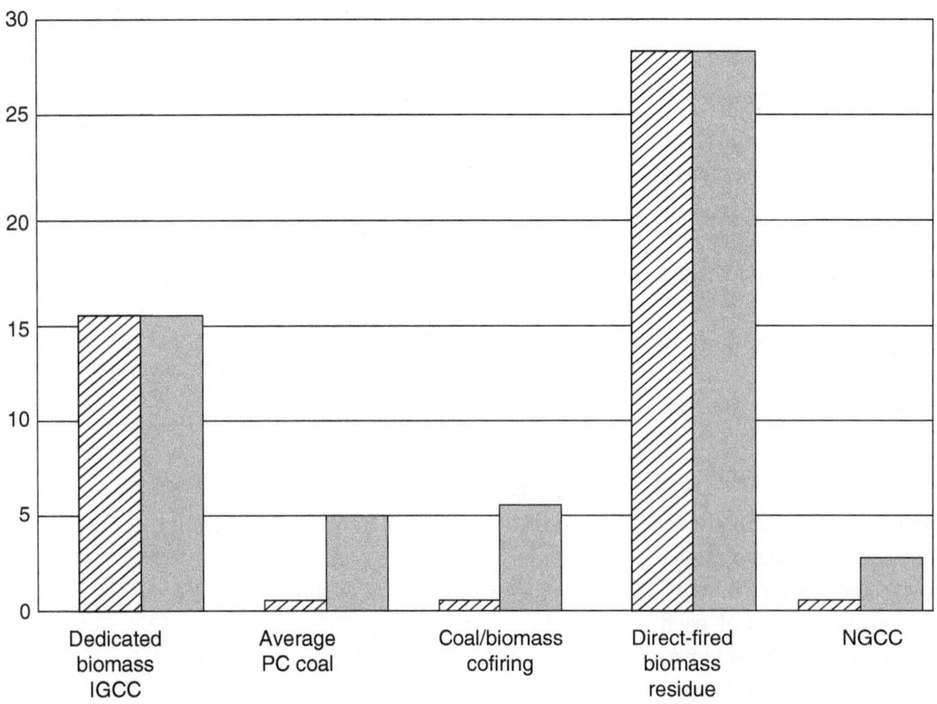

Figure 6.13 Life cycle energy balance.

Mining the coal consumes 25% of this energy, while transporting the coal is responsible for 32%. Greater than 97% of the upstream energy consumption related to the natural gas IGCC system is due to natural gas extraction and pipeline transport steps, including fugitive losses. Although upstream processes in the biomass systems also consume energy, shorter transportation distances and the fact that FGD is not required reduce the total energy burden.

6.6.1.2 Global Warming Potential

Figure 6.14 shows the net emissions of the three GHGs quantified for these studies: carbon dioxide (CO_2), methane (CH_4), and nitrous oxide (N_2O). The biomass IGCC system has a much lower GWP than the fossil systems because of the absorption of CO_2 during the biomass growth cycle. Sensitivity analyses demonstrated that even moderate amounts of soil carbon sequestration (1900 kg/ha/7-year rotation) would result in the biomass IGCC system having a zero-net GHG balance. Sequestration amounts greater than this would result in a negative release of GHGs, and a system that removes carbon from the atmosphere overall.

The base case presented here assumes no net change in soil carbon, since actual gains and losses will be very site specific. The direct-fired biomass system in which the residue is used at the power plants has a highly negative rate of GHG emissions because the generation of methane associated with biomass decomposition is avoided. Based on current disposal practices, it was assumed that 46% of the residue biomass used in the direct-fired and cofiring cases would have been sent to a landfill and that the remainder would end up as mulch and other low-value products. Decomposition studies reported in the literature were used to determine that if the biomass residue had not been used at the power plant, approximately 9% of the carbon would have ended up as CH_4 and 61% as CO_2. The remaining carbon is resistant to decomposition in the landfill. Had all the residue biomass been decomposed aerobically, the CO_2 produced would have been 1.85 kg/kg biomass. If the biomass residue was not used at the power plant, the decomposition pathways described above would have resulted in total GHG emissions of 2.48 kg CO_2-equiv/kg biomass (1.117 kg CO_2 + 0.065 kg CH_4). The net difference is the reason for the negative GHG emissions associated with the direct-fired system.

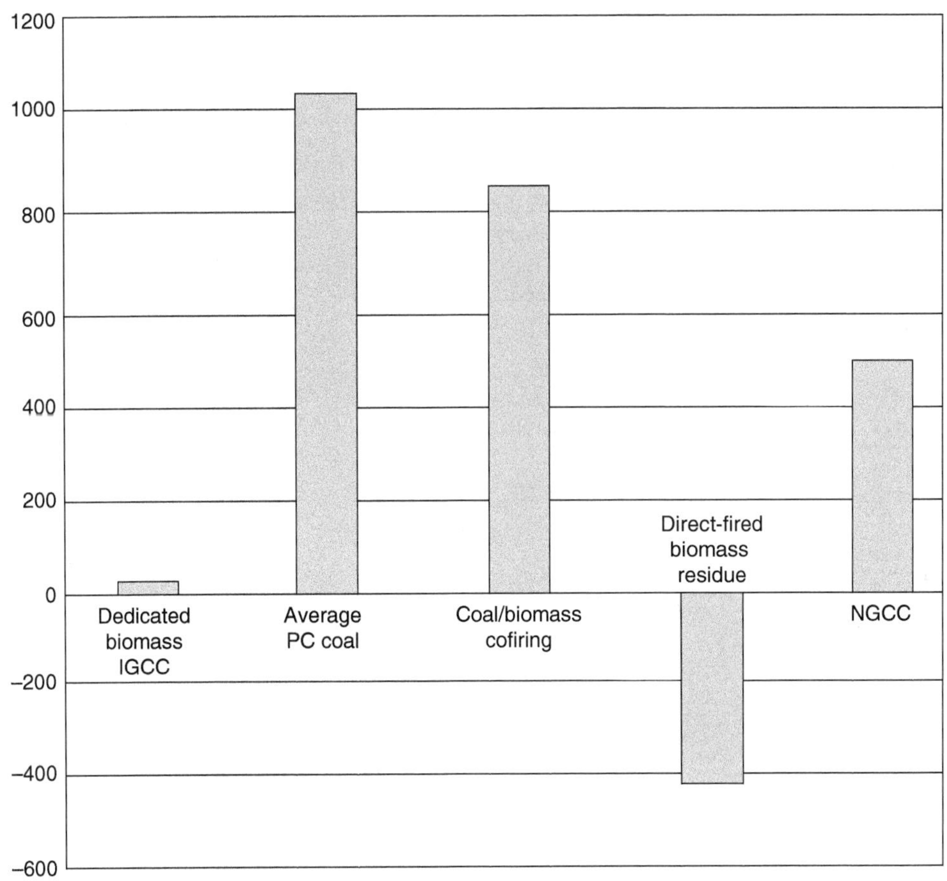

Figure 6.14 Net GHG emission.

The NGCC plant has the lowest GWP of all fossil systems because of its higher efficiency, despite natural gas losses that increase net CH_4 emissions. Natural gas losses during extraction and delivery were assumed to be 1.4% of the gross amount extracted. Because of the potency of methane as a GHG, nearly one-quarter of the total GWP of this system is due to these losses. Cofiring biomass with coal at 15% by heat input reduces the GWP of the average coal-fired power plant by 18%. The reduction in GHGs is greater than the rate at which biomass is cofired because of the avoidance of methane emissions associated with decomposition that would have occurred had the biomass not been used at the power plant. Biomass disposal and decomposition emissions for this scenario are the same as those used in the direct-fired case.

6.6.1.3 Air Emissions

Figure 6.15 charts the following emissions: particulates, oxides of sulfur (SO_x), oxides of nitrogen (NO_x), CH_4, CO_2, and non-methane hydrocarbons (NMHCs). Methane emissions are high for the natural gas case due to natural gas losses during extraction and delivery. The direct-fired biomass and coal/biomass cofiring cases have negative methane emissions due to avoided decomposition processes (landfilling and mulching). CO and NMHCs are higher for the biomass case because of upstream diesel combustion during biomass growth and preparation. Cofiring reduces the coal system air emissions by approximately the rate of cofiring, with the exception of particulates, which are generated during biomass chipping and handling.

6.6.1.4 Resource Consumption

Figure 6.16 shows the total amount of nonrenewable resources consumed by the systems investigated. Limestone is used in significant quantities by the coal-fired power plants for FGD. The natural gas IGCC plant consumes almost negligible quantities of resources, with the exception of the feedstock itself, including that lost during extraction and delivery.

6.6.2 Sensitivity Analysis

A sensitivity analysis was conducted on each system to determine which parameters had the most influence on the results and to pinpoint opportunities for reducing the environmental burden of the system. In general, parameters associated with increasing the system efficiency and reducing the fossil fuel usage had the largest effects. Additionally, for the biomass systems, variables associated with growing a dedicated feedstock and factors affecting how much CO_2 and CH_4 are avoided by using biomass residue significantly affected the GWP of the system. Overall, however, the sensitivity analyses demonstrated that the conclusions that can be drawn from these studies remain relatively constant as different parameters are varied.

6.6.3 Summary and Conclusions

It is evident that biomass power systems reduce the environmental burden associated with power generation. The key comparative results can be summarized as follows:

- The GWP of generating electricity using a dedicated energy crop in an IGCC system is 4.7% of that of an average US coal power system.

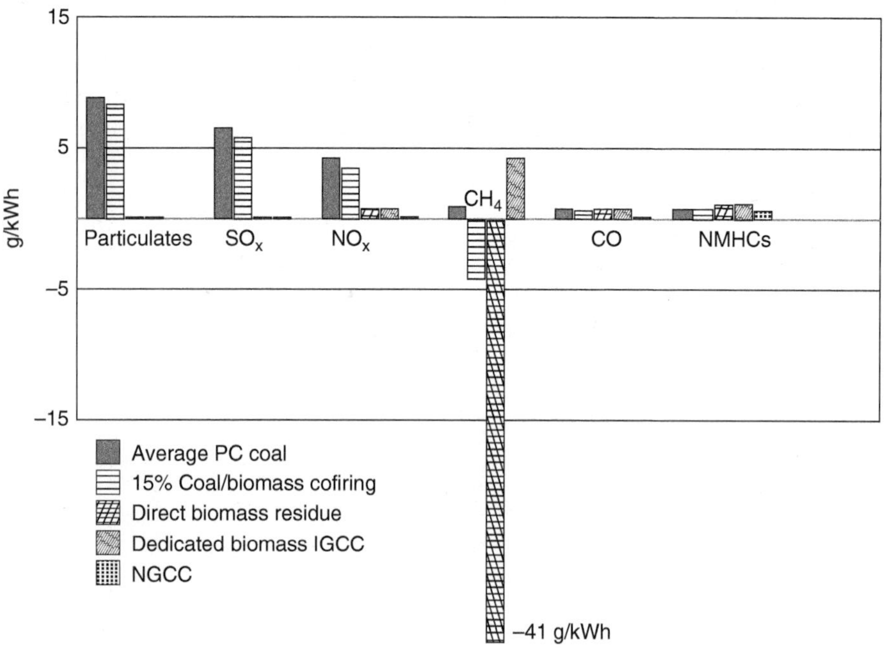

Figure 6.15 Emissions of PM, SO_x, NO_x, CH_4, CO, and NMHCs.

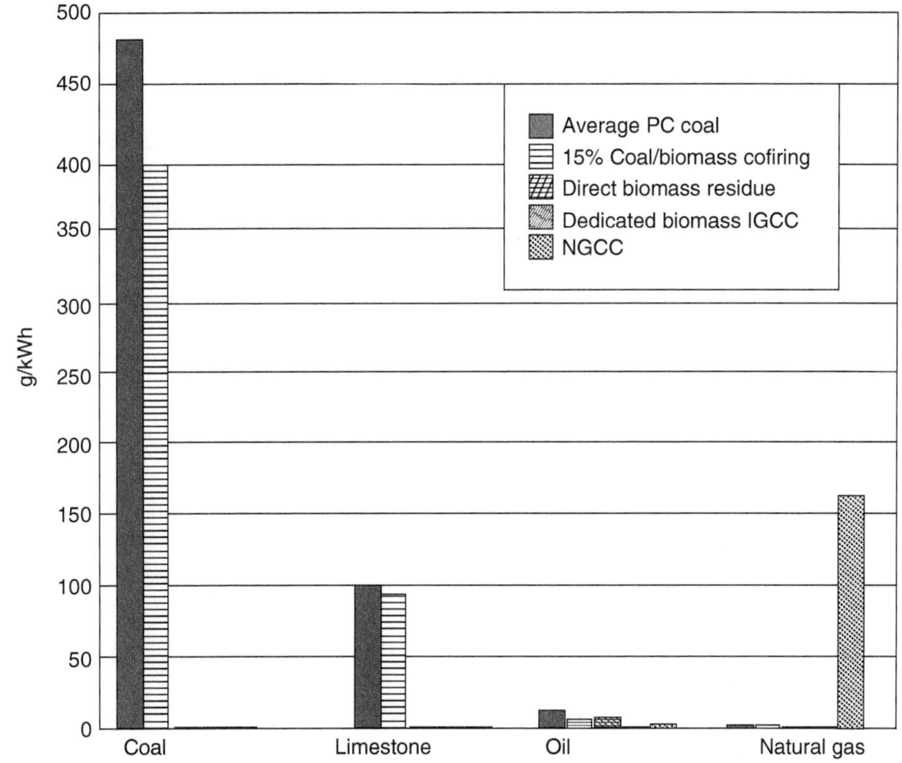

Figure 6.16 Total amount of nonrenewable resources consumed by the systems.

- Cofiring residue biomass at 15% by heat input reduces the GHG emissions and net energy consumption of the average coal system by 18 and 12%, respectively.
- The life cycle energy consumption of the coal and natural gas systems are significantly lower than those of the biomass systems because of the consumption of nonrenewable resources.
- Not counting the coal and natural gas consumed at the power plants in these systems, the net energy consumption is still lower than that of the biomass systems because of energy used in processes related to flue-gas cleanup, transportation, and natural gas extraction and coal mining.
- The biomass systems produce very low levels of particulates, NO*x*, and SO*x*, compared to the fossil systems.
- System methane emissions are negative when residue biomass is used because of avoided decomposition emissions.
- The biomass systems consume very small quantities of natural resources compared to the fossil systems.
- Other than natural gas, the natural gas IGCC consumes small amounts of resources.

These results demonstrate that overall, biomass power provides significant environmental benefits over conventional fossil-based power systems. In particular, biomass systems can significantly reduce the amount of GHGs that are produced, per kilowatt-hour of electricity generated. Additionally, because the biomass systems use renewable energy instead of nonrenewable fossil fuels, they consume very small quantities of natural resources and have a positive net energy balance. Cofiring biomass with coal offers an opportunity to reduce the environmental burdens associated with the coal-fired power systems that currently generate over half of the electricity in the United States. Finally, by reducing NO_x, SO_x, and particulates, biomass power can improve local air quality over coal-fired power generation.

6.7 Best Available Control Technology (for Environmental Remediation)

6.7.1 What Is "Best Available Control Technology"?

Best available control technology (BACT) is a pollution control standard mandated by the Clean Air Act:

> ...an emission limitation based on the maximum degree of reduction of each pollutant subject to regulation under this Act emitted from or which results from any major emitting facility, which the permitting authority, on a case-by-case basis, taking into account energy, environmental, and economic impacts and other costs, determines is achievable for such facility through

application of production processes and available methods, systems, and techniques, including fuel cleaning, clean fuels, or treatment or innovative fuel combustion techniques for control of each such pollutant.

The USEPA determines what air pollution control technology will be used to control a specific pollutant to a specified limit. When a BACT is determined, factors such as energy consumption, total source emission, regional environmental impact, and economic costs are taken into account. It is the current EPA standard for all polluting sources that fall under the New Source Review (NSR) and PSD permitting guidelines, and is determined on a case-by-case basis. The BACT standard is significantly more stringent than the reasonably available control technology standard but much less stringent than the lowest achievable emissions rate standard (U.S. EPA, 2016a).

6.7.1.1 Reasonably Available Control Technology

Reasonably available control technology (RACT) is a pollution control standard created by the EPA and is used to determine what air pollution control technology will be used to control a specific pollutant to a specified limit. RACT applies to existing sources in areas that are not meeting national ambient air quality standards on controlled air pollutants and is required on all sources that meet these criteria. The RACT standard is less stringent than the BACT.

6.7.1.2 Lowest Achievable Emissions Rate

The lowest achievable emissions rate (LAER) is used by the EPA to determine if emissions from a new or modified major stationary source are acceptable under SIP guidelines. LAER standards are required when a new stationary source is located in a non-attainment air-quality region (*see* Figure 4.14). It is the most stringent air pollution standard above the best available control technology and reasonably available control technology standards.

6.7.1.3 Life Cycle-Based Environmental Law

Life cycle concepts are increasingly becoming a part of environmental regulation, especially those involving GHG emissions. Now LCA is often used in writing about carbon accounting. In these times of heightened concern over climate change, individuals, organizations, and governmental agencies alike are eager to measure the release and impact of GHGs. For example, a variety of governmental agencies are developing approaches for regulating the emissions of GHGs associated with the production and use of fuels. As an example, the Energy Independence and Security Act (HR 6) of 2007 (EISA 2007) aims to: move the United States toward greater energy independence and security; increase the production of clean renewable fuels; protect consumers; increase the efficiency of products, buildings, and vehicles. States are also implementing GHGs emission regulation (USEPA 2010). For example, California Global warming Solutions Act of 2006 resulted in regulations that establish a limit for life cycle GHG emissions of transportation fuels. Both the California low-carbon fuel standard and Section 526 of EISA require a life cycle evaluation of the GHG emissions of transportation fuels and this is becoming a common approach to considering GHG emissions (Allen et al. 2009).

6.7.1.4 Life Cycle Best Control Technology

A number of quantitative tools are emerging that enable life cycle assessments and analyses, and subsequently provide guidance on selecting best available control technologies for PSD, NPDES, and other environmental permitting processes. Many of our environmental laws are based on risk frameworks. Now, the life cycle–based best control technology decisions in environmental permitting processes are made that incorporate not only health and environmental but also the economic and societal impacts. Next, we present a case study on BACT for a power-generating facility.

6.8 BACT: Applications to Gas Turbine Power Plants

In the United States, BACT comprises the equipment and methods needed to achieve the maximum degree of reduction of pollutants subject to federal regulations that are emitted from any proposed major stationary source or major modification of such a source. The pollutants subject to review under the PSD regulations, and for which a BACT analysis is required, include nitrogen oxides (NO_x), carbon monoxide (CO), particulates less than or equal to 10 μm in diameter (PM_{10}), and VOC. All PM is assumed to be PM_{10}. The BACT review follows the "top-down" approach recommended by the USEPA. The top-down process calls for all available control technologies for a given pollutant to be identified and ranked in descending order of control effectiveness. The permit applicant should first examine the highest-ranked ("top") option. The top-ranked options should be established as BACT unless the permit applicant demonstrates to the satisfaction of the permitting authority that technical consideration, or energy, environmental, or economic impacts justify a conclusion that the top-ranked technology is not "achievable" in that case. If the most effective control strategy is eliminated in this fashion, then the next most effective alternative should be evaluated, and so on, until an option is selected as BACT.

The USEPA has broken down this analytical process into the following five steps:

Step 1: Identify all available control technologies
Step 2: Eliminate technically infeasible options
Step 3: Rank remaining control technologies
Step 4: Evaluate most effective controls and document results
Step 5: Select the BACT

If there is no single BACT for any industry, it is determined on a case-by-case basis, taking into account cost-effectiveness, economic, energy, environmental, and other effects of proposed solutions. Three important terms in BACT analysis are PSD, "significant emission rate" (SER), and ambient air quality analysis and national ambient air quality standards (NAAQS). The purpose of the PSD program is to implement the Federal Clean Air Act requirements for the prevention of "significant" deterioration of air quality. These insure that the permitting of proposed new industrial facilities and the associated economic growth will occur in a manner consistent with the preservation of clean air resources. The program provides for special emphasis on implementation of BACT, protecting of scenic areas such as national park, and informed public participation. SER refers to net emissions increase or the potential of a source to emit pollutants, a SER equal to or greater than the rates listed in Title 40 CFR 51. An ambient air quality analysis must be carried out for each regulated pollutant and must demonstrate that the source will not cause nor contribute to a violation of any applicable NAAQS.

As industries age and expand, they acquire new emissions units for pollutants and modify old ones. Federal law calls for a BACT analysis of each such new or altered unit of "major stationary sources." Major stationary source, as defined in Title 40 CFR Part 51, Subpart 1, review of new sources and modification – any stationary source that (1) emits, or has the potential to emit (PTE), 250 T/Y or more of any pollutant under the CAA; or (2) emits, or has the PTE, 100 tpy or more of a regulated pollutant within one of the 28 sources categories (Title 40 CFR).

The USEPA has consistently interpreted the statutory and regulatory BACT definitions as containing two core requirements that the agency believes must be met by any BACT determination. First, the BACT analysis must *consider* the most stringent available technologies (i.e. those which provide the "maximum degree of emissions reduction"). Second, any decision to require a lesser degree of emissions reduction must be justified by an objective analysis of "energy, environmental, and economic impacts."

BACT must be at least as stringent as any New Source Performance Standard (NSPS) applicable to the emissions source. With this set of data as a baseline, a BACT analysis often proceed by way of the so-called top-down approach recommended by the EPA. The first step is to determine for the emission unit in question the most stringent control available for a similar or identical source or source category. If it can be shown that this level of control is technically infeasible for the unit in question, the next most stringent level of control is determined and similarly evaluated. This process continues until the BACT level under consideration cannot be eliminated by any substantial or unique technical or environmental concerns. The remaining technologies are evaluated on the basis of operational and economic effectiveness (USEPA 1990).

6.8.1 Importance of Energy Efficiency

USEPA believes that it is important in BACT reviews for permitting authorities to consider options that improve the overall energy efficiency of the source or modification – through technologies, processes, and practices at the emitting unit. In general, a more energy efficient technology burns less fuel than a less energy efficient technology on a per unit of output basis. For example, coal-fired boilers operating at supercritical steam conditions consume approximately 5% less fuel per megawatt hour produced than boilers operating at subcritical steam conditions (US DOE 2007). Thus, considering the most energy efficient technologies in the BACT analysis helps reduce the products of combustion, which includes not only GHGs but other regulated NSR pollutants (e.g. NO_x, SO_2, $PM/PM_{10}/PM_{2.5}$, CO, etc.). Thus, it is also important to emphasize that energy efficiency should be considered in BACT determinations for all regulated NSR pollutants (not just GHGs).

The emission units of a gas turbine power plant for which a BACT analysis is required include the combustion turbines, duct burners, the auxiliary boiler, and the cooling towers. Due to their status as emergency/backup units and/or very limited run time, the emergency diesel generator and the diesel fire water pump are not included in the BACT analysis. The top-down BACT approach analysis must look not only at the most stringent emission control technology previously approved but also evaluate all demonstrated and potentially applicable technologies, including innovative controls, lower polluting processes, etc. For a gas turbine power plant, these technologies and emissions data can be identified through a review of sources made available by the EPA. Foremost among these is RBLC, an umbrella term standing for RACT/BACT/LAER clearinghouse, where RACT is reasonably available control technology and LAER is lowest achievable emission rates. Other sources are EPA's NSR and Clean Air Technology Center websites, the EPA Technology Transfer Network, STAPPA/ALAPCO (the State and Territorial Air Pollution

Program Administrators and the Association of Local Air Pollution Control Officials), and Clean Air World.

Table 6.15 presents the technologies we shall consider in the sections that follow and their approximate control efficiencies.

6.8.2 NO_x BACT Review

6.8.2.1 Combustion Turbines and Duct Burners

NO_x is produced through two mechanisms: high temperature processes, which create thermal NO_x (products of the reaction of nitrogen and oxygen gases in the air) and combustion of nitrogen-containing materials, which produces fuel NO_x. Table 6.16 lists the technologies that were identified for controlling NO_x emissions from gas turbines and their effective emission levels.

6.8.2.2 $SCONO_X$

$SCONO_X$ is an emerging proprietary catalytic and absorption technology that has shown some promise for turbine applications. Unlike selective catalytic reductions (SCRs), which requires ammonia injection, this system does not require ammonia as a reagent; its parallel catalyst beds are alternately taken off line through means of mechanical dampers for regeneration.

Despite its advantages, however, the process $SCONO_X$ catalyst is subject to the same fouling or masking degradation that is experienced by any catalyst operating in a turbine exhaust stream. There is also a small energy loss from the performance loss due to the pressure drop across the catalyst.

The vendor of $SCONO_X$ guarantees performance to all owners and operators of natural gas-fired combustion turbines, regardless of size or gas turbine supplier. The system is designed to reduce both CO and NO_x emissions from natural gas-fired power plants to levels below ambient concentrations. Indeed, the EPA considers $SCONO_X$ a technically feasible and commercially available air pollution control technology and expects its emission levels for criteria pollutants such as NO_x, CO, and VOC to be comparable or superior to previously applied technologies for large combined cycle turbine applications.

$SCONO_X$ has been demonstrated successfully on smaller power plants, including a 32 MW combined-cycle General Electric LM2500 gas turbine in Los Angeles (Das 2003). This facility uses water injection in conjunction with $SCONO_X$ to achieve a NO_x emissions rate of 0.75 ppm on a 15-minute rolling average. The $SCONO_X$ technology has also been successfully demonstrated on a 5 MW Solar Turbine Model Taurus 50 at the Genetics Institute in Andover, Massachusetts. The system is reducing NO_x down to 0.5 ppm NO_x, on a one-hour rolling average. The permit for the power plant was originally issued for 2.5 ppm NO_x.

The manufacturer guarantees CO emissions of 1 ppm and NO_x emissions of 2 ppm. According to one set of figures, when NO_x is reduced from 12.18 ppm (gas turbine with duct

Table 6.15 Technologies and their approximate control efficiencies.

Pollutant	Technology	Potential control efficiency (%)
NO_x	$SCONO_X$	70–95
	Selective catalytic reduction (SCR)	50–95
	Dry low NO_x combustors	40–60
	Selective non-catalytic reduction (SNCR)	40–60
	Water/steam injection	30–50
	Good combustion practices	Base case
CO	Catalytic oxidation	60–80
	Good combustion practices	Base case
PM_{10}	*Good combustion practices*	10–30
	Fuel specification: clean-burning Fuels	Base case
VOC	Catalytic oxidation	60–80
	Good combustion practices	Base case

Source: From USEPA (2016b).

Table 6.16 NO_x control technologies and effective emission levels.

Technology	Typical control range (% removal)	Typical emission level (ppm)
$SCONO_X$	90–95	2–2.5
$XONON_{TM}$ flameless combustion	80–90	3–5
Selective catalytic reduction (SCR) with low-NO_x combustor or SCR with water injection	50–95	2–6
SCR with water/steam injection or advanced low-NO_x combustor	50–95	6–9
Dry low-NO_x combustor and/or aggressive water injection	30–70	9–25
Water/steam injection or low-NO_x burners	30–70	25–35

Source: From USEPA (2016b), Catalytica Energy Systems Inc. (2004), and California Air Resources Board (2000).

burner firing) to 2 ppm, the cost effectiveness is $13 627/T of NO_x removed (Catalytica Energy Systems Inc. 2004).

6.8.2.3 XONON

Several companies are reported to be working on a second technology for the control of NO_x. Introduced commercially by Catalytica Combustion Systems, it is being marketed under the name XONON. This technology replaces traditional flame combustion with flameless catalytic combustion. NO_x control is accomplished through the combustion process using a catalyst to limit the temperature in the combustor below the temperature where NO_x is formed. The XONON demonstrated to achieve near-zero emissions. The XONON combustion system consists of four sections: (i) the preburner, for start-up, acceleration of the turbine engine, and adjusting catalyst inlet temperature if needed; (ii) the fuel injection and fuel-air mixing system, which achieves a uniform fuel-air mixture to the catalyst; (iii) the flameless catalyst module, where a portion of the fuel is combusted flamelessly; and (iv) the burnout zone, where the remainder of the fuel is combusted.

The single field installation of the XONON technology at a municipal power company is being used to perform engineering studies of the technology at Silicon Valley Power, in Santa Clara, California. NO_x emissions are well below 2.5 ppm on the 1.5 MW Kawasaki M1A-13A gas turbine. Catalytica has a collaborative commercialization agreement with General Electric Power Systems, committing to the development of XONON. In conjunction with General Electric Power systems, the XONON system was specified for use with the GE 7FA turbines at the proposed 750 MW natural gas-fired Pastoria Energy Facility, near Bakersfield, California. The project entered commercial operations in 2003. Because the NO_x emissions limitations of 2.5 ppm have been demonstrated in practice by a commercial facility, this technology is considered commercially available at this time (Catalytica Energy Systems Inc. 2004).

6.8.2.4 Selective Catalytic Reduction

SCR systems selectively reduce NO_x by injecting ammonia (NH_3) into the exhaust gas stream upstream of a catalyst. NO_x, ammonia, and oxygen react on the surface to form molecular nitrogen (N_2) and water. The overall chemical reaction can be expressed as

$$4NO + 4NH_3 + O_2 \rightarrow 4N_2 + 6H_2O$$

Parallel plates or honeycomb structures, permeated with the catalyst, are installed in the form of rectangular modules, downstream of the gas turbine in simple-cycle configurations, and into the heat recovery steam generator (HRSG) portion of the gas turbine downstream of the superheater in combined-cycle and cogeneration configurations.

The turbine exhaust gas must contain a minimum amount of oxygen and be within a somewhat narrow temperature range in order for the SCR system to operate properly. The temperature range is dictated by the catalyst: if it is too low, the reaction efficiency drops and increased amounts of NO_x and ammonia are released from the stack; if it becomes too high, the catalyst may begin to decompose. Turbine exhaust gas is generally too hot to be passed through the catalyst, so it is cooled by the HRSG, which extracts energy from the hot turbine exhaust gases and creates steam for use in other industrial processes or to turn a steam turbine. In simple-cycle power plants where no heat recovery is accomplished, high temperature catalysts (e.g. zeolite) are an option. SCR can typically achieve NO_x emission reductions in the range of about 80–95%.

SCR is the most widely applied post-combustion control technology in turbine applications and is currently accepted as LAER for new facilities located in ozone non-attainment regions. It can reduce NO_x emissions to as low as 4.5 ppmvd for standard combustion turbines without duct burner firing and as low as 2–2.5 ppmvd when combined with lean-premix combustion (again without duct burner firing). SCR uses ammonia as a reducing agent in controlling NO_x emissions from gas turbines. The portion of the unreacted ammonia passing through the catalyst and emitted from the stack is called ammonia slip. The ammonia is injected into the exhaust gases prior to passage through the catalyst bed. There is also a potential for increased particulate emissions from formation of ammonia salts. SCR may also results in the generation of spent vanadium pentoxide catalyst, which is classified as a hazardous waste. In addition, there is an energy loss from the performance loss due to the pressure drop across the SCR catalyst.

Gas turbines using SCR typically have been limited to 10 ppmvd ammonia slip (emissions of ammonia that has not reacted with nitrogen) at 15% oxygen. However, levels as low as 2 ppmvd at 15% oxygen have been proposed and guaranteed by control equipment vendors. In addition, Massachusetts and Rhode Island have established ammonia slip LAER levels of 2 ppmvd. Massachusetts has permitted at least two large gas turbine power plants using SCR reduction with 2 ppmvd ammonia slip limits. California recommended that the establishment of ammonia slip levels below 5 ppmvd at 15% oxygen on the basis of guarantees from control equipment vendors of single-digit levels for ammonia slip.

Data supplied by the vendor show that when NO_x is reduced from 15 ppm (gas turbine with duct burner firing) to 3.5 ppm, the cost effectiveness is $9473/T of NO_x removed.

6.8.2.5 Lean-Premix Technology or Dry-Low NO_x

Processes that use air as a diluent to reduce combustion flame temperatures achieved reduce NO_x by premixing the fuel and air before they enter the combustor. This type of process is called lean-premix combustion and goes by a variety of names, including the dry-low NO_x (DLN) process of General Electric, the dry-low emissions process of Rolls-Royce, and the SoLoNO$_x$ process of Solar Turbines.

Lean premixed designs reduce combustion temperatures, thereby reducing thermal NO_x. In a conventional turbine combustor, the air and fuel are introduced at an approximately stoichiometric ratio and air/fuel mixing occurs at the flame front where diffusion of fuel and air reaches the combustible limit. A lean premixed combustor design premixes the fuel and air prior to combustion. Premixing results in a homogeneous air/fuel mixture, which minimizes localized fuel-rich pockets that produce elevated combustion temperatures and higher NO_x emissions. A lean air-to-fuel ratio approaching the lean flammability limit is maintained, and the excess air serves as a heat sink to lower combustion temperatures, which lowers thermal NO_x formation. A pilot flame is used to maintain combustion stability in this fuel-lean environment. Lean-premix combustors can achieve emissions of about 9 ppmvd NO_x at 15% oxygen (~94% control).

To achieve low NO_x emission levels, the mixture of fuel and air introduced into the combustor must be maintained near the lean flammability limit of the mixture. Lean-premix combustors are designed to maintain this air/fuel ratio at rated load. At reduced load conditions, the fuel input requirement decreases. To avoid combustion instability and excessive CO emissions that occur as the air/fuel ratio reaches the lean flammability limit, lean-premix combustors switch to diffusion combustion mode at reduced load conditions. This switch to diffusion mode means that the NO_x emissions in this mode are essentially uncontrolled.

Lean-premix technology is the most widely applied pre-combustion control technology in natural gas turbine applications. It has been demonstrated to achieve emissions of approximately 9 ppmvd NO_x at 15% oxygen (Catalytica Energy Systems Inc. 2004).

6.8.2.6 Steam/Water Injection

In steam/water injection, the technology commonly chosen to reduce the NO_x emissions in natural gas turbine, higher combustion temperatures are used to achieve greater thermodynamic efficiency. In turn, more work is generated by the gas turbine at a lower cost. However, the higher the gas turbine inlet temperature, the more NO_x that is produced. Diluent injection, or wet controls, can be used to reduce NO_x emissions from gas turbines. Diluent injection involves the injection of a small amount of water or steam via a nozzle into the immediate vicinity of the combustor burner flame. NO_x emissions are reduced by instantaneous cooling of combustion temperatures from the injection of water or steam into the combustion zone. The effect of the water or steam injection is to increase the thermal mass by mass dilution and thereby reduce the peak flame temperature in the NO_x forming regions of the combustor. Water injection typically results in a NO_x reduction efficiency of about 70%, with emissions below 42 ppmvd NO_x at 15% oxygen. Steam injection has generally been more successful in reducing NO_x emissions and can achieve emissions less than 25 ppmvd NO_x at 15% oxygen (~82% control).

6.8.2.7 Summary of NO_x BACT for Turbines and Duct Burners

Table 6.17 provides information on the emissions, control effectiveness, economics, and environmental impacts measures for the control of NO_x discussed in Sections 6.8.2.2, 6.8.2.4, and 6.8.2.5. The analysis was performed on a unit (turbine and duct burner) basis.

SCONO$_X$ provides the highest level of NO_x reduction. However, this very new technology has yet to prove itself for long-term commercial operation on large-scale combined-cycle plants. It is the relatively high cost per emission reduction of this control technology ($13 627/T of NO_x removed) that rules out SCONOx as a control option. The next most effective control technology for NO_x is a combination of DLN combustors and SCR. The adverse environmental impact of SCR is primarily from the emissions of ammonia which is on the EPA's list of extremely hazardous substances. Although these adverse impacts of the SCR process can be minimized with proper system design and operation, it is ruled out as a control operation because the cost is prohibitive ($10 191/T of NO_x removed).

The next most effective control technology is DLN combustors. At reduced loads, combustion instability requires

Table 6.17 NO_x emissions, control effectiveness, economics, and environmental impacts.

Technology effectiveness	NO_x emissions reduction (TPY)	Capital cost ($)	Annualized cost ($)	Cost effectiveness ($/T)	Adverse environmental impacts
SCONO$_X$ (2 ppm)	399.5	14 922 733	5 444 139	13 627	Yes
DLN + SCR (3.5 ppm)	366.5	3 476 578	3 471 362	9 473	Yes

that a switch to diffusion combustion mode, in which NO_x emissions are essentially uncontrolled. However, this can be minimized with proper system design and operation.

6.8.3 CO BACT Review: Combustion Turbines and Duct Burners

$SCONO_X$ reduces CO emissions by oxidizing CO to CO_2. When CO is reduced from 11.51 ppm (gas turbine with duct burner firing) to 1 ppm, the cost effectiveness is \$21 706/T of CO removed.

In the United States, combustion turbines and duct burners are subject to the federal NSPS, but the regulations provide no CO emission limits. The following sections assess the control strategies that are potentially feasible for decreasing CO emissions from the facility.

6.8.3.1 Catalytic Oxidation

The rate of formation of CO during natural gas combustion depends primarily on the efficiency of combustion. The formation of CO occurs in small, localized areas around the burner where oxygen levels cannot support the complete oxidation of carbon to CO_2. Efficient burners can minimize the formation of CO by providing excess oxygen or by mixing the fuel thoroughly with air. CO emissions resulting from natural gas combustion can be decreased via catalytic oxidation. The oxidation is carried out by the well-known overall reaction:

$$CO + \frac{1}{2}O_2 \rightarrow CO_2$$

Several noble metal-enriched catalysts at high temperatures promote this reaction. Under ideal operating conditions, this technology can achieve an 80% reduction in CO emissions. Prior to entering the catalyst bed where the oxidation reaction occurs, the exhaust gas must be preheated to about 400–800 °F. Below this temperature range, the reaction rate drops sharply, and effective oxidation of CO is no longer feasible. Moreover, there is an energy loss because of the reduction in performance due to the pressure drop across the CO catalyst.

Sulfur and other compounds in the exhaust may foul the catalyst, leading to decreased activity. Catalyst fouling occurs slowly under normal operating conditions and may be accelerated by even moderate sulfur concentrations in the exhaust gas. The catalyst can be chemically washed to restore its effectiveness, but eventually, irreversible degradation occurs. Catalyst replacement is usually necessary every 5–10 years depending on the type and operating conditions.

An economic analysis for the catalytic oxidation of CO emissions based on vendor information estimates the cost at \$5084/T of CO removed. This cost level is considered to be economically infeasible for BACT. In addition to cost, catalytic oxidation would lead to increased downtime for catalyst washing and would present hazardous waste concerns during catalyst disposal. Due to the high cost and concerns with downtime and hazardous material disposal, catalytic oxidation is not selected as BACT for control of CO emissions from the turbines and duct burners.

6.8.3.2 Good Combustion Practices

Clearinghouse (RBLC) data show that the majority of BACT determinations for CO relied on the use of good combustion practices. Since add-on controls for CO were shown to be economically infeasible, the proposed BACT for CO emissions is the use of good combustion practices.

6.8.3.3 Combustion Control

Because CO results from the incomplete combustion of fuel, combustion control is an inherent design feature of combustion turbines and duct burner. Control is accomplished by providing adequate fuel residence time and high temperature in the combustion zone to ensure complete combustion. These control methods, however, also result in increased emissions of NO_x. Conversely, a low NO_x emission rate achieved through flame temperature can result in higher levels of CO emissions. Thus, a compromise is needed to set the flame temperature at the level that will achieve the lowest NO_x emission rate possible while keeping CO emissions rates at acceptable levels.

6.8.3.4 Summary of CO BACT for Turbines and Duct Burners

Table 6.18 provides information on the emissions, control effectiveness, economics, and environmental impacts associated with control of CO. The analysis was performed on a unit (turbine and duct burner) basis.

Table 6.18 CO emissions, control effectiveness, economics, and environmental impacts.

Technology effectiveness	CO emissions reduction (TPY)	Capital cost (\$)	Annualized cost (\$)	Cost effectiveness (\$/T)	Adverse environmental impacts
$SCONO_X$ (1 ppm)	251	14 922 733	5 444 139	21 706	Yes
DLN + CO catalyst (3.4 ppm)	194	1 302 514	2 208 461	5084	Yes

$SCONO_X$ provides a higher level of CO reduction than a combination of DLN combustors and CO oxidation. The adverse impacts include an energy loss from the performance loss due to the pressure drop across the CO catalyst and emissions of sulfates condense as additional PM_{10} or $PM_{2.5}$. However, at a cost of \$5084/T of CO removed, and the second option is more than four times as cost-effective as a control option.

Combustion control is selected as BACT for CO control with the limit of 9 ppm at 15% O_2 (annual average) without duct burners firing and 11.5 ppm at 15% O_2 with duct burner firing (annual average).

6.8.4 BACT Evaluation for PM/PM_{10} Emissions

Although no specific PM/PM_{10} emission limits are established in the NSPS for combustion turbines, BACT must be considered for the control of particulate.

6.8.4.1 Step 1: Identify Potential Control Technologies

The first step in the BACT analysis is to identify all potential control technologies available for the control of PM/PM_{10} emissions. The technologies identified for the control of PM/PM_{10} emissions are as follows:

- Electrostatic precipitation (ESP)
- Fabric filter
- Good combustion practices
- Fuel specification: clean burning fuels

6.8.4.2 Step 2: Eliminate Technically Infeasible Options

The second step in the BACT analysis is to eliminate any technically infeasible control technologies. Each control technology for each pollutant is considered, and those that are clearly technically infeasible are eliminated. The following is a list of control options deemed technically infeasible for the control of PM/PM_{10} emissions.

6.8.4.3 Electrostatic Precipitators

ESP technology removes particulate from an exhaust stream by electrically charging the particles and collecting the charged particles on plates. ESP performance is greatly affected by the particles' ability to accept and maintain an electrical charge. Because of the resistivity of gas turbine exhaust particles, ESP technology is ineffective for the control of PM/PM_{10} emitted from the proposed PSE (Puget Sound Energy) turbine.

Electrostatic precipitators are also not considered technically feasible options for combustion turbines due to the high exhaust flow rate and the low concentration of particulate in the turbine exhaust.

6.8.4.4 Fabric Filters

Fabric filters remove PM/PM_{10} from an exhaust stream by collecting the particulate on the filter as the air stream passes through the filter. Fabric filters typically cannot withstand high exhaust temperatures (>500 °F) and would be damaged by the high temperature of turbine exhaust. Thus, fabric filters are not technically feasible for the proposed PSE turbine.

6.8.4.5 Step 3: Ranking of Remaining Control Technologies by Control Effectiveness

The third step in the BACT analysis is to rank the remaining control technologies in order of control effectiveness. Table 6.19 presents the control technologies and their approximate control efficiencies.

6.8.4.6 Step 4: Evaluation of the Most Effective Emissions Controls

Because ESP technology and fabric filters have been deemed technically infeasible, the implementation of good combustion practices and the firing of clean burning fuels are the only remaining options for the control of PM/PM_{10} emissions. The combination of implementing good combustion practices and the firing of clean burning fuels is the most effective emission control option.

6.8.4.7 Step 5: Select BACT for the Control of PM/PM_{10} Emissions

Properly tuned burners firing natural gas and light oils inherently emit low levels of particulate matter. The RBLC database indicates that good combustion control is widely accepted as BACT for turbines firing natural gas. Thus, it is selected to use of clean burning fuels (natural gas and fuel oil with a sulfur content <0.05%) and good combustion practices as BACT for PM/PM_{10} emissions.

Table 6.19 Ranking of feasible control technologies by effectiveness.

Pollutant	Control technology	Potential control efficiency (%)
PM/PM_{10}	Good combustion practices	Base case
	Fuel specification: clean burning fuels	Base case

6.8.5 VOC Control Technologies

The NSPS regulations for combustion turbines and duct burners provide no VOC emission limits. The following sections assess the control strategies that are potentially feasible for decreasing VOC emissions from the facility.

6.8.5.1 Catalytic Oxidation

The formation of VOC in combustion units depends primarily on the efficiency of combustion. Inefficient combustion leads to the formation of aldehydes, aromatic carbon compounds, and various other organic compounds by several mechanisms. Catalytic oxidation decreases VOC emissions by facilitating the complete combustion of organic compounds to water and carbon dioxide. Prior to entering the catalyst bed where the oxidation reaction occurs, the exhaust gas must be preheated to about 400–800 °F.

The RBLC database shows few instances of catalytic oxidation being selected as BACT for VOC at any gas-fired turbine power plant nationwide. An economic analysis for the catalytic oxidation of VOC emissions based on vendor information estimates the cost at $30 811/T of VOC removed. This cost level is economically infeasible for VOC removal. Thus, catalytic oxidation would not be selected as BACT for VOC emissions from turbines and duct burners proposed in the present time frame.

6.8.5.2 Good Combustion Practices

All the RBLC database BACT determinations for VOC outside California and New York show the use of combustion control or good combustion practices. Thus, in most areas good combustion practices are acceptable as BACT for VOC, with limits of 7.25 ppmvd at 15% O_2, as methane.

6.8.6 BACT Evaluation for SO_2 and H_2SO_4 Emissions

Control techniques available to reduce SO_2 and H_2SO_4 emissions include FGD systems and the use of low-sulfur fuels. Although FGD systems are common in boiler application, the RBLC database shows no known FGD systems on combustion turbines. Thus, the use of an FGD system is not warranted and an FGD system should be rejected as a BACT control alternative.

Another available technique is the use of low-sulfur fuels. There are no adverse environmental or energy impacts associated with the properly specified use of pipeline natural gas with low-sulfur content, and this control alternative should be acceptable as BACT. NSPS Subpart GG requires that either SO_2 emissions from a gas turbine be limited to 0.015% by volume (or 150 ppmvd) at 15% O_2, or that the turbine fuel contain less than 0.8% sulfur by weight.

The power plant proposes to combust only low-sulfur fuel (i.e. pipeline quality natural gas or diesel with a sulfur content < 0.05% by weight) in the proposed units, which is a widely accepted BACT determination for SO_2. By firing low-sulfur fuels, the turbines will meet the NSPS emission standard of 150 ppm at 15% O_2.

6.8.6.1 Step 1: Identify Potential Control Technologies

The first step in the BACT analysis is to identify all potential control technologies available for the control of SO_2 emissions. The technologies identified for the control of SO_2 emissions are as follows:

- FSD
- Spray dryers
- Fuel specification: low-sulfur fuels

6.8.6.2 Step 2: Eliminate Technically Infeasible Options

Neither FGD systems nor spray dryers have been applied to the exhaust gases from turbines, and significant technological difficulties are envisioned to apply to either of these technologies with high exhaust temperatures. The high temperature of the turbine exhaust deems FGD units and spray dryers infeasible. Thus, both FGD systems and spray dryers are eliminated for the control of SO_2 emissions from the proposed PSE turbines.

6.8.6.3 Step 3: Ranking of Remaining Control Technologies by Control Effectiveness

All add-on control options for the control of SO_2 emissions have been eliminated due to technical unfeasibility. For the control of SO_2 emissions, the use of low-sulfur fuels is the only control method determined to be technically feasible.

6.8.6.4 Step 4: Evaluation of the Most Effective Emissions Controls

As stated above, all add-on control options for the control of SO_2 emissions have been eliminated due to technical unfeasibility. Thus, the use of low-sulfur fuels containing less than 0.05% by weight of sulfur is the most effective means for reducing SO_2 emissions.

6.8.6.5 Step 5: Select BACT for the Control of SO_2 Emissions

The low SO_2 emissions levels inherent with low-sulfur fuel in a turbine constitutes BACT. Thus, the permittee proposes that BACT for the proposed turbines is the firing of low-sulfur fuels.

Problems

6.1 Write about LCA centers or societies in any country or within a continent of your choice.

6.2 Why LCA? Describe the four major phases involved in an LCA study.

6.3 Describe all benefits of conducting an LCA. Give an example.

6.4 Describe a methodological approach in which life cycle principles can be used to design and develop a process or product as a tool for innovation.

6.5 Give an example of LCA environmental impact trade-off among competitive products or manufacturing processes.

6.6 Describe two approaches that can be used to infuse life cycle design changes into manufacturing.

6.7 What are the advantages and disadvantages of the mechanical and kraft pulping processes in context of an LCA?

6.8 Give an example of a product that would be a good candidate for a remanufacture at the end of its normal life-cycle. Why would it be a good candidate?

6.9 Name a few LCA and LCI software tools that you can use in your senior design project. Provide a brief introduction to each software and their availability online including the URL.

6.10 Name all six categories that are incorporated in an LCIA of a product or process.

6.11 Explain how an LCA study helps decision-makers select the product or process that causes the least impact to the environment and public health.

6.12 Review the EIOLCA model developed by Carnegie-Mellon University and how you could incorporate this model in your capstone design and/or senior research projects on a sustainable product design and development. This model also estimates the materials and energy resources required for, and the environmental emissions resulting from, activities in our economy.

References

Allen, D.T., Allport, C., Atkins, K. et al. (2009). The Aviation Fuel Life Cycle Assessment Working Group, Framework and Guidance for Estimating Greenhouse Gas Footprints of Aviation Fuels (Final Report). Prepared for Universal Technology Corporation and the Air Force Research Laboratory. *April Interim Report. AFRL-RZ-WP-TR-2009-2206.*

Azapagic, A. (2000). Life cycle assessment: a tool for innovation and improved environmental performance, Part VI35. In: *Science, Technology and Innovation Policy: Opportunities and Challenges for the Knowledge Economy*, 519–530. Westport, CT: Quorum Books.

Bare, J.C. (2003). The tool for the reduction and assessment of chemical and other environmental impacts (TRACI), USEPA, 2003. TRACI is available on the Internet at no cost at http://epa.gov/ORD/NRMRL/std/sab/iam_traci.htm (accessed 12 February 2019).

Brungs, W.A. (1973). Literature review of the effects of residual chlorine on aquatic life. *Journal of the Water Pollution Control Federation* 45 (10): 2180–2193.

California Air Resources Board (2000) Best Available Control Technology Guidelines, South Coast Air Quality Management District, 17 August 2000. http://www.arb.ca.gov/bact/bact.htm.

Catalytica Energy Systems Inc. (2004). Mountain View, CA: Catalytica Energy's Xonon Technology. http://www.catalyticaenergy.com/xonon/technology.html (accessed 12 January 2019).

Cooper, J.S. (2003). Life cycle assessment and sustainable development indicators. *Journal of Industrial Ecology* 7 (1): 12–15.

Curran, M.A. (2003). Do bio-based products move us toward sustainability? A look at three USEPA case studies. *Environmental Progress* 22 (4): 277–292.

Curran, M.A. (ed.) (2012). *Life Cycle Assessment Handbook: A Guide for Environmentally Sustainable Product.* Salem, MA/Hoboken, NJ: Scrivener Publishing LLC/Wiley.

Curran, M.A. (ed.) (2015). *Life Cycle Assessment Student Handbook.* Salem, MA/Hoboken, NJ: Scrivener Publishing LLC/Wiley.

Dandy J.W. (1967). The effects of chemical characteristics of the environment on the activity of an aquatic organism. Thesis. University of Toronto, Ontario, Canada.

Dandy, J.W. (1972). Activity response to chlorine in the Brook Trout, Salvelinus fontinalis (Mitchill). *Canadian Journal of Zoology* 50: 405.

Das, T.K. (2002). Evaluating the life cycle environmental performance of chlorine disinfection and ultraviolet technologies. *Clean Technology and Environmental Policy* 4: 32–43.

Das, T.K. (2003). Hog-Fuel Boiler's Reasonably Available Control Technology (RACT) Determination. *Air Quality Program, Washington Department of Ecology, Olympia, Washington, Report #03-02-009*, 137 pages. http://www.ecy.wa.gov/biblio/0302009.html (accessed 14 February 2019).

Das, T. K. (2004). Disinfection. In the R.E. Kirk and D. F. Othmer *Encyclopedia of Chemical Technology*, 5, Vol. 8, pp. 605–672, Hoboken, NJ: Wiley.

Das, T.K. (2005). *Toward Zero Discharge: Innovative Methodology and Technologies for Process Pollution Prevention*. Hoboken, NJ: Wiley.

Das, T.K. and Ekstrom, L.P. (1999). UV application to a major wastewater treatment plant on the west coast. *Paper No. 100d, American Institute of Chemical Engineers Spring Annual Meeting*, Houston (April).

Das, T.K. and Houtman, C.J. (2004). "Evaluating chemical-, mechanical-, and bio-pulping processes and its sustainability characterization using life-cycle assessment," Invited paper in the special topical issue of environmental progress – sustainability in chemical engineering. *Environmental Progress and Sustainable Energy* 23 (4): 347–357.

Das, T.K. and Jain, A.K. (2001). Pollution prevention advances in pulp and paper processing. *Environmental Progress and Sustainable Energy* 20: 87–92.

Das, T.K. and Scott, G.M. (2003). Life cycle assessment comparisons of chemical-, mechanical-, and bio-pulping processes. Presented at the AIChE Topical Conference on Sustainability and Life-Cycle, San Francisco, CA (16–21 November 2003).

Dondoroff, P. and Katz, M. (1950). Critical review of literature on the toxicity of industrial wastes and their components to fish. *Sewage and Industrial Wastes* 22: 1432.

Ecology, Washington Department of Ecology (1998). Ecology's criteria for sewage works design. Chapter T5 (December).

EISA (2007). Energy Independence and Security Act of 2007. http://frwebgate.Access.gpo.gov/egi-bin/getdoc.cgi?

ENDS (1996). Dow Europe and the challenge of eco-efficiency. *Company Report No. 170. ENDS Report, 252* (16–20 January).

Energy Ideas Clearinghouse (2018). Washington State University Cooperative Extension Energy Program, Olympia, Washington. http://www.energy.wsu.edu/nwalliance-eic (accessed 4 April 2018).

Geissdoerfer, M., Savaget, P., Bocken, N., and Hultink, E.J. (2017). The circular economy: a new sustainability paradigm? *Journal of Cleaner Production* 143: 757–768.

Hart, J.L. (1973). *Pacific Fishes of Canada*. Ottawa: Fisheries Research Board of Canada.

Heltz, G.R. and Nweke, A.C. (1995). Incompleteness of wastewater dechlorination. *Environmental Science and Technology* 29 (4): 1018–1022.

Holland, G.A. (1960). Toxic effects of organic and inorganic pollutants on young salmon and trout. *State of Washington Department of Fisheries, Res. Bull., No. 5*, pp. 198.

Hynninen, P. (ed.) (1998). *Papermaking Science and Technology. Book 19: Environmental Control*. Jyvaskyla, Finland: Fapet Oy/Enviro Data Oy.

International Organization of Standardization (ISO) (2000). *Environmental Management – Life Cycle Assessment – Examples of Application of ISO 14041 to Goal and Scope Definition and Inventory Analysis*. Geneva, Switzerland: International Organization for Standardization.

ISO (1997). Environmental Management – Life Cycle Assessment – Principles and Framework (ISO/FDIS 14040), ISO TC 207.

ISO (1998a). Life Cycle Assessment – Impact Assessment (ISO 14042), ISO TC 207/SC5/WG4.

ISO (1998b). *Environmental Management: Life Cycle Assessment – Life Cycle Interpretation (ISO/DIS 14043)*, ISO TC 207. Charlottenlund: Dansk Standard.

Kirk, T.K., Akhtar, M., and Blanchette, R.A. (1994). Biopulping: Seven years of consortia research. *Proceedings of Tappi Biological Sciences Symposium*, Atlanta, GA, pp. 57–66.

Kirk, T.K., Koning, J.W. Jr., Burgess, R.R., and Akhtar, M. (1993). *Biopulping: A glimpse of the future?* Research Paper FPL-RP-523. Madison, WI: U.S. Department of Agriculture, Forest Service, Forest Products Laboratory.

Kozloff, E.N. (1974). *Keys to the Marine Invertebrates of Puget Sound, the San Juan Archipelago, and Adjacent Regional*. Seattle: University of Washington Press.

Lacey, Olympia, Tumwater, and Thurston (LOTT) (1994). Pilot Study Final Report on UV Disinfection of Wastewater, The City of Olympia, Washington.

Lienig, J. and Bruemmer, H. (2017). *Fundamentals of Electronic Systems Design*. Springer.

Loge, F.J., Darby, J.L., and Tchobanoglous, G. (1996). UV disinfection of wastewater: probabilistic approach to design. *Journal of Environmental Engineering* 122 (12): 1078.

Merkens, J.C. (1958). Studies on the toxicity of chlorine and chloramines to the rainbow trout. *Water and Waste Treatment Journal* 7: 150.

Metcalf & Eddy Inc (2003). *Wastewater Engineering: Treatment and Reuse*, 4e. New York, NY: McGraw-Hill.

National Council of the Paper Industry for Air and Stream Improvement (NCASI) (1983). *TGNMO Emissions from the Thermomechanical Pulping Process*, NCASI Technical Bulletin, vol. 410. Research Triangle Park, NC: NCASI.

National Renewable Energy Laboratory (NREL) (2011). U.S. National Life Cycle Inventory Database. www.nrel.gov/lci (accessed 14 February 2019).

Pauli, G. (1996). *Breakthroughs*. Haslemere: Epsilon Press Limited.

Pike, D.J. (1971). Toxicity of chlorine to brown trout, *New Zealand Wildlife*, No 33.

Pineda-Henson, R., Culaba, A.B., and Mendoza, G.A. (2002). Evaluating environmental performance of pulp and paper manufacturing using the analytical hierarchy process and life-cycle assessment. *Journal of Industrial Ecology* 6: 15–28.

Scheible, O.K. (1987). Development of a rationally based design protocol for the ultraviolet light disinfection process. *Journal (Water Pollution Control Federation)* 59: 25.

Scheible, O.K. and Forndran, A. (1986). *Ultraviolet Disinfection of Wastewaters from Secondary Effluent and Combined Sewer Overflows*, EPA-600/2-86/005. Cincinnati, OH: USEPA.

Scott, G.M., Akhtar, M., Lentz, M.J. et al. (1998a). Engineering aspects of fungal pretreatment for wood chips. In: *Environmentally Friendly Pulping and Bleaching Methods*. New York, NY: Wiley.

Scott, G.M., Akhtar, M., Lentz, M.J. et al. (1998b). New technology for papermaking commercialization biopulping. *Tappi Journal* 81 (12): 153–157.

SETAC (1991). *A Technical Framework for Life Cycle Assessment, Workshop Report from the Smugglers Notch, Vermont, USA*. Sandestin: The Society of Environmental Toxicology and Chemistry.

SETAC (1993). *Guidelines for Life-Cycle Assessment: A Code of Practice*. Pensacola, FL: The Society of Environmental Toxicology and Chemistry Sesimbra, Portugal (31 March to 3 April).

Smook, G.A. (1992). *Handbook for Pulp and Paper Technologies*, 2e. Bellingham, WA: Angus Wilde Publications.

Someshwar, A.V. and Pinkerton, J.E. (1992). Mechanical pulping. In: *Air Pollution Engineering Manual* (eds. A.J. Buonicore and W.T. Davis), 848–849. New York, NY: Van Nostrand Reinhold.

Sprague, J.B. and Drury, D.E. (1969). Avoidance reactions of salmonid fish to representative pollutants. In: *Advances in Water Pollution Research, Proceedings of the Fourth International Conference on Water Pollution Research Held in Prague* (ed. S.H. Jenkins), 169. London, UK: Pergamon Press.

Sundholm, J. (ed.) (1999). *Papermaking Science and Technology, Book 5: Mechanical Pulping*. Helsinki, Finland: Fapet Oy/Finish Pulp and Paper Research Institute.

Sykes, M. (1994). Environmental compatibility of effluents of aspen biomechanical pulps. *Tappi Journal* 77: 160–164.

Temmer, B., Soroushian, F., and Noesen, M. (2000). UV emerges bright spot for wastewater disinfection. *Pollution Engineering* 32: 46–49.

Thatcher, T.O. (1977). The relative sensitivity of Pacific Northwest fishes and invertebrates to chlorinated sea water. Battelle, Pacific Northwest Laboratories, Sequim, Washington. In: *Proceedings of the Second Conference on Water Chlorination Environ, Impact and Health Effects*, vol. 2 (eds. R.L. Jolly, H. Gorchov and D.H. Hamilton), 341–350. Richland, WA: Battelle, Pacific Northwest National Laboratory (31 October to 4 November).

Trojan Technologies Inc. (2010). Overview of UV disinfection. 3020 Gore Road, London, Ontario, Canada N5V 4 TS.

U.S. Pulp and Paper Industry (1988). *An energy perspective*, Report Number DOE/RL/01830T57. Richland, WA: Pacific Northwest Laboratory.

UNEP-SETAC (2000). United Nations Environmental Program–Society of Environmental Toxicology and Chemistry *Letter of Intent: Best Available Practice of Life Cycle Assessment with Generic Application Dependency*. Signed by the International Council of SETAC, and UNEP, Division Technology, Industry, and Economics. www.uneptie.org/pc/sustain/lca/letter-of-intent.htm (accessed 29 March 2018).

UNEP-SETAC (2011). *Global Guidance Principles for Life Cycle Assessment Databases: A Basis for Greener Processes and Products*. Paris, France: United Nations Environmental Programme – Society of Environmental Toxicology and Chemistry.

USDOE (2007). Cost and Performance Baseline for Fossil Energy Plants – Volume 1: Bituminous Coal and Natural Gas to Electricity. DOE/NETL-2007/1281.

USEPA (1985). *Design Manual, Odor and Corrosion Control in Sanitary Sewage Systems and Treatment Plant*, EPA/625/1-85/018, (October). Washington, DC: Office of Water.

USEPA (1986). *Design Manual: Municipal Wastewater Disinfection*, EPA/625/1-86/021. Cincinnati, OH: Office of Research and Development, Water Engineering Research Laboratory.

USEPA (1989a). *EPA Quality Assurance Management Staff, Development of Data Quality Objectives: Description of Stages I and II*, EPA QA/G-4. Washington, DC: USEPA, Office of Research and Development.

USEPA (1989b). *Risk Assessment Guidance for Superfund, Vol. 2, Environmental Evaluation Manual*, EPA/540/1-89/001. Washington, DC: USEPA.

USEPA (1989c). *Exposure Factors Handbook*, EPA/600/889/043. Washington, DC: Office of Health and Environmental Assessment.

USEPA (1990). *Workshop Manual: Prevention of Significant Deterioration and Nonattainment Area Permitting*. Research Triangle Park, NC: Office of Air Quality, Planning and Standards, USEPA.

USEPA (1992a). *Facility Pollution Prevention Guide*, EPA/600/R-92/088. Washington, DC: U.S. EPA, Office of Research and Development (May).

USEPA (1992b). *Users Manual for UVDIS, Version 3.1 UV Disinfection Process Design Manual*, EPAG0703, Contract No. 68-C8-0023. Cincinnati, OH: USEPA.

USEPA (1993). *Office of Research and Development, Life Cycle Assessment: Inventory Guidelines and Principals*, EPA/600/R-92/245. Washington, DC: USEPA.

USEPA (1995). *Office of Solid Waste, Guidelines for Assessing the Quality of Life Cycle Inventory Analysis*, EPA/530/R-95/010. Washington, DC: USEPA.

USEPA (1996). Chemical Accidental Prevention and Risk Management Programs. *40 CFR 68 Subpart B* (June). http://response.restoration.noaa.gov/chemaids/rmp/rmp.html (accessed 27 February 2019).

USEPA (1998). EPA's Risk Management Program Guidance for Wastewater Treatment Plants, *40CFR68, EPA 550-B-98-010*, Washington, DC (June).

USEPA (2003). The tool for the reduction and assessment of chemical and other environmental impacts (TRACI), USEPA, 2003. http://epa.gov/ORD/NRMRL/std/sab/iam_traci.htm (accessed 27 February 2019).

USEPA (2006). *Life Cycle Assessment: Principles and Practices*, EPA/600/R-06/066. Washington, DC: Office of Research and Development.

USEPA (2010).Life Cycle Assessment of Greenhouse Gas Emissions from Renewable Fuels. http://www.epa.gov/wastes/otaq/renewablefuels/420f09024.htm (accessed 29 March 2018).

USEPA (2012). Sustainable materials management: non-hazardous materials and waste management hierarchy. https://www.epa.gov/smm/sustainable-materials-management-non-hazardous-materials-and-waste-management-hierarchy (accessed 24 Septemer 2019).

USEPA (2016a). Technology Transfer Network, Clean Air Technology Transfer, RACT/BACT/LAER Clearinghouse, http://cfpub1.epa.gov/rblc/htm/b102.cfm (accessed 6 September 2019).

USEPA (2016b), STAPPA/ALAPCO Clean Air World, http://cleanairworld.org/scripts/us.temp.asp?id-307 (accessed 6 September 2019).

White, G.C. (2010). *White's Handbook of Chlorination and Alternative Disinfectants*, 5e. Hoboken, NJ: Wiley.

7

Economics of Manufacturing Pollution Prevention

Toward an Environmentally Sustainable Industrial Economy

7.1 Introduction

The failure of many pollution-prevention programs can be traced to the inability of the engineers and scientists to convince business leadership to change manufacturing processes unfavorable to the environment. Often, this reluctance to change is not because the recommended process improvements were technically unsound, but because the engineering team failed to speak the language of business, that is dollars and cents. The role of economics in pollution prevention is very important, even as important as the ability to identify technologies changes to the process, new and emerging technologies, process and/or product modification, Zero Discharge technologies, technologies for biobased engineered chemicals, products, renewable energy sources, and its associated costs.

Market acceptance of new technologies, products, processes, and services depend upon the complex interplay of cost, physical properties, environmental performance, public policy, cultural prejudices, and other factors. Accurate forecasting is a difficult and time-consuming activities best left to the experts. However, cost estimating is a valuable skill that allows an engineer to obtain "ballpark" approximation of project costs. The goal is to obtain an estimate that is within ±30% of the actual cost if the enterprise were pursued. Such estimates are relatively easy to develop. Engineers use a relatively small set of financial tools to assess alternative capital investments and to justify their selections to upper management. Most often, the engineer's purpose is to show how a recommended investment will improve the company's profitability. Spending on capital improvements is largely voluntary. Adding an assembly line or changing from one type of gasket material to another can be postponed, or the proposal, or rejected outright. This is not the case with pollution control devices, which are necessary for compliance with state and federal pollution standards and generally must be installed by or before a mandated deadline. Consequently, a decision to install device X may not originate with the engineer. Instead, the need for pollution control equipment might be identified by a company's environmental manager who then passes the decision to acquire device X down to the engineer. This chapter gives some cost-estimating methods that can be used to assess the costs of implementing pollution prevention technologies; these methods are useful, as well, in cost comparisons, and in considerations of cost-effectiveness, best available technologies and emerging technologies. Also provided is some information about the cost of renewable resources and the cost of manufacturing biobased products from such feedstocks. We begin, however, with a few remarks about the economic matrix in which profitable pollution control technology and best control technology are implemented.

7.2 Economic Evaluation of Pollution Prevention

The economic evaluation of engineering projects typically involves estimation of equipment, installation, raw material, energy, and maintenance cost. Disposal and pollution control costs are often factored into these calculations in determining economic rates of return, but other regulatory and social costs are not. In this chapter, total cost assessment of waste management alternatives is described, and the hidden costs, future liabilities, and less tangible costs associated with waste generation are discussed.

7.2.1 Total Cost Assessment of Pollution Control and Prevention Strategies

Traditional economic measure of engineering projects evaluate equipment, raw-material, energy, operating, and maintenance costs. These evaluations generally overlook some of the costs of waste generation. A more complete accounting of environmental costs is referred to as total

Industrial Environmental Management: Engineering, Science, and Policy, First Edition. Tapas K. Das.

Companion website: www.wiley.com/go/Das/IEM_1e

cost assessment. Four types of costs are identified, which are labeled as tier 0, tier 1, tier 2, and tier 3. Tier 0 costs are the "usual" costs that are included in a conventional analysis of a project. Tier 1 costs are included in a conventional analysis of a project. Tier 1 costs include permitting, reporting, monitoring, manifesting, and insurance costs and are often referred to as "hidden" costs because they are usually treated as overhead costs and are not directly charged to a project. Waste disposal costs are sometimes treated as overhead costs as well. Tier 2 costs include future liabilities, which are extremely difficult to accurately evaluate. Even more difficult to evaluate are tier 3 costs, which include consumer responses, employee relations, and public image. All four tiers of costs provide information for financial analysis methods, where measures such as ROR and payback periods (PPs) are evaluated.

7.2.2 Economics of Pollution Control Technology

In the present economic system, the goal of a sustainable development process is to maintain intergenerational equity by ensuring quality of life for future generations, which essentially requires the stopping of further damage to the environment. From this point of view the necessity of an integrated system consisting of modern manufacturing production and controlling technologies is being gradually recognized. To this end, various regulatory measures are being implemented, mainly in the industrial sector, which is the major contributor of pollution, for adopting such technologies.

7.3 Cost Estimates

The costs and estimating methodology in this section are directed toward the "study" estimate with a nominal accuracy of ±30%. According to Perry's *Chemical Engineer's Handbook*, a study estimate is "used to estimate the economic feasibility of a project before expending significant funds for piloting, marketing, land surveys, and acquisition. However, it can be prepared at relatively low cost with minimum data" (Perry and Green 1997). Specifically, the development of a study estimate calls for knowledge of the following:

- Location of the source within the plant
- Rough sketch of the process flow sheet (i.e. the relative locations of the equipment in the system)
- Preliminary sizes of, and material specifications for, the system equipment items
- Approximate sizes and types of construction of any buildings required to house the control system
- Rough estimates of utility requirements (e.g. electricity)
- Preliminary flow sheet and specifications for ducts and piping
- Approximate sizes of motors required

In addition, since the accuracy of an estimate (study or otherwise) depends on the amount of engineering work expended on the project, the user will need an estimate of the labor hours required for engineering and drafting activities. There are four other types of estimate.

The order-of-magnitude estimate, a rule-of-thumb procedure, can be used only for plant installations of the repetitive type for which there exists good cost history. Its error bounds are greater than ±30%. (However, Perry states that "no limits of accuracy can safely be applied to [an order-of-magnitude estimate].") The sole input required for making an order-of-magnitude estimate is the control system's capacity (often measured by the maximum volumetric flow rate of the gas passing through the system).

The other three types of estimate, listed next, are preferable.

- Scope, budget authorization, or preliminary. This estimate, nominally of ±20% accuracy, requires more detailed knowledge than the study estimate regarding the site, flow sheet, equipment, buildings, etc. In addition, rough specifications for the insulation and instrumentation are also needed.
- Project control or definitive. These estimates, accurate to within ±10%, require yet more information than the scope estimates, especially concerning the site, equipment, and electrical requirements.
- Firm or contractor's or detailed. This is the most accurate (±5%) of the estimate types, requiring complete drawings, specifications, and site surveys. Consequently, detailed cost estimates are typically not available until right before construction. Common sense suggests that there is seldom enough time to prepare such estimate before approval to proceed with the project has been obtained.

7.3.1 Elements of Total Capital Investment

Total capital investment (TCI) includes all costs required to purchase equipment needed for the control system (purchased equipment costs), the costs of labor and materials for installing that equipment (direct installation costs), costs for site preparation and buildings, and certain other costs (indirect installation costs). TCI also includes costs for land, working capital, and off-site facilities.

Direct installation costs include costs for foundations and supports, erecting and handling the equipment, electrical work, piping, insulation, and painting. Typical indirect installation costs are engineering costs, construction and field expenses (i.e. costs for construction supervisory personnel, office personnel, rental of temporary offices, etc.), contractor fees (for construction and engineering firms involved in the project), start-up and performance test costs (to get the control system running and to verify that it meets performance guarantees), and associated costs with contingencies. "Contingencies" is a catch-all category that covers unforeseen costs that may arise if, for example, it becomes necessary to redesign, modify equipment, to pay higher costs of equipment or field labor costs, or to compensate for delays in start-up. Contingency costs are not the same as uncertainty and retrofit factor costs.

The elements of TCI are displayed in Figure 7.1. Note item "battery limits" cost, which is the sum of the purchased equipment cost, direct and indirect installation costs, and costs of site preparation and buildings. By definition, this is the total estimate for a specific job; any support facilities that may be needed (e.g. control systems) are assumed to already exist at the plant and are not included in the battery limits. For systems installed in new plants, off-site facilities (special facilities for supporting the control system) also might be excluded from the battery limits. Off-site facilities are exemplified by units to produce steam, electricity, and treated water; laboratory buildings; and railroad spurs, roads, and other transportation infrastructure items. Pollution control devices rarely consume enough energy to warrant dedicated off-site capital units. However, it may be necessary – especially in the case of control systems installed in new or "grassroots" plants – to build extra capacity into the site's generating plant to service the system. (A Venturi scrubber, which often requires large amounts of electricity, is a good example.) Nevertheless, that the capital cost of a device does not include utility costs, even if the device were to require an offsite facility. Utility costs are charged to the project as operating costs at a rate that covers both the investment and operating and maintenance costs for the utility.

As Figure 7.1 shows, the installation of pollution control equipment may also require land, but since most add-on control systems take up very little space (a quarter-acre or less), this cost tends to be relatively small. Certain control

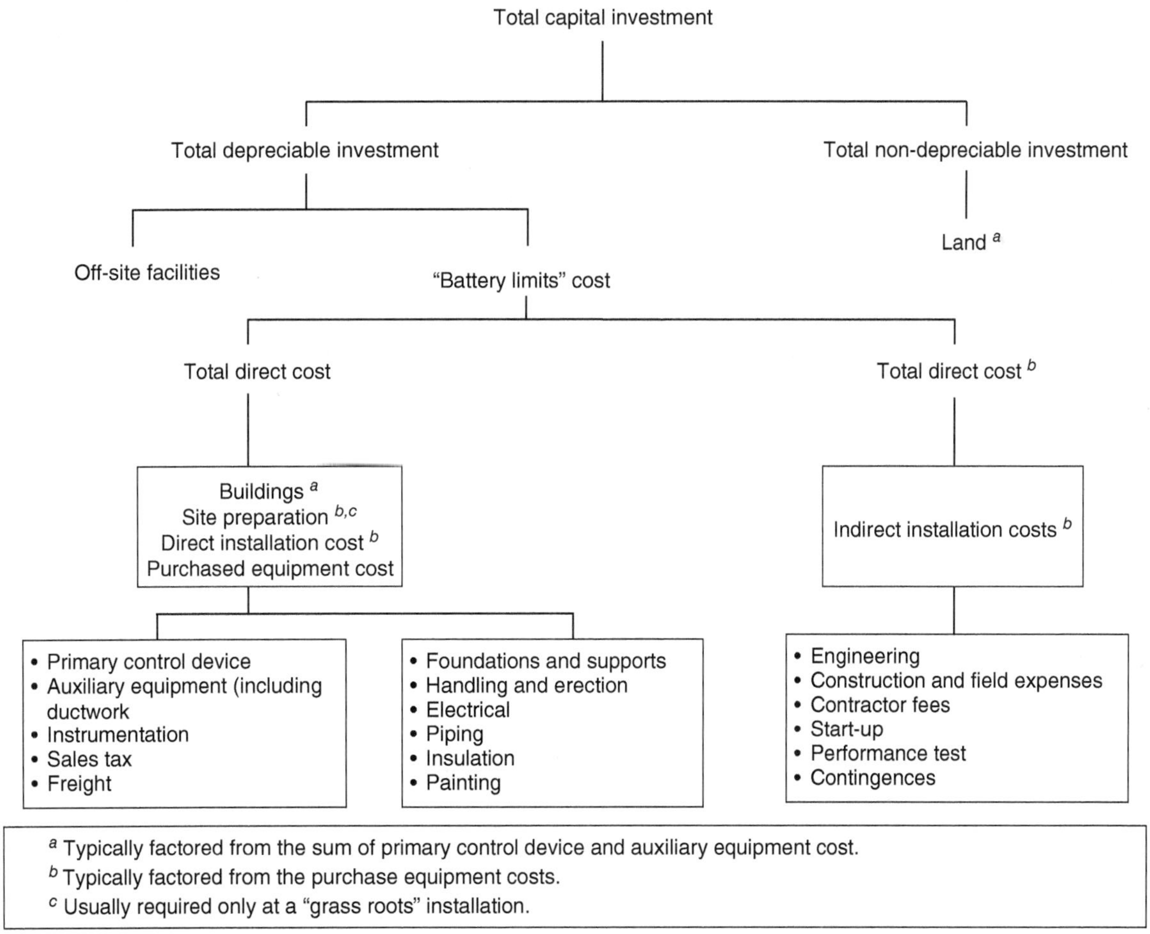

Figure 7.1 Elements of TCI.

systems, such as those used for flue-gas desulfurization or selective catalytic reduction (SCR), require larger quantities of land for the equipment, chemicals storage, and waste disposal. In these cases, especially when a retrofit installation must be performed, space constraints can significantly influence the cost of installation, and the purchase of additional land may be a significant factor in the development of the project's capital costs. Land retains its value over time; however, land is not treated the same as other capital investments, retains its value over time, the purchase price of new land needed for sitting a pollution control device can be added to the TCI, but it must not be depreciated. Instead, if the firm plans to dismantle the device at some future time, either the land should be excluded from the analysis or its value included at the disposal point as an "income" to the project, to net it out of the cash flow (CF) analysis.

One might expect to include initial operational costs (the initial costs of fuel, chemicals, and other materials, as well as labor and maintenance related to start-up) in the operating cost section of the cost analysis instead of in the capital component, but such an allocation would be inappropriate. Routine operation of the control does not begin until the system has been tested, balanced, and adjusted to work within its design parameters. Until then, all utilities consumed, all labor expended, and all maintenance and repairs performed are a part of the construction phase of the project and are included in the TCI in the "start-up" component of the indirect installation costs.

7.3.2 Elements of Total Annual Cost

Total annual cost (TAC) has three elements: direct costs (DC), indirect costs (IC), and recovery credits (RC), which are related by the following equation:

$$\text{TAC} = \text{DC} + \text{IC} - \text{RC} \tag{7.1}$$

Clearly, the basis of these costs is one year, a period that allows for seasonal variations in production (and emissions generation) and is directly usable in financial analyses. The various annual costs and their interrelationships are displayed in Figure 7.2.

DC are those that tend to be directly proportional (variable costs) or partially proportional (semi-variable costs) to some measure of productivity – generally the company's productive output, but for purposes of pollution control, the proper metric may be the quantity of exhaust gas processed by the control system per unit time. Conceptually, a variable cost can be graphed in cost/output space as a positive sloped straight line that passes through the origin. The slope of the line is the factor by which output is multiplied to derive the total variable cost of the system. Semi-variable costs can be graphed as a positive sloped straight line that passes through the cost axis at a value greater than zero – that value being the "fixed" portion of the semi-variable cost and the slope of the line being analogous to that of the variable cost line discussed above.

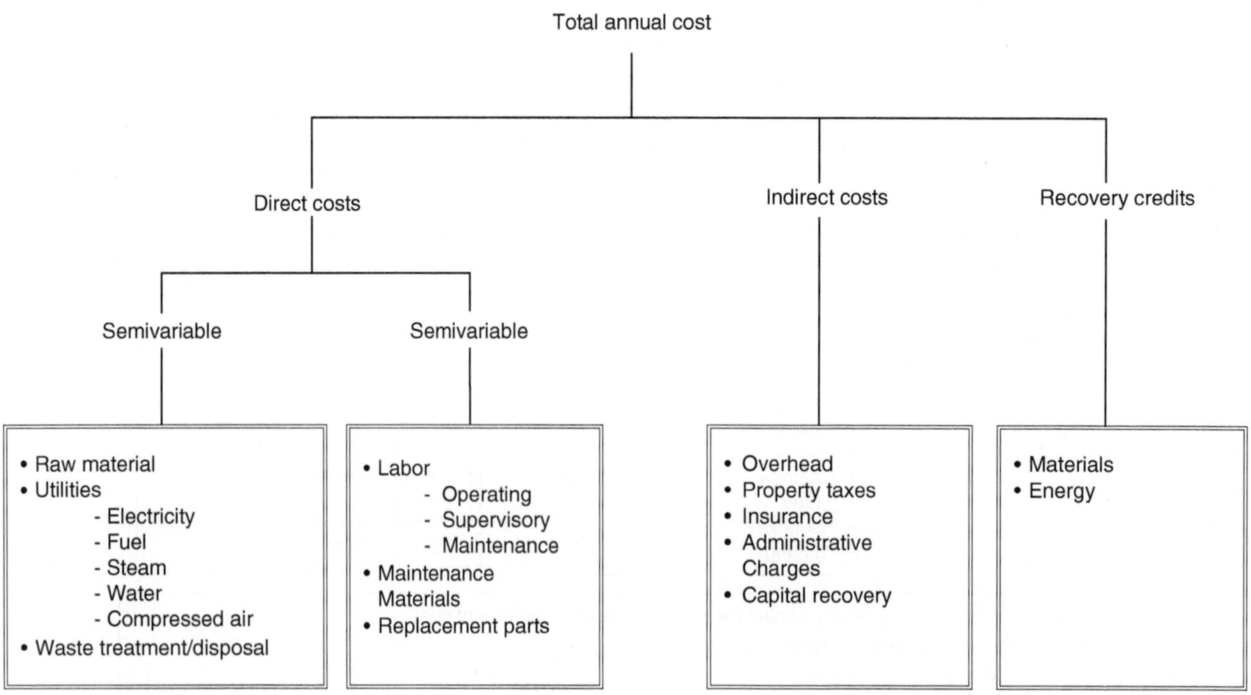

Figure 7.2 Elements of TAC.

7.4 Economic Criteria for Technology Comparisons

Technology evaluation encompasses not only technical feasibility of a particular technology but also the economics of its implementation. While there are many measures of economic merit, two measures – investment and net present value (NPV) – provide a complete set of information on which to base an informed, economic decision. "Investment" refers to the money that must be spent initially to design and build the new facilities. NPV is the after-tax worth in today's dollars of all the future cash that an alternative will either consume or generate; it includes the effects of new investment, working capital, operating costs, revenues, and income taxes. NPV is the most popular singular measure of the economic merit of an alternative, and many spreadsheet programs can easily calculate it.

To evaluate alternative pollution control devices, the analyst must be able to compare them in a meaningful manner. Since different controls have different expected useful lives and will result in different CFs, the first step is to use the principle of the time value of money to normalize the returns of the alternatives being compared. The process through which future CFs are translated into current dollars is called present value (PV) analysis. When the CFs involve income and expenses, it is also commonly referred to as NPV analysis. In either case, the calculation is the same: adjust the value of future money to values based on the same (generally year zero of the project), employing an appropriate interest (discount) rate and then add them together. The decision rule for NPV analysis is that projects with negative NPVs should not be undertaken; and for projects with positive NPVs, the larger the NPV, the more attractive the project. Derivation of a CF's NPV involves the following steps:

- Identification of alternatives – for example, the choice between a fabric filter baghouse and an electrostatic precipitator for removing particulate matter from a flue-gas stream.
- Determination of costs and CFs over the life of each alternative.

The relationship of the future value of money to the PV of a sum of money compounded annually at an interest rate, i, for a total of n years is shown by the following equation:

$$\text{FV} = \text{PV}\left(1+i\right)^n \tag{7.2}$$

where

FV = future value
PV = present value
i = annual earnings rate
n = years in the future

This equation is used to determine how many future dollars will be realized by investing present dollars at the interest rate i for n years. For example, \$100 invested today at 5% for 12 years will yield the following return:

$$\text{FV} = \$100\left(1+0.05\right)^{12} = \$180$$

Similarly, inflation at 5% per year makes \$100 worth of merchandise today cost \$180 in 12 years. Discounting is the opposite of compounding: the PV of a sum of money to be spent or received in the future is reduced according to

$$\text{PV} = \frac{\text{FV}}{\left(1+i\right)^n}$$

where i is now referred to as the discount rate. Thus, if the discount rate is 5% per year, we calculate the PV of \$100 to be spent 12 years as follows:

$$\text{PV} = \frac{\$100}{\left(1+0.05\right)^{12}} = \$56$$

Since the PV of money to be spent in the future declines with increasing interest rate and number of years into the future, the importance of accounting for the time at which money is spent or received is evident.

7.5 Calculating CF

A complete financial analysis requires the inclusion of CF. On a year-by-year basis, the CF from a manufacturing investment is determined through calculations of cash costs, total costs, and cost-effectiveness.

Cash costs, those associated with operating and overhead, include raw materials (chemicals, catalysts, solvent, etc.) utilities (steam, electricity, natural gas, water, etc.), maintenance materials and labor, operations labor, technical support, startup costs, taxes and insurance, and administrative costs.

Total costs include the sum of cash costs as just defined and depreciated. Note that depreciation is not a cash cost and is used only to calculate income taxes. If the investment generates revenues or savings from which the total costs may be subtracted, the resulting amount is known as the "pretax earnings." Income taxes (calculated on pretax earnings) are next subtracted to yield the after-tax earnings. A year-by-year CF can then be determined by summing the four real CFs: investment, cash costs, revenues, and income taxes.

NPV is the sum of the PVs of a series of CFs:

$$NPV = CF_0 + (CF_1 \times D_1) + (CF_2 \times D_2) + (CF_3 \times D_3) + \cdots \quad (7.3)$$

where, CF_0, CF_1, CF_2 represent CF in year 0, 1, 2, and so on, and the D terms are discount factor for the respective years.

In the context of pollution control, cost-effectiveness is defined as the annualized cost of the control option divided by the annual emission reductions resulting from the control option. The following information is required to calculate the cost-effectiveness of a proposed control option: (i) the capital cost of purchasing and installing the control equipment or making a process modification, (ii) the annual operating costs of the control option, and (iii) an estimate of the emissions before and after application of the control option.

The capital costs of purchasing the control option should be determined from actual vendor price quotes for each proposed control option. Installation costs should also be based on vendor price quotes. If vendor price quotes are unavailable, elements of the installation cost may be estimated (Vatavuk 2002). To illustrate cost-effectiveness calculations, let us consider a facility that is a major source of NO_x emissions.

Example 7.1 Total Capital Cost

As part of a company's ongoing pollution prevention program, an existing appliance coating operation is to be replaced by a minibell automatic electrostatic spray system. The company estimates that the coating operation is presently operating at a 50% transfer efficiency, and it has received a guarantee of 90% transfer efficiency with a new unit. The company operates 250 days/year and processes 200 appliance units per day. An average of 2 gal of paint are presently required to coat each unit. The coating material has a volatile organic compound (VOC) content of 5.0 lb VOC/gal solids. Information on the minibell system is given below. If the paint worth \$5/gal shows that an annual profit will be realized by replacing the existing spray system with the more efficient minibell unit. Cost information on the proposed minibell system is given below.

Installed cost = \$1 125 000
Operation and maintenance cost = \$70 000/year
Lifetime guarantee = 15 years
Interest rate per purchase cost = 8%

Solution

4) Obtain the annual operating and maintenance cost. From the data given,

$$\text{Annual O\&M cost} = \$70\,000/\text{year}$$

5) Calculate the annualized capital cost, ACC.

$$\begin{aligned} ACC &= (IC)(CRF) \\ &= (IC)(i)(1+i)^n / \left[(1+i)^n - 1\right] \\ &= (1\,125\,000)(0.08)(1.08)^{15} / \left[(1.08)^{15} - 1\right] \\ &= \$131\,444 \end{aligned}$$

6) Calculate the TAC, if the minibell system is installed

$$\begin{aligned} TAC &= O\&M + ACC \\ &= 70\,000 + 131\,444 \\ &= \$201\,500 \end{aligned}$$

7) Calculate the total annual operating savings, TAS, if the minibell system is installed. Since the sprat system operates more efficiently, a savings will be realized because of the reduced paint usage. At 50% transfer efficiency only 1 gal of the 2 gal of paint used per unit is actually applied. The paint usage for the minibell system, operating at 90% transfer efficiency, is (1.0/0.9) or 1.11 gal. Thus, the gallons saved per unit is 2.0 − 1.11 = 0.89 gal. The total annual savings, TAS, becomes

$$\begin{aligned} TAS &= (\text{gas used/number units/day})(\text{operating days})(\text{cost}) \\ &= (0.89\,\text{gal})(299\,\text{units/day})(250\,\text{days/year}) \\ &\quad (5.0\text{ lb VOC/gal}) \\ &= \$222\,5000 \end{aligned}$$

8) Calculate the annual savings.
The savings, S, can be calculated by subtracting the result of step (3) from that of step (4).

$$\begin{aligned} S &= \$222\,500 - \$201\,500 \\ &= \$21\,000 \end{aligned}$$

Once a particular process scheme has been selected, it is common practice to optimize the process from a capital cost and O&M standpoint. There are many optimization procedures available, most of them too detailed for meaningful application to simple pollution prevention projects. These sophisticated optimization techniques, some of which are routinely used in the design of conventional chemical and petrochemical plants, invariably involve computer calculations.

Example 7.2 Cost-Effectiveness Calculation

For a boiler operating at 100% of its operating capacity ($50\,\text{Btu} \times 10^6$), the following information was determined for the purposes of calculating the cost-effectiveness of installing SCR as a control option. The facility is a major source of NO_x emissions.

Given:

Capital cost of control option = \$1 500 000
Capital recovery factor (CRF) = 0.1627 (assuming 10% interest for 10 years)
Annual operating and maintenance costs = \$98 000
Uncontrolled emissions = 21 T of NO_x per year
Capture efficiency = 100%
Control efficiency = 80%

Solution

We insert the data into the following equations:

$$\text{Cost-effectiveness}\left(\$/\text{T}\right) = \text{annualized costs/T of pollutant reduced}$$

$$\text{Annualized costs}\left(\$\right) = \left(\text{capital cost} \times \text{CRF}\right) + \text{annual operating costs}$$

$$\text{Pollutant reduced}\left(\text{T}\right) = (\text{uncontrolled emissions} \times \text{capture efficiency} \times \text{control efficiency})$$

The result is

$$\begin{aligned}\text{Cost-effectiveness} &= \left(\$1\ 500\ 000 \times 0.162\ 7 + \$98\ 000\right) / \left(21\,\text{tons} \times 1.0 \times 0.80\right) \\ &= \$20\ 400\ /\ \text{ton}\end{aligned}$$

Inducing industries to adopt effective new production methods and pollution control technologies in timely fashion is a challenge for regulators everywhere. Environmental protection costs money, and the charges that will be incurred by a given enterprise depend on the nature and extent of the pollution it generates. These factors, in turn, are functions of the type and scale of production, the plant's input, the technology used, and the size of the plant.

Generators of pollution range from private individuals and small business to municipalities, large corporations, and the government itself. Each polluting entity must decide how to invest in the socially desirable goal of a cleaner environment; influencing such decisions, inevitably, will be the benefit the entity expects to obtain from the investment.

When business owners consider how to implement the installation of modern pollution control technology, their decisions incorporate the results of comparative analyses of the costs and benefits of the investments contemplated. Later we shall present a case study focused largely on financial costs and benefits associated with biobased energy and products (Section 7.4). To provide the necessary background for this material, we turn now to some basics of cost estimation and capital investments.

7.5.1 Achieving a Responsible Balance

In the real worlds, taking socially desirable action for pollution prevention may not be in the best interest of an industry in the short run, since environmental protection costs money. The type of production, input used, technology employed, space used as well as plant size and the scale of production determine the nature and extent of pollution which, in turn, influence the abatement cost. Hence, the decisions of the profit maximizing private investor for implementation of pollution abatement measure are influenced by the benefit expected as a result of the investment required. On the other hand, the task of government planners and policy makers is to encourage industry to adopt modern production and pollution control technologies to provide a better environment for the society. In upgrading their facilities to meet the near antipollution regulations, industries turn for guidance to the principles of best available control technology (BACT).

7.6 From Pollution Control to Profitable Pollution Prevention

As companies incorporate pollution prevention approaches in their strategic planning, capital investment priorities, and process design decisions, it is vital that they understand both the quantitative and qualitative dimensions of assessing pollution prevention projects. These projects tend to reduce or eliminate costs that may not be captured in cursory financial analyses due to the way costs are categorized and allocated by conventional management accounting systems. Additionally, pollution prevention projects often have impacts on a broad range of issues, such as market share and public impact that are difficult to quantify but that can be of strategic importance. Identifying and analyzing all costs and less tangible items is an important step in an evaluation of the potential benefits of pollution prevention projects.

Pollution prevention can take many forms – from simple "housekeeping improvements," which cost little to carry out, to installation of expensive capital equipment. Although many pollution prevention projects, such as material substitution or process redesign, do not require large outlays for purchase of equipment, they may require extensive qualitative assessments related to such issues as product quality or employee health and safety. The analytical tools described here are applicable to the assessment of most pollution prevention initiatives that fit under the umbrella of the capital budgeting process. Pollution prevention projects involving capital budgeting generally include

1) new manufacturing equipment
2) replacement equipment
3) plant expansion and construction

Capital budgeting is a process of evaluating capital investment options based on a company's needs and analyzing the impact of an investment on a company's CF over time. Pollution prevention and other capital projects are justified by showing how the project will increase the company's earnings as well. Financial tools demonstrate the importance of the pollution investment on a life cycle or total cost basis in terms of revenue, expenses, and profits. Key concepts and factors used in capital budgeting are described next.

Life cycle costing (LCC): Also referred to as total cost accounting, this method analyzes the costs and benefits associated with a piece of equipment or a procedure over the entire time the equipment or procedure is to be used. LCC is also a tool to determine the most cost-effective option among different competing alternatives to purchase, own, operate, maintain and, finally, dispose of an object or process, when each is equally appropriate to be implemented on technical grounds. For example, for a highway pavement, in addition to the initial construction cost, LCC takes into account all the user costs (e.g. reduced capacity at work zones), and agency costs related to future activities, including future periodic maintenance and rehabilitation. All the costs are usually discounted and total to a present-day value known as NPV. This example can be generalized on any type of material, product, or system.

In order to perform a LCC scoping is critical, what aspects are to be included and what not? If the scope becomes too large, the tool may become impractical to use and of limited ability to help in decision-making and consideration of alternatives; if the scope is too small, then the results may be skewed by the choice of factors considered such that the output becomes unreliable or partisan. LCC has a broader scope, including environmental costs. To help building and facility managers make sound decisions, the US Federal Energy Management Program provides guidance and resources on applying LCC that permits the cost-effectiveness of energy and water efficiency investments to be evaluated. LCC can be conducted in two approaches: deterministic and probabilistic method.

Present value (PV): The importance of PV, or present worth, lies in the fact time is money. The preference between a dollar now and a dollar a year from now is driven by the fact that dollar in-hand can earn interest. Mathematically, this relationship is shown earlier in Eq. (7.2), with an example on present and future values.

Comparative factors for financial analysis: The more common methods for comparing investment options all utilize the PV equation presented in Eq. (7.2). Generally, one of the following four factors is used:

1) *Payback period (PP):* This factor measures how long it takes to return the initial investment capital. Conceptually, the project with the quickest return is the best investment.
2) *Internal ROR:* This factor is also called *return on investment* or ROI. It is the interest rate that would produce a return on the invested capital equivalent to the project's return. For example, a pollution prevention project with an internal ROR of 20% would indicate that pursing the project would be equivalent to investing the money in a bank and receiving 20% interest.
3) *Benefits cost ratio:* This factor is a ratio determined by taking the total PV of all financial benefits of a pollution prevention project and dividing by the total PV of all costs of the project. If the ratio is greater than 1.0, the benefits outweigh the costs and the project is economically worthwhile to undertake.
4) *Present value of net benefits:* This factor shows the worth of a pollution prevention project as a PV sum. It is determined by calculating the PVs of all benefits, doing the same for all costs and subtracting the two totals. The new result would be an amount of money that would represent the tangible value of undertaking the project.

7.6.1 Life Cycle Costing

While firms can use any of these factors, the importance of the LCC (for total cost analysis) makes the PV of net benefits the preferred method. Additional details of LCC and TCA follows.

LCC tool and the total cost assessment (TCA) tool are introduced as concept overviews. Both tools can be used to establish economic criteria to justify pollution prevention projects. TCA is often used to describe the internal costs and savings, including environmental criteria. LCC includes all internal costs plus external costs incurred throughout the entire life cycle of a process, product, or activity.

LCC has been used for many years by both public and private sector. It associates economic criteria and societal (external) costs with pollution prevention opportunities. The purpose of LCC is to quantify a series of time varying costs for a given opportunity over an extended time horizon, and to represent these costs as a single value. These time varying costs usually include the following:

1) *Capital expenditures.* Cost for large, infrequent investment with long economic lives (e.g. new structures, major renovation and equipment replacement).
2) *Nonrecurring operations and maintenance (O&M).* Costs reflecting items that occur on a less frequent than annual basis that are not capital expenditures (e.g. repair or replacement of parts in a solvent distillation unit).
3) *Recurring O&M.* Costs for items that occur on an annual or more frequent basis (e.g. oil and hydraulic fluid changes).

4) *Energy*. All energy or power generation related costs. Although energy costs can be included as a recurring O&M cost, they are usually itemized because of their economic magnitude and sensitivity to both market prices and building utilization.
5) *Residual value*. Costs reflecting the value of equipment at the end of the LCC analysis period. This considers the effects of depreciation and service improvements.

By considering all costs, a LCC analysis can quantify relationships that exist between cost categories. For example, certain types of capital improvements will reduce operations, maintenance, and energy costs while increasing the equipment's residual value at the end of the analysis period.

Societal (external) costs include those resulting from health and ecological damages, such as those related to unregulated emissions, wetland loss, or deforestation, can also be reflected in a LCC analysis either in a qualitative or in a quantitative manner. LCC includes the following cost components:

1) *Extraction of natural resources*. The costs of extracting the materials for use and any direct or indirect environmental cost for the process.
2) *Production of raw materials*. All of the costs of processing the raw materials.
3) *Making the basic components and products*. The total cost of material fabrication and product manufacturing.
4) *Internal storage*. The cost of packaging and storage of the product before it is shipped to distributors and/or retail stores.
5) *Distribution and retail storage*. The cost of distributing the products to retail stores including transportation costs, and the cost of retail storage before purchase by the consumers.
6) *Product use*. The cost of consumer use of the product. This could include any fuels, oils, maintenance, and repairs which must be made to the equipment.
7) *Product disposal or reuse and/or recycling*. The costs of disposal or recycling of the product.

7.6.2 Total Cost Assessment

The TCA tool is especially interesting because it usually employs both economic and environmental criteria. As with the LCC analysis, the TCA study is usually focused on a particular process as it affects the bottom-line costs to the user. Environmental criteria are not explicit, i.e. success is not measured by waste reduction or resource conservation, but by cost savings. However, since the purpose of TCA is to change accounting practices by including environmental costs, environmental goals are met through cost reductions.

Because of its focus on cost and cost effectiveness, TCA shares many of the features of LCC analysis by tracking DC, such as capital expenditures and O&M expenses/revenues. However, TCA also includes IC, liability costs, and less tangible benefits – subjects that are not customarily included in both economic and environmental goals, because of its private-sector orientation, as well as other economic computation methods.

7.6.3 Economic Consideration Associated with Pollution Prevention

The greatest driving force behind any pollution prevention (P^2) plan is the promise of economic opportunities and cost savings over long-term. Pollution prevention has been recognized as one of the lowest-cost options for waste/pollutant management. Hence, an understanding of the economics involved in P^2 programs/options is important in making decisions at both the engineering and management levels. Every engineer should be able to exercise an economic evaluation of a proposed project. If the project cannot be justified economically after all factors – include those discussed earlier have been taken into account – it should obviously not be pursued. The earlier such a project is identified, the fewer resources will be wasted.

Before the true cost or profit of a P^2 program can be evaluated, the factors contributing to the economics must be recognized. As discussed earlier, there are two traditional contributing factors (capital costs, and operating and maintenance costs), but there are also other important costs and benefits associated with P^2 that need to be quantified if a meaningful economic analysis is going to be performed. The TCA aims to quantify not only the economic aspect of P^2 but also the social costs associated with the process, product, or a service from cradle-to-grave (i.e. life cycle). The TCA attempts to quantify less tangible benefits such as the reduced risk derived from not using a hazardous substance. The future is certain to see more emphasis placed on the TCA approach in any P^2 program. For example, a utility considering the option of converting from a gas-fired boiler to coal-firing is usually not concerned with the environmental effects and implications associated with such activities as mining, transporting, and storing the coal prior to its usage as an energy feedstock. Pollution prevention approaches will become more aware of this kind of need.

7.7 Resource Recovery and Reuse

A critical element of an interim strategy is enhanced recovery. This can be approached from two directions: reuse of products and recycling of materials. Reuse of products includes return, reconditioning, and remanufacturing. The energy required for reuse and recycling is one of the key factors determining recoverability of a product. The closer the recovered product is to the form it needs to be in for recycling, the less energy is required to make that transformation. From the standpoint of economic development it is worth pointing out that the reconditioning or remanufacturing cycle is relative; it requires roughly half the energy and twice the labor per physical unit of output.

Recycling materials means closing the loop between the supply of postconsumer waste and the demand for resources for production. Recycling of materials will be the business of the Zero Emissions engineer; reuse of products will also involve the Zero Emissions engineer, but will have lots of front-end work from another professional, the concurrent engineer. Concurrent engineering, which incorporates aspects of industrial engineering, product design, and product manufacturing, is an integrated approach that seeks to optimize materials, assembly, and factory operation. These engineers examine the broader context of a product, including technology for managing the environmental impacts of its transport, intended use, recyclability, and disposal, as well as the environmental consequences of the extraction of the raw material used in its production.

The ultimate fate of all materials is thus dissipation, being discarded, or recycling and recovery. With 94% of materials extracted from the environment being converted to wastes, current levels of recovery are clearly not sufficient. Recovery of materials from wastes will reduce the extent of resource extraction (but will not slow the speed of material flows through the economy). Aluminum and lead are two resources currently being heavily recycled, but evidence shows that there is potential for a lot more resources to be recovered from wastes.

Sherwood plots are diagrams that permit the graphic comparison of concentrations of materials in nature against their commodity cost. The sample plot (see Figure 1.7) shows that the price for a commodity depends on its concentration in nature before extraction and refining. In a similar plot (see Figure 1.8) for concentrations of materials in wastes, it again shows a consistent line.

Together, the Sherwood plots demonstrate the recovery potential of materials. The elements plotted above the line shown in Figure 1.7 should be vigorously recycled because they are present in individual by-products in relatively high concentrations. Lead, zinc, copper, nickel, mercury, arsenic, silver, selenium, antimony, and thallium are more economical to recover from waste than from nature. Extensive waste trading could significantly reduce the quantity of material requiring disposal because resource extraction uses from wastes, not virgin feedstocks (Allen and Rosselot 1997).

7.8 Profitable Pollution Prevention in the Metal-Finishing Industry

Surface finishing of metal products is a major manufacturing operation performed in thousands of production shops to provide weather-resistant, wear-resistant, and aesthetically pleasing finishes for thousands of manufactured products.[1] Surface-finishing technology involves direct atom-to-atom bonding between a base material (such as steel, aluminum, brass, or plastics) and a metal or organic surface top coating that provides the desired material performance and appearance properties. Surface pretreatment is crucial for proper performance and durability of the produced part. Cleaning and oxide removal is critical, so pretreatment processes usually involve many tanks with various purposes. Multistep surface preparation processes are generally employed to remove oils, soiling and dirty materials, old coatings, corrosion products, residual cutting fluids, brazing residuals, pickling acid residuals, cleaner residuals, etc. The surface preparation process removes contaminants, preserves the cleaned surface, and/or modifies the surface for the next coating. After the finishing process, the parts require rinsing/cleaning to remove residual plating solution. These baths, both plating and cleaning, ultimately are exhausted because of depletion of strength or buildup of impurities and, as a result, become major waste streams (Burckle and Ferguson 2005; USEPA 2001).

In the past, pollution control in the metal-finishing industry has been achieved primarily through end-of-pipe treatment. Facilities added waste treatment systems as a final process step to meet regulatory discharge limits to municipal sewers or as direct dischargers. This approach results in the production of residual sludges contaminated with heavy metals that require appropriate pretreatment for environmentally safe disposal (the so-called "Hammer Provisions" of the Resource Conservation and Recovery Act that apply to hazardous waste disposal in the United States). In the mid-1980s, various facilities began integrating "pollution prevention" into their operations to reduce the amount of wastewater generated and treated. However, the speed of adoption has been relatively slow, most noticeably in smaller facilities (USEPA 1982, 1985).

The most logical and efficient approach to effectively achieve environmental protection is by preventing pollution. Prevention is achieved using techniques that reduce, eliminate, or recycle/reuse waste materials so that the generation of waste is eliminated, avoiding the need for waste disposal, and wastes that are generated are handled responsibly, either through recovery and recycling to displace other material needs and/or to conserve energy, or minimized and made

1 Much of this material was derived from the Section 9.2.3 Burckle and Ferguson (2005).

safe for disposal into the environment. The metal-finishing industry began to apply pollution prevention principles to reduce process waste emissions in the mid-1980s, but there were no public statistics generated of the impact on pollution avoided or economic benefit accrued. In discussing the progress of domestic metal-finishing industry in transitioning to operations based upon pollution prevention, we begin by describing the National Metal Finishing Strategic Goals Program (SGP) initiated by the US Environmental Protection Agency (USEPA) in 1998 in cooperation with industry and public interest stakeholders (USEPA 1998, 2001).

7.8.1 National Metal Finishing Strategic Goals Program

This was the first nationwide program initiated by the USEPA to build voluntary participation in a compliance program by providing incentives in return for a real commitment to reducing pollution to levels below those required for compliance. The SGP was created to address environmental problems and institutional barriers to achieving compliance in a number of industries. It was designed to provide a working relation that facilitates and encourages companies to go beyond environmental compliance to pollution prevention–based operations. SGP member companies were offered incentives, resources, and a means for removing regulatory and policy barriers as they work to achieve specific environmental goals.

The SGP program brought together stakeholders to identify important issues, conduct demonstration projects, and develop consensus policy recommendations that offered opportunities for reducing pollution. The stakeholders included industry, labor, environmental groups, state and local government, and other federal agencies. The metal-finishing industry was represented by the National Association of Metal Finishers, American Electroplaters and Surface Finishers Society, Metal Finishing Suppliers Association, and Surface Finishing Industry Council. Through the SGP (USEPA 2001), the participants worked together to improve environmental performance and the bottom line.

Participation required top management commitment to the implementation and maintenance of an approved Environmental Management System (EMS). The EMS has caused management to take careful note of the true costs of pollution and institute policies and procedures to eliminate this waste at its source (USEPA 2003a). The results of these efforts have provided the first large-scale quantification of the power of pollution prevention for achieving significant reductions in pollution and the resulting economic benefits in the metal-finishing industry. The SGP program was designed to help a metal-finishing company achieve its goals – both environmental and economic. A wide variety of state and local SGP resources (USEPA 2001) were available to provide a company with the tools need to get the job done. These include the following:

- Free, non-regulatory environmental audits
- Funding for environmental technologies
- On-site technical assistance in (evaluation and planning) achieving compliance and improving pollution prevention, safety, and health measures
- Free assistance from interns to help fill out the SGP data worksheets
- Free workshops on energy, water, and waste reduction
- Regulatory flexibility
- EMS training
- Public recognition

All parties benefited from the use of these tools. A few of the advantages listed in the project report are listed here.

- Companies received the incentives and resources needed to take the risk of going beyond compliance requirements.
- As the participating companies employ less-polluting technologies, waste is reduced, less pollution is discharged to the environment, and the plant saves money, becoming more cost efficient and competitive.
- As the results demonstrated in the participating plants are employed by other plants to reduce pollution and improve competitiveness, the practices of pollution prevention are spread throughout the industry.
- The industry as a whole benefits from the positive action of SGP member companies.
- As the metal-finishing industry becomes increasingly more self-regulated, government regulators – from the EPA to the local publicly owned treatment works – save time and money.

The SGP has seven environmental performance goals (USEPA 2001) that form the core of the program (see Table 7.1). Through the end of 2000, many participants had already made significant progress toward meeting the goals, which translated into real environmental gains:

- 380 million gal of water conservation
- 120 million lb of hazardous waste not sent to landfills
- 665 000 lb of organic chemicals not released to the environment.

A comparison of the reductions achieved through 2000 and 2003 shows continuing and significant improvements in pollution prevention. This continued progress of the participants is a positive indicator of the benefits of a program driven by the employment of the EMS. The SGP participants were provided free, non-regulatory environmental audits and access to technical support in achieving compliance

Table 7.1 Progress towards SGP goals.

SGP goal	Average achievement by SGP metal finishers (over all projects)	
	2000	2003
1) 50% reduction in water usage	41% reduction	56% reduction
2) 25% reduction in energy use	14% reduction	41% reduction
3) 90% reduction in organic toxic release inventory (TRI) releases	77% reduction	84% reduction
4) 50% reduction in metals released to water and air	58% reduction	65% reduction
5) 50% reduction in land disposal of hazardous sludge	36% reduction	48% reduction
6) 98% metals utilization	17% utilization factor	64% utilization factor
7) Reduction in human exposure to toxic materials in the facility and surrounding community	51% of activities accomplished	85% of activities accomplished

and improving pollution prevention. Data like this readily demonstrates the power of the EMS management tool to reduce pollution. But to see how this progress was achieved, we must examine the role of pollution prevention (P^2) technologies in enabling the reductions.

7.8.2 The Role of Pollution Prevention Technologies

7.8.2.1 Moving Toward the Zero Discharge Goal

A highly desirable "state" for environmental protection would be Zero Discharge of pollutants to the air, water, and land. Today, this is a goal not yet realized. However, "approaching" Zero Discharge has been found to be realistic, as demonstrated by the continual improvement in environmental performance achieved in the SGP for the metal-finishing sector. Approaching Zero Discharge can be defined as reducing wastes emitted from a process by a significant amount, with significant reduction ranging from a low defined by the regulatory standard and the high defined by the technology employed.

There are two key elements to achieving movement toward the state of Zero Discharge – implementation of an EMS and deployment of certain technologies based upon pollution prevention. The framework of an EMS is the *management tool* that provides a state of "continuous assessment and compliance of plant operations," while P^2 technologies provide the *technological tools* needed to achieve significant improvements in performance and reductions in generation of waste.

7.8.2.2 Planning and Implementation

As part of the planning phase, a compliance assessment of a participating plant is necessary to establish a baseline against which progress and savings in incremental costs can be measured. The goal of the plant assessment is to identify the root causes of the most significant problems, to identify areas in which P^2 options could save the most money, and to assign priorities to address the most significant problems. The plant assessment must include an inventory of all chemicals, wastes, bath chemistries; the overall plant layout; and a site inspection. Once the problems are identified and prioritized, various solutions can be proposed and the effects of their deployment evaluated. A comprehensive plan addresses housekeeping and maintenance issues in order to sustain any P^2 efforts and also to establish good standard practice.

This planning approach is most efficiently implemented through an EMS. In addition, the imperatives of "total quality management" apply. Management must buy into the process and be willing to provide the necessary resources to achieve success. Experience has shown that often the best solutions come from those working most closely to the problem. It is important that employees be included in the improvement program and kept well informed so that they will become stakeholders in the process.

Implementation through an EMS has several advantages. First it is more effective because it provides a tool for involving management through its provision for continuous improvement in both environmental performance and worker health and safety. Second, it provides a mechanism to integrate process and product quality issues that influence reduction of waste and the improvement of productivity and profitability. The review of these considerations should be incorporated into the process and reviewed over time as a pathway for identifying future opportunities and establishing priorities. This process leads to the identification of the power of pollution prevention technologies to achieve these savings offered by waste reduction. The Agency provides an EMS template tool (USEPA 2004; Stander and Theodore 2008) to assist those who are interested.

Many of the same elements that must be defined for the compliance assessment are also needed to establish an

EMS. Certification under the ISO 14001 – an EMS that is standardized worldwide – is being increasingly required of those companies engaged in the manufacture of products for export, including components in the supply chain of such products. Also the EPA offers Compliance Incentives – "policies and programs that eliminate, reduce, or waive penalties under certain conditions for business, industry, and government facilities which voluntarily discover, promptly disclose, and expeditiously correct environmental problems" and is including a requirement for the implementation of EMSs in settlements. Information about these incentives, programs, and the environmental benefits achieved by such programs may be found on the Internet (http://www.epa.gov/compliance/incentives/index.html).

7.8.2.3 Practicing Pollution Prevention

Technologies based on the precept of Pollution Prevention serve to eliminate the generation of wastes or to reduce the disposal of wastes through recycle/reuse. Pollution in the metal-finishing industry is basically the discharge of some unwanted form of material or other resource (energy, labor, time, etc.). Loss of these resources equates to the loss of profit and economic productivity. It stands to reason that the more of a resource used above the minimum required by the process, the more this incremental use (read: "waste") adds to the "unnecessary" component of the total costs of the operation. In the case of water use, for example, this unnecessary cost is not just the cost of the excess water wasted. The cost of this waste is magnified by the costs for its overall management for treatment and disposal, including the capital and operating costs for moving the excess water through the process, its cleaning, and the disposal of the treated water and any residual wastes. The treatment and disposal costs are usually more costly than the initial raw material. Often these costs (process operations and compliance costs) are not tracked by a company because the accounting process used is inadequate to perform such tracking.

Companies participating in the SGP have found significant cost savings by implementing P^2 practices (USEPA 2001). Significant cost savings result in improved economic efficiency and an improved bottom line on the balance sheet, making the operation more competitive. Other advantages include the protection of employees, reduction of liabilities, and, at the same time, enhancement of the company's business image.

Success in achieving the goals of the SGP for the metal-finishing sector has been attributed largely to the employment of pollution prevention techniques. A principal purpose of the program was to demonstrate the attractive environmental and economic advantages offered by P^2-based approaches to waste reduction. The outcome anticipated was to "stimulate a keen awareness and appreciation" of these extremely valuable tools so that they would be adopted into everyday practice on a sector-wide basis. In addition, the adoption of an EMS and integration into company operations ensures continued improvements (both in environmental performance and in cost savings) within the entire fabric of the participating companies.

There are many approaches to reduce pollution in various unit processes that were investigated and documented by USEPA sponsored research in the 1980s, including housekeeping and maintenance methods and management of process chemicals and rinse water. These methods are described in the EPA Capsule Report entitled *Approaching Zero Discharge in Surface Finishing* (USEPA 2000) and the training course document entitled *P2 Concepts & Practices for Metal Plating & Finishing* (AEFS, American Electroplaters and Surface Finishers Society). These learning aids address a number of process design areas such as those listed in Table 7.2.

Transitions of manufacturing operations to P^2-based systems have been demonstrated using a range of technology options from relatively simple improvements in existing process technology to more sophisticated approaches based upon significant process changes. There are numerous examples of the power of pollution prevention technologies, ranging from relatively simple improvements of water management to more sophisticated recovery processes, as documented in the results of the SGP and other programs.

Mini-Case Study (USEPA 1998): *A Water Management P^2 Example*

A California company implemented a number of relatively simple and low-cost P^2 operations: flow restrictors in rinse tanks to manage rinse water, drag-out reduction using drip boards between process tanks and hang bars above plating tanks. Between 1997 and 2000, the plant achieved a 50% reduction in water consumption, a 25% reduction in energy consumption, and 98% metals utilization. Since October 2001, the plant has increased savings in energy and water consumption by improving reductions to 62.5 and 70% less process water from the original baseline, respectively.

For other case studies, go to access on-line at:

http://es.epa.gov/cooperative/topics/metcasestudies.html

http://www.epa.gov/ORD/NRMRL/std/mtb/metal_finishing.htm#nmmsfp

Table 7.2 Pollution prevention technologies for surface finishers.

To improve in these operations	Employ these technologies and practices
Extend bath life	
• reduce chemical consumption • reduce bath dumping • maximize on-line time • maintain product quality	• improve bath chemical solution management • reduce contamination by improved housekeeping • add new continuous bath purification technologies
Reduce water consumption	
• improve water utilization to decrease water demand and sewer charges • increase in-plant recycling • decrease water discharges and water intake	• employ a rinse method that uses less water • reduce drag-in and drag-out • employ new process analysis and planning methods for water conservation • use advanced technologies for removal of chemicals to permit rinse water recycling
Minimize waste	
• decrease usage of all chemicals and water • decrease waste generation • improve recovery and recycling in-plant	• improve process management and oversight • implement an environmental management system • improve rinsing operations as described above • increase in-plant recycling of water
Reduce the use of hazardous chemicals	
• reduce environmental risks and costs • improve worker safety • maintain or improve productivity and quality • acquire new or maintain existing markets where clients are "greening the supply chain"	• move to less toxic chemistries • use new finishing processes and coatings • reduce the number of stages in a line

7.8.2.4 Case Study: An Emerging Profitable Pollution Prevention Technology

The USEPA recently completed a study of the proprietary Picklex process. Picklex® is a "'non-polluting' pretreatment/conversion coat which replaces chromate conversion coating and zinc & iron phosphatizing in powder coating, paint and other organic finish applications" (Ferguson and Monzyk 2003; Ferguson et al. 2001).

Description Metal pretreatment is crucial for cleaning and oxide removal to obtain proper performance and durability of the produced part, and it usually involves many tanks with various purposes. For example, aluminum anodizing may require eight pretreatment tanks; in conventional chromate conversion coating, there are twelve stages, while conventional zinc phosphatizing requires six. All of these baths contain heavy metals and must be dumped periodically.

The volume of hazardous/toxic waste streams produced from metal surface–finishing operations is significant (USEPA 1995a). The elimination of any of the surface-processing steps is beneficial to the manufacturing process because it reduces processing costs, waste production, and energy consumption. Strong acids used in the pretreatment of metals pose a great health and safety risk to workers. With this in mind, a no-waste or waste-reducing surface-finishing agent designed to lower processing costs for metal-finishing operations would be of great benefit.

The Picklex process (a proprietary process) is one such alternative to conventional metal surface pretreatment that offers significant reductions in waste generation. The process was developed by International Chemical Products, Inc. with assistance from the USEPA. The Picklex process works in a completely different way than conventional processes. It incorporates the corrosion products from the metal surface into the protective coating that is applied to that same metal surface. This new P^2 approach "thinks outside the box" by solving the common environmental problem of metal buildup in the processing tanks while forming a protective surface coating on the work piece. A two-phase program was undertaken to evaluate the ability of Picklex to perform technically and economically as a pollution prevention–based replacement for conventional metal pretreatment or pretreatment/conversion coat in finishing operations. The goal was to demonstrate its potential to eliminate or significantly reduce the amount of hazardous and toxic chemicals consumed and the amount of pollution generated while maintaining equal or better product performance properties.

A broad laboratory evaluation of Picklex was undertaken in which full multistep, bench-scale, batch operation tests using side-by-side test lines of seven conventional processes and of Picklex were performed. The results of the laboratory tests were quite favorable, and field tests (Ferguson and Monzyk 2003) were conducted to evaluate in-plant performance in side-by-side tests with conventional preparation technologies. Only the most promising applications – the replacing of chromate conversion coating and zinc phosphatizing – were taken to field evaluation. In these focused field tests, Picklex was used to prepare aluminum and steel substrates for powder coating only. Although the testing scope was narrowed, it was more detailed. A total of 41 different combinations of substrate, degreaser, pretreatment, conversion coat, and powder coat were tested in the field evaluation. Only uncontaminated panels and components, without corrosion products, were used. The results for all test cases demonstrated that the quality of the final product was equal in all respects to conventional practice. Product quality was targeted as the single most important acceptance criterion within all the comprehensive results, because if the final product quality was not acceptable, the pollution reductions would not matter. The field test results replicated those obtained earlier, indicating that the lab testing may be used to accurately predict results achievable in practice.

7.8.2.5 Environmental Benefit

Two commercial processes, chromate conversion coating and zinc phosphatizing, were employed in the field tests as the experimental control baseline (Figures 7.3 and 7.4, respectively). Metal buildup in these conventional processes ruined the baths, and the baths had to be dumped. In the metal-finishing sector, the discarding of such metal-containing waste streams is a major contributor to pollution. The Picklex chemical used in the process tank is used up on the product; however, with filtering, the bath does not become contaminated. As a result, it is not necessary to dump the bath to control impurity buildup, and the chemical make-up is simply added to the bath as needed. This process was demonstrated to have a strong potential as a leading technology for pollution prevention in surface preparation operations.

As can be seen in the comparison of the conventional practices to this emerging technology, the Picklex process eliminates up to eleven steps in metal-finishing processes. The reductions of waste at the source were very significant. These reductions were accomplished by the elimination of the hexavalent chromium tank and subsequent rinsing operations; the elimination of the acid baths used for etching and cleaning along with subsequent rinsing operations; the elimination of the zinc phosphatizing baths and subsequent rinsing operations; and the elimination of the deoxidizing bath and subsequent rinses. All of the steps eliminated are steps that emit pollution. In addition, the

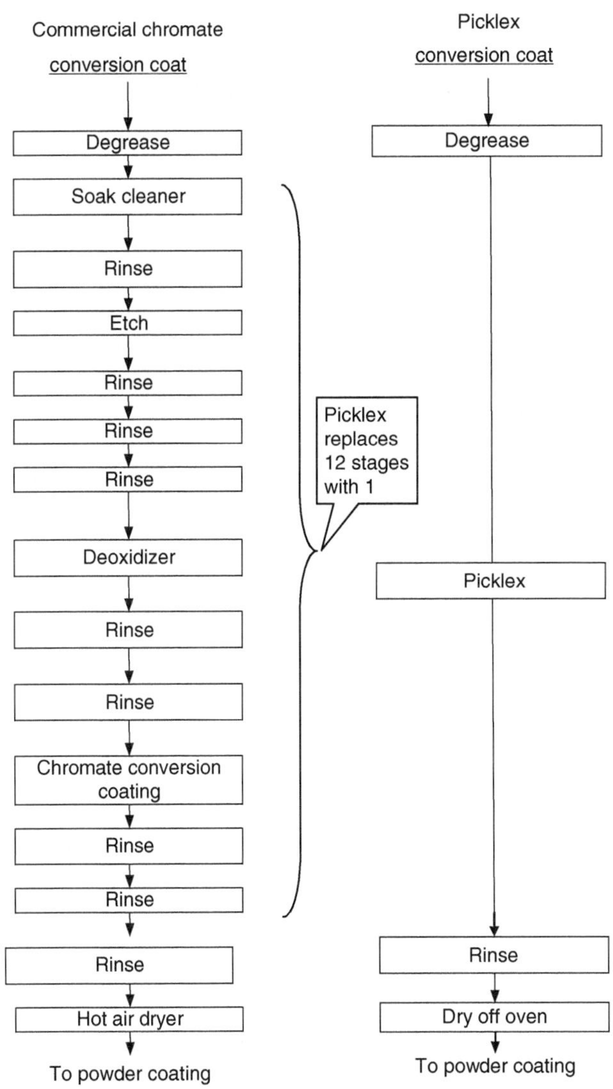

Figure 7.3 Comparison of Picklex to conventional chromate conversion coating.

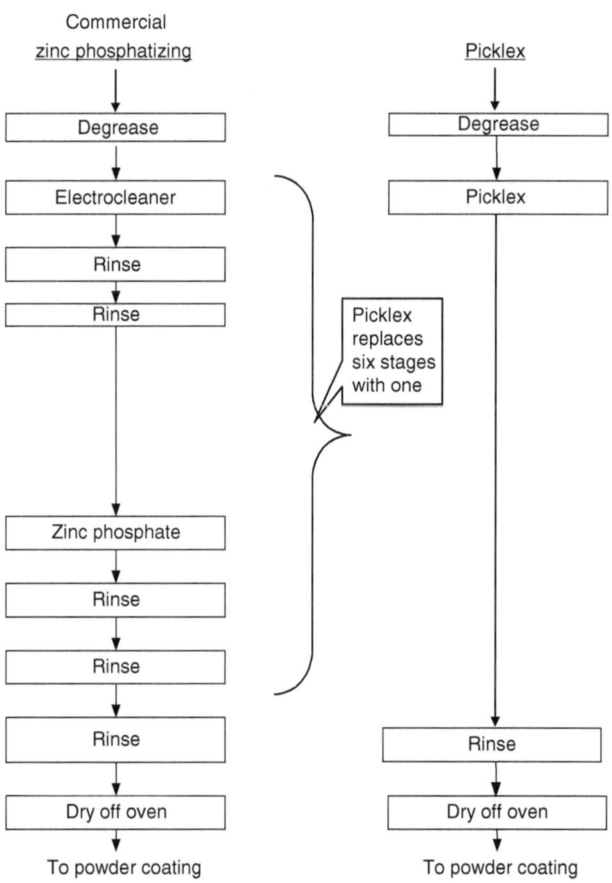

Figure 7.4 Comparison of Picklex to conventional zinc phosphatizing.

Picklex process significantly reduces the pollution from the remaining steps. It does this while providing the same high-quality finish and wear resistant capabilities as the conventional processes.

Although normally the Picklex bath does not need to be dumped, but, as part of the study, the bath was evaluated for waste treatment issues in the event of a spill or accidental contamination. Fresh, spent, and impurity-spiked spent Picklex samples were treated using the conventional pH 9 precipitation industrial wastewater treatment method to produce samples for a waste disposal assessment for Picklex. The waste solids were assessed for toxic leachability. The treated water supernatants for discharge were also examined and found as most likely dischargeable. Waste treatment may not be frequent or necessary in many processes. Actual discharge limits from industrial waste treatment plant operations are site-specific and are determined on a case-by-case basis with local, state (and federal only where necessary) regulatory agencies. Hence, no exact classification of these potential waste solutions is possible until a specific location is known. All leachates are passed with respect to the applicable federal regulations (USEPA 1995b). Therefore, Picklex does not appear to present unusual waste treatment issues.

7.8.2.6 Economic Benefit

A cost comparison was made for two most promising processes – replacing chromate conversion coating and zinc phosphatizing. These were chosen based on the potential cost savings that industry could achieve. There would appear to be a significant economic incentive to migrate existing practice to this new technology, based on the significant capital and annual operating costs that is projected for the Picklex technology, summarized in Table 7.3. In fact, a new shop adopting the process in place of chromate conversion coat would save capital costs of \$254 000, with the reduction of the process from 12 to 1 tank (Figure 7.3). For zinc phosphatizing on steel, the savings is estimated to be \$230 000.

Table 7.3 Potential cost savings.

	Savings of Picklex over conventional process	
Cost type	**Chromate conversion coat on aluminum (\$)**	**Zinc phosphatizing on steel (\$)**
Capital cost savings	254 000	230 000
Annual operating cost savings	46 000	32 000

Source: From Ferguson and Wilmoth (2000) and Ferguson et al. (2001).

7.8.2.7 Technology Transferability

The Picklex technology is very robust and flexible (Ferguson and Monzyk 2003). It is applicable from the very large to very small operations. It can be used in a new installation, in a retrofit for process modernization, or as a drop in replacement for any existing facility. Picklex can be applied by dipping the part in tanks or by spraying or brushing. Cost savings are very significant and are a function of the size of the facility and the number of processing lines that are required. Transferability is enhanced by the prospective significance of the overall facility cost savings and productivity improvements offered.

7.8.3 Value-Added Chemicals from Pulp Mill Waste Gases

Methanol, formed during the pulping of wood and is contaminated with reduced sulfur compounds and terpenes, is the largest single source of VOC emissions from kraft pulp mills, accounting for 70–80% of total emissions. The Cluster Rule (Cluster Rule Regulation: 40CFR Part 63 1998) limits methanol emissions for all pulp mills in the United States. Canada faces similar legislation (Das 1997, 1999a, b; Das and Houtman 2004; Das and Jain 2001).

Methanol emissions can be reduced by collecting condensate streams from the digesters, evaporators, and other sources in the mill. The collected condensate streams can then be steam-stripped to concentrate the methanol for incineration. A few mills are "hard-piped" to send this methanol-laden stream to bio-treatment plant, thus avoiding incineration.

Stripper overhead gas (SOG) contains roughly equal part of methanol and water (40–50 wt%) and roughly equal parts of TRS and terpenes (1–5 wt%).

In the United States, the SOG is likely to be burned in an incinerator, kiln, or boiler. Recently, some mills have found it advantageous to rectify the SOG to about 80% methanol and collect it as a liquid, which has a higher fuel value. With this concentrated (70–80%) methanol stream available, plants can cut down on the amount of natural gas that must be bought.

A process has been developed at Georgia-Pacific that converts the methanol and TRS (mostly mercaptans) in the rectified SOG into formaldehyde. Based on work by Wachs (1999a, b), this patented catalytic process (Burgess et al. 2002) has achieved commercially viable yields of formaldehyde (70–80%) from a typical pulp mill SOG feedstock containing methanol, water, and TRS compounds.

The conversion of methanol and TRS to formaldehyde presents kraft mills with a more profitable alternative for SOG than incinerating the gas as a fuel. The formaldehyde produced can be used by resin manufacturers to produce

Table 7.4 Annualized income, costs, and earnings 2000 ADTPD mill, 14 lb methanol/T SOG.

Income 15 000 000 lb/year formaldehyde (50%) at $0.06/lb, FOB mill	$900 000
Operating cost	
Direct labor	$110 000
Utilities	$25 000
Methanol in feed at $2.25/MMBTU fuel value	$260 000
Misc., catalyst, supplies, etc.	$70 000
Total operating cost	$465 000
Gross margin	$435 000
Depreciation	$210 000
Earning before tax	$225 000
Credit for steam generated (depending on the mill's steam balance)	$95 000
Credit for terpene recovered and sold	$110 000
Credit for SO_2 recycled	$45 000
Total by-product credits	$250 000
Total earnings, including by-product credits	$475 000

Source: From Burgess et al. (2002).

thermosetting resins commonly used in plywood and other structural panels. A typical pulp mill of 2000 air-dried tons per day (ADTPD) output may achieve a payout of 2–4 years, depending on the price of methanol and local economics.

The process also produces two levels of low-pressure stream, 60 and 70 psig, usable within a paper mill, by reducing most of the methanol to formaldehyde, rather than CO_2. The reduction in CO_2 emissions is about 80–85% of the amount otherwise generated by incineration or by the bio-treatment plant. For a 2000 APTDP mill, this equates to 28 T/Y of CO_2 emissions that are avoided.

A typical itemization of income, costs, and earnings is given in Table 7.4. For a 2000 ADTPD mill, a payout of 3–4 years is calculated based on formaldehyde prices of $0.06/lb (50%) basis. This assumes a customer-shipping radius of 500 miles from the mill producing the formaldehyde. A heat value credit to the mill is included (equivalent of natural gas fuel) for the methanol that would otherwise have been incinerated and used to generate steam via heat recovery exchangers.

7.8.4 Recovery and Control of Sulfur Emissions

Sulfur is often considered one of the four basic raw materials in the chemical industry. It can be recovered as a by-product from sulfur removal and recovery processes (Kirk and Othmer 2004). Historically, sulfur recovery processes focus on the removal and conversion of hydrogen sulfide (H_2S) and sulfur dioxide (SO_2) to elemental sulfur, as these species represent significant emission from pulping process. Various processes for the removal of SO_2 in the combustion gases are available. Direct catalytic oxidation of SO_2 to SO_3, and subsequent absorption of SO_3 in water to produce sulfuric acid, is an alternative method (Paik and Chung 1996). Total mill TRS emissions of 10–20% are contributed by vent streams from brownstock washers, foam tanks, black-liquor filters, oxidation tanks, and storage tanks that are not typically collected in the noncondensable gas system (Pinkerton 1999). The emissions from these sources can be collected and combusted for energy recovery, reducing the atmospheric emissions of TRS and VOCs.

7.8.4.1 Freshwater Use Reduction and Chemical Recovery and Reuse Save Million at Pulp Mill

Effluent Discharges Before pulp mill effluents can be released to the environment they must be treated. Primary treatment involves the use of settling ponds or tanks in which suspended solids settle out of the liquid effluent. Solids can be composted and spread on land, converted to other useful products, or incinerated. Secondary effluent treatment includes oxidation and aeration in shallow basins having wide areas or in smaller areas using mechanical agitators and spargers to oxygenate fluids before release. Biological filter systems can be used to remove organic compounds and heavy metals, and often the process can be accelerated by adding nutrients and by using oxygen rather than air. In some areas, natural wetland systems have been designed to achieve this. Other means of treatment include lime coagulation and the elevation of pH to precipitate organic color bodies as calcium lignates. Once precipitated, the sludge is dewatered and incinerated to destroy organics.

It has been reported by NCASI that there was a 30% reduction in effluent flow from mills between 1975 and 1988. During the same period, final effluent 5-day BOD and TSS decreased by 75 and 45% (NCASI 1991). Data for 1975, 1985, and 1988 are presented in Table 7.5. These reductions will continue as pulp and paper mills implement the so-called best management plans (BMPs) required

Table 7.5 Effluent discharges from pulp and paper mills.

	1975	1985	1995
Effluent flow (gal/T)	22 800	17 200	16 000
BOD (lb/T)	18.0	4.8	4.4
TSS (lb/T)	13.0	8.3	7.1

Source: From Das (1997).

by the Cluster Rule. BMPs will require better management of process losses and spills and are expected to reduce effluent discharges. For example, in roughly 10 years, Simpson Tacoma kraft mill reduced its freshwater consumption by approximately 50% through various pollution prevention methods (K. Schumacher and M. Mammenga, personal communication; USEPA 1992a, b).

Simpson Tacoma now uses about 18 million gal of freshwater per day (mgd) vs. 32 mgd in 1990s. This reduction in freshwater usage saves about $1.92 million/year (350/MG, city of Tacoma charges per average fixed and variable cost). The plant also saves money through reducing sodium hydroxide (NaOH) losses to sewers and to product fiber, by stopping overflowing weak washing dilute NaOH solution and through recycling processed water. Otherwise, NaOH would be needed to the process to make up for soda (sodium) lost. The saving is about 3.4 million/year (based on $375/T of NaOH and 25 T of NaOH saving per day). Simpson also saves $0.18 million/year by reducing losses of black liquor sulfur to sewer and stack (K. Schumacher and M. Mammenga, personal communication).

Other firms save water by installing savealls (devices which separate fiber from process water), heat exchangers, and other equipment which permit more reuse of process water. Internal water cleaning systems make it possible to substitute filtered white water for clean water. Separating process cooling and clean water is often necessary to achieve balance in operations and water use.

7.8.4.2 Brine Concentrator for Recycling Wastewater

Brine concentrators are vapor compression evaporator systems that produce distilled water and a very small salt concentrate stream. These are ideal for water recycling because the concentrate stream is so low that wastewater can be treated economically with a very high recovery and with no liquid discharge (Dalan 2005).

Mini-Case Study

Saving the Colorado River

The market for brine concentrators initially arose because of federal clean water regulations. The Colorado River, a major source of drinking and irrigation water for the southwestern United States, had been growing increasingly saline as a result of human activities. It was to control such damage to the environment here and elsewhere that the USEPA, in the 1970s, promulgated regulations curtailing discharges to the Colorado and forbidding construction of new plants that could not achieve Zero Discharge of water into the river.

To comply with the regulations, both new and existing power plants that were using river water had to recycle their water wastes. Although vapor compression was not new to industry as an energy source, the technological fit with power plants was a natural because it allowed these facilities to use electricity they were generating as the source of the mechanical energy needed in recycling water.

Federal law was not the only factor prompting industry to turn to the use of brine concentrators for recycling wastewater. All over the country, there were local siting regulations, as well. Thus, with the advent of the private power industry in the early 1990s, entrepreneurs turned to Zero Discharge water systems, which allowed them to use sites with a limited water supply, far away from discharge points. Similarly, the move to clean-burning natural gas diminished the importance of locating power plants near sources of fuel or water.

As of 2004, there were approximately 60 brine concentrators in the Unites States. They are sold as package plants, designed and constructed with energy conservation principles. All the original units were at coal-fired power plants, but as metal smelters, manufacturers of chemicals and semiconductors, and other enterprises began to recognize the need to eliminate water discharges, the use of brine concentration has spread.

7.8.4.3 Economics of Brine Concentrator Systems

Based on the parametric cost information on brine concentrator systems, we will now overview some economic aspects of these systems.

Brine concentrators, and by extension, integrated zero liquid discharge systems, only make sense in grassroots facilities only when

- there is a shortage of surface or well water
- the facility needs water to operate
- there is no water discharge option, such as a source of potable water or a river
- a discharge permit is unobtainable

Zero Liquid Discharge takes away the siting constraint; that is, it is no longer necessary to locate the plant near a large usable water source or suitable discharge point. The economic advantage of such an effect is hard to generalize, being very specific to the actual circumstance.

The NPV of a Zero Discharge facility is negative. Dalan and Rosain (1992) found that a 265 gpm brine concentrator had operating costs of $8.94/1000 gal of water treated (total of $1 252 600/year), while the avoided cost of extra demineralizer regeneration chemicals was $216 000/year. The avoided cost amounts to 1.54/1000 gal. Since any dollar amount here is an operating expense, the CF is negative.

Table 7.6 Operating cost breakdown for a 265 gpm system resulting in a cost of $9.62/1000 gal of feed.

Item	Consumption	Unit cost ($)	Annual cost ($)
Operating labor	1 mhr (maximum hourly rate)/h	50/mhr	438 000
Maintenance (labor)	2 mhr/h	50/mhr	87 600
Maintenance (materials, including spare parts)			80 000
Electricity	1 617 kWh	0.05/kWh	707 000
Chemicals			
Sulfuric acid	293 lb/day	0.06/lb	6 416
Polymers	40 lb/day	1.50/lb	21 900
Total chemicals		28 316	
Total annual operating cost		1 340 916	
$/1000 gal of feed		9.62	

Source: From Dalan and Rosain (1992), updated by Dalan (2005).

A typical specific operating expense is in the $5–7/1000 gal treated range. In this estimate is the price of electricity at a retail price of 0.05/kWh. In grassroots power plant planning, the electricity is many times considered a parasitic load on the power plant (electricity needed to produce the power). In this accounting method, the cost of electricity is zero. The elimination of electricity as an operating cost brings the total treatment cost down to the $2–3/1000 gal range. The typical energy load for a brine concentrator is 100 kWh/1000 gal of water produced.

The determination of water economics is very geography specific. For example, in western Washington, water and sewer bills typically are in the $1–2/1000 gal range, whereas in eastern Washington, water costs more, if indeed it is economically available.

For non-power plant applications, local high prices for electricity can be overcome by seeking out alternate energy sources. Compressors (the majority energy user) in brine concentrator plants have been installed that are steam driven or natural gas (via a natural gas engine) driven. Table 7.6 presents breakdown of typical operating costs (Dalan 2005; Haussman and Rosain 1996).

For existing facilities, installing a Zero Discharge plant makes sense when

- the discharge permit conditions change, with the result that the existing facility can no longer be operated economically
- the cost of water rights plus treatment fees exceed the operating costs of a Zero Discharge facility

This last point is illustrated by a mini-case study.

Mini-Case Study

Calculating Payback for a Zero Discharge System

A power plant has a 300 gpm blow downstream from the plant. Preliminary estimates show that a Zero Discharge plant would cost $4.55/1000 gal to operate. The local utility raises the water and disposal fees to $3.00/1000 gal, at a total cost to the plant of $6.00/1000 gal.

The capital cost of a 300 gpm of distillate flow rate is approximately $7.5 million (Dalan 2005). The annual saving is ($1.45 × 300 × 1 440 × 365)/1 000, or $228 636/year. This represents a project with a simple ROI of 3% and a simple payback of 32 years.

7.9 Use of Treated Municipal Wastewater as Power Plant Cooling System Makeup Water: Tertiary Treatment vs. Expanded Chemical Regimen for Recirculating Water Quality Management

7.9.1 Introduction

Every day, water-cooled thermoelectric power plants in the United States withdraw from 60 billion to 170 billion gal of freshwater from rivers, lakes, streams, and aquifers, and consumes from 2.8 billion to 5.9 billion gal of that water. Freshwater withdrawals for cooling in thermoelectric power production account for about 40% of all withdrawals, essentially the same amount as withdrawals for agricultural irrigation, as documented by the U.S. Geological Survey. Sustained droughts nationwide underscore the critical need to think about using treated municipal wastewater (MWW) for use in cooling in electric power generation. It needs a great deal of water for electric power production, to condense stream in the power plant stream cycle. Air cooling is possible, but it is more costly and less efficient. Water will continue to be the preferred coolant for new thermoelectric power plants (Dzombak 2013).

Motivation for the project: Increase in demand for electricity brings with it an increase in water needed for cooling. The cooling of thermoelectric power plants accounts for 41% of all freshwater withdrawal in the United States, i.e. approximately the same amount of water as is withdrawn for agricultural irrigation. Some areas of the United States have little or no freshwater available for use. Alternative sources of water are needed for new electric power production. The U.S. Department of Energy has been conducting and sponsoring research to investigate the feasibility and costs of using alternative sources of water for power plant cooling, especially in recirculating cooling systems which are required for most new power production in the United States.

Goals and highlights of the project: Treated MWW is a common, widely available alternative source of cooling water for thermoelectric power plants across the United States, as determined in a predecessor DOE project (2006–2009) by the project team. However, the biodegradable organic matter, ammonia–nitrogen, carbonate, and phosphates in the treated wastewater pose challenges with respect to enhanced biofouling, corrosion, and scaling, respectively. The overall objective of this study was to evaluate the benefits and LCC of implementing tertiary treatment of secondary treated MWW prior to use in recirculating cooling systems.

The study comprised bench-and pilot-scale experimental studies with three different tertiary treated MWWs, and LCC and environmental analyses of various tertiary treatment schemes. Sustainability factors and metrics for reuse of treated wastewater in power plant cooling systems were also evaluated. The three tertiary treated wastewaters studied were secondary treated MWW subjected to acid addition for pH control (MWW_pH); secondary treated MWW subjected to nitrification and sand filtration (MWW_SF); and secondary treated MWW subjected to nitrification, sand filtration, and GAC adsorption (MWW_NFG).

Tertiary treatment was determined to be essential to achieve appropriate corrosion, scaling, and biofouling control for use of secondary treated MWW in power plant cooling systems. The ability to control scaling, in particular, was found to be significantly enhanced with tertiary treated wastewater compared to secondary treated wastewater. MWW_pH treated water (adjustment to pH 7.8) was effective in reducing scale formation, but it increased corrosion and the amount of biocide required to achieve appropriate biofouling control. Corrosion could be adequately controlled with tolytriazole addition (4–5 ppm TTA), however, which was the case for all of the tertiary treated waters. For MWW_NF treated water, the removal of ammonia by nitrification helped to reduce the corrosivity and biocide demand. Additional GAC adsorption treatment, MWW_NFG, yielded no net benefit. For all of the tertiary treatments, biofouling control was achievable, and most effectively with preformed monochloramine (2–3 ppm) in comparison with NaOCl and ClO_2.

LCC analyses were performed for the tertiary treatment systems studied experimentally and for several other treatment options. A public domain conceptual costing tool (LC3 model) was developed for this purpose. MWW_SF (lime softening and sand filtration) and MWW_NF were the most cost-effective treatment options among the tertiary treatment alternatives considered because of the higher effluent quality with moderate infrastructure costs and the relatively low doses of conditioning chemicals required (Dzombak 2013).

Life cycle inventory (LCI) analysis along with integration of external costs of emissions with DC was performed to evaluate relative emissions to the environment and external costs associated with construction and operation of tertiary treatment alternatives. Integrated LCI and LCC analysis indicated that MWW_NF and MWW_SF alternatives exhibited moderate external impact costs with moderate infrastructure and chemical conditioner dosing, which makes them (especially MWW_NF) better treatment alternatives from the environmental sustainability perspective since they exhibited minimal contribution to environmental damage from emissions (Dzombak 2013).

7.9.2 Key Points

1) This study undertook a comprehensive, integrated approach by looking at all aspects of the water quality control problem in a recirculating cooling system employing treated MWW as makeup water, and we determined optimal approaches for tertiary treatment considering both direct economic costs and environmental impacts of alternative water treatment/management approaches in an integrated manner.
2) The work included pilot-scale demonstrations in the field, in addition to laboratory and modeling work, in partnership with a MWW treatment facility.
3) In regard to originality and innovation, this study was the first comprehensive research study in the public

domain of the use of treated MWW as makeup water in recirculating cooling systems of electric power plants.

4) The challenges of using treated MWW in power plant cooling systems are many and include the technical challenges of controlling corrosion, scaling, and biofouling in a recirculating cooling system employing a low-quality makeup water; the challenge of determining capital and operating costs in a complex operational environment in which the water is being concentrated four times or more in the recirculating cooling system; and the challenge of assessing environmental and social risks and benefits for use of treated wastewater as cooling system makeup water. It was a complex problem, and further complicated by the economic, social acceptance, and sustainability issues involved.
5) Alternative sources of water are needed for new power production in regions without new sources of freshwater available. Our research will help advance the use of treated MWW in electric power production and thus help contribute to economic development. Further, in this research they evaluated explicitly the relative sustainability of various water treatment/management alternatives by considering environmental impact and social acceptance factors in addition to direct economic costs of the alternatives.

7.9.3 The World's First Zero Effluent Pulp Mill at Meadow Lake: The Closed-Loop Concept

The \$250 million Millar Western Meadow Lake Mill is located on a 247-acre site about 200 miles northwest of Saskatoon, Saskatchewan. It uses mechanical action supplemented by mild chemicals to turn aspen wood chips into bleached chemi-thermomechanical pulp (BCTMP), about 240 000 MT/Y.[2] More efficient than the kraft process, this approach uses half the trees to make the same amount of pulp, producing almost one ton of pulp for each ton of wood on a water-free basis. The Millar Western BCTMP process also eliminates chlorine compounds and odorous sulfur-based impregnation chemicals. This environmentally friendly mill uses hydrogen peroxide to increase the brightness of the pulp, making it suitable for printing and writing grades of paper as well as for tissue and paper towels.

The plant is the first pulp mill in the world to operate a successful zero liquid discharge system. Effluent from the thermomechanical pulping process is concentrated from 2% solids to 35% solids by three falling film vapor compression evaporators, followed by two steam-driven concentrators which further concentrate the effluent to about 70% solids. Of the 1760 gal/min of effluent sent to the system, 1720 is recovered as high-purity water for reuse in the pulping process. Solids are burned in the boiler; the smelt is cast into ingots and stored on site for future chemical recovery.

In early1990s, when Millar Western Pulp (Meadow Lake) Ltd. announced plans to build a mill in northern Saskatchewan, the community was concerned about the pollution it would generate, especially effluent discharged to the Beaver River. Though a biological treatment system planned at the mill would have made the effluent cleaner than river water, Millar Western decided to go one step further and eliminate all effluent discharge from the pulp mill. The zero effluent system at Meadow Lake is the first of its kind in the world. The evaporator system, the key equipment in the water recovery process, was designed and supplied by Resources Conservation Company (Fosberg 1992). All effluent coming out of the mill is treated in the water recovery plant. As a result, the mill only needs about 300 gpm of makeup water to replace water lost to the atmosphere by evaporation. The same type of pulp mill without a water recovery plant would need about 2500 gpm of raw makeup water. The effluent treatment system started up in January 1992, when the mill went on line.

Millar Western's Meadow Lake BCTMP mill is an example of successful closure of the water cycle in a mechanical pulp mill. An earlier attempt in Canada to close a kraft mill by recycling bleach plant effluents through the kraft chemical recovery process had failed on account of the buildup of corrosive materials (Smook 1992). Thus, it is useful to study in detail the advanced system in place at Meadow Lake.

The effluent produced by the BCTMP process at Meadow Lake is discharged at a rate of almost 1800 gpm. It has a temperature of 150 °F, a pH of about 8 and contains about 20 000 ppm dissolved solids. Figure 7.5 shows a more detailed view of the water recovery portion of the system, consisting of five stages: clarification, evaporation, concentration, stripping, and incineration.

7.9.3.1 Clarification

The first unit operation to receive pulp mill wastewater is the floatation clarifiers. Since removal of fiber is very important to the performance of the evaporators, the mill decided to install two clarifiers instead of one. This allows for maximum removal efficiency and flexibility. Chemicals are added to aid in flocculation and floatation of the solids.

To ensure that upsets in the pulp mill do not directly affect the evaporators, an on-line meter measures suspended solids in the clarifier accepts stream. When the suspended solids exceeds 900 ppm, the clarifier accepts are directed to the settling ponds. Clarifier accepts normally go directly to the evaporators in the winter to conserve heat.

2 MT/Y = metric tons per year.

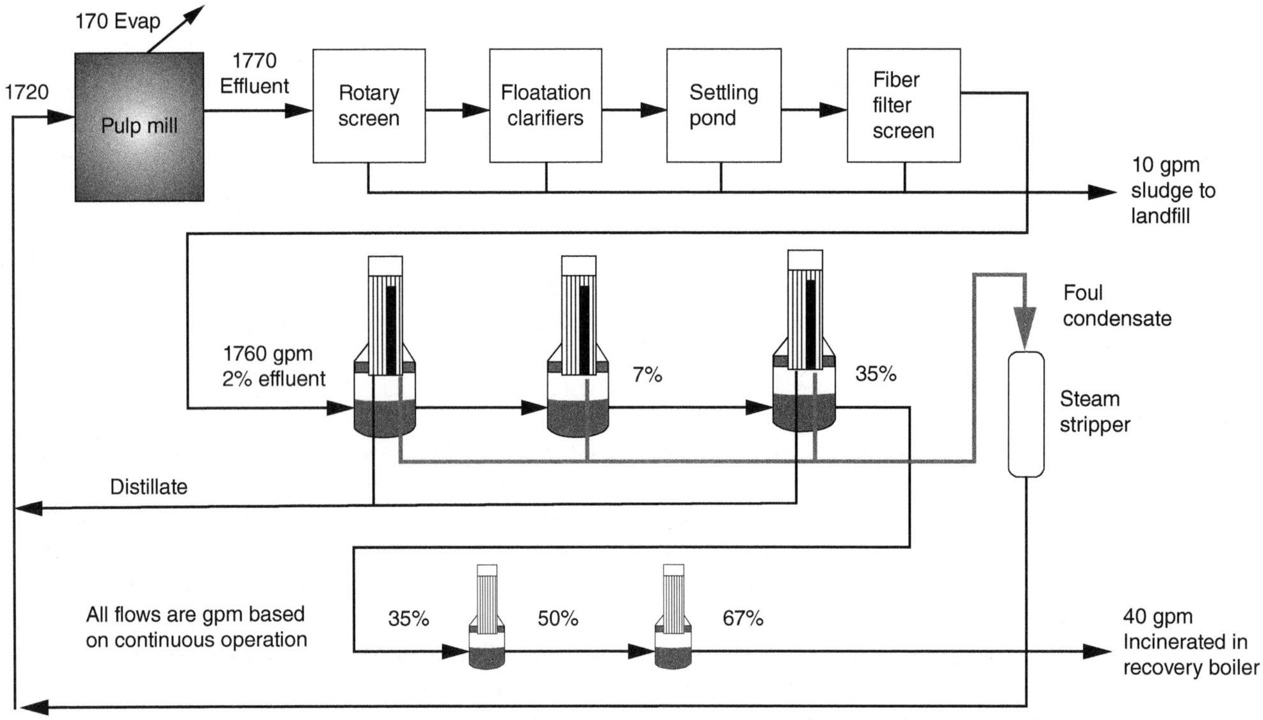

Figure 7.5 Effluent treatment system in Meadow Lake BCTMP mill.

In the summer the accepts go preferentially to the settling pond to dump heat since the heat balance changes from season to season.

7.9.3.2 Evaporation

The heart of the zero effluent system is three vertical-tube, falling-film vapor compression evaporators which operate as explained (Das 2005). At 100 ft tall, and with thousands of square feet of heat transfer surface, this is the largest train of mechanical vapor recompression evaporators in the world. The evaporators concentrate effluent from 2% solids to 35% solids, by means of an energy-efficient mechanical vapor-compression process that recovers distilled water from the effluent. The evaporator consists principally of a heating element, vapor body, recirculation pump, and a vapor compressor.

The effluent is pumped from the vapor body sump to the top of the heating element (tube bundle). A distributor is installed in the top of each tube, causing the effluent to flow down the inside of each tube in a thin film. The distributor helps prevent fouling of the heat transfer tubes by keeping them evenly and constantly wet. It also allows the mill to operate at reduced capacity if desired, since the heating surfaces will remain wet regardless of the amount of effluent being processed. (The evaporators are also capable of handling 1.2 times more than design flow rates from the pulp mill which gives the mill a significant amount of catch-up ability). When the effluent reaches the bottom of the tubes, the recirculation pump sends it back to the top for further evaporation.

As the effluent flows through the heated tubes, a small portion evaporates. The vapor flows down with the liquid. When it reaches the bottom of the tube bundle, the vapor flows out of the vapor body through a mist eliminator and then to the compressor. The compressed steam (at a few psi) is then ducted to the shell side of the tube bundle, where it condenses on the outside of the tubes. As it does so, it gives up heat to the tubes, resulting in further evaporation of the liquid inside. A large amount of heat transfer surface is provided, which minimizes the amount of energy consumed in the evaporation process. Operation of the vapor compression evaporator system requires only 65 kWh/1000 gal of feed.

As the vapor loses heat to the tubes, it condenses into distilled water, which flows down the outside of the tubes. Because the water that first condenses out of the steam is cleaner than water condensing later, baffles are provided within the heating element to create two separate regions for condensing. Steam flows first through the clean condensate region where most condenses. The remaining vapor, which is rich in volatile organics such as methanol, condenses in the foul condensate region of the heating element.

A major portion (70%) of the clean condensate is sent directly to the pulp mill for use as hot wash water at the back end of the mill. The balance of the clean condensate goes to the distillate equalization pond where it is combined with makeup water from Meadow Lake and serves as the cold water supply to the mill. The foul condensate, which contains the volatile organic materials, is reused after stripping in a steam stripper. The steam stripper top product (which contains the concentrated organics) is incinerated.

7.9.3.3 Concentration

Like the three evaporators, the two concentrators are vertical-tube, falling-film design. Rather than using a vapor compressor to drive the system, the concentrator is operated with steam generated by the recovery boiler. The evaporation process in the concentrators is essentially the same as in the evaporators, but the effluent is concentrated further, to about 67% solids. The concentrated effluent is incinerated in the recovery boiler. The lead concentrator takes the liquor from 35 to 50%, while the lag concentrator goes from 50 to 67% solids.

7.9.3.4 Stripping

The foul condensate, only about 10% of the total condensate, is stripped of volatile organic compounds in a packed column stripper. The VOCs are selectively concentrated in the foul condensate because of the condensate segregation features built into the evaporator heating elements. Process steam from the concentrator is sent to a reboiler, which generates stripping steam from a portion of the stripped condensate. The stripped condensate is combined with the clean condensate and reused in the mill. The concentrated VOCs are incinerated in the recovery boiler as a concentrated vapor.

7.9.3.5 Incineration

At the recovery boiler, the organic components of the effluent are incinerated, a process that also generates steam to operate the concentrators. Inorganic chemicals in the effluent are recovered in the smelt from the boiler, which is cast into ingots and stored on site. The mill is considering recovering the sodium carbonate, which would then be converted to sodium hydroxide, a major chemical used in the BCTMP process.

7.9.4 Successful Implementation of a Zero Discharge Program

In July 1996, a paper company located on the west bank of the Mississippi River undertook a program to eliminate the discharge of industrial wastewater to the river. A wastewater recycling system consisting of pumps, surge tank, and filtration system reduced discharges by 99%. The successful pollution prevention includes the annual elimination of 562 million gal of wastewater, 149 000 lb of total suspended solids, and 57 000 lb of biochemical oxygen demand. The plant primarily manufactures colored construction-grade paper from a mixture of secondary fiber, stone groundwood pulp, and kraft pulp.

The company initiated a Zero Discharge program, described in detail by Klinker (1996), having two goals:

1) Eliminate the discharge of wastewater into the Mississippi River.
2) Improve the efficiency of water use in manufacturing to reduce the mill's dependence on river water.

Recycling treated wastewater into the mill's water supply system would accomplish these goals.

Besides the regulatory motivation for reusing wastewater, the company had concerns about periodic interruptions in the flow of water to the mill. A local power company's hydroelectric plant located immediately upstream caused these interruptions. On occasion, the utility lowered the river level by halting water flow through a canal that also feeds water from the Mississippi to the paper mill. When this occurred, the mill had to stop its manufacturing process until there was sufficient volume of water to run the mill. By reusing wastewater, the company could reduce its dependency on river water and avoid this disruption to production.

The nucleus of the Zero Discharge program, a closed-loop wastewater recycling system in the mill, not only offers environmental benefits but also generates difficulties because of the increase in the volume of recycled wastewater used in manufacturing and the expenses associated with addressing these problems.

7.9.4.1 Closing the Loop

Figure 7.6 is a diagram of the company's closed-loop wastewater recycling system. Before the Zero Discharge program began, the Mississippi River supplied all the process and cooling water for the mill. Freshwater from the river entered the mill, passed over a fine mesh screen, and entered a 2700-gal freshwater tank. The house pump directed it to process and cooling water demand points in the mill.

The resulting process wastewater underwent treatment in the company-owned, activated sludge, wastewater treatment plant. Discharge was through a process-wastewater outfall designated outfall 01. Cooling wastewater discharge was at outfall 02. Wastewater sludge underwent dewatering on a belt filter press followed by land application on company owned agricultural land.

Before the Zero Discharge program, the plant discharged an average of 607 000 gal of process wastewater and 1.14 million gal of cooling wastewater into the river each day. By

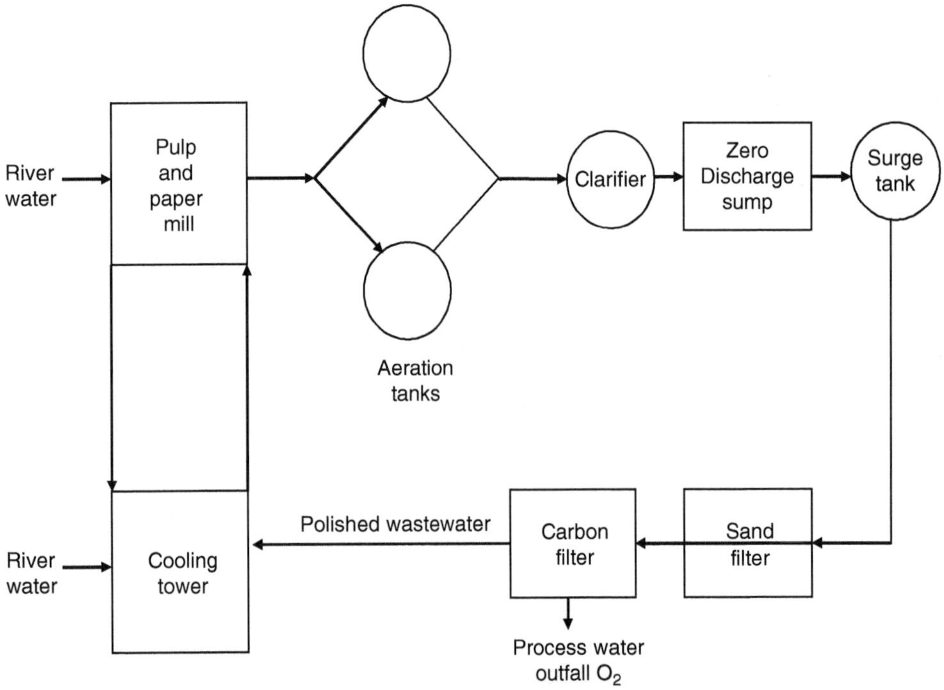

Figure 7.6 Schematic of a pulp and paper mill closed-loop recycling system for Zero Discharge system.

using the closed-loop system to pump increased amounts of treated wastewater into the freshwater tank, the mill was able to state that its process and cooling water consisted of nearly 100% recycled wastewater. The total volume of wastewater discharged to the river decreased by 82% (Klinker 1996).

7.9.5 Conclusions

The US pulp and paper industry has made significant progress towards reducing water consumption and increased water recycling and reuse through innovative technologies and process modification. Some mills have implemented processes that "close the loop" and proved to be successful Zero Discharge bleach plant systems. The effects of the EPA's Cluster Rule, which became applicable on 15 April 2001, have yet to be fully felt. It is reasonable to expect, however, that the new linkage of federal regulations aimed at reducing air and water pollution will result in an even higher level of processed water recycling, reuse within the mills, as well as greater pollution prevention and Zero Discharge in water, air, and solid waste areas the next decade or two. As technologies to reduce facility water, chemical, and energy use have advanced, other chemical industries, like pulp and paper and power generating industry, have increasingly embraced the use of reclaimed water for a wide-ranging suite of purposes: from process water, boiler feedwater, and cooling tower use to finishing toilets and site irrigation. Current technologies produce reclaimed water that can provide the same performance as more-expensive potable water. As water resources become increasingly valued around the world, industrial water reuse is expected to expand (DaSilva and Goodman 2014).

7.10 Consequences of Dirty Air: Costs–Benefits

Emissions of nitrogen oxides (NO_x), sulfur dioxide (SO_2), VOCs, and mercury, as well as carbon dioxide (CO_2), and hotter climate undermine public health, the environment, and the overall state economy. The worst air quality in the United States is in California, a state known for its efforts to raise environmental standards, cut greenhouse gas emissions, and combat climate change (California Air Resource Board Best Available Control Technology Guidelines 2000).

California has the highest ozone levels of any state, according to the American Lung Association's "State of the Air 2017" report on air pollution, which analyzed counties across the country based on levels of ozone and particle pollution. The situation in California is steadily improving, but the state still lags behind the rest of the United States. Three California counties – Los Angeles-Long Beach, Bakersfield, and Fresno-Madera – had the worst smog levels in the entire country. More than 90% of California residents live in counties with unhealthy air, according to the American Lung Association (see Figure 5.2).

Why is California's ozone problem so extreme?

California's population is rapidly growing, from 15.8 million in 1960 to 39.2 million in 2016, according to the U.S. Census Bureau. Major urban areas like Los Angeles are designed around car travel. In addition, ozone is naturally produced when the sun's rays split oxygen molecules – meaning California's sunshine exacerbates its existing smog problem.

7.10.1 Public Health

Nationwide, the report found nearly four in ten Americans live where pollution levels are often dangerous to breathe. But overall, the number of Americans exposed to unhealthy levels of air pollution dropped to about 125 million people, down from 166 million in last year's report (which covered the period from 2012 to 2014).

Air pollution can cause asthma attacks, heart attacks, lung cancer, reproductive harm, and premature death. Globally, the World Health Organization says three million people die prematurely each year from breathing polluted air, and children are especially at risk. Airborne mercury falls into the rivers and estuaries, contaminating freshwater and saltwater fish populations. Mercury compounds bioaccumulate in the food chain, making king mackerel, bass (in some areas), and bowfin unfit for human consumption by children and women of childbearing age.

7.10.2 Visibility

Visibility in the southeast has declined by 75% from natural levels. One should be able to see out 93 miles on an average day, but now air pollution has reduced this to an average of 20 miles. On any given summer day, due to regional haze in the regional, there is a good chance that views may be entirely obscured by pollution.

7.10.3 Ecosystems

Air pollution causes acid rain and nitrogen deposition, which make vegetation more susceptible to disease and pests, contributing to stunted growth and significant declines in populations of redwood, dogwood, spruce, fir, beech, and other tree species. Rainfall in the regions is more acidic than normal rainfall. In the west, airborne nitrogen adds to nutrient pollution in sensitive coastal watersheds, contributing to algal blooms and fish kills.

7.10.4 Economic Consequences

California's dirty air threatens the vitality of the state's economy. Dirty air is estimated to cost the state over $20.0 billion annually in morbidity and mortality costs. Ozone's effect on plants is "significant stress factor in agricultural production". Air pollution reduces crop yields, which causes California farmers to lose estimated more than $300 million each year. Frequent smog alerts in ozone-smog non-attainment areas discourage hiking and other outdoor activities. The impairment of visibility undermines California's $10 billion tourism industry. The loss in economic activity in the state's National Parks is estimated to cost more than $500 million each year.

7.10.5 Global Climate Change

Carbon dioxide pollution from mobile and point sources is one of the primary pollutants that contributes to global warming, which over the next 30 years is expected to raise sea levels off the California coast lines.

7.10.6 Quality of Life

A loss of air quality clearly diminishes the quality of life for all Californians, but putting a price tag on it is difficult. Nonetheless, some indirect measures are possible. For example, quality of life has long been a major factor in persuading new businesses to locate in the state, which implies that as air quality declines, it will be harder for the state to attract new investment and jobs.

7.10.7 Costs–Benefits Analysis

Section 812 of the Clean Air Act Amendments of 1990 provides a case study of a status that requires a particularly detailed risk assessment (or cost–benefits). The goal of this section of the Clean Air Act is to assess the costs of the Clean Air Act. It requires the USEPA to estimate the hazardous associated with the air pollutants covered under the act (USEPA 2011).

The statistics just given are not comprehensive: the economic effects of acid rain, eutrophication, mercury exposure, and other environmental problems are not included. Thus, the actual costs of dirty air are much higher than those reflected by the dollar amounts cited. Nor can any no economic calculation quantify the pain experienced by a child who suffers from asthma or a grandparent with cardiopulmonary disease whose death is hastened by exposure to high levels of fine particulate matter.

7.11 Some On-Going Pollution Prevention Technologies

Industrial air pollution prevention (P^2) efforts have focused on both source and waste reduction, and on reuse and recycling. A key approach is preventing air pollution within a company's manufacturing processes, particularly in chemical process industries. Frequently, we fail to consider that control systems themselves are industrial processes that consume energy and can emit significant pollutants into the atmosphere. There has been a tendency to pursue the control of each target pollutant independently while ignoring secondary pollutants. Opportunities to reduce energy consumption and secondary pollutants simply by selecting sustainable pollution control technologies are increasing.

Air pollution arises from a number of sources: stationary such as factories and other manufacturing processes; mobiles such as automobiles, recreational vehicles, snowmobiles and watercraft; and area sources which are all other emissions associated with human activities. Air quality problems are closely associated with combustion processes occurring in the industrial and transportation sectors of the economy.

Air pollution control is an area where P^2 and sustainability concepts are relatively new and just beginning to be applied, and the promise of significant environmental and economic benefits is strong. However, the regulatory development processes and framework have not fully embraced them.

Since large industries could reasonably assimilate the cost of controls into their cost of doing business, the United States' air quality regulatory approach began with these sources having the most potential for immediate environmental improvement. The regulatory requirements have progressively tightened standards and applied limits to smaller sources too. The regulations began targeting priority pollutants as separate and distinct entities to be controlled. These tendencies are part of the problem Congress was addressing through the *Pollution Prevention Act of 1990*, as stated in finding number 3:

> The opportunities for source reduction are often not realized because existing regulations, and the industrial resources they require for compliance, focus upon treatment and disposal, rather than source reduction; existing regulations do not emphasize multi-media management of pollution; and businesses need information and technical assistance to overcome institutional barriers to the adoption of source reduction practices. (Pollution Prevention 1997; State of California January 2001; United State Congress, Pollution Prevention Act of 1990; CFR 2019)

However, some P^2 concepts and the lack of cost-effective treatment technologies for smaller sources have also guided the regulatory approach. For example, many paint and coatings industry regulations have set limits on VOCs in the product formulations as the primary compliance mode, with end-of-pipe treatment as an alternative compliance option (State of California 2001). On the other hand, current regulations also have requirements such as BACT, requiring large sources to achieve a high level of treatment for each target pollutant with only limited consideration of the cost, in either money or secondary environmental impacts.

An area of particular opportunity is the control of VOCs and odors. Traditional control technologies, such as thermal and catalytic oxidation, are effective in reducing VOC emissions by more than 95%. They have become the standard by which other VOC control technologies are measured for acceptability through regulatory concepts such as BACT. However, these technologies frequently demand significant energy and produce secondary wastes such as NO_x, SO_x, and greenhouse gases. It is not uncommon for a facility to have challenges meeting NO_x emission standards due to the secondary emissions from their VOC removal systems. VOC control technologies frequently combust large quantities of natural gas to maintain the required temperature for destruction of the VOCs, resulting in significant emissions of greenhouse gases.

Example 7.3 Solvent Emission from a Dry Cleaning Shop

A dry cleaning shop in Los Angeles is trying to estimate if it will need to apply Reasonably Available Control Technology (RACT) to its operations. The Clean Air Act threshold for a major source in Los Angeles is 10 T (1 T = 2000 lb) of VOC emissions per year. The shop operates its cleaning facilities 8 h/day, 6 days/week, and 51 weeks/year. The shop's new vapor recovery equipment has a 250 gal reservoir and loses an average of 0.70 gal of solvent/hour during operations. The solvent is perchloethylene, which has a liquid density of 1483 kg/m^3 (12.17 lb/gal).

1) Does this shop need to apply RACT?
2) If, instead of Los Angeles, this shop were to be located in New York City, would it need to apply RACT? The VOC emission threshold for New York City is 25 T/Y.

Solution

To calculate the shop's hourly mass emissions, the following expression is used:

$$\begin{aligned}\text{Mass rate} &= \left(\text{volumetric solvent loss rate}\right)\left(\text{solvent density}\right)\\ &= \left(0.70\,\text{gal}/\text{h}\right)\left(12.37\,\text{lb/gal}\right) = 8.66\,\text{lb/h}\end{aligned}$$

The mass emission rate in lb/day is then:

$$= \text{lb/day} = \left(8.66\,\text{lb/h}\right)8\,\text{h/day} = 69.3\,\text{lb/day}$$

The mass emission rate is then converted to annual emissions in lb/year as

$$\text{lb/year} = (69.3\ \text{lb/day})(6\ \text{day/week})(51\ \text{week /year}) = 21\ 206\ \text{lb/year}$$

The emission rate is then converted to T/Y as

$$\text{ton/year} = (21\ 206\ \text{lb/year})(1\ \text{ton/2 000 lb}) = 10.6\ \text{ton/year.}$$

The actual annual emission rate is then compared to the major source threshold for Los Angeles of 10 T/Y, i.e. 10.6 T/Y actual > 10 T/Y. Since the result marginally exceeds the 10 T/Y threshold, the shop must therefore apply RACT.

The major source threshold for New York City is 25 T/Y. Since this shop emits 10.6 T/Y, it is below the 25 T/Y threshold for New York City, and therefore does not need to apply RACT.

As a rule of thumb, the release of 1 gal/h of solvent with specific gravity 1.0 is equivalent to approximately 10 T/Y for a 52 week, 8-hour work schedule, and 30 T/Y for a 24-hour work schedule.

Example 7.4 Cost of a Catalytic/Thermal Incinerator
A catalytic incinerator is investigated to replace the thermal incinerator operating at a rate of 2.53 lb/min fuel gas flow rate at 750 °C. Because of the lower required temperature, the required fuel gas flow rate drops to 1.44 lb/min. At a cost of $4.00/MBtu for fuel gas, calculate the maximum justifiable cost of the catalytic system. Assume that the costs of the shells of both systems are the same, so the only additional cost is that of the catalysts itself. Assume the catalyst must be replaced every five years, and the incinerator operates 24 h/day, 300 days/year.

Solution
Calculate the amount of fuel saved.

$$\text{Fuel}_{\text{saved}} = (2.53 - 1.44)\frac{\text{lb}}{\text{min}} = 1.09\frac{\text{lb}}{\text{min}}$$

Calculate the yearly fuel saved.

$$\text{Energy}_{\text{saved}} = \left(1.09\frac{\text{lb}}{\text{min}}\right)\left(21\ 560\frac{\text{Btu}}{\text{lb}}\right) = 23\ 500\frac{\text{Btu}}{\text{min}}$$

Calculate the yearly fuel savings knowing when the incinerator operates.

$$\text{Yearly}_{\text{savings}} = \left(23\ 500\frac{\text{Btu}}{\text{min}}\right)\left(\frac{\$4.00}{\text{MBtu}}\right)\left(\frac{1\ \text{MBtu}}{10^6\ \text{Btu}}\right)\left(60\frac{\text{min}}{\text{h}}\right)\left(24\frac{\text{h}}{\text{day}}\right)\left(300\frac{\text{day}}{\text{year}}\right) = \$40\ 608\,/\,\text{year.}$$

Calculate how much money could be saved in 5 years' time.

$$\text{Total}_{\text{saved}} = \left(\frac{40\ 608}{\text{year}}\right)(5\ \text{year}) = \$203\ 040$$

With this considered, the maximum that can be spent is roughly $203 000 for the catalyst to make the costs of the thermal and catalytic incinerators equal.

$$\text{Cost} = \$203\ 000\quad(\text{answer}).$$

7.11.1 Economic Performance Indicators

Costs associated with poor environmental and societal performance can be very large. Waste disposal feed, permitting costs, and liability costs can all be substantial. Wasted raw material, waste energy, and reduced manufacturing throughput are also consequences of wastes and emissions. Corporate image and relationships with workers and communities can suffer if performance is substandard. But how can these costs be quantified? This section will review the tools available for estimating environmental and societal costs and benefits. These include traditional concepts such as the time value of money, PV, PP, internal ROR, and other financial evaluation calculations. Nontraditional tools include methods for monetizing environmental costs that are hidden from normal accounting procedures. We'll also touch upon less tangible costs and benefits that can still be monetized.

In general, traditional accounting practices have acted as a barrier to implementation of sustainable engineering projects because they hide the costs of poor environmental and societal performance. Many organizations are now giving more consideration to all significant sources of environmental and societal costs. The principle is that if costs are properly accounted for, management practices that foster good economic and societal performance will also foster superior environmental and societal performance (USEPA 1995c).

7.11.2 Estimates of Environmental Costs

It is generally recognized that in environmental and societal cost accounting, words like *full* (e.g. full-cost accounting), *total* (e.g. total cost assessment), *true*, and *life cycle* (e.g. life cycle costing) are used to indicate that not all costs are captured in traditional accounting and capital budgeting practices. Among the easiest environmental costs to track are the costs associated with treating emissions and disposing of wastes. DC of pollution abatement are tracked by the U.S. Census Bureau and have been increasing steadily. Expenditures in 1972 totaled $52 billion (in 1990 dollars)

Table 7.7 Pollutant abatement expenditures by US manufacturing industries.

Industry sector	Pollution control expenditure (as % of sales)	Pollution control expenditure (as % of value added)	Capital expenditure (as % of total capital expenditure)
Petroleum	2.25	15.42	25.7
Primary metals	1.68	4.79	11.6
Pulp mill	5.70	12.39	17.2
Paper	1.87	4.13	13.8
Chemical manufacturing	1.88	3.54	13.4
Stone product	0.93	1.77	7.2
Lumber	0.63	1.67	11.1
Leather products	0.65	1.37	16.2
Fabricated materials	0.65	1.34	4.6
Food	0.42	1.11	5.3
Rubber	0.49	0.98	2.0
Textile	0.38	0.93	3.3
Electric product	0.49	0.91	2.9
Transportation	0.33	0.80	3.0
Furniture	0.38	0.73	3.4
Machinery	0.25	0.57	1.9

Source: Data reported by US Congress (1994); original data collected by US Census Bureau. http://www.census.gov

and were projected to grow to approximately $140 billion (in 1990 dollars), or 2.0–2.2% of gross national product, in the year 2000 (for a review and analysis of these data, see United State Congress 1994).

These expenditures are not distributed uniformly among industry sectors. As shown in Table 7.7, sectors such as petroleum refining and chemical manufacturing spend much higher fractions of their net sales and capital expenditures on pollution abatement than do other industrial sectors. Therefore, in these industrial sectors, minimizing costs by preventing wastes and emissions will be far more strategic an issue than in other sectors.

Example 7.5 Potential Pollution Control Costs for Greenhouse Gases

Estimate control costs for greenhouse gases for an electricity generating unit (EGU) or power plant) as a percentage of sales revenues. Assume that (i) the EGU uses coal as a fuel and converts the heat of combustion of coal into electricity with 35% efficiency, (ii) the heating value of coal is 10 000 BTU/lb, (iii) coal is 85% carbon, (iv) carbon costs $20/T to capture and sequester, and (v) electricity can be sold for $0.10/kWh.

Solution

Basis:

$$1\,\text{T coal} = 1\,\text{T coal} \times 0.85\,\text{T C} \times 2000\,\text{lb/T} \times 10\,000\,\text{Btu/lb} \times \text{kWh}/3413\,\text{Btu} \times 0.35 = 1743\,\text{kWh/T of C.}$$

The control costs are $20 for 1743 kWh electricity generation leads to 1743 kWh × $0.10/kWh = $174 in electricity sale. So the control cost is about 11% of sale.

Pollution abatement costs reported by individual companies both reflect these general trends and provide more detail about the magnitude and the distribution of environmental expenditures. For example, Tables 7.8 and 7.9 show the distribution of environmental costs reported by the Amoco Yorktown refinery and DuPont's LaPorte chemical manufacturing facility (Heller et al. 1995; Shields et al. 1995). In the case of Amoco refinery only about a quarter of the quantified environmental costs are associated with waste treatment and disposal; the costs are summarized in Table 7.7 associated with removing sulfur from fuels, meeting other environmentally based fuel requirements, and maintaining environmental equipment were greater than the costs associated with waste treatment and disposal. This indicates that the magnitude of environmental costs is

Table 7.8 Summary of environmental costs at the Amoco Yorktown Refinery.

Cost category	Percentage of annual non-crude operating costs
Waste treatment	4.9
Maintenance	3.3
Product requirements	2.7
Depreciation	2.5
Administration, compliance	2.4
Sulfur recovery	1.1
Waste disposal	0.7
Fees, fines, penalties	0.2
Total costs	17.8

Source: From Heller et al. (1995).

Table 7.9 Summary of environmental costs at the DuPont LaPorte Chemical Manufacturing Facility.

Cost category	Percentage of annual non-crude operating costs
Taxes, fees, training, legal	4.0
Depreciation	3.2
Operations	2.6
Contract waste disposal	2.4
Utilities	2.3
Salaries	1.8
Maintenance	1.6
Engineering services	1.1
Total costs	19.0

Source: From Shields et al. (1995).

substantially greater than that reported in Table 7.7 and that these costs may be hard to identify.

7.11.3 Total Annualized Cost for BACT

In Chapter 6, top-down BACT analyses for criteria pollutants were presented for a gas turbine facility (Section 6.6). As an example, here we present the TACs and costs per ton pollutant of particulate matter removed against exhaust gas rate and boiler steam capacity for four control options: (i) multi-cyclones (MC), (ii) Venturi scrubber (VS), (iii) electrostatic precipitator (ESP), and (iv) fabric filter (FF). The TACs were computed using the EPA's air pollution control cost manual and the control-cost spreadsheets (APTI 1995; USEPA 1996). The USEPA control-cost spreadsheets consider most aspects of cost evaluation, using operating parameters, design parameters, capital costs, and annual cost. The TAC spreadsheets for control options, e.g. MC, FF, ESP, and VS, are given in the report by Das (2003).

The TAC for a chosen boiler exhaust flow rate at 18 685 dscfm (Table 7.10). Cost effectiveness, or dollar per ton of pollutant removal, is one of the key economic criterion used to determine if a control option is acceptable for permitting use. Cost effectiveness is calculated as the TAC of the control option being considered divided by the baseline emissions minus the control option emissions rate (Figures 7.7 and 7.8) (Das 2001, 2003; Peters et al. 2002).

7.11.4 Cost Per Ton (T) of Pollutant Removal

The following is the calculation of ton per year (T/Y) and dollar per ton ($/T) using Figure 7.7 for four control technologies (MS, VS, ESP, and) for particulate matter (PM) removal.

Example 7.6 Multi-Cyclone
Basis: 1 minute

Inlet PM concentration per minute = 2.0 gr/dscfm

(where dscfm = dry standard cubic feet per minute)

$$= \frac{2\times 60\times 24\times 365}{7000}\left(\frac{\text{lb}}{\text{dscfm}\cdot\text{year}}\right)$$

(where 7000 grains (gr) = 1 lb)

$$= \frac{2\times 60\times 24\times 365}{7000\times 2000}\left(\frac{\text{T}}{\text{dscfm}\cdot\text{Y}}\right)$$

(where 1 T = 2000 lb)

$$= 0.0751\left(\frac{\text{T}}{\text{dscfm}\cdot\text{Y}}\right)$$

For 90% removal efficiency

$$= 0.0751\times 0.9$$

$$= 0.0676\left(\frac{\text{T}}{\text{dscfm}\cdot\text{Y}}\right)$$

For an inlet exhaust gas flow rate = 18 685 dscfm

$$\text{T/Y} = 0.067\,6\times 18\,685 = 1\,263$$

TAC for MC at 18 685 dscfm

$$= \$82\,292\,(\text{Figure 7.7})$$

$$\therefore \$/\text{T} = \frac{82\,292}{1\,263} \simeq 65$$

Table 7.10 TAC and cost per ton of PM removals VS flow rate and boiler steam capacity.

Flow rate (dscfm)	Steam (lb/h)	Cost, MC	Cost, VS	Cost, ESP	Cost, FF	T/Y,MC 90%	$/TY, MC	T/Y,VS 90%	$/TY, VS	T/Y,ESP 99.5%	$/TY,ESP	T/Y,FF 99.5%	$/TY, FF
1 500	3 500	18 203	124 592	66 563	107 370	101	180	101	1 229	112	594	112	958
6 744	15 000	37 960	139 485	147 139	140 647	456	83	456	306	504	292	504	279
18 685	40 000	82 292	173 398	296 329	216 415	1 263	65	1 263	137	1 396	212	1 396	155
21 118	45 000	91 283	179 983	320 471	231 852	1 428	64	1 428	126	1 578	203	1 578	147
67 540	137 776	264 428	298 057	711 959	595 016	4 566	58	4 566	65	5 045	141	5 045	118
99 478	200 000	381 768	376 049	967 426	797 667	6 725	57	6 725	56	7 431	130	7 431	107
141 109	280 000	534 321	475 904	1 316 887	1 061 828	9 539	56	9 539	50	10 541	125	10 541	101
223 023	435 000	833 661	668 958	1 996 247	1 581 600	15 076	55	15 076	44	16 660	120	16 660	95
268 464	520 000	999 392	774 805	2 355 281	1 869 935	18 148	55	18 148	43	20 054	117	20 054	93
311 500	600 000	1 156 191	874 473	2 690 738	2 143 013	21 057	55	21 057	42	23 269	116	23 269	92
365 608	700 000	1 353 146	999 146	3 107 276	2 486 341	24 715	55	24 715	40	27 311	114	27 311	91

Source: From Das (2003).
Assuming grain loading in the boiler exhaust gas stream = 2.0 gr/dscfm.
Multi-cyclone (MC) efficiency = 90%.
Venturi scrubber (VS) efficiency = 90%.
Electrostatic precipitator (ESP) efficiency = 99.5%.
Fabric filter (FF) efficiency = 99.5%.

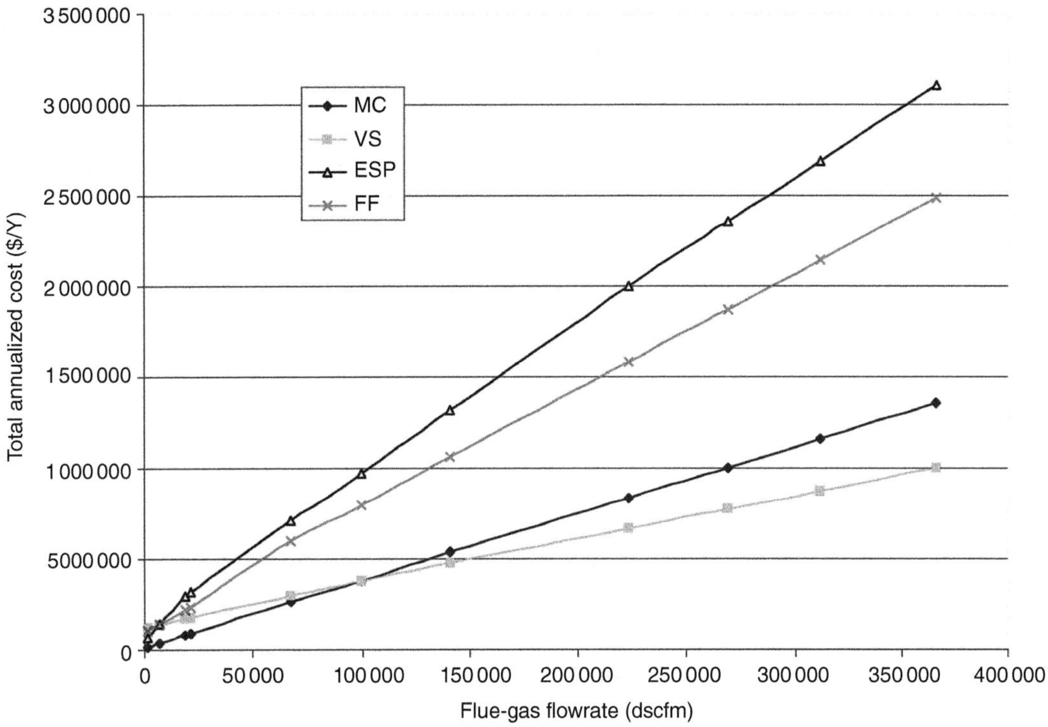

Figure 7.7 Flue-gas flow rate vs. annualized control cost for PM.

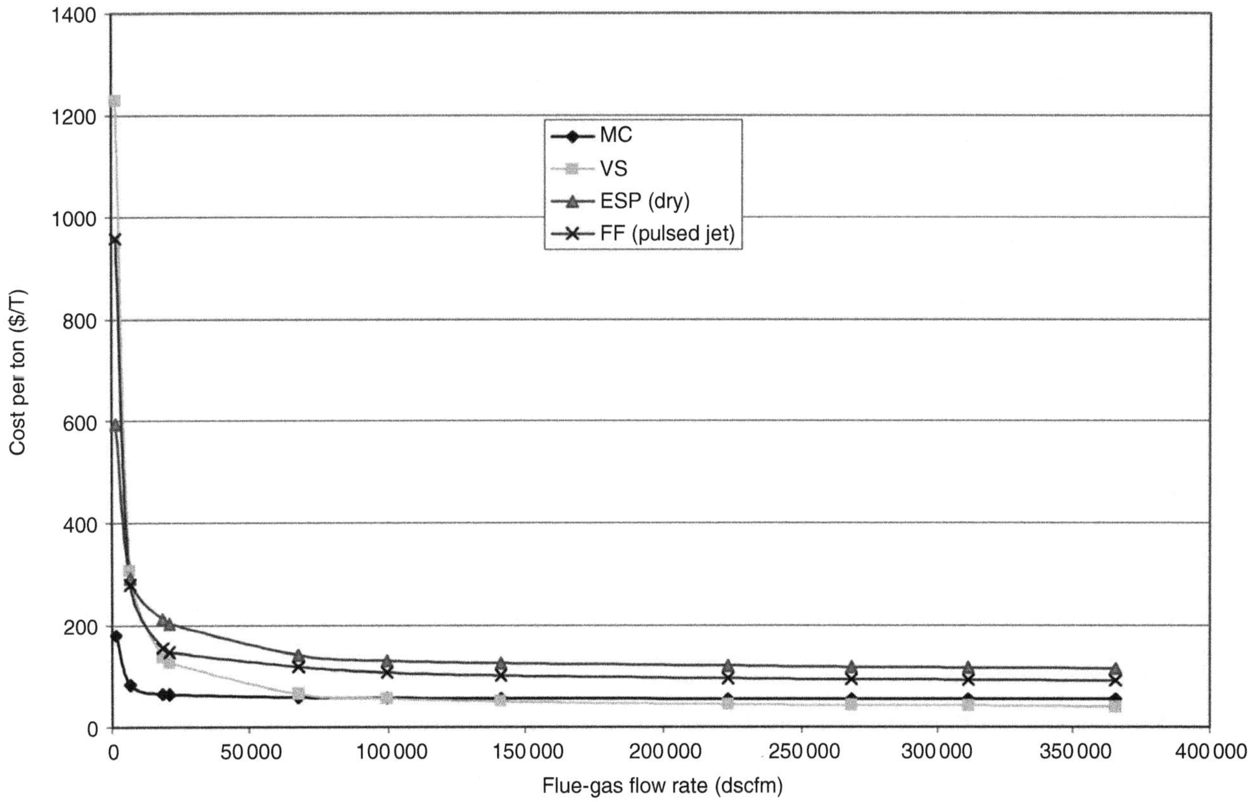

Figure 7.8 Flue-gas flow rate vs. cost effectiveness ($/T) for PM using four control technologies.

Example 7.7 Venturi Scrubber
Input PM loading to VS per minute = 2.0 gr/dscfm

$$\therefore \text{T/Y} = \left(\frac{2\times60\times24\times365}{7000\times2000}\right)\text{dscfm} = 0.0751\,\text{dscfm}$$

For 18 685 dscfm

$$\begin{aligned}\text{T / Y} &= 18\,685\times0.075\,1\\ &= 1\,403\end{aligned}$$

For 90% removal efficiency

$$\text{T/Y} = 1403\times0.9 \simeq 1263$$

Annualized cost at 18 685 dscfm = \$173 398 (Figure 7.7)

$$\therefore \$/\text{T} = \frac{173\,398}{1\,263} = 137$$

Example 7.8 Electrostatic Precipitator
Input PM loading to ESP and FF per minute = 2.0 gr/dscfm

$$\therefore \text{T/Y} = \left(\frac{2\times60\times24\times365}{7000\times2000}\right)\text{dscfm} = (0.0751)\text{dscfm}$$

For 18 685 dscfm

$$\begin{aligned}\text{T/Y} &= 18\,685\times0.075\,1\\ &= 1\,403\end{aligned}$$

For 99.5% removal efficiency for both ESP and FF

$$\text{T/Y} = 1403\times0.995 \simeq 1396$$

Annualized cost at 18 685 dscfm = \$296 329 (Figure 7.7)

$$\therefore \$/\text{T} = \frac{296\,329}{1\,396} = 212$$

Example 7.9 Fabric Filter
Input PM loading to FF per minute = 2.0 gr/dscfm

$$\therefore \text{T/Y}\left(\frac{2\times60\times24\times365}{7000\times2000}\right)\text{dscfm} = (0.0751)\text{dscfm}$$

For 18 685 dscfm

$$\begin{aligned}\text{T/Y} &= 18\,685\times0.075\,1\\ &= 1\,403\end{aligned}$$

For 99.5% removal efficiency for FF

$$\text{T/Y} = 1403\times0.995 \simeq 1396$$

Annualized cost at 18 685 dscfm = \$216 415 (Table 7.10)

$$\therefore \$/\text{T} = \frac{216\,415}{1\,396} = 155$$

Conclusion: For the boiler exhaust flow rate at 18 685 dscfm, MC could operate at least expensive rate (\$65/T) having 90% efficiency to remove PM_{10}; on the other hand, ESP could operate at most expensive rate (\$212/T) having 99.5% efficiency to remove PM_{10}.

7.12 Cost Indices and Estimating Cost of Equipment

Most cost data that are available for making a preliminary or predesign estimation are only valid at that time they were developed. Because prices may have changed considerably with time due to changes in economic conditions, some methods must be used for updating cost data applicable at a past date to costs that are representative of conditions at a later time. This can be done by the use of cost indices.

A cost index is a ratio of the cost of an item or equipment group at a specific time to the cost of the item at some base time in the past. Indices are published by the government for labor and materials (both for retail and for wholesale). The most familiar index the Commerce Price Index provides a measure of inflation each year. In equivalent cost estimation, an equipment or process cost index is used to update equipment cost as follows:

$$\text{Cost}_{\text{present}} = \text{Cost}_{\text{past}}\left(\text{present index/past index}\right) \quad (7.4)$$

Two indices frequently used in estimating the cost of process and air pollution control equipment are shown in Table 7.11. Brief descriptions of the Marshall-Swift and Chemical Engineering Cost Indices are given in the *Chemical Engineering* journal.

Example 7.10 The purchase cost of a 0.2 m^3 glass-lined jacketed reactor (without drive) was \$10 000 in 1991. Estimate the purchased cost of a similar 1.2 m^3, glass-lined, jacketed reactor (without drive) in 1996. Use the annual average Chemical Engineering Plant Cost Index to update the purchase cost of the reactor. The glass-lined jacketed reactor cost vs. capacity exponent is given as 0.54.

Solution
The Chemical Engineering Plant Cost Index in 1991 was 361.3 and in 1996 was 381.7 (Table 7.11). Therefore using Eq. (7.4), we get the cost of reactor in 1996

$$= (\$10\,000)(381.7/361.3)(1.2/0.2)^{0.54} = \$27\,800$$

7.12.1 Equipment Costs

Process equipment costs can be correlated with size or capacity by the following relationship.

$$\begin{aligned}\text{Cost}_\text{B} &= \text{Cost}_\text{A}\left(\text{capacity}_\text{B}/\text{capacity}_\text{A}\right)^{b}\\ &= \text{cost}_\text{A}\left(\text{size}_\text{B}/\text{size}_\text{A}\right)^{b'}\end{aligned} \quad (7.5)$$

Table 7.11 Cost indices as annual averages.

Base year	Marshall-Swift (1926 = 100) all industries	Chemical engineering plant cost index (1958 = 100)
1984	780.4	322.7
1985	789.6	325.3
1986	797.6	318.4
1987	813.6	323.8
1988	852.0	342.5
1989	895.1	355.4
1990	915.1	357.6
1991	930.6	361.3
1992	943.1	358.2
1993	964.2	359.2
1994	993.4	328.1
1995	1027.5	381.1
1996	1039.1	381.7
1997	1056.8	386.5
1998	1061.9	389.5
1999	1068.3	390.6
2000	1080.6	392.6
2001	1093.9	394.3
2002	1104.2	395.6
2003	1123.6	402.0
2004	1178.5	444.2
2005	1244.5	468.2
2006	1302.3	499.6
2007	1373.3	525.4
2008	1449.3	575.4
2009	1486.6	521.9
2010	—	550.8
2011	—	585.7

Source: Adopted from indices published monthly by *Chemical Engineering*.

where

capacity_B = equipment throughput, in standard units of cfm, gpm, etc.
size_B = equipment size, in standard units of 2 ft^2, ft^3, etc.
b, b' = constant

The exponents b and b' vary from 0.5 to 0.8 and will average between 0.6 and 0.7 for many types of equipment. Typical exponent values for various types of air pollution control equipment are shown in Table 7.12.

Example 7.11 A carbon-steel cyclone designed for a throughput of 5 000 scfm was purchased in 2003 for 15 000.

Table 7.12 Cost exponents for air pollution control equipment.

Equipment type	Size range (cfm)	Cost exponent
Dry cyclones	5 000–100 000	0.65
Multi-cyclones	10 000–200 000	0.65
Scrubbers		
Impingement	5 000–900 000	0.80
Gravity spray	5 000–200 000	0.62
Centrifugal	5 000–100 000	0.76
Venture, low energy	5 000–100 000	0.76
Venture, high energy	5 000–100 000	0.72
Electrostatic		
Precipitators	10 000–1 000 000	0.62
Reverse pulse		
Filters	1 000–100 000	0.62
Adsorbers, fixed bed	2 000–20 000	0.72

Source: Adopted from Peters and Timmerhaus (2002).

Estimate the cost of a 10 000 scfm carbon = steel cyclone purchased in 2007.

Solution

We use the Marshall-Swift (M-S) equipment cost index from Table 7.11 to extrapolate the cost of a 5000 scfm unit to 2007. We find the cost exponent for cyclones in Table 7.12.

M-S index in 2003 = 1123.6
M-S index in 2007 = 1373.3
Cost exponent = 0.65; using the Eq. (7.5), we get
$\text{Cost}_{10\,000\,\text{scfm},\,2007} = \$15\,000(10\,000/5\,000)^{0.65}(1\,373.3/1\,123.6) =$ \$28 768 = \$28 800 (or possibly \$29 000. Using the Chemical Engineering Index results in an estimate of \$31 000.)

Example 7.12 Given that a wet scrubber had a total installed capital cost (TIC) of \$800 000 in 2001, estimate the 2008 TIC using the two cost indices presented in Table 7.11. Use a reasonable method to estimate the TIC in the year 2013.

Solution

c) $\text{M-S}_{2001} = 1093.9;\ \text{M-S}_{2008} = 1449.3$

$$\text{M-S}_{\text{factor}} = \text{M-S}_{2008}/\text{M-S}_{2001} = 1.325$$

$$\text{TIC}_{2001} = \$800\ 000$$

$$\text{TIC}_{2008} = \text{TIC}_{2001} \times \text{M-S}_{\text{factor}} = \$800\ 000 \times 1.325 = \$1\ 060\ 000$$

Assume inflation = 2.5%

$$\text{TIC}_{2013} = \text{TIC}_{2008}(1.0 + \text{Inflation})^5 = \$1\ 060\ 000(1.0 + 0.025)^5 = \$1\ 199\ 292 \text{ rounded to } \$1\ 200\ 000$$

d) $CE_{2001} = 394.3$; $CE_{2008} = 575.4$

$$CE_{factor} = CE_{2008}/CE_{2001} = 1.459$$

$$TIC_{2001} = \$800\ 000$$

$$TIC_{2008} = TIC_{2001} \times CE_{factor} = \$1\ 167\ 435$$

Assume inflation = 2.5%

$TIC_{2008} = TIC_{2001}(1.0 + 0.025)^5 = \$1\,320\,845$, rounded to $1 321 000

7.13 Waste-to-Energy

Waste-to-energy (WtE) or energy-from-waste (EfW) is the process of generating energy in the form of electricity and/or heat from the primary treatment of waste. WtE is a form of energy recovery. Most WtE processes produce electricity and/or heat directly through combustion, or produce a combustible fuel commodity, such as methane, methanol, ethanol, or synthetic fuels.

Gasification and pyrolysis processes have been known and used for centuries and for coal as early as the eighteenth century. Development technologies for processing (residual solid mixed waste) has only become a focus of attention in recent years stimulated by the search for more efficient energy recovery (Fichtner Consulting Engineers 2004).

Municipal solid waste (MSW) to a large extent is of biological origin (biogenic), e.g. paper, cardboard, wood, cloth, food scraps. Typically, half of the energy content in MSW is from biogenic material. Consequently, this energy is often recognized as renewable energy according to the waste input.

7.13.1 Methods

7.13.1.1 Incineration

Incineration, the combustion of organic material such as waste with energy recovery, is the most common WtE implementation. All new WtE plants in OECD countries incinerating waste (residual MSW, commercial, industrial, or refuse-derived fuel [RDF]) must meet strict emission standards, including those on nitrogen oxides (NO_x), sulfur dioxide (SO_2), heavy metals, and dioxins. Hence, modern incineration plants are vastly different from old types, some of which neither recovered energy nor materials. Modern incinerators reduce the volume of the original waste by 95–96%, depending upon composition and degree of recovery of materials such as metals from the ash for recycling.

Incinerators may emit fine particulate, heavy metals, trace dioxin, and acid gas, even though these emissions are relatively low from modern incinerators. Other concerns include proper management of residues: toxic fly ash, which must be handled in hazardous waste disposal installation as well as incinerator bottom ash, which must be reused properly.

Critics argue that incinerators destroy valuable resources and they may reduce incentives for recycling. The question, however, is an open one, as European countries which recycle the most (up to 70%) also incinerate to avoid landfilling.

Incinerators have electric efficiencies of 14–28%. In order to avoid losing the rest of the energy, it can be used for, e.g., district heating (cogeneration). The total efficiencies of cogeneration incinerators are typically higher than 80% (based on the lower heating value of the waste).

The method of incineration to convert MSW is a relatively old method of WtE production. Incineration generally entails burning waste (residual MSW, commercial, industrial, and RDF) to boil water which powers steam generators that make electric energy and heat to be used in homes, businesses, institutions, and industries. One problem associated is the potential for pollutants to enter the atmosphere with the flue gases from the boiler. These pollutants can be acidic and in the 1980s were reported to cause environmental damage by turning rain into acid rain. Since then, the industry has removed this problem by the use of lime scrubbers and ESPs on smokestacks. By passing the smoke through the basic lime scrubbers, any acids that might be in the smoke are neutralized which prevents the acid from reaching the atmosphere and hurting the environment. Many other devices, such as FFs, reactors, and catalysts destroy or capture other regulated pollutants. According to the *New York Times*, modern incineration plants are so clean that "many times more dioxin is now released from home fireplaces and backyard barbecues than from incineration." According to the German Environmental Ministry, "because of stringent regulations, waste incineration plants are no longer significant in terms of emissions of dioxins, dust, and heavy metals."

7.13.2 Other Technologies

There are a number of other new and emerging technologies that are able to produce energy from waste and other fuels without direct combustion. Many of these technologies have the potential to produce more electric power from the same amount of fuel than would be possible by direct combustion. This is mainly due to the separation of corrosive components (ash) from the converted fuel, thereby allowing higher combustion temperatures in, e.g., boilers, gas turbines, internal combustion engines, and fuel cells. Some are able to efficiently convert the energy into liquid or gaseous fuels.

Thermal technologies:

- Gasification: produces combustible gas, hydrogen, synthetic fuels
- Thermal depolymerization: produces synthetic crude oil, which can be further refined
- Pyrolysis: produces combustible tar/bio-oil and chars
- Plasma arc gasification or plasma gasification process (PGP): produces rich syngas including hydrogen and carbon monoxide usable for fuel cells or generating electricity to drive the plasma arch, usable vitrified silicate and metal ingots, salt, and sulfur

Non-thermal technologies:

- Anaerobic digestion: biogas rich in methane
- Fermentation production: examples are ethanol, lactic acid, hydrogen
- Mechanical biological treatment (MBT)
  - MBT + anaerobic digestion
  - MBT to refuse derived fuel

7.13.3 Global Developments

During the 2001–2007 period, the WtE capacity increased by about 4 million MT/Y. Japan and China each built several plants based on direct smelting or on fluidized bed combustion of solid waste. Japan is the largest user in thermal treatment of MSW in the world, with 40 million T. Some of the newest plants use stoker technology and others use the advanced oxygen enrichment technology. There are also over one hundred thermal treatment plants using relatively novel processes such as direct smelting, the Ebara fluidization process, and the thermoselect JFE gasification and melting technology process. In India its first energy bio-science center was developed to reduce the country's greenhouse gases and its dependency on fossil fuel. As of June 2014, Indonesia had a total of 93.5 MW installed capacity of WtE, with a pipeline of projects in different preparation phases together amounting to another 373 MW of capacity.

Biofuel Energy Corporation of Denver, CO, opened two new biofuel plants in Wood River, Nebraska, and Fairmont, Minnesota, in July 2008. These plants use distillation to make ethanol for use in motor vehicles and other engines. Both plants are currently reported to be working at over 90% capacity. Fulcrum BioEnergy Incorporation located in Pleasanton, California, is building a WtE plant near Reno, Nevada. The plant is scheduled to open in 2020. https://fulcrum-bioenergy.com/facilities. BioEnergy Incorporation predicts that the plant will produce approximately 10.5 MG/year of ethanol from nearly 200 000 T/Y of MSW.

WtE technology includes fermentation, which can take biomass and create ethanol, using waste cellulosic or organic material. In the fermentation process, the sugar in the waste is changed to carbon dioxide and alcohol, in the same general process that is used to make wine. Normally, fermentation occurs with no air present. Esterification can also be done using WtE technologies, and the result of this process is biodiesel. The cost effectiveness of esterification will depend on the feedstock being used, and all the other relevant factors such as transportation distance, amount of oil present in the feedstock, and others. Gasification and pyrolysis by now can reach gross thermal conversion efficiencies (fuel to gas) up to 75%; however, a complete combustion is superior in terms of fuel conversion efficiency. Some pyrolysis processes need an outside heat source which may be supplied by the gasification process, making the combined process self-sustaining.

7.13.3.1 Carbon Dioxide Emissions

In thermal WtE technologies, nearly all of the carbon content in the waste is emitted as carbon dioxide (CO_2) to the atmosphere (when including final combustion of the products from pyrolysis and gasification; except when producing bio-char for fertilizer). MSW contain approximately the same mass fraction of carbon as CO_2 itself (27%), so treatment of 1 MT (1.1 short tons) of MSW produce approximately 1 MT (1.1 short tons) of CO_2.

In the event that the waste was landfilled, 1 MT (1.1 short tons) of MSW would produce approximately 62 m^3 (2200 ft^3) methane via the anaerobic decomposition of the biodegradable part of the waste. This amount of methane has more than twice the global warming potential than the 1 MT (1.1 short tons) of CO_2, which would have been produced by combustion. In some countries, large amounts of landfill gas are collected, but still the global warming potential of the landfill gas emitted to atmosphere in, e.g., the United States in 1999 was approximately 32% higher than the amount of CO_2 that would have been emitted by combustion.

In addition, nearly all biodegradable waste is biomass. That is, it has biological origin. This material has been formed by plants using atmospheric CO_2 typically within the last growing season. If these plants are regrown, the CO_2 emitted from their combustion will be taken out from the atmosphere once more.

Such considerations are the main reason why several countries administrate WtE of the biomass part of waste as renewable energy. The rest – mainly plastics and other oil and gas derived products – is generally treated as nonrenewables.

7.13.4 Examples of WtE Plants

According to ISWA there are 431 WtE plants in Europe (2005) and 89 in the United States (2004). The following are some examples of WtE plants.

Waste incineration WtE plants:

- Lee County Solid Waste Resource Recovery Facility, Fort Myers, Florida, USA (1994)
- Montgomery County Resource Recovery Facility in Dickerson, Maryland, USA (1995)
- Essex County Resource Recovery Facility, Newark, New Jersey, USA (1996)
- Spokane Waste-to-Energy Facility, Washington, USA (1996)
- Spittelau (1971), and Flötzersteig Vienna, (Wien Energie), Austria (1963)
- SYSAV Waste-to-Energy Plant, Malmö, Sweden (2003 and 2008)
- Stoke Incinerator, Stoke-on-Trent, England, UK (1989)
- Teesside EfW plant near Middlesbrough, North East, England, UK (1998)
- Edmonton Incinerator in Greater London, England, UK (1974)
- Burnaby Waste-to-Energy Facility, Metro Vancouver, Canada (1988)
- Timarpur-Okhla Waste to Energy Plant, New Delhi, India (2001)
- Algonquin Power, Brampton, Ontario, Canada (1990)

7.13.5 Case Study: Energy Recovery from Municipal Solid Waste: Profitable Pollution Prevention at the City of Spokane, Washington (see Appendix G)

Example 7.13 Waste-to-Energy

The City of Newcastle-Upon-Tyne, UK, incinerates some of its municipal RDF for electricity production. The city operates two 13 MW RDF boilers, which have 88% overall efficiency. The RDF has an average heat content 1650 Btu/lb. Calculate the following:

a) The daily mass of RDF (T/day) required to burn to generate the same power.
b) If coal were used in place of RDF, how much coal (in T/Y) that would be needed to burn to produce the same energy. The coal has an average heat content of 33 516 kJ/kg.

Solution

This is essentially a mass and energy balance problem, in which solid wastes with significant heat content are being collected and burned to generate heat and electricity.

a) The daily mass of RDF in T/day
First, find input power needed.

$$\frac{2(13\text{ MW})}{0.88} = 30.58\text{ MW} = 30.58\frac{\text{MJ}}{\text{s}}$$

Then, mass of RDF required to produce same power

$$= \frac{30.58\frac{\text{MJ}}{\text{s}}\times 10^3\frac{\text{KJ}}{\text{MJ}}\times 0.948\frac{\text{Btu}}{\text{KJ}}\times 86\ 400\frac{\text{s}}{\text{day}}}{1\ 650\frac{\text{Btu}}{\text{lb}}}$$

$$= 1.518\times 10^3\frac{\text{lb}}{\text{day}} = 760\frac{\text{ton}}{\text{day}}\text{RDF}$$

b) Mass of coal required to produce same power

$$= \frac{\left(30.58\frac{\text{MJ}}{\text{s}}\times 0.948\frac{\text{Btu}}{\text{KJ}}\times 86\ 400\frac{\text{s}}{\text{day}}\right)}{\left(33\ 516\frac{\text{KJ}}{\text{Kg}}\right)\left(\frac{\text{MJ}}{10^3}\text{KJ}\right)}\left(\frac{2.205\text{ lb}}{\text{Kg}}\right)$$

$$= 74\ 732\frac{\text{lb}}{\text{day}} = 37.37\frac{\text{ton}}{\text{day}}\text{coal}$$

Example 7.14 Waste-to-Energy

Assume the proximate and ultimate analysis of a solid waste.

Proximate analysis	(wt%)	Ultimate analysis	(wt%)
Moisture	20	Moisture	20
Volatile matter	60	Carbon	30
Ash	20	Hydrogen	4
Total	100	Oxygen	25.5
		Nitrogen	0.37
		Sulfur	0.13
		Ash and metal (inerts)	20
		Total	100

Also assume the following conditions are applicable:

1) The as-fired heating values of the solid waste are represented by the equations:

$$\text{HHV} = 145.7(\text{C}) + 619(\text{H}) - 59.8(\text{O}) + 45.1(\text{S})\text{Btu/lb}, \quad (7.6)$$

$$\text{LHV} = \text{HHV} - 10.50(\text{W} + 9\text{H})\text{Btu/lb} \quad (7.7)$$

where HHV is the high-heating value and LHV is the low-heating value; C, H, O, S, and W are the weight percentages of carbon, hydrogen, oxygen, sulfur, and W is the weight of the moisture, respectively. And the answer is in terms of Btu/lb.

2) The grate residue contains 10% unburned carbon.
3) The entering air is at 77 °F; the residue is at 700 °F.
4) The radiation loss is 0.006 Btu/Btu.
5) The specific heat of residue is 0.3 Btu/lb °F.
6) The latent heat of water is 1040 Btu/lb.
7) All the oxygen in the waste is bound as water.
8) The theoretical air requirements based on stoichiometry are as follows:

Element	Reaction with oxygen	Air requirement (lb/lb)
C	$C+O \rightarrow CO$	11.52
H	$2H_2+O_2 \rightarrow 2H_2O$	34.56
S	$S+O_2 \rightarrow SO_2$	4.31

9) The moisture in combustion air is negligible.
10) The net hydrogen available for combustion is equal to the fraction $(H-\frac{O}{8})$.
    In other words, this is the percentage of hydrogen minus ⅛th the percentage oxygen. This accounts for "bound water" in the dry combustible material.
11) The heating value for carbon is 14 000 Btu/lb.
    Find:
    a) Estimate the HHV and LHV of this waste using the Eqs. (7.6) and (7.7).
    b) Calculate the air requirement for complete combustion of 100 T of this waste.
    c) Determine the heat available in the exhaust gases from combustion of 100 T/day waste by a heat balance, both in million Btu and kWh units.

Solution

a) From Eq. (7.6),

$$\begin{aligned}HHV &= 145.7(C)+619(H)-59.8(O)+45.1(S)\\ &=145.7\times(30)+619\times4-59.8\times25.5+45.1\times0.13\\ &=4371+2476-1525+5.86\\ &=5327.86\ \text{Btu/lb}=5328\ \text{Btu/lb}\end{aligned}$$

and from Eq. (7.7),

$$\begin{aligned}LHV &= HHV-10.50(W+9H)\\ &=5328-10.50(20+9\times4)=5328-588\\ &=4740\ \text{Btu/lb}\end{aligned}$$

b) First, compute the weights of the elements of the waste:

Elements	Computation (lb)	lb/day
Carbon	0.30 × 200 000	60 000
Hydrogen	0.04 × 200 000	8 000
Oxygen	0.255 × 200 000	51 000
Nitrogen	0.0037 × 200 000	740
Sulfur	0.0013 × 200 000	260
Ash and metals (inert)	0.20 × 200 000	40 000
Moisture	0.20 × 200 000	40 000

Second, compute the residue:

$$\text{Inerts}=40\ 000\ \text{lb/day}$$

Because the grate residue contains 10% unburnt carbon (an assumption made or this problem),

$$\text{Total residue}=\frac{40\ 000}{0.90}=44\ 444\ \text{lb/day}$$

$$\text{Carbon in residue}=44\ 444-40\ 000=4\ 444\ \text{lb/day}$$

Third, compute the available hydrogen and bound water:

$$\begin{aligned}\text{Net available hydrogen} &= \left(4.0\%-\frac{25.5\%}{8}\right)\\ &=0.812\ 5\%\\ &=0.008\ 125\times200\ 000\ \text{lb}\\ &=1\ 625\ \text{lb/day}\end{aligned}$$

$$\begin{aligned}\text{Hydrogen in bound water} &= 4\%-0.812\ 5\%=3.18\%\\ &=0.031\ 875\times200\ 000\ \text{lb/day}\\ &=6\ 375\ \text{lb/day}\end{aligned}$$

Therefore,

$$\begin{aligned}\text{Bound water} &= \text{oxygen}+\text{hydrogen in bound water}\\ &=51\ 000+6\ 375=57\ 375\ \text{lb/day}\end{aligned}$$

Fourth, calculate the air required:

Element	Air requirement (lb/day)
Carbon = (60 000 − 4 444) × 11.52	640 005
Hydrogen = 1.625 × 34.56	56.160
Sulfur = 260 × 4.31	1 120
Total dry theoretical air (C + H + S)	697 285
Total dry air (100-percent excess)	1 394 570

c) First, determine the amount of water produced by combustion of the available hydrogen.

$$H_2O=\frac{18\ \text{lb of}\ H_2O}{2\ \text{lb of H}}\left(1\ 625\frac{\text{lb}}{\text{day}}=14\ 625\frac{\text{lb}}{\text{day}}\right)$$

The remaining computations of heat balance for the process are as follows:

Item	Heat value (10^6 Btu/day)
Gross heat input	
$2\times10^5\frac{\text{lb}}{\text{d}}\left(5328\frac{\text{Btu}}{\text{lb}}\right)$	1065.6
Heat lost in unburned carbon:	
$4\ 444\frac{\text{lb}}{\text{day}}\left(14\ 000\frac{\text{Btu}}{\text{day}}\right)$	62.21

(continued)

Item	Heat value (10^6 Btu/day)
Radiation lost:	
$0.006\frac{\text{Btu}}{\text{Btu}}\left(1.0655\times10^{6}\frac{\text{Btu}}{\text{day}}\right)$	6.39
Inherent moisture:	
$40\,000\frac{\text{lb}}{\text{day}}\left(1.040\frac{\text{Btu}}{\text{lb}}\right)$	41.60
Moisture in bound water:	
$57\,375\frac{\text{lb}}{\text{day}}\left(1.040\frac{\text{Btu}}{\text{lb}}\right)$	59.65
Moisture from the combustion of available hydrogen:	
$14\,625\frac{\text{lb}}{\text{day}}\left(1.040\frac{\text{Btu}}{\text{lb}}\right)$	15.21
Sensible heat in residue:	
$44\,444\frac{\text{lb}}{\text{day}}\left(0.3\frac{\text{Btu}}{\text{lb}}\right)\left(700°\text{F}-77°\text{F}\right)$	8.30
Total losses	193.38
Net heat available in flue gases:	
$\left(1065.6-193.38\right)\times10^{6}\frac{\text{Btu}}{\text{day}}$	877.22
Therefore, net energy available in kWh $= 877.22\times10^{6}$ Btu $= 877.22\times10^{6}\,\text{Btu}\times2.93\times10^{-4}\,\text{kWh/Btu}$ $= 257\ 000\ \text{kWh}$	
Combustion efficiency:	
$\left(\dfrac{877.22\times10^{6}\,\frac{\text{Btu}}{\text{day}}}{1065.5\times10^{6}\,\frac{\text{Btu}}{\text{day}}}\right)\times100\%$	81.86%

7.14 Sustainable Economy and the Earth

Evidence that the economy is in conflict with Earth's natural systems can be seen in the daily news reports of collapsing fisheries, shrinking forests, eroding soils, deteriorating rangelands, expanding deserts, rising carbon dioxide levels, falling water tables, rising temperatures, more destructive storms, melting glaciers, rising sea level, dying coral reefs, and disappearing species. These trends, which mark an increasingly stressed relationship between the economy and the Earth's ecosystem, are taking a growing economic toll. At some point, this could overwhelm the worldwide forces of progress, leading to economic decline. The challenge for our generation is to reverse these trends before environmental deterioration leads to long-term economic decline, as it did for so many earlier civilizations.

7.14.1 What Is a Sustainable Economy?

An environmentally sustainable economy – an eco-economy – requires that the principles of ecology be observed in establishing the framework for formulating economic policy and that economists understand that all economic activity, indeed all life, depends on the Earth's ecosystems – the complex of individual species living together, interacting with each other, and with their physical habitat. Millions of plant and animal species exist in an intricate balance, woven together along the food chain by nutrient cycles, the hydrological cycle, and global climate systems. Economists know how to translate goals into policy. Economists and ecologists, along with engineers, scientists, and policy makers, working together are being challenged to design and build an eco-economy, one that can sustain progress (Brown 2001a, b; Western Governors' Association, 3 December 2004).

Through industrial ecology, eco-industrial parks, eco-efficiency, Zero Discharge manufacturing, and engineering new products from agricultural materials (biorenewable resources), an eco-economy that facilitates sustainable economic progress offer the potential for the achievements, over the long-term, of a sustainable economy and a healthy planet. Humankind has gone outside the biotic environment for the majority of its material needs only recently. Plant-based resources were the predominant source of energy, organic chemicals, and fibers in the West as recently as 150 years ago, and they continue to play important roles in many developing countries.

The transition to nonbiological (or nonrenewable) sources of energy and materials was relatively swift and recent in the history of the world. Biorenewable resources are by definition sustainable natural resources; that is, they are self-renewing at a rate that ensures their availability for use by future generations. Biorenewable resources can be converted into either bioenergy or biobased engineering products. Bioenergy, also known as biomass energy, is the result of the conversion of the chemical energy of a biorenewable resource into heat and stationary power. Biobased products include transportation fuels, chemicals, and natural fibers derived from biorenewable resources. Transportation fuels are generally liquid fuels, such as ethanol or biodiesel, but compressed hydrogen and methane have also been proposed and evaluated for use in vehicular propulsion.

Chemicals may include pharmaceuticals, nutraceuticals, and other fine chemicals, but the emphasis here is on

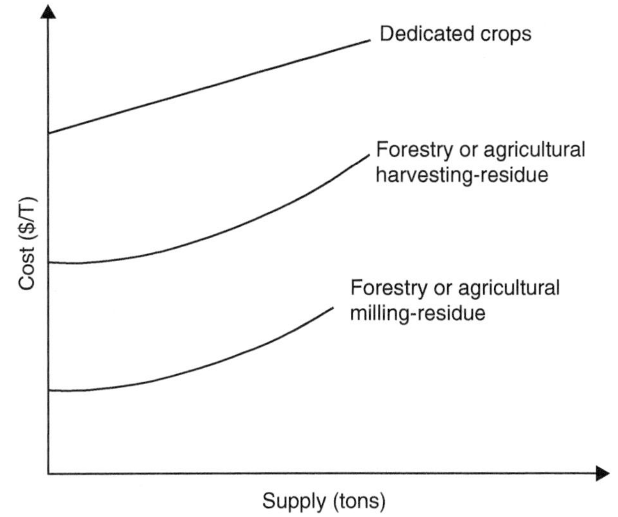

Figure 7.9 Example of supply curves to different kinds of biorenewable resources.

commodity chemicals, which produce high demands for biorenewable resources. An example is polylactic acid, which is derived from the fermentation of sugars hydrolyzed from cornstarch and can be converted into biobased polymers used in a variety of consumer products such as carpets.

The cost of a biorenewable resource is related to the demand for the resource by a supply curve. Figure 7.9 is a generalized representation of a supply curve for the three kinds of biorenewable resources. The cost of biorenewable resources is highly variable and dependent on local conditions of supply and demand. This is particularly true for the costs involved in processing residue and wastes. Table 7.13 provides an estimate of the availability and cost of several kinds of residues and wastes along with a comparison of the costs of a few crop. The cultivation and harvesting of dedicated energy crops, on the other hand, is amenable to standardized cost estimating since information on "unit operation," such as planting, fertilizing, and harvesting, can be readily obtained from knowledgeable sources (Brown 2001b).

7.14.2 Costs of Manufacturing Various Biobased Products and Energy

Over the next decades, a much larger fraction of fuels, chemicals, and materials will be produced from renewable plant materials. These biobased industrial products offer the potential for a much more sustainable economy based on environmentally superior products. This section briefly describes the associated costs of producing electricity, fuels, and chemicals from various feedstocks using biorenewable resources.

Table 7.13 Availability and cost of potential feedstocks.

Feedstocks	Production (106 T/Y)	Price (194 $/kg)
Corn	191	0.09
Potato	17	0.16
Sorgham	16	0.09
Beet molasses	1	0.09
Cane molasses	1	0.03
Sugar cane	25	0.03
Agricultural residues		
Low cost	4	12.9
Mid cost	36	38.8
High cost	50	47.4
Forest residue-logging		
Low cost	3	12.9
Mid cost	3	25.9
High cost	3	43.1
Forest residue mill	3	17.2
Municipal solid waste		
Mixed paper	26	0–19
Packaging	14	0–5.2
Urban wood	3.5	12.9–25.9
Yard waste	11	0.12.9

Source: Crop data from Polman (1994) and Waste and residue data from Lynd (1996).

7.14.2.1 Electricity from Combustion of Biomass

The capital and operating costs for steam power plants fired with biomass are relatively well known because of significant operating experience with these systems. The capital cost for a new plant ranges between $1400 and $1800/kW capacity. Accordingly, a 50-MW biomass plant based on direct-combustion would cost approximately $80 million. On the basis of a target price of $41.90/GJ for biomass, the cost of production for direct-fired biomass power is about $0.06/kWh (Environmental Law & Policy Center 2001).

7.14.2.2 Electricity from Gasification of Biomass

The capital cost for a gasification plant, including fuel feeding and gas cleanup, is dependent on both the size and the operating pressure of the system. In the United States, an atmospheric-pressure gasifier producing 50 MW thermal energy would cost about $15 million. A gasification/gas-turbine power plant producing 50 MW of electricity would have total capital cost of between $75 and $138 million (between $1500 and $2750/kW), the smaller number reflecting improved technical know-how after building at least ten plants. Electricity production costs would range from $0.05 to $0.09/kWh if fuel is available at an optimistic price range of $1.00–$1.50/GJ.

Capital costs for high-temperature fuel cells suitable for integrated gasification/fuel cell power plants currently cost $3000/kW. Molten carbonate fuel cells are expected to be $1500/kW at the time of market entry, decreasing to about $1000/kW for a commercially mature unit. The cost of electricity from a mature unit operating on natural gas is projected to be between $0.049 and $0.085/kWh. More attractive economics result if less expensive fuel is available. The cost of electricity generated from landfill gas using mature fuel-cell technology is expected to be comparable to that for an internal combustion engine/electric generator set, i.e., about $0.05/kWh (Bridgewater 1995; Hirschenhoofer et al. 1994; William and Larson 1993).

7.14.2.3 Biogas from Anaerobic Digestion

Anaerobic digestion is commercially developed for the purpose of treating wastewater. Power production from anaerobic digestion is in its infancy. The methane generated is often fired in internal combustion engines to produce electricity in an effort to help offset costs of waste treatment, but there are no immediate prospects for it to replace natural gas. Capital costs for anaerobic digestion facilities processing more than 200 T/day of volatile solids are estimated to be between $44 000 and $132 000 for each T/day of volatile solids processed. Methane yields will be approximately 0.38 m^3/kg of volatile solids. Thus, a 200 T/day anaerobic digestion plant could produce 28 million m^3 of methane per year, representing almost 2900 GJ/day of chemical energy. Projected operating costs for producing methane from dedicated energy crops were in the range of $5–6/GJ in 1986 dollars. In comparison, the cost of natural gas in the United States, which shows large seasonal and geographical variations, ranges between $1.90 and $4/GJ. In niche markets, where the feedstock is inexpensive and natural gas is not available, biogas can be a viable alternative energy source (Benson et al. 1986). A similar biogas manufacturing using poultry WtE by catalytic steam gasification process is described in Chapter 10.

7.14.2.4 Ethanol from Biomass

The cost of producing ethanol from biomass varies tremendously depending on the feedstock employed, the size and management of the facility and the market value of coproducts generated as part of some conversion processes. Cost information for ethanol plants to be built in the United States is most reliable for those using cornstarch, the basis of the US ethanol industry. A 5000 barrel/day plant (about 265 million l/day) built from the ground up will have a capital cost of about $140 million in 1987 dollars, or $0.53/l of annual capacity. Smaller facilities can have capital costs as high as $0.79/l of annual capacity, and poorly designed facilities of any size may cost $1.06/l of annual capacity. On the other hand, ethanol plants that are built from existing facilities, such as refineries or chemical plants, or ethanol plants integrated into a larger industrial facility, can have substantially lower capital costs, often in the range of $0.26–0.40/l of annual capacity.

Low-end production costs are about $0.26/l. However, the volumetric heating value of ethanol is only 66% that of gasoline. This production cost, therefore, is equivalent to gasoline selling for $0.39/l before tax, transportation, or profit. In contrast, refinery price for gasoline in 1990 dollars was $0.20/l. Currently, the economics of fermentation are such the commercial viability of ethanol is entirely dependent on government incentives in the form of tax credit, currently $0.16 for each liter of ethanol used for fuel blending. Also, a strong market for fermentation by-products is key factor in the economic viability of ethanol-from-corn.

Technology to convert lignocellulose to sugar is expected to reduce the cost of fuel ethanol, although dedicated economic information is not currently available. Capital cost for a 5000 barrel/day plant to produce ethanol from lignocellulose using simultaneous saccharification and fermentation is estimated to be $175 million (1994 dollars). Assuming wood costs $42/dry ton, ethanol can be produced for about $0.31/l. Combining economies of scale with advances in processing technology are projected to decrease production costs to 40.31/l. However, some reports have suggested that ethanol from cellulose will have to cost as little as $0.08–0.11/l to be competitive with the gasoline prices anticipated early in the twenty-first century (Lynd et al. 1996; National Advisory Panel 1987).

7.14.2.5 Methanol from Biomass-Derived Syngas

Capital investment for a 7500 barrel/day plant to produce methanol from biomass would be about $280 million in 1991 dollars. The cost of methanol from $40/dry ton of wood is projected to be about $0.27/l. Since the volumetric heating value of methanol is only 49% that of gasoline, the production cost from this plant is equivalent to gasoline selling for $0.55/l. Methanol from natural gas can be produced at significantly lower cost, but this assumes much larger plant capacities to capture economies of scale. Such large plants are not feasible for widely dispersed biomass feedstocks. New methanol-synthesis technologies may be able to significantly reduce this price. The U.S. Department of Energy's methanol from biomass program has a goal of $0.15/l ($7.90/GJ) based on feedstock cost of $1.90/GJ (Klass 1998).

7.14.2.6 Bio-Oil from Fast Pyrolysis

Capital investment for a 5000 barrel/day plant to produce pyrolysis liquids would be \$63 million in 1987 dollars. Assuming biomass feedstock was available at \$1.70/GJ, this size plant could produce pyrolysis liquids for \$0.18/l, which has an energy value of \$6.70/GJ (Elliott et al. 1990).

7.14.2.7 Biodiesel from Vegetable Oils

Capital costs for a biodiesel facility are relatively modest, costing about \$250 000 for a 50 barrel/day (3.2 million l/year). However, feedstock costs for production of biodiesel are relatively higher than feedstocks for production of other kinds of fuel, ranging from \$0.16 to \$0.26/l for waste fats to \$0.53–0.79/l for vegetable oils. Under the best scenarios, a biodiesel plant might produce fuel for \$0.44/l. Diesel fuel produced from petroleum typically sells for less than \$0.25/l (Gavett et al. 1993).

7.14.2.8 Succinic Acid

Succinic acid is used in producing food and pharmaceutical products, surfactants and detergents, biodegradable solvents and plastics, and ingredients to stimulate animal and plant growth. Although it is a common metabolite formed by plants, animals, and microorganisms, its current commercial production of 15 000 T/Y is from petroleum. However, the recently discovered rumen organism *Actinobacillus succinogenes* produces succinic acid with yields as high as 110 g/l, offering prospects for producing this chemical from biorenewable resources. In contrast to most other commercial fermentations, the process consumes CO_2 and, integrated with a process like ethanol fermentation, succinic acid production could contribute to reduction in greenhouse gas emissions.

Optimum yields occur under pH conditions where succinate salt rather than free acid is produced. Thus, recovery entails concentration of the salt, conversion back to free acid, and polishing of the acid to the desired purity. Downstream purification can account for 60–70% of the product cost.

7.14.2.9 Lactic Acid

Lactic acid, a three-carbon molecule, is used in the production of *polylactide* resin, a biodegradable polymer expected to compete with polyethelene and polystyrene in the synthetic fibers and plastic markets. Lactic acid is currently produced by milling corn, separating the starch, hydrolyzing the starch to glucose, and anaerobically fermenting the glucose to lactic acid with *Bacillus dextrolactius* or *Lactobacillus delbrueckii*. Esterification with ethanol produces ethyl lactate, which can be polymerized to polylactate resin.

Problems

7.1 A manufacturing process has present operating costs of \$10.9 million annually and waste disposal costs of 3.3 million. The plant manager wishes to upgrade the process by installing new equipment at a capital cost of \$23.3 million. This will reduce the annual operating and waste disposal costs to \$7.1 and \$1.4 million respectively. Compare the two processes on a present worth basis, assuming no salvage value. Use 12% interest for 10 years as a basis for calculation.

7.2 (a) Should a project with $\frac{\Sigma \text{ benefit}}{\Sigma \text{ cost}} = 0.85$ be undertaken? Explain your answer; (b) What are the drawbacks of relying on the benefit/cost ratio for justifying a project?

7.3 A community plans to build a landfill to dispose of wastes for next 20 years. The cost estimates and expected revenue are as follows:

Landfill construction cost	\$7 600 000
Start-up costs	\$500 000
Operation and maintenance costs per year	\$1 000 000
Tipping fees received per year	\$4 500 000

Assume that citizens will only support the new landfill facility if the benefit/cost ratio is greater than 1.5. At a 7% interest rate over the 20 years life, determine if the landfill should be built.

7.4 A county wants to build a toll-free bridge over a river in order to take time off of a route that goes through the mountains. Benefits are decreased travel time, fewer accidents, reduced gas usage, less pollution, etc. Should be bridge be built? Benefit and cost data are given below.

	Millions (\$)
Initial cost	40
Annual maintenance	12
Annual user benefits	49
Residual value	0

7.5 Given that a wet scrubber had a TIC of \$800 000 in 2001, estimate the 2008 TIC using the two cost indices (Marshall-Swift & Chemical Engineering Plant) presented in Table 7.11. Use a reasonable method to estimate the TIC in the year 2013.

7.6 A company uses a 50% solution of NaOH to maintain the pH of water circulated in one of its furnace exhaust scrubbers. The cost of the solution is \$200/T free-on-board. The supplier says it would prefer to sell the company a 65% NaOH solution and the company says the water content is immaterial as long as the equivalent costs are the same including freight. If the freight is \$15/T, what is the maximum price per ton the company should pay for the 65% solution?

7.7 An engineer has compiled the data below for two different project options – one that includes a comprehensive pollution prevention program/option and one that does not (data are given in a table below). From an economic point of view, which project should the engineer select? The lifetime of the equipment is 10 years and the interest rate is 10%.

Pollution prevention cost data:

	Project with pollution prevention (W) (\$)	Project without pollution prevention (WO) (\$)
Equipment cost	1 294 000	1 081 000
Installation cost	786 000	659 000
Operating labor (per year)	39 900	8 500
Maintenance (per year)	43 000	17 000
Utilities (per year)	958 000	821 000
Overhead (per year)	51 300	13 900
Taxes, insurance and administr. (per year)	86 200	72 600
Credits (per year)	380 000	0

7.8 A small rural community with a population of 850 people generates recyclable solid waste, at an average rate of 2.0 lb/day/capita. The chemical composition of the wastes is $C_{89.10}H_{138.61}O_{51.98}N_1$. The approximate Btu values for individual materials in RDF can be calculated using the following equation:

$$\text{Btu / lb} = 145\text{C} + 610\left(\text{H}_2 - \tfrac{1}{8}\text{O}_2\right) + 40\text{S} + 10\text{N};$$

where C, H, O, S, and N values are in wt%.
Determine

a) the percentage distribution by weight of elements in the waste, round off the coefficients and set up a table
b) the daily energy content of the waste that is received by a local WtE facility.

7.9 In the City of Aberdeen, Scotland, a landfill with a gas collection system is in operation and serves a population of 250 000. At a rate of (4.4 lb)/(person-day) RDF from the city's solid waste is generated. Gas is produced at an annual rate of 2.8 l/lb at the city's landfill. The gas contains 60% methane. The gas recovery is about 20% of that generated. The heat content of the landfill gas is approximately 480 Btu/ft^3 (this value is lower than the theoretical value because of dilution of methane with air during gas recovery). The landfill operator and the city have proposed to supply the methane to homes that would be built in the vicinity of the landfill. The homes are estimated to use an average of 95×10^6 Btu/year for heating and cooking. Peak usage during winter is 1.3 times the average usage. How many homes could be served by the energy generated at the landfill?

7.10 Recently, a cast iron leaf pressure filter with 100 ft^2 was purchased for clarifying an inorganic liquid stream for \$15 000. In a similar application, the company will need a 450 ft^2 cast iron leaf pressure filter. The size exponent for this filter is 0.6. Estimate the purchased price of the 450 ft^2 unit.

7.11 A stainless steel centrifuge costs \$83 000 in 1990. What is the cost of that same centrifuge in 2001? Use the CE Index.

7.12 The delivered equipment cost (DEC) for a fixed-bed carbon adsorption system with a capacity of 5000 acfm was \$100 000 in 1994. Estimate the total installation cost of a similar system with a capacity of 7500 acfm in 2000. Assuming the size exponent for a carbon-adsorption system is 0.7 and TIC is 1.5 times the DEC.

7.13 Search the literature on mercury emissions from coal-fired power plants and make a brief report. In your report, address the following: total mercury emissions in the United States, range of mercury content in coal, US regulations dealing with mercury from coal-fired power plants, and methods for controlling mercury (with approximate control efficiencies) in power plant exhaust gases.

7.14 A water resources project has produced benefits that equal \$40 000 at the end of 20 years. Determine the PV of these benefits using a 6% interest rate? Determine the PV of the benefits at the end of the 20th year.

7.15 A water resources project has produced benefits that equal \$20 000 at the end of the first year and increase on a uniform gradient series to \$100 000 at the end of the 5th year. The benefits remain constant at \$100 000 each year until the end of the year 25. Determine the PV of these benefits using a 6% interest rate?

7.16 Exhaust gas from a gas turbine facility (2×10^5 lb/h steam capacity) flows at a rate of 100 000 dscfm at 1200 °F with a PM loading of 1.5 gr/dscfm. As part of PSD/Title V permitting process, a top-down BACT analysis is needed in which the TAC and cost per ton of PM removal (\$/T) to be evaluated for four control options: (i) MC, (ii) VS, (iii) ESPs, and (iv) FF. Calculate the \$/T of PM removal by four control technologies using the TACs for four control options presented in Figure 7.7. Assume MC and VS have 90% and ESP and FF have 99.5% removal efficiencies, respectively.

References

Air Pollution Training Institute (APTI) (1995). *APTI Course 427 Lecture Note on Control of Gaseous Emissions*. Research Triangle Park, NC: USEPA.

Allen, D.T. and Rosselot, K.S. (1997). *Pollution Prevention for Chemical Processes*, 19–32. New York, NY, Chapter 2: Wiley.

Benson, P.H., Hayes, T.D., and Isaacson, R. (1986). Regional and community approaches to methane from biomass and waste: an industry perspective. *Proceedings Energy from Biomass and Waste X*, Orlando, FL (April 1986).

Bridgewater, A.V. (1995). The technical and economic feasibility of biomass gasification for power generation. *Fuel* 74: 631–653.

Brown, L.R. (2001a). *Eco-Economy: Building an Economy for the Earth*. New York, NY: W. W. Norton.

Brown, R.C. (2001b). *Biorenewable Resources: Engineering New Products from Agriculture*. Iowa State Press.

Burckle, J.O. and Ferguson, T.D. (2005). Advances in pollution prevention in the metal finishing industry. In: *Toward Zero Discharge: Innovative Methodology and Technologies for Process Pollution Prevention* (ed. T.K. Das). Hoboken, NJ: Wiley.

Burgess, T.L., Gibson, A.G., Furstein, S.J., and Wachs, I.E. (2002). Converting waste gases from pulp mills into value-added chemicals. *Environmental Progress* 21 (3): 137–141.

California Air Resource Board (2000). Statewide Best Available Control Technology (BACT) Clearinghouse. South Coast Air Quality Management District (August 17). http://www.arb.ca.gov/bact/bact.htm (accessed 11 January 2019).

Cluster Rule Regulation (1998). 40CFR Part 63, Subpart S (foul condensates), Section 63:446.

Dalan, J. (2005). Brine concentrators for recycling wastewater. In: *Toward Zero Discharge: Innovative Methodology and Technologies for Process Pollution Prevention* (ed. T.K. Das). Hoboken, NJ: Wiley.

Dalan, J. and Rosain, R. (1992). *Zero Discharge Wastewater Treatment Facility for a 900 MW GCC Power Plant*, Report No. EPRI TR-100375. Palo Alto, CA: Electric Power Research Institute.

Das, T.K. (1997). A pollution prevention strategy in the kraft pulp and paper manufacturing process technologies. Paper 188c, American Institute of Chemical Engineers National Meeting, Los Angeles, CA.

Das, T.K. (1999a). Process pollution prevention advances in the pulp and paper industry. Paper 274b, American Institute of Chemical Engineers National Meeting, Dallas, TX.

Das, T.K. (1999b). Process technology advances in the pulp and paper industry. Paper 273e, American Institute of Chemical Engineers National Meeting, Dallas, TX.

Das, T.K. (2001). Qualitative Analysis of Control Technologies, Costs, and Environmental Impacts Analyses-Wood-fired Boiler RACT Determination. Air Quality Program, Washington Department of Ecology, Draft Report.

Das, T.K. (2003). Hog-Fuel Boiler's Reasonably Available Control Technology (RACT) Determination. Air Quality Program, Washington Department of Ecology, Olympia, Washington, Report #03-02-009, 137 pages. http://www.ecy.wa.gov/biblio/0302009.html (accessed 14 February 2019).

Das, T.K. (2005). *Toward Zero Discharge: Innovative Methodology and Technologies for Process Pollution Prevention*. Hoboken, NJ: Wiley-InterScience.

Das, T.K. and Houtman, C. (2004). Evaluating chemical-, mechanical-, and bio-pulping processes and its sustainability characterization using life cycle assessment. *Environmental Progress* Special Issue on Sustainability 23 (4): 347–357.

Das, T.K. and Jain, A.K. (2001). Pollution prevention advances in pulp and paper processing. *Environmental Progress* 22 (2): 87–92.

DaSilva, A. and Goodman, A. (2014). Reduce water consumption through recycling. *Chemical Engineering Progress* 110: 29–37.

Dzombak, D.A. (2013). Use of treated municipal wastewater as power plant cooling system makeup water: tertiary treatment versus expanded chemical regimen for recirculating water quality management. *Environmental Engineers and Scientists* 49 (2): 30.

Elliott, D.C., Osman, A., Borje Gevett, S. et al. (1990). Biomass for renewable energy and fuels. In: *Energy from Biomass and Wastes XIII* (ed. D.L. Klass). Chicago, IL: Institute of Gas Technology.

Environmental Law and Policy Center (2001). Repowering the Midwest: the clean energy development plan for the Heartland. www.repowermidwest.org (accessed 18 February 2019).

Ferguson, D., Hindin, B., Chen, A., and Monzyk, B. (2001). Picklex as a non-polluting metal surface finishing pretreatment and pretreatment/conversion coating. *Clean Products and Processes* 2: 220–227.

Ferguson, D. and Monzyk, B. (2003). Non-polluting metal surface finishing pretreatment and pretreatment/conversion coating. *Plating and Surface Finishing* 90: 66–75.

Ferguson, D. and Wilmoth, R. (2000). Cost-effective control of hexavalent chromium air emissions from functional chromium electroplating. *Environmental Health and Safety Solutions* 1 (3): 10–19.

Fichtner Consulting Engineers Ltd. (2004). *The Viability of Advanced Thermal Treatment of MSW in the UK*. London: ESTET.

Fosberg, T. (1992). Case study: water recovery systems at the world's first zero effluent pulp mill. *Proceedings of the International Water Conference-51st Annual Meeting*, Pittsburgh, USA (19–21 October 1992).

Gavett, E.E., van Dyne, D., and Blasé, M. (1993). Fuel cells for distributed generation. In: *Energy from Biomass and Waste XIII* (ed. D.L. Klass). Chicago, IL: Institute of Gas Technology.

Haussman, C. and Rosain, R. (1996). *Power Plant Wastewater Treatment Technology Review Report*, Report No. EPRI TR-10781. Palo Alto, CA: Electric Power Research Institute.

Heller, M., Shields, P.D., and Beloff, B. (1995). Environmental accounting case study: Amoco Yorktown refinery. In: *Green Ledgers: Case Studies in Corporate Environmental Accounting* (eds. D. Ditz, J. Ranganathan and D. Banks). Washington, DC: World Resources.

Hirschenhoofer, J. H., Stauffer, D. B., and Engleman, R. R. (1994). *Fuel Cells: A Handbook*, Department of Energy Technical Report DOE/METC-94/1006 (DE94004072). Washington, DC: United States Department of Energy. https://www.annualreviews.org/doi/abs/10.1146/annurev.energy.21.1.403 (accessed 24 September 2019).

Kirk, R.E. and Othmer, D.F. (2004). *Encyclopedia of Chemical Technology*, 5e. Hoboken, NJ: Wiley.

Klass, D.L. (ed.) (1998). Thermal conversion. In: *Biomass for Renewable Energy, Fuels, and Chemicals*. San Diego: Academic Press.

Klinker, R. (1996). Successful implementation of a Zero Discharge program. *TAPPI Journal* 79 (1): 97–102.

Lynd, L.R. (1996). Overview and evaluation of fuel ethanol from cellulosic biomass: technology, economics, the environment, and policy. *Annual Review of Energy and the Environment* 21: 403–465.

Lynd, L.R., Elander, R.T., and Wyman, C.E. (1996). Likely features and costs of manure biomass ethanol technology. *Applied Biochemistry and Biotechnology* 57/58: 741–761.

National Advisory Panel on Cost Effectiveness of Fuel Ethanol Production (1987). Fuel Ethanol Cost-Effectiveness Study Final Report.

National Council of the Paper Industry for Air and Stream Improvement, Inc. (NCASI) (1991). *Progress in Reducing Water Use and Wastewater Loads in the U.S. Paper Industry*, Technical Bulletin No. 603. Research Triangle Park, NC: NCASI.

Paik, S.C. and Chung, J.S. (1996). Selective hydrogenation of sulfur dioxide with hydrogen to elemental sulfur over transition metal sulfide supported on AL_2O_3. *Applied Catalysis B: Environment* 8: 267–279.

Perry, R.H. and Green, D.W. (1997). *Perry's Chemical Engineers' Handbook*, 7e, 9–10. New York, NY: McGraw-Hill and 9–16.

Peters, M.S. and Timmerhaus, K. (2002). *Plant Design and Economics for Chemical Engineers*, 5e. New York: McGraw-Hill.

Peters, M.S., Timmerhaus, K., and West, R.E. (2002). *Plant Design and Economics for Chemical Engineers*, 5e. New York, NY: McGraw-Hill.

Pinkerton, J.E. (1999). Trends in U.S. kraft mill TRS emissions. *Tappi Journal* 82 (4): 166–169.

Pollution Prevention 1997: A National Progress Report. http://www.epa.gov/opptintr/p297/foreword.pdf (accessed 4 March 2019).

Polman, K. (1994). Review and analysis of renewable feedstocks for the production of commodity chemicals. *Applied Biochemistry and Biotechnology* 45 (/46): 709–722.

Shields, P., Heller, M., Kite, D., and Bellof, B. (1995). Environmental accounting case study: DuPont. In: *Green Ledgers: Case Studies in Corporate Environmental Accounting* (eds. D. Ditz, J. Ranganathan and D. Banks). Washington, DC: World Resources.

Smook, G.A. (1992). *Handbook for Pulp and Paper Technologies*, 2e. Bellingham, WA: Angus Wilde Publications.

Stander, L. and Theodore, L. (2008). *Environmental Regulatory Calculations Handbook*. Hoboken, NJ: Wiley.

State of California (2001). South Coast Air Quality Management District Rule 1151 (amended December 1998) and Rule 1132 (adopted January 2001).

Title 40 Code of Federal Regulation (CFR) (2019). Part 51, Subpart 1, Review of New Sources and Modifications.

United State Congress (1990). Pollution Prevention Act of 1990. http://www.epa.gov/opptintr/p2home/p2policy/act1990.htm ().

United State Congress (1994). Office of Technology Assessment.

USEPA (1982). *Control and Treatment Technologies for the Metal Finishing Industry: In-Plant Changes*, EPA 625/ 8-82-008. USEPA, Office of Research and Development.

USEPA (1985). *Environmental Regulations and Technology: The Electroplating Industry*, EPA 625/ 10-85-001. USEPA, Office of Research and Development.

USEPA (1992a). *Facility Pollution Prevention Guide*, EPA/600/R-92/088. Washington, DC: Office of Research and Development.

USEPA (1992b). *Model Pollution Prevention Plan for the Kraft Segment of the Pulp and Paper industry*, EPA/910/9-92-030. Seattle, WA: EPA Region 10.

USEPA (1995a). Toxic Release Inventory (TRI) release data for fabricated metals facilities SIC 340. http://www.epa.gov/triexplorer (accessed 7 March 2019).

USEPA (1995b). 40CFR261 Identification and Listing of Hazardous Wastes. *Code of Federal Regulations*, US GPO. http://www.gpoaccess.gov/cfr/index.html (accessed 19 March 2019).

USEPA (1995c). *An Introduction to Environmental Accounting as a Business Management Tool: Key Concepts and Terms*, EPA 742-R-95-001. London: ACCA.

USEPA (1996). *Office of Air Quality Planning and Standards (OAQPS), Control Cost Manual*, EPA 453/B-96-001, 5e. Research Triangle Park, NC: USEPA, Office of Air Quality Planning and Standards.

USEPA (1998). Metal finishing/electroplating industry: case studies. http://es.epa.gov/cooperative/topics/metcasestudies.html (accessed 20 March 2019).

USEPA (2000). *Approaching Zero Discharge in Surface Finishing*, EPA/625/R-99/008. Research Triangle Park, NC: USEPA, Office of Research and Development.

USEPA (2001). *Living the Vision: Accomplishments of the National Metal Finishing Strategic Goals Program*, Research Triangle Park, NC. USEPA, Office of Research and Development. EPA 240-R-00-007. www.strategicgoals.org (accessed 20 March 2019).

USEPA (2003a). Environmental management systems: systematically improving your performance. *MetFin_BizCase_15*. http://www.epa.gov/ispd/metalfinishing/metfin_pdf/metfin_bizcase.pdf (accessed 22 March 2019).

USEPA (2004). *A Guide to Developing an Environmental Management System for Metal Finishing Facilities*. Washington, DC: USEPA, Sector Strategies. http://www.epa.gov/sectors/metalfinishing/metfin_ems.html (accessed 27 March 2019)

USEPA (2011). Clean Air Act, Section 812 Analyses on the Benefits and Costs of the Clean Air Act from 1990 to 2020. www.epa.gov/ora/sec812 (accessed 27 March 2019).

Vatavuk, W.M. (2002). *Control Cost Manual, Emissions Standards Division of the Office of Air Quality, Planning and Standards*, EPA/452/B-02-001. USEPA.

Wachs, I.E. (1999a). Treating methanol-containing waste gas streams. US Patent 5,907,066.

Wachs, I.E. (1999b). Production of formaldehyde from methyl mercaptans. US Patent 5,969,191.

Western Governors' Association (2004). Principles for environmental management in the west (3 December 2003). http://www.westgov.org/wga/policy (accessed 22 February 2004).

William, R.H. and Larson, E.D. (1993). Fossil energy. In: *Advances in Gasification-Based Biomass Generation, Renewable Energy: Source for Fuels and Electricity* (eds. T.B. Johnson, H. Kelly, A.K. Reddy and R.H. Williams). Washington, DC: Island Press.

8

Lean Manufacturing

Zero Defect and Zero Effect: Environmentally Conscious Manufacturing

8.1 Introduction

In this chapter our major focus is on lean manufacturing of various products while applying techniques and methodologies to achieve zero defect in products, and significantly eliminate waste and discharges to environment from the manufacturing process.

The quality of products, processes, and services has become a major decision factors in most industries and businesses today. Regardless of whether the consumer is an individual, a corporation, a military defense program, or a retail store, when the consumer is making purchase decisions, he or she is likely to consider quality to be equal in importance to cost and schedule. Consequently, *quality improvement* has become a major concern to many US industries and businesses. *Quality control and quality assurance* is a critical process of producing high-quality products that should be a top priority of any manufacturing firm in today's competitive environment. This also means designing and building products that conform to specifications and satisfy or exceed customer expectation. The American Society for Quality Control (ASQC) defines quality as "the totality of features and characteristics of a product or process or service that bear on its ability to satisfy given needs (Evans and Lindsay 1992). *Quality control (QC)* is a process by which entities review the quality of all factors involved in production. ISO 9000 defines quality control as "A part of quality management focused on fulfilling quality requirements" (ISO 2018a).

Quality means *fitness for use* with *zero defects or zero effects* in environmentally conscious manufacturing. For example, you or I may purchase automobiles that we expect to be free of manufacturing defects and that should provide reliable and economical transportation, a retailer buys finished goods with the expectation that they are properly packaged and arranged for easy storage and display, and a manufacturer buys raw material and expects to process it with no rework or scrap. In other words, all consumers expect that the products and services they buy will meet their requirements. Those requirements define fitness for use.

Quality or fitness for use is determined through the interaction of *quality of design* and *quality of conformance with quality control*. By quality of design for environment and other aspects, we mean the different grades or levels of performance, reliability, serviceability, and function that are the result of deliberate engineering and *environmental management* decisions. By quality of conformance, we mean systematic *reduction of variability and elimination of defects* until every, unit, batch, and product produced is identical in physical and chemical properties (*zero defect and zero effect*).

Some confusion exists in our society and manufacturing industry about quality improvement; some people still think that it means gold-plating a product or spending more money to develop a process or/and product. This thinking is wrong. Quality improvement means the *systematic elimination of waste*. Examples of waste include scrap and rework in manufacturing, inspection and test, errors on documents (such as engineering drawings, checks, purchase orders, and plans), customer complaint hot lines, warranty costs, and the time required to do things over again that could have been done right the first time. A successful quality-improvement effort can eliminate much of this waste and lead to lower costs, higher productivity, increased customer satisfaction, increased business reputation, higher market share, and ultimately higher profits for the company. One vice president of DuPont has said that

Industrial Environmental Management: Engineering, Science, and Policy, First Edition. Tapas K. Das.

Companion website: www.wiley.com/go/Das/IEM_1e

whenever the company eliminates a pound of waste, the material most likely ends up in a product. At DuPont, titanium dioxide wastes are converted to high-purity table salt, fertilizer, and food-grade carbon dioxide.

The lean manufacturing is a system whose techniques aim to significantly eliminate waste in the manufacturing process. The Toyota way or system is about Toyota culture in auto industry which focuses on the management principle of Toyota in a 4P model: Philosophy, Process, People, and Problem Solving – the foundation of which is a long-term philosophy that emphasizes on adding value to customers and society. Building on that is Toyota's investment in *lean processes*, which concentrates on shortening lead time by *eliminating waste*. Eliminating waste is done by people and partners using rigorous problem-solving methods, the two top layers of the 4P model. More on the Toyota management system and *lean manufacturing*, or simply lean, is described in Section 8.2.

8.2 Engineering Data Summary and Presentation

Before moving into the theory and practice of *statistical process control* (SPC), it is important that engineers get involved with exchanging ideas, and at each step of the way colleagues in engineering will be sharing information and data with one another, and engineers will be making presentations to decision makers who have no background in engineering. In those situations, the quality and clarity of displayed data will play a large role in statistical analysis and process control. Well-constructed graphics and data summaries are essential to good statistical thinking because they focus the engineer on important features of the data. They help the engineer make sense of the data and can provide insight about potential problem-solving approaches or the type of model should be used.

The computer has become an important tool in the presentation and analysis of data. Although many statistical techniques require only a handheld calculator, this approach can be time-consuming. Computers can perform the tasks more efficiently. Most statistical analysis is done using a prewritten library of statistical programs. The user enters the data and then selects the types of analysis and output displays that are of interest. Statistical software packages are available for both mainframe machines and personal computers. Among the most popular and widely used packages are Statistical Analysis System for both servers and personal computers (PCs) and Minitab for the PC. We can describe data feature numerically. For example, we can characterize the location or central tendency in the data by ordinary arithmetic average or mean. Because we almost always think of our data as a sample, we will refer to the arithmetic mean as the *sample mean* (Montgomery et al. 2011).

8.2.1 Sample Mean

If the n observations in a sample are denoted by $x_1, x_2, \ldots, x_n$, the sample mean is

$$\overline{x} = \frac{x_1 + x_2 + \cdots + x_n}{n} = \textstyle\sum_{i=1}^{n} x_i \tag{8.1}$$

The sample mean is the average value of all the observations in a given data set. Usually, these data are a sample of observations that have been selected from some large *population* of observations. We could also think of calculating the average value of all the observations in a population. This average is called the population mean, and it is denoted by the Greek letter μ (mu). When there is a finite number of observation (say, N) in the population, the population mean is

$$\mu = \frac{\sum_{i=1}^{N} x_i}{N} \tag{8.2}$$

Although the sample mean is useful, it does not convey all of the information about a sample of data. The variability or scatter in the data may be described by the *sample variance* or the *sample standard deviation*. If the n observations in a sample are denoted by x_1, x_2, ..., x_n, then the sample variance is

$$S^2 = \frac{\sum_{i=1}^{n} \left(x_i - \overline{x}\right)^2}{n-1} \tag{8.3}$$

Analogous to the sample variance S^2, there is a measure of variability in the population called *population variance*. We will use the Greek letter σ^2 (sigma squared) to denote the population variance. The positive square root of σ^2, or σ, will denote *population standard deviation*. When the population is finite and consists of N values, we may define the population variance as

$$\sigma^2 = \frac{\sum_{i=1}^{n} \left(x_i - \mu\right)^2}{N} \tag{8.4}$$

Example 8.1 An important quality characteristic of water is the concentration of suspended solid particle in mg/l. Twelve measurements on suspended solids from Wisconsin River by an industrial site are as follows: 42.4, 65.7, 29.8, 58.7, 52.1, 55.8, 57.0, 68.7, 67.3, 67.3, 54.3, and 54.0. Calculate the sample average and sample standard deviation.

Solution

Sample average: $\bar{x} = \dfrac{\sum^{n} x_i}{i=1} = \dfrac{\sum_{i=1}^{12} x_i}{12} = \dfrac{673.1}{12} = 56.09\ \text{mg/l}$

Sample standard deviation: $\sum_{i=1}^{12} x_i = 673.10; \sum_{i=1}^{12} x_i^2 = 39\ 178$

$$S = \sqrt{\frac{\sum_{i=1}^{n} x_i^2 - \frac{\left(\sum_{i=1}^{n} x_i\right)^2}{n}}{n-1}}$$

$$S = \sqrt{\frac{39\ 168 - \frac{(673.10)^2}{12}}{12-1}} = \sqrt{\frac{1\ 412.70}{11}}$$
$$= \sqrt{128.43} = 11.33 mg/1$$

Example 8.2 The following data are direct solar intensity measurements (W/m^2) on different days at a location in the State of Rajasthan, India: 562, 869, 708, 775, 775, 704, 809, 856, 655, 806, 878, 909, 918, 558, 768, 870, 918, 940, 946, 661, 820, 898, 935, 952, 957, 693, 835, 905, 939, 955, 960, 498, 653, 730, and 753.

Calculate the sample mean and sample standard deviation. Prepare a dot diagram of these data. Indicate where the sample mean falls on this diagram. Provide a practical interpretation of the sample mean.

Solution

Sample average:

$$\bar{x} = \frac{\sum_{i=1}^{35} x_i}{35} = \frac{28\ 368}{35} = 810.51\ \text{W/m}^2$$

Sample variance:

$$\bar{x} = \frac{\sum_{i=1}^{35} x_i}{35} = 28\ 368$$

$$\sum_{i=1}^{35} x_i^2 = 23\ 552\ 500$$

Sample standard deviation:

$$S^2 = \frac{\sum_{i=1}^{n} x_i^2 - \left(\sum_{i=1}^{n} x_i\right)^2}{n-1};$$

$$S = \sqrt{\frac{\sum_{i=1}^{n} x_i^2 - \frac{\left(\sum_{i=1}^{n} x_i\right)^2}{n}}{n-1}}$$

$$= \frac{23\ 552\ 500 - \frac{(28\ 368)^2}{35}}{35-1} = \frac{559\ 830.74}{34}$$
$$= 16\ 465.62\ \text{W}/\text{m}^2$$

Therefore, sample deviation, $S = \sqrt{16\ 465.62} = 128.32\ \text{W/m}^2$

The sample standard deviation could also be found using

$$S = \sqrt{\frac{\sum_{i=1}^{n} (x_i - \bar{x})^2}{n-1}}, \quad \text{where} \quad \sum_{i=1}^{35} (x_1 - \bar{x})^2 = 559\ 830.74$$

Dot diagram (rounding of the data is used to create the dot diagram)

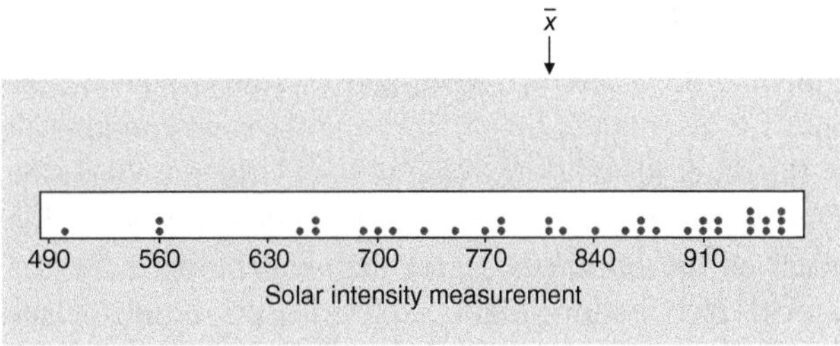

The sample mean is the point at which the data would balance if it were on a scale.

8.2.2 Stem-and-Leaf Diagram

The dot diagram is a useful data display for small samples, up to (say) about 20 observations. However, when the number of observations is moderately large, other graphical displays may be more useful. A *stem-and-leaf display* or *stem-and-leaf plot, also called a stemplot*, is a device for presenting quantitative data in a graphical format, similar to a histogram, to assist in visualizing the shape of a distribution. They evolved from Arthur Bowley's work in the early 1900s and are useful tools in exploratory data analysis. Stemplots became more commonly used in the 1980s after the publication of John Tukey's book on *exploratory data analysis* in 1977 (Tukey 1977).

A *stem-and-leaf display* often refers to another chart type. A simple stemplot may refer to plotting a matrix of y values onto a common x-axis, and identifying the common x value with a vertical line, and the individual y values with symbols on the line (Wild and Seber 2000).

Unlike histograms, stem-and-leaf displays retain the original data to at least two significant digits, and put the data in order, thereby easing the move to order-based inference and nonparametric statistics.

A basic stem-and-leaf display contains two columns separated by a vertical line. The left column contains the *stems* and the right column contains the *leaves*.

8.2.3 Constructing a Stem-and-Leaf Display

Example 8.3 To construct a stem-and-leaf display, the observations must first be sorted in ascending order: this can be done most easily if working by hand by constructing a draft of the stem-and-leaf display with the leaves unsorted, then sorting the leaves to produce the final stem-and-leaf display. Here is the sorted set of data values that will be used in the following example:

$$44, 46, 47, 49, 63, 64, 66, 68, 68, 72, 72, 75, 76, 81, 84, 88, 106$$

Next, it must be determined what the stems will represent and what the leaves will represent. Typically, the leaf contains the last digit of the number and the stem contains all of the other digits. In the case of very large numbers, the data values may be rounded to a particular place value (such as the hundreds place) that will be used for the leaves. The remaining digits to the left of the rounded place value are used as the stem. In this example, the leaf represents the ones place and the stem will represent the rest of the number (tens place and higher).

The stem-and-leaf display is drawn with two columns separated by a vertical line. The stems are listed to the left of the vertical line. It is important that each stem is listed only once and that no numbers are skipped, even if it means that some stems have no leaves. The leaves are listed in increasing order in a row to the right of each stem.

It is important to note that when there is a repeated number in the data (such as two 72s) then the plot must reflect such (so the plot would look like 7 | 2 2 5 6 7 when it has the numbers 72 72 75 76 77).

Stem	Leaf	Frequency
4	4 6 7 9	4
5		0
6	3 4 6 8 8	5
7	2 2 5 6	4
8	1 4 8	3
9		0
10	6	1

$$\text{Key}: 4\,|\,4 = 44$$

Rounding may be needed to create a stem-and-leaf display. Based on the following set of data, the stemplot below would be created:

$$-23.678\,758, -12.45, -3.4, 4.43, 5.5, 5.678, 16.87, 24.7, 56.8$$

For negative numbers, a negative is placed in front of the stem unit, which is still the value $X/10$. Non-integers are rounded. This allowed the stem-and-leaf plot to retain its shape, even for more complicated data sets as in this example below:

Stem	Leaf	Frequency
−2	3	1
−1	2	1
−0	3	1
0	4 6 6	3
1	7	1
2	5	1
3		0
4		0
5	7	1

$$\text{Key}: -1\,|\,2 = -12$$

The stem-and-leaf diagram displays make it relatively easy to find data features such as percentiles, quartiles, and median. The *median* is a measure of central tendency that divides the data into two equal parts, half below the median and half above. If the number of observations is even, the median is halfway between the two central values.

8.2.4 Application

Stem-and-leaf displays are useful for displaying the relative density and shape of the data, giving the reader a quick overview of distribution. They retain (most of) the raw numerical data, often with perfect integrity. They are also useful for highlighting outliers and finding the mode. However, stem-and-leaf displays are only useful for moderately sized data sets (around 15–150 data points). With very small data sets a stem-and-leaf displays can be of little use, as a reasonable number of data points are required to establish definitive distribution properties. A dot plot may be better suited for such data (Example 8.1). With very large data sets, a stem-and-leaf display will become very cluttered, since each data point must be represented numerically. A histogram or boxplot may become more appropriate as the data size increases.

8.2.5 Histogram

A *histogram* is a more compact summary of data than a stemplot. A *histogram* is an accurate representation of the distribution of numerical data. It is an estimate of the probability distribution of a continuous variable (quantitative variable) and was first introduced by Karl Pearson

(1895). It differs from a bar graph, in the sense that a bar graph relates two variables, but a histogram relates only one. To construct a histogram, the first step is to "bin" the range of values – that is, divide the entire range of values into a series of intervals – and then count how many values fall into each interval. The bins are usually specified as consecutive, nonoverlapping intervals of a variable. The bins (intervals) must be adjacent and are often (but are not required to be) of equal size (Howitt and Cramer 2008).

If the bins are of equal size, a rectangle is erected over the bin with height proportional to the frequency – the number of cases in each bin. A histogram may also be normalized to display "relative" frequencies. It then shows the proportion of cases that fall into each of several categories, with the sum of the heights equaling 1.

However, bins need not be of equal width; in that case, the erected rectangle is defined to have its *area* proportional to the frequency of cases in the bin (Freedman et al. 1997). The vertical axis is then not the frequency but *frequency density* – the number of cases per unit of the variable on the horizontal axis. Examples of variable bin width are displayed on Census bureau data (Freedman et al. 1997). As the adjacent bins leave no gaps, the rectangles of a histogram touch each other to indicate that the original variable is continuous (Stangor 2011).

Histograms give a rough sense of the density of the underlying distribution of the data, and often for density estimation: estimating the probability density function of the underlying variable. The total area of a histogram used for probability density is always normalized to 1. If the lengths of the intervals on the x-axis are all 1, then a histogram is identical to a relative frequency plot.

Once the number of bins and the lower and upper boundary of each bin have been determined, the data are sorted into the bins and a count is made of the number of observations in each bin. To construct the histogram, use the horizontal axis to represent the measurement scale for the data and the vertical scale to represent the counts, or frequencies. Sometimes the frequencies in each bin are divided by the total number of observations (n), and then the vertical scale of the histogram represents relative frequencies. Rectangles are drawn over each bin, and the height of each rectangle is proportional to frequency (or relative frequency). Most statistics packages construct histograms.

Example 8.4 Construct a cumulative frequency plot and histogram for the solar intensity data in Example 8.2. Use six bins.

Solution
For six bins

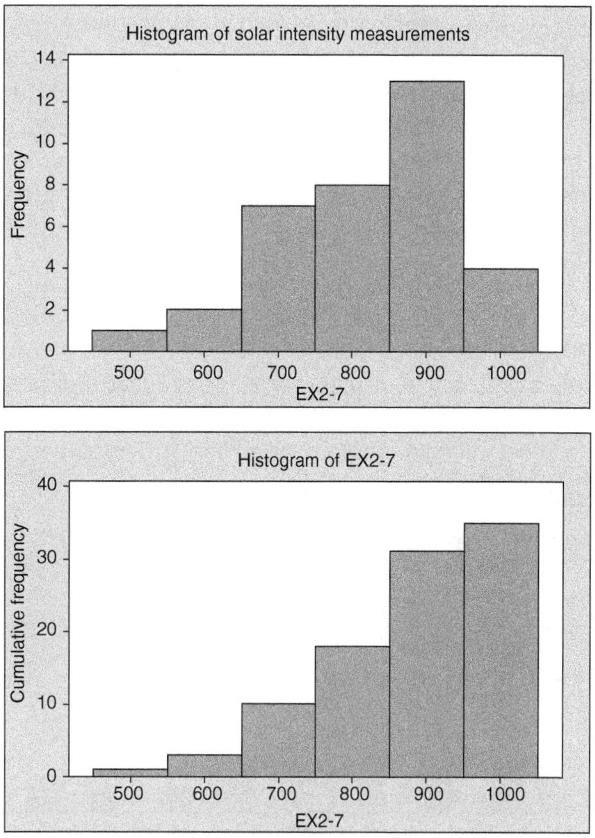

Histogram

Histogram

8.2.6 Pareto Diagram

A Pareto Diagram is an important variation of the histogram. A *Pareto Diagram* is named after the Italian economist Vilfredo Pareto, and they usually exhibit *Pareto principle*, that is, most of the defects can usually be accounted for by a few of the categories. The Pareto principle essentially states that 80% of the effect is caused by 20% of the causes (commonly called 80/20 rule or "the vital few and the insignificant many"). The Pareto chart organizes data to show which items or issues have the biggest impact on the process or system. Then, on the chart, we stratify the data to show the groups, starting with the largest and working down to the lowest number of items in each group. The idea is that by organizing the data in this format, we can develop a plan to work to work on problems that will give us the biggest return for our process improvement efforts.

The sole purpose of the Pareto chart is to highlight the most important among a (typically large) set of factors. Using the chart, one can identify the vital 20% factors causing the majority of the defects. In quality control, it often represents the most common sources of defects, the highest occurring type of defect, or the most frequent reasons for customer complaints, and so on (Montgomery 1985; Tague 2004). Wilkinson (2006) devised an algorithm for producing statistically based acceptance limits (similar to confidence intervals) for each bar in the Pareto chart. Pareto chart can also help in the SPC of zeroing the cause and effects in a manufacturing process toward achieving the ideal of "zero defect and zero effect."

Example 8.5 The following information on structural defects in automobile doors is obtained: dents, 4; pits, 4; parts assembled out of sequence, 6; parts under-trimmed, 21; missing holes/slots, 8; parts not lubricated, 5; parts out of contour, 30; and parts not deburred, 3. Construct and interpret a Pareto chart.

Solution

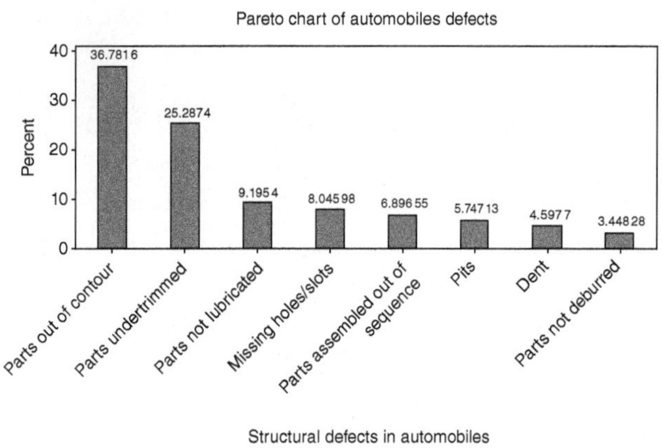

Figure 8.1 Pareto chart.

Roughly 63% of defects are described by parts out of contour and parts under trimmed.

8.2.7 Boxplots

A display that further summarizes information about the distribution of the values is the *boxplot*. Instead of plotting the actual values, a boxplot displays summary statistics for the distribution. It is a plot of the 25th, 50th, and 75th percentiles, as well as values far removed from the rest.

Figure 8.2 shows an annotated sketch of a boxplot. The lowest boundary of the box is the 25th percentile. Tukey (1977) refers to the 25th and 75th percentile "hinges." Note that the 50th percentile is the median of the overall data set, the 25th percentile is the median of those values below the median, and the 75th percentile is the median of those values above the median. The horizontal line inside the box

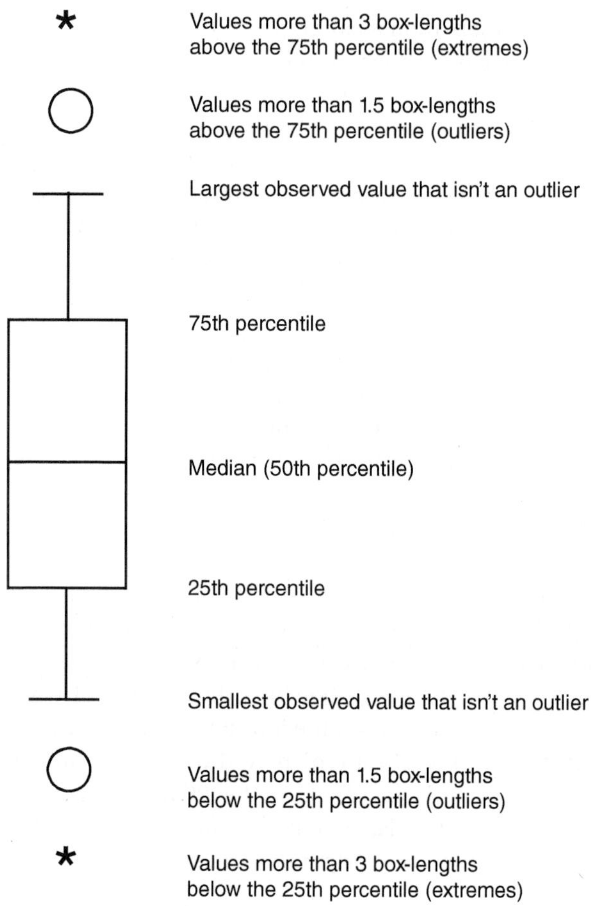

Figure 8.2 Annotated boxplot.

represents the median. Of the cases, 50% are included within the box. The box length corresponds to the interquartile range, which is the difference between the 25th and 75th percentiles.

The boxplot includes two categories of cases with outlying values. Cases with values that are more than 3 box lengths from the upper or lower edge of the box are called *extreme values*. On the boxplot, these are designed with an asterisk (*). Cases with values that are between 1.5 and 3 box-lengths from the upper or lower edge of the box are called *outliers* and are designed with a circle. The largest and smallest observed values that aren't outliers are also shown. Lines are drawn from the ends of the box to these values. (These lines are sometimes called *whiskers* and the plot is then called a *box-and-whiskers plot.*)

Despite its simplicity, the boxplot contains an impressive amount of information. From the median you can determine the central tendency, or location. From the length of the box, you can determine the spread, or variability, of your observations. If the median is not in the center of the box, you know that the observed values are skewed. If the median is closer to the top of the box than to the bottom, the opposite is true: the distribution is negatively skewed. The length of the tail is shown by the whiskers and the outlying and extreme points.

Example 8.6 Eighteen measurements of the disbursement rate (cm^3/s) of a chemical disbursement system are recorded and sorted:

6.50	6.77	6.91	7.38	7.64	7.74	7.90	7.91	8.21
8.26	8.30	8.31	8.42	8.53	8.55	9.04	9.33	9.36

a) Compute the sample mean and sample variance.
b) Find the sample upper and lower quartiles.
c) Find the sample median.
d) Construct a boxplot of the data.
e) Find the 5th and 95th percentiles of the inside diameter.

Solution

a) Sample mean: 8.059; Sample variance: 0.661
b) Q_1: 7.575, Q_3: 8.535
c) Sample median: 8.235

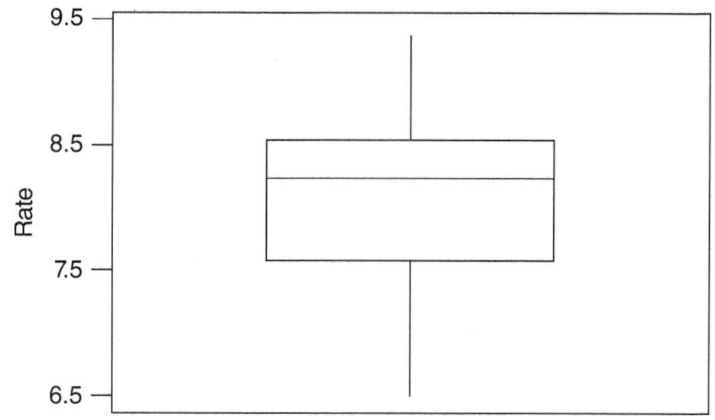

d) 5th percentile: 6.175; 95th percentile: 9.331.

8.2.8 Statistical Tools for Experimental Design: Process and Product Development

This section is about *SPC*, a collection of tools that are essential in quality-improvement activities. Statistical methods play a vital role in quality improvement. Some applications are outlined next:

1) In product and/or process design development, statistical methods, including designed experiments, can be used to compare different materials, components, or ingredients and to help determine both system and component tolerances. This application can significantly lower development costs and reduce development time.
2) Statistical method can be used to determine the capability of a manufacturing process. SPC can be used to systematically improve a process by reducing variability.
3) Experimental design methods can be used to investigate improvements in the process. These improvements can lead to higher yields and lower manufacturing costs.
4) Life testing provides reliability and other performance data about the product. This can lead to new and improved designs and products that have longer useful lives and lower operating and maintenance costs.

It is essential that engineers, scientists, managers, and policy makers have an in-depth understanding of these statistical tools in any industry or business that wants to be a high-quality, low-cost producer. In this section we provide an introduction to the basic methods of statistical quality control that, along with experimental design, form the basis of a successful quality-improvement effort.

SPC has its origins in the 1920s. Dr. Walter A. Shewhart of the Bell Telephone Laboratories was one of the early pioneers of the field. In 1924 he wrote a memorandum illustrating a control chart, one of the basic SPC tools. World War II saw the widespread dissemination of these methods to US industry. Dr. Edward Deming and Dr. Joseph Juan were instrumental in spreading the methodology after World War II.

SPC is a set of problem-solving tools that may be applied to any process. The following are the major tools of SPC:

1) Stem-and-leaf diagram
2) Histogram
3) Pareto chart
4) Cause-and-effect diagram
5) Defect-concentration diagram
6) Control chart
7) Scatter diagram
8) Check sheet

Although these tools are an important part of SPC, they comprise only the technical aspect of the subject. An equally important element of SPC is attitude – a desire of all individuals in the organization for continuous improvement in quality and productivity through the systematic reduction of variability. The control chart is the most powerful of the SPC tools. For complete discussion of these methods, see Montgomery (2009).

8.3 Time Series: Process over Time

Data are often collected in a manufacturing process over time in many practical situations. Phenomena that might affect the process or system often become more visible in a time-oriented plot of the data, and the stability of the process can be better judged. For example, the control chart is a technique that displays data over time and permits the engineer to assess the stability of a process.

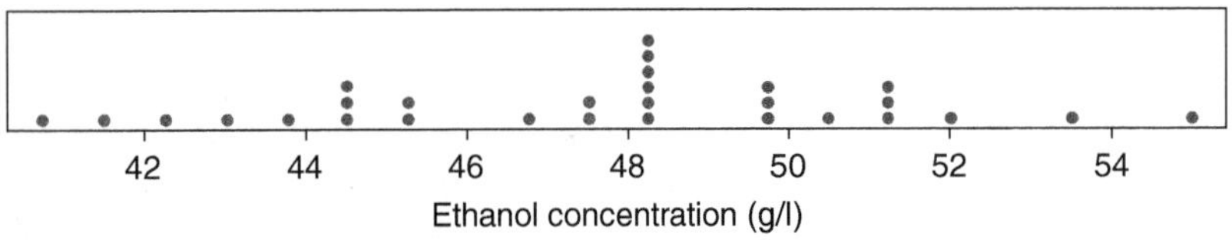

Figure 8.3 A dot diagram illustrates concentration variations but does not identify the problem.

Figure 8.3 is a dot diagram of ethanol concentration readings taken hourly from a binary distillation column in an industrial complex. The large variation displayed on the dot diagram indicates a possible problem, but the chart does not help explain the reason for the variation. Because the data are collected over time, they are called a time series. A graph of the data versus time, as shown in Figure 8.4, is called a time series plot. A possible downward shift in the process mean level is visible in the plot, and an estimate of the time of the shift can be obtained.

8.3.1 Basic Principles

In any production process, regardless of how well designed or carefully maintained it is, a certain amount of inherent or natural variability will always exist. This natural variability, or "background noise," is the cumulative effect of many small, essentially unavoidable causes. When the background noise in a process is relatively small, we usually consider it an acceptable level of process performance. In this framework of SPC, this natural variability is often called a "stable system of chance cause." A process that is operating with only *chance causes* of variation present is said to be in statistical control. In other words, the chance causes are an inherent part of the process.

Other kinds of variability may occasionally be present in the output of a process. This variability in key quality characteristics usually arises from these sources: improperly adjusted machines, operator errors, or defective raw materials. Such variability is generally large when compared to the background noise and it usually represents an unacceptable level of process performance. We refer to these sources of variability that are not part of the chance cause pattern as *assignable causes.* A process that is operating in the presence of assignable causes is said to be out of control. The terminology *chance and assignable causes* was developed by Dr. Walter Shewhart (Montgomery et al. 2011).

The control chart is a device for describing exactly what is meant by statistical control; as such, it may be used in a variety of ways. In many applications, it is used for online process monitoring. That is sample data are collected and used to construct the control chart, and if the sample values of $\bar{x}$ (say) fall within the control limits and do not exhibit any systematic pattern, we say the process is in control at the levels indicated by the chart. Note that we may be interested here in determining both whether the past data came from a process that was in control and whether future samples from this process indicate statistical control. The most important use of a control chart is to *improve* the process. This process-improvement activities using the control chart is illustrated in Figure 8.5. Note that (i) most processes do not operate in a state of statistical control; (ii) consequently, the routine and attentive use of control charts will identify assignable causes. If these causes can

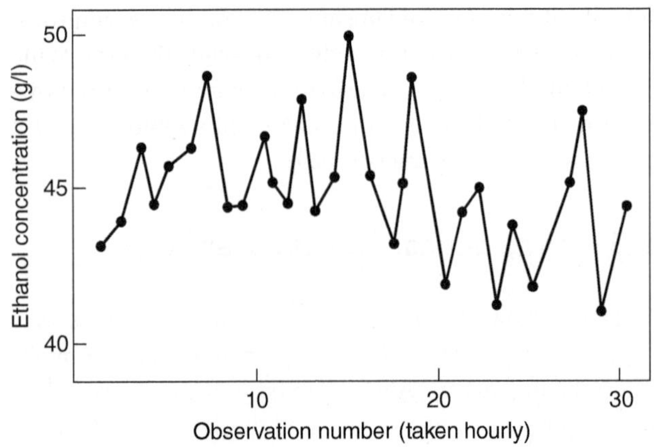

Figure 8.4 A time series plot of ethanol concentration provides more information than the dot diagram in Figure 8.3.

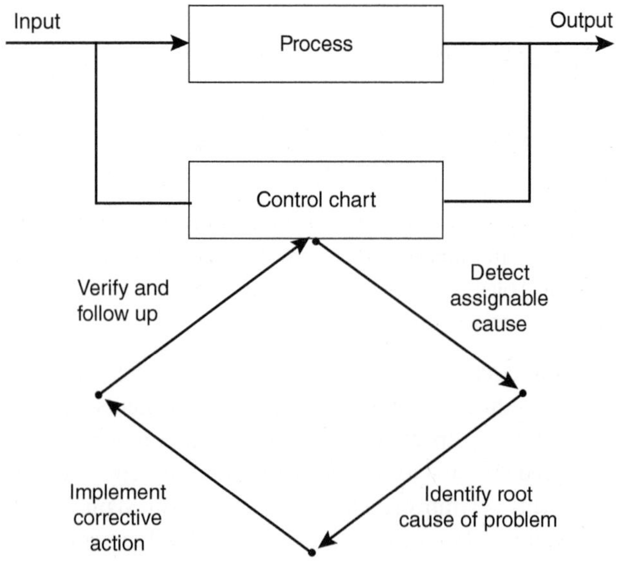

Figure 8.5 Schematic diagram of process improvement using the control chart.

be eliminated from the process, variability will be reduced and the process will be improved; (iii) the control chart will only detect assignable causes. Management, operator, and engineering action will usually be necessary to eliminate the assignable cause. An action plan for responding to control chart signals is vital (Montgomery et al. 2011).

In identifying and eliminating assignable causes, it is important to find the underlying *root cause* of the problem and to attack it. A cosmetic solution will not result in any real, long-term process improvement. Developing an effective system for corrective action is an essential component of an effective SPC implementation. The control chart may also be used as an *estimating device*. That is, from a control chart that exhibits statistical control, we may estimate certain parameters, such as the mean, standard deviation, and fraction nonconforming or fallout. These estimates may then be used to determine the *capability* of the process to produce acceptable products. Such *process capability (PC) studies* have considerable impact on many management decision problems that occur over the product cycle, including make-or-buy decisions, plant and process improvements that reduce process variability, and contractual agreement with customers or suppliers regarding product quality.

Control charts for quality characteristics are called *attributes control charts*. Control charts have had a long history of use in industry. There are at least five reasons for their popularity:

1) *Control charts are a proven technique for improving productivity*. A successful control chart program will reduce scrap and rework, which are the primary productivity killers in any operation. If you reduce scrap and rework, productivity increases, cost decreases, and production capacity (measured in the number of *good parts per hour*) increases.
2) *Control charts are effective in defect prevention*. The control chart helps keep the process in control, which is consistent with the "do it right the first time" philosophy. It is never cheaper to sort out the "good" units from the "bad" later on than it is to build it right initially. If you do not have effective process control, you are paying someone to make a nonconforming product.
3) *Control charts prevent unnecessary process adjustments*. A control chart can distinguish between background noise and abnormal variation; no other device, including a human operator, is as effective in making this distinction. If process operators adjust the process based on periodic tests unrelated to a control chart program, they will often overreact to the background noise and make unneeded adjustments. These unnecessary adjustments can usually result in a deterioration of process performance.
4) *Control charts provide diagnostic information*. Frequently, the pattern of points on the control chart will contain information that is of diagnostic value to an experienced operator or engineer. This information allows the operator to implement a change in the process that will improve the performance.
5) *Control charts provide information about PC*. The control chart provides information about the value of important process parameters and their stability over time, which allows an estimate of PC to be made. This information is of tremendous use to product and process designers.

Control charts are among the most effective management control tools, and they are as important as cost controls and material controls. Modern computer technology has made it easy to implement control charts in any type of process because data collection and analysis can be performed on a microcomputer or a local area network terminal in real time, online at work center (Montgomery et al. 2011).

8.4 Process Capability

In any manufacturing process and operation, variability exists in the process output. In a machining operation, which is one of the most accurate processes, the machined parts may appear to be identical, but close inspection reveals dimensional differences from one part to the next. Manufacturing variations can be divided into two types: *random and assignable*.

Random variations are caused by many factors: human variability with each operation cycle, variations in raw material, machine vibration, and so on. Individually, these factors may not amount to much, but collectively the errors can be significant enough to cause trouble unless they are within the tolerances for the part. Random variations typically form a normal statistical distribution. The output of the process tends to cluster about the mean value, in terms of the product's quality characteristic of interest (e.g., length, or diameter, or weight). A large proportion of the population of parts is centered around the mean, with fewer parts away from the mean. When the only variations in the process are of this type, the process is said to be in *statistical control*. This kind of variability will continue as long as the process is operating normally. It is when the process deviates from this normal operating condition that variations of the second type appear.

Assignable variations indicate an exception from normal operating conditions. Something has occurred in the process that is not accounted for by random variations. Reasons for assignable variations include operator mistakes, defective raw materials, tool failures, machine malfunctions, and so on. Assignable variations in manufacturing usually betray themselves by causing the output to deviate from the normal distribution. The process is no longer in statistical control.

PC relates to the normal variations inherent in the output when the process is in statistical control. By definition, *PC* equals ± 3 standard deviations about the mean output value (a total of six standard deviations):

$$PC = \mu \pm k\sigma \tag{8.5}$$

where PC, process capability; μ, process mean, which is set at the nominal value of the product characteristic; k is the "distance" of control limits from the center line, expressed in standard deviation units. A common choice is $k = 3$ and σ, standard deviation of the process. Assumptions underlying this definition are (i) the output is normally distributed and (ii) steady-state operation has been achieved and the process is in statistical control. Under these assumptions, 99.73% of the parts or products produced will have output values that fall with ± 3.0 of the mean. This general theory of control charts was first proposed by Dr. Walter A. Shewhart, and control charts developed according to these principles are often called *Shewhart control charts.*

8.4.1 Statistical Process Control

SPC involves the use of various statistical methods to assess and analyze variations in a process. SPC methods include simply keeping records of the production data, histograms, *PC*, and control charts. Control charts are the most widely used SPC method, and our discussion in this section will focus on them.

The underlying principle in control charts is that the variations in any process divide into two types (Section 8.4): (i) random variations, which are the variations present if the is in statistical control and (ii) assignable variations, which indicate a departure from statistical control. It is the objective of a control chart to identify when the process has gone out of statistical control, thus signaling that some corrective action should be taken.

A *control chart* is a graphical technique in which statistics computed from measured values of a certain process characteristic are plotted over time to determine if the process remains in statistical control. The general form of the control chart is illustrated in Figure 8.6. The chart consists of three horizontal lines that remain constant over time: a center, a lower control limit (LCL), and an upper control limit (UCL). The center is usually set at the normal design value. The UCL and LCL are generally set at ± 3 standard deviation of the sample means.

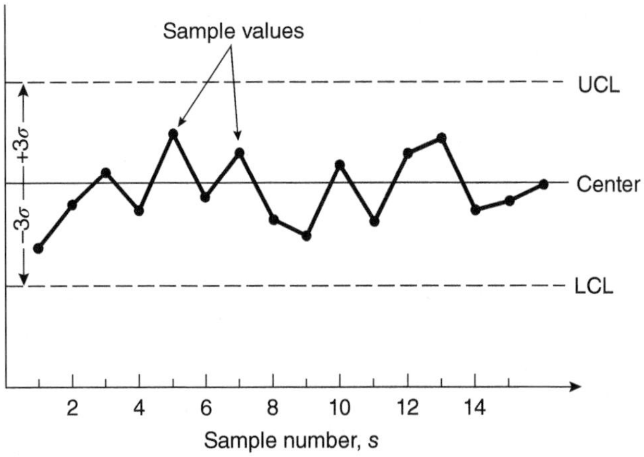

Figure 8.6 A control chart set at ±3 standard deviations of the sample means.

It is highly unlikely that a sample drawn from the process lies outside the UCL or LCL while the process is in statistical control. Thus, if it happens that a sample value does fall outside these limits, it is interpreted to mean that the process is out of control. Therefore, an investigation is undertaken to determine the reason for the out-of-control condition, with appropriate corrective action to eliminate the condition. By similar reasoning, if the process is found to be in statistical control and there is no evidence of undesirable trends in the data, then no adjustments should be made, since they would introduce an assignable variation to the process. The philosophy "if it ain't broken, don't fix it" is applicable in control charts.

There are two basic types of control charts: (i) control charts for variables and (ii) control charts for attributes. Control charts for variables require a measurement of the quality characteristic of interest. Control charts for attributes simply require a determination of whether a part is defective or how many defects there are in the sample.

8.4.2 Control Charts for Variables

A process that is out of statistical control manifests this condition in the form of significant changes in (i) process mean and/or (ii) process variability. Corresponding to these possibilities, there are two principal types of control charts for variables: (i) $\bar{x}$-chart and (ii) R-chart. The $\bar{x}$-chart (call it "x-bar chart") is used to plot the average measured value of a certain quality characteristics for each of a series of samples taken from the production process. It indicates how the process mean changes over time. The R-chart plots the range of each sample, thus monitoring the variability of the process and indicating whether it changes over time.

A suitable quality characteristic of the process must be selected as the variable to be monitored on the $\bar{x}$ and $\bar{R}$ charts. In a mechanical process, this might be a shaft diameter or other critical dimension. Measurements of the process itself must be used to construct the two control charts. With the process operating smoothly and absent of assignable variations, a series of samples (e.g. $m = 20$ or more is generally recommended) of small size (e.g., $n = 5$ parts per sample) are collected and the characteristics of interest is measured for each part. The following procedure is used to construct the center, LCL, and UCL, for each chart:

1) Compute the mean $\bar{x}$ and range R for each of the m samples.
2) Compute the grand mean, $\bar{\bar{x}}$, which is the $\bar{x}$ values for the m samples; this will be the center for the $\bar{x}$ chart.
3) Compute $\bar{R}$, which is the mean of the R values for the m samples; this will be the center for the R chart.
4) Determine the UCL and LCL for the $\bar{x}$ and R charts. Values of standard deviation could be estimated from the sample data and used to compute these control limits. However, an easier approach is based on the statistical factors tabulated in Table H.1 that have been derived specifically for these control charts. Values of the factors depends on sample size n. For the $\bar{x}$ chart.

$$\mathrm{LCL} = \bar{\bar{x}} - A_2\bar{R} \tag{8.6}$$

$$\mathrm{UCL} = \bar{\bar{x}} + A_2\bar{R} \tag{8.7}$$

And for the $\bar{R}$ chart,

$$\mathrm{LCL} = D_3\bar{R} \tag{8.8}$$

$$\mathrm{UCL} = D_4\bar{R} \tag{8.9}$$

Example 8.7 $\bar{x}$ and $\bar{R}$ Charts

Eight samples ($m = 8$) of size 4 ($n = 4$) have been collected from a manufacturing process that is in statistical control, and the dimension of interest has been measured for each part. It is desired to determine the values of the center, LCL, and UCL to construct the $\bar{x}$ and $\bar{R}$ charts. The calculated values of $\bar{x}$ and $\bar{R}$ for each sample are given below (measured values are in inches), which is step 1 in our procedure.

s	1	2	3	4	5	6	7	8
$\bar{x}$	2.008	1.998	1.993	2.002	2.001	1.995	2.004	1.999
$\bar{R}$	0.027	0.011	0.017	0.009	0.014	0.020	0.024	0.018

Solution

In step 1, we present constants for the $\bar{x}$ and $\bar{R}$ in Table 8.1, derived from Table H.1.

Table 8.1 Constants for the $\bar{x}$ and $\bar{R}$.

	$\bar{x}$ chart		$\bar{R}$ chart
Sample size, n	A_2	D_3	D_4
3	1.023	0	2.575
4	0.729	0	2.282
5	0.577	0	2.114
6	0.483	0	2.004
7	0.419	0.076	1.924
8	0.373	0.136	1.864
9	0.337	0.184	1.816
10	0.308	0.223	1.777

In step 2, we compute the grand mean of the sample averages.

$$\bar{\bar{x}} = (2.008 + 1.998 + \cdots + 1.999)/(8) = 2.000$$

In step 3, the mean value of $\bar{R}$ is computed.

$$\bar{R} = (0.027 + 0.011 + \cdots + 0.018)/(8) = 0.0175$$

In step 4, the values of LCL and UCL are determined based on factors in Table 8.1. First, using Eqs. (8.6) and (8.7) for the $\bar{x}$ chart,

$$\mathrm{LCL} = 2.000 - 0.729(0.0175) = 1.9872$$

$$\mathrm{ULC} = 2.000 + 0.729(0.0175) = 2.0128$$

And for the $\bar{R}$ chart using Eqs. (8.8) and (8.9)

$$\mathrm{LCL} = 0(0.0175) = 0$$

$$\mathrm{UCL} = 2.282(0.0175) = 0.0399$$

The two control charts are constructed in Figure 8.7 with the sample data plotted in the charts.

If the mean and standard deviation for the process were known, an alternative way to calculate the center, UCL, and LCL for the $\bar{x}$ chart would be as follows:

$$\mathrm{LCL} = \mu - 3\sigma/(\sqrt{n}) \tag{8.10}$$

$$\mathrm{UCL} = \mu + 3\sigma/(\sqrt{n}) \tag{8.11}$$

where μ, process mean; σ, standard deviation of the process, and n, sample size. The LCL and UCL values given by Eqs. (8.10) and (8.11) are theoretically the same values as those calculated by Eqs. (8.6)–(8.9). However, when first setting up the $\bar{x}$ chart for a process, the mean and standard deviation for the process variable of interest are generally not known. Accordingly, Eqs. (8.6)–(8.9) based on measured $\bar{x}$ and $\bar{R}$ values can be conveniently used to compute the control chart parameters. With the control limits set at

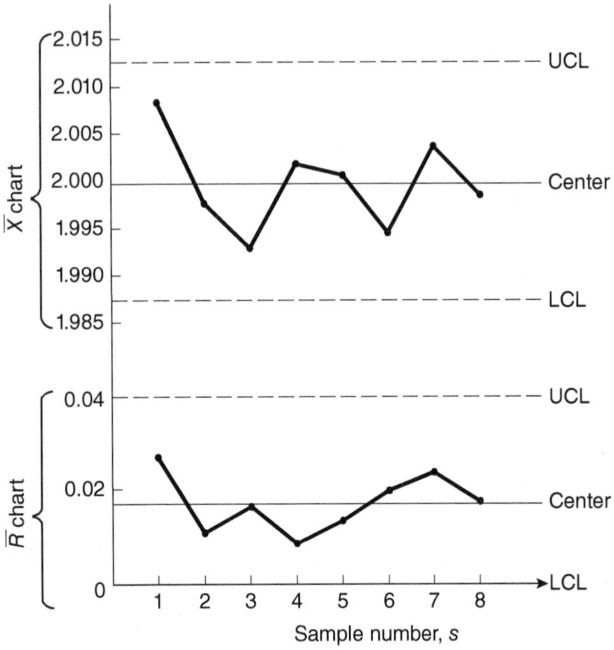

Figure 8.7 Control charts for Example 8.7.

the values defined in Eqs. (8.6)–(8.9) or Eqs. (8.10) and (8.11), 99.73% of the random samples drawn from a process that is in statistical control lie inside the control limits.

Note that the standard deviation of the sample means is related to the population standard deviation by the reciprocal of the square root of n, the number of units in the sample, that is,

$$\sigma_x = \sigma / \left(\sqrt{n}\right) \tag{8.12}$$

where σ_x, standard deviation of the sample mean and the other terms are as defined above.

8.4.3 PC Analysis

It is usually necessary to obtain some information about analysis of the PC – that is, the performance of the process when it is operating in control. PC is also a mathematical calculation to determine how the manufacturing process (*voice of the process*) compares with the engineering specifications. The intention is that the engineering specification will match the pressing needs and desires of the customers who use the products and services we produce (*voice of the customer*). There are several different calculations that have been developed over the years with the most popular ones being the *PC ratio* (C_p). The C_p can be expressed as an index that is defined as follows:

$$C_p = \left(\text{USL} - \text{LSL}\right) / 6\sigma \tag{8.13}$$

The numerator of C_p is the width of the specifications. The three-sigma limits on either side of the process mean are sometimes called natural tolerance limits because these represent limits that an in-control process should meet with most of the units produced. Consequently, six sigma is often referred to as the width of the process. C_p is the process potential calculation. C_p looks at the engineering specification limits as the numerator and the manufacturing process spread over six standard deviation value. If variable process behavior charts are being used in the manufacturing process, then the range value can be used to estimate the six standard deviations value and the engineering prints will contain the specifications that manufacturing is supposed to work with. The numerical value can never be negative. A value of $C_p = 1.0$ indicates that the tails of the six standard deviation calculation meet the engineering specification width.

8.5 Lean Manufacturing

Lean manufacturing or *lean production*, often simply *lean's* origins date back to the post World War II era in Japan. It was developed by Taiichi Ohno, a Toyota production executive, in response to a number of problems that plagued Japanese industry. The main problem was that of high-variety production required to serve the domestic Japanese market. Mass production techniques, which were developed by Henry Ford to economically produce long runs of identical product, were ill-suited to the situation faced by Toyota. Those conditions faced by Toyota in the late 1940s are common throughout industry and *lean* is being adopted by business all over the world as a way to improve efficiency and to serve customers better.

The Lean approach [the term Lean was first coined in the late 1980s by MIT researchers, John F. Krafcik (1988)] systematically minimizes waste – called *muda* – in the value stream. Krafcik's research was continued by the International Motor Vehicle Program at MIT, which produced the international best-selling book coauthored by Womack, Jones, and Roos entitled *The Machine That Changed the World* (1990).

Muda includes all types of defective work, not just defective products. Wasted time, motion, and materials are all *muda*. Ohno (1988) identified the following types of *muda* in business:

1) Defects
2) Overproduction
3) Inventories (in process or finished goods)
4) Unnecessary processing
5) Unnecessary movement of people
6) Unnecessary transport of goods
7) Waiting

Womack and Jones (2003) added another type of *muda*:

8) Designing goods and services that don't meet customers' needs.

To "think lean" is to declare war on *muda*. It is to focus on muda's opposite value. Defining values mean answering the questions:

- What are customers willing to pay for?
- By what processes are these values created?
- How does each activity in the process help meet the wants and needs of the customers?
- How can we make the value creation processes flow more efficiently?
- How can we be sure that we're producing only what is needed, when it's needed?
- How can we become perfect at creating value?

The Lean answers to these questions can be grouped into five categories: value, the vale stream, flow, pull, and perfection.

8.5.1 Overview

For many, lean is *the set of "tools" that assist in the identification and steady elimination of waste*. As waste is eliminated, quality improves while production time and cost are reduced. There is a second approach to lean manufacturing, which is promoted by Toyota, called the Toyota Production System (TPS) in which the focus is upon improving the "flow" or smoothness of work, thereby steadily eliminating *mura* ("unevenness") through the system and not upon "waste reduction" per se. Techniques to improve flow include production leveling, "pull" production (by means of *kanban*), and the *Heijunka box*. This is a fundamentally different approach from most improvement methodologies and requires considerably more persistence than basic application of the tools, which may partially account for its lack of popularity (Liker and Hoseus 2008). The difference between these two approaches is not the goal itself, but rather the prime approach to achieving it. The implementation of smooth flow exposes quality problems that already existed, and thus waste reduction naturally happens as a consequence. The advantage claimed for this approach is that it naturally takes a system-wide perspective, whereas a waste focus sometimes wrongly assumes this perspective.

Both lean and TPS can be seen as a closely connected set of potentially competing principles whose goal is cost reduction by the elimination of waste (Ohno 1988). These principles include pull processing, perfect first-time quality, waste minimization, continuous improvement, flexibility, building and maintaining a long-term relationship with suppliers, autonomation (smart automation), load leveling and production flow, and visual control. The disconnected nature of some of these principles perhaps springs from the fact that the TPS has grown pragmatically since 1948 as it responded to the problems it saw within its own production facilities. Thus, what one sees today is the result of a "need"-driven learning to improve where each step has built on previous ideas and not something based upon a theoretical framework.

Toyota's view is that the main method of lean is not the tools, but the reduction of three types of waste: *muda* (non-value-adding work), *muri* (overburden), and *mura* (unevenness), to expose problems systematically and to use the tools where the ideal cannot be achieved. From this perspective, the tools are workarounds adapted to different situations, which explains any apparent incoherence of the principles above (Ohno 1988).

Also known as the flexible mass production, the TPS has two pillar concepts: *Just-in-time* (JIT) or "flow," and "autonomation" (Ohno 1988). Adherents of the Toyota approach would say that the smooth flowing delivery of value achieves all the other improvements as side-effects. If production flows perfectly (meaning it is both "pull" and with no interruptions) then there is no inventory; if customer-valued features are the only ones produced, then product design is simplified and effort is only expended on features the customer values. The other of the two TPS pillars is the very human aspect of automation, whereby automation is achieved with a human touch. In this instance, the "human touch" means to automate so that the machines/systems are designed to aid humans in focusing on what the humans do best (Ohno 1988).

Lean implementation emphasizes the importance of optimizing work flow through strategic operational procedures while minimizing waste and being adaptable. Flexibility is required to allow production leveling (*Heijunka*) using tools such as *research and development* (R&D). However, adaptability is often constrained, and therefore may not require significant investment. More importantly, all of these concepts have to be acknowledged by employees who develop the products and initiate processes that deliver value. The cultural and managerial aspects of lean are arguably more important than the actual tools or methodologies of production itself. There are many examples of lean tool implementation without sustained benefit, and these are often blamed on weak understanding of lean throughout the whole organization.

Lean aims to enhance productivity by simplifying the operational structure enough to understand, perform, and manage the work environment. To achieve these three goals simultaneously, one of Toyota's mentoring methodologies (loosely called *Senpai* and *Kohai* which is Japanese for senior and junior) can be used to foster lean thinking throughout the organizational structure from the ground up. The closest equivalent to Toyota's mentoring process is the concept of *Lean Sensei*, which encourages companies, organizations, and teams to seek third-party experts that can provide advice and coaching (Womack and Jones 2003). Spear and Bowen (1999) identified four rules that characterizes the "Toyota DNA":

1) All work shall be highly specified as to content, sequence, timing, and outcome.
2) Every customer–supplier connection must be direct, and there must be an unambiguous yes or no way to send requests and receive responses.
3) The pathway for every product and service must be simple and direct.
4) Any improvement must be made in accordance with the scientific method, under the guidance of a teacher, at the lowest possible level in the organization.

8.5.2 History: Pre-Twentieth Century

8.5.2.1 Benjamin Franklin

Benjamin Franklin (Figure 8.8) contributed greatly to waste reduction thinking. Most of the basic goals of lean manufacturing and waste reduction were derived from Benjamin Franklin through documented examples. *Poor Richard's Almanac* says of wasted time, "He that idly loses 5s. worth of time, loses 5s., and might as prudently throw 5s. into the river." He added that avoiding unnecessary costs could be more profitable than increasing sales: "A penny saved is two pence clear. A pin a-day is a groat a-year. Save and have."

Again Franklin's *The Way to Wealth* says the following about carrying unnecessary inventory. "You call them goods; but, if you do not take care, they will prove evils to some of you. You expect they will be sold cheap, and, perhaps, they may [be bought] for less than they cost; but, if you have no occasion for them, they must be dear to you. Remember what Poor Richard says, 'Buy what thou hast no need of, and ere long thou shalt sell thy necessaries'. In another place he says, 'Many have been ruined by buying good penny worth'." Henry Ford cited Franklin as a major influence on his own business practices, which included *JIT* manufacturing.

The accumulation of waste and energy within the work environment was noticed by motion efficiency expert Frank Gilbreth, who witnessed the inefficient practices of masons who often bend over to gather bricks from the ground. The introduction of a non-stooping scaffold, which delivered the bricks at waist level, allowed masons to work about three times as quickly, and with the least amount of effort.

8.5.2.2 William Boeing

William E. Boeing (Figure 8.9) was born in Detroit, Michigan on 1 October 1881. Boeing moved to the Pacific Northwest at Hoquiam, Washington, and purchased extensive timberlands around Grays Harbor on the Pacific (west) side of the Olympic Peninsula and also bought into lumber operations. He made a success of the venture, in part by shipping lumber to the East Coast via the new Panama Canal, generating funds that he would later apply to a very

Figure 8.8 Benjamin Franklin. *Source:* https://commons.wikimedia.org/wiki/File:Benjamin-Franklin-U.S.-$100-bill.jpg. Public domain.

Figure 8.9 William (Bill) Boeing. *Source:* https://en.wikipedia.org/wiki/William_Boeing.

different business. William Boeing left Yale University in 1903 to take advantage of opportunities in the risky and cyclical, but financially rewarding, Northwest timber industry. That experience would serve him well in aviation.

In 1916, Boeing went into business with George Westervelt as "B & W" and founded the Pacific Aero Products Co. The company's first plane was the Boeing Model 1 (B & W Seaplane). When America entered the First World War on 8 April 1917, a little more than a month later, Boeing changed the name from Pacific Aero Products Co. to Boeing Airplane Company and obtained orders from the US Navy for 50 planes. At the end of the War, Boeing began to concentrate on commercial aircraft. He secured contracts to supply airmail service and built a successful airmail operation and later passenger service that evolved into United Airlines.

Under his guidance, a tiny airplane-manufacturing company grew into a huge corporation of related industries. When post-Depression legislation in 1934 mandated the dispersion of the corporation, Boeing sold his interests in the Boeing Airplane Co., but he continued to work on other business ventures.

He became one of America's most successful breeders of thoroughbred horses. He never lost his interest in aviation, and during World War II he volunteered as a consultant to the company. He lived until 1956, long enough to see the company he started enter the jet age.

William E. Boeing was a private person, a visionary, a perfectionist, and a stickler for the facts. The wall of his outer office bore a placard that reads: "2329 Hippocrates said: 1. There is no authority except facts. 2. Facts are obtained by accurate observation. 3. Deductions are to be made only from facts. 4. Experience has proved the truth of these rules." According to his son, William Boeing, Jr., Boeing was a fast and avid reader and remembered everything he read. He was also a perfectionist. While visiting his airplane building shop at the Duwamish shipyard in 1916, Boeing saw a set of improperly sawed spruce ribs. He brushed them to the floor and walked all over them until they were broken. A frayed aileron cable caused him to remark, "I, for one, will close up shop rather than send out work of this kind."

8.5.2.3 Frederick Winslow Taylor

Frederick Winslow Taylor (Figure 8.10), the father of scientific management, introduced what are now called standardization and best practice deployment. In *Principles of Scientific Management* (Taylor 1911), Taylor said: "And whenever a workman proposes an improvement, it should be the policy of the management to make a careful analysis of the new method, and if necessary conduct a series of experiments to determine accurately the relative merit of the new suggestion and of the old standard. And whenever the new method is found to be markedly superior to the old, it should be adopted as the standard for the whole establishment."

Figure 8.10 Frederick Winslow Taylor. *Source:* https://commons.wikimedia.org/wiki/File:Frederick_Winslow_Taylor.jpg. Public domain.

Taylor also warned explicitly against cutting piece rates (or, by implication, cutting wages or discharging workers) when efficiency improvements reduce the need for fresh labor: "...after a workman has had the price per piece of the work he is doing lowered two or three times as a result of his having worked harder and increased his output, he is likely entirely to lose sight of his employer's side of the case and become imbued with a grim determination to have no more cuts if soldiering [marking time, just doing what he is told] can prevent it."

Shigeo Shingo, the best-known exponent of single minute exchange of die and error-proofing or poka-yoke, cites *Principles of Scientific Management* as his inspiration (Shigeo Shingo 1987). American industrialists recognized the threat of cheap offshore labor to American workers during the 1910s, and explicitly stated the goal of what is now called lean manufacturing as a countermeasure. Henry Towne, past President of the American Society of Mechanical Engineers, wrote in the Foreword to Frederick Winslow Taylor's *Shop Management* (1911): "We are justly proud of the high wage rates which prevail throughout our country, and jealous of any interference with them by the products of the cheaper labor of other countries." To maintain this condition, to strengthen our control of home markets, and, above all, to broaden our opportunities in foreign markets where we must compete with the products of other industrial nations, we should welcome and encourage every influence tending to increase the efficiency of our productive processes.

8.5.2.4 Henry Ford

Henry Ford (Figure 8.11) initially ignored the impact of waste accumulation while developing his mass assembly manufacturing system. Charles Buxton Going wrote in 1915: "Ford's success has startled the country, almost the world, financially, industrially, mechanically. It exhibits in higher degree than most persons would have thought possible the seemingly contradictory requirements of true efficiency, which are: constant increase of quality, great increase of pay to the workers, repeated reduction in cost to the consumer. And with these appears, as at once cause and effect, an absolutely incredible enlargement of output reaching something like one hundredfold in less than ten years, and an enormous profit to the manufacturer" (Arnold and Faurote 1915).

Ford, in *My Life and Work* (Ford and Crowther 1922), provided a single-paragraph description that encompasses the entire concept of waste:

Figure 8.11 Henry Ford. *Source:* https://simple.wikipedia.org/wiki/Henry_Ford. Public domain.

> I believe that the average farmer puts to a really useful purpose only about 5% of the energy he expends. ... Not only is everything done by hand, but seldom is a thought given to a logical arrangement. A farmer doing his chores will walk up and down a rickety ladder a dozen times. He will carry water for years instead of putting in a few lengths of pipe. His whole idea, when there is extra work to do, is to hire extra men. He thinks of putting money into improvements as an expense.... It is waste motion—waste effort—that makes farm prices high and profits low.

Poor arrangement of the workplace – a major focus of the modern kaizen – and doing a job inefficiently out of habit – are major forms of waste even in modern workplaces.

Ford also pointed out how easy it was to overlook material waste. A former employee, Harry Bennett, wrote: "One day when Mr. Ford and I were together he spotted some rust in the slag that ballasted the right of way of the D. T. & I [railroad]. This slag had been dumped there from our own furnaces. 'You know,' Mr. Ford said to me, 'there's iron in that slag'. You make the crane crews who put it out there sort it over, and take it back to the plant" (Bennett and Marcus 1951).

In other words, Ford saw the rust and realized that the steel plant was not recovering all of the iron.

Ford's early success, however, was not sustainable. As James Womack and Daniel Jones pointed out in "Lean Thinking," what Ford accomplished represented the "special case" rather than a robust lean solution (Womack and Jones 2003). The major challenge that Ford faced was that his methods were built for a steady-state environment, rather than for the dynamic conditions firms increasingly face today (Ruffa 2008). Although his rigid, top-down controls made it possible to hold variation in work activities down to very low levels, his approach did not respond well to uncertain, dynamic business conditions; they responded particularly badly to the need for new product innovation. This was made clear by Ford's precipitous decline when the company was forced to finally introduce a follow-on to the Model T.

Design for manufacture (DFM) is a concept derived from Ford which emphasizes the importance of standardizing individual parts as well as eliminating redundant components in *My Life and Work* (Ford and Crowther 1922). This standardization was central to Ford's concept of mass production, and the manufacturing "tolerances," or upper and lower dimensional limits that ensured interchangeability of parts became widely applied across manufacturing. Decades later, the renowned Japanese quality guru, Genichi Taguchi, demonstrated that this "goal post" method of measuring was inadequate. He showed that "loss" in capabilities did not begin only after exceeding these tolerances, but increased as described by the Taguchi Loss Function at any condition exceeding the nominal condition. This became an important part of W. Edwards Deming's quality movement of the 1980s, later helping to develop improved understanding of key areas of focus such as cycle time variation in improving manufacturing quality and efficiencies in aerospace and other industries.

While Ford is renowned for his production line, it is often not recognized how much effort he put into removing the fitters' work to make the production line possible. Previous to the use, Ford's car's components were fitted and reshaped by a skilled engineer at the point of use, so that they would connect properly. By enforcing very strict specification and quality criteria on component manufacture, he eliminated this work almost entirely, reducing manufacturing effort by between 60 and 90% (Hounshell 1984). However, Ford's mass production system failed to incorporate the notion of "pull production" and thus often suffered from overproduction.

8.5.3 Toyota Develops TPS

8.5.3.1 Sakichi Toyoda

Toyota's development of ideas that later became lean may have started at the turn of the twentieth century with Sakichi Toyoda (Figure 8.12), in a textile factory with looms that stopped themselves when a thread broke. This became the seed of autonomation and *Jidoka*. Toyota's journey with *JIT* may have started back in 1934 when it moved from textiles to produce its first car. Kiichiro Toyoda, founder of Toyota Motor Corporation, directed the engine casting work and discovered many problems in their manufacturing. He decided he must stop the repairing of poor quality by intense study of each stage of the process. In 1936, when Toyota won its first truck contract with the Japanese government, his processes hit new problems and he developed the "Kaizen" improvement teams.

Levels of demand in the postwar economy of Japan were low and the focus of mass production on lowest cost per item via economies of scale therefore had little application. Having visited and seen supermarkets in the United States, Taiichi Ohno (1988) recognized the scheduling of work should not be driven by sales or production targets but by actual sales. Given the financial situation during this period, overproduction had to be avoided and thus the notion of pull (build to order rather than target driven push) came to underpin production scheduling.

It was with Taiichi Ohno at Toyota that these themes came together. He built on the already existing internal schools of thought and spread their breadth and use into

Figure 8.12 Sakichi Toyoda. *Source:* https://commons.wikimedia.org/wiki/File:Sakichi_TOYODA.jpg.

what has now become the *Toyota Production System* (TPS). It is principally from the TPS (which was widely referred to in the 1980s as *JIT manufacturing*), but now including many other sources, that lean production is developing. Norman Bodek wrote the following in his foreword to a reprint of Ford's *Today and Tomorrow*.

Toyoda was first introduced to the concepts of JIT and the TPS in 1980. Subsequently, Toyoda had the opportunity to witness its actual application at Toyota on one of our numerous Japanese study missions. There he met Mr. Taiichi Ohno, the system's creator. When bombarded with questions from our group on what inspired his thinking, he just laughed and said he learned it all from Henry Ford's book. The scale, rigor, and continuous learning aspects of TPS have made it a core concept of lean.

8.5.4 Tata Group

Jamsetji Tata, in full *Jamsetji Nusserwanji Tata* (born 3 March 1839, Navsari, Gujarat, India – died 19 May 1904, Bad Nauheim, Germany), Indian philanthropist and entrepreneur founded the Tata Group (Figure 8.13). His ambitious endeavors helped catapult India into the league of industrialized countries.

Figure 8.13 Jamsetji Tata. *Source:* https://en.wikipedia.org/wiki/File:JNTata.jpg. Public domain.

Born into a Parsi family, Jamsetji was the first child and only son of Nusserwanji Tata. After graduating from Elphinstone College, Bombay (now Mumbai), in 1858, he joined his father's export-trading firm and helped establish its branches in Japan, China, Europe, and the United States. In 1868 Jamsetji founded a trading company that later evolved into the Tata Group. In 1872 he focused on cotton manufacturing and subsequently founded mills at Nagpur, Bombay, and Coorla. His enterprises were noted for efficiency, for improved labor-protection policies, and for the introduction of finer grades of fiber. He also planned for the Bombay-area hydroelectric power plants that became the Tata Power Company in 1906.

In 1901 Jamsetji began organizing India's first large-scale ironworks, and six years later these were incorporated as the Tata Iron and Steel Company (now Tata Steel). Under the direction of his sons, Sir Dorabji Jamsetji Tata (1859–1932) and Sir Ratanji Tata (born in 1937 – retired in 2012), the Tata Iron and Steel Company became the largest privately owned steelmaker in India and the nucleus of a group of companies producing not only textiles, steel, and hydroelectric power but also chemicals, agricultural equipment, trucks, locomotives, and cement. Jamsetji's other commercial ventures included the Taj Mahal Palace, the first luxury hotel in India. After Jamsetji's death in 1904, his family retained control of the Tata Group and built it into a global conglomerate that by the early twenty-first century included more than 100 companies.

A noted philanthropist, Jamsetji established the J.N. Tata Endowment in 1892, which encouraged Indian students to pursue higher education. In 1898 he donated land for a research institute in Bangalore (now Bengaluru), and his sons ultimately established (1911) the Indian Institute of Science there. The Tata family went on to become perhaps the most important private funder of technical education and scientific research in India.

8.6 Types of Waste

Although the elimination of waste may seem like a simple and clear subject, it is noticeable that waste is often very conservatively identified. This then hugely reduces the potential of such an aim. The elimination of waste is the goal of lean, and Toyota defined three broad types of waste: *muda*, *muri,* and *mura*; for many lean implementations this list shrinks to the first waste type only with reduced corresponding benefits.

To illustrate the state of this thinking, Shigeo Shingo observed that only the last turn of a bolt tightens it – the rest is just movement. This ever finer clarification of waste is key to establishing distinctions between value-adding activity, waste, and non-value-adding work (Ohno 1988). Non-value adding work is waste that must be done under the present work conditions. One key is to measure, or estimate, the size of these wastes, to demonstrate the effect of the changes achieved and therefore the movement toward the goal.

The "flow" (or smoothness) based approach aims to achieve JIT, by removing the variation caused by work scheduling and thereby provide a driver, rationale, or target and priorities for implementation, using a variety of techniques. The effort to achieve JIT exposes many quality problems that are hidden by buffer stocks; by forcing smooth flow of only value-adding steps, these problems become visible and must be dealt with explicitly.

Muri is all the unreasonable work that management imposes on workers and machines because of poor organization, such as carrying heavy weights, moving things around, dangerous tasks, even working significantly faster than usual. It is pushing a person or a machine beyond its natural limits. This may simply be asking a greater level of performance from a process than it can handle without taking shortcuts and informally modifying decision criteria. Unreasonable work is almost always a cause of multiple variations.

To link these three concepts is simple in TPS and thus lean. Firstly, *muri* focuses on the preparation and planning of the process, or what work can be avoided proactively by design. Next, *mura* then focuses on how the work design is implemented and the elimination of fluctuation at the scheduling or operations level, such as quality and volume. *Muda* is then discovered after the process is in place and is dealt with reactively. It is seen through variation in output. It is the role of management to examine the *muda*, in the processes and eliminate the deeper causes by considering the connections to the *muri* and *mura* of the system. The *muda* and *mura* inconsistencies must be fed back to the *muri*, or planning, stage for the next project.

A typical example of the interplay of these wastes is the corporate behavior of "making the numbers" as the end of a reporting period approaches. Demand is raised to "make plan," increasing (*mura*), when the "numbers" are low, which causes production to try to squeeze extra capacity from the process, which causes routines and standards to be modified or stretched. This stretch and improvisation leads to *muri*-style waste, which leads to downtime, mistakes and back flows, and waiting, thus the muda of waiting, correction, and movement.

The original seven *mudas* (TIM WOOD) are as follows:

- **T**ransport (moving products that are not actually required to perform the processing)
- **I**nventory (all components, work in process, and finished product not being processed)
- **M**otion (people or equipment moving or walking more than is required to perform the processing)
- **W**aiting (waiting for the next production step, interruptions of production during shift change)
- **O**verproduction (production ahead of demand)
- **O**ver processing (resulting from poor tool or product design creating activity)
- **D**efects (the effort involved in inspecting for and fixing defects) (Womack and Jones 2003).

Eventually, an eighth "muda" was defined by Womack et al. (1990); it was described as manufacturing goods or services that do not meet customer demand or specifications. Many others have added the "waste of unused human talent" to the original seven wastes. For example, six sigma includes the waste of skills, referred to as "underutilizing capabilities and delegating tasks with inadequate training." Other additional wastes added were, for example, "space." These wastes were not originally a part of the seven deadly wastes defined by Taiichi Ohno in TPS, but they were found to be useful additions in practice. In 1999 Geoffrey Mika in his book, "Kaizen Event Implementation Manual" added three more forms of waste that are now universally accepted: the waste associated with working to the wrong metrics or no metrics, the waste associated with not utilizing a complete worker by not allowing them to contribute ideas and suggestions and be part of Participative Management, and lastly the waste attributable to improper use of computers; not having the proper software, training on use, and time spent surfing, playing games, or just wasting time. For a complete listing of the "old" and "new" wastes, see Bicheno and Holweg (2009).

The identification of non-value-adding work, as distinct from wasted work, is critical to identifying the assumptions behind the current work process and to challenging them in due course (Ohno 1988). Breakthroughs in SMED and other process changing techniques rely upon clear identification of where untapped opportunities may lie if the processing assumptions are challenged.

8.7 Six Sigma in Industry

Sigma (σ) denotes the standard deviation of a population. Six Sigma is the term used in industry to describe a process that has no more than 3.4 defects out of a million. The reference to Six Sigma suggests six standard deviations away from the center of a normal distribution, but the assumption of a perfectly stable process is placed with the assumption of a process that drifts slightly, so the defects per million. This defect rate is so low that defects are considered to be almost eliminated. Adopting the Six Sigma goal of quality, Motorola saved more than \$950 million in years. AlliedSignal reported a savings of \$1.5 billion. GE, Polaroid, and Texas Instruments are other major companies that have adopted the Six Sigma goal.

8.8 Lean Implementation Develops from TPS

8.8.1 Lean Leadership

The role of the leaders within the organization is the fundamental element of sustaining the progress of lean thinking. Experienced kaizen members at Toyota, for example, often bring up the concepts of *Senpai*, *Kohai*, and *Sensei*, because they strongly feel that transferring of Toyota culture down and across Toyota can only happen when more experienced Toyota Sensei continuously coach and guide the less experienced lean champions.

One of the dislocation effects of lean is in the area of *key performance indicators* (KPIs). The KPIs by which a plant/facility are judged will often be driving behavior, because the KPIs themselves assume a particular approach to the work being done. This can be an issue where, for example a truly lean, *Fixed Repeating Schedule* and JIT approach is adopted, because these KPIs will no longer reflect performance, as the assumptions on which they are based become invalid. It is a key leadership challenge to manage the impact of this KPI chaos within the organization.

Similarly, commonly used accounting systems developed to support mass production are no longer appropriate for companies pursuing lean. Lean accounting provides truly lean approaches to business management and financial reporting.

After formulating the guiding principles of its lean manufacturing approach in the TPS, Toyota formalized in 2001 the basis of its lean management: the key managerial values and attitudes needed to sustain continuous improvement in the long run. These core management principles are articulated around the twin pillars of Continuous Improvement (relentless elimination of waste) and Respect for People (engagement in long-term relationships based on continuous improvement and mutual trust).

This formalization stems from problem solving. As Toyota expanded beyond its home base for the past 20 years, it hit the same problems in getting TPS properly applied that other Western companies have had in copying TPS. Like any other problem, it has been working on trying a series of countermeasures to solve this particular concern. These countermeasures have focused on culture: how people behave, which is the most difficult challenge of all. Without the proper behavioral principles and values, TPS can be totally misapplied and fail to deliver results. As with TPS, the values had originally been passed down in a master-disciple manner, from boss to subordinate, without any written statement on the way. Just as with TPS, it was internally argued that formalizing the values would stifle them and lead to further misunderstanding. However, as Toyota veterans eventually wrote down the basic principles of TPS, Toyota set to put the Toyota Way into writing to educate new joiners.

Continuous Improvement breaks down into three basic principles:

1) *Challenge*: Having a long-term vision of the challenges, one needs to face to realize one's ambition (what we need to learn rather than what we want to do and then having the spirit to face that challenge). To do so, we have to challenge ourselves every day to see if we are achieving our goals.
2) *Kaizen*: Good enough never is, no process can ever be thought perfect, so operations must be improved continuously, striving for innovation and evolution.
3) *Genchi Genbutsu*: Going to the source to see the facts for oneself and make the right decisions, create consensus, and make sure goals are attained at the best possible speed.

Respect for People is less known outside of Toyota and essentially involves two defining principles:

1) *Respect*: Taking every stakeholders' problems seriously, and making every effort to build mutual trust. Taking responsibility for other people reaching their objectives.
2) *Teamwork*: This is about developing individuals through team problem-solving. The idea is to develop and engage people through their contribution to team performance. Shop floor teams, the whole site as team, and team Toyota at the outset.

8.8.2 Differences from TPS

While lean is seen by many as a generalization of the TPS into other industries and contexts, there are some acknowledged differences that seem to have developed in implementation.

1) *Seeking profit* is a relentless focus for Toyota exemplified by the profit maximization principle (price − cost = profit) and the need, therefore, to practice systematic cost reduction (through TPS or otherwise) to realize benefit. Lean implementations can tend to de-emphasize this key measure and thus become fixated with the implementation of improvement concepts of "flow" or "pull." However, the emergence of the "value curve analysis" promises to directly tie lean improvements to bottom-line performance measurements.
2) *Tool orientation* is a tendency in many programs to elevate mere tools [standardized work, value stream mapping (VSM), visual control, etc.] to an unhealthy status beyond their pragmatic intent. The tools are just different ways to work around certain types of problems, but they do not solve them for you or always highlight the underlying cause of many types of problems. The tools employed at Toyota are often used to expose particular problems that are then dealt with, as each tool's limitations or blind spots are perhaps better understood. So, for example, value stream mapping focuses upon material and information flow problems (a title built into the Toyota title for this activity) but is not strong on Metrics, Man, or Method. Internally, they well know the limits of the tool and understood that it was never intended as the best way to see and analyze every waste or every problem related to quality, downtime, personnel development, cross training–related issues, capacity bottlenecks, or anything to do with profits, safety, metrics or morale, etc. No one tool can do all of that. For surfacing these issues, other tools are much more widely and effectively used.
3) *Management technique rather than change agents* has been a principle in Toyota from the early 1950s when they started emphasizing the development of the production manager's and supervisors' skills set in guiding natural work teams and did not rely upon staff-level change agents to drive improvements. This can manifest itself as a "Push" implementation of lean rather than "Pull" by the team itself. This area of skills development is not that of the change agent specialist but that of the

natural operations work team leader. Although less prestigious than the TPS specialists, development of work team supervisors in Toyota is considered an equally, if not, more important topic merely because there are tens of thousands of these individuals. Specifically, it is these manufacturing leaders that are the main focus of training efforts in Toyota since they lead the daily work areas, and they directly and dramatically affect quality, cost, productivity, safety, and morale of the team environment. In many companies implementing lean, the reverse set of priorities is true. Emphasis is put on developing the specialist, while the supervisor skill level is expected to somehow develop over time on its own.

4) *Lack of understanding* is one of the key reasons that a large share of lean manufacturing projects in the West fail to bring any benefit. In *Factory Physics*, Hopp and Spearman describe this as *romantic JIT*, where the belief in the methods is more important than the actual understanding and results. In this aspect, lean manufacturing is more of a religion than a science. Others have compared it to cargo cult science.

8.8.3 Lean Services

Lean principles have been successfully applied to various sectors and services. For example, call center services use these principles to improve live agent call handling. By combining agent-assisted automation and lean's waste reduction practices, a company reduced handle time, reduced between agent variability, reduced accent barriers, and attained near perfect process adherence (Padmane et al. 2013).

Lean principles also have applications to software development and maintenance as well as other sectors of *information technology* (IT) (Hanna 2007). More generally, the use of lean in IT has become known as Lean IT. Lean methods are also applicable to the public sector, but most results have been achieved using a much more restricted range of techniques than lean provides (Radnor et al. 2006).

The challenge in moving lean to services is the lack of widely available reference implementations to allow people to see how directly applying lean manufacturing tools and practices can work and the impact it does have. This makes it more difficult to build the level of belief seen as necessary for strong implementation. However, some research does relate widely recognized examples of success in retail and even airlines to the underlying principles of lean (Ruffa 2008). Despite this, it remains the case that the direct manufacturing examples of "techniques" or "tools" need to be better "translated" into a service context to support the more prominent approaches of implementation, which has not yet received the level of work or publicity that would give starting points for implementers. The upshot of this is that each implementation often "feels its way" along as must the early industrial engineering practices of Toyota. This places huge importance upon sponsorship to encourage and protect these experimental developments.

Lean management is nowadays implemented also in nonmanufacturing processes and administrative processes. In nonmanufacturing processes, lean management is still a huge potential for optimization and efficiency increase (Januska and Stastna 2013).

8.8.4 Goal and Strategy

The espoused goals of lean manufacturing systems differ between various authors. While some maintain an internal focus, e.g., to increase profit for the organization, others claim that improvements should be done for the sake of the customer (Liker and Hoseus 2008; Wormack et al. 1990).

Some commonly mentioned goals are as follows:

- Improve quality: To stay competitive in today's marketplace, a company must understand its customers' wants and needs and design processes to meet their expectations and requirements.
- Eliminate waste: Waste is any activity that consumes time, resources, or space but does not add any value to the product or service.
- Reduce time: Reducing the time it takes to finish an activity from start to finish is one of the most effective ways to eliminate waste and lower costs.
- Reduce total costs: To minimize cost, a company must produce only to customer demand. Overproduction increases a company's inventory costs because of storage needs.

The strategic elements of lean can be quite complex and comprise multiple elements. Four different notions of lean have been identified (Pettersen 2009):

1) Lean as a fixed state or goal (being lean)
2) Lean as a continuous change process (becoming lean)
3) Lean as a set of tools or methods (doing lean/toolbox lean)
4) Lean as a philosophy (lean thinking)

8.8.5 Examples: Lean Strategy in the Global Supply Chain and Its Crisis

8.8.5.1 Strategy

Lean production has been adopted into other industries to promote productivity and efficiency in an ever-changing market. In global supply chain and outsource scale, IT is necessary and can deal with most of hard lean practices to synchronize pull system in supply chains and value system.

The manufacturing industry can renew and change strategy of production just in time.

The supply chains take changes in deploying second factory or warehouse near their major markets in order to react to the consumers' need promptly instead of investing manufacturing factories on the lost-cost countries. For instance, Dell sells computers directly from their website, cutting franchised dealers out of their supply chains. Then, the firm uses outsourced partners to produce its components and deliver components to their assembly plants on these main markets around the world.

Zara made decision of speeding their fashion to the consumers market by fast-producing cloths within five weeks with their local partners in Spain and never involved in mass production to pursue new styles and keep products fresh.

The other way to avoid market risk and control the supply efficiently is to cut down in stock. P&G has done the goal to cooperate with Walmart and other wholesales companies by building the response system of stocks directly to the suppliers companies (Martin 2005). With the improvement of global scale supply chains, firms apply lean practices (JIT, supplier partnership, and customer involvement) built between global firms and suppliers intensively to connect with consumers markets efficiently.

8.8.5.2 Crisis

After years of success of Toyota's Lean Production, the consolidation of supply chain networks has brought Toyota to the position of being the world's biggest carmaker in the rapid expansion. In 2010, the crisis of safety-related problems in Toyota made other carmakers that duplicated Toyota's supply chain system wary that the same recall issue might happen to them.

James Womack had warned Toyota that cooperating with single outsourced suppliers might bring unexpected problems (Hopp 2011; Ohno 1988). Indeed, the crisis cost a great fortune and left Toyota thinking whether the *JIT* practice has problems about outsourced suppliers without enough experience and senior engineers could not achieve the monitoring job close to their suppliers out of Japan. That is proven as the economy of scale becomes global, the soft-learn practices become more important in their outsourced suppliers, if they could keep good Sensei relationship with their partners and constantly modify production process to perfection. Otherwise, Toyota begins to consider whether to have more choices of suppliers of producing the same component; it might bring more safety on risk-control and reduce the huge cost that might happen in the future.

The appliance of *JIT* in supply chain system is the key issue of lean implementation in global scale. How do the supply partners avoid causing production flow? Global firms should make more suppliers who can compete with each other in order to get the best quality and lower the risk of production flow at the same time.

8.8.6 Steps to Achieve Lean Systems

The following steps should be implemented to create the ideal lean manufacturing system:

- Design a simple manufacturing system
- Recognize that there is always room for improvement
- Continuously improve the lean manufacturing system design (Ohno 1988)

8.8.6.1 Design a Simple Manufacturing System

A fundamental principle of lean manufacturing is demand-based flow manufacturing. In this type of production setting, inventory is only pulled through each production center when it is needed to meet a customer's order. The benefits of this goal include the following:

- Decreased cycle time
- Less inventory
- Increased productivity
- Increased capital equipment utilization

8.8.6.2 Continuous Improvement

A continuous improvement mindset is essential to reach the company's goals. The term "continuous improvement" means incremental improvement of products, processes, or services over time, with the goal of reducing waste to improve workplace functionality, customer service, or product performance. Lean is founded on the concept of continuous and incremental improvements on product and process while eliminating redundant activities. "The value of adding activities are simply only those things the customer is willing to pay for, everything else is waste, and should be eliminated, simplified, reduced, or integrated." Improving the flow of material through new ideal system layouts at the customer's required rate would reduce waste in material movement and inventory (Bicheno and Holweg 2009).

8.8.7 Measure

Overall equipment effectiveness is a set of performance metrics that fit well in a lean environment. Also, principal members of technical staff (PMTS), methods-time measurement, cost analysis, and perhaps time study can be used to evaluate the wastes and IT effectiveness in the operational processes. For example, Jun-Ing Ker and Yichuan Wang analyze two prescribing technologies, namely no carbon required and digital scanning technologies to quantify the advantages of the medication ordering, transcribing, and dispensing process in a multi-hospital health system. With comparison between these two technologies, the statistical analysis results show a significant reduction on process times by adopting digital scanning technology. The results indicated a reduction of 54.5% in queue time, 32.4% in order entry time, 76.9% in outgoing delay time, and 67.7% in outgoing transit time with the use of digital scanning technology (Ker et al. 2014).

8.8.8 Implementation Dilemma

One criticism of lean is that its practitioners may focus on tools and methodologies rather than on the philosophy and culture of lean. Consequently, adequate management is needed in order to avoid failed implementation of lean methodologies (Hopp and Spearman 2008). Another pitfall is that management decides what solution to use without understanding the true problem and without consulting shop floor personnel. As a result, lean implementations often look good to the manager but fail to improve the situation (Hopp and Spearman 2008).

In addition, many of the popular lean initiatives, coming from the TPS, are solutions to specific problems that Toyota was facing. Toyota, having an undesired current condition, determined what the end state would look like. Through much study, the gap was closed, which resulted in many of the tools in place today. Often, when a tool is implemented outside of TPS, a company believes that the solution lay specifically within one of the popular lean initiatives. The tools which were the solution to a specific problem for a specific company may not be able to be applied in exactly the same manner as designed. Thus, the solution does not fit the problem and a temporary solution is created vs. the actual root cause (Ohno 1988).

The lean philosophy aims to reduce costs while optimizing and improving performance. *VSM* and 5S are the most common approaches companies take on their first steps toward making their organization leaner. Lean actions can be focused on the specific logistics processes, or cover the entire supply chain. Conducting a gap analysis determines the company's "must take" steps to improve the value stream and achieve the objective. Based on that evaluation, the improvement group conducts the *failure mode effects analysis*, in order to identify and prevent risk factors. It is crucial for front-line workers to be involved in VSM activities since they understood the process and can directly increase the efficiency. Although the impact may be small and limited for each lean activity, implementing a series of small improvements incrementally along the supply chain can bring forth enhanced productivity (Merrill 2013).

After adopting the lean approach, both managers and employees experience change. Therefore, decisive leaders are needed when starting on a lean journey. There are several requirements to control the lean journey. First and most importantly, experts recommend that the organization have its own lean plan, developed by the lean leadership. In other words, the lean team provides suggestions for the leader who then makes the actual decisions about what to implement. Second, coaching is recommended when the organization starts off on its lean journey. They will impart their knowledge and skills to shop floor staff and the lean implementation will be much more efficient. Third, the metrics or measurements used for measuring lean and improvements are extremely important. It will enable collection of the data required for informed decision-making by a leader. One cannot successfully implement lean without sufficient aptitude at measuring the process and outputs. To control and improve results going forward, one must see and measure, i.e. map, what is happening now (Ohno 1988).

8.9 Manufacturing System Characteristics: Process Planning Basics

Investments in Computer Integrated Manufacturing Systems are driven by the desire for high-quality components, small batch sixes, agility, and short lead times. In the early stages of part specification, the creation of an integrated CAD/CAM environment is the key to rapid off-line simulation and verification of new part designs. Design, process planning, and manufacturing integration occurs at several levels depending upon the objective of the exercise and the degree of "flexibility" (that is to say the degree to which the design or process are fixed). The level of integration attainable between design, manufacturing, and finishing is dependent upon several factors (Stein and Dornfeld 1996):

1) Nature of the design, manufacturing, or finishing task
2) Environment in which the task is performed
3) Tools available to assist the designer with the task

Four distinct levels of integration between the tasks of design, manufacturing, and finishing have been identified in present day and future production environment (Stein and Dornfeld 1997). Levels of integration can be described in terms of ability at each level to predict, influence, and optimize part production objectives at various stages of the total part production process. The objectives encompass a variety of process metrics which influence the goals of maintaining the tolerance of shape, tolerance of form, desired surface characteristics, and subsurface damage/residual stress. These four levels of integration from a process planning point of view are illustrated in Table 8.2, using feature quality as a process metric.

Table 8.2 Four levels of integration in the design to fabrication cycle.

Integration level	Process planning software expert and agent tasks	Degree of freedom for adjustment
Level I	Feature prediction, control, and optimization in an iterative design and process planning environment	Design: High Manufacturing: High Finishing: High
Level II	Feature prediction, control, and optimization through the selection of a manufacturing plan in an "over-the-wall" design-to-manufacturing environment	Design: Low Manufacturing: High Finishing: High > Low
Level III	Feature prediction and control through limited adjustments to a preestablished manufacturing process	Design: Low Manufacturing: Limited Finishing: High
Level IV	Feature prediction for finishing process planning, finishing tool trajectories, and sensor-feedback strategies	Design: Low Manufacturing: Low Finishing: High

Source: From Stein and Dornfeld (1997).

At Level I, the highest level of integration, the designer is contemplating the design of the component. At this level, any information that correlates the design, the process plan, and manufacturing can be utilized to improve the overall "manufacturability" of the component. At a slightly lower level of integration, Level II, the design may be fixed, but it is still possible to develop a process plan and a manufacturing configuration to insure that the part specifications are met. At a still lower level, Level III, the design and process plan, as well as the machinery for manufacture, may be fixed. Even at this level, however, it is still possible and useful to consider optimization and fine-tuning of the manufacturing process to accommodate unexpected problems through changes in total geometry or level of integration, and it may be of interest only to assist in insuring that subsequent manufacturing processes, such as finishing, for example, are efficiently and accurately carried out. As the level of integration between design, manufacturing, and finishing increases from Level IV to Level I, the ability of the software tools (and the designer) to influence and optimize the process metric of feature size, shape, or quality increases. The result of allowing the designer to work within the higher level of integration is to increase the sensitivity of design decisions for one process metric, such as burr formation, and to enable the designer to optimize and balance opposing process metric through analysis, simulation, and decision evaluation.

8.10 Design for Life Cycle

Design for life cycle refers to the product after it has been manufactured. In many cases, a product can involve a significant cost to the customer beyond the purchase price. These costs include installation, maintenance and repair, spare parts, future upgrading of the product, safety during operation, and disposition of the product at the end of its useful life. Table 8.3 lists most of the factors associated with life cycle design of the product.

The cost of a product is a major factor in determining its commercial success. Cost affects the price charged for the

Table 8.3 Factors in design for life cycle.

Factor	Typical issues and concerns
Delivery	Transport cost, time to deliver, storage and distribution of mass produced items, type of carrier required (truck, railway, air transport)
Installability	Utility requirements (electric power, air pressure, and others), construction costs, field assembly, support during installation
Reliability	Service life of product, failure rate, reliability testing requirements, materials used in the product, tolerances
Maintainability	Design modularity, types of fasteners used in assembly, preventive maintenance requirements, ease of serving by customer
Serviceability	Product complexity, diagnostics techniques, training of field service staff, access to internal working of product, tools required, availability of spare parts
Human factors	Ease and convenience of use, complexity of controls, potential hazards, risk of injuries during operation
Upgradeability	Compatibility of current design with future modules and software, cost of upgrades
Disposability	Materials used in the product, recycling of components, waste hazards

Source: Adapted from Groover (1999).

product and the profit made on it. *Design for product cost* refers to the efforts of a company to identify the impact of design decisions on overall product costs and to control those costs through optimal design. Many of the DFM guidelines are directed at reducing product costs. It is often useful for a company to develop a product cost model to predict how design alternatives might affect costs of materials, manufacturing, and inspection.

To the customer, the price paid for the product may be a small portion of its total cost when life cycle costs (LCCs) are included. Some customers (e.g. the federal government) consider LCCs in purchasing decisions. The manufacturing must often include service contracts that limit customer vulnerability to out-of-control maintenance and service costs. Here, accurate estimates of these LCCs must be included in the total product cost (Groover 1999).

8.11 Sustainable Design and Environmentally Conscious Design and Manufacturing

In this section, we will dwell on discussing energy and, its close compliment, the environment with respect to precision manufacturing technology. These two – energy and the environment – will drive the technology and practice of precision manufacturing in the future. For sure, the more incremental development, following Moore's Law – higher density data storage, advanced optics, more complex medical devices, nanoscale devices, sophisticated telecommunications and IT advances, etc. – will be present and continually push us. But it is the double header of energy and environmental impacts of product design, manufacturing, use and reuse, or disposal that will dominate our future. This is expected to compel us toward:

- technology for advanced energy sources (e.g. batteries, photovoltaic/solar, hydrogen fueled motors, photosynthesis, hybrid vehicle, high-efficiency engines, hydropower, wind power, advance bio-fuel manufacture)
- sustainable manufacturing [e.g. green machines, green factories, green processes, and systems (including supply chain), minimizing or eliminating waste, consumables, energy, green products (*zero defect and zero effect*), reuse, or remanufacturing].

For sustainable manufacturing, we need to rethink our approach to product design and manufacturing. For example, how can we provide incremental solutions to the large problems of global warming, excessive energy and resource consumption, manufacturing related atmospheric contamination, etc.?

8.11.1 Technologies for Sustainable Manufacturing

Increasingly, concerns about the availability and cost of energy and raw materials, the impacts of industrial activity, and the efficient reuse of consumer products at the end of their life are driving efforts toward *green manufacturing*. This is part of a much larger international concern over the overall accomplishment of sustainability in design and manufacturing. Sustainability, as defined here, implies a level of resource utilization that is very comprehensive in its scope. It must be in accord with a level necessary to ensure that, over time, the resource will (i) not only be available, but (ii) demand is reduced to approach a level of what is naturally sustainable. That is, the "gap" between current use and sustainable use must be understood. This is a hard measure to meet but it is, ultimately, what is necessary (Dornfeld and Lee 2008).

The topic of "green business" is very popular today in the business world. Companies from GE (with its "eco-friendly" campaign) to Ford (with its sustainable "supply chain" for some products) are finding it both profitable and responsible to promote green products and green economy. Automaker like Toyota emphasizes the need for the entire manufacturing process, including the supply chain, to be green. Others, like Cummins Engine, are engaged in "remanufacturing" for engine components that have life after first use.

A short introduction to green manufacturing is presented followed by a review of the elements and scope of the system that should logically be included in any discussion of green manufacturing. The concept of "technology wedges" and how they might be applied to identifying opportunities for environmentally benign manufacturing and integrating solutions with existing processes or supply chains is discussed.

8.11.2 Green Manufacturing Pipeline

Integration of green components and analysis into manufacturing systems was considered early on by Sheng et al. (1995). Some researchers have addressed specific parts of the manufacturing process in more detail with success, e.g., minimum quantity lubrication/dry machining, which is now being implemented in a number of manufacturing operations internationally (Zimmerman et al. 2003). These specific accomplishments are often referred to as "technology wedges" that form part of a comprehensive solution. The wedges are explained in more detail in the next section. Ishii has looked at incorporating end-of-life considerations when products are designed (1999).

A schematic of the green elements of design and manufacturing is shown in Figure 8.14. The "life cycle" of the

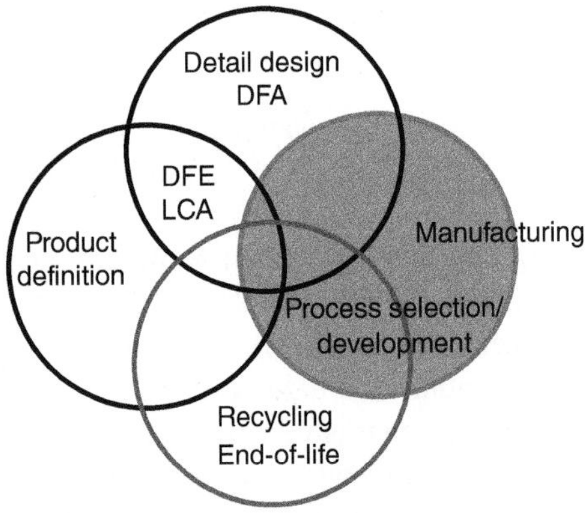

Figure 8.14 Schematic showing relationship of green elements of design and manufacturing from product definition to end-of-life.

product from design to recycling or reuse at end of life must be considered. The figure refers to life cycle analysis (LCA) tools (see Section 6.3) (Das 2002, 2005; Das and Houtman 2004), design for assembly (DFA), and design for environment (DFE). The figure distinguishes between the design of a product and the processes or systems that make the product. Of importance is that both the design of the process and how the process operates to manufacture the product are considered. Substantial opportunities exist in the design of machines that make products (Taniguchi et al. 2005), as well as in their productive operation [macro and microplanning, for example, see Srinivasan and Sheng (1999a, b)].

Regarding green manufacturing practices, it is necessary to consider the environmental impact of all the elements in a product's life cycle as shown in Figure 8.15. The figure covers the broadest scope possible from the original source of raw materials (including mining costs) to eventual disposal of the used product. Each element includes waste (or non-utilized resources and by-product). The energy and resources consumed in each of these stages of the process, and the waste output, will "count against" any benefits of the new product or machine resulting from the process. In fact, to be complete, we could consider manufacturing to include elements of most of the stages of the product's life cycle since mining, conversion, and transport are arguably part of manufacturing. At a more detailed level, sustainable manufacturing can be addressed by improving the individual components to manufacturing (Figure 8.16). Here we see contributors to sustainable manufacturing including power sources and processes.

To be effective, there must be a net energy, resource, or environmental impact savings over the life of any product or process produced in an "environmentally benign" manner to meet the sustainability definition. Colloquially, it can be said: "if you consume as much resources in manufacturing a product as you save in use of the product you have not succeeded!" This requires careful accounting for the impact of all the process chain elements and that is often not easy to do.

8.11.3 Sustainable Manufacturing: Is Green Equivalent to Sustainable?

If we look at consumption of resources (any resource) or the impact of that consumption over time we can identify a trend, generally increasing over time. The sustainable rate of that consumption, or impact, is often much less. Here we can discuss sustainability in the context of a well cited

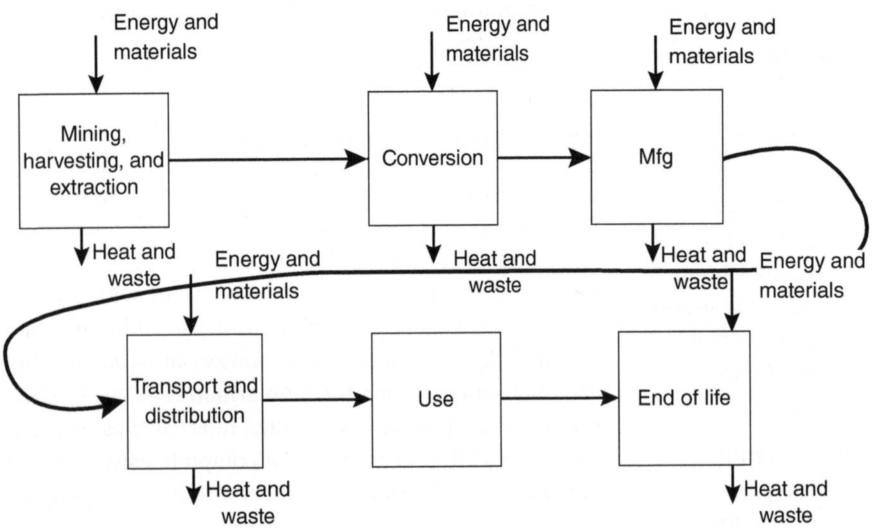

Figure 8.15 Schematic showing the elements to consider the environmental impacts in green manufacturing practices throughout a product's life-cycle.

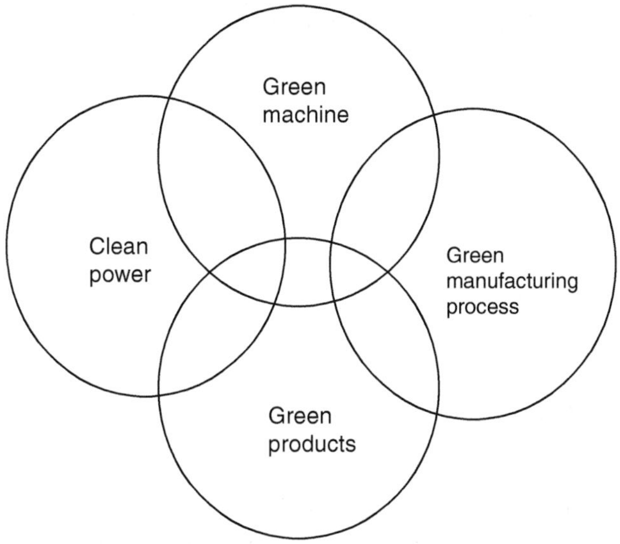

Figure 8.16 Components of green and sustainable manufacturing including machine and processes and power sources.

of the Bruntland Commission, formally the World Commission on Environment and Development, as part of a study to look at conditions for fostering sustainable development in the world (Our Common Future 1987). They defined sustainable development is "development that meets the needs of the present without compromising the ability of future generation to meet their own needs."

This is a lofty goal to meet when designing products and systems of production. If we define a level of sustainability so that we don't "compromise the...future," paraphrasing here from Brundtland, it means our factories and processes are at or below the level of sustainable impact or consumption. Figure 8.17 illustrates the challenge. The "wedge" between our rate of consumption or its impact and the sustainable rate of consumption or impact is often large and growing. In order to return to a sustainable level of

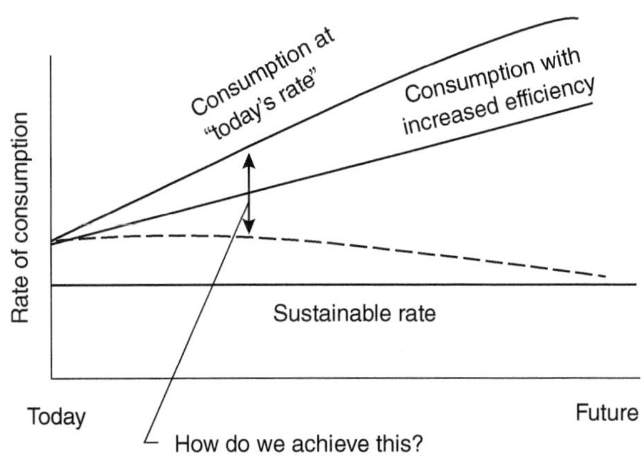

Figure 8.17 Sustainability frame of reference.

consumption, especially from the perspective of manufacturing, we will need to reduce the consumption, or impact, at a rate larger than its natural growth (think population increase times per capita consumption or impact). Just reducing consumption or impact, although preferable than not, is insufficient. That is a challenge! One way to look at this is from the standpoint of "technology wedges."

8.11.4 Manufacturing Technology Wedges

Pacala and Socolow (2004) describe the concept of "stabilization wedges" with respect to technology that could solve the climate problem (with respect to fossil fuels) for the next 50 years using current technology. The idea is to close the gap between current rate of impact and the sustainable level with "wedges" of technology – each making a small but measureable improvement. They introduce the idea that focuses on the "gap" between the current trends in fossil fuel emissions (the impact or consumption) relative to the atmosphere's capability to accommodate emissions (the sustainable level). This is a similar gap to that (normal consumption vs. sustainable consumption) in Figure 8.17. Each individual wedge represents the ability of some existing current technology to reduce, on its own, some portion of fossil emission. Then, summed together, these wedges provide the necessary reduction in emissions to achieve an overall "sustainable" situation. Their goal was to achieve CO_2 stabilization at 500 ppm by 2025, a sustainable level.

The question is whether or not this strategy could be employed in manufacturing to accomplish environmentally benign manufacturing processes. There is a renewed interest in developing process enhancements that contribute to reducing an environmental impact. As yet, there is no strategy to coordinate a set of enhancements and new capabilities that will, combined, render a process "sustainable" as in the case of the fossil fuel emission example (Dornfeld and Lee 2008).

There are some interesting possibilities. With the development of new manufacturing technologies for microscale and nanoscale manufacturing, as well as the various alternative energy sources being pursued (from fuel cell to photovoltaics), attention will need to be paid to ensure that any new processes will have some damaging effects on environment and health (see more details in Section 9.2.6). That is, wedges must be designed to be "net-positive." An improvement in one element of a process or system of manufacture cannot be at the expense of another segment of the cycle (Hawken 1993). This is specially complicated with the complex supply chains employed today. Continuous improvement is as valid here as in other areas of manufacturing. We strive to remove "wasted time and effort." Why not also try to do this for energy/consumables/waste? Here it is proposed that "technology wedges" – analogous to "stabilization wedges" – offer a framework and potential metric for addressing these energy challenges.

There are a number of fundamental rules which govern how wedge technology can be employed in manufacturing. In no particular order, they are as follows (Dornfeld and Wright 2007):

Rule 1. The cost of materials and manufacturing [in terms of energy consumption and Green House Gas (GHG) emissions, etc.] associated with the wedge cannot exceed the savings generated by the implementation of the wedges over the life of the process or system or product in which it is employed (see Section 8.11.4).

Rule 2. The technology must be applied at the lowest level in the process chain.

Rule 3. The cost and impact of the technology must be calculated in terms of the basic metrics of the manufacturing system and the environment. That is, cost and impact must be expressible in units of dollars (or euros, yen, etc.), carbon equivalent, global warming gas creation or reduction, joules, cycle time and production rate, quality measures, lead time, working capital, and so on relative to present levels of consumption, use, time, etc.

Rule 4. The technology must take into consideration societal concerns along with business and economy (Hawken 1993).

Rule 5. There must an accompanying analytical means or design tool so that it can be evaluated at the design stage of the process or system. It must be an integrated approach.

These rules, when applied, will insure a balanced and accurate appraisal of the impact of the technology from an environmental perspective and also insure the rules are feasible from a business perspective. This will avoid creation of anomalies in the supply chain caused by a gain which yields a global net loss. Many examples of a net loss currently exist in society, such as with so-called "high tech/low cost" products, some of which operate efficiently, but which have short useful lives. The consumer "learns" to expect this and, thus, expects to replace the product in a short time. This trend destroys a tremendous amount of product value and resources and, worse, it encourages increased environmental damage. The life cycle of such "through-away" products is often not considered in the design, fabrication, use, or reuse.

8.12 Lean Six Sigma

8.12.1 Introduction

Lean Six Sigma is a methodology that relies on a collaborative team effort to improve performance by systematically removing waste and reducing variation. It combines lean manufacturing/lean enterprise and Six Sigma to eliminate the eight kinds of waste (muda):

1) Transporting
2) Inventory
3) Motion
4) Waiting
5) Overprocessing
6) Overproduction
7) Defects
8) Under utilization of employees

Six Sigma seeks to improve the quality of process outputs by identifying and removing the causes of defects (errors) and minimizing variability in (manufacturing and business) processes. Synergistically, Lean aims to achieve continuous flow by tightening the linkages between process steps, while Six Sigma focuses on reducing process variation (in all its forms) for the process steps, thereby enabling a tightening of those linkages. In short, Lean exposes sources of process variation and Six Sigma aims to reduce that variation, enabling a virtuous cycle of iterative improvements toward the goal of continuous flow.

Six Sigma is a business process for improving the quality, reducing costs, and increasing customer satisfaction. This name comes from statistics, where the high level of quality represents no more than 3.4 defects per million. The goal of Six Sigma is to identify and remove causes of defects. Plan-do-check-act (PDCA) or plan-do-study-act (PDSA) is an iterative method of continuous improvement. It is also related to the improvement cycle DMAIC (define, measure, analyze, improve, and control). The DMAIC cycle is used for existing products, related to DMADV (define, measure, analyze, design, verify) which is used during product development.

8.12.2 The History of Six Sigma: 1980s–2000s

8.12.2.1 Motorola Story

The origin of Six Sigma is found at Motorola. In a study of the field life of a product, a Motorola engineer observed that products that were defect-free at manufacturing rarely failed in early use by a customer. However, products which had failed during production and undergone rework were more susceptible to early failures in the hands of the customer. It was as if those products that were originally made well worked well, while those that originally had a defect might have other defects as well. Catching and fixing one defect did not assure a good product.

It was not enough to find and fix defects, they had to be prevented.

Preventing defects is possible only when the processes that produce the product are set up so that defect can't or don't happen. Motorola focused on "how" the work was done in each and every process. If defects were prevented, then there would be no need for rework and quality would be higher. Since there were lots of processes, Motorola developed a measure for Six Sigma work that could be applied to every process.

The measure Motorola developed was called the *sigma level or sigma value* for each *critical-to-quality* (CTQ) associated with a process. A CTQ analysis is a way of studying the flowchart of a process to find problems. The analysis studies inputs and outputs and identifies the steps that influence quality of the process and its outputs. If a process is in an ideal state, it's stable and predictable and meets the CTQs. If it's stable and predictable but doesn't meet the CTQs, it's in a threshold state. If it's free from special cause(s), it should be relatively straightforward to move it into an ideal state, perhaps through a DMAIC project (see Section 8.12.2.3).

The sigma level is based on a statistical analysis linking the *defects per million opportunities* (DPMO) and the capability of the CTO with respect to customer requirements. A CTQ with 3.4 DPMO is associated with capability value of $C_p = 2.0$ and $C_{pk} = 1.5$ and is considered to have achieved the Six Sigma level. Motorola successfully applied the concepts, philosophy, and techniques of Six Sigma to the design, developments, and production of the Bandit pager. The quality level of this pager was unsurpassed. Motorola showed that the traditional approach of detecting and fixing defects resulted in CTQs at four sigma level of quality. Six Sigma quality, or 3.4 DPMO, led to eliminate of costly inspection and rework, which in turn led to decrease in manufacturing time and increase in customer satisfaction. Customers were happy and Motorola reaped staggering financial savings.

8.12.2.2 General Electric Story

In 1995, Larry Bossidy of AlliedSignal introduced General Electric (GE) top management to Six Sigma. Bossidy's account of the benefits realized at AlliedSignal led Jack Welch, then CEO of GE, to launch a Six Sigma initiative which would bring all of GE's products to a Six Sigma level by 2000. At the time, GE's processes were operating between three and four sigma, which would mean an average of 35 000 DPMO. To go from three or four sigma to Six Sigma would require a monumental training and education effort. GE rolled out a rigorous training in the Six Sigma methodology with a heavy emphasis on statistical methods. From 1996 through 1998, GE's training investment approached $1 billion and the returns on that investment were already keeping pace. By 1999 they expected $1.5 billion in savings. GE's businesses range from a variety of products to various services including GE capital. Throughout GE, Six Sigma initiative returned benefits to the organization's bottom line and its ability to serve its customers.

8.12.2.3 DMAIC Model

The five phases of the DMAIC model are *define, measure, analyze, improve, and control* (Figure 8.18). While this model suggests a linear progression with each phase leading to the next, there will always be some iterative work between the phases. Each phase involves a certain amount of work to accomplish which is integral to the successful completion of a Six Sigma project (Ward and Poling 2004).

The Phases of DMAIC The *define phase* is for focusing the Six Sigma project and creating a roadmap with timeline to guide all project activities toward a successful conclusion. Under the champion leadership, the project team works on mapping the process associated with the project and setting up communication with all stakeholders of the process. Logistical decisions about team members and meeting times are decided. The charter is finalized and signed off by everyone. If questions arise about availability of resources such as time or people, the champion helps resolves these issues.

The *measure phase* concentrates on collecting data to assess current process performance. It involves deciding all of the measures that are pertinent to the project. These include the scorecard metrics that tie to the scoreboard of the organization as well as all CTQ characteristics for the process. All measure must be clearly identified and data collection plans for each one setup. Then data can be collected regularly for the analyze phase.

In the *analyze phase*, the various techniques are used to determine how the current process performance compares to the performance goals in the project charter. Data are analyzed with the express purpose of understanding why the current process works the way it does. Specifically, it is important to determine if the process has a predictable or unpredictable behavior. Predictable processes have routine sources of variation that are always present and impact the process day in and day out. Unpredictable processes have both routine sources of variation plus special, unpredictable sources of variation that can knock a process off track.

Activities in the *improve phase* are directed toward finding out how to improve the process. This phase involves investigations to see what changes to the process will make

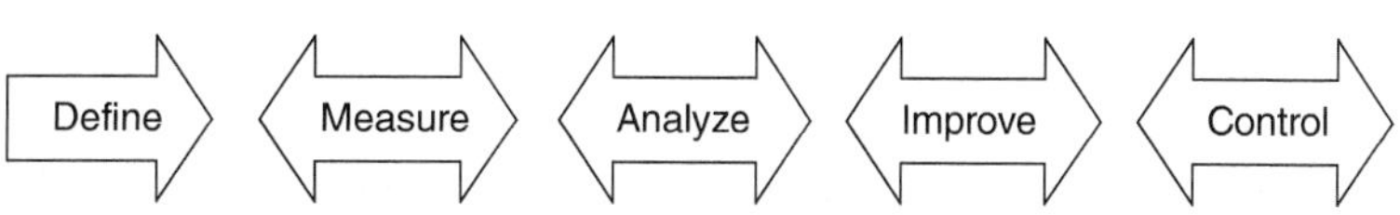

Figure 8.18 Five major components of the DMAIC model.

the process better to meet the project goals. The project team may conduct experiments to see what can be achieved and the associated costs. Once several solutions are identified, the project team can evaluate and pilot them for their potential benefits. Before a final improvement solution is selected and implemented, the project team needs to assess any potential problems with all solutions. Resolution of all potential problems will assure that the solution will not be undermined in the future.

Finally, the *control phase* in the DMAIC model must focus on making the changes permanent for sustained process improvement. This means that the project team will work on those activities that are critical for the project team to turn over an improved process to the process owners. To reap the benefits of any Six Sigma project, the improvements must be sustainable by the organization, not by the project team. Elements included in the control phase are training of employees, control plans, and process measures that will keep attention on the process. The transition from the control phase of the DMAIC model used by the project team back to the organization comes with the standardized and institutionalized phases.

8.12.2.4 Calculating Process Sigma Values

Process sigma values provide a way of comparing performance of different processes, which can help you to prioritize the projects. The process sigma value represents the population of cases that meet the CTQs right first time. Sigma values are often expressed as *defects per million opportunities*, rather than per hundred or per thousand, to emphasize the need for world-class performance. Not all organizations using Six Sigma calculate process sigma values. Some organizations just use the number of defects or the percentage of orders meeting CTQs to show their performance. Either way, if benchmarking is to be meaningful, the calculations must be made in a consistent manner. Table H.2 includes "yield" percentages, sigma levels, DPMOs – the right first time data. You can see that Six Sigma performance equates to only 3.4 DPMO (Morgan and Brenig-Jones 2016).

Lean Six Sigma uses the DMAIC phases similar to that of Six Sigma. Lean Six Sigma projects comprise aspects of Lean's waste elimination and the Six Sigma focus on reducing defects, based on critical to quality characteristics. The DMAIC toolkit of Lean Six Sigma comprises all the Lean and Six Sigma tools. The training for Lean Six Sigma is provided through the belt-based training system similar to that of Six Sigma. The belt personnel are designated as white belts, yellow belts, green belts, black belts, and master black belts, similar to judo.

8.12.2.5 Lean Six Sigma Organization Structure

For each of these belt levels skill sets are available that describe which of the overall Lean Six Sigma tools are expected to be part at a certain belt level. These skill sets provide a detailed description of the learning elements that a participant will have acquired after completing a training program. The level upon which these learning elements may be applied is also described. The skill sets reflect elements from Six Sigma, Lean, and other process improvement methods like the theory of constraints and total productive maintenance.

8.12.3 5S

5S is a lean practice used to keep production workspace orderly and keep the workforce committed to maintaining order.

Japanese terms

1) *Seiri* – Put things in order
2) *Seiton* – The proper arrangement
3) *Seiso* – Clean (keep polished)
4) *Seiketsu* – Purity (maintain clean)
5) *Shitsuke* – Commitment instilling attitude/atmosphere to maintain 5S

English terms

1) *Sort* – Get rid of what is not necessary
2) *Straighten* – Everything has a place and is in it
3) *Shine* – Keep clean machine/workplace
4) *Standardize* – Systems and procedures to maintain 1–3
5) *Sustain* – Maintain systems/procedures (1–4)

8.13 Six Sigma and Lean Manufacturing

There are many parallels between the Six Sigma and lean philosophies. Both are focused on the process and process improvements. The techniques of each one complement the other and provide enhanced results.

8.13.1 Comparing the Two Methodologies

Six Sigma focuses on improving existing processes to give results that are essentially defect-free. A CTQ characteristic or CTQ that reaches the Six Sigma level will have only 3.4 DPMO. There are two ideas at work here: first, preventing defects is more cost effective and reduces the need for inspection and rework; second, all products and services have multiple CTQ characteristics. If each CTQ reaches the six sigma level, then the product or service will work as the customer expects.

The reality is that Six Sigma can lead to an improved process with mechanisms to maintain the improvements, but there are still inefficiencies in the process. All of the benefits of eliminating non-value-added steps may not yet be evaluated or addressed. Also, the flow throughout an organization that is integral to the pull systems of lean may not have been considered. What we know today as lean manufacturing began with the concepts that Ford Motor Company pioneers at the Rouge facilities in Michigan. Based on the idea of raw steel coming in and finished cars going out, the Rouge facility had taken production to new heights. Then Toyota took this idea a step further. The brilliance of lean was to move beyond craft and mass production techniques to short runs with essentially no setup time required. With short runs and small batch sizes, quality problems would be noticed immediately and there would be no backlog of interior parts or materials. Small batches made quality immediate visible and led to quality improvements (see Table 8.4).

Waste of time, waste of materials, and the waste of money – all drive the lean concepts. A number of offshoots of the original lean ideas have emerged. Eliminating wasted efforts by removing non-value-added steps in a process has led to *VSM*. Eliminating excess inventory and small batches has led to a "pull" system of production. All of the lean efforts are directed toward removing wastes of various types in the production systems. A lean approach opens up the opportunity to achieve reductions in time to market and in costs while improving product quality.

Both concepts focus on the process. Lean techniques may well be the ones to use to achieve the improvement in quality and cost reduction that are the objectives of the Six Sigma initiative. Six Sigma may well provide the statistical techniques that will take the lean initiatives to the next level. They both are critical and if orchestrated in concert can be much more effective than either one individually (Snee and Hoerl 2003).

8.14 Cost vs. Quality Analysis

A cost–quality analysis is a way of studying the flowchart of a process to identify problems. The analysis helps a team look critically at individual steps of a process to find opportunities for improvement. Cost of quality means those costs incurred because of poor quality, costs that would not be incurred if things were done right the first time and every time. In a cost–quality analysis, one looks for activities that incur costs of quality.

When to use:

- When flowcharting a process, to be sure that cost-of-quality activities are included.
- After flowcharting a process, to identify problems, potential causes, and areas to concentrate improvement efforts.

Procedure:

1) Draw a detailed flowchart of the process.
2) Identify all process steps (including recycle loops) that incur costs of quality: inspection, fix, and damage control. (Later in this section, see "Considerations" for definitions of these costs.) Draw a red circle (red for stop) around those steps or recycles.
3) If no or few steps have red circles, ask "what can go wrong? How do we tell if things go wrong? How does the process handle things going wrong?" If necessary, add steps to the flowchart.
4) For each red circle, ask what process step, done perfectly, would allow us to eliminate this red-circled step? Draw a green circle (green for Go) around each step identified here.
5) The green circles show areas to look for ways to prevent problems and in general to seek improvement. Green circles will contain the root causes of problems identified by the red circles.

Example:

Figure 8.19 shows a cost–quality analysis for the flowchart for filling an order. Imagine the dotted circles are red: inspections, fixes (including rework), and damage control. They were drawn first. Note that on this flowchart, all the decision diamonds are checking, or inspection, steps. Also, the two recycle loops are fix steps.

Table 8.4 Comparing the two methodologies: Six Sigma and Lean Manufacturing.

	Objectives	Focus	What's missing
Six Sigma	• Improved customer satisfaction • Improved quality • Increased profitability	• Reducing variation in products and services • Eliminating defects to achieve a six sigma level of quality	• Speed and efficiency • Streamlined process
Lean Manufacturing	• Reduced waste • Decreased inventory • Reduced costs • Increased speed	• Streamlining processes • Improving efficiency	• Reduction in variation • Prevention of defects

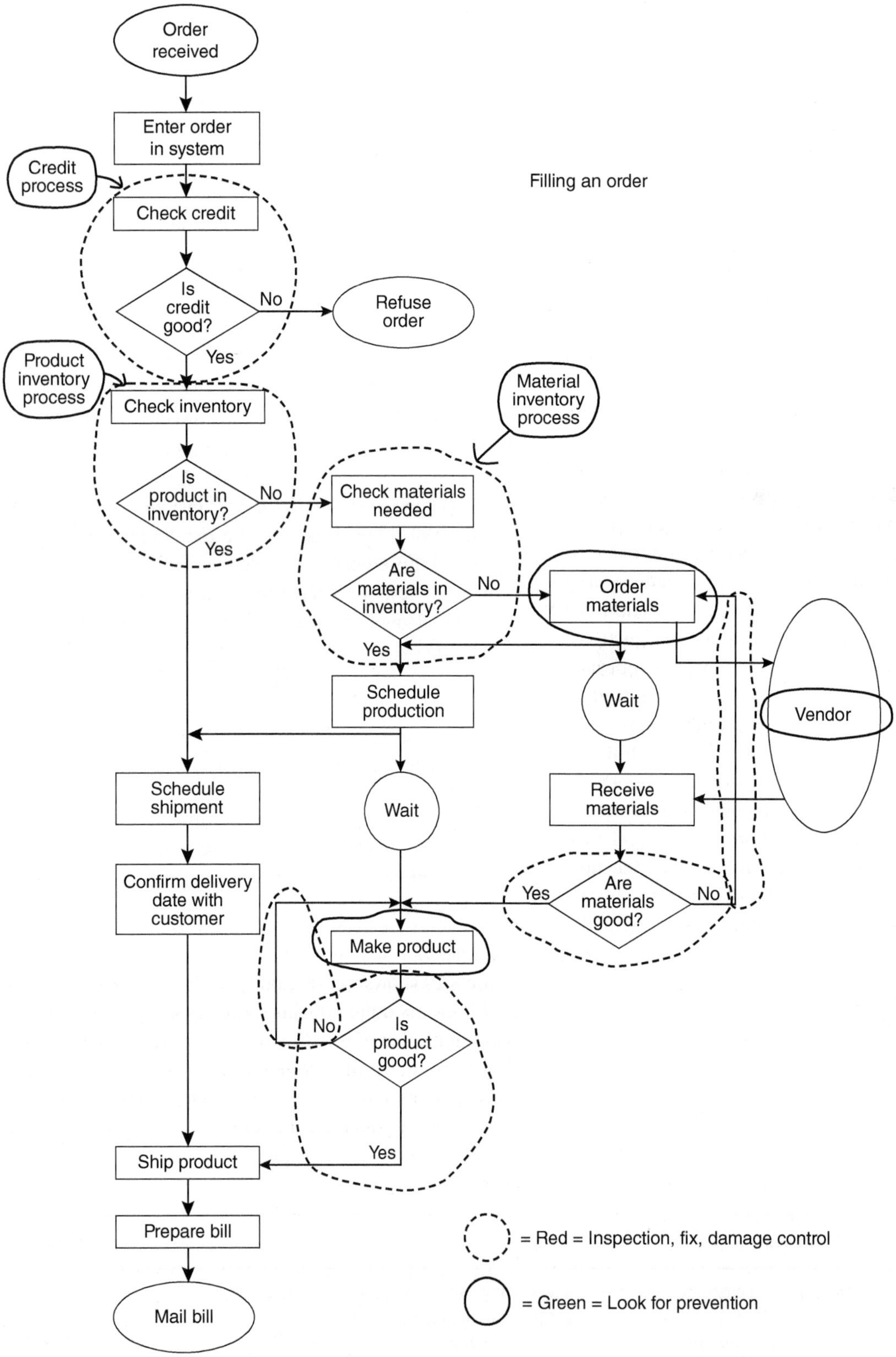

Figure 8.19 Cost vs. quality analysis example.

There are no damage control steps because this flowchart does not show what happens when things go wrong with the external customer. What if a good customer is mistakenly refused credit? Or the customer is unhappy with the delivery date? Or the wrong product is shipped? Or the bill is incorrect? If those possibilities were shown, they would be circled in red.

At the next step in the analysis, preventions for each dotted circle are sought in previous boxes on the flowchart. For the inspection "is product good?" and the recycle loop flowing from the "No" answer, the prevention is "make product." If product were made perfectly each time, the question "is product good?" would be unnecessary. Imagine that the solid circle around "make product" is green. For "are materials good?" and the recycle loop flowing from that decision diamond, the preventions are "order materials" and "vendor." Notice that although we don't control what the vendors do, their processes cause us to add an inspection step. That is why it is important to involve suppliers in your quality-improvement efforts.

If problems with the external customer had been flowcharted and circled in red, then green circles would be drawn around "check credit," "schedule shipment," "ship product," and prepare bill." Doing this right prevents later damage control. For three red (dotted) circles, no opportunities for prevention can be found on this flowchart. "Are materials in inventory?" depends on the material inventory process; "Is product in inventory?'" depends on the product inventory process; "Is credit good?" depends on the credit process (and also on the customer). Names of those processes have been written in and circled with green. The team may decide that to improve the order-filling process, the material inventory process that links with it should be studied.

8.14.1 Considerations

- There are three types of activities that incur costs to be eliminated:
  1) *Inspection*: Testing, inspecting, or checking to determine whether a failure or error has occurred. For example, proofreading, countersigning a document.
  2) *Fix*: Correcting, reworking, or disposing of errors or poor output that are identified before the customer is affected. For example, redoing a report when incorrect information is discovered in it.
  3) *Damage control*: Correcting, reworking, or disposing of errors or poor output that are identified after the customer is affected. Also, handling problems that arise because the customer has been affected. For example, explaining to the customer how we are solving the problem.
- There is a fourth type of cost of quality: prevention. It usually is preferable (cheaper and easier) to handle problems through prevention rather than waiting until inspection, fix, or damage control are needed. Quality improvement is based on preventing problems from occurring. Through a cost-of-quality analysis, you can identify process steps that are inspection, fix, or damage control (red circles) in order to determine prevention measures (green circles) that can reduce or even eliminate those steps.
- The process steps circled in red probably will never actually disappear. We are all human, so we will never be perfect. However, thinking about the ideal, perfect world helps us to see where to focus improvement efforts. For example, we may always want to do a final inspection to check whether the product is good. But to improve the process, we should work on getting better at making product, rather than on better checking.
- If a flowchart has a few or no red circles, it may have been drawn with very little detail, as a big picture. This analysis can be done only with flowcharts that include enough detail to show the problems and fixes.
- Often, opportunities for improvement will be found beyond the boundaries of your flowchart. A narrow scope may have been the best way to draw the part of the process in which you were interested. However, do not limit yourself to that scope when you identify opportunities for improvement (Tague 2004).

8.15 Assessing and Reducing Risk in Design: Cost to Manufacturer

The design process that follows a spiral approach is preferable to the current "loops," which represent feedback. That is, the client can infer that the design is progressing, but in order to incorporate various viewpoints, it is losing ground (and costing more money and time). Often, however, a synergetic and innovative design never goes backward. In fact, better and, frequent, more cost-effective features are being integrated into the project continuously. This goes beyond the "pay now versus pay later" decision, although consideration of the entire life cycle will save time and money, not to mention liabilities down the road, after completion (see Figure 8.20) (Vallero and Brasier 2008). Increased safety and sustainability can be gained by considering secondary costs in product and system design. This is the beginning of the life cycle perspective, especially when cost and impacts are considered in addition to financial measures.

Any improvement in safety as it relates to an engineered product is often accompanied by an increase in the cost of that product. On the other hand, products that are not safe incur secondary costs to the manufacturer beyond the

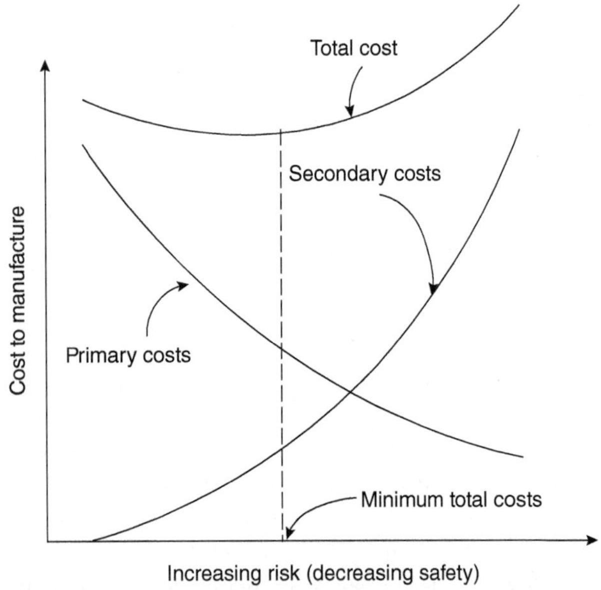

Figure 8.20 Safety and environmental risks associated with primary and secondary costs. Increasing risk and lower manufacturing and product costs cause liabilities down the road (this goes beyond saying: "pay now or pay later"). *Source:* From Martin and Schinzinger (2004).

primary (production) costs that must also be taken into account – costs associated with warranty expenses, loss of customer goodwill, and even loss of customers because of injuries sustained from the use of the product, litigation, possible downtime in the manufacturing process, and so forth (see Figure 8.20). It is therefore important for manufacturers and users alike to reach some understanding of the risks connected with any given product and know what it might cost to reduce those risks (or not reduce them).

8.16 The Heart and Soul of the Toyota Way: Lean Processes

The *Toyota Way* summarized the management principles of Toyota in a 4P model: Philosophy, Process, People, and Problem Solving (see Figure 8.21). The 4Ps form a pyramid, the foundation of which is a long-term *philosophy* that focuses on adding value to customers and society. Building on that is Toyota's investment in *lean processes*, which concentrates on shortening lead time by eliminating waste. Eliminating waste is done by people using rigorous problem-solving methods – the top two layers of the pyramid. Toyota's management system is described by 14 principles within these 4 levels (Figure 8.21).

8.16.1 Fourteen Principles of the Toyota Way

8.16.1.1 Long-Term Philosophy

Principle 1. Base your management decisions on a long-term philosophy, even at the expense of short-term financial goals.

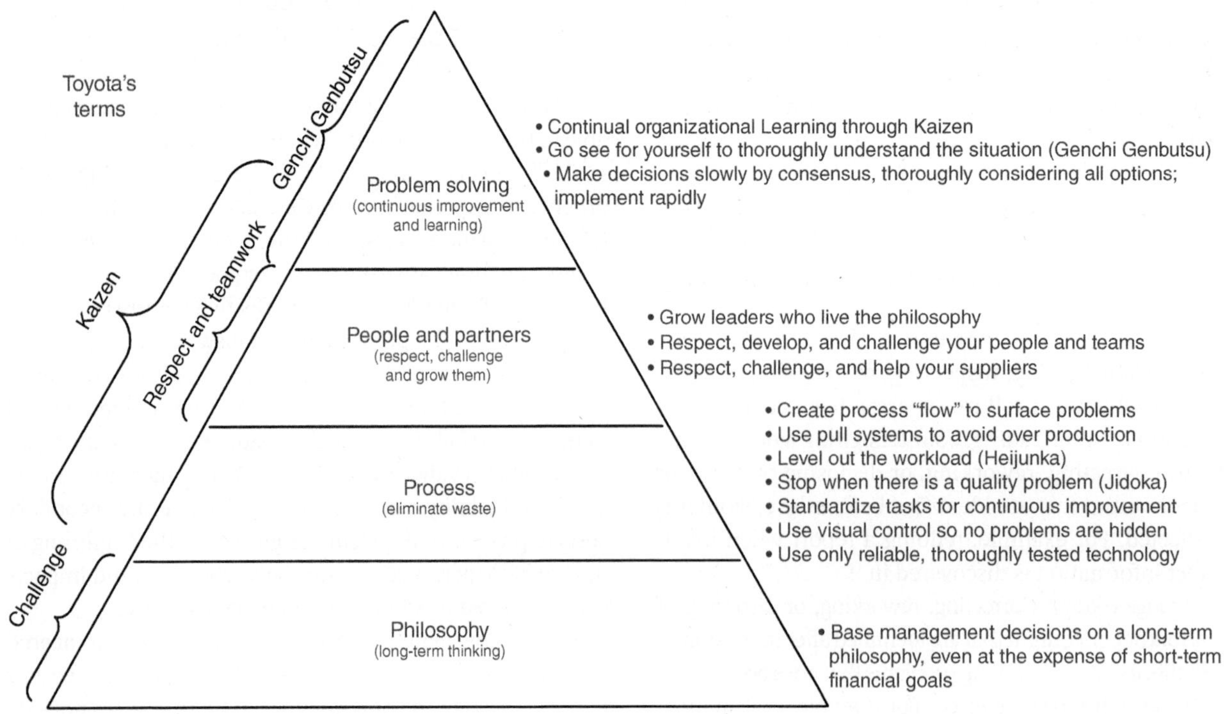

Figure 8.21 4P model of the Toyota way.

8.16.1.2 Lean Processes: The Right Process will Produce the Right Results

Principle 2. Create a continuous process "flow" to bring problems to the surface.
Principle 3. Use "pull" systems to avoid overproduction.
Principle 4. Level out the workload – like the tortoise, not the hare.
Principle 5. Build a culture of stopping to fix problems, to get quality right the first time.
Principle 6. Standardized tasks and processes are the foundation for continuous improvement and employee empowerment.
Principle 7. Use visual controls so no problems are hidden.
Principle 8. Use only reliable, thoroughly tested technology that serves your people and processes.

8.16.1.3 Develop and Challenge Your People and Partners Through Long-term Relationships

Principle 9. Grow leaders who thoroughly understand the work, live the philosophy, and teach it to others.
Principle 10. Develop exceptional people and teams who follow your company's philosophy.
Principle 11. Respect your suppliers by challenging them and helping them improve.

8.16.1.4 Problem Solving and Continuous Improvement Drive Organizational Learning

Principle 12. Go and see for yourself to thoroughly understand the situation.
Principle 13. Make decisions slowly by consensus, thoroughly considering all opinions; implement decisions rapidly.
Principle 14. Become a learning organization through relentless reflection and continuous improvement (Liker 2004).

If we carefully look at these 14 principles, we find they are all statements of beliefs and values. They are about Toyota's culture.

Philosophy is about Toyota's purpose and why they exist.

Process is about what Toyota believes leads to operational excellence – constantly eliminating waste.

People are what drive the company forward and culture is what teaches the people how to act, think, and feel to work together toward a common goal.

Problem solving is the way Toyota people focus their efforts to continually improve.

8.16.2 Life Cycle Cost Analysis (LCCA)

Life cycle cost analysis (LCCA) is a tool to determine the most cost-effective option among different competing alternatives to manufacturing process, purchase, own, operate, maintain and, finally, dispose of an object or process, when each is equally appropriate to be implemented on technical grounds. For example, for a highway pavement, in addition to the initial construction cost, LCCA takes into account all the user costs (e.g. reduced capacity at work zones) and agency costs related to future activities, including future periodic maintenance and rehabilitation. All the costs are usually discounted and total to a present-day value known as net present value. This example can be generalized on any type of material, product, or system.

In order to perform a LCCA, scoping is critical – what aspects are to be included and what not? If the scope becomes too large, the tool may become impractical to use and of limited ability to help in decision-making and consideration of alternatives; if the scope is too small then the results may be skewed by the choice of factors considered such that the output becomes unreliable or partisan. Usually, the LCCA term implies that environmental costs are not included, whereas the similar whole-life costing, or just LCA, generally has a broader scope, including environmental costs

8.16.3 Cost of Quality: Poor vs. Good Quality

When calculating the business case for a Six Sigma project, the cost of poor quality (COPQ), which is the cost caused through producing defects, is a commonly used concept. Within the total amount of quality cost, however, COPQ represents only a certain proportion. Costs do not result from only producing and fixing failures; a high amount of costs comes from ensuring that good products are produced. This section explains the cost of quality as a more comprehensive concept covering the COPQ and the cost of good quality. In short, any cost that would not have been expended if quality were perfect contributes to the cost of quality.

8.16.4 Cost of Quality: Not Only Failure Cost

The cost of quality has two main components: the cost of good quality (or the cost of conformance) and the COPQ (or the cost of nonconformance). Figure 8.22 shows the following.

- The COPQ affects:
- Internal and external costs resulting from failing to meet requirements.
- The cost of good quality affects:
- Costs for investing in the prevention of nonconformance to requirements.
- Costs for appraising a product or service for conformance to requirements.

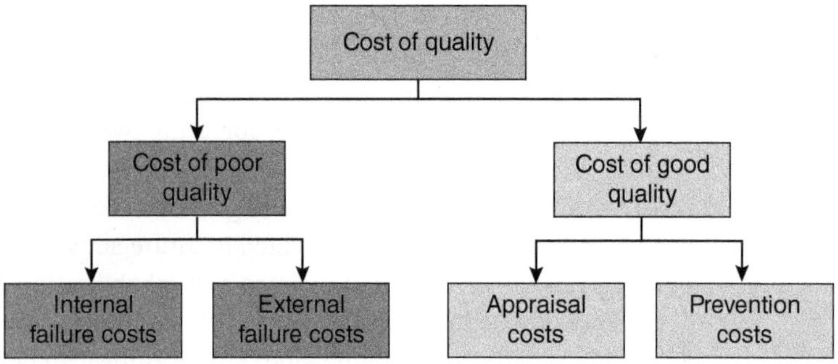

Figure 8.22 Cost of quality.

8.16.5 COPQ: Internal Failure Costs

Internal failure costs are costs that are caused by products or services not conforming to requirements or customer/user needs and are found before delivery of products and services to external customers. They would have otherwise led to the customer not being satisfied. Deficiencies are caused both by errors in products and by inefficiencies in processes. Examples include the costs for

- rework
- delays
- redesigning
- shortages
- failure analysis
- retesting
- downgrading
- downtime
- lack of flexibility and adaptability

8.16.6 COPQ: External Failure Costs

External failure costs are costs that are caused by deficiencies found after delivery of products and services to external customers, which lead to customer dissatisfaction. Examples include the costs for

- complaints
- repairing goods and redoing services
- warranties
- customers' bad will
- losses due to sales reductions
- environmental costs

8.16.7 Cost of Good Quality: Prevention Costs

Prevention costs are costs of all activities that are designed to prevent poor quality from arising in products or services. Examples include the costs for

- quality planning
- supplier evaluation
- new product review
- error proofing
- capability evaluations
- quality-improvement team meetings
- quality-improvement projects
- quality education and training

8.16.8 Cost of Good Quality: Appraisal Costs

Appraisal costs are costs that occur because of the need to control products and services to ensure a high-quality level in all stages, conformance to quality standards, and performance requirements. Examples include the costs for

- checking and testing purchased goods and services
- in-process and final inspection/test
- field testing
- product, process, or service audits
- calibration of measuring and test equipment

The total quality costs are then the sum of these costs. They represent the difference between the actual cost of a product or service and the potential (reduced) cost given no substandard service or no defective products. Many of the costs of quality are hidden and difficult to identify by formal measurement systems. The iceberg model (Figure 8.23) is very often used to illustrate this matter: only a minority of the costs of poor and good quality are obvious – appear above the surface of the water. But there is a huge potential for reducing costs under the water. Identifying and improving these costs will significantly reduce the costs of doing business.

8.16.9 The Six Sigma Philosophy of Cost of Quality

What is the relation between the cost of good quality and the COPQ? The traditional view would be to conclude that if a company wants to reduce defects and by this reduce the COPQ, the cost of good quality would have to be increased, meaning higher investments in any kind of checking, testing, evaluation, training of operators, etc. Following the Six Sigma philosophy, however, of building quality into process, service

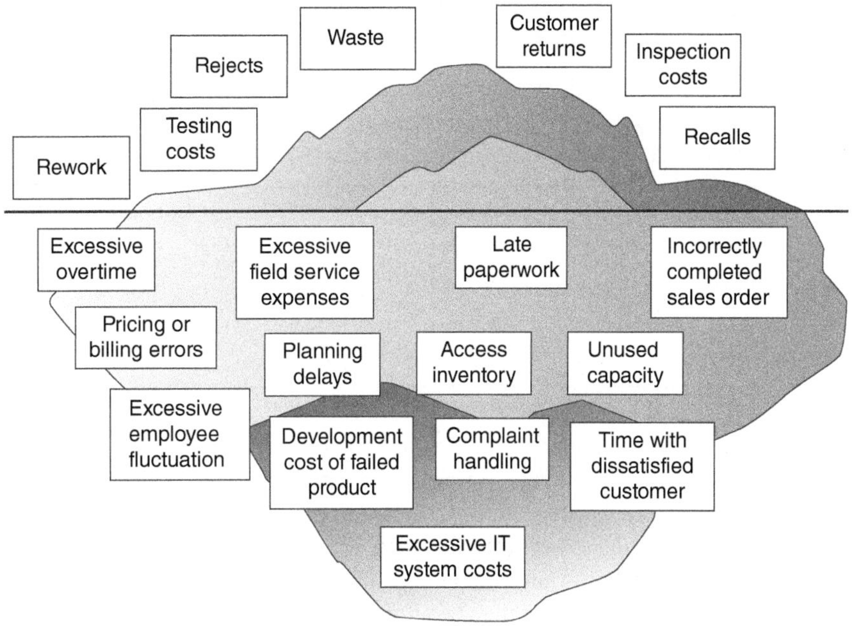

Figure 8.23 Iceberg model of cost of quality.

and products and doing things right the first time, the increase of the cost of good quality, while striving for zero defect performance, can be smoothed if processes get better.

As Figure 8.24 shows, business processes with better process sigma will have significantly lower prevention and appraisal costs. Although you will never fully eliminate appraisal and prevention costs (as opposed to failure costs that in an ideal zero defect world would also be zero), their reduction due to better process performance will be significant.

Table 8.5 shows how dramatically the cost of quality as a percentage of sales decreases if the process sigma improves. Assuming that the average performance of a company is 3 sigma, 25–40% of its annual revenue gets chewed up by the cost of quality. Thus, if a company can improve its quality by 1 sigma level, its net income will increase hugely. Therefore, Six Sigma quality, or 3.4 DPMO, led to elimination of costly inspection and rework, which in turn led to decrease in manufacturing time and increase in customer satisfaction that yield much higher financial savings (Morgan and Brenig-Jones 2016).

8.16.10 Energy-Efficiency Plan for Lean Manufacturing

Manufacturing accounts for almost 25% of all US energy use. Industry spends almost $230 billion on energy each year, but these costs could be significantly reduced through energy-efficiency efforts. More than $40 billion could go back into company pockets each year by leveraging proven energy-efficiency technologies and practices (Lung 2018).

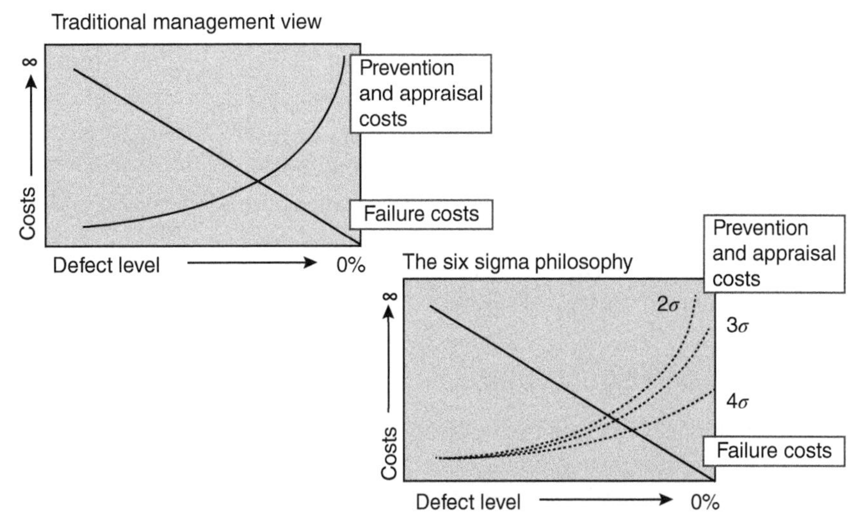

Figure 8.24 Traditional management view vs. Six Sigma philosophy.

Table 8.5 Dramatic improvement in the cost of quality as a percentage of sales decreases if the process sigma improves.

Sigma level	Defects per million opportunities (DPMO)	Cost of quality as percentage of sales (%)
2	308 000	>40
3	66 800	25–40
4	6 210	15–25
5	230	5–15
6	3.4	<1

In industrial process, the goal is to produce a product at a high level of quality and safety, for the least amount of money – i.e. efficiently. *Energy efficiency* is an *integral part of lean and sustainable manufacturing*, and energy efficiency plays a key role in this effort. We should consider *energy a raw material* – it would then be prudent to manage this resource – using less raw materials to make the same amount of product at the same quality (0 defect) – also less CO_2 emission.

The first step in managing anything – finances, health, emissions etc. – is to measure it. Energy intensity is a useful metric for quantifying energy efficiency because it not only accounts for the amount of energy consumed but also the amount of product produced. Many facilities pursuing an energy efficiency program have an *energy manager* who is accountable for the energy use decisions. No matter how efficient the new pump you install is, it won't be very efficient if it is over designed and let run under capacity, e.g. a 5 hp pump is only delivering fluids at 0.5 hp. Therefore, designing and selecting an optimum pump size is important to energy-saving process.

The U.S. Department of Energy's (DOE's) Better Plants program has been partnering with US-based manufacturers and industrial-scale energy users to improve energy performance. The US DOE's Better Plants program also provides technical assistance, training, software, and tools to help manufacturers improve their energy performance. This voluntary program provides participants with a broad range of tools and resources to overcome barriers and identify opportunities to save money. More than 200 organizations are Better Plants partners who collectively saved more than 1 quadrillion Btu of energy in total, which translates to about $5.3 billion in energy cost savings (Lung 2018).

8.16.11 Become ISO 50001 Ready

The International Organization for Standardization (ISO) created ISO 50001 as a voluntary energy management standard to help organizations use energy efficiently and cut cost. The US DOE developed the 50001 Ready recognition program as a way to guide organizations in setting up an energy management system using the ISO 50001 structure. 50001 Ready allows organization to self-attest to these efforts and can be a useful first step in pursuing ISO 50001 certification (Lung 2018). Achieving 50001 Ready recognition involves three steps:

1) Complete the 25 tasks of the 50001 Ready Navigator, a free online tool that helps guide and track progress.
2) Self-attest to the completion of the 25 Navigator tool tasks.
3) Measure and improve site energy performance over time.

8.16.12 A Ten-Step Outline for Energy Analysis: Understand the Energy Used to Transform Raw Material into Finished Product to Enhance Energy Efficiency (Stowe 2018)

Step 1. Identify the raw materials
Step 2. Characterize the final product(s)
Step3. Tour the plant
Step 4. Develop the process flow (or block) diagram
Step 5. Catalog energy inputs
Step 6. Identify energy wastes
Step 7. Identify energy-recovery possibilities
Step 8. Pinpoint energy-efficiency opportunities
Step 9. Identify new technology opportunities
Step 10. Implement solutions; energy analysis for energy-efficiency program should be an ongoing process

8.17 Essential Roles of Industrial Environmental Managers

Industrial environmental managers (IEMs) working in manufacturing facility environmental programs are responsible for overall environmental management system (EMS) in a comprehensive, systematic, planned, and documented manner regarding *waste minimization or reduction* and *compliance* with environmental regulations and standards. These managers act as stewards of environmental ethics. The IEMs work closely within an environmental management process (EMP) team, including process operation engineers, quality control engineers with a goal of high-quality manufactured products while optimizing their use of resources and reducing wastes. Their tasks include but are not limited to support training fresh environmental engineers and personnel: safety, sampling, monitoring, analyzing, summarizing, and managing environmental data; preparing various permits for compliance assurance, and reporting of specialized environmental performances information to internal and external stakeholders of the industry. These activities facilitate the EMP team, organizational structure, planning, and resources for developing, implementing, and maintaining policy for environmental protection (Sroufe 2003).

8.18 Goals of IEMs

The major goals of IEMs are to continuously increase industry and corporate regulatory compliances and concurrently reduce wastes. Typically, an environmental manager follows a PDCA cycle (Figure 8.25). The diagram shows the process of first developing an environmental policy, planning the environmental management process, and then implementing it. The process also includes checking the system and acting on it. The model process continuously guides improvement through organizational review and revision of the system (ISO 2018b). This is a model that can be used by a wide range of organizations – from manufacturing facilities to service industries to government agencies.

8.19 Environmental Compliance and Compliance Assurances

Environmental Compliance and Its Assurances means conforming to environmental laws, regulations, standards, and other requirements such as site permits to operate. In recent years, environmental concerns have led to a significant increase in the number and scope of compliance imperatives across all global regulatory environments. Being closely related, environmental concerns and compliance activities are increasingly being aligned with

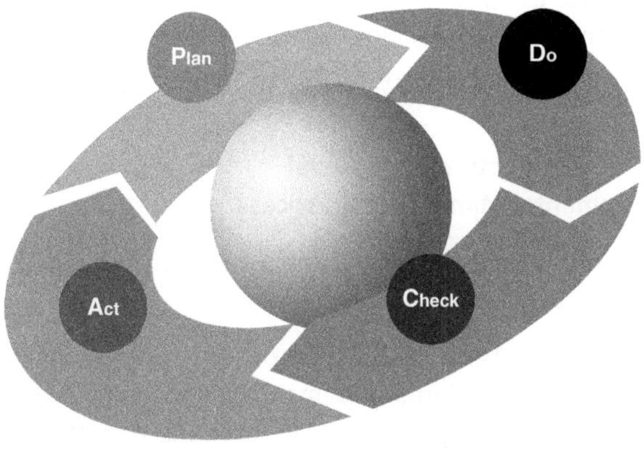

Figure 8.25 PDCA cycle for environmental management process.

corporate performance goals and being integrated to some extent to avoid conflicts, wasteful overlaps, and gaps (Tarantino 2008).

Compliance with the above requirements and obligations requires meeting certain conditions. Typically, these include

- developing comprehensive *spreadsheets and compliance assurance matrices*
- managing monitoring programs or schedules, ensuring that the monitoring required in various permits has been done at the correct locations, for the correct parameters and at the correct frequency
- preprocessing, performing calculations, and validating the data for compliance with any alert or reporting levels
- generating routine compliance reports for state and federal regulatory authorities

The management of the above can be complex and time-consuming, leading to an increasing uptake of software systems designed to manage environmental compliance. These are often referred to as *environmental data management systems*. Criteria must be considered when selecting environmental compliance software: proven capability, high performance, transparent, traceable data handling, a robust calculation engine, advanced factor handling, simple integration, automated workflows and quality assurance, and flexible reporting and data extraction.

8.20 Waste Reduction

Waste reduction or minimization is a set of processes and practices intended to reduce the amount of waste produced. By reducing or eliminating the generation of harmful and persistent wastes, waste minimization supports efforts to promote a more sustainable society. Waste reduction involves redesigning products and processes and/or changing societal patterns of consumption and production.

The most environmentally resourceful, economically efficient, and cost effective way to manage waste often is to not have to address the problem in the first place. Managers see waste reduction as a primary focus for most waste management strategies. Proper waste treatment and disposal can require a significant amount of time and resources; therefore, the benefits of waste minimization or reduction can be considerable if carried out in an effective, safe and sustainable manner.

Traditional waste management focuses on processing waste after it is created, concentrating on reuse, recycling, and waste-to-energy conversion (also see Appendix G).

Waste reduction involves efforts to avoid creating the waste during manufacturing. To effectively implement waste reduction the IEM requires knowledge of the production process, cradle-to-grave analysis (the tracking of materials from their extraction to their return to earth), and details of the composition of the wastes.

The main sources of waste vary from country to country. In industrialized and developing countries, most waste comes from various manufacturing industries, followed by mining, construction and demolition of buildings, and commerce. Industrial waste is often tied to requirements in the supply chain. In the 1980s, the term supply-chain management (SCM) was developed to express the need to integrate the key business processes, from end user through original suppliers (Oliver and Webber 1992). Original suppliers are those that provide products, services, and information that add value for customers and other stakeholders. The basic idea behind SCM is that companies and corporations involve themselves in a supply chain by exchanging information about market fluctuations and production capabilities.

For example, a German paper factory receives its daily supply of 100 tons of recyclable paper as its raw material. A company handling a product may insist that it should be shipped using particular packing because it fits downstream needs. Another example, a Brazilian scrap metal recycling company receives over 50 tons of used aluminum cans to reprocess and recycle. Brazil recycles 90.4% of its aluminum can production, equivalent to 14.5 billion beverage cans per year, ranking first in the world, more than Japan's 83.2% recovery rate. Brazil has topped the aluminum can recycling charts eight years in a row (AFG 2018).

8.20.1 Reuse and Recycling Processes

- Reuse of scrap material
  Scraps can be immediately reincorporated at the beginning of the manufacturing line so that they do not become a waste product. Many industries routinely do this; for example, paper mills return any damaged rolls to the beginning of the production line, and in the manufacture of plastic items, off-cuts and scrap are reincorporated into new products.
- Improved quality control and process monitoring
  Steps can be taken to ensure that the number of reject batches is kept to a minimum. This is achieved by increasing the frequency of inspection and the number of points of inspection. For example, installing automated continuous monitoring equipment can help to identify production problems at an early stage.
- Waste exchanges
  This is where the waste product of one process becomes the raw material for a second process. Waste exchanges represent another way of reducing waste disposal volumes for waste that cannot be eliminated.
- Ship to point of use
  This involves making deliveries of incoming raw materials or components direct to the point where they are assembled or used in the manufacturing process to minimize handling and the use of protective wrappings or enclosures.
- Zero waste
  This is a whole systems approach that aims to eliminate waste at the source and at all points down the supply chain, with the intention of producing no waste. It is a design philosophy which emphasizes waste prevention as opposed to end-of-pipe waste management (Das 2005).

8.20.2 Benefits of Waste Minimization

Waste minimization can protect the environment and often turns out to have positive economic benefits. Waste minimization can improve the following:

- Efficient production practices. Waste minimization can achieve more output of product per unit of input of raw materials.
- Economic returns. More efficient use of products means reduced costs of purchasing new materials improving the financial performance of a company.
- Public image. The environmental profile of a company is an important part of its overall reputation and waste minimization reflects a proactive movement toward environmental protection.
- Quality of products produced. New innovation and technological practices can reduce waste generation and improve the quality of the inputs in the production phase.
- Environmental responsibility. Minimizing or eliminating waste generation makes it easier to meet targets of environmental regulations, policies, and standards. The environmental impact of waste will be reduced.

8.20.3 Key Features: Industrial Environmental Management Process

An industrial environmental management process (IEMP) and roles of environmental managers:

- Serves as a tool or process to improve environmental performance and information mainly "design, pollution control and waste minimization, training, reporting to top management, and the setting of goals."

- Provides a systematic way of managing the organization's environmental affairs.
- Is the aspect of the industry's overall management structure that addresses immediate and long-term impacts of its products, services, and processes on the environment. IEMP assists with planning, controlling, and monitoring policies in the organization. Develops some detailed spreadsheets and compliance matrices.
- Gives order and consistency for organizations to address environmental concerns through the allocation of resources, assignment of responsibility and ongoing evaluation of practices, procedures, and processes.
- Creates environmental buy-in from upper management and employees and assigns accountability and responsibility.
- Sets framework for training to achieve objectives and desired performance.
- Helps all who are involved in the processes to better understand legislative requirements to determine a product's or service's impact, significance, priorities, and objectives.
- Focuses on continual improvement of the system and a way to implement policies and objectives to meet a desired result. This also helps with reviewing and auditing the IEMP to find future opportunities.
- Encourages contractors and suppliers to establish their own IEMP.
- Facilitates record keeping and reporting/e-reporting to federal, state, and local governmental environmental agencies through direct upload.

Problems

8.1 The College of Engineering at a major university wants to develop an empirical model to predict the success of its undergraduate students in completing their degrees.

A What would you use as the response or outcome variable in this study?

B Discuss how you would collect data to build this model. Is this is a retrospective study or an observational study?

8.2 An engineer wants to obtain a random sample of the output of a process manufacturing digital cameras. She samples on three different days, and on each day she selects five cameras at random at 3 p.m. and 5 p.m. from the production line output. Is this a random sample?

8.3 An engineer draws a simple random sample of 50 bearings from a large production run and measures their inside diameter. He finds that three of them are outside the diameter specifications. Since 6% of the sample units are defective, he reports to the manufacturing manager that exactly 6% of the production is defective. Is this the correct conclusion to draw from these data?

8.4 A student in a laboratory course on quality control methods measure the length of a bolt several times with a micrometer. When would it be reasonable to consider these measurements a random sample? What is the population?

8.5 The pH of a solution is measured eight times by one operator using the same instrument. She obtains the following data: 7.15, 7.20, 7.18, 7.19, 7.21, 7.20, 7.16, and 7.18.

A Calculate the sample mean. Suppose that the desirable value for this solution was specified to be 7.20. Do you think that the sample mean value computed here is close enough to the target value to accept the solution as conforming to target? Explain your reasoning?

B Calculate the sample variance and sample standard deviation. What do you think are the major sources of variability in this experiment? Why is it desirable to have a small variance of these measurements?

8.6 Draw a dot diagram that illustrates the variation and range of data but doesn't identify the problem. The data are 40.75, 41.5, 42.25, 43, 43.75, 44.5, 44.5, 45.25, 45.25, 46.75, 47.5, 47.5, 48.25, 48.25, 48.25, 48.25, 48.25, 49.75, 49.75, 50.5, 51.25, 51.25, 51.25, 52, 53.5, 55.

8.7 Draw a Pareto chart in analyzing the occurrence of chemical tank overflows due to specific causes, such as sensor, pump, electrical contacts, operator, maintenance, and repair.

Data are given for recurrence failures due to various causes as follows:

Sensor	45
Electrical contacts	35
Pump	11
Operator	5
Repair person	5

8.8 (a) What is the basic concept of 80/20 rule concerning cause and effect in manufacturing? (b) Draw a Pareto chart using the data given in the table below and show which items or issues have the biggest impact on the process or system.

Defect	Count
Element 7	56
Element 6	13
Element 3	7
Element 1	5
Element 8	3
Element 2	1
Others	1

8.9 Suppose daily inspection of a product results in rejection due to scratches, dents, smudge, failure to meet specifications, and other causes. Suppose the data shown in a table below was gathered during a given shift at a large company. Draw (a) Excel pie chart; (b) Excel bar chart; and (c) a Pareto chart and show which items or issues have the biggest impact on the process or system.

Category	Frequency	Percent
Dent	15	30
Other	7	14
Scratch	10	20
Smudge	4	8
Spec	14	28

8.10 The reasons due to which aircraft tanks were classified as defective were collected over several months and are shown as the causes and counts in the table below. Form a Pareto chart for the defective aircraft tanks.

Cause	Count	Cause	Count	Cause	Count
Adhesive	6	Dimensions	36	Paint dam	1
Alignment	2	Fairing	3	Paint spec	2
Alodine	1	Film	5	Primer dam	1
Cast voids	2	Machining	29	Procedure	1
Damaged	34	Masking	17	Rusted	13
Dalam comp	2	Out order	4	Salt spray	4
				Wrong part	3

8.11 Explain how to construct a Pareto chart for the data in Problem 8.10.

8.12 Name two scientists who pioneered the SPC techniques and briefly mention their contributions.

8.13 A factory quality control center measured moisture contents in 160 packages and recorded sample mean, $\overline{X}$, sample range, $\overline{R}$, and standard deviation, $\overline{s}$, respectively. Compute (a) the $3\sigma\overline{R}$ chart UCL and LCL for the moisture sample mean and range values; (b) the $3\sigma\overline{X}$ chart UCL and LCLs the moisture sample mean and range values.

8.14 Define five clauses mentioned in the ISO 9000 series pertinent to quality management systems (QMS). Also, define ISO 14000 for an EMS and environmental policy pertinent to QMS and industrial waste minimization.

8.15 What is quality?

8.16 What are major differences between quality assurance, quality control, and inspection?

8.17 What is the cost of quality?

8.18 What are the concepts of "Lean Manufacturing"?

8.19 What is "KANBANs"?

8.20 Describe briefly "Toyota's Seven Wastes."

8.21 Describe the "value-added and waste elimination" initiatives as part of "Lean Principles."

8.22 Describe DMAIC model which is integral to the success of a Six Sigma project.

8.23 Compare the two methodologies: Six Sigma and Lean Manufacturing.

8.24 Name four risk categories encountered by manufacturers and briefly describe each.

8.25 Engineering is an important and learned profession. As members of this profession, the engineers are expected to exhibit the highest standards of ethics. Soon to be a member of your engineering profession, share ideas and ideals about "ethics in engineering" in your own words.

References

Aluminum for Future Generation (AFG) (2018). http://recycling.world-aluminium.org/regional-reports/brazil.html (accessed 16 December 2018).

Arnold, H.L. and Faurote, F.L. (1915). *Ford Methods and the Ford Shops*. New York: The Engineering Magazine Company.

Bennett, H. and Marcus, P. (1951). *We Never called Him Henry*. New York: Fawcett Publications.

Bicheno, J. and Holweg, M. (2009). *The Lean Toolbox: Essential Guide to Lean Transformation*, 4e. Buckingham: PICSIE Books.

Brundtland, H. (1987). *Our Common Future: The World Commission on Environment and Development*. Oxford: Oxford University Press.

Das, T.K. (2002). Evaluating the life cycle environmental performance of chlorine disinfection and ultraviolet technologies. *Clean Technology and Environmental Policy* 4: 32–43.

Das, T.K. (2005). *Toward Zero Discharge: Innovative Methodology and Technologies for Process Pollution Prevention*. Hoboken, NJ: Wiley.

Das, T.K. and Houtman, C.J. (2004). Evaluating chemical-, mechanical-, and bio-pulping processes and its sustainability characterization using life-cycle assessment. *Environmental Progress and Sustainable Energy* 23 (4): 347–357.

Dornfeld, D. and Lee, D.-E. (2008). *Precision Manufacturing*. New York: Springer.

Dornfeld, D. and Wright, P. (2007). Technology wedges for implementing green manufacturing. *Transactions of North American Manufacturing Research Institute* 35: 193–200.

Evans, J.R. and Lindsay, W.M. (1992). *The Measurement and Control of Quality*, 2e. Minneapolis/St. Paul, MN: West Publishing Company.

Ford, H. and Crowther, S. (1922). *My Life and Work*. New York: Garden City Publishing Company.

Freedman, D., Pisani, R., and Purves, R. (1997). *Statistics*, 3e. W. W. Norton.

Groover, M.P. (1999). *Fundamentals of Modern Manufacturing: Materials, Processes, and Systems*. New York: Wiley.

Hanna, J. (2007). Bringing lean principles to service industries. HBS Working Knowledge. (Summary article based on published research of Professor David Upton of Harvard Business School).

Hawken, P. (1993). *The Ecology of Commerce*. New York: Harper Collins.

Hopp, W.J. (2011). *Supply Chain Science*. Long Grove, IL: Waveland Press.

Hopp, W.J. and Spearman, M.L. (2008). *Factory Physics*, 3e. Long Grove, IL: Waveland Press.

Hounshell, D.A. (1984). *From the American System to Mass Production, 1800–1932: The Development of Manufacturing Technology in the United States*. Baltimore, MD: Johns Hopkins University Press.

Howitt, D. and Cramer, D. (2008). *Introduction to Statistics in Psychology*, 4e. Upper Saddle River, NJ: Prentice Hall.

International Organization for Standardization (ISO) (2018a). ISO 9000 quality management. www.iso.org (accessed 24 september 2019).

International Organization for Standardization (ISO) (2018b). Environmental management software: the better way to manage and maintain ISO 14001 compliance. www.iso.org (accessed 16 December 2018).

Ishii, K. (1999). Incorporating end-of-life strategy in product definition. *Eco-Design '99: First International Symposium on Environmentally Conscious Design and Inverse Manufacturing*, Tokyo, Japan (February 1999).

Januska, M. and Štastna, L. (2013). Industrial engineering in the non-manufacturing processes. *Proceedings of the 22nd International Business Information Management Association Conference: International Business Information Management Association (IBIMA)*, Rome, Italy (13–14 November 2013), 747–766.

Ker, J.I., Wang, Y., Hajli, M.N. et al. (2014). Deploying lean in healthcare: evaluating information technology effectiveness in US Hospital Pharmacies. *International Journal of Information Management* 34 (4): 556–560.

Krafcik, J.F. (1988). Triumph of the lean production system. *Sloan Management Review* 30 (1): 41–52.

Liker, J.K. (2004). *The Toyota Way*. New York: McGraw-Hill.

Liker, J.K. and Hoseus, M. (2008). *Toyota Culture: The Heart and Soul of The Toyota Way*. New York: McGraw-Hill.

Lung, R. (2018). Better plants: working with manufacturers to save money. *Chemical Engineering Progress*: 31–34.

Martin, C. (2005). *Logistics and Supply Chain Management: Creating Value-Added Networks*, 3e. Upper Saddle, NJ: Prentice-Hall.

Martin, M. and Schinzinger, R. (2004). *Ethics in Engineering*, 4e. New York: McGraw-Hill.

Merrill, D. (2013). The lean supply chain: watch your waste line. *Inbound Logistics: The Magazine for Demand-Driven Logistics* 33.

Montgomery, D.C. (1985). *Statistical Quality Control*. New York: Wiley.

Montgomery, D.C. (2009). *Introduction to Statistical Quality Control*, 6e. Hoboken, NJ: Wiley.

Montgomery, D.C., Runger, G.C., and Hubele, N.F. (2011). *Engineering Statistics*, 5e. Hoboken, NJ: Wiley.

Morgan, J. and Brenig-Jones, M. (2016). *Lean Six Sigma for Dummies*, 3e. Chichester: Wiley.

Ohno, T. (1988). *Toyota Production System: Beyond Large Scale Production*. Portland, OR: Productivity Press.

Oliver, R.K. and Webber, M.D. (1992). Supply-chain management: logistics catches up with strategy. In: *Logistics: The Strategic Issues* (ed. M. Christopher). London: Chapman Hall.

Pacala, S. and Socolow, R. (2004). Stabilization wedges: solving the climate problem for the next 50 years with current technologies. *Science* 305: 968–972.

Padmane, G.T., Sohel, M.T., and Patil, P.D. (2013). Role of IT in lean manufacturing: a brief scenario. *Journal of Modern Engineering Research (IJMER)* 3 (5): 3026–3031.

Pearson, K. (1895). Contributions to the mathematical theory of evolution. II. Skew variation in homogeneous material. *Philosophical Transactions of the Royal Society A: Mathematical, Physical and Engineering Sciences* 186: 343–414.

Pettersen, J. (2009). Defining lean production: some conceptual and practical issues. *The TQM Journal* 21 (2): 127–142.

Radnor, Z., Walley, P., Stephens, A., and Giovanni, B. (2006). *Evaluation of the Lean Approach to Business Management and ITs Use in the Public Sector*. Edinburgh: Warwick Business School, Scottish Executive Social Research.

Ruffa, S.A. (2008). *Going Lean: How the Best Companies Apply Lean Manufacturing Principles to Shatter Uncertainty, Drive Innovation, and Maximize Profits*. New York: AMACOM.

Sheng, P., Dornfeld, D., and Worhach, P. (1995). Integration issues in green design and manufacturing. *Manufacturing Review* 8 (2): 95–105.

Shingo, S. (1987). *The Sayings of Shigeo Shingo: Key Strategies for Plant Improvement*, 1e. Nikkan Kōgyō Shimbun (English Edition by Productivity Press).

Snee, R. and Hoerl, R. (2003). *Leading Six Sigma: A Step-by-Step Guide Based on Experience with GE and Other Six Sigma Companies*. Upper Saddle River, NJ: Financial Times Prentice Hall.

Spear, S. and Bowen, H.K. (1999). Decoding the DNA of the Toyota production system. *Harvard Business Review* 77 (5): 1–12.

Srinivasan, M. and Sheng, P. (1999a). Feature-based process planning for environmentally conscious machining. Part 1: Macro-planning. *Robotics and Computer Integrated Manufacturing* 15: 257–270.

Srinivasan, M. and Sheng, P. (1999b). Feature-based process planning for environmentally conscious machining. Part 2: Macro-planning. *Robotics and Computer Integrated Manufacturing* 15: 271–281.

Sroufe, R. (2003). Effects of environmental management systems on environmental management practices and operations. *Production and Operations Management* 12 (3): 416–431.

Stangor, C. (2011). *Research Methods for the Behavioral Sciences*. Wadsworth: Cengage Learning.

Stein, J. and Dornfeld, D.A. (1996). Integrated design and manufacturing for precision mechanical components. *Proceedings of the First IDMME Conference*, Nantes, France (April 1996).

Stein, J. and Dornfeld, D.A. (1997). An architecture for integrated design and manufacturing for precision mechanical components. *Transactions of North American Manufacturing Research Institute* 25: 249–254.

Stowe, M. (2018). A ten-step process for energy analysis. *Chemical Engineering Progress* 114: 36–41.

Tague, N.R. (2004). *Seven Basic Quality Tools: The Quality Toolbox*. Milwaukee, WI: American Society for Quality.

Taniguchi, M. Kakinuma, Y., Aoyama, T. and Inasaki, I. (2005). Influences of downsized machine tools on the environment. *Proceeding MTTRF Meeting*, San Francisco, CA.

Tarantino, A. (2008). *Governance, Risk, and Compliance Handbook: Technology, Finance, Environmental, and International Guidance and Best Practices*. Hoboken, NJ: Wiley.

Taylor, F.W. (1911). *The Principles of Scientific Management*. New York and London: Harper & Brothers.

Tukey, J.W. (1977). *Exploratory Data Analysis*. Reading, MA: Addison, Wesley.

Vallero, D. and Brasier, C. (2008). *Sustainable Design: The Science of Sustainability and Green Engineering*. Hoboken, NJ: Wiley.

Ward, S. and Poling, S.R. (2004). *Manufacturing Engineering Handbook* (ed. H. Geng), Chapter 56, 56.1–56.11. New York: McGraw-Hill.

Wild, C. and Seber, G. (2000). *Chance Encounters: A First Course in Data Analysis and Inference*. Hoboken, NJ: Wiley.

Wilkinson, L. (2006). Revising the Pareto Chart. *The American Statistician* 60: 332–334.

Womack, J.P. and Jones, D.T. (2003). *Lean Thinking: Banish Waste and Create Wealth in Your Corporation*. New York: Simon & Shuster.

Womack, J.P., Jones, D.T., and Roos, D. (1990). *The Machine That Changed the World*. New York: Free Press.

Zimmerman, J., Hayes, K., and Skerlos, S. (2003). Influence of ion type and concentration on the emulsion stability and machining performance of two semi-synthetic metalworking fluids. *Environmental Science and Technology* 38: 2483–2490.

9

Industrial Waste Minimization Methodology

Industrial Ecology, Eco-Industrial Park and Manufacturing Process Intensification and Integration

9.1 Introduction

In the United States, roughly 12 billion T of nonhazardous waste is generated and disposed of by the US industries. That amount is about is over 200 lb of industrial waste per person per day. The largest industrial contributors to nonhazardous waste are manufacturing industries (~7600 million T/Y), oil and gas production (2100–3600 million T/Y), and the mining industry (>1400 million T/Y). Contributors of lower amounts are electricity generators (fly ash and flue-gas desulfurization waste), construction waste, hospital infectious waste, and waste tires. The rate of industrial hazardous waste generation in the United States is approximately 750 million T/Y (Allen and Shonnard 2012; Allen and Rosselot 1997). Once these materials are designated as hazardous, the costs of managing, treating, storing, and disposing of it increase dramatically. This chapter describes some specific industrial waste minimization processes and technologies that have been successfully operating and provides other methodologies, including industrial ecology, eco-industrial park (EIP), manufacturing process intensification, and integration. The wastes (in air, water, or as solid) or by-products generated during manufacturing process are recovered. The materials and energy recovered from waste streams either are reused in the plant or are sold to another plant as feedstock. It is in practice, as well as in theory, possible to isolate some industrial facilities almost completely from the environment by recycling all wastes into materials that can then be manufactured into consumer products. An example of such a facility is a coal-fired power plant. An electron beam–ammonia conversion unit adds ammonia to the effluent gases, which is then irradiated electronically, producing ammonium nitrate and ammonium sulfate that are sold as feedstock to fertilizer manufacturing; enhanced recovery of mercury from flue gas by adsorption and mercury recovery from a coal-fired plant. The details of these two processes are given as case studies later in this chapter. Also, two separate case studies have been presented that highlight a profitable industrial "by-product-to-energy" recovery generating electricity and heat, and making chemicals and energy from gasification of black liquor, as by-product of pulping process.

Our goal is to modify industrial processes so that services and manufactured goods can be produced without waste. But it is important to understand that some manufacturing processes inherently produce wastes, even after all reasonable efforts at pollution prevention. Thus, in some cases, the use of a conversion technology may be more appropriate than a program of pollution prevention: many industrial wastes can be processed to render them viable as material inputs to another industry or to part of an industrial cluster of several connected industries – as part of the movement of "industrial ecology."

9.2 Industrial Ecology

Industrial ecology (IE) is the study of material and energy flows through industrial systems. The global industrial economy can be modeled as a network of industrial processes that extract resources from the Earth and transform those resources into commodities which can be bought and sold to meet the needs of humanity. IE seeks to quantify the material flows and document the industrial processes that make modern society function. Industrial ecologists are often concerned with the impacts that industrial activities have on the environment, with use of the planet's supply of natural resources, and with problems of waste disposal. IE is a growing multidisciplinary field of research which combines aspects of engineering, economics, sociology, toxicology, and the natural sciences (Allenby 2006; Ashton 2009; Jensen 2011).

IE provides the theoretical scientific basis upon which understanding, and reasoned improvement, of current practices can be based. It incorporates, among other things, research involving energy supply and use, new materials,

Industrial Environmental Management: Engineering, Science, and Policy, First Edition. Tapas K. Das.

Companion website: www.wiley.com/go/Das/IEM_1e

new technologies, basic sciences, economics, law, management, and social sciences. It encompasses concurrent engineering, design for the environment (DFE), dematerialization, process pollution prevention, waste conversion, waste exchange, by-product utilization to a new product, waste minimization, and recycling, with *Zero Emissions* as an important subset. IE can also be a policy tool. It is a view of a system in which one seeks to optimize the total materials cycle from virgin material to finished material, to component, to product, to obsolete product, and to ultimate disposal. Factors to be optimized include resources, energy, and capital.

IE is a dynamic, systems-based framework that enables management of human activity on a sustainable basis by minimizing energy and materials usage, ensuring acceptable quality of life for people, minimizing ecological impacts of human activity to levels natural systems can sustain, and maintaining economic viability of systems for industry, trade, and commerce. The IE approach involves (i) application of science to industrial systems, (ii) defining the system boundary to incorporate the natural world, and (iii) seeking to optimize that system. In this context, industrial systems apply not just to private sector manufacturing and service but also to government operations, including provision of infrastructure.

IE is a framework for designing and operating industries as living systems interdependent with natural systems, and therefore it is essential to grasp the concept of industrial metabolism. Whereas the biosphere's metabolism is a near-perfect recycling system because so many of its components are capable of biological regeneration, in industry, metabolism depends largely on the combustion of fossil fuels that are not regenerated within the system.

Thus, students of industrial metabolism investigate the mass flows of key industrial materials of environmental significance and their associated emissions. With this information in hand, industrial ecologists are better able to balance environmental concerns and economic performance, factoring in the real value of nonrenewable resources and the real costs of environmental pollution. These efforts are enhanced by improvements in understanding of local and global ecological constraints.

IE supports coordination of design over the life cycle of products and processes. IE enables creation of short-term innovations in a long-term context. While much of the initial work in IE has focused on manufacturing, a full definition of industrial systems includes service, agricultural, manufacturing, military, and other public operations, as well as infrastructure such as landfills, water and sewage systems, and transportation systems.

The concept of an "industrial ecosystem" received wide attention when *Scientific American* published an article by two General Motors researchers who suggested that the days of finding an "open space beyond the village gates" for the by-products of industrial activity were quickly fading (Frosch and Gallopoulos 1989). The concept of IE has spawned an ever increasing amount of research and activities. At the most basic level, IE describes a system where one industry's wastes (outputs) become another's raw materials (inputs). Within this "closed loop, illustrated in Figure 9.1," fewer materials would be wasted. Thus, if businesses were able to turn waste into food, they could sharply reduce pollution and the need for raw materials (Van der Ryn and Cowan 1996).

Many industries have long had symbiotic relationships where wastes and materials are transformed internally or by others. For example, metal industries use scrap materials in the production process; the advent of the electric arc furnace (EAF) increased the ability of steel manufacturers to use scrap materials. Petrochemical and chemical companies are adept at finding new production uses or markets for waste materials (Richards et al. 1994). The growth in rubber, plastics, paper, and glass recycling has generated new uses for previously discarded materials. As Ernest Lowe suggests, IE is a broad holistic framework for guiding the transformation of the industrial systems. The shift from the linear model (mine pit to producer, to consumer, to dump) to a closed-loop model, more closely resembling the cyclical flows of ecosystems, has stimulated new ways of thinking in forward-looking companies, in a number of universities, and in governmental agencies like the Environmental Protection Agency and the Department of Energy in the United States (Low and Warren 1996; Lowe 1995, 2001).

9.2.1 What Is EIP?

According to Ernest Lowe, John Warren, Andreas Hein, and others, an EIP is a community of manufacturing and service businesses seeking enhanced environmental and economic performance through collaboration in managing environmental and resource issues, including energy, water, and materials (Lowe and Warren 1996). By working together, the community of businesses seeks a collective benefit that is greater than the sum of the individual benefits each company would realize if it optimized its individual performance only. The goal of an EIP is to improve economic performance of the participating companies while minimizing their environmental impact.

Industrial activity releases wastes into local, regional, and eventually global ecosystems. Since the 1970s, myriad laws and regulations have been promulgated to limit emissions to air and water, and to regulate solid and hazardous waste disposal. Further, industrial activity results in nonpoint-source

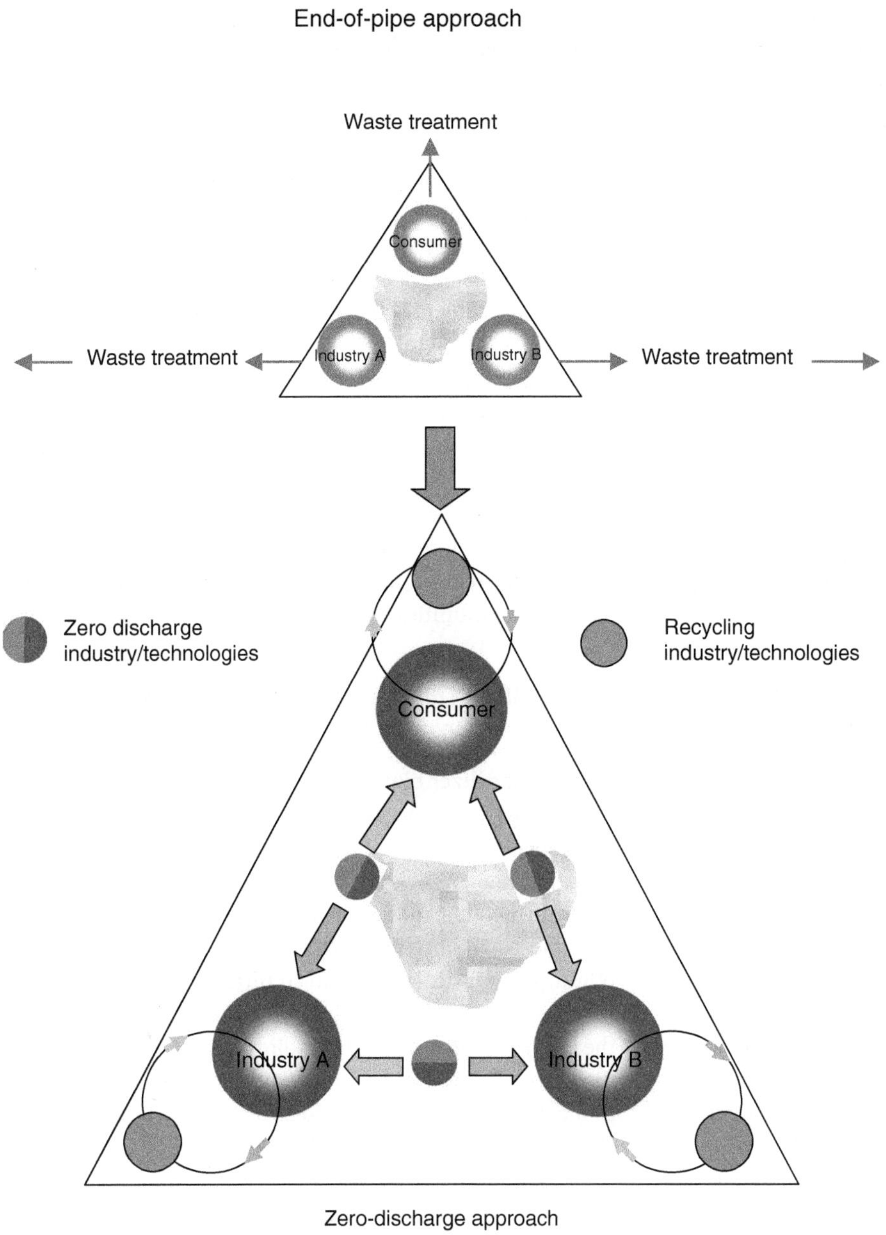

Figure 9.1 Schematic diagram of end-of-pipe and Zero-Emission approaches. *Source:* From Ebara Corporation, Tokyo, Japan (2010) and Das (2005).

pollution caused by general runoff, spills, or illegal dumping. Business and the environment have traditionally been considered natural enemies. The assumption has long been "more environmental protection corresponds to higher costs for business"; however, new developments in research and business operations are challenging this assumption. Many companies such as 3 M have long realized the economic benefits of applying environmental principles to business operations. The goal of an EIP is to minimize the ecological impact of industrial activity and to improve business performance. In the United States and Canada, several projects are testing this idea, which has already demonstrated its effectiveness in Europe, most spectacularly perhaps, in Kalundborg, Denmark, as discussed by Hein et al. (2015); EIO 2012; Lowe (2001), and others.

An EIP is a set of businesses that share resources in order to increase profitability and reduce environmental impact. The implementation of EIPs may significantly contribute to the creation of a sustainable economy. Despite this prospect, the actual development of EIPs is challenging, as a variety of factors must be considered. Not only technical, economic, and environmental factors are relevant but numerous stakeholder relationships as well, such as between firms, governmental bodies, and local communities. Hein et al. (2015) presents a conceptual framework that is used to capture these diverse aspects and the relationships between them. The Unified Modeling Language is used for modeling its concepts and relationships. First, based on a literature survey, relevant concepts of EIPs are identified. One central concept is "industrial symbiosis." A novel value-based interpretation of industrial symbiosis is presented. Second, the park's economic, local, and regional development context, as well as its internal technical components and their relationships, are modeled. Finally, the framework is used for modeling a concrete EIP, in this case part of the Kalundborg EIP (Hein et al. 2015).

9.2.2 EIP Development

EIP development (EIPD) is a new paradigm for achieving excellence in business and environmental performance. It opens up innovative new avenues for managing businesses and conducting economic development. By creating linkages among local "resources," including businesses, nonprofit groups, governments, unions, and educational institutions, communities can creatively foster dynamic and responsible growth. Antiquated business strategies, based on isolated enterprises, are no longer responsive enough to market, environmental, and community requirements.

Economic development is a never-ending challenge for communities. As the global marketplace has become increasingly competitive, municipalities, counties, states, and regions seek new strategies for attracting good investments with good jobs. Moreover, communities everywhere are demanding improvements in local ecosystems. In the past, economic development and environmental protection were seen as mutually exclusive. However, new practices and activities are challenging that assumption. One broad category of activities falls under the umbrella of EIPD.

Sustainable EIPD looks systematically at development, business, and the environment attempting to stretch the boundaries of current practice. On one level, it is as directly practical as making the right connections between wastes and resources needed for production. At another level, it is a whole new way of thinking about doing business and interacting with communities. The eco-industrial approach has many ways of being applied. At a most basic level, each organization seeks higher performance internally. However, most eco-industrial activity is moving ahead by increasing interconnections between companies.

Just as in nature interconnected systems work together to ensure survivability and efficient use of resources and energy, in the business world, strategic partnerships, networked manufacturing, and preferred supplier arrangements assist companies to grow, to contain costs, and to reach for new opportunities. Eco-industrial development can help to achieve these goals by offering businesses access to cost-effective, quality resources for producing products or delivering services.

The vital task of securing community support for eco-industrial projects calls for multi-stakeholder engagement in setting a vision for local development. Companies that offer attractive models in their community outreach will have an edge in obtaining popular support. Attention to getting wide backing for a project in the early planning stages can reduce delays due to the reflexive community opposition that often crops up when a major change is proposed. A positive and proactive stance in anticipating and overcoming residents' concerns can be most helpful.

The three mini-case studies presented in Section 9.2.3 illustrate the range of results that are possible when an attempt is made to husband the resources and improve the environment of communities dominated by a single industry.

9.2.3 EIPs – The Ebara Process: Mini Case Study 9.1 in Japan

The first fertilizer plant using the Ebara process using SO_2 and NO_x gases from a nearby coal-fired power station was developed and marketed by the Tokyo-based Ebara International Corporation (Das 2005; Ebara 2010). Electron beam irradiation of a flue gas containing both SO_2 and NO_x (about 95% NO and 5% NO_2) acts on these pollutants by oxidizing them to nitrogen dioxide and sulfur trioxide anions. Addition of water and ammonia to these ionized gases yields solid products, ammonium nitrate (NH_4NO_3), and ammonium sulfate ($(NH_4)_2SO_4$) that can be separated and sold as fertilizers. The simplified chemical reactions are as follows:

$$NO + O \rightarrow NO_2$$

$$NO_2 + OH \rightarrow HNO_3$$

$$SO_2 + O \rightarrow SO_3$$

$$NH_3 + HNO_3 \rightarrow \left(NH_4NO_3\right)$$

$$2NH_3 + H_2O + SO_3 \rightarrow \left(NH_4\right)_2 SO_4$$

The main attraction of the Ebara process (Figure 9.2) is the simultaneous removal of both SO_2 and NO_x. The complex reaction mechanism entails ionization, the formation of excited electronic states, the transfer of excitation energy between molecules, molecular dissociation, electron capture, neutralization, and radical reactions. The flue gas is first cooled and humidified, then irradiated by means of a high-intensity electron beam. The flue gas is routed first through electrostatic precipitators (ESPs) or

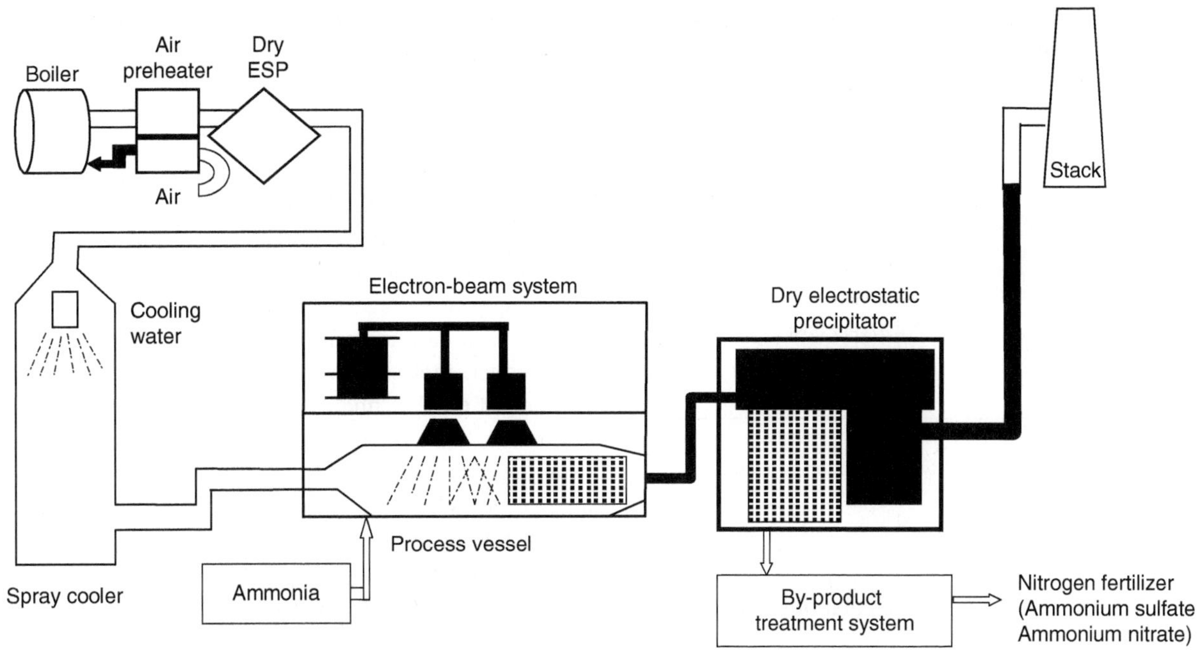

Figure 9.2 Schematic diagram of the electron beam–ammonia conversion of SO_2 and NO_x to fertilizer (Das 2005).

other mechanical collectors to remove particulates, such as fly ash. The gas is next piped to a spray cooler and humidified and cooled to a temperature in the 160–200 °F range, the optimal temperature. Anhydrous ammonia is introduced in stoichiometric amounts at this point. The ammoniated flue gas is passed to an electron-beam reactor chamber, where it is irradiated (1.5 Mrad dose), producing nitrate and sulfate anions. These species react with ammonia to yield a dry, particulate mixture of NH_4NO_3 and $(NH_4)_2SO_4$.

9.2.3.1 Advantages and Disadvantages

There are at least five advantages of the electron beam process:

1) SO_2 and NO_x are removed together in the same piece of equipment by the addition of one chemical. The method can meet all New Source Performance Standards (NSPS) for SO_2 removal because it has the potential to remove 98% of all SO_2 and 70–90% NO_x depending on the process condition (CFR-40 1999). High SO_2 removal can be achieved with very low power consumption (1% of the block). Most of the energy is consumed for NO_x reduction. NSPS standards are complex, but in the case of 3–6% sulfur coal, a 90% reduction is required.
2) Volatile organic compounds are removed in the process as well.
3) Since no waste is produced, there is no need to devise a means of disposal.
4) Since the process produces a dry product, high maintenance costs associated with abrasive and corrosive slurries are avoided.
5) A salable product results.

Among the corresponding disadvantages and challenges associated with this process is the need to keep ammonia injection close to the stoichiometric point to alleviate any ammonia slip. In addition, large electron-beam generators are needed for a large-scale plant because the current electron beam has a limited range. Technical concerns about this process include questions about the ability to scale up the electron accelerators for a full-scale application and the associated power requirements for the technology. To accelerate radiation, a combined microwave-electron-beam process has been developed and patented, and this hybrid radiation system will be suitable for a large-scale plant (Das 2005; Ebara 2010).

9.2.3.2 Cost Analysis

Table 9.1 presents an annualized cost analysis, including a projected overall return on investment for the installation and operation of a system of E-beam–ammonia conversion of SO_2 and NO_x. The cost analysis shows that the coal-fired power-generating industry can earn about $11 million/year from this process. The system itself is much cheaper than conventional flue-gas treatments – with 25% less construction costs and 20% less running costs – and requires considerably less space.

Table 9.1 Return on investment calculations: electron beam–ammonia SO_2 and NO_x conversion.

Sample calculation	Total profit/cost ($ million)
Profit from product sales	
Fertilizer: 110 000 T/Y (400 T/day) × $100/T	11/year profit
Operating cost	−0.9/year
Personnel expenses, etc.	−3.8/year
Ammonia: 29 000 T/Y (105 T/day) × $130/T	−4.7/year cost
Facility construction cost	−38.1 cost

Source: From Ebara Corporation (2010).
At an annual interest rate of 12%, facility construction cost would have been paid back within 15 years.

But there are caveats in all such calculations. For starters, the Ebara process would be sensitive to the cost of anhydrous ammonia. To control this variable, research is being conducted to discover a way to extract ammonia from municipal wastewater (Ebara 2010). If ammonia were available from a nearby municipal wastewater treatment plant, it would be even more economical to run this process. In any event, the fertilizer-producing variant of the Ebara process would be economically beneficial only in locations well sited with respect to markets for the fertilizer materials produced as by-products, and variability in capital and electricity costs.

9.2.4 Mini-Case Study 9.2: Seshasayee Paper and Board Ltd. in India

As a success example, consider one where a corporate group devised and adopted a growth strategy whereby a majority of the wastes from one activity were converted to feedstock for another activity.

Seshasayee's original enterprise was a paper mill. To ensure a regular supply of raw material, the company then set up a sugar mill. Bagasse, a waste from the sugar mill, was used as a raw material for papermaking. Another waste product, molasses, was used in a distillery for the production of ethyl alcohol. To guarantee a regular supply of sugarcane, the company took interest in the cultivation of this crop by organizing the farmers in the region. Seshasayee struck long-term agreements with the farmers to buy back their produce and, in turn, took the responsibility of supplying them with water. Part of the water for cultivation was treated wastewater from the paper manufacturing operations. The company also used bagasse pith (a waste after the papermaking) and other combustible agricultural wastes in the region as energy sources. The agro industrial eco-complex is illustrated in Figure 9.3.

This is particularly important to the theme of a process integration and intensification work because it involves the production of a good (paper) and a human food (sugar), and the sustainable generation of energy (from bagasse – e.g. waste-to-energy) and sustainable use of water. In this case, the discussion of the mini-case study shows that it is possible to combine the production of a

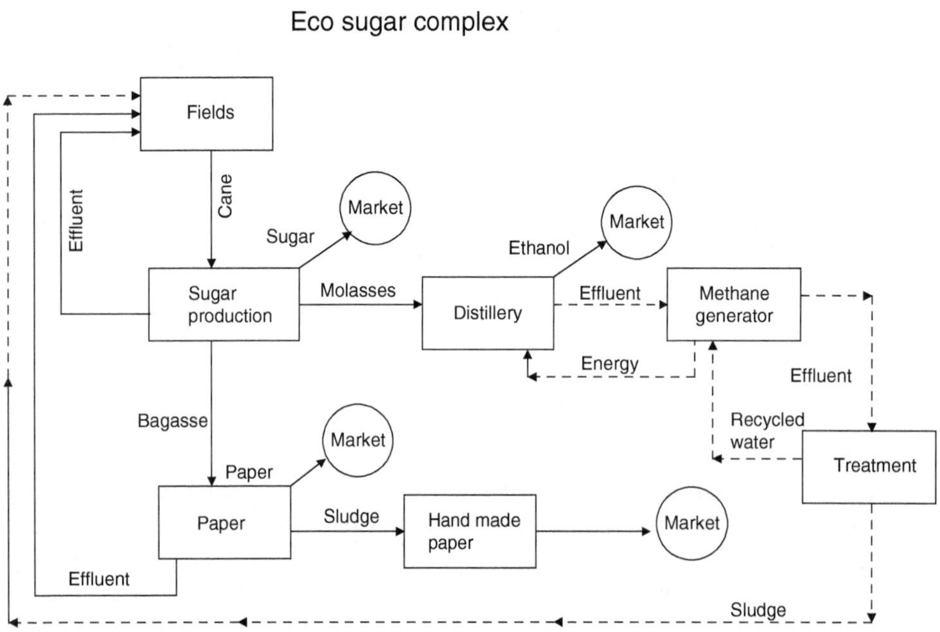

Figure 9.3 An agroindustrial eco-complex. *Source:* From Das (2005) and Das and Cabezas (2018).

food and the generation of energy with the efficient use of water resources. This is important because the production of a commercial food (sugar) is economically and socially important, especially if it can be produced as sustainably as possible.

9.2.5 Mini-Case Study 9.3: Materials and Energy Flow in an EIP in North Texas, USA

Consider, for example, an EIP in North Texas where the central facility is a steel mill. This facility, shown conceptually in Figure 9.4, utilizes scrap cars as the primary feed material. The steel from the vehicles goes to an electric arc furnace (EAF), producing a variety of steel products. The furnace also produces a significant quantity of EAF dust, which contains significant quantities of zinc, lead, and other metals. In the North Texas facility, the EAF dust is sent to a cement kiln where the trace metals (copper, chromium, manganese, nickel, lead, and other metals) have values. Automobile shredder residue (ASR) can be burned for energy recovery, or some of the plastics in the residue can be separated.

Another alternative for EAF dust, currently being exploited in Europe, is as a feed for zinc and lead recovery operations. The recovered zinc can then be used in producing galvanized steel products and batteries can be used as an alternative source of zinc.

These case studies illustrate the basic principle of EIP – integrating flows of materials, energy, and water in diverse industrial operations, increasing mass and energy efficiency. The three case studies examined in this section involved exchanges between facilities that are located adjacent to each other; however, co-location of facilities in an EIP is not always necessary.

9.2.6 Mini-Case Study 9.4: EIP Including Numerous Symbiotic Factories for Manufacturing Very Large Scale Photovoltaic System

The global climate destabilization underway is primarily due to human combustion of fossil fuels for energy and the resultant greenhouse gas (GHG) emissions (IPCC 2007). There is a large consensus among scientists that if current trends continue in climate destabilization, the earth will reach a point of no return. The challenge of reducing atmospheric GHG like carbon dioxide (CO_2) emissions is significant – with the Kyoto Protocol and the recent Paris Agreement, an entire order of magnitude below those values necessary to stabilize the global climate. To obtain a stable climate, GHG emissions would have to be reduced to a point equal to the natural absorption of CO_2. After this stabilization occurs, the level of natural absorption will gradually fall as the vegetation sink is exhausted so GHG emissions would need to fall to the level of ocean uptake alone. This level is not well quantified, but it may demand emissions reductions to 5 Gt CO_2 equiv/year (more than 80% below current levels) by the second half of the next century (Hansen and Sato 2004; Prentice et al. 2001).

In order to increase both the economic and environmental performance of the manufacturing sector, the government could introduce a second set of policies with the large-scale PV (photovoltaic) manufacturing in order to facilitate the widespread adoption of industrial symbiosis. In industrial symbiosis, traditionally separate industries are considered collectively to gain competitive advantage by instituting the mutually beneficial physical exchange of materials, energy, water, and/or by-products. Such a system collectively optimizes material and energy use at efficiencies beyond those achievable by any individual process alone. The key benefits of industrial

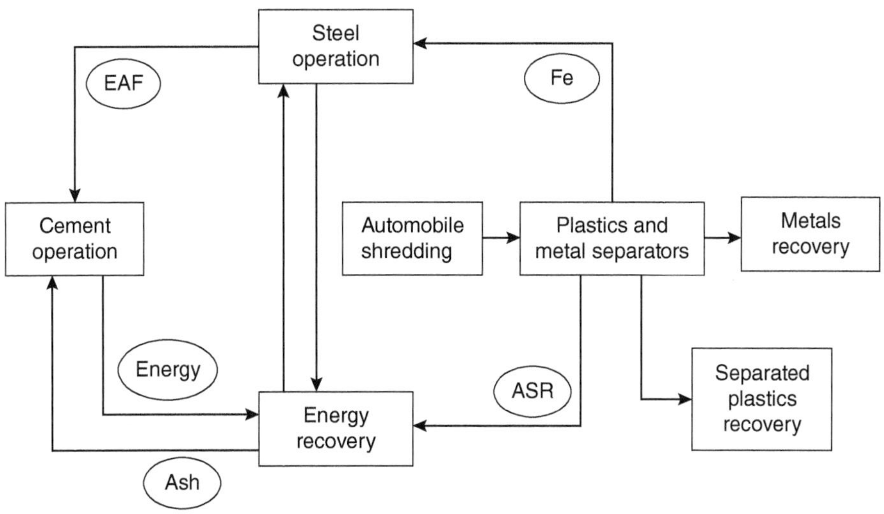

Figure 9.4 Materials and energy flow in an EIP in North Texas, USA.

symbiosis are collaboration and the synergies offered by geographic proximity (Chertow 2000). Industrial symbiotic systems such as the now-classic network of companies in Kalundborg, Denmark, have spontaneously evolved from a series of micro-innovations over a long time scale. In order to accelerate the process and demonstrate the possibilities for mutually beneficial collaboration, the government could engineer the design of the new solar PV manufacturing plant using industrial symbiosis.

The multi-GW PV factory could sit at the center of a next generation EIP. The EIP would be made up of at least eight symbiotic factories as seen in Figure 9.5. These could be located outside of a major population center to provide raw materials, labor, and a ready market. The first factory would be a conventional recycling facility (1). In this way, the glass and aluminum needed to fabricate the solar cell could be recovered from recycled materials and thus have a lower embodied energy (95% lower for aluminum and 20% for glass) (Milne and Readon 2005).

The raw glass from the recycling plant will be fed to a sheet glass factory (2) and melted using natural gas. Generally, the high-quality requirements of flat glass prohibit the use of post-consumer waste glass. However, the glass industry is exploring a method to eliminate problems with color contamination by using thin plastic coatings, which can be made with a variety of colors and that would vaporize during remelting without affecting the quality of the new glass. This would benefit the industry considerably because using recycled glass, called cullet, has several important benefits that include (i) lowering the consumption of raw materials, (ii) reducing the release of CO_2 formed in the chemical reaction of raw materials, (iii) increasing the life of the furnace by up to 30% due to lower melting temperatures, (iv) reducing energy use during the melting stage of production and thus reducing additional GHGs and operating costs, and (v) reducing the costs associated with pollution abatement due to lower emissions of NO_x, SO_2, and particulates (Pearce 2008). The factory will output cut sheets of 3 mm thick glass with seamed edges and low-iron content in order to obtain a high-solar transparency. Finally, the glass will be tempered for mechanical strength and coated

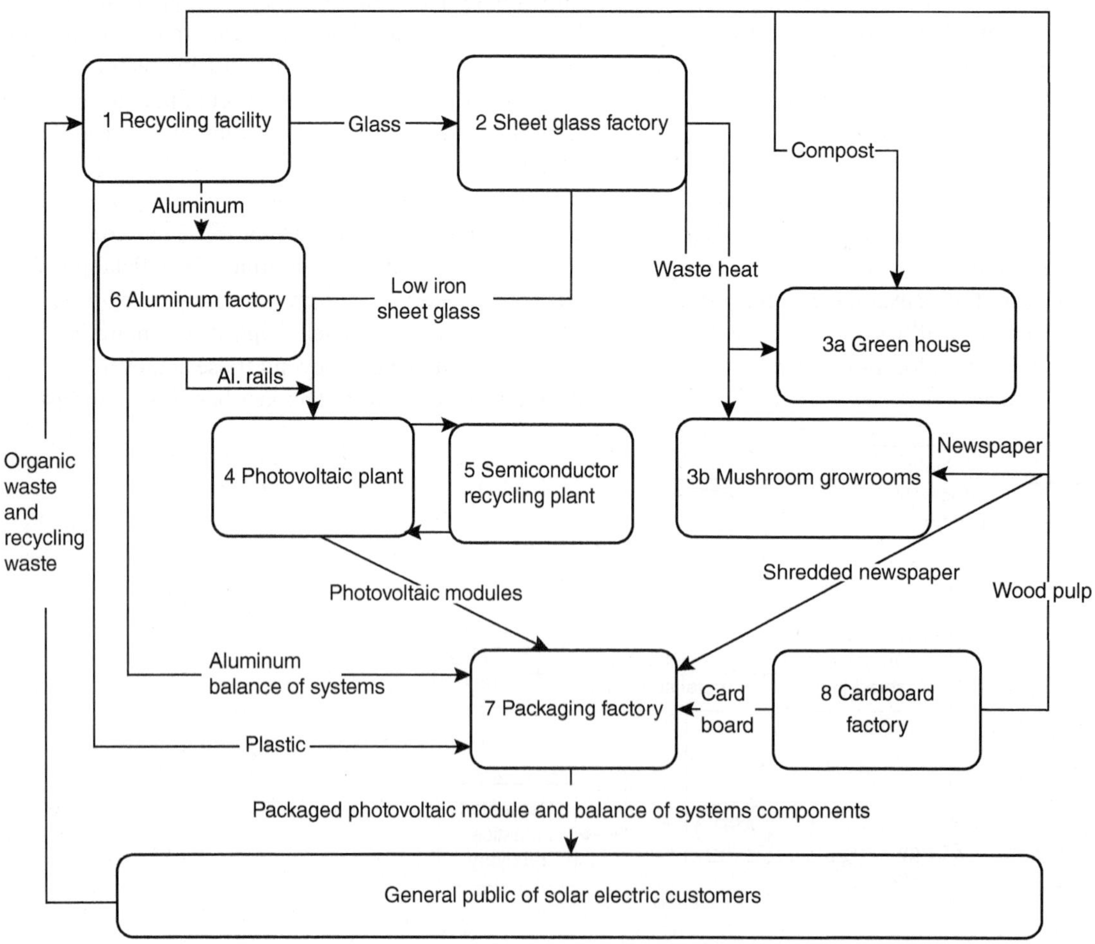

Figure 9.5 Schematic diagram of an EIP to produce giga-watt solar PV. *Source:* From Pearce (2008).

with a transparent conductor such as tin oxide, zinc oxide, or indium tin oxide to be used as the substrates.

The production stages in the glass factory that utilize large amounts of heat will have integrated thermal recovery to provide lower grade heat for the other facilities and a greenhouse complex (3a). In the greenhouse complex, exotic plants can be grown year round in even northern climates utilizing the waste heat from the manufacturing plants in the EIP. Similarly, waste heat could be utilized to provide grow rooms for mushrooms (3b). In both agricultural plants, the food or other agricultural products will be sold outside of the park and the growing medium will be provided by the recycling facility (1), compost for the greenhouse (3a), and wood pulp or compost for the mushroom-growing facility (3b). In warmer climates, the waste heat could be used to drive absorption chillers perhaps providing cooling for an office park.

The substrates will then be fed directly into the PV module plant (4). Then a group of semiconductor and metal thin-film deposition systems will create and pattern the active layers of the solar cells. All waste semiconductors and metals will be captured and returned to a semiconductor recycling plant (5) to supplement the incoming and generally expensive high-purity materials going into the deposition systems. The output of the PV deposition and patterning lines will be PV solar panels ready for protective coatings and packaging.

The aluminum extracted from common drinking cans in the recycling center (1) will be fed to an aluminum fabrication factory (6) that will produce coated aluminum rails for holding the glass solar panels. The aluminum rails will be extruded and used to provide a simple and inexpensive means of attachment to rooftops, ground mounted systems, or building integrated PV. In addition, the extruded aluminum rails could be designed into ground and flat roof mounting balance of system components. Similar to the glass manufacturing plant (2), waste heat will be recovered and used in the symbiotic collective or to heat the greenhouse (3a) or mushroom grow rooms (3b).

Next, in the packaging factory (7), the solar panels are interconnected if necessary and are sprayed with a protective polymer coating to seal them to the environment. The primary constituents of the polymer coating could again be acquired from the recycling plant (1) and common plastics. The panels would then be wired with quick connects so they can be easily installed in the field by connecting to each other an inverter or battery bank. Finally, the panels would be packaged for shipment to prevent damage in cardboard boxes and cushioned with shredded newspaper. The newspapers would again come from the recycling plant (1) and the cardboard could come from a cardboard plant (8), which would gain its raw materials from the recycling factory (1).

By co-locating these factories in the EIP, the transportation costs and energy between them can be minimized and many of the inputs for the solar PV plant can literally come from waste products in the surrounding population centers. It should be noted that each factory will be scaled appropriately for the symbiotic system and should be individually profitable so that independent businesses can replicate this model by co-locating and benefit from industrial symbiosis in future facilities.

9.3 Water–Energy Nexus

Present day water and energy systems are tightly intertwined. Water is used in all phases of energy production and electricity generation. Energy is required to extract, convey, and deliver water of appropriate quality for diverse human uses, and then again to treat wastewaters prior to their return to the environment. Historically, interactions between energy and water have been considered on a regional or technology-by-technology basis. At the national and international levels, energy and water systems have been developed, managed, and regulated independently. Water and energy are critical, mutually dependent resources – the production of energy requires large volumes of water and water infrastructure requires large amounts of energy (Das 2017; Das and Cabezas 2018; EPRI 2011; USDOE 2014a, 2013a, b).

Water is required to generate energy: Thermoelectric cooling, hydropower, energy mineral extraction and mining, fuel production (including fossil fuels, biofuels, and other nonconventional fuels), and emission controls all rely on large amounts of water. In the United States, the thermoelectric generating industry is the largest withdrawal user of water. According to USGS (United State Geological Survey), 349 billion gallons of freshwater were withdrawn per day in the United States in the year 2005. The largest use, thermoelectric, accounted for 41% of freshwater withdrawn at 143 billion gal/day. However, freshwater consumption for thermoelectric purposes is low (only 3%) when compared to other use categories such as irrigation, which was responsible for 81% of water consumed.

- *Water withdrawal*: The total volume removed from a water source such as a lake or river. Often, a large portion of this water is returned to the source and is available to be used again.
- *Water consumption*: The amount of water removed for use and not returned to its source.

Water supply also requires energy use: A large amount of energy is needed to extract, convey, treat, and deliver potable water. Additionally, energy is required to collect, treat, and dispose of wastewater. In 2010, the US water system consumed over 600 billion kWh, or approximately 12.6% of the nation's energy according to a study by researchers at the University of Texas at Austin. The study found water systems use about 25% more energy than is used for residential or commercial lighting in the United States.

Water and energy are both multifaceted issues with many variables impacting their supply, demand, and management. Lawmakers should consider the following variables which add complexity to the management of water and energy:

- *Growing population*: According to a 2012 United States Census Bureau projection, the US population could reach 400 million people by 2051. Population growth affects energy use through increases in housing, commercial floor space, transportation, and economic activity. The US Energy Information Administration (EIA) estimates that total electricity consumption will grow from 3841 billion kWh in 2011 to 4930 billion kWh in 2040, an average annual rate of 0.9%. With a higher generating capacity, the United States will require additional water withdrawals.
- *Agriculture*: Feeding a growing population may require greater agricultural water use. Agriculture accounts for approximately 37% of total freshwater withdrawals in the United States, and 81% of water consumption.
- *Geographical water demand*: Water supply and demand are not geographically linked. From 1990 to 2010, the second largest regional population growth, 13.8%, occurred in the west, which is one of the most water-deficient regions in the United States. Additionally, water consumption in the western United States is much higher than other regions due to agricultural demands. It is estimated that it takes over one million gallons of water a year to irrigate one acre of farmland in arid conditions. In other words about 86% of irrigation water withdrawals were in western states in 2000.
- *Climate change*: Climate change could affect water supply and electricity use. Warmer or colder weather patterns could result in increases or decreases in energy use. Changes in precipitation in a region could increase or decrease the ability to store water, agricultural production and water use, and overall water supply.

States are beginning to assess their energy options and promote policies that allocate financial support to a diverse range of technologies to encourage responsible, sustainable energy production. States are also becoming aware of the limitations to accessible water, and as our energy demands grow, competition for water among municipalities, farmers, industrial, and power suppliers will increase. Water and energy are linked at both the supply side (electric generation and water/wastewater facilities) and the end-use side (residential, commercial, industrial, and agriculture sectors) (Figure 9.6). In order to sustain energy production and a dependable water supply, the United States must gain a

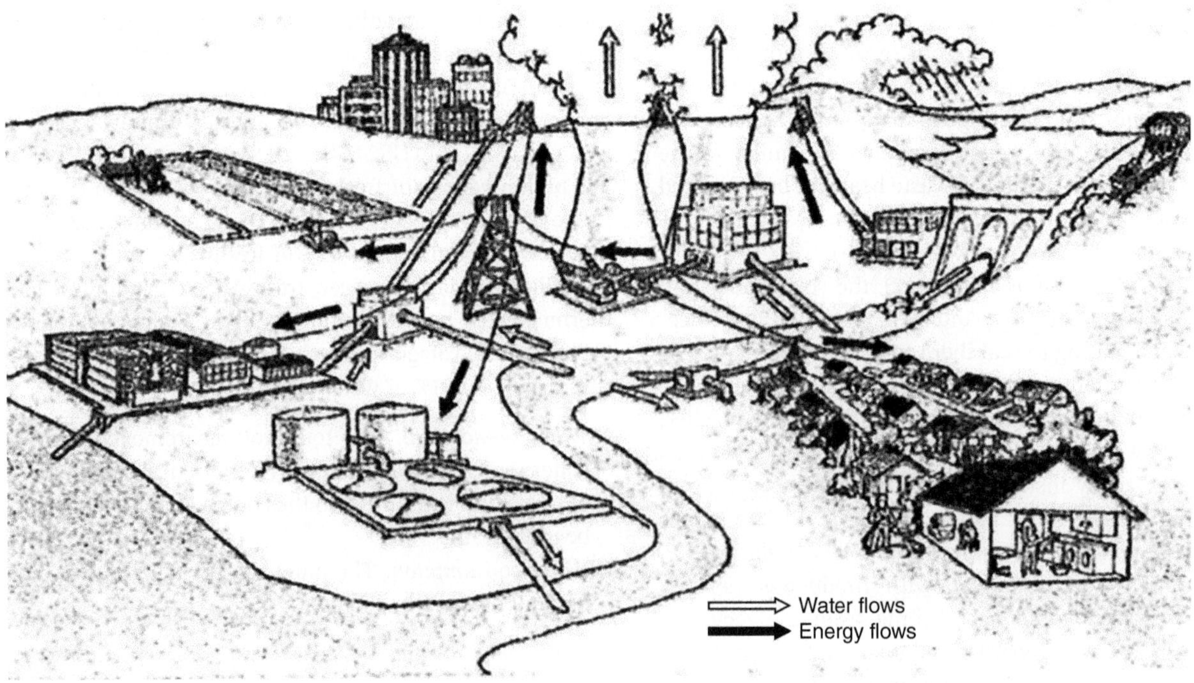

Figure 9.6 Examples of interrelationships between water and energy. *Source:* Adopted from USDOE (2014b).

detailed understanding of the interdependencies of water and energy systems and balance the needs of all users. State lawmakers and constituents will be critical in this process given their responsibility formulating policy, convening stakeholders, facilitating negotiations, and ratifying reached agreements.

Flows of energy and water are intrinsically interconnected, in large part due to the characteristics and properties of water that make it so useful for producing energy and the energy requirements to treat and distribute water for human use. This interconnectivity is illustrated in the Sankey diagram in Figure 9.7, which captures the magnitude of energy and water flows in the United States on a national scale. As shown in the diagram, thermoelectric power generation withdraws large quantities of water for cooling and dissipates tremendous quantities of primary energy due to inefficiencies in converting thermal energy to electricity. The intensity of water use and energy dissipated varies with generation and cooling technology.

As the largest single consumer of water, agriculture competes directly with the energy sector for water resources. However, agriculture also contributes indirectly to the energy sector via production of biofuels. Both connections will be strained by increasing concerns over water availability and quality. In addition, water treatment and distribution for drinking water supply and municipal wastewater also require energy.

Significant aspects of water and energy flows do not appear in Figure 9.6. First, flows will change over time, and anticipated changes in flows are important to consider when prioritizing investment in technology and other solutions. Increased deployment of some energy technologies in the future, such as carbon capture and sequestration, could lead to increases in the energy system's water intensity, whereas deployment of other technologies, such as wind and solar photovoltaics, could lower it. In addition, there is significant regional variability in the water and energy systems, their interactions, and resulting vulnerabilities. For example, producing oil and natural gas through horizontal drilling and hydraulic fracturing has the potential for localized water quantity and quality impacts that can be mitigated through fluid life cycle management. Large volumes of water produced from oil and gas operations in general present both localized management challenges and potential opportunities for beneficial reuse.

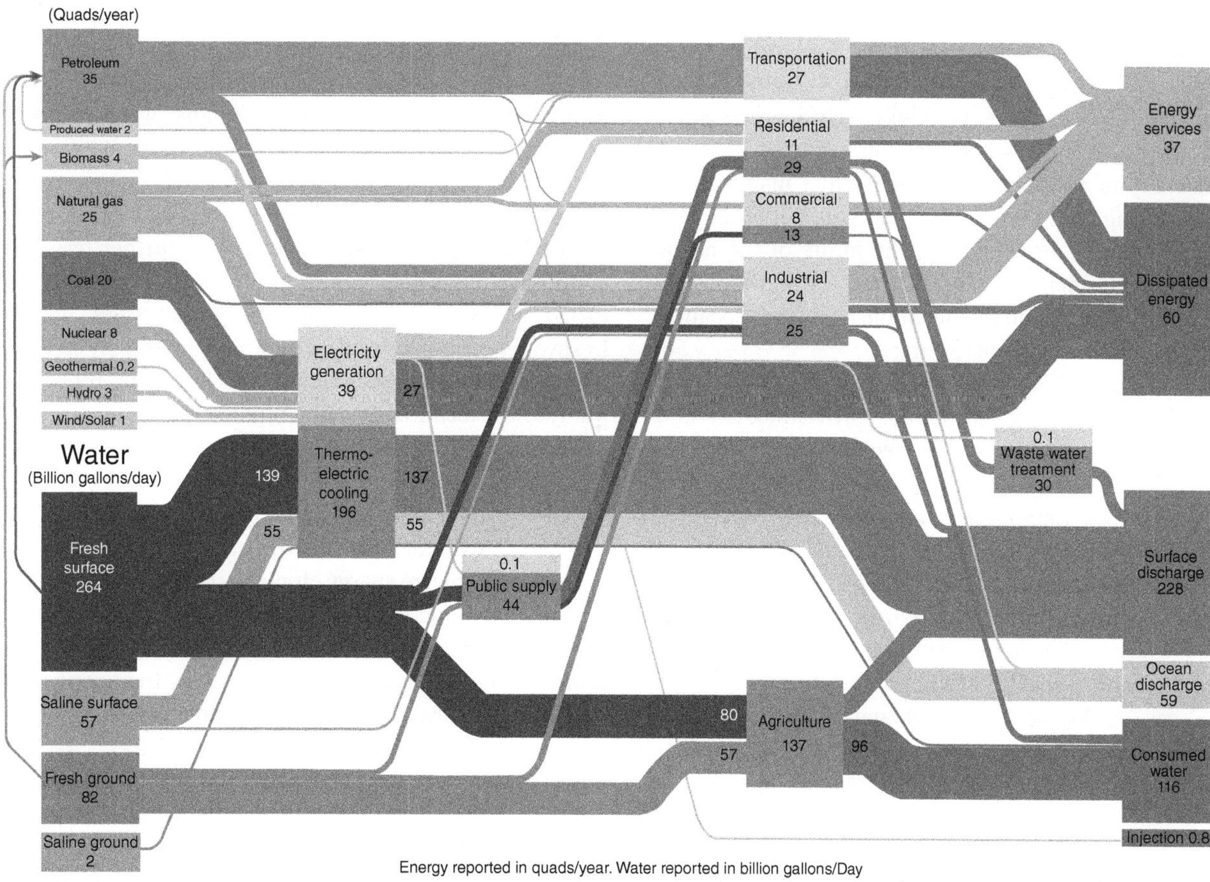

Figure 9.7 Hybrid Sankey diagram of 2011 US interconnected water and energy flows. *Source:* From USDOE (2014a).

The energy requirements for water systems also have regional variability, based on the quality of water sources and pumping needs.

Water availability will affect the future of the water–energy nexus. While there is significant uncertainty regarding the magnitude of effects, water availability and predictability may be altered by changing temperatures, shifting precipitation patterns, increasing variability, and more extreme weather. Shifts in precipitation and temperature patterns – including changes in snowmelt – will likely lead to more regional variation in water availability for hydropower, biofeedstock production, thermoelectric generation, and other energy needs. Rising temperatures have the potential to increase the demand for electricity for cooling and decrease the efficiency of thermoelectric generation, as well as increase water consumption for agricultural crops and domestic use. These changes and variations pose challenges for energy infrastructure resilience.

Water and energy needs will also be shaped by population growth and migration patterns, as well as changes in fuels used and energy technologies deployed. For example, projected population growth in the arid southwest will amplify pressure on water and energy systems in that region. Increased production of oil and gas may increase both localized demand for water and generation of produced water that requires management. According to EIA data, planned retirements and additions of electricity generation units and cooling systems will likely decrease water withdrawals, increase water consumption, and increase the diversity of water sources used. While many of the forces affecting the water–energy nexus are out of the federal government's direct control, the future of the nexus hinges on a number of factors that are within the DOE's scope of influence, including technology options, location of energy activities, and energy mix.

The decision-making landscape for the nexus is shaped by political, regulatory, economic, environmental, and social factors, as well as available technologies. The landscape is fragmented, complex, and changing; the incentive structures are overlapping but not necessarily consistent. Water is inherently a multi-jurisdictional management issue and is primarily a state and local responsibility. States and localities vary in philosophies regarding water rights. There is also variation across states in relevant energy policies, including renewable portfolio standards, regulation of oil and gas development activities, and regulation of thermoelectric water intake and discharge. Regulations for both oil and gas development and thermoelectric water use are currently undergoing substantial change. Energy for water is also the subject of policy activity at multiple scales, from appliance standards to municipal water treatment funding mechanisms. A more integrated approach to the interconnected energy and water challenges could stimulate the development and deployment of solutions that address objectives in both domains (Table 9.2; AGU 2012; Clark and Veil 2009; DOE 2013a, b; EPRI 2011).

9.3.1 Technology Roadmaps and R&D

There are a number of technologies that support water-efficient energy systems or energy-efficient water systems. These technologies are at various stages of research, development, demonstration, and deployment. Figure 9.8 illustrates a range of technologies optimizing water use for energy in waste heat recovery, cooling, alternate fluids, and process water efficiency.

Cooling for thermoelectric generation is an important target for water efficiency because it withdraws large quantities of water for cooling and dissipates tremendous amounts of primary energy. One approach to reduce thermoelectric and other cooling requirements, along with associated water use, is to reduce the generation of waste heat through more efficient power cycles (e.g. the recompression closed-loop Brayton cycle). Another option is to increase the productive use of the waste heat, such as through thermoelectric materials, enhancements in heat exchanger technologies, or low-temperature co-produced geothermal power. A third approach to improve the water efficiency of cooling systems is through advancements in technologies, including air flow designs, water recovery systems, hybrid or dry cooling, and treatment of water from blowdown.

Opportunities to optimize water use also exist in other parts of the overall energy system. With further research, alternative fluids may replace freshwater in hydraulic fracturing, geothermal operations, and power cycles. Process freshwater efficiency can be improved in carbon capture, bioenergy feedstock production, and industrial processes. Many of the technologies that improve water efficiency are enhanced by advances in materials, including thermoelectric properties, heat-driven state change, scaling/fouling resistance, and temperature and pressure tolerance.

Figure 9.9 shows water treatment technologies that can potentially enhance energy efficiency of water systems and enable the productive, economical, and safe use of nontraditional water resources for energy and nonenergy applications. Such improvements in water treatment and management have particular use for treating oil- and gas-produced waters, as well as saline aquifers, brackish groundwater, brines, seawater, and municipal wastewater. For saline sources, promising water treatment technologies include membrane distillation, forward osmosis, evaporation, nanomembranes, and capacitive deionization. For municipal wastewater, treatment technologies include

Table 9.2 Comparison of the water withdrawal and water consumption factors (in gal/MWh) for fuel-based electricity-generating technologies.

Fuel type	Cooling	Technology	Median withdrawal	Median consumption
Nuclear	Tower	Generic	1101	672
	Once-through	Generic	44350	269
	Pond	Generic	7050	610
Natural gas	Tower	Combined cycle	225	205
		Steam	1203	826
		Combined cycle with CCS	506	393
	Once-through	Combined cycle	11380	100
		Steam	35000	240
	Pond	Combined cycle	5950	240
	Dry	Combined cycle	2	2
Coal	Tower	Generic	1005	687
		Supercritical	634	493
		IGCC	393	380
		Supercritical with CCS	1147	846
		IGCC with CCS	642	549
	Once-through	Generic	36350	250
		Supercritical	15046	103
	Pond	Generic	12225	545
		Supercritical	15046	42
Biopower	Tower	Steam	878	553

Source: From National Renewable Energy Laboratory (NREL) (2011).

anammox systems, anaerobic pretreatments, and anaerobic membrane bioreactors. In addition, the biosolids contained in wastewater can be a source of methane energy.

9.3.2 Circular Economy

There are various definitions of a circular economy (CE): An idea for a truly sustainable future that works without waste, in symbiosis with our environment and resources. A future where every product is designed for multiple cycles of use, and different material or manufacturing cycles are carefully aligned, so that the output of one process always feeds the input of another. Rather than seeing emissions, manufacturing by-products, or damaged and unwanted goods as "waste," in the CE they become raw material, nutrients for a new production cycle (Figure 9.10).

Waste Resources Action Program in the United Kingdom defines it as an alternative to a traditional linear economy (make, use, dispose) in which we keep resources in use for as long as possible, extract the maximum value from them while in use, then recover and regenerate products and materials at the end of each service life. It is also described as a regenerative system in which resource input and waste, emission, and energy leakage are minimized by slowing, closing, and narrowing energy and material loops. This can be achieved through long-lasting design, maintenance, repair, reuse, remanufacturing, refurbishing, and closed recycling loops (Geissdoerfer et al. 2017). This is in contrast to a linear economy which is a "take, make, dispose" model of production. The Ellen MacArthur Foundation (2012) works with business, governments, and education to help explain the concepts and benefits of a CE. The Foundation's "butterfly diagram" is often used to illustrate a CE, but for those new to the concept, it can be difficult to understand.

9.3.2.1 Origins

As early as 1966 Kenneth Boulding already raised awareness of an "open economy" with unlimited input resources and output sinks in contrast with a "closed economy," in which resources and sinks are tied and remain as long as possible a part of the economy. The concept of a CE was raised by two British environmental economists David W. Pearce and R. Kerry Turner in 1989. In *Economics of*

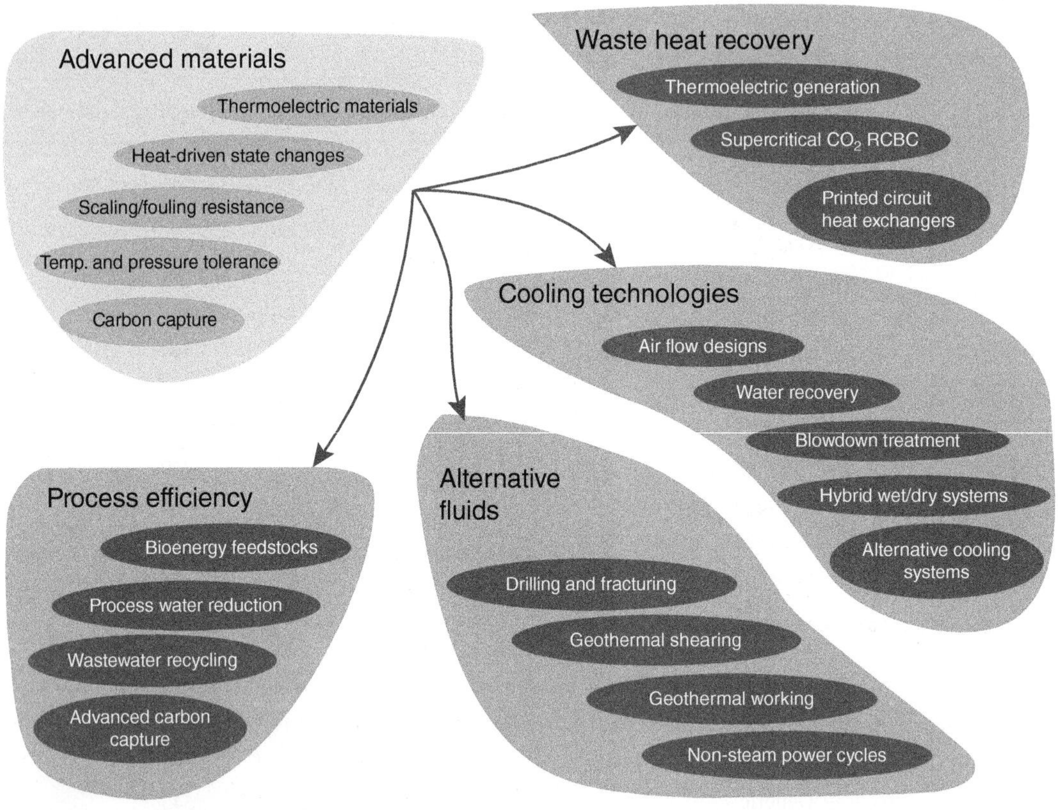

Figure 9.8 Representative problem/opportunity spaces in water for energy. *Source:* Adapted from USDOE (2014a).

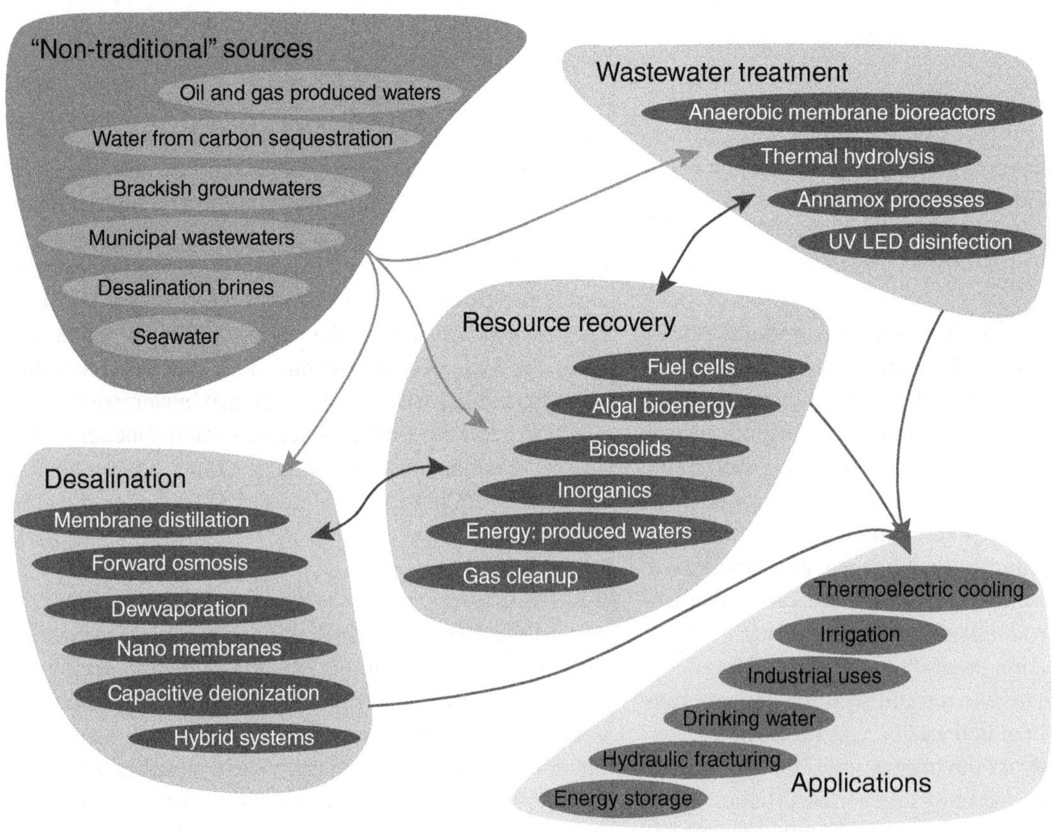

Figure 9.9 Representative problem/opportunity spaces in energy for and from water. *Source:* From USDOE (2014a).

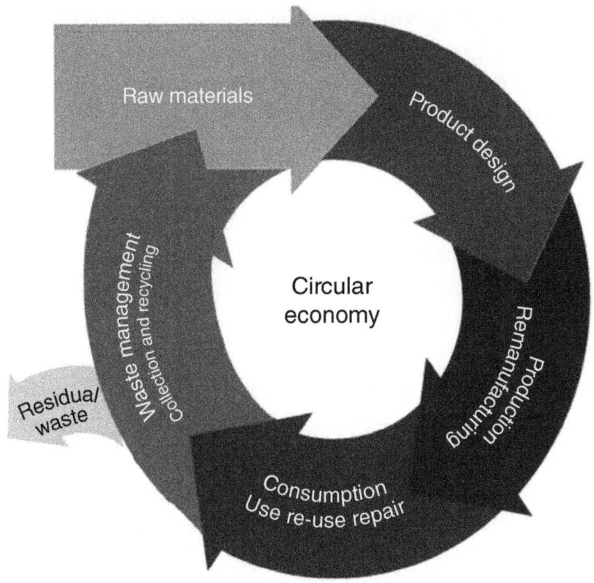

Figure 9.10 A schematic depicting the "take, make, waste" linear approach to the CE.

Natural Resources and the Environment (Pearce and Turner 1989), they pointed out that a traditional open-ended economy was developed with no built-in tendency to recycle, which was reflected by treating the environment as a waste reservoir. Other early schools of thought include Professor Walter Stahel, Gunter Pauli, William McDonough, and Michael Braungart, and complementary approaches such as IE, Permaculture, and The Natural Step.

9.3.2.2 Moving Away from the Linear Model

Linear "take, make, dispose" industrial processes and the lifestyles that feed on them deplete finite reserves to create products that end up in landfills or in incinerators. This realization triggered the thought process of a few scientists and thinkers, including Walter Stahel, an architect, economist, and a founding father of industrial sustainability. Credited with having coined the expression "cradle to cradle" (in contrast with "cradle to grave," illustrating our "resource to waste" way of functioning), in the late 1970s, Stahel worked on developing a "closed-loop" approach to production processes, co-founding the Product-Life Institute in Geneva more than 25 years ago. In the United Kingdom, Steve D. Parker researched waste as a resource in the UK agricultural sector in 1982, developing novel closed-loop production systems mimicking, and integrated with, the symbiotic biological ecosystems they exploited.

9.3.2.3 Emergence of the Idea

In their 1976 Hannah Reekman research report to the European Commission, "The Potential for Substituting Manpower for Energy," Walter Stahel and Genevieve Reday sketched the vision of an economy in loops (or CE) and its impact on job creation, economic competitiveness, resource savings, and waste prevention. The report was published in 1982 as the book *Jobs for Tomorrow: The Potential for Substituting Manpower for Energy*.

Considered as one of the first pragmatic and credible sustainability think tanks, the main goals of Stahel's institute are product-life extension, long-life goods, reconditioning activities, and waste prevention. It also insists on the importance of selling services rather than products, an idea referred to as the "functional service economy" and sometimes put under the wider notion of "performance economy," which also advocates "more localization of economic activity" (Clift and Allwood 2011).

In broader terms, the circular approach is a framework that takes insights from living systems. It considers that our systems should work like organisms, processing nutrients that can be fed back into the cycle – whether biological or technical – hence the "closed loop" or "regenerative" terms usually associated with it. The Ellen MacArthur Foundation, an independent charity established in 2010, has more recently outlined the economic opportunity of a CE. As part of its educational mission, the Foundation has worked to bring together complementary schools of thought and create a coherent framework, thus giving the concept a wide exposure and appeal (Ellen MacArthur Foundation 2012).

Most frequently described as a framework for thinking, its supporters claim it is a coherent model that has value as part of a response to the end of the era of cheap oil and materials and can contribute to the transition to a low-carbon economy. In line with this, a CE can contribute to meet the COP 21 Paris Agreement. The emissions reduction commitments made by 195 countries at the COP 21 Paris Agreement are not sufficient to limit global warming to 1.5 °C. To reach the 1.5 °C ambition, it is estimated that additional emissions reductions of 15 billion T CO_2/year need to be achieved by 2030. Circle Economy and Ecofys estimated that CE strategies may deliver emissions reductions that could basically bridge the gap by half (Blok et al. 2018). However, we have to keep in mind that economic and business goals and environmental goals are two distinct goal sets (see Appendix I).

9.3.2.4 Sustainability

The CE seems intuitively to be more sustainable than the current linear economic system. The reduction of resource inputs into and waste and emission leakage out of the system reduces resource depletion and environmental pollution. However, these simple assumptions are not sufficient to deal with the involved systemic complexity and disregards potential trade-offs. For example, the social dimension of sustainability seems to be only marginally addressed in many publications on the CE, and there are cases that require different or additional strategies, like purchasing new, more energy efficient equipment. By reviewing the literature, a team of researchers from Cambridge and TU Delft could show that there are at least eight different relationship types between sustainability and the CE (Geissdoerfer et al. 2017)

1) Conditional relation
2) Strong conditional relation
3) Necessary but not sufficient conditional relation
4) Beneficial relationship
5) Subset relation (structured and unstructured)
6) Degree relation
7) Cost-benefit/trade-off relation
8) Selective relation

9.3.2.5 Use Waste as a Resource

The second element aims to utilize waste streams as a source of secondary resources and recover waste for reuse and recycling and is grounded on the idea that waste does not exist. It is necessary here to design out waste, meaning that both the biological and technical components (nutrients) of a product are designed intentionally in such a way that waste streams are minimalized. Closed recycling loops are key here, one for manufacture (production-waste recycling) and two for disposal of the product (product and material recycling) (Lienig and Bruemmer 2017). This follows a cradle-to-cradle design rather than a cradle-to-grave process.

9.3.2.6 Design for the Future

Account for the systems perspective during the design process, to use the right materials, to design for appropriate lifetime, and to design for extended future use. Meaning that a product is designed to fit within a materials cycle, can easily be dissembled, and can easily be used with a different purpose. In addition to strategies like emotionally durable design, it involves anticipating product recycling (the reuse and further use of the product), and material recycling (the reuse and further use of its constituent materials) during the design process (Lienig and Bruemmer 2017). It should be stressed that there is not something like one ideal blueprint for future design. Modularity, versatility, and adaptiveness are to be prioritized in an uncertain and fast-evolving world, meaning that diverse products, materials, and systems, with many connections and scales, are more resilient in the face of external shocks, than monotone systems built simply for efficiency.

9.3.2.7 Preserve and Extend What's Already Made

While resources are in use, maintain, repair, and upgrade them to maximize their lifetime and give them a second life through take back strategies when applicable. This could mean that a product is accompanied with a pre-thought maintenance program to maximize its lifetime, including a buyback program and supporting logistics system. Leasing programs ("purchase the usage" instead of "purchase the product"), secondhand sales, or product recycling also falls within this element (Lienig and Bruemmer 2017).

9.3.2.8 Collaborate to Create Joint Value

Within a CE, one should work together throughout the supply chain, internally within organizations and with the public sector to increase transparency and create joint value. For the business sector this calls for collaboration within the supply chain and cross-sectoral, recognizing the interdependence between the different market players. Governments can support this by creating the right incentives, for example via common standards within a regulatory framework and provide business support.

9.3.2.9 Incorporate Digital Technology

Track and optimize resource use and strengthen connections between supply chain actors through digital, online platforms, and technologies that provide insights. It also encompasses virtualized value creation and delivering, for example via 3D printers, and communicating with customers virtually.

9.3.3 Rethink the Business Model

Consider opportunities to create greater value and align incentives through business models that build on the interaction between products and services. Do we need to own products or is one satisfied by using it as service? Can the current business model, for example, be modified to lease its products? This also includes rethinking circular production processes (McDonough and Braungart 2002)

9.3.3.1 Framework

The various approaches to "circular" business and economic models have slightly different emphasis on the key components. They share several common principles aiming to

- extend the life of materials and products, where possible over multiple "use cycles";
- use a "waste = food" approach to help recover materials and ensure those biological materials returned to earth are benign, not toxic;
- retain the embedded energy, water, and other process inputs in the product and the material for as long as possible;
- use systems-thinking approaches in designing solutions;
- regenerate or at least conserve nature and living systems;
- push for policies, taxes and market mechanisms that encourage product stewardship, for example "polluter pays" regulations (Weetman 2016).

The many different "understandings" of the CE are evidenced by a recent review of 114 different publications. Some of the key aspects are outlined in the following.

9.3.3.2 Systems Thinking

System thinking is the ability to understand how things influence one another within a whole. Elements are considered as "fitting in" their infrastructure, environment, and social context. While a machine is also a system, systems thinking usually refers to nonlinear systems: systems where through feedback and imprecise starting conditions, the outcome is not necessarily proportional to the input and where evolution of the system is possible – the system can display emergent properties. Examples of these systems are all living systems and any open system such as meteorological systems or ocean currents, even the orbits of the planets have nonlinear characteristics.

Understanding a system is crucial when trying to decide and plan (corrections) in a system. Missing or misinterpreting the trends, flows, functions of, and human influences on, our socio-ecological systems can result in disastrous results. In order to prevent errors in planning or design, an understanding of the system should be applied to the whole and also to the details of the plan or design. The Natural Step created a set of systems conditions (or sustainability principles) that can be applied when designing for (parts of) a CE to ensure alignment with functions of the socio-ecological system.

The concept of the CE has previously been expressed as the circulation of money versus goods, services, access rights, valuable documents, etc., in macroeconomics. This situation has been illustrated in many diagrams for money and goods circulation associated with social systems. As a system, various agencies or entities are connected by paths through which the various goods etc., pass in exchange for money. However, this situation is different from the CE described above, where the flow is unilinear – in only one direction, that is until the recycled goods are again spread over the world.

9.3.4 Biomimicry

Janine Benyus, author of *Biomimicry: Innovation Inspired by Nature*, defines her approach as "a new discipline that studies nature's best ideas and then imitates these designs and processes to solve human problems. Studying a leaf to invent a better solar cell is an example. I think of it as 'innovation inspired by nature'" (Benyus 2003). Biomimicry relies on three key principles:

- Nature as model: Biomimicry studies nature's models and emulates these forms, processes, systems, and strategies to solve human problems.
- Nature as measure: Biomimicry uses an ecological standard to judge the sustainability of our innovations.
- Nature as mentor: Biomimicry is a way of viewing and valuing nature. It introduces an era based not on what we can extract from the natural world but on what we can learn from it.

9.3.4.1 Cradle to Cradle

Created by Walter R. Stahel, a Swiss architect who graduated from the Swiss Federal Institute of Technology Zürich in 1971. He has been influential in developing the field of sustainability by advocating philosophies of "service-life extension of goods – reuse, repair, remanufacture, upgrade technologically" as they apply to industrialized economies. He cofounded the Product Life Institute in Geneva, Switzerland, a consultancy devoted to developing sustainable strategies and policies, after receiving recognition for his prize-winning paper "The Product Life Factor" in 1982. His ideas and those of similar theorists led to what is now known as the *CE*, in which industry adopts the reuse and service-life extension of goods as a strategy of waste prevention, regional job creation, and resource efficiency in order to decouple wealth from resource consumption; in other words, to dematerialize the industrial economy. Recent technical developments such as the recyclebot, which profitably converts postconsumer plastic waste to 3D printing feedstock to make higher-value products, (upcycling) provide financial incentives to tighten the loop of the CE (McDonough and Braungart 2002; Zhong 2018).

9.3.4.2 Toward the CE

In January 2012, a report was released entitled *Towards the Circular Economy: Economic and business rationale for an accelerated transition*. The report, commissioned by the Ellen MacArthur Foundation and developed by McKinsey & Company, was the first of its kind to consider the economic and business opportunity for the transition to a restorative, circular model. Using product case studies and economy-wide analysis, the report details the potential for significant benefits across the EU. It argues that a subset of the EU manufacturing sector could realize net materials cost savings worth up to \$630 billion annually toward 2025 – stimulating economic activity in the areas of product development, remanufacturing, and refurbishment. *Towards the Circular Economy* also identified the key building blocks in making the transition to a CE, namely in skills in circular design and production, new business models, skills in building cascades and reverse cycles, and cross-cycle/cross-sector collaboration (Ellen MacArther Foundation 2012).

In January 2015 a *Definitive Guide to The Circular Economy* (Ellen MacArthur Foundation 2015) was published by Cowes with the specific aim to raise awareness among the general population of the environmental problems already being caused by our "throwaway culture." Waste Electrical and Electronic Equipment, in particular, is contributing to excessive use of landfill sites across the globe in which society is both discarding valuable metals and dumping toxic compounds that are polluting the surrounding land and water supplies. Mobile devices and computer hard drives typically contain valuable metals such as silver and copper but also hazardous chemicals such as lead, mercury, and cadmium. Consumers are unaware of the environmental significance of upgrading their mobile phones, for instance, on such a frequent basis but could do much to encourage manufacturers to start to move away from the wasteful, polluting linear economy toward are sustainable CE.

9.3.4.3 Circular Business Models

A CE calls upon chances to make more prominent esteem and adjust motivating forces through plans of action that expand on the connection among items and administrations. Essentially, this implies a round plan of action isn't centered simply around offering an item, yet it incorporates a move in contemplating incentive, presenting an entire scope of various plans of action to be utilized. This involves both the motivators and advantages offered to clients for bringing back utilized items and an adjustment in income streams, including installments for a round item or administration, or installments for conveyed accessibility, utilization, or execution identified with the item-based administration advertised. These better approaches for working together expect organizations to make an alluring plan of action for lenders, and agents to change the way they see the dangers and openings related with these models (Lewandowski 2016).

9.4 CE Indicators in Relation to Eco-Innovation

Nowadays, one of the most significant challenges in environmental management across the world is ensuring that our activities conform to the principles of sustainable development (Allen and Shonnard 2012; Clift and Allwood 2011; Das 2005; Das et al. 2001). Approaches to sustainable development have focused on "top-down" quantitative indicators based on scientific expertise and have a tendency to measure progress at global, national, or regional levels (Thomé et al. 2016). In recent years, the strategy of adopting a CE has gained increasing currency as a concept for the pursuit of global sustainability (Staniškis 2012). The most important benefit in moving to a more CE-based approach is the possibility of retaining the added value in products for as long as possible (Smol et al. 2015), extracting their maximum value and eliminating waste. CE-based systems keep resources within the economy. When a product has reached the end of its life, products can be efficiently reused again and again and hence create further value (COM, No. 398 2014). One of the factors determining the possibility of moving toward a CE is the implementation of innovation technologies, with a particular emphasis on eco-innovation. Despite the fact that these two issues are linked, a uniform methodology designed to compare the degree to which economies have moved toward a CE, which also includes eco-innovation, has not currently been established.

9.4.1 Development of the Concept of the CE

The CE is a relatively new concept, although the idea behind the CE has existed for a long time (Murray et al. 2015). "The better a real factory makes use of its waste, the closer it gets to its ideal, the bigger is the profit" (Lancaster 2002). The CE model is actually the opposite of a leaner one and is based on closed loops like a biological life cycle. It was not widely debated in the academic and scientific literature on sustainability, but it has become more popular with recent research (Pitt and Heinemeyer 2015; Stahel 2015).

In the twenty-first century the "preventive approach" has been replaced with the "restorative approach" both in Europe and across the whole world. The way of thinking in the twenty-first century has started to be more global, holistic, and systematic. Society and government have begun to introduce one more element – "restore" – into the "reduce-recycle-reuse concept." The CE model is based on concepts such as "cradle-to cradle™," where industry, by being waste-free, operates with no impact upon the environment (McDonough and Braungart 2002).

As for the development of the CE approach at the non-governmental level, the experience of the United Kingdom should be mentioned. A leading follower of the CE in the United Kingdom is the Ellen Macarthur Foundation, an NGO which has produced three reports on the concept (Ellen MacArthur Foundation 2012, 2013, 2014). The reports examine the potential of the CE as a new concept for development. The Foundation is very active and for the time being they already have obtained support for global innovation.

At the same time, existing CE approaches are valuable and have a tendency to develop further. They are strongly focused on resource efficient production. This can be proved by the main principles of the concept presented in the analytical report "Towards a Circular Economy: Business Rationale for an Accelerated Transition" (Ellen MacArthur Foundation 2015; Ellen MacArthur Foundation 2016; Smol et al. 2017):

- Principle 1: Preserve and enhance natural capital by controlling finite stocks and balancing renewable resource flows.
- Principle 2: Optimize resource yields by circulating products, components, and materials at the highest utility at all times in both technical and biological cycles.
- Principle 3: Foster system effectiveness by revealing and designing out negative externalities.

It can be observed that an innovative approach is in fact needed for each aspect of the CE concept. All strategic EU documents on the CE and the reports presented above see innovation as the heart of any transition to a CE. It is also obvious that a special role should be given to eco-innovations and that they should be key drivers, because the CE concept is all about economic growth, creating jobs, and at the same time reducing environmental impacts, including carbon emissions (Smol et al. 2015, 2017).

9.5 Process Intensification and Integration Potential in Manufacturing

Process intensification (PI) aims to dramatically improve manufacturing processes through the application of novel process systems and equipment. The novel approaches can be used to overcome bottlenecks, such as those imposed by thermodynamics, or to combine processing phenomena into fewer processing units with a concurrent reduction of capital and operation and maintenance costs and energy, water and materials intensity. PI approach goes beyond the incremental improvements achieved through optimizing existing equipment and process systems and achieves step changes in energy and materials efficiency, total life-cycle cost reduction, and environmental impact by minimizing wastes at the sources via various hierarchy pollution prevention techniques (Bielenberg and Bryner 2018).

9.5.1 What Is PI?

PI is not well defined like many multidimensional concepts. Although PI may be best explained through examples, here are a few potential definitions to consider when thinking about PI (Bielenberg and Bryner 2018).

Well-known experts in PI Stankiewicz and Moulijin (2000), both at Delft University of Technology, defined PI as the development of innovative apparatuses and technologies that bring dramatic improvements in chemical and allied manufacturing and processing, substantially reducing equipment volume, energy consumption, or waste minimization, less environmental and health impacts, and ultimately yields to cheaper, safer, sustainable technologies. They added four guiding principles to that definition (EFCE 2015):

1) Maximize the effectiveness of intramolecular and intermolecular events
2) Provide all molecules the same process experience
3) Optimize driving forces at all scales and maximize the specific surface areas to which they apply
4) Maximize synergistic effects from partial processes.

PI targets dramatic improvements in manufacturing and processing by rethinking existing operation schemes into ones that are both more precise and efficient than existing operations. There are a series of technologies that enable equipment sizes to be radically reduced. PI and microreaction technology, as well as reports of experimental results in the use of novel PI systems, including the static mixers, high-gravity (HiGee) technology, cyclic and reactive distillation, compact high specific surface heat exchanger, multifunctional reactors, microchannel reaction systems, microengineering, microtechnology, the catalytic plate reactor, and a chemical microsystem for pervaporation. Such technologies enable to plant sizes to be correspondingly reduced. The very low inventories have environmental benefits and there are also claimed cost benefits (Reay et al. 2013). An incidental benefit is that the processes may be economic at a smaller scale (mostly batch processes), and that partly contributes to economic sustainability.

The European Roadmap on Process Intensification describes PI as providing "radically innovative principles (paradigm shift) in process and equipment design, which can benefit (often with more than a factor of two) process and chain efficiency, capital and operating expenses, quality, wastes, process safety, process integration and more" (EFCE 2015).

Reay et al. (2013) describe PI as a "chemical and process design approach that leads to substantially smaller, cleaner, safer, and more energy-efficient process technology." The common thread among these definitions is a focus on new schemes and equipment that create improved processes by combining, controlling, and/or enhancing the chemistry and transport phenomena in a chemical process.

A classic example of PI equipment is the static mixers. Although there are many different designs for static mixers, the basic concepts are the same. Stationary mixing elements placed in the path of fluid flow create locally highly mixed channels for the fluid to move through. Homogeneous mixing occurs quickly, with no external energy input other than that associated with the small pressure drop, at typically low capital costs. Static mixers can be incorporated into other unit operations (e.g. reactors) to enable the combination of processes and can be tailored to match mixing scales and times to optimize overall process efficiency. For example, static mixers can be placed in a tubular reactor for a two-phase reaction system – creating a high level of mixing while maintaining a largely plug-flow profile (typically found at a much smaller scale) at the larger reactor scale. Such an approach could offer many advantages over the alternative of operating a large continuous stirred-tank reactor to maintain high levels of mixing.

There are many other examples of PI equipment, including microchannel reactors, spinning-disc reactors, centrifugal contractors, and dividing-wall column (Figure 9.11) to name a few. Each of these relies on a novel driving force (e.g. rotation) or nonstandard configuration (e.g. microchannels) to enable increased control over mixing, reaction, and heat, mass, and momentum transfer to bring about step changes in the reduction of energy consumption and capital costs.

The dividing-wall column (Figure 9.11) is one form of process intensification that enables the separation of a three-phase system in a single distillation tower. The internal wall splits the column into two halves. The three-phase system is pumped into one side of the column and is reflected by the wall. The lightest component drops to the bottom and is withdrawn. The intermediate component is initially entrained in both streams; the intermediate component that flows upward subsequently separates and falls down on the opposite side of the wall, while the component that is entrained in the heavy component separates and flows up the back side of the column, where the entire intermediate stream is recovered through a side port.

9.5.2 Case Study 9.5: Elimination of Dioxin and Furans by Alternative Chemical PI

In early 1988, a study of effluents from bleached pulp mills showed significant levels of dioxins and furans. As a result of these findings, the industry implemented a series of process changes including (i) eliminating the use of certain defoamers which contained dioxin and furan precursors, (ii) decreasing the use of chlorine as bleaching chemical, and (iii) increasing the use of chlorine dioxide for pulp bleaching. Between 1988 and 1996, there was a very significant reduction in effluent 2,3,7,8-tetrachloro dibenzo-*p*-dioxin (TCDD) and 2,3,7,8-tetrachloro dibenzo-*p*-furan (TCDF) concentrations. For example, in 1988 40% of the mill effluent samples contained less than 10 parts per quadrillion (ppq) of 2,3,7,8-TCDD. In 1990, 70% of the samples contained less than or equal to 10 ppq of 2,3,7,8-TCDD. By 1996, all but two mill effluent samples showed 2,3,7,8-TCDD levels below 10 ppq. This record exemplifies how pollution prevention leading to Zero Discharge of dangerous environmental contaminants can be achieved at the source through process changes and substituting and intensifying chemical processes. Figure 9.12 shows a steady reduction of dioxin/furan formation to below detection level by substituting elemental chlorine to chlorine dioxide as a bleaching agent (Das and Jain 2001; NCASI 1993).

9.5.3 Mini-Case Study 9.2: Multi-Pollutants Capture and Recovery of SO_x, NO_x, and Mercury in Coal-Fired Power Plant

9.5.3.1 Advanced Multi-Pollutant Control: Intensified Regenerative Activated Coke Technology

Regenerative activated coke technology (ReACT™) is an advanced intensified multi-pollutants technology that achieves simultaneous capture of SO_x, NO_x, and mercury in one vessel. The process was first developed in Germany, and it was subsequently advanced and commercialized by J-Power EnTech in Japan where ReACT™ has been implemented at three large-scale coal-fired power plants and at several steel mills, petrochemicals, refineries, and waste incinerator facilities, as a high-efficiency emission control system.

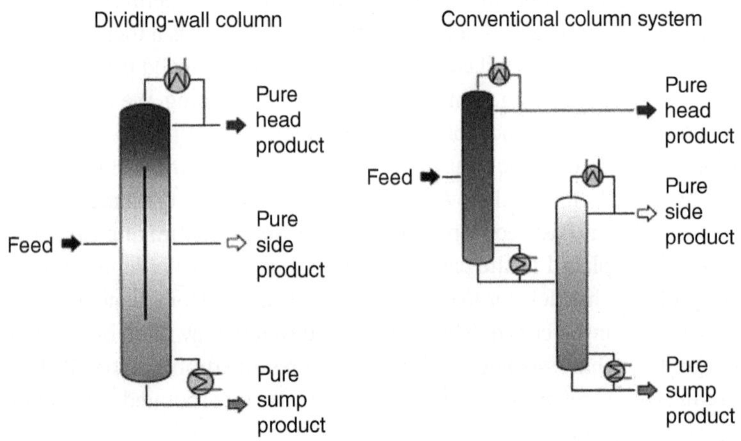

Figure 9.11 Dividing-wall column for three-phase separation in distillation tower. *Source:* From Bielenberg and Bryner (2018).

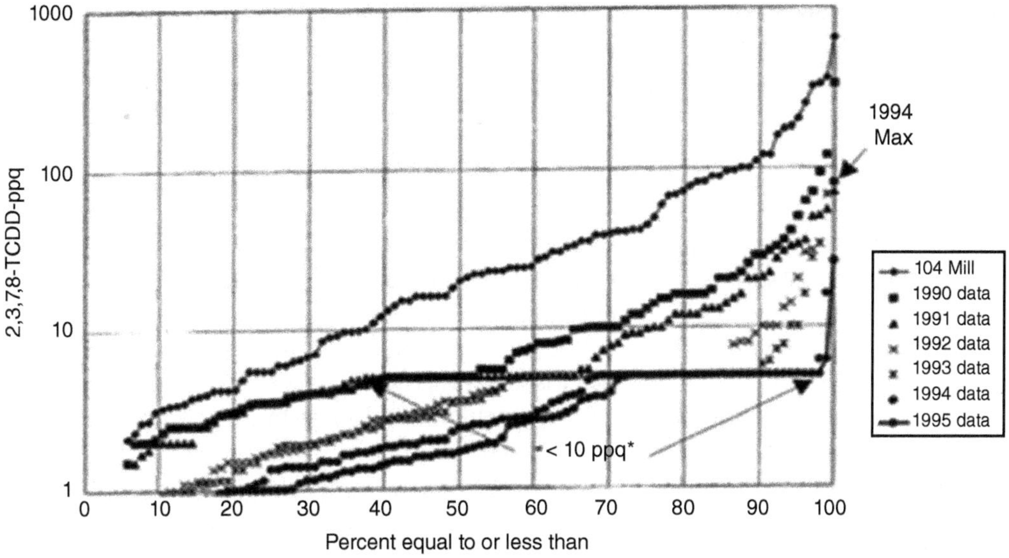

Figure 9.12 Elimination of dioxin/furans formation in bleaching pulp by intensifying process and substituting chemicals. *Source:* From Das (2005).

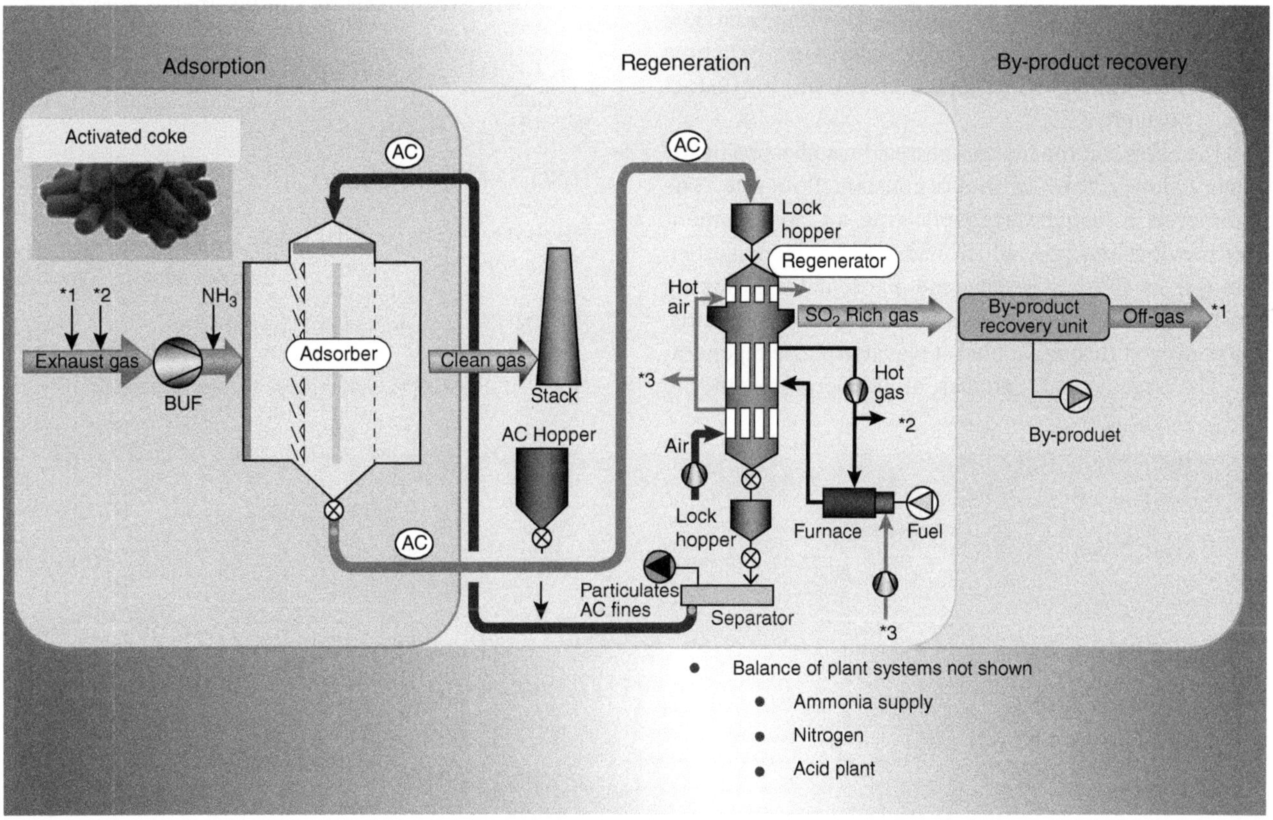

Figure 9.13 Schematic diagram of three-stage ReACT™ process. *Source:* From Reynolds (2018).

Figure 9.13 shows the three-stage ReACT™ process, including adsorption by activated coke (AC) followed by regeneration of SO_x and mercury, and then by-product recovery of sulfur-rich gas converting into a marketable sulfuric acid.

ReACT™ technology is based on adsorption of SO_2 and NO_x on AC in a moving bed, with subsequent regeneration of the coke and production of saleable by-product sulfuric acid. This technology is a completely dry scrubbing event that does not evaporate any water into the flue-gas stream, can produce extremely low level of SO_2 emissions, and can also provide additional control for NO_x, PM, and Hg as co-benefits. The moving bed dry scrubber operates downstream of a primary particulate control device, which for many plants is traditional ESP technology. For special cases where extremely low levels of particulate are permitted, a downstream polishing ESP can also be applied.

A mix of custom and traditional material handling equipment transports the AC to and from the absorber and regenerator. Careful material handling and the robust nature of the AC allow the coke to be recycled with minimal makeup requirements.

Simultaneous capture of SO_2, NO_x, and Hg is accomplished in one vessel along with a further reduction in particulate emissions. High removal efficiencies include up to 99.9% SO_2, 50–80% NO_x, and above 90% Hg (including elemental). Particulate is also reduced by about 50%. Advantages of this technology include high-emission reduction efficiencies, cost-effectiveness, low power consumption, minimum installation space, minimal water requirements (dry process), no gas reheating, and low maintenance requirements. The AC process involves three steps: (i) adsorption, (ii) regeneration, and (iii) by-product recovery (Figure 9.13).

In the absorber, the flue gas passes through a bed of AC moving slowly downward at a constant flow rate. The adsorber is a single stage, reflecting an improvement over previous designs, which had a two-stage adsorber. Removal of SO_2, NO_x, Hg, and particulates is taking place in one step. As the AC is saturated with pollutants, it is conveyed through a bucket elevator to the regenerator. The regenerator operates at temperatures in the 399–499 °C range. Simultaneously, sulfuric acid or ammonium salts in the AC are decomposed to N_2, SO_2, SO_3, and H_2O. Hg is retained in the AC and removed at some period of time, depending on the Hg concentrations at the inlet of ReACT™.

After cooling, the regenerated AC passes through a separator (vibrating screen), which separates particulates and small particles of AC, from the larger AC, which is returned to the adsorber. The AC removed from the separator can be returned to the boiler for burning or sold and used in industrial applications, such as dioxin adsorption – SO_2-rich gas from the generator is converted to a saleable product, such as sulfuric acid, in the by-product recovery facility (Figure 9.14).

- Adsorption and reaction on AC: SO_2 and SO_3 are adsorbed at high efficiency on the AC surfaces as sulfuric acid, ammonium sulfate, and ammonium bisulfate.
- The catalytic effect of AC also leads to reduction reactions to reduce NO_x in the presence of ammonia. The NO_x control activity is completed in the adsorber as follows:

$$SO_2 + H_2O + \frac{1}{2}O_2 \rightarrow H_2SO_4^*$$

$$SO_3 + H_2O \rightarrow H_2SO_4^*$$

$$H_2SO_4^* + NH_3 \rightarrow NH_4HSO_4^*$$

$$NH_4HSO_4^* + NH_3 \rightarrow (NH_4)_2SO_4^*$$

(* denotes adsorbed species)

$$NO + NH_3 + \frac{1}{2}O \rightarrow N_2 + 3/2H_2O$$

$$NO + C\ldots R \rightarrow N_2\ (C\ldots R \text{ represents a surface functional site})$$

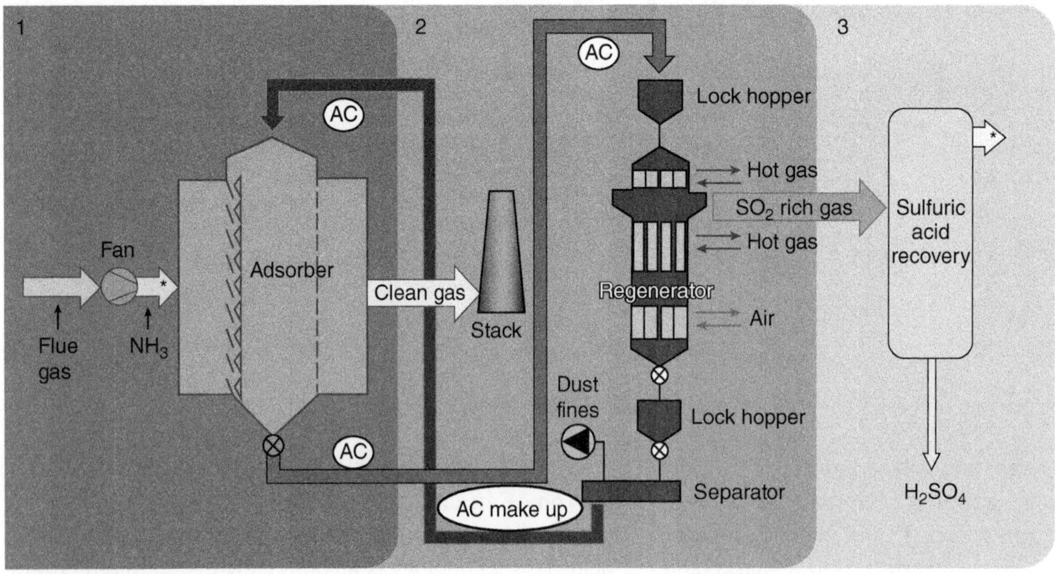

Figure 9.14 SO_x recovery and converting it to H_2SO_4 in acid plant. *Source:* From Reynolds (2018).

9.5.3.2 AC and Regenerated Coke

The AC is a carbonaceous material produced by steam activation (at ~900 °C). It has high mechanical strength against abrasion and crushing. Its surface area is 150–300 m^2/g, less than the conventional activated carbon, but not much higher than the metallurgical coal. AC comes in tablet or almond-type shape and is generally much larger than the activated carbon. AC is easier to produce and less expensive than activated carbon. Figure 9.15 shows the internal model of AC and its pore structure. Figure 9.16 shows mercury vapor removal efficiency versus time.

9.5.3.3 Thermal Regeneration of AC

At the temperatures in the regenerator heating section adsorbed sulfuric acid, ammonium sulfate, and ammonium bisulfate are decomposed to SO_2. AC pellets enter via lock hoppers and is in gravity counter-flow against desorbed gases and N_2 carrier gas. Sulfur-rich gas containing SO_2, CO_2, N_2, and H_2O exits the regenerator to an acid plant as follows:

$$H_2SO_4^* + \tfrac{1}{2}C \rightarrow SO_2 + \tfrac{1}{2}CO_2 + H_2O$$

$$(NH_4)_2SO_4^* \rightarrow SO_3 + NH_3 + 2H_2O$$

$$NH_4HSO_4^* \rightarrow SO_3 + 2NH_3 + 2H_2O$$

$$3SO_3 + 2NH_3 \rightarrow 3SO_2 + 3H_2O + N_2$$

(*denotes adsorbed species; non-isothermal desorption kinetics reach peak rates at 300 °C and completion by 450 °C).

9.5.3.4 Performance and Benefits of ReACT™

Table 9.3 presents a typical performance at J-Power EnTech 2x600MW ISOGO coal-fired power plant in Japan (Reynolds, personal communication). The benefits of the advanced multi-pollutants control by the ReACT™ can be summarized as follows:

- *By-product revenue* – Sulfuric acid is the world's number one commodity chemical with a market value of \$50 200/T. Instead of producing disposal gypsum or fly ash/gypsum waste, ReACT™ produces saleable by-product.
- *Avoided disposal costs* – For every ton of SO_2 controlled in conventional flue-gas desulfurization (FGD), about three tons of solid waste is generated. More if fly ash is part of the FGD waste stream.
- *Near zero water use* – ReACT™ uses minimal water, in significant contrast to FGD systems. For a 500 MW plant, a WFGD (wet flue gas desulfurization) system would require 275 000 000 gal/year while ReACT™ would use near zero.

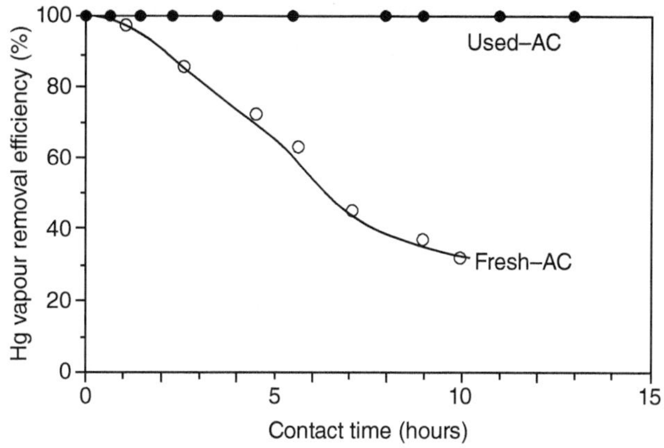

Figure 9.16 Mercury vapor removal efficiency (%) versus time.

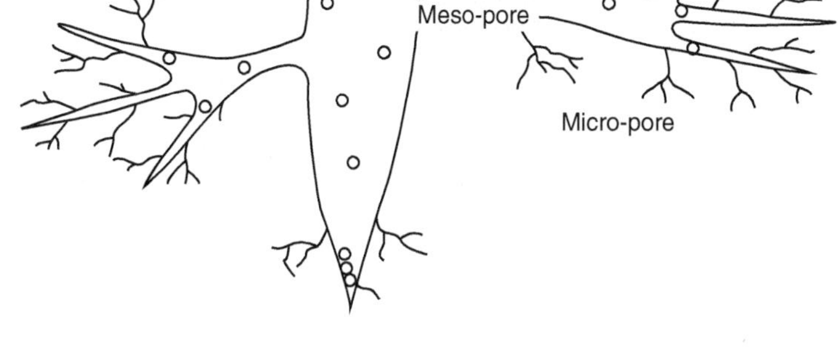

Figure 9.15 Internal model of AC – pore structure.

Table 9.3 Typical performance of advanced multi-pollutant removal.

		Operating results		
Pollutants	**Emissions permit**	**Efficiency (%)**	**Inlet concentration**	**Outlet concentration**
SO_x	10 ppm (0.025 lb/MMbtu)	>98	<410 ppm <0.85 lb/MMbtu	<1 ppm <0.002 lb/MMBtu
NO_x	13 ppm (0.02 lb/MMbtu)	10–50	<20 ppm 0.03 lb/MMbtu	<7 ppm <0.011 lb/MMbtu
Particulate	5 mg/Nm^3 (0.004 lb/MMbtu)	>95 (>99.9 w/ESPs)	<100 mg/Nm^3 <0.1 lb/MMbtu	<3 mg/Nm^3 <0.002 lb/MMbtu (w/backend ESP)
Hg	--	>90	2.5 μg/Nm^3	<0.25 μg/Nm^3

- *Minimal plant modifications* – ReACT™ can flow to the existing stacks with no change in liner materials. ReACT™ is located downstream of existing equipment and does not necessitate modifications to upstream equipment.
- *NO_x performance options* – ReACT™ systems may be designed for a range of NO_x reduction options – from co-benefit levels of 30% through alternative designs reaching 80%.

9.6 Manufacturing Process Integration

In response to the staggering environmental and energy problems associated with manufacturing facilities, the chemical and allied process industries have dedicated much attention and many resources to mitigating the detrimental impact on the environment, conserving resources, and reducing the intensity of energy usage. These efforts have gradually shifted from a unit-based approach to a systems-level paradigm. The past decade has seen significant industrial and academic efforts devoted to the development of holistic process design methodologies that target energy conservation and waste reduction from a system's perspective. Implicit in the holistic approach, however, is the need to realize that changes in a unit or a stream often propagate throughout the process and can have significant effects on the operability and profitability of the process. Furthermore, the various process objectives (e.g. technical, economic, environmental, and safety) must be integrated and reconciled. These challenges call for the development and application of a systematic approach that transcends the specific circumstances of a process and views the environmental, energy, and resource-conservation problems from a holistic perspective. This approach is called *process integration* (Parthasarathy and Dunn 2005; Smith 2016). It emphasizes the unity of the process and it is broadly divisible into the categories of *mass integration* and *energy integration*. Parthasarathy and Dunn (2005) provide an energy and heat exchange networks with a mini-case-study on heat exchange network, mass integration is briefly introduced, and the concept of mass integration is explored in detail.

9.6.1 Process Integration Technique Has Few Possible Applications

1) A holistic approach to process design which emphasizes the unity of the process and considers the interactions between different unit operations from the outset, rather than optimizing them separately. This can also be called *integrated process design* or *process synthesis* (Smith 2016) describe the approach well. An important first step is often *product design* which develops the specification for the product to fulfill its required purpose.
2) *Pinch analysis*, a technique for designing a process to minimize energy consumption and maximize heat recovery, also known as *heat integration*, *energy integration*, or *pinch technology*. The technique calculates thermodynamically attainable *energy targets* for a given process and identifies how to achieve them. A key insight is the pinch temperature, which is the most constrained point in the process. A detailed explanation of the techniques is given by Parthasarathy and Dunn (2005). This definition reflects the fact that the first major success for process integration was the thermal pinch analysis addressing energy problems. Other pinch analyses were developed for several applications such as mass-exchange networks, water minimization, and material recycle. A very successful extension was "Hydrogen Pinch," which was applied to refinery hydrogen management. This allowed refiners to minimize the

capital and operating costs of hydrogen supply to meet ever stricter environmental regulations and also increase hydrotreater yields (Hallale 2001).

In the context of chemical engineering, process integration can be defined as a holistic approach to process design and optimization, which exploits the interactions between different units in order to employ resources effectively and minimize costs. Process integration is not limited to the design of new plants, but it also covers retrofit design (e.g. new units to be installed in an old plant) and the operation of existing systems. Hallale (2001) explains that with process integration, industries are making more money from their raw materials and capital assets while becoming cleaner and more sustainable.

The main advantage of process integration is to consider a system as a whole (i.e. integrated or holistic approach) in order to improve their design and/or operation. In contrast, an analytical approach would attempt to improve or optimize process units separately without necessarily taking advantage of potential interactions among them.

For instance, by using process integration techniques it might be possible to identify that a process can use the heat rejected by another unit and reduce the overall energy consumption, even if the units are not running at optimum conditions on their own. Such an opportunity would be missed with an analytical approach, as it would seek to optimize each unit, and thereafter it wouldn't be possible to reuse the heat internally.

Typically, process integration techniques are employed at the beginning of a project (e.g. a new plant or the improvement of an existing one) to screen out promising options to optimize the design and/or operation of a process plant. Also it is often employed, in conjunction with simulation and mathematical optimization tools to identify opportunities in order to better integrate a system (new or existing) and reduce capital and/or operating costs.

Most process integration techniques employ Pinch analysis or Pinch tools to evaluate several processes as a whole system. Therefore, strictly speaking, both concepts are not the same, even if in certain contexts they are used interchangeably. The review by Hallale (2001) explains that in the future, several trends are to be expected in the field. In the future, it seems probable that the boundary between targets and design will be blurred and that these will be based on more structural information regarding the process network. Second, it is likely that we will see a much wider range of applications of process integration. There is still much work to be carried out in the area of separation, not only in complex distillation systems but also in mixed types of separation systems. This includes processes involving solids, such as flotation and crystallization. The use of process integration techniques for reactor design has seen rapid progress, but it is still in its early stages. Third, a new generation of software tools is expected. The emergence of commercial software for process integration is fundamental to its wider application in process design.

9.7 New Sustainable Chemicals and Energy from Black Liquor Gasification Using Process Integration and Intensification

In the areas of chemicals and allied products and energy process engineering, it is of particular interest to design processes with low-energy requirements (energy intensive) (preferably renewable materials, e.g. biomass) and the use of minimum raw materials or use of a by-product as feedstock (materials intensive) and minimum number of equipment (i.e. lean manufacturing). This is the basis from which the area of PI arises. The main objective of PI is to propose production and manufacturing processes in plants as small as possible and with low total energy, materials, water, and other resources. To reduce the size of plants, the scope is reducing the size of equipment or the number of operations in the processes, or using multifunction within the process is must. Since energy requirements of industrial processes are usually satisfied with steam, for which production is necessary to burn some kind of fuel (usually fossil fuels), by reducing energy requirements a direct reduction on environmental impact (e.g. greenhouse gases, oxides of carbon, nitrogen and sulfur, and particulate matter). In this extended case study, black liquor, a by-product of wood-pulping process, is utilized to make chemicals and biofuels, in particular methanol, dimethyl ether, and syngas. Different feedstock, use of by-product and alternatives and principles of PI are commented (Gómez-Castro et al. 2016).

9.7.1 Introduction

The US pulp and paper industry is the largest producer and user of biomass energy in the United States today, nearly all derived from sustainably grown trees. Renewable resources used at pulp mills include bark, wood wastes, and black liquor, the lignin-rich by-product of cellulose-fiber extraction.

With substantial renewable energy resources at its immediate disposal and with potentially much more extensive resources available in the long term, the US pulp and paper industry has the potential to contribute significantly to addressing climate change and US energy security concerns, while also improving its global competitiveness. A key requirement for achieving these goals is the commercialization of breakthrough technologies, especially gasification, to enable the clean and efficient conversion of biomass to useful energy forms, including electricity and transportation fuels.

Gasification technology enables low-quality solid fuels like biomass to be converted with low pollution into a fuel gas (synthesis gas or "syngas") consisting largely of hydrogen (H_2) and carbon monoxide (CO). Syngas can be burned cleanly and efficiently in a gas turbine to generate electricity. It can be passed over appropriate catalysts to synthesize clean liquid transportation fuels or chemicals. It can also be converted efficiently into pure H_2 fuel.

While most pulp and paper manufacturing facilities in the United States today do not export electricity and none export transportation fuels, their established infrastructure for collecting and processing biomass resources provides a strong foundation for future gasification-based "biorefineries" that might produce a variety of renewable fuels, electricity, and chemicals in conjunction with pulp and paper products (Figure 9.17) (Perlack et al. 2005).

If the biomass resources from which energy carriers are produced at such biorefineries are sustainably grown and harvested, there would be few net lifecycle emissions of CO_2 associated with biorefineries and their products. To the extent that the biorefinery products replace fuels or chemicals that would otherwise have come from fossil fuels, there would be net reductions in CO_2 emissions from the energy system as a whole. The reductions would be even more significant if by-product CO_2 generated at biorefineries were to be captured and sent for long-term underground storage (Larson et al. 2006b). Carbon capture and storage with fossil fuels is of wide interest today. Several large-scale CO_2 storage projects (storing >1 million T/Y of CO_2) are operating and more are under development worldwide to demonstrate feasibility (USDOE 2006).

Coupled with the potential to address national energy security and global warming concerns is the looming need in the US pulp and paper industry for major capital investments to replace the aging fleet of Tomlinson recovery boilers used today to recover energy and pulping chemicals from black liquor. The majority of Tomlinson boilers operating in the United States were built beginning in the late 1960s through the 1970s. With serviceable lifetimes of 30–40 years, the Tomlinson fleet began undergoing a wave of life-extension rebuilds in the mid-1980s. Within the next 10–20 years, rebuilt boilers will be approaching the age at which they will need to be replaced, the capital investment for which at a typical mill will be between \$100 and \$200 million. A similar situation exists in the European pulp industry. This situation provides an unusual window of economic opportunity for introducing black liquor gasifiers as replacements for Tomlinson boilers. Concerted efforts are ongoing in the United States and Sweden to develop commercial black liquor gasification technologies (Ekbom et al. 2005; Larson et al. 2006a; Swedish Forest Agency 2008).

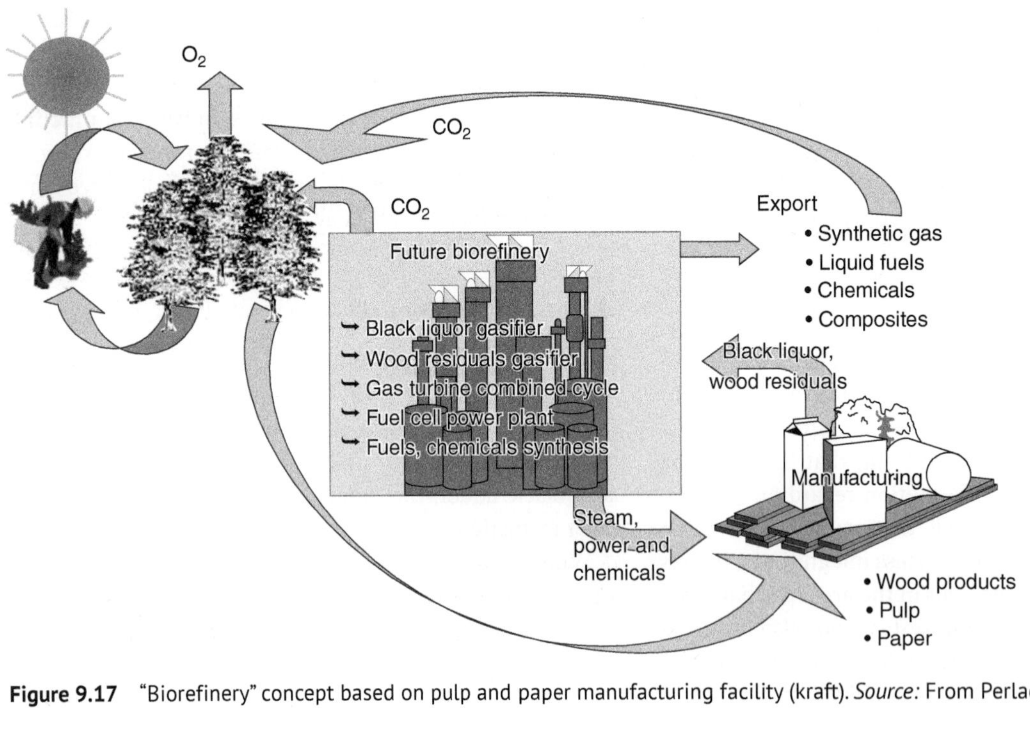

Figure 9.17 "Biorefinery" concept based on pulp and paper manufacturing facility (kraft). *Source:* From Perlack et al. (2005).

9.7.2 Black Liquor Gasification (BLG): Introduction

Black liquor is a biomass feedstock with unique properties suitable for gasification (Ådahl et al. 2004; Andersson and Harvey 2004; Bajpai, 2008, 2013; Berglin et al. 2002; Dahlquist et al. 2009; Grigoray 2009; Marklund 2006; Salomonsson, 2013; Sricharoenchaikul 2009, 2001; Waldner and Vogel 2005; Wikipedia 2016). First of all, it is available at existing industrial sites in large quantities. Second it is liquid. This makes it possibly to easily feed it by pumping into the pressurized gasifier. The liquid state also makes the black liquor easy to atomize into a fine mist that reacts very fast in the gasifier. Third, the gasification of black liquor char is more rapid than for any other feedstock as the inherently high sodium and potassium content of black liquor acts as a catalyst. These properties make it possible to apply the high temperature, entrained flow gasification principle to black liquor. This type of gasification process provides many advantages over alternative gasification technologies. It is very rapid, single-stage gasification process with low reactor volume, and low emissions (including particulate matters and odorous TRS gases) and low power consumption.

9.7.2.1 Black Liquor Properties

The black liquor is an aqueous solution of lignin residues, hemicellulose, and the inorganic chemicals used in the process. The black liquor is sent through a series of evaporators to increase the solid content in order to make it suitable for combustion. In the evaporators, the BLS content is increased from about 15 wt% to about 70–75 wt%. The concentrated black liquor that goes into the recovery boiler consists of about 30% water, 30% valuable inorganic cooking chemicals, and 40% lignin and other organic matter separated from the wood. The higher heating value is about 14 000 kJ/kg solids. Typically, a large pulp-mill will produce about 200 T/day of black liquor corresponding to about 300 MWh (Bajpai 2014).

The organic matter in the black liquor is made up of water/alkali soluble degradation components from the wood. Lignin is degraded to shorter fragments with sulfur content at 1–2% and sodium content at about 6% of the dry solids. Cellulose and hemicellulose is degraded to aliphatic carboxylic acid soaps and hemicellulose fragments. The extractives gives tall oil soap and crude turpentine. The soaps contain about 20% sodium. Table 9.4 presents some properties of black liquor.

9.7.2.2 BLG Technologies

BLG may be performed either at low temperatures or at high temperature based on whether the process is conducted above or below the melting temperature range (650–850 °C) of the spent pulping chemicals (Sricharoenchaikul 2001).

Low-temperature gasification – Low-temperature gasifier operates at 600–850 °C, below the melting point of inorganics, thus avoiding smelt–water explosions.

High-temperature gasification – High-temperature gasification units generally operate in the 900–1000 °C range and produce a molten smelt.

In low-temperature gasification, the alkali salts in the condensed phase remain as solid products, while molten salts are produced in high-temperature gasification. Low-temperature gasification is advantageous over high-temperature gasification because gasification at low temperature yields improved sodium and sulfur separation. Additionally, low-temperature gasification requires fewer constraints for materials of construction because of the solid product. However, syngas of low-temperature gasification may contain larger amounts of tars, which can contaminate gas cleanup operations in addition to contaminating gas turbines upstream of the gasifier. These contamination problems can result in a loss of fuel product from the gasifier (Patrick and Siedel 2003; Sricharoenchaikul 2001).

Several companies have performed trials to develop a commercially feasible process for BLG. History of BLG development is very well described by Whitty and Baxter (2001) and Whitty and Verrill (2004). In the search for alternative ways to recovering the cooking chemicals, gasification techniques have been thoroughly examined several times. In total, more than 20 different technologies have been investigated over the years (Swedish Energy Agency 2008). The most interesting attempts of accomplishing low-temperature black liquor gasification (LTBLG) and high-temperature black liquor gasification (HTBLG) processes are presented in Table 9.5. Only two technologies are currently being commercially pursued: the MTCI (Manufacturing and Technology Conversion International Company, low temperature) and Chemrec (high temperature) technologies. A pressurized, oxygen-blown, HTBLG technology, shown in Figure 9.18, is being developed by Chemrec, a Swedish company (Chemrec 2006; Whitty 2006; Whitty and Nilsson 2001). Table 9.6 gives the difference between these two technologies.

Table 9.4 Properties of black liquor.

Density (50% solids) at 93 °C	1.26 g/cm^3
Viscosity (50% solids at 121 °C) Viscosity (70% solids at 125 °C)	5.0 and 100.0 cP
Specific heat (40% solids)	2.0 KJ/(kg·K)
Heating value (35% lignin and 70% solids)	~14 000 kJ/kg

Table 9.5 Black liquor gasification processes.

SCA-Billerud process
Copeland process
Weyerhaeuser's "dry pyrolysis" process
St. Regis hydropyrolysis process
The Texaco coking process
VIT circulating fluidized gasification process
Babcock & Wilcox's bubbling fluidized bed gasification process
NSP or Ny Sodahus process
DARS process (Direct Alkali Recovery System)
ABB circulating fluidized bed gasification process
MTCI/TRI's fluidized bed steam reforming process
Kellogg, Brown, and Root's spouting and transport fluidized bed gasifiers
Paprican's AST process
University of California's "pyrolysis, gasification, combustion" process
The Champion–Rockwell molten salt gasification process
The SKF plasma black liquor gasifier
Ahlstrom's suspension gasifier
Tampella's entrained flow gasifier
Noell's entrained flow gasifier
Champion–Rockwell molten salt gasification process
Chemrec entrained flow gasification process
Catalytic hydrothermal gasification of black liquor

Source: From Whitty and Verrill (2004), Grigoray (2009), Naqvi et al. (2010), and Swedish Energy Agency (2008).

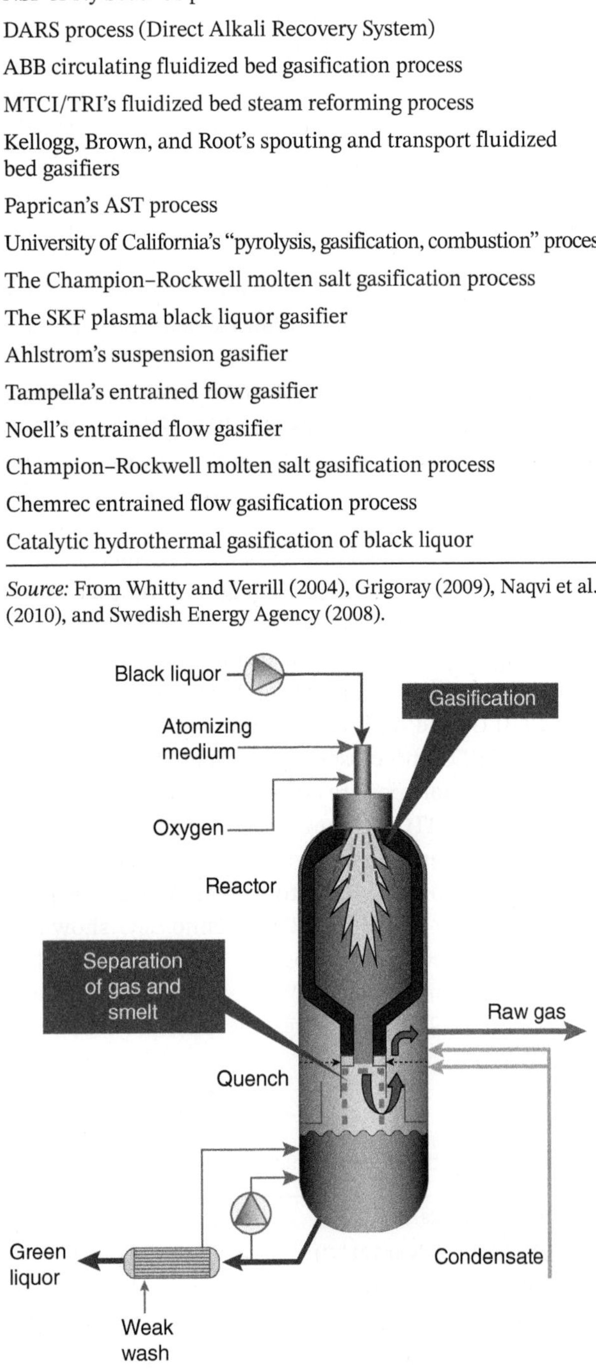

Figure 9.18 Pressurized, oxygen-blown, HTBLG technology being developed by Chemrec, a Swedish Company. *Source:* From Chemrec (2006), Whitty (2006) and Whitty and Nilsson (2001).

Table 9.6 Difference between LTBLG and HTBLG.

Property	LT (TRI)	HT (Chemrec)
Heating	Indirectly (syngas)	Directly (black liquor)
Chemical recovered	Solid phase	Smelt
Sulfur split (gas/smelt)	90/10	50/50
Syngas composition	High H_2 concentration	Moderate H_2 concentration
Syngas energy content	High	Low

Source: From Swedish Energy Agency (2008).

9.8 Chemical Recovery and Power/Steam Cogeneration at Pulp and Paper Mills

9.8.1 The Pulp and Paper Industry

The pulp and paper industry is one of the largest industries in the world. It is also an important source of employment in many countries. A sustainably managed pulp and paper industry can bring several benefits to the local economy and people particularly in rural areas. The industry is dominated by North American, Northern European, and East Asian countries like Japan. Latin America and Australasia also have significant pulp and paper industries. Over the next few years, it is expected that both India and China will become the key in the industry's growth.

Pulp and paper are primarily made out of wood fibers originating from natural forests or pulpwood plantations. Recycled fiber and other fiber sources such as agricultural residue are also utilized and recycled fiber is becoming more commonly used in pulp and papermaking. There are two significant pulping technologies available that differ greatly in terms of process, i.e. mechanical and chemical pulping. Approximately 30% of the total pulp production in European Union is from mechanical pulping, while the rest is produced by means of chemical pulping (Das 2005; Das and Houtman 2004; Swedish Forest Agency 2008). North America has major pulp and paper industry, about 21% of the total pulp produced is from mechanical pulping and rest is produced chemically.

A pulp mill that produces bleached kraft pulp generates 1.7–1.8 T of black liquor (measured as dry content) per ton of pulp. Black liquor, thus, represents a potential energy source of 250–500 MW/mill. As modern kraft pulp mills have a surplus of energy, they could become key suppliers of renewable fuels in the future energy systems. Today, black liquor is the most important source of energy from biomass. It is, thus, of great interest to convert the primary energy in the black liquor to an energy carrier of high value. Table 9.7 gives the relative annual production of some major fuels (Farmer and Sinquefield 2009). A key advantage of black liquor

Table 9.7 Annual production of selected fuels.

Production	Million T/Y
Black liquor solids	200
Paper and board	180
Crude oil	4000
Hard coal	500

Source: From Farmer and Sinquefield (2009).

compared to biofuels and fossil fuels is that it is already at the mill; the handling infrastructure already exists and there are no collection and transport costs.

Worldwide, the pulp and paper industry currently processes about 170 million T of black liquor (measured as dry solids) per year, with a total energy content of about 2 EJ, making black liquor a very significant biomass source (IEA Bioenergy Report 2009). In comparison with other potential biomass sources for chemical production, black liquor has the great advantage that it is already partially processed and exists in a pumpable, liquid form. Using black liquor as a raw material for liquid or gaseous biofuel production in a biorefinery has many advantages: biomass logistics are extremely simplified as the raw material for fuel production is handled within the boundaries of pulp and paper plant; the process is easily pressurized, which increases fuel production efficiency; due to the processing of wood to pulp, the produced syngas has a low methane content, which optimizes fuel yield; pulp mill economics become less sensitive to pulp prices when diversified with another product; and gasification capital cost is shared between recovery of inorganic chemicals, steam production, and syngas production. Overall, if the global production of black liquor was to be used for transportation biofuel production, then this would correspond to about 48 million T of methanol, compared with current world production from fossil fuels of about 32 million T, a significant impact.

9.8.2 Black Liquor Gasification Combined Cycle Power/Recovery

Gasifying black liquor enables it to be used as fuel in a gas turbine combined cycle, a much more efficient electricity generating option than the Tomlinson boiler steam turbine technology. In a typical "mill scale" black liquor gasification combined cycle (BLGCC), shown in Figure 9.19, the black liquor is gasified, and the syngas product is cooled, cleaned, stripped of H_2S (using a Selexol® unit), and then burned in the gas turbine. The turbine exhaust passes through a "duct burner" to the heat recovery steam generator (HRSG), where steam is raised to drive a steam turbine. A small amount of natural gas is burned in the duct burner to enable production of the requisite process steam needed to run the mill. Steam is extracted at two different pressures from the steam turbine. The HRSG steam production is augmented by steam delivered from hog fuel boilers assumed to be pre-existing at the reference mill. One design constraint in the BLGCC analysis was the size of the existing hog fuel boilers, which limited the available steam delivered from these boilers (This limitation is the reason a small quantity of natural gas is used in the duct burner (Black Liquor, Wikipedia, 2016).

9.8.3 Biorefinery

Figure 9.20 presents a schematic of the dimethyl ether (DME) biorefinery. In the DME biorefinery, the black liquor gasifier, supplied with oxygen from an air separation unit, provides all of the synthesis gas to a liquid-phase DME reactor. Unconverted synthesis gas is separated from product DME and 97% of it is recycled to the synthesis reactor to increase DME production. The 3% purge stream taken from the recycle loop prevents excessive buildup of inert gases. The purge gas is sent to the hog fuel boiler, where it burns with an amount of wood residues selected such that the amount of steam generated is sufficient to

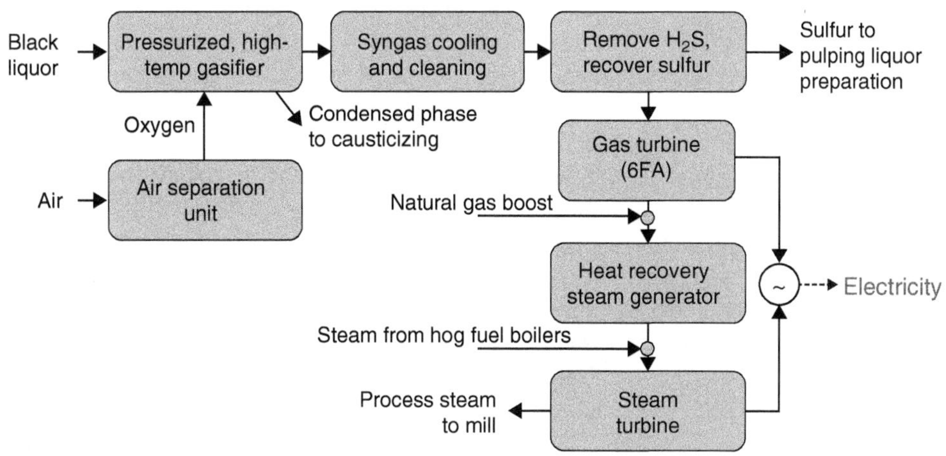

Figure 9.19 Simplified schematic representative "mill-scale" BLGCC (BLGCC). *Source:* From Larson et al. (2006a).

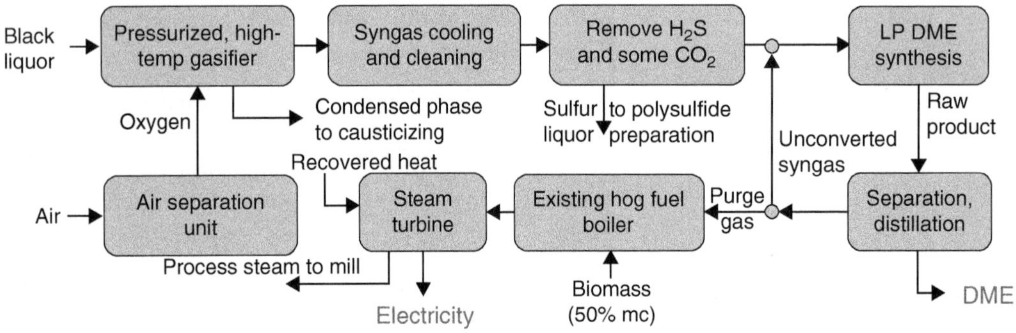

Figure 9.20 Schematic of biorefinery of DME. Key features include recycling of unconverted syngas to increase DME production and use of steam power. *Source:* From Larson et al. (2006a).

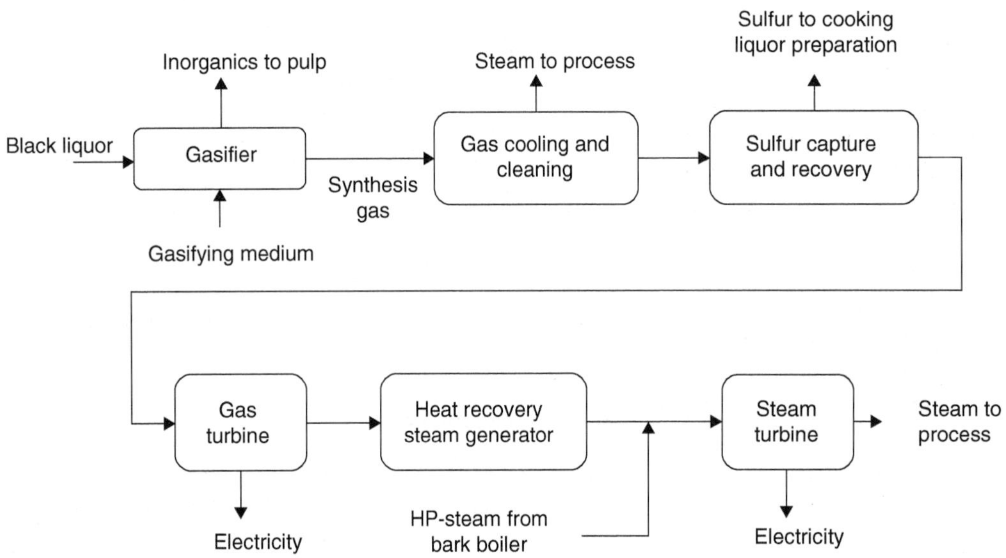

Figure 9.21 Flow diagram of power/recovery system in pulp mill BLGCC. *Source:* From Naqvi et al. (2010).

meet all of the mill's process steam needs. Heat recovered from the black liquor and syngas processing areas are integrated into the boiler to increase steam production and minimize the amount of woody residues needed. The steam is expanded through a back-pressure turbine to generate some electricity which goes toward meeting the mill's process electricity needs. To fully meet the mill's electricity need, some electricity must be imported from the grid. Because black liquor is being converted primarily into liquid fuel and not electricity, the amount of electricity imported is larger than with a conventional Tomlinson system (ETC 2011; Larson et al. 2006a).

Figure 9.21 gives a simplified representation of the basic design of the BLGCC system (Naqvi et al. 2010). The electricity output will be increased especially with pressurized gasification and a gas turbine–based cogeneration system. A number of studies were performed to estimate potential electricity generation in the pulp industry by implementing a BLGCC technology (Eriksson and Harvey 2004; KAM Report 2003; Larson et al. 2003).

Hydrogen production from gasified black liquor has also been studied. Andersson and Harvey (2004) evaluated energy and net CO_2 emissions consequences of integrated hydrogen production from BLG in a chemical pulp mill. The base capacity of the pulp mill was 2000 ADt/day of pulp production. Hydrogen production at the reference KAM pulp mill required import of 210 000 T of biomass for steam production and 0.43 TWh/year of electricity. If black liquor from all pulp mills in Sweden were gasified to produce hydrogen, the estimated import of biomass was 4.9 and 2.3 TWh/year of electricity. The results showed a large potential of hydrogen production, i.e. 188 T/day of hydrogen fuel which was equivalent to 261 MW [lower heating value (LHV), 120 MJ/kg] (KAM Report 2003).

Larson et al. (2006a, b) made a "well-to-wheels" (WTW) comparative environmental assessment of BLG for liquid fuel production with BLGCC electricity production at kraft pulp mills in the United States. The introduction of BLGCC systems in the United States had potential to displace up to 35 million T net CO_2, 16 000 T net SO_2, and 100 000 T net NO_x within 25 years (Naqvi et al. 2010).

9.8.4 Liquid Fuels Synthesis

The conversion of clean synthesis gas into a liquid fuel involves passing the syngas over a catalyst that promotes the desired synthesis reactions and then refining the raw product to obtain the final desired liquid fuel. Two basic designs for commercial synthesis reactors have been developed: gas-phase (or fixed-bed) and liquid-phase (or slurry-bed). Fixed-bed reactors have a long commercial history, but liquid phase reactors have been gaining popularity in commercial applications because of attractive performance attributes and lower cost. Liquid phase reactors are now commercially offered for Fischer–Tropsch liquids (FTL), methanol, and DME synthesis. Liquid phase reactors for mixed-alcohol synthesis are still under development.

Fixed-bed and liquid-phase reactor designs differ primarily in their handling of reactor temperature control. Synthesis reactions are exothermic, such that the reactor temperature increases as the reactions proceed if no heat is removed. Higher temperatures promote faster reactions, but maximum (equilibrium) conversion is favored by lower temperatures. Also, catalysts are deactivated when overheated. Thus, the temperature rise in a synthesis reactor must be controlled. In commercial practice, a reactor-operating temperature of 250–280 °C for methanol, DME, or FTL synthesis balances kinetic, equilibrium, and catalyst activity considerations. For mixed-alcohols synthesis, which is not yet a commercially established technology, higher reaction temperatures (300–400 °C) have been indicated with catalysts identified to date (Aden et al. 2005; Air Products and Chemicals, Inc. 2001).

A gas-phase reactor incorporates the flow of syngas over a fixed-bed of catalyst pellets. With this design it is difficult to maintain isothermal conditions by direct heat exchange (due to low gas-phase heat transfer coefficients). To limit temperature rise, the synthesis reactions are typically staged, with cooling between reactor stages. Also, by limiting the initial concentration of CO entering the reactor (to 10–15 vol%) the extent of the exothermic reactions can be controlled. Control of the CO fraction is achieved in practice by maintaining a sufficiently high recycle of unconverted H_2-rich syngas back to the reactor.

In a liquid-phase reactor, syngas is bubbled through an inert mineral oil containing powdered catalyst in suspension (Figure 9.22). Much higher heat release rates (i.e. extents of reaction) can be accommodated without excessive temperature rise as compared to a gas-phase reactor because of more effective reactor cooling by boiler tubes immersed in the fluid. The vigorous mixing, intimate gas-catalyst contact, and uniform temperature distribution enable a high conversion of feed gas to liquids in a relatively small reactor volume. Conversion by liquid-phase FT synthesis is especially high. A single-pass fractional conversion of CO of about 80% can be achieved (Bechtel Group, Inc. 1990), compared to less than 40% for conversion with traditional fixed-bed FT reactors. For the FT reactor conditions it is assumed in simulations, the single-pass CO conversion is about 65%.

9.8.5 Dimethyl Ether

Single-step DME synthesis reactors typically utilize a mix of two catalysts, one promoting the synthesis of methanol from syngas ($CO + 2H_2 \rightarrow CH_3OH$) and one promoting the dehydration of the methanol to DME ($2CH_3OH \rightarrow CH_3OCH_3 + H_2O$). Both liquid-phase and fixed-bed reactors are offered commercially. Leading developers of liquid-phase DME synthesis reactors are DME Development, Inc.,

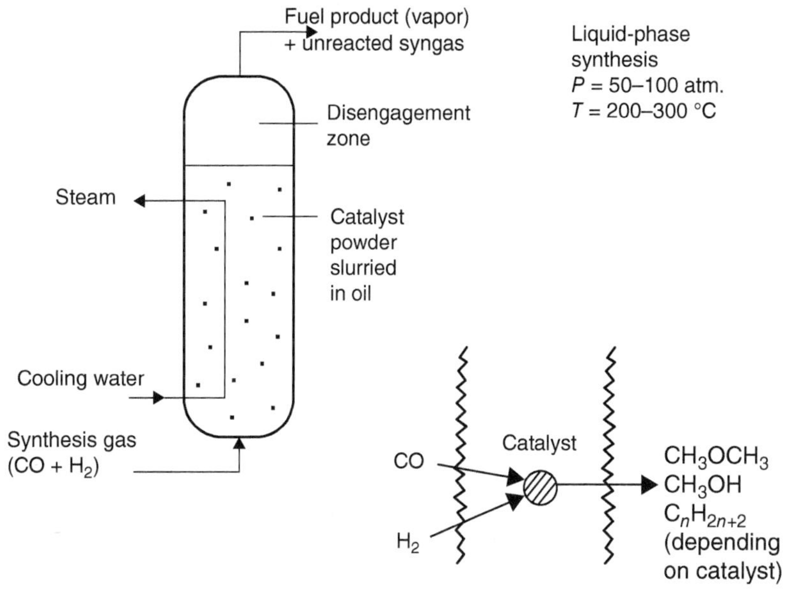

Figure 9.22 Schematic of liquid-phase synthesis reactor. *Source:* From Larson et al. (2006a).

a Japanese consortium of nine companies led by NKK and Nippon Sanso (Adachi et al. 2000; Air Products and Chemicals 2001; Brown et al. 1991; Fujimoto et al. 1995; Ohno 2006).

The DME reactor design of Air Products is derived from its liquid-phase methanol (LPMEOH) synthesis process developed in the 1980s. A commercial-scale LPMEOH demonstration plant (250 T/day methanol capacity) has been operating since 1997 with gas produced by the Eastman Chemical Company's coal gasification facility in Kingsport, Tennessee (Eastman Chemical and Air Products and Chemicals 2003). The construction of this facility was preceded by extensive testing in a 10 tpd process development unit (PDU) in LaPorte, Texas. The PDU was operated in 1999 to generate test data on direct DME synthesis (Air Products and Chemicals 1993, 2001, 2002; Weyerhaeuser Company 2000).

9.8.6 Pressurized Chemrec BLG

In 2010, a black liquor gasification DME and motor fuel (MF) demonstration plant was inaugurated in Piteå in Sweden (Landälv et al. 2010). Figure 9.23 shows a schematic of the Piteå DME and MF production plant (Landälv et al. 2010; Larson et al. 2006a). When used as a fuel in a truck diesel engine, BioDME gives an equally high-efficiency rating and a lower noise level compared with a conventional engine. Compared with diesel, BioDME generates an impressive lower carbon dioxide emission (CO_2). In addition, its combustion produces very low emissions of particulate matters (PM) and nitrogen oxides (NO_x). This makes BioDME an ideal fuel for diesel engines. DME is a gas at ambient conditions but is liquefied at low pressure, just five bars and is handled as a liquid during distribution and use. It is simple to handle, similar to propane. DME can be produced from natural gas and also from various types of biomass. When it is produced from biomass, it is known as BioDME (Larson et al. 2006a; Salomonsson 2011).

The BioDME project aims to demonstrate production of environmentally optimized synthetic biofuel from lignocellulosic biomass at an industrial scale (Landävl 2010; Lindblom 2012; Salomonsson 2013). The project involves a consortium of Chemrec, Haldor Topsøe, Volvo, Preem, Total, Delphi, and ETC. The project is supported by the Swedish Energy Agency and the EU's Seventh Framework Program. The output of this demonstration is DME produced from black liquor through the production of clean syngas and a final fuel synthesis step. The overall possible chemicals that can be produced from the syngas are hydrogen, methanol, DME, FT fuels, ethanol, and methyl tertiary butyl ether (Tampier et al. 2004).

9.8.7 Catalytic Hydrothermal Gasification of Black Liquor

The process is also known as supercritical water oxidation. This is a novel technology to produce methane-rich synthesis gas as an alternative to replace the conventional recovery system (Naqvi et al. 2010). Heat demand for bringing water to supercritical conditions is less than that for evaporating at subcritical pressure. This phenomena leads to better energy savings as compared to conventional recovery system (Sricharoenchaikul 2009; Vogel et al. 2005). High-water content in the black liquor – nearly 80% under supercritical condition – would increase gasification reactions that lead to high organics conversion to the synthesis gas (Calzavara et al. 2005; Williams and Onwudili 2006). This phenomenon results in direct introduction of black liquor to hydrothermal gasifier, removing energy demanding evaporation unit in conventional process, thus decreasing the overall stream demand of the pulp mill. The multi-effect evaporation unit represents nearly 37% of total heat demand of the pulp mill and can be removed in this technique. Unlike other gasification processes, black liquor from the digestion unit is directly introduced to catalytic hydrothermal gasifier at supercritical water conditions

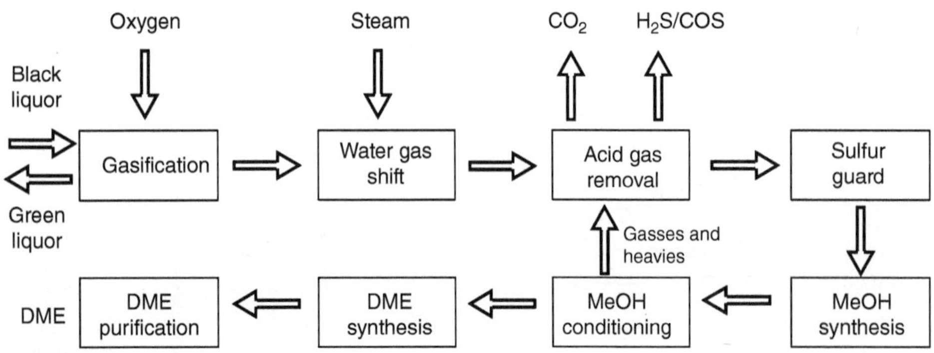

Figure 9.23 Schematic of the Pitéa DME production plant. *Source:* Adapted from Landälv et al. (2010) and Larson et al. (2006a).

(600 °C, 300 bar). The catalytic hydrothermal gasification process involves three steps:

1) *Heat up phase* (decomposition unit): During this phase, large molecules in black liquor hydrolyze to form alcohols due to the presence of lignin. Cellulose present in lignin decomposes rapidly in water at about 250 °C.
2) *Salt separation:* In salt separator, inorganic salts present in black liquor precipitate and are returned to the pulp mill.
3) *Catalytic reactor:* This is the actual methane synthesis unit where smaller organic molecules, such as carboxylic acid, alcohols, and aldehydes, are converted to CH_4, CO_2, H_2, and CO.

Carboxylic acid is converted to methane, carbon dioxide, and hydrogen. The tar formation is avoided due to supercritical conditions and presence of catalyst.

Sricharoenchaikul (2009) studied the feasibility of supercritical water oxidation technique to convert black liquor to value-added fuel products and recovery of pulping chemicals. The experiments were conducted in a quartz capillary heated in a fluidized bed reactor based on important operating parameters which were pressure, temperature, feed concentration, and reaction time. The investigator found that pressure between 220 and 400 bar had insignificant influence on the syngas and carbon conversion. But increasing temperature and residence time between 375–650 °C and 5–120 seconds resulted in high-carbon conversion, greater synthesis gas production, and energy efficiency.

9.8.8 Fischer–Tropsch Liquids

Commercial Fischer–Tropsch catalysts include iron- and cobalt-based materials. Cobalt catalysts produce a large heavy-wax fraction that can be easily and with high selectivity refined into desired lighter products by subsequent hydrocracking (breaking up of the large hydrocarbon molecules into desired final products in a hydrogen-rich environment). Hydrocracking of the large straight-chain hydrocarbons formed by FT synthesis can be done under much less severe temperature conditions (350–400 °C for cracking to C5–C18 range hydrocarbons) than is required for hydrocracking of aromatic molecules found in conventional petrochemical refining. Iron-based catalysts produce a broader product mix that requires a greater level of refining than with cobalt catalysts. Also, unlike cobalt catalysts, iron catalysts promote water-gas shift activity ($CO + H_2O \rightarrow H_2 + CO_2$), making them well suited for use with syngas characterized by H_2/CO ratios below the stoichiometric value of 2.2 for FT synthesis.

9.8.9 Mixed Alcohols

Compared with FTL or DME synthesis, the technology for synthesis of methyl-, ethyl-, and propanol-mixed alcohols (MA) is considerably less commercially advanced, and there is sparse published literature on which to base detailed reactor performance estimates. Catalysts that have been examined in the past can be divided into four categories (Liu et al. 1997): ruthenium-based catalysts, modified methanol catalysts, modified Fischer–Tropsch catalysts, and molybdenum sulfide–based catalysts. Among these, the MoS_2-based catalysts (originally discovered by researchers at Dow and Union Carbide in the 1980s) have received considerable recent attention due to their high tolerance for sulfur-contaminated syngas, their water–gas shift activity, and their high activity and selectivity for linear alcohols. Selectivity is an especially important characteristic because if all possible chemical reactions between CO and H_2 are allowed to compete without constraints, reactions other than those for synthesis of higher alcohols will thermodynamically out-compete reactions for synthesis of higher alcohols. In particular, the formation of Fischer–Tropsch hydrocarbons (α-olefins and *n*-paraffins) from CO and H_2 is thermodynamically favored over the formation of higher alcohols. For this reason, to maximize performance of alcohol synthesis catalysts, high selectivity is an essential feature.

The simplified reactions scheme adopted for the higher alcohols are as follows:

$$CO + H_2O \leftrightarrow CO_2 + H_2$$

$$CO + 2H_2 \leftrightarrow CH_3OH$$

$$CH_3OH + 2H_2 \rightarrow C_2H_5OH + H_2O$$

$$2CH_3OH + CO + 2H_2 \rightarrow C_3H_7OH + 2H_2O$$

9.8.10 "WTW" Environmental Impact of Black Liquor Gasification

In addition to energy aspects of the biorefinery systems discussed earlier, we have also examined environmental attributes. Water effluents, air emissions, and solid wastes are all of potential concern. In assessing the impact that biorefinery systems would have on these effluents relative to levels found with Tomlinson power/recovery systems, one may consider changes both in direct effluents and in effluents associated with the displacement of grid electricity generation and conventional petroleum-based motor fuels. In particular, to effectively estimate the full environmental impacts of biorefineries, the current analysis involves estimating the emissions impacts from resource extraction to end use. This so-called WTW analysis is a common approach for making meaningful comparisons between different alternative and conventional fuels. This approach is necessary because of the different upstream production and conversion processes, different downstream vehicle/engine types for different fuels and significant differences in fuel properties and combustion characteristics.

Biofuels which are produced from BLG process excel in terms of WTW carbon dioxide emission reduction and energy efficiency. Also, synthetic diesel and DME produced from forest harvest residues over the BLG route both showed among the highest WTW GHG reduction and energy efficiency. The total available black liquor volume in the United States with the conversion efficiency of this process is equivalent to approximately 5 billion gal/year as ethanol. The renewable fuels standard calls for 16 billion gal of cellulosic biofuels by 2022. Therefore, this route can give a significant contribution to meeting this target.

The pulp and paper companies in the United States today are meeting severe competition from low-cost producers overseas and from alternative solutions in both packaging and printed media. Mill operators and their investors now option to transform mills into biorefineries that use this fuels-from-the-forest process. This transformation completely changes a pulp mill's competitive position by adding 30–50% of profitable revenue with the typical 25–40% internal rate of return. It also makes needed reinvestment possible by replacing aged recovery boilers with high-maintenance costs and low performance. The fuel plant investment can also be used to provide additional recovery capacity, allowing for higher pulp production in many cases. Mills producing only 500 T of BLS per day are viable as fuels-from-the-forest biorefineries are using this method. Most mills are significantly larger. Such a biorefinery mill would produce upward of 8 million gal a year of green motor fuel calculated as gasoline equivalents at the minimum capacity size. In a mill investing in second-generation biofuels technology, jobs are not only preserved but also additional jobs are created, mainly for extraction of biomass from the forest as well as to operate and maintain the biofuels plant. Other economics and social benefits are also significant such as possible tax benefits and air emissions reductions (Bajpai 2014).

Typical capital investment for a biorefinery project that uses fuels from the forest is \$200–400 million, depending on plant size and the cost to interconnect to the mill. The BLG industry is vigorously pursuing federal and state grants and loan guarantees to ramp up this technology as soon as possible to large-scale commercial capacity. As a source of ultra-clean, renewable motor fuels, the BLG route that transform pulp and paper mills into refineries is standing up to a critical scrutiny as a viable and practical way of producing alternative, renewable energy, while making good use of the land and being gentle to the environment (Bajpai 2014).

Forest biorefinery utilizing gasification in a BLGCC configuration rather than a Tomlinson boiler is predicted to produce significantly fewer pollutant emissions due to the intrinsic characteristics of the BLGCC technology. Syngas cleanup conditioning removes a considerable amount of contaminants and gas-turbine combustion is more efficient and complete than boiler combustion. There could also be reductions in pollutant emissions and hazardous wastes resulting from cleaner production of chemicals and fuels that are now manufactured using fossil energy resources. In addition, it is generally accepted that production of power, fuels, chemicals, and other products from biomass resources creates a net zero generation of carbon dioxide, as plants are renewable carbon sinks. A key component of the forest biorefinery concepts is sustainable forestry. The forest biorefinery concept utilizes advanced technologies to convert sustainable woody biomass to electricity and other valuable products and would support the sustainable management of forest lands (Farmer 2005). In addition, the forest biorefinery offers a productive value-added use for renewable resources such as wood thinning and forestry residues and also urban waste (Mabee et al. 2005; Miller et al. 2005).

BLG whether conducted at high or low temperatures is still superior to the current recovery boiler combustion technology (Bajpai 2008, 2012). The thermal efficiency of gasifiers is estimated to be 74% compared to 64% in modern recovery boilers, and the integrated gasification and combined cycle (IGCC) power plant could potentially generate same amount of fuel (Dance 2005; Farmer and Sinquefield 2003). While the electrical production ratio of conventional recovery boiler power plants is 0.025–0.10 MWe/MWt, the IGCC power plant can produce an estimated 0.20–0.22 MWe/MWt (Farmer and Sinquefield 2003; Sricharoenchaikul 2001). This increase in electrical efficiency is significant enough to make pulp and paper mills potential exporters of renewable electric power. Alternatively, pulp mills could become manufacturers of bio-based products by becoming biorefineries. Additionally, the new technology could potentially save more than 100 trillion Btu of energy consumption annually, and within 25 years of implementation, it could save up to 360 trillion Btu/year of fossil fuel energy (Larson et al. 2003). The new technology also offers the benefits of improved pulp yields if alternative pulping chemistries are included and reductions in solid waste discharges. Also, the process is inherently safer because the gasifier does not contain a bed of char smelt unlike in recovery boilers, which reduces the risk of deadly smelt–water explosions (Argonne National Laboratory 2006; Sricharoenchaikul 2001).

9.8.11 Water and Solid Waste

Water quality, temperature, and consumption are all potential concerns with biorefineries. Over time, as demand rises for limited freshwater supplies, these issues are likely to only become more important. Briefly, the issues are as follows:

- *Water quantity*: Any water savings results in a direct financial benefit to a mill and also addresses growing concerns over the availability of freshwater for other purposes (e.g. agriculture, human consumption).
- *Water quality*: It is of major concern for rare and endangered species, recreation, and for its effects on other users downstream (e.g. municipalities).
- *Thermal discharge*: The temperature of the cooling water discharge is also of concern for its effect on flora and fauna.

Depending on the configuration (amount of fuels and electricity produced), the biorefinery will have different effects on water quantity and thermal discharge at a mill, but overall the conversion to biorefineries is not expected to significantly impact water quality, especially when considering the impacts on displaced grid power (Larson et al. 2003). Also, IGCC power plant will reduce cooling water and make-up water discharges locally at the mill, and because the efficient gasifiers will cause grid power reductions, substantial reductions in cooling water requirement at power plant will also occur (Larson et al. 2003). Overall, the implementation of IGCC power plant will cause net savings in cooling water requirements and net reductions in wastewater discharges. An additional benefit is also the avoided water usage in conventional fuel production, which has not been quantified. Moreover, the consequences of spills from petroleum and petroleum product transportation are also reduced. Also, some of the biofuels, namely DME and MA, pose much lower risks of groundwater contamination in the event of a fuel leak or spill (e.g. at refueling stations). FT liquids, since they contain very low aromatics, should also pose a lower risk than conventional diesel and gasoline.

Solid waste issues relate to the quantity and toxicity of any solids that must be disposed of. In this regard, biorefineries are not expected to result in significant changes at the mill, in part because the solids produced (mainly ash from biomass) are not problematic to deal with. There will be the need to periodically replace catalysts and guard beds, such as zinc oxide (for H_2S) and activated carbon (for other trace contaminants). Nevertheless, as with water usage, the impacts of displaced grid power (particularly for the coal component of that grid power) and conventional transportation fuel use would likely result in important reductions of solid waste generation (Larson et al. 2003).

9.8.12 Mill-Related Air Emissions

The most significant effluent differences between biorefineries and Tomlinson systems are expected to be in air emissions. This is particularly expected to be the case in a WTW context. As discussed below, air emissions were estimated in detail for both the biorefineries and the Tomlinson power/recovery systems. For comparison, the BLGCC case from our earlier study is also shown, with the updated assumptions consistent with the current analysis (e.g. grid power emissions). Actual air emissions data are available for modern Tomlinson systems. Since emissions data do not exist for BLGCC or biorefinery systems, estimates were made starting with data for coal IGCC and natural-gas combined cycle power systems and adjusting appropriately. Note that relative to the BLGCC configuration, sources of air emissions in a biorefinery are expected to be similar, namely the power island. The production of the biofuel itself does not lead to significant additional sources of air emissions at the biorefinery.

The air emissions analysis presented below is not intended to serve as a complete lifecycle analysis of biorefinery emissions. Rather the estimates provide indicative results of the potential impacts of biorefinery options relative to "business as usual" in the pulp and paper industry. For example, upstream emissions for grid power (i.e. fuel production and transportation to the power plant) are not included, but these are relatively small compared to the power plant emissions themselves and to the total emissions from conventional motor fuel chains. To the extent that most of the biorefinery configurations result in more displaced grid power than the Tomlinson case, the emissions benefits estimated in this study can be considered conservative because they do not also factor in emissions reductions related to fuel supply for power plants.

Air emissions fall into three basic categories: criteria pollutants, hazardous air pollutants (HAPs), and GHGs. This study includes quantitative estimates for the criteria pollutants: sulfur dioxide (SO_2), nitrogen oxides (NO_x), carbon monoxide (CO), volatile organic compounds (VOCs), PM, and total reduced sulfur (TRS). Estimates are also made for carbon dioxide (CO_2), the major greenhouse gas. HAPs and other emissions issues are discussed qualitatively.

9.8.13 Tomlinson Boiler Air Emissions

Modern Tomlinson boilers are characterized by emissions of criteria pollutants that are similar overall to grid power (some are higher, like CO and PM, while others are lower, like SO_2 and NO_x). The most significant pollutants, in terms of both environmental impacts and relative emissions rates from Tomlinson boilers, are NO_x and particulates. While many furnaces already have particulate controls in place, there is no effective form of NO_x after treatment. Furnace rebuilds and replacements trigger the New Source Review (NSR) process, which generally results in process modifications being made to reduce TRS emissions. Installation of more efficient particulate control is also common following a NSR, and generally modern furnaces have better design and controls than older ones, which results in lower overall emissions. Table 9.8 presents a qualitative analysis of relative environmental impact of different mill-level emissions, together with relative emission rates for controlled and uncontrolled Tomlinson furnaces and with biorefinery technology (Argonne National Laboratory 2006; USEPA 2002).

9.8.14 Economic Development Opportunities

Biorefineries could have important economic development benefits, stemming from the enhancement of the competitiveness of the pulp/paper industry. The financial analysis illustrated the potential for attractive financial returns and significantly increased cash flows relative to Tomlinson systems. The related economic development benefits include preserving and growing employment in the industry and potentially adding to rural and semi-rural employment by creating increased demand for raw materials for paper production and biomass supply and, in the longer term, energy and other products derived from biomass. On a national scale these impacts are likely to be modest, but in certain regions or states (especially southeast, upper mid-west), the impacts could be very significant. However, if biorefineries help catalyze a new, larger, bio-energy industry, the economic impacts would be more substantial at the national level as well.

A kraft mill that adopts BLG exclusively to displace the boiler and to produce more electricity can be profitable (Larson et al. 2003, 2006a). But at present, a much developed biorefinery with a diverse product range adds many new options that may become economically attractive if BLG technology can solve a few technical challenges (Bajpai 2008). Current BLG technology has excited new interest in biorefinery and has the ability to leverage BLG into a wide portfolio of products, retrofitted to different types of mill, including medium-sized mills facing shutdown. The Chemrec development plant in Piteå, Sweden, is the most developed BLG model with strong industry momentum. Volvo Technology Transfer AB and affiliates new hold a majority share. The biorefinery aims to produce pulp (and paper) and to use the BLG syngas to produce DME fuel. DME is clean-burning and has advantage over diesel. Chemrec has successfully implemented a DME demonstration in Piteå. The Chemrec 2020 plan anticipates five large and three small BLGDME and BLGMF plants.

Table 9.8 Qualitative indication of relative environmental impact of different mill-level emissions, together with relative emission rates for controlled and uncontrolled Tomlinson furnaces and with biorefinery technology.

Pollutant/ discharge	Relative environmental impact of pollutant	Relative emissions rates from Tomlinson furnace (uncontrolled)	Relative emissions rates with control on Tomlinson	Relative emission rates expected with biorefineries
SO_2	H	L	L (not required)	VL
NO_x	H	M	M[d]	VL
CO	L	M (can be highly variable)	M[d]	VL
VOC	H	L	L[d]	VL
PM[b]	H	H[c]	L–M	VL
CH_4	L–M	L	L[d]	VL
HAPs	M–H	L[c]	L[c]	VL
TRS[e]	L	L	L[d]	VL
Waste Water[f]	M–H	L	L	VL–L
Solids[a]	L	L	L	L

Source: From Larson et al. (2006b).
VL, very low; L, low; M, moderate; H, high.
[a] General importance, not specifically for the P&P industry.
[b] PM, particulate matter. Of greatest concern with PM emissions are fine particulates smaller than 10 and 2.5 μm in diameter (PM10 and PM2.5 respectively).
[c] Current MACTII rules are expected to result in about a 10% reduction of HAPs and a modest reduction in PM.
[d] Not generally practiced other than by maintaining good combustion efficiency.
[e] Total reduced sulfur.
[f] For power systems, the issue is mainly one of the cooling water (quantity and discharge temperature).

Economic projections are attractive. This is largely due to return from DME (Ekbom et al. 2003, 2005; Connor 2007).

If biorefineries were to penetrate slowly rather than rapidly into the market, the cumulative (25-year) energy savings would be roughly 2–12 quads less. Assuming a rough average fossil fuel price range of \$5–10/MMBtu (which corresponds to \$29–58/barrel of crude oil or 1.7–3.4 ¢/kWh of electricity), the corresponding added energy costs would be \$10–120 billion over this period (Larson et al. 2006a, b).

For certain emissions, it is also possible to estimate a market value since there are existing cap-and-trade markets. At \$625/T (the recent price for SO_2 allowances), and assuming prices remain at this level in real terms, SO_2 reductions have a cumulative value of up to \$301 million over the 25-year period following commercial introduction of biorefineries. (In some of the configurations, the net SO_2 benefit is negative because of the large decreases expected in grid power SO_2 emissions discussed earlier.) NO_x, if valued at \$2100/T over the same period, has a market value as high as of \$1.5–2.6 billion in the *Aggressive* market penetration scenario. If a system for trading CO_2 is put in place, the CO_2 value could be as high as \$37 billion in the *Aggressive* market penetration scenario at a price of \$25/MT of CO_2. While it will likely be difficult for biorefineries to capture all of these additional revenue streams, these estimates provide an indication of the value to the nation of emissions reductions that biorefineries could enable. Thus, in addition to energy costs savings, the value of lost SO_2, NO_x, and CO_2 emissions reductions due to slower market penetration could also be in billions of dollars (Larson et al. 2006b).

9.8.15 Cost-Benefit Analysis

A cost-benefit analysis of BLG-based bio-refining in the kraft pulp and paper mill was studied by Larson et al. (2006a) and Consonni et al. (2009). The study assumed that BLG systems are technologically comparable with Tomlinson recovery boilers. Hence, BLGCC or bio-refinery systems can be integrated within kraft pulp mill replacing the conventional recovery technology.

The reference mill with a base capacity of 2458 T BLS/day was an integrated pulp mill producing uncoated free sheet paper from a 65/35 mix of hardwood and softwood. DME production from LTBLG was compared to HTBLG with identical black liquor input. DME energy flow of 137.4 MW was estimated in the LTBLG integration and 168 MW for the HTBLG case. The power boiler consumed larger fuel (unconverted syngas and wood residues) in the LTBLG case than in the HTBLG case but resulted in 64% more back-pressure steam turbine electricity production. From the preliminary calculations, it appeared that the LTBLG in DME pulp mill bio-refinery integration would produce 15–20% less DME than in the HTBLG bio-refinery, with both configurations requiring electricity import. In addition, the LTBLG case required about double biomass import than the HTBLG, and fuel oil import for the lime kiln was about 25% higher. Table 9.9 compares potential electricity or fuel production from various studies.

9.9 Conclusions

One may consider a modern pulp and paper mill as a first-generation forest biorefinery, with steam, power, and other products being produced alongside the wide range of paper products we normally associate with the industry. Black liquor and biomass gasification are key technology platforms for realizing the forest biorefinery of the future. Current research and development has shown that gasification-based pulp mill biorefinery technologies, once fully commercialized, offer the potential for attractive investment returns. They also offer the potential for important contributions toward national petroleum savings, emissions reductions, improved energy security, and rural economic development – contributions that could be two times or larger the size of contributions from the existing US corn–ethanol industry. The introduction of BLGCC systems in the United States had potential to displace up to 35 million T net CO_2, 16 000 T net SO_2, and 100 000 T net NO_x within 25 years.

These potential private and public benefits arise, fundamentally, because of the integration of biorefining with pulp and paper production, such that the biorefinery is providing chemical recovery services, process steam, and process electricity in addition to exporting liquid fuels (FTL, DME, MA) and perhaps some electricity. We have analyzed in detail a variety of integrated pulp/paper mill biorefinery designs encompassing a broad range of product slates. An overarching finding is that integration can effectively enable more efficient use of biomass resources for liquid biofuel production compared to nonintegrated biofuel production. Integration also can effectively reduce the capital investment required per unit of biofuel production to levels comparable to investments needed for coal-to-liquids facilities that are an order of magnitude or more larger than prospective pulp mill biorefineries. Finally, integration can effectively reduce the cost of producing gasification-based biofuels to approximately \$1/gal of ethanol equivalent, which would make them competitive with the current cost target developed by analysts at the National Renewable Energy Laboratory for ethanol made from lignocellulosic biomass by enzymatic hydrolysis/fermentation processes.

Table 9.9 Biorefinery estimates based on BLG studies, e.g. BLGCC and BLG for biofuel production.

Parameter	BLGCC		BLG for biofuel production						
Reference	Larson et al. (2003)	Eriksson and Harvey (2004)	Ekbom et al. (2003)	Andersson and Harvey (2004)		Larson et al. (2006b)		Naqvi et al. (2010)	
Product	Electricity	Electricity	MeOH	DME	H_2	DME	FTL	MA	CH_4
Pulp production (ADt/day)	1600	2000	2000	2000	2000	1600	1600	1600	1000
BLS flow (tDS/day)	2724	3420	3420	3420	3420	2724	2724	2724	1700
BLS flow (MW)	350.7	487	487	487	487	350	350	350	243.5
Biomass import (MW)	27.1	21.3	129	125	123.5	77.4	102	89.2	107
Electricity import/export (+/−) (MW)	15.2	86.5	45.9	−48.7	−56.7	99.6	12.4	8.2	1.1
Fuel production (MW)	—	—	272	275	261	168	112	60	240.2

Source: From Naqvi et al. (2010).
Fuel values are based on LHV.

9.9.1 Summary

- The forest biorefinery offers a business strategy that potential forestry companies are seriously considering for improving the overall financial performance of the sector. However, there are considerable technology and business risks related to its implementation. These risks can be mitigated to a great extent by using systematic product and process design tools for analyzing biorefinery strategies.
- Industry leaders, investors, policy-makers, and others are now beginning to better understand the vital role to be played by biorefineries as we move from a fossil fuel-based energy economy toward a bio-based one (Connor 2007).
- When properly located and operated, the potential of an integrated forest biorefinery is believed to be huge: a very attractive and synergistic business and economic opportunities for both co-located pulp and paper mill and for the biorefinery itself.
- Biorefineries are a key pathway to our biofuture, displacing fossil fuels and supplying clean, renewable, and carbon neutral energy. Biorefineries fit very well at pulp and paper mills because of their inherent ability to gather and process biomass and create energy from biomass.
- BLG for liquid fuel production with BLGCC electricity production at kraft mills has potential to significantly reduce air emissions, including CO_2, SO_2, NO_x, CO, VOCs, PM, and TRS.
- Forest biorefinery has significant impact on the society, our economy, environment, and sustainable forestry and pulp and paper industry.

Problems

9.1 Which of the following are steps needed to achieve zero waste?

- **A** Some specialized or specific waste minimization process and hazardous waste designation technology.
- **B** Old technologies are intensified with integration processes.
- **C** New products are created using current products

9.2 Why can all wastes reach a zero waste "grave"?

9.3 Biorefineries have current and future speculative production of renewable fuels, electricity, and chemicals in conjunction with pulp and paper mills.

- **A** Where do national energy security and global warming concerns enter this picture?
- **B** Black liquor gasification processes and properties parallel in value, and liquid fuels synthesis have potential to save fossil fuel energy and reduce solid waste discharges. What steps are used in black liquor gasification and liquid fuel synthesis?
- **C** What five changes can you anticipate for the near future in use of biorefineries?

9.4 Define industrial ecology (IE). Give your perspective about the IE pertaining to industrial environmental management.

9.5 What is an "eco-industrial park"? Give examples with a case study.

9.6 Define "cradle to cradle."

9.7 Describe circular economy.

9.8 What are process intensification and process integration in manufacturing? Describe the differences between the two approaches.

9.9 Describe how (a) process redesign and (b) process modifications could achieve greater pollution prevention for chemical production industries.

9.10 Describe the water and energy nexus in industrial sector. Describe how climate change could affect water supply and electricity use.

9.11 Highlight deinking process in recovery fiber from recycled fiber. A benefit/cost analysis of the process.

References

Adachi, Y., Komoto, M., Watanabe, I. et al. (2000). Effective Utilization of Remote Coal Through Dimethyl Ether Synthesis. *Fuel* 79: 229–234.

Ådahl, A., Harvey, S., and Berntson, T. (2004). Process industry energy retrofits: the importance of emission baselines for greenhouse gas reduction. *Energy Policy* 32: 1375–1388.

Aden, A., Spath, P., and Atherton, B. (2005). The potential of thermochemical ethanol via mixed alcohols production. Milestone completion report, National Renewable Energy Laboratory, Golden, CO, 31 October.

AGU, American Geophysical Union (2012). *Water-Energy Nexus: Solutions to Meet a Growing Demand*. Washington, DC: AGU.

Air Products and Chemicals, Inc. (1993). *Synthesis of Dimethyl Ether and Alternative Fuels in the Liquid Phase from Coal-Derived Synthesis Gas*, Report DOE/PC/89865-T8. Pittsburgh, PA: US Department of Energy, Pittsburgh Energy Technology Center.

Air Products and Chemicals, Inc. (2001). Liquid Phase Dimethyl Ether Demonstration in the LaPorte Alternative Fuels Development Unit. Prepared for US Department of Energy by APCI, Allentown, Pennsylvania, January.

Air Products and Chemicals, Inc. (2002). Market Outlook for Dimethyl Ether (DME). Prepared for US Department of Energy by APCI, Allentown, Pennsylvania, April.

Allen, D.T. and Rosselot, K.S. (1997). *Pollution Prevention for Chemical Processes*. Hoboken, NJ: Wiley.

Allen, D.T. and Shonnard, D.R. (2012). *Sustainable Engineering: Concepts, Design, and Case Studies*. Hoboken, NJ: Wiley.

Allenby, B. (2006). The ontologies of industrial ecology. *Progress in Industrial Ecology: An International Journal* 3 (1/2): 28–40.

Andersson, E. and Harvey, S. (2004). Pulp-mill integrated bio-refineries: a framework for assessing net CO_2 emission consequences. *Proceedings, AIChE 2004 Fall Annual Meeting*, Austin, TX (7–12 November 2004), 203–208.

Argonne National Laboratory (2006). Greenhouse Gases, Regulated Emissions, and Energy Use in Transportation (GREET) Model, version 1.7 (beta), released January 18.

Ashton, W. (2009). The structure, function, and evolution of a regional industrial ecosystem. *Journal of Industrial Ecology* 13 (2): 228–246.

Bajpai, P. (2008). *Chemical Recovery in Pulp and Paper Making*, 166 pp. Leatherhead: PIRA International.

Bajpai, P. (2012). *Biotechnology in Pulp and Paper Processing*. New York, NY: Springer-Verlag.

Bajpai, P. (2013). *Biorefinery in the Pulp and Paper Industry*, 114 pp. London: Elsevier Inc.

Bajpai, P. (2014). *Black Liquor Gasification*, 92 pp. London: Elsevier Inc.

Bechtel Group, Inc. (1990). *Slurry Reactor Design Studies. Slurry vs. Fixed-Bed Reactors for Fischer-Tropsch and Methanol: Final Report*, US Department of Energy Project No. DE-AC22-89PC89867. Pittsburgh: Pittsburgh Energy Technology Center.

Benyus, J. (2003). *Biomimicry: Innovation Inspired by Nature*. New York: HarperCollins.

Berglin, N., Lindblom, M., and Ekbom, T. (2002). Efficient production of methanol from biomass via black liquor gasification. *TAPPI Engineering Conference*, San Diego, CA (7–11 September 2002).

Bielenberg, J. and Bryner, M. (2018). Realize the potential of process intensification. *Chemical Engineering Progress* 114 (3): 41–45.

Blok, K., Hoogzaad, J., Ramkumar, S. et al. (2018). Implementing Circular Economy Globally Makes Paris Targets Achievable. *Circle Economy. Circle Economy, Ecofys* (accessed 12 April 2018).

Brown, D.M., Bhatt, B.L., Hsiung, T.H. et al. (1991). Novel technology for the synthesis of dimethyl ether from syngas. *Catalysis Today* 8: 279–304.

Calzavara, Y., Joussot-Dubien, C., Boissonnnet, G., and Sarrade, S. (2005). Evaluation of biomass gasification in supercritical water process for hydrogen production. *Energy Conversion and Management* 46: 615–631.

CFR-40 (1999) Protection of Environment.

Chemrec (2006). Chemrec pressurized BLG (black liquor gasification): status and future plans. Paper presented at the 7th International Colloquium on Black Liquor Combustion and Gasification, Jyväskylä, Finland (2 August 2006).

Chertow, M.R. (2000). Industrial symbiosis: literature and taxonomy. *Annual Review of Energy and the Environment* 25: 313–337.

Clark, C.E. and Veil, J.A. (2009). *Produced Water Volumes and Management Practices in the United States*. Argonne, IL: Argonne National Laboratory.

Clift, R. and Allwood, J. (2011). Rethinking the Economy. *The Chemical Engineer*, March.

Commission of European Communities (COM) (2014). *Towards a Circular Economy: A Zero Waste Program for Europe*, Communication No. 398. Brussels: Commission of European Communities.

Connor, E. (2007). The integrated forest biorefinery: the pathway to our bio-future. *Proceedings of the International Chemical Recovery Conference: Efficiency and Energy Management*, Quebec City, QC (29 May to 1 June 2007), pp. 323–327.

Consonni, S., Katofsky, R., and Larson, E. (2009). A gasification-based bio-refinery for the pulp and paper industry. *Chemical Engineering Research and Design* 87 (9): 1293–1317.

Dahlquist, E., Avelin, A., and Yan, J. (2009). Black liquor gasification in a CFB gasifier: system solutions. *Proceedings of the First International Conference on Applied Energy (ICAE09)*, Hong Kong (5–7 January 2009).

Dance, M. (2005). Hydroxide formation and carbon species distribution during high-temperature kraft black liquor gasification. MSc thesis. Georgia Institute of Technology.

Das, T.K. (2005). Process pollution prevention in pulp and paper industry. In: *Toward Zero Discharge: Innovative Methodology and Technologies for Process Pollution Prevention*. Hoboken, NJ: Wiley-InterScience.

Das, T.K. (2017). Optimum, sustainable and integrated water management. In: *Sustainable Water Management Technologies*, vol. I (ed. D. Chen), 397–414. Boca Raton, FL: CRC Press.

Das, T.K. and Cabezas, H. (2018). Tools and concepts for environmental sustainability in the food-energy-water nexus: chemical engineering perspective. *Environmental Progress and Sustainable Energy* 37 (1): 73–81.

Das, T.K. and Houtman, C.J. (2004). Evaluating chemical-, mechanical-, and bio-pulping processes and its sustainability characterization using life-cycle assessment. *Environmental Progress and Sustainable Energy* 23 (4): 347–357.

Das, T.K. and Jain, A.K. (2001). Pollution prevention advances in pulp and paper processing. *Environmental Progress* 20 (2): 87–92.

Das, T. K., Sharma, M. P., and Butner, S. (2001). The concept of an Eco-Industrial Park: a solution to industrial ecology. Paper presented at the Topical Conference: Industrial Ecology and Zero-Discharge Manufacturing, American Institute of Chemical Engineers Spring Meeting, Paper No. 18c, Houston, TX (April 2001).

Eastman Chemical and Air Products and Chemicals (2003). Project Data on Eastman Chemical Company's Chemical-from-Coal Complex in Kingsport, TN. Prepared for US Department of Energy (under cooperative agreement DE-FC22-92PC90543).

Ebara Corporation, Tokyo, Japan (2010). Progress on Our Zero Emission Challenge. Summary of Annual Reports.

Eco-Innovation Observatory (EIO) (2012). Methodological Report. Eco-Innovation Observatory. Funded by the European Commission, DG, Environment, Brussels.

Ekbom, T., Berglin, N., and Logdberg, S. (2005). Black liquor gasification with motor fuel production – BLGMF-II: a techno-economic feasibility study on catalytic Fischer-Tropsch synthesis diesel production in comparison with methanol and DME as transport fuels. P21384-1, Nykomb Synergestics AB, Stockholm.

Ekbom, T., Lindblom, M., Berglin, N., and Ahlvik, P. (2003). Technical and commercial feasibility study of black liquor gasification with methanol/DME production as motor fuels for automotive uses: BLGMF. Report for Contract No. 4.1030/Z/01-087/2001.

Ellen Macarthur Foundation (2012). *Towards the Circular Economy*, Economic and Business Rationale for a Circular Economy, vol. 1. Cowes: Ellen Macarthur Foundation.

Ellen Macarthur Foundation (2013). *Towards the Circular Economy*, Opportunities for the consumer goods sector, vol. 2. Cowes: Ellen Macarthur Foundation.

Ellen Macarthur Foundation (2014). *Towards the Circular Economy*, Accelerating the Scale-Up Across Global Supply Chains, vol. 3. Cowes: Ellen Macarthur Foundation.

Ellen Macarthur Foundation (2015). *Towards a Circular Economy: Business Rationale for an Accelerated Transition*. Ellen Macarthur Foundation.

Ellen Macarthur Foundation (2016). The New Plastics Economy: Rethinking the Future of Plastic.

Ellen Macarthur Foundation, SUN, McKinsey and Co (2015). *Growth Within: Circular Economy Vision for a Competitive Europe*. London: McKinsey Center for Business and Environment.

EPRI, Electric Power Research Institute (2011). *Water Use for Electricity Generation and Other Sectors: Recent Changes (1985–2005) and Future Projections (2005–2030)*. Palo Alto, CA: EPRI.

Eriksson, H. and Harvey, S. (2004). Black liquor gasification: consequences for both industry and society. *Energy* 29: 581–612.

ETC, (2011). Gasification of Black Liquor. Popular Science Report from the BLG II Program 2007–2010.

European Federation Chemical Engineering (EFCE) (2015). European Roadmap on Process Intensification.

Farmer, M.C. (2005). The adaptable integrated biorefinery for existing pulp mills. Presentation at TAPPI Engineering, Pulping, and Environmental Conference, Philadelphia, PA (28–31 August 2005).

Farmer, M.C. and Sinquefield, S. (2003). *An External Benefits Study of Black Liquor Gasification. Final Report*. Atlanta: Georgia Institute of Technology, June 15.

Farmer, M.C. and Sinquefield, S. (2009). *Black Liquor Gasification: Paths and Obstacles to Commercialization*. Surrey, UK: Pira Industry Insight.

Frosch, R. and Gallopoulos, N. (1989). Strategies for manufacturing. *Scientific American* (September 1989), p. 144–152.

Fujimoto, K., Shikada, T., and Yamaoka, Y. (1995). Method of producing dimethyl ether. US Patent 5,466,720, 14 November 1995.

Geissdoerfer, M., Savaget, P., Bocken, N., and Hultink, E.J. (2017). The circular economy: a new sustainability paradigm? *Journal of Cleaner Production* 143: 757–768.

Gómez-Castro, F.I., Cano-Rodriguez, I.C., and Gamiño-Arroyo, Z. (2016). Process intensification in the production of liquid biofuels: strategies to minimize environmental impact. In: *Process Intensification in Chemical Engineering: Design Optimization and Control* (eds. J.D. Segovia-Hernandez and A. Bonilla-Petriciolet). Cham: Springer.

Grigoray, O. (2009). Gasification of black liquor as a way to increase power production at kraft pulp mills. Master thesis. Lappeenrantea University of Technology Faculty of Technology Degree Program of Chemical Technology.

Hallale, N. (2001). Burning bright: tends in process integration. *Chemical Engineering Progress* 97 (7): 30–41.

Hansen, J. and Sato, M. (2004). Greenhouse gas growth rates. *Proceedings of the National Academy of Sciences* 101 (46): 16109–16114.

Hein, A.M., Jankovic, M., Farel, R., and Yannou, B. (2015). A conceptual framework for Eco-Industrial Parks. *Proceedings of the ASME 2015 International Design Engineering Technical Conferences & Computers and Information in Engineering Conference IDETC/CIE*, Boston, MA (2–5 August 2015).

Intergovernmental Panel on Climate Change (IPCC) (2007). *Climate Change 2007: The Physical Science Basis Summary for Policymakers*. Geneva: U.N. Environment Program and World Meteorological Organization.

International Energy Agency (2009). Black liquor gasification summary and conclusions from the IEA Bioenergy ExCo54 Workshop. Bioenergy Report.

Jensen, P.D. (2011). Reinterpreting industrial ecology. *Journal of Industrial Ecology* 15 (5): 680–692.

KAM Final Report (2003). Eco-cyclic pulp mill: KAM project. Report no. A100, Stockholm, Sweden: STFI. www.stfi-packforsk.se

Lancaster, M. (2002). Principles of sustainable and green chemistry. In: *Handbook of Green Chemistry and Technology* (eds. J. Clark and D. Manqueary). Oxford: Blackwell Science.

Landälv, I., Granberg, F., and Nelving, H. (2010). Production of di methyl ether from kraft black liquor for use in heavy duty trucks. *Proceedings of International Chemical Recovery Conference*, Williamsburg, VA (29 March 29 to 1 April), vol. 2, p. 179–186.

Landävl, I. (2010). Woods to wheel: Chemrec's BioDME demonstration plant at the Smurfit Kappa mill. Pulp Paper Int.

Larson, E.D., Consonni, S., and Katofsky, R.E. (2003). *A Cost-Benefit Assessment of Biomass Gasification Power Generation in the Pulp and Paper Industry, Final Report*. Princeton, NJ: Princeton Environmental Institute.

Larson, E.D., Consonni, S., Katofsky, R.E. et al. (2006a). *A Cost-Benefit Assessment of Biomass Gasification Power Generation in the Pulp and Paper Industry, Final Report*. Princeton, NJ: Princeton Environmental Institute.

Larson, E.D., Williams, R.H., and Jin, H. (2006b). Electricity and fuels from biomass with carbon capture and storage. *Proceedings of the 8th International Conference on Greenhouse Gas Control Technologies*, Trondheim, Norway (19–22 June 2006).

Lewandowski, M. (2016). Designing the business models for circular economy: towards the conceptual framework. *Sustainability* 8 (1): 43.

Lienig, J. and Bruemmer, H. (2017). Recycling requirements and design for environmental compliance. In: *Fundamentals of Electronic Systems Design*, 193–218. New York: Springer.

Lindblom, M. (2012). Chemrec, history and future plants Chemrec sales presentation to ETF. *Nenet* http://www.nenet.se/.../Chemrec_biodme_motor_fuels_from_the_forest_matsli... (accessed 27 February 2019).

Liu, Z., Li, X., Close, M.R. et al. (1997). Screening of alkali-promoted vapor-phase-synthesized molybdenum sulfide catalysts for the production of alcohols from synthesis gas. *Industrial and Engineering Chemistry Research* 36: 3085–3093.

Lowe, E. and Warren, J. (1996). *The Source of Value: An Executive Briefing and Sourcebook on Industrial Ecology*. Washington, DC: Pacific Northwest National Laboratory.

Lowe, E.A. (1995). The Eco-Industrial Park: a business environment for a sustainable future. Paper presented at Designing, Financing and Building the Industrial Park of the Future Workshop, San Diego, CA.

Lowe, E.A. (2001). Eco-industrial Park Handbook for Asian Developing Countries. A Report to Asian Development Bank. Environment Department, Indigo Development, Oakland, CA.

Mabee, W.E., Gregg, D.J., and Saddler, J.N. (2005). Assessing the emerging biorefinery sector in Canada. *Applied Biochemistry and Biotechnology* 121–124: 765–777.

Marklund, M. (2006). Pressurized entrained-flow high temperature black liquor gasification CFD based reactor scale-up method and spray burner characterization. PhD thesis. Luleå University of Technology.

McDonough, W. and Braungart, M. (2002). *Cradle to Cradle: Remaking the Way We Make Things*. North Point Press.

Miller, M., Justiniano, M., and McQueen, S. (2005). *Energy and Environmental Profile of the U.S. Pulp and Paper Industry*. Columbia, MD: Energetics Incorporated.

Milne, G. and Readon, C. (2005). Embodied energy. In: *Your Home Technical Manual: Australia's Guide to Environmentally Sustainable Homes*. Adelaide, Australia: Commonwealth of Australia.

Murray, A., Skene, K., and Haynes, K. (2015). The circular economy: an interdisciplinary exploration of the concept and application in a global contex. *Journal of Business Ethics* 127 (3): 525–536.

Naqvi, M., Yan, J., and Dahlquist, E. (2010). Black liquor gasification integrated in pulp and paper mill: a critical review. *Bioresource and Technology* 101: 8001–8015.

NCASI (1993). *Summary of Data Reflective of Pulp and Paper Industry Progress in Reducing the TCDD/TCDF Content of Effluents, Pulps, and Wastewater Treatment Sludges*. New York: NCASI, June.

NREL (2011). *A Review of Operational Water Consumption and Withdrawal Factors for Electricity-Generating Technologies*, TP-6A20-50900. Golden, CO: NREL.

Ohno, Y. (2006). The role of DME in the World: a perspective. *Proceedings of the Second International DME Conference*, London (15–17 May 2006).

Parthasarathy, G. and Dunn, R. (2005). Minimization of environmental discharge through process integration. In: *Toward Zero Discharge: Innovative Methodology and Technologies for Process Pollution Prevention* (ed. T.K. Das), 597–645. Hoboken, NJ: Wiley-InterScience.

Patrick, K. and Siedel, B. (2003). Gasification edges closer to commercial reality with three new N.A. mills startups. *PaperAge* 119 (Part 7): 30–33.

Pearce, D.W. and Turner, R.K. (1989). *Economics of Natural Resources and the Environment*. Baltimore MD: Johns Hopkins University Press.

Pearce, J.M. (2008). Industrial symbiosis for very large scale photovoltaic manufacturing. *Renewable Energy* 33: 1101–1108.

Perlack, R.D., Wright, L.L., Turhallow, A. et al. (2005). *Biomass as Feedstocks for a Bioenergy and Bioproducts Industry: The Technical Feasibility of a Billion-Ton Annual Supply*, ORNL/TM-2005/66 for USDA and USDOE, April. Oak Ridge, TN: Oak Ridge National Laboratory.

Pitt, J. and Heinemeyer, C. (2015). Introducing ideas of a circular economy. In *Environment, Ethics and Cultures: Design and Technology Education's Contribution to Sustainable Global Futures*. London, UK: University of London. ScienceDirect.com.

Prentice, I.C., Farquhar, G.D., Fasham, M.J. et al. (2001). The carbon cycle and atmospheric carbon dioxide. In: *Climate Change 2001: The Scientific Basis. Contribution of Working Group I to the Third Assessment Report of the Intergovernmental Panel on Climate Change* (eds. J. Houghton et al.). Cambridge: Cambridge University Press.

Reay, D., Ramshaw, C., and Harvey, A. (2013). *Process Intensification: Engineering for Efficiency, Sustainability and Flexibility*, 2e. Oxford: Butterworth-Heinemann.

Richards, D.J., Allenby, B.R., and Frosch, R.A. (1994). The greening of industrial ecosystems: overview and perspective. In: *The Greening of Industrial Ecosystems* (eds. B.R. Allenby and D.J. Richards), 1–19. Washington, DC: National Academy Press.

Salomonsson, P. (2011). BIOMED project. *Proceedings of the Seventh Asian DME Conference*, Nigata, Japan (November 2011).

Salomonsson, P. (2013). Final report of the European BioDME project. *Proceedings of the Fifth International DME Conference*, Ann Arbor, MI (18 April 2013).

Smith, R. (2016). *Chemical Process Design and Integration*, 2e. Hoboken, NJ: Wiley.

Smol, M., Kulczycka, J., and Avdiushchenko, A. (2017). Circular economy indicators in relation to eco-innovation in European regions. *Clean Technologies and Environmental Policy* 19 (3): 669–678.

Smol, M., Kulczycka, J., Henclic, K. et al. (2015). The possible use of sewage sludge ash (SSA) in the construction industry as a way towards a circular economy. *Journal of Cleaner Production* 95: 45–54.

Sricharoenchaikul, V. (2001). Fate of carbon containing compounds from gasification of kraft black liquor with subsequent catalytic conditioning of condensable organics. PhD dissertation. Georgia Institute of Technology.

Sricharoenchaikul, V. (2009). Assessment of black liquor gasification in supercritical water. *Bioresource Technology* 100: 638–643.

Stahel, W.R. (2015). *The Business Angle of a Circular Economy: Higher Competitiveness, Higher Resource Security and Material Efficiency*. Stockholm: EMF.

Staniškis, J.K. (2012). Sustainable consumption and production: how to make it possible. *Clean Technologies and Environmental Policy* 14 (6): 1015–1022.

Stankiewicz, A.I. and Moulijn, J.A. (2000). Process intensification: transforming chemical engineering. *Chemical Engineering Progress* 96 (1): 22–34.

Swedish Energy Agency (2008). *System Aspects of Black Liquor Gasification: A Review of Existing Reports*. Eskilstuna: Swedish Energy Agency ISSN 1403-1892.

Swedish Forest Agency (2008). *Forestry Statistics*. Jönköping, Sweden: Swedish Forest Agency.

Tampier, M., Smith, D., Bibeau, E., and Beauchemin, P.A. (2004). Identifying Environmentally Preferable Uses for Biomass Resources. Stage I Report: Identification of Feedstocks-to-Product Threads. Prepared for Natural Resources Canada, Commission for Environmental Co-operation, and National Research Council of Canada. Envirochem Services Inc., Vancouver, BC.

Thomé, A.M., Scavarda, A., Ceryno, P.S., and Remmen, A. (2016). Sustainable new product development: a longitudinal review. *Clean Technologies and Environmental Policy* 18 (7): 2195–2208.

USDOE (2006). *Proceeding of the 5th Annual Conference on Carbon Capture and Sequestration*, Alexandria, VA (8–11 May 2006) and *Proceedings of the 8th International*

Conference on Greenhouse Gas Control Technologies, Trondheim, Norway (19–22 June 2006).

USDOE (2013a). *Effects of Climate Change on Federal Hydropower: Report to Congress*. Washington, DC: DOE. http://www1.eere.energy.gov/water/pdfs/hydro_climate_change_report.pdf (accessed 10 April 2018).

USDOE (2013b). *U.S. Energy Sector Vulnerabilities to Climate Change and Extreme Weather*. Washington, DC: DOE. http://energy.gov/sites/prod/files/2013/07/f2/20130716-Energy%20Sector%20Vulnerabilities%20Report.pdf (accessed 28 February 2019).

USDOE (2014a). *The Water-Energy Nexus: Challenges and Opportunities*. USDOE, Pacific Northwest National Laboratory.

USDOE (2014b). Energy Demands on Water Resources. USDOE Report to Congress on the Interdependency of Energy and Water.

USEPA (2002). *AP-42, Compilation of Air Pollutant Emission Factors*, Section 1.6: Wood Residue Combustion in Boilers. Research Triangle Park, NC: U.S. Environmental Protection Agency, March.

Van der Ryn, S. and Cowan, S. (1996). *Ecological Design*. Washington, DC: Island Press.

Vogel, F., DiNaro, J., Marrone, P. et al. (2005). Critical review of kinetic data for the oxidation of methanol in supercritical water. *Journal of Supercritical Fluids* 34: 249–286.

Waldner, M. and Vogel, F. (2005). Renewable production of methane from woody biomass by catalytic hydrothermal gasification. *Industrial and Engineering Chemistry Research* 44 (13): 4543–4551.

Weetman, C. (2016). *A Circular Economy Handbook for Business and Supply Chains: Repair, Remake, Redesign, Rethink*. London: Kogan Page.

Weyerhaeuser Company (2000). Biomass Gasification Combined Cycle. Final Report Under Contract DE-FC36-96GO10173 to U.S. Department of Energy, Federal Way, WA, 160 pages.

Whitty, K. (2006). Conversion of black liquor in a fluidized bed steam reformer. Paper presented at the 7th International Colloquium on Black Liquor Combustion and Gasification, Jyvaskyla, Finland (2 August 2006).

Whitty, K. and Baxter, L. (2001). State of the art in black liquor gasification technology. *Proceedings of the Joint International Combustion Symposium*, Kauai, Hawaii (9–12 September 2001).

Whitty, K. and Nilsson, A. (2001). Experience from a high temperature, pressurized black liquor gasification pilot plant. *Proceedings of the Tappi Chemical International Recovery Conference*, Whistler, BC (11–14 June 2001). Atlanta: Tappi Press, pp. 655–662.

Whitty, K. and Verrill, C.L. (2004). A historical look at the development of alternative black liquor recovery technologies and the evaluation of black liquor gasifier designs. *Proceedings of the International Chemical Recovery Conference*, Charleston, SC (6–10 June 2004).

Wikipedia (2016). Black liquor, 2 February 2016. https://en.wikipedia.org/wiki/Black_liquor (accessed 6 August 2019).

Williams, P. and Onwudili, J. (2006). Subcritical and supercritical water gasification of cellulose, starch, glucose, and biomass waste. *Energy Fuels* 20: 1259–1265.

Zhong (2018). Tightening the loop on the circular economy: coupled distributed recycling and manufacturing with recyclebot and RepRap 3-D printing. *Resources, Conservation and Recycling* 128: 48–58.

10

Quality Industrial Environmental Management

Sustainable Engineering in Manufacturing

10.1 Introduction: Industry and the Global Environmental Issues

Environmental issues are commanding considerable attention internationally. Climate change, water availability, pollution, and waste generation and disposal are among the leading challenges in this regard. As a major user of raw materials and energy and a major source of pollutants and waste, industry is an important player. Progress in dealing more systematically with issues has been made on several fronts during the past several decades, with industry taking a more active role in addressing current and emerging issues. During the late 1980s and 1990s, institutions for addressing environmental matters were developed, international co-operation and public awareness increased, and private-sector actions became more common (UNEP 1997). At the same time, legal frameworks and economic instruments were introduced, while technological advances helped to boost cleaner production. Finally, governments became more circumspect in pursuing developmental objectives with mandated environmental impact assessments becoming more common worldwide (IGEL 2012; Western Governors' Association 2004).

Despite the progress that has been made in developing institutions, methods, and techniques, environmental degradation remains an important issue. In its recent analysis of the global environment, the United Nations points out that full-scale emergencies exist in a number of areas, with environmental stewardship lagging behind economic and social development (UNEP 1999). Air pollution, for example, is at a crisis point in many major cities. Moreover, increased greenhouse gas emissions are seen as making it impossible to prevent global warming, despite the efforts undertaken in the Kyoto Protocol and Paris Agreement (see Figure 5.18). There are also important issues to be addressed concerning resource use. Natural capital is declining in a number of key areas, with the implication that efficiency in the use of resources will have to increase, or substitutes will have to be developed in order to meet future consumer demands (Hohen 2000).

10.1.1 Industry Role and Trends

The positive role that industry can play in meeting environmental and interrelated social and economic challenges has been recognized for some time. It was highlighted in *Agenda 21*, the global plan for sustainable development adopted at the United Nations Conference on Environment and Development (1992). More efficient production processes, preventive strategies, cleaner technologies, and procedures throughout the product life cycle, for example, can be important mechanisms for reducing impacts on resource use and the environment. In practical terms, this means, in the first instance, identifying and promoting sustainable practices. This, in fact, is already being done increasingly in industry, reflecting changes in government policies and heightened industry involvement in identifying and addressing environmental challenges. As discussed at the Forum, industry is moving from an "end-of-pipe" reactive strategy that focused on capturing and disposing of pollutants generated during manufacturing to a more holistic approach that integrates environmental considerations more effectively over a broader range of firm functions – including product design and procurement practices, as well as production processes.

The UN, in its *Global Environmental Outook-2000* report, points out that industry has responded in several areas with impressive results (UNEP 1999). The voluntary actions taken to reduce resource use and waste were viewed as a positive step that boded well for the environment. Industry was encouraged to pursue these initiatives, paying particular attention to ways in which these efforts could be expanded to include a broader range of small and medium-sized enterprises.

Industrial Environmental Management: Engineering, Science, and Policy, First Edition. Tapas K. Das.

Companion website: www.wiley.com/go/Das/IEM_1e

Pursuing environmental initiatives, however, does not occur in a vacuum. Such efforts can compete and conflict with other objectives – notably those concerned with financial performance. Some, in fact, have gone so far as to question the viability of "corporate sustainability" given the nature and scope of the potential conflicts (Box 10.1). If financial performance takes precedence, objectives in the environmental and social fields are likely to be compromised or, at the very least, be subject to important constraints. As discussed at the Forum, the trade-offs become most explicit when performance in different domains is evaluated discretely, using a *triple-bottom-line* (i.e. financial, environmental, and social) approach.

10.1.2 Code of Ethics for Engineers

10.1.2.1 Engineers Play a Key Role in Manufacturing

Engineers play a key role in global sustainable manufacturing and development by designing production systems for materials, minerals, chemicals, energy, water, electricity generation and distribution, transportation, buildings and other structures, and consumer products (see Appendix J). These designs have impact on the environment, economics, and societal benefits at spatial scale that vary from local to global and temporal scales that vary from minutes to decades. As engineers create designs, they do not only evaluate their designs at multiple spatial and temporal scales, they also embed their designs in complex systems.

The field of transportation provides an illustration of the multiple layers of systems in which engineers create designs. Among the most visible products designed by engineers are automobiles. Engineers design engines, and improvements to the design of a fossil-fuel-powered engine for an automobile can increase fuel efficiency and reduce environmental impacts of emissions associated with burning fuels, while simultaneously reducing the cost of operating the vehicle. The size, power, and fuel efficiency of the engine must be balanced with the weight of the vehicle system. Further reductions in emissions and operating costs might be possible by lowering the weight of the vehicle. The use of materials and fuels by automobiles are embedded in complex fuel and material supply systems. Developing systems to recycle the materials that make up the automobile at the end of its useful life might improve the environmental and economic performance of global materials flows. Use of alternative power sources, such as electricity or biofuels can impact flows of fuels, which, in turn, might impact global flows of materials such as water. Finally, the design of cities that reduce the need for personal transportation could dramatically reduce the environmental impacts of transportation systems and would also transform social structures.

10.1.2.2 Sustainable Engineering

Sustainable engineering is the design, commercialization, and use of processes and products that are feasible and economical while minimizing both the generation of pollution at the source and the risk to human health and the environment. The discipline embraces the concept that decisions to protect human health and the environment can have the greatest impact and cost effectiveness when applied early in the "design and development phase of a process or product." Sustainable engineering transforms existing engineering disciplines and practices to those that promote

Box 10.1 Corporate Sustainability

The precise meaning of the term "corporate sustainability" is much disputed. For some, a company that is truly pursuing sustainable practices must be phasing out its use of substances that are systematically at odds with a sustainable world, while measuring what share of the world's limited regenerative capacity is occupied by its use of renewable resources. Similar to this is the view that corporate sustainability means establishing boundary conditions, limits, metabolic ratios, and footprints.[1] Others intellectually tied to the social auditing movement, assign more weight to the processes that lead to sustainable practices (as distinct from the results of those practices). They see the involvement of stakeholders – and indicators defined by those stakeholders – as a precondition for truly sustainable corporate activities. Still others, led by the World Business Council on Sustainable Development, define sustainability largely in terms of eco-efficiency, i.e. as producing goods and services of greater and greater value relative to the burden they impose on the environment. Finally, some commentators view corporate sustainability as an oxymoron. That is, they see today's global corporations – which measure their success primarily in terms of short-term increases in shareholder value and which make few commitments to their communities or workers – as inherently at odds with the intergenerational and social equity aspects of sustainability.

Source: From White (1999).

[1] Boundary conditions, limits, and metabolic ratios refer to the assimilative capacity of air, water, and land relative to the resource use and wastes associated with a firm's products and services.

sustainability. This new discipline incorporates the development and implementation of technologically and economically viable products, processes, and systems that promote human welfare while protecting human health and elevating the protection of the biosphere as a criterion in engineering solutions. To fully implement sustainable green engineering solutions, engineers use the following principles (Das 2005b).

Sandestin Sustainable Engineering Principles
Principle 1: Engineer processes and products holistically, use systems analysis, and integrate environmental impact assessment tools.
Principle 2. Conserve and improve natural ecosystems while protecting human health and well-being.
Principle 3. Use life-cycle thinking in all engineering activities.
Principle 4. Ensure that all material and energy inputs and outputs are as inherently safe and benign as possible.
Principle 5. Minimize depletion of natural resources.
Principle 6. Strive to prevent waste.
Principle 7. Develop and apply engineering solutions that take into consideration local geography, aspirations, and cultures.
Principle 8. Create engineering solutions that go beyond current or dominant technologies; improve, innovate, and invent technologies to achieve sustainability.
Principle 9. Actively engage communities and stakeholders in development of engineering solutions, recognizing green engineering's duty to inform society about its practices.
There is a duty to inform society of the practice of sustainable engineering.

10.1.3 Sustainable Engineering Design Principles

Starting from the introduction of sustainable development concepts in the "Brundtland Report" (WCED 1987), there have been many attempts to incorporate sustainability principles into engineering design. For example, the "Hannover Principles" express the view that human systems must be designed to coexist with natural systems, renewable resources should be used, safe and long-lived products are desired, and elimination of waste is a priority (Das 2005c; McDonough and Braungart 2002). The Augsburg Materials Declaration Research (2002) identifies eight factors that must be considered to achieve sustainable production, including integration of environmentally benign, materials, and manufacturing over all stages of the life cycle; optimization and exploitation of raw materials and natural resources; energy-efficient production technologies and product distribution; regenerative energy sources; and durability, recyclability, and closed loops. The "'12' Principles of Green Engineering" (Anastas and Zimmerman 2003) include design to be inherently safe and benign, design for recycle or a commercial afterlife, energy and mass efficiency, and integration with existing energy and material flows. The Sandestin Green Engineering Principles, developed as an outcome of a multidisciplinary engineering conference, emphasize the need for holistic thinking and the use of environmental impact and integrative analysis tools such as life cycle assessment (LCA; Abraham and Nguyen 2003; Allen and Shonnard 2012; Shonnard et al. 2007). The Sandestin Green Engineering Principles (also referred to as "Sustainable Engineering Principles"; Abraham 2006) were developed based on a starting list of principles complied from a literature review of available sustainability or green-related principles and declaratory statements, including the Hannover Principles, CERES, the Augsburg Materials Declaration, the Twelve Principles of Green Chemistry, Ahwahnee Principles (1991), and Earth Charter Principles. These sustainability concepts and engineering design principles can be summarized in the statement: *The goal of sustainable engineering design is to create products that meet the needs of today in an equitable fashion while maintaining healthy ecosystems and without compromising the ability of future generations to meet their resource needs.*

The *Sandestin Sustainable Engineering Principles* and the 12 *Principles of Green Engineering* are illustrative of the multiple sets of principles available. They capture similar but also complementary elements of sustainability and engineering design; therefore, these two sets of principles will be described in more detail. Here are the Sandestin Green Engineering Principles (Sustainable Engineering Principles) (Abraham and Nguyen 2003; Shonnard et al. 2007):

Principle 1: Engineer processes and products holistically use system analysis and integrate environmental impact assessment tools. These concepts resonate in a number of Green and Sustainable Engineering principles and are addressed at length in various textbooks, including *Green Engineering: Environmentally Conscious Design of Chemical Processes* (Allen and Shonnard 2002). The principle points out the importance of systematic evaluation and reduction of human health and environmental impacts of designs, products, technologies, processes, and systems. The use of system-based techniques such as heat and mass integration techniques is essential to minimize human health and environmental impacts of designs through material and energy optimization. The principle also conveys the importance of not shifting risk (e.g. reducing releases to one environmental medium may increase risk to another medium and/or increase the likelihood of worker exposures and jeopardize worker safety) (Abraham 2006; Shonnard et al. 2007). A well-known example of the consequences of shifting risk is the use of methyl-tert butyl ether (MTBE) as a gasoline line additive. MTBE was added to gasoline to reduce emissions of carbon monoxide from automobile tailpipe, thereby protecting human health. Its greater mobility in soil and water environments, however, meant that spills of MTBE could more readily migrate to and disperse in water supplies than spills of gasoline (for more details see http://www.epa.gov/otaq/consumer/fuels/oxypanel/blueribb.htm).

Principle 2: Conserve and improve natural ecosystems while protecting human health and well-being. This principle expresses the importance of understanding environmental processes for engineers involved in design of chemicals, automobiles, building, and other manufactured goods. There are many examples where a lack of understanding caused severe environmental harm and raised the level of health risk to humans and other forms of life. For example, chlorofluorocarbons (CFCs) were thought to be ideal refrigerants. They replaced dangerous refrigerant fluids like ammonia and made storage of food and building eliminate control far safer. Their benefits to human health and well-being were clear; however, once the role of CFC in stratospheric ozone destruction chemistry was worked out, it became clear that there were hazards to human health associated with CFC use. Not every engineer needs to be an expert in environmental processes and health effects, but designers should be aware of the potential harm that can be caused and work with multidisciplinary experts to achieve more sustainable solutions (Allen and Shonnard 2012; Das 2005e).

Principle 3: Use of life-cycle thinking in all engineering activities. This principle complements Principle 1. Every engineered product is created, functions over a useful life, and eventually disposed of to the environment. Life-cycle thinking can help avoid a narrow outlook on environmental, social, and economic concerns and help make informed decisions. Life-cycle thinking that gets incorporated into design will help identify design alternatives that minimize environmental impacts at the various life stages. This same kind of thinking can also consider economic and social aspects. The importance of life-cycle thinking can be illustrated (see Chapter 6).

Principle 4: Ensure that all material and energy inputs and outputs are as inherently safe and benign as possible. This principle complements Principle 3. These characteristics of materials and chemicals must be applied to all stages of a product's life, from extraction to use and disposal. The following are some questions that must be asked in each stage of the life cycle: Are the materials toxic? Are these inherently benign (in terms of toxicity) materials that can be used as substitutes? Will exposure during manufacturing be a health problem to workers? Does the product pose minimal impact during recycle and disposal? Will an unintentional release of materials quickly degrade in the environment? Properties of materials relevant to safety, beyond toxicity, must also be considered, such as flammability, explosivity, and corrosivity.

Principle 5: Minimize depletion of natural resources. As the world's population continues to grow and becomes more affluent, natural resources will be used at ever greater rates, and the importance of this principle is raised. An overview of materials, energy, and water use is provided in Chapter 1. Efficient use of nonrenewable energy resources is of primary interest, and development of renewable alternatives for energy and materials should be given a high priority in engineering design.

Principle 6. Strive to prevent waste. Waste not only represents material that takes up space in landfills but more importantly represents a loss of efficiency in a production system that includes many input materials and energy sources. When waste is avoided through design, the environmental impacts associated with the input materials and the energy that went into producing the discarded product are also avoided. The following are some questions that must be asked during design: Are there ways (e.g. procedures, engineering) to improve the yield or efficiency of raw materials? How can the releases or wastes be recycled and reused? Can the product be reused after its normal commercial life, hence minimizing the raw materials needed to manufacture new products? Methods for minimizing wastes, particularly in chemical processes, are discussed at length in various textbooks, including *Green Engineering: Environmentally Conscious Design of Chemical Processes* (Allen and Shonnard 2002).

Principle 7: Develop and apply engineering solutions, while being cognizant of local geography, aspirations, and cultures. Engineering design are directed toward meeting individual human and societal needs, and in order to better achieve this goal, awareness of the societal context of the design is crucial. An engineering design in one society, such as rapid public transportation systems, may not meet the aspirations and needs in another society, even though the design achieves environmental objectives. The main point is to move each society, though engineering design, toward more sustainable utilization of resources in a way that achieves that society's or individual's aspirations (Allen and Shonnard 2012).

Principle 8: Create engineering solutions beyond current or dominant technologies: improve, innovate, and invent (technologies) to achieve sustainability. Sustainability can be a powerful motivation for change in engineering design, technologies, process, and products. This principle emphasizes the importance of being innovative (i.e. "out-the-box" thinking) in the development of new technologies. The knowledge gained through considering the many dimensions of sustainability should be reflected in how engineering designs accomplish societal objectives.

Principle 9: Actively engage communities and stakeholders in the development of engineering solutions. There are many examples of stakeholder and community engagement in the development of engineered solutions in a wide range of activities, including city planning, infrastructure development, and production of manufactured goods. One illustrative example is in the Mining Minerals Sustainable Development North America project (IISD 2002) that has been adopted by key members of this industry. Central to these mining project questions is community engagement from project inception to mine closure. During this engagement, the communities surrounding the proposed mine development express their wishes with regard to managing the economic development and any concerns over local environmental consequences.

A second set of engineering design principles, the 12 Principles of Green Engineering, from Anastas and Zimmerman (2003) and Allen and Shonnard (2012), is presented here as follows.

Principles of Green Engineering

Principle 1: Designers need to strive to ensure that all material and energy inputs and outputs are as inherently nonhazardous as possible.

Principle 2: It is better to prevent waste than to treat or clean up waste after it is formed.

Principle 3: Separation and purification operations should be designed to minimize energy consumption and materials use.

Principle 4: Products, processes, and systems should be designed to minimize mass, energy, space, and time efficiency.

Principle 5: Products, processes, and systems should be "output pulled" rather than "input pulled" through the use of material and energy.

Principle 6: Embedded entropy and complexity must be viewed as an investment when making design choices on recycle, reuse, or beneficial disposition.

Principle 7: Targeted durability, not immortality, should be a design goal.

Principle 8: Design for unnecessary capacity or capability (e.g. "one-size-fits-all") solutions should be considered a design flaw.

Principle 9: Material diversity in multicomponent products should be minimized to promote disassembly and value retention.

Principle 10: Design of products, processes, and systems must include integration and interconnectivity with available energy and materials flows.

Principle 11: Products, processes, and systems should be designed for performance in a commercial "afterlife."

Principle 12: Material and energy inputs should be renewable rather than depleting.

Principle 1: Designers need to strive to ensure that all material and energy inputs and outputs are as inherently nonhazardous as possible. This principle recognizes that significant costs and hazards result from the selection of sources of materials and energy. Additional control systems are required to capture and destroy hazardous materials during production, use, and disposal, all of which add to the cost of the design. If inputs to the system are inherently less hazardous, the risks of failure will be reduced and the amount of resources expended on control, monitoring, and containment will be less.

Principle 2: It is better to prevent waste than to treat or clean up waste after it is formed. The creation of waste in engineered systems adds to the complexity, efforts, and expense of the design. This is especially true for hazardous wastes, which require extraordinary measures for their control, monitoring, transport, and disposal. To reduce waste generation, the design must strive to incorporate as much of the input materials as possible into final products. This strategy can be applied at many scales, for example, at the molecular level in the design of chemical reactions, at larger scales such as in machining of parts, and further in assembly of discrete parts. Any waste that is generated should be considered as raw material to be used again in the current product system or as input to a separate product system.

Principle 3: Separation and purification operation should be designed to minimize energy consumption and materials use. In the chemical and mineral-processing industries, large-scale separation processes are among the largest energy-consuming units and generates a significant proportion of emissions and wastes. Even in industry sectors where the mass of the products produced is not large, separation processes can be significant. In electronics manufacturing, the generation of ultra-pure water and the creation of ultra-clean work environments require separation processes with significant costs and energy demands. Design for efficient separation is very important for these industries, and several approaches can be investigated during design. Gains in *energy efficiency* can be attempted through heat integration by considering all streams needing to gain or lose energy in the process and also outside the process if in close proximity to other facilities. Similarly, pollution can be prevented by considering mass integration, taking waste streams from one process and using them as raw materials for another. In certain instances, products can be induced to self-separate by adjusting conditions to take advantage of physical and chemical properties of the chemicals. As another example, in mechanical systems, reversible fasteners can be used to encourage the disassembly of manufactured parts at the end of life.

Principle 4: Products, processes, and systems should be designed to maximize mass, energy, space, and time efficiency. Energy and mass efficiency were dealt with in Principle 3, so this discussion will focus on space and time efficiency. Space and time are interrelated in many engineered systems, but most obviously in transport of raw material and products. Reducing the distance between points of use for materials can save time and reduce pollution. Close proximity can facilitate exchanges of waste heat and materials in highly integrated production systems. Colocation of manufacturing and recycle facilities can also lead to efficiency gains in many production systems. Industrial parks near residential areas can lead to sharing of excess heat with communities. However, these proximity opportunities that take advantage of space and time factors also must consider safety concerns due to potential exposure to emissions and industrial accidents. The excess time that a product sits in inventory can run up against storage stability limits and could lead to excess waste generation.

Principle 5: Products, processes, and systems should be "output pulled" rather than "input pushed" through the use of materials and energy. It is well known that some chemical reactions can be "pulled" to completion by removing certain co-products from the reaction mixture. This chemical phenomenon, which is termed *Le Châtelier's* principle, can be applied to engineering design across scales of production. "Just-in-time" manufacturing is an example of this principle where only the necessary units are produced in the necessary quantities at the necessary time by bringing production rates exactly in line with demand. Planning manufacturing systems for final output eliminates the wastes associated with over-production, waiting time, processing, inventory, and resource inputs.

Principle 6: Embedded entropy and complexity must be viewed as an investment when making design choices on recycle, reuse, or beneficial disposition. Entropy and complexity are related concepts when considering engineered systems. Products having a high degree of order and structure are at a low state of entropy. The higher the degree of complexity and structure in a product, the greater is the amount of energy invested to create such structure and complexity. When considering end-of-life options for products, the degree of complexity and structure should point the way to proper reuse, recycle, and remanufacturing options. Highly complex parts should be reused if at all possible in order to avoid the investment required to create a replacement part from virgin (newly extracted from nature) resources.

Principle 7: Targeted durability, not immortality, should be a design goal. Many products last much longer than the expected commercial life. There can be multiple impacts from this extended durability. For example, buildings with inefficient energy systems, designed when energy was relatively inexpensive, may become inoperable in times of expensive energy. At a different scale, the challenge in the design of molecules and of manufactured parts is to create products that are durable yet do not persist indefinitely in the environment. Durability means that the products for the intended commercial life and are therefore readily reconfigured or degraded at the end of life into harmless substances that assimilate easily into natural cycles.

Principle 8: Design for unnecessary capacity or capability (e.g. "one size fits all") solution should be considered a design flaw. Most products are overdesigned to cover a wide application range and settings. Automobiles must be designed to function not only in warm temperatures but also in extremes of cold. However, there are instances where design for "one size fits all" does not make the most sense and is potentially wasteful. For example, a jacket could be designed for the coldest possible climate, but this garment would not be much use for a stroll along the beach on a breezy evening in Florida in the winter. Likewise, lighting of a large classroom, office space, or a room at home with a single light switch would not make as much sense if only one person were in the room at a given time. In such a case, district lighting would save on energy when only one or a few occupants are present in a large space.

Principle 9: Material diversity in multicomponent products should be minimized to promote disassembly and value retention. This design principle has elements in common with Principle 6. Increasing material diversity in products has the effect of making recycling more difficult and expensive because the number of recycling options and their complexity increase as material diversity increases. Different kinds of material and the use of different additives have a strong influence on recycling methods and costs.

Principle 10: Design of products, processes, and systems must include integration and interconnectivity with available materials and energy flows. This design principle states that products, processes, and entire engineered systems should be designed to use the existing infrastructure of material and energy flows. Integration with existing infrastructure can occur at the scale of a unit operation, production line, manufacturing facility, or industrial park. Taking advantage of existing material and energy flows will minimize the need to generate energy and/or acquire and process raw materials. Applications of this principle include the recovery and use of heat from exothermic chemical reactions, the cogeneration of heat and power, and the recovery of electrical energy by regenerative braking in hybrid vehicles.

Principle 11: Products, processes, and systems should be designed for performance in a commercial "afterlife." Designing components for a second, third, or even longer life is an important strategy in product design. When components are recovered and reused in next-generation products, the environmental impacts of raw material acquisition from virgin resources and conversion are eliminated. And the overall life cycle environmental impacts are reduced. This strategy is especially important for products that become obsolete prior to component failure, such as cell phones and other electronic devices. Important examples of this principle also include the recovery and recycle of spent copy toner cartridges, the renovation of industrial building for housing and reuse of beverage containers as practiced in Germany, where bottles are more substantial in their construction to allow for collection, washing, sterilization, refilling, and relabeling.

Principle 12: Material and energy inputs should be renewable rather than depleting. The use of nonrenewable raw materials from nature in the design of engineered systems moves the Earth system incrementally toward depletion of finite resources and is therefore unsustainable by definition. All renewable resources (also see Figure 7.9) derive their usefulness and energy from the sun, and as a result these system inputs can be sustainable, if used at a level consistent with their rate of renewable resources in that it can serve not only as an energy source but also as a feedstock for design of material products (AIChE, CWRT 2000; Das 2005g). One important form of a biomass product is liquid transportation fuels. Biofuels are renewable on relatively short timescales, and the cycling of biofuel carbon between the atmosphere, biomass/biofuel, and back to the atmosphere again is readily integrated into natural cycles in a way that might not cause accumulation of CO_2 in the atmosphere. An offshoot of this principle is the importance of design for products and systems that integrate well with natural cycles of elements across the life cycle, from raw material acquisition to end-of-life process.

These engineering design principles, in addition to the others mentioned in the beginning of this section, establish a framework for designing more sustainable products and processes. At first, changes in engineering design are likely to be improvements to inherently unsustainable products and systems, but over time it is hoped that these principles will move industry and consumers toward inherently sustainable products and production systems. However, there will be tensions in applying these principles. What if making a process inherently safer requires more energy? What if minimizing water use requires more energy? Engineers are accustomed to addressing trade-offs between objectives, but methods for doing so require measures of performance. In most engineering designs the measure is cost. In delivering a specified level of performance of a product, technology, or service, the goal is to minimize cost.

10.1.4 Design for Environmental Practices

Resource consumption and mitigation strategies for product production which minimizes environmental and health impacts. These design for the environment (DfE) practices have been adapted from USEPA (2017), Spangenberg et al. (2010), Allen and Shonnard (2012), and Allenby (2012).

The following are the five main concepts that fall under the DfE umbrella.

1) *Design for environmental processing and manufacturing*: This ensures that raw material extraction (mining, drilling, etc.), processing (processing reusable materials, metal melting, etc.), and manufacturing are done using materials and processes which are not dangerous to the environment or to the employees working on said processes. This includes the minimization of waste and hazardous by-products, air pollution, energy expenditure, and other factors.
2) *Design for reliability (DFR):* Is it reliable and safe given the range of environments in which the product is likely to be used, and how it will be used? For example, an electronic device used where there's high risk of static electricity must be designed to avoid failure when

exposed to static discharge. From a sustainable engineering perspective, DFR can extend product life, which is generally, although not always, a good thing. Note that in such cases sustainable engineering aligns environmental values with social values: customers like reliable and safe product, and the environment is benefitted by the reduction in waste while product assures reliability and safety.

3) *Design for environmental packaging:* This ensures that the materials used in packaging are environmentally friendly, which can be achieved through the reuse of shipping products, elimination of unnecessary paper and packaging products, efficient use of materials and space, use of recycled and/or recyclable materials.
4) *Design for disposal or reuse:* The end-of-life of a product is very important, because some products emit dangerous chemicals into the air, ground, and water after they are disposed of in a landfill. Planning for the reuse or refurbishing of a product will change the types of materials that would be used, how they could later be disassembled and reused, and the environmental impacts such materials have.
5) *Design for energy efficiency:* The design of products to reduce overall energy consumption throughout the product's life.

LCA is employed to forecast the impacts of different (production) alternatives of the product in question, thus being able to choose the most environmentally friendly. A life-cycle analysis can serve as a tool when determining the environmental impact of a product or process. Proper LCAs can help a designer compare several different products according to several categories, such as energy use, toxicity, acidification, CO_2 emissions, ozone depletion, resource depletion, and many others. By comparing different products, designers can make decisions about which environmental hazard to focus on in order to make the product more environmentally friendly.

10.1.5 Why Do Firms Want to Design for the Environment?

Modern day businesses all aim to produce goods at a low cost while maintaining quality, staying competitive in the global marketplace, and meeting consumer preferences for more environmentally friendly products. To help businesses meet these challenges, EPA encourages businesses to incorporate environmental considerations into the design process. The benefits of incorporating DfE include cost savings, reduced business and environmental risks, expanded business and market opportunities, and to meet environmental regulations (USEPA 2001).

10.1.6 How Does a Business Design for the Environment?

A business can design for the environment by the following ways:

- Evaluating the human health and environmental impacts of its processes and products.
- Identifying what information is needed to make human health and environment decisions.
- Conducting an assessment of alternatives.
- Considering cross-media impacts and the benefits of substituting chemicals.
- Reducing the use and release of toxic chemicals through the innovation of cleaner technologies that use safer chemicals.
- Implementing pollution prevention, energy efficiency, and other resource conservation measures.
- Making products that can be reused and recycled.
- Monitoring the environmental impacts and costs associated with each product or process.
- Recognizing that although change can be rapid, in many cases a cycle of evaluation and continuous improvement is needed (USEPA 2017).

10.1.7 Design for Environment

DfE is a design approach to reduce the overall human health and environmental impact of a product, process, or service, where impacts are considered across its life cycle. Different software tools have been developed to assist designers in finding optimized products or processes/services. DfE is also the original name of a USEPA program, created in 1992, that works to prevent pollution and the risk pollution presents to humans and the environment. The program provides information regarding safer chemical formulations for cleaning and other products. The USEPA renamed its program "Safer Choice" in 2017 (USEPA 2017).

Initial guidelines for a DfE approach were written in 1990 by East Meets West, a New York–based nongovernmental organization. It became a global movement targeting design initiatives and incorporating environmental motives to improve product design in order to minimize health and environmental impacts by incorporating it from design stage all the way to the manufacturing process. The DfE strategy aims to improve technology and design tactics to expand the scope of products. By incorporating eco-efficiency into design tactics, DfE takes into consideration the entire life cycle of the product, while still making products usable but minimizing resource use. The key focus of DfE is to minimize the environmental-economic cost to consumers while still focusing on the life-cycle framework of the product. By balancing both customer needs as well as environmental and social impacts, DfE aims to "improve the product use experience both for consumers and producers, while minimally impacting the environment" (Luttropp and Lagerstedt 2006).

10.1.8 Design for Regulatory Compliance

Purposeful, cost effective, safe product design supporting sustainable environment works as a catalyst for viability in engineering multiple applications with continued attention to reuse, flexibility in maintenance and environmental constraints. Engineering research analysis expands product applications within known legal parameters. Modest to major product engineering changes include interchangeable parts, moveable manufacture locations, use of by-products, and alternate packaging. All current methods of manufacture that are compliant and that keep appraised of anticipated regulatory changes use integrity in their engineering design and design support which lessens collateral damages.

10.1.9 Design for Testability

Product design quantitative analysis and qualitative analysis is integral in determining if a product is created. Knowledge of sources of raw materials, recycled materials, and shared production facilities play a part in production design testing and quality of product. Anticipation of competition gives rise to refining of methodology of production and new types of testing needed over time. Manufacture flexibility application tests for pilots and of extrapolated uses help redefine products. The worst case scenario is creating a product of such complexity that there are no competent tests available and thus creating a total waste product.

10.1.10 Design and Test for Service and Maintenance

Human use, manufacture use, and equipment use factors combine many different tests. Cultural tests are important from the name of a product to ease of use and repair. Reliability of product includes promised and desirability of length of service. If reuse probability is high and application is broad, then tests include upgrade factors, resource reuse factors, and development factors. An oil and filter change, or a car brake repair, or a flushing toilet repair package is likely to find a larger market if it is easily accessible, will fit unilaterally and has simple, correct directions for removal and installation. Large products must consider movement and workers available for maintenance specific to the product such as tower cranes.

It is desirable that a product is designed so that when the product breaks down, it is cheaper for consumers to fix it rather than to buy a new one. This would also contribute to a reduction in solid waste stream.

10.1.11 Design for Manufacturing

Processes used in the manufacture of products follow known production guidelines and adhere to environmental rules and regulations and the ease at which this is done includes the ability of the designer to incorporate already existing designs, refine and adapt product mixes, and to manipulate products to use designs in ingenious manners to create complimentary products using the same or improved tooling. Products meet strict monetary guidelines dedicated to responsible selection of materials, clean production; monitoring processes to avoid waste within local and global boundaries.

Design for manufacturing (also sometimes known as *design for manufacturability* or *DFM*) is the general engineering practice of designing products in such a way that they are easy to manufacture. The concept exists in almost all engineering disciplines, but the implementation differs widely depending on the manufacturing technology. DFM describes the process of designing or engineering a product in order to facilitate the manufacturing process in order to reduce its manufacturing costs. DFM will allow potential problems to be fixed in the design phase which is the least expensive place to address them. Other factors may affect the manufacturability such as the type of raw material, the form of the raw material, dimensional tolerances, and secondary processing such as finishing.

DFM process asks typical questions: Is the product easy to manufacture? Does it facilitate environmentally and socially responsible manufacturing by, for example, having components that can be cleaned without the use of toxic materials? Does it reduce production of products that fail quality standards, thus reducing manufacturing waste? From a sustainable engineering perspective, much of the environmental and social impacts of manufacturing can be ameliorated not just by proper manufacturing line design, but through good product design as well (Allenby 2012).

Depending on the various types of manufacturing processes, there are a set of guidelines for DFM practices. These DFM guidelines help to precisely define various tolerances, rules, and common manufacturing checks related to DFM. While DFM is applicable to the design process, a similar concept called *Design for Six Sigma* is also practiced in many organizations.

10.1.12 Design for Assembly

Manufacture for assembly can pay homage to work done in vehicle creation that originally used all three methods of assembly to produce high-quality products; human, mechanical, and a combination of human and mechanical assembly. The goal of production is to use thoughtful creation to produce a product with speed and without error which includes the use of human ergonomics, technical oversight of mechanical creation combined with regular maintenance. Assembly that doesn't overheat, freeze, or breakdown easily gets high marks in service longevity and higher profits.

10.1.13 Design for Disassembly

What is put together needs to be made such that it can come apart without harm. The areas of concern range from upgrades, to rebuilds of product; modifications and adaptations of machinery used to create products; flexibility of product manufacture to meet changing standards; and improved reduction of waste and cost of manufacture by being able to reverse processes.

When decisions are made to initially make a product, mock ups are made and pilots are put into production and the product is put to rigorous testing, engineering also takes on the job to refuse to duplicate one-case only scenarios, reliability includes the unexpected and how to fix it.

10.1.14 Design for Sustainable Manufacturing

Sustainable manufacturing can be defined as the creation of a manufactured product through a concurrent improvement in the resulting effect on factory and product sustainability. The concept of sustainable manufacturing demands a renewed design of production systems in order to condition the related sustainability on product life cycle and factory operations.

Designing sustainable production systems imply, on the one hand, the analysis and optimization of intra-factory aspects that are related to manufacturing plants. Such aspects can regard the resource consumption restrain, reuse resources discarded from previous products rather than to continue using virgin materials to improve the process efficiency, the ergonomics for the factory workers, the elimination of hazardous substances, the minimization of factory emissions and waste as well as internal emissions, the integrated management of information in the production facilities, and the technological updating of machines and plants.

Other inter-factories aspects concern the sustainable design of manufactured products, product chain dematerialization, management of the background and foreground supply chains, support of circular economy paradigm, and the labeling for sustainability.

10.1.15 Design for Sustainability

Design for sustainability (also called environmentally sustainable design, environmentally conscious design, etc.) is the philosophy of designing physical objects, the built environment, and services to comply with the principles of social, economic, and ecological sustainability.

The intention of sustainable design is to "eliminate negative environmental impact completely through skillful, sensitive design." Manifestations of sustainable design require renewable resources, impact the environment minimally, and connect people with the natural environment.

Great design is sustainable design: rather than considering green building design as an externality, architects must think about it as a set of principles for great design. This includes better user experience and comfort, doing more with less to enable the building to easily achieve peak performance, and maximizing the effectives of durable, quality materials. Rather than the well-known edict "form follows function," it's time to think "form follows environment."

Beyond the "elimination of negative environmental impact," sustainable design must create projects that are meaningful innovations that can shift behavior. A dynamic balance between economy and society intended to generate long-term relationships between user and object/service and finally to be respectful and mindful of the environmental and social differences (McLennan 2004).

Many more modules might be used, depending in part on the particular product and the specific design issues associated with it. Moreover, these sorts of product design processes require expertise in very different areas, which is one reason modern product design is usually done by teams (AIChE, IFS 2004; Allenby 2012; Sikdar 2003). For example, the law department would provide input on applicable regulations and potential liabilities that should be addressed, while the environment, health, and safety group (or corporate equivalent) would provide the technical expertise on environmental implications of product design and manufacturing options. Among engineering team members, it is likely that some would be experts in design of manufacturing processes, some in product design, some in materials science, and, increasingly, some experts in lifecycle assessment of process and product impact and performance. Note, however, that design is always a balancing process, with many different constraints and objectives; what emerges must reflect real-world conditions, rather than wishful thinking. Thus, for example, sustainability activities are free to demand that all toxics be eliminated from a complex product, but it is rare design indeed where engineers can do that and at the same time meet all other relevant conditions. Similarly, it is often the case that activist groups do not reflect the characteristics of all consumer segments in the marketplace; firms can lead, but at least in a free market they must also be responsive to consumer demand (Allenby 2012).

So what is the process that a product might go through before it actually gets to the market? Figure 10.1 maps out a typical route, with decision points, usually called "gates," noted as numbers (Allenby 2012; Graedel and Allenby 2003). At each gate, either a formal or informal review by the team responsible for the product, which might include not just the design team members, but people from marketing, sourcing, logistics, external affairs (an organization responsible for both representing the firm to external audiences and informing the firm about looming developments, such as material bans, that might affect the product), is held. The continued commercial viability of the product is assessed, given the product characteristics that are increasingly locked in as the design proceeds. The reason for such continued assessments is clear: the earlier an uncompetitive product is identified and killed, the less time, money, and resources are wasted. Thus, each gate not only informs the next stage of the design process, but is an important governance mechanism assuring robust design. From a sustainable engineering perspective, these gates become even more important, as they are existing mechanisms for managing the ever increasing demands of stakeholders and NGOs on firms, their products, and their activities around the world. A good gate process can be an important part of institutionalizing sustainable engineering. Conversely, an inadequate one – say, one that ignores the requirements of important markets, or of powerful activist stakeholders, or fails to identify important sustainability issues so that they may be addresses early in the design stage – can make a design less efficient, more expensive, and perhaps inadequate (Allenby 2012).

10.2 Integrating LCA in Sustainable Product Design and Development

One of the internal applications of LCA is in process or product design and development. Recently, an LCA-related tool called Life Cycle Product/Process Design (LCPPD) has started to emerge as an extension of life-cycle thinking and as an aid in sustainable product and process design and development. The LCPPD procedure is outlined in Figure 10.2.

LCA is used throughout the development procedure, initially with reference to an existing process or product. This holistic approach is dynamic, involving a continuous exchange of information within and outside the design team to explore systematically the possibilities for improvement. LCA, green chemistry, green engineering, design-for-environment, industrial ecology, and other emerging areas of chemical engineering and environmental engineering provide greater opportunities for developing sustainable and innovative processes and products (Azapagic 2000; Bare 2003; Bhander et al. 2003; Biswas et al. 1998; Brandão et al. 2010; Cooper 2003; Curran 2003, 2012, 2015; Das 2002; Das et al. 2001, 2003; Gonzalez and Smith 2003; ISO 1997; Linke et al. 2003; SETAC 1991, 1993; UNEP-SETAC 2000).

Once the main environmental impacts have been quantified, potential improvements are identified and the subsequent design focuses on these. The improvements are achieved through the selection of materials and technologies so as to minimize the environmental impacts but still satisfy such other parameters as efficiency, technical performance, costs, legislation, and the needs of customers and suppliers. When all of these requirements have been met, LCA is performed again, to identify and quantify the improvements that have been made.

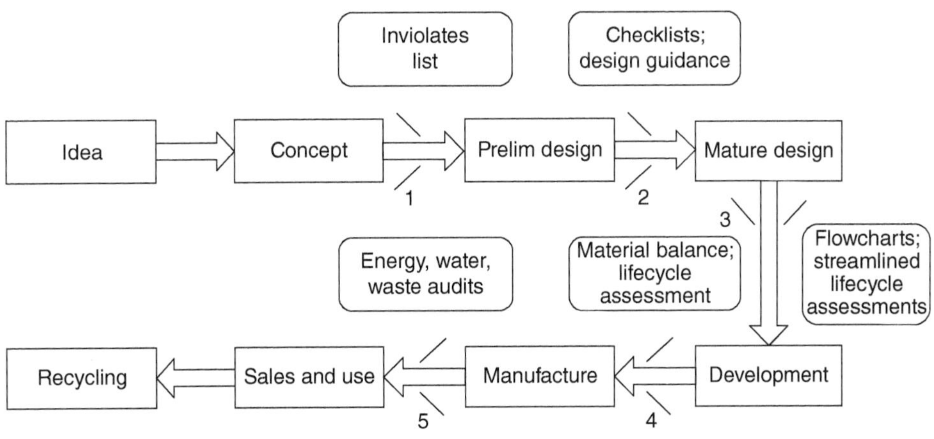

Figure 10.1 Environment and sustainability in the product realization process.

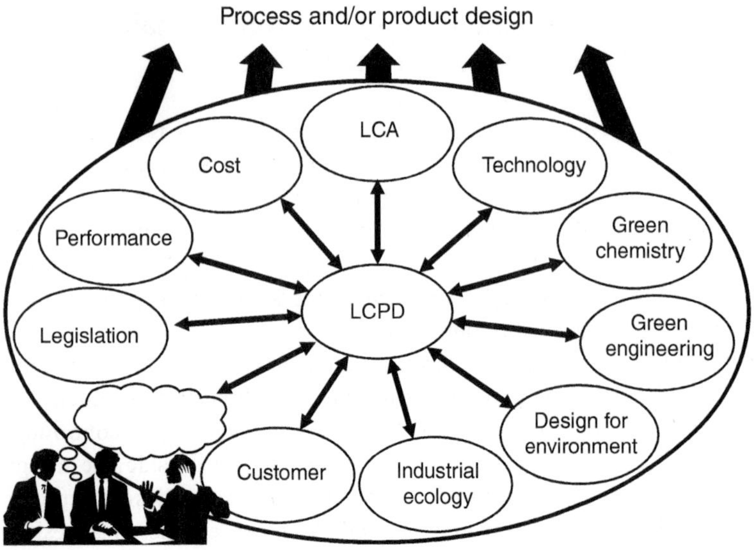

Figure 10.2 Methodological framework for life cycle process or product design as a tool for innovation.

Clearly, the LCA process is iterative, with a continuous exchange of information between the stakeholders, and yields a number of possibilities for improvements. Thus, LCPPD offers the potential for technological innovation in the product or process concept and structure through selecting the best material and process alternatives over the whole cycle. This can be of particular importance in the context of International Standards Organization (ISO) 14000, which require companies to maintain full awareness of the environmental consequences of their actions, both on and off site (Azapagic 2000).

Other applications are related to identifying possibilities for improvements in the environmental performance of an existing process or product or for the design of new ones. Studies made pursuant to ISO 14000, which are usually industry-specific, are mainly used for product or process innovation as well as to demonstrate the environmental progress of a company. Another use of LCA is for public policy making by governments.

Because of its holistic approach to system analysis, LCA is becoming an increasingly important decision-making tool in environmental system management. Its main advantage over other, site-specific methods for environmental analysis, such as environmental impact assessment or environmental audits, lies in broadening of system boundaries to include all burdens and impacts in the life cycle of a product or a process, not merely focusing on the emissions and wastes generated by a plant or manufacturing site.

LCA can be used both internally by a company or externally by industry, policy makers, planners, educators, and others with an interest in the outcome. If the results of LCA are to be used internally by a company, the possible areas where it can be useful include, but are not limited to, the following:

- Strategic planning or environmental strategy development
- Problem solving in the system
- Environmental reporting, auditing, and marketing
- Identification of opportunities for, and the tracking of, environmental improvements
- Process and product design, innovation, improvement, and optimization

The external applications of LCA include its use as a marketing tool, to support environmental labeling or claims, for educational or informational purposes, and to support policy decisions. LCA, like design for environment, industrial ecology, and other emerging areas of chemical engineering and environmental engineering provides greater opportunities for developing sustainable and innovative processes and products (Azapagic 2000; Bhander et al. 2003; Cooper 2003; Curran 2003, 2012, 2015; Das 2005a; Gonzalez and Smith 2003; Linke et al. 2003).

10.3 Green Chemistry: The Twelve Principles of Green Chemistry

Green chemistry is the process of designing of chemical products and chemical processes that reduces or eliminate the use and/or generation of hazardous substances. This process involves definition of the feed materials, reaction pathways, products, and reactor conditions to minimize the impact on the environment, including workers.

Green chemistry uses all the chemical principles and techniques at its disposal to prevent pollution at its source. Green Chemical Engineering does likewise but also includes the design and operation of processes as well as the design and manufacture of products to minimize pollution and the risks to human health and the environment. Progress has been made in the search for processes using fewer toxic chemicals and producing less waste while requiring less energy (Dewulf et al. 2000; Matlack 2003).

Green Chemistry and Green Chemical Engineering promote the goal of achieving the so-called triple bottom line of economic prosperity and continuity for corporations, social well-being for communities and employees, and environmental protection and resource conservation. Economic, financial, and environmental accounting and risk management tools are being developed to assess green technologies to include LCA (see Chapter 6), total cost assessment (TCA; see Chapter 5), and ecological risk assessment (ERA; see Chapter 4). The goal of achieving green and sustainable chemical manufacturing has been embraced by a number of government and nongovernmental organizations (Mathew 2003).

10.3.1 The Principles of Green Chemistry

The design and implementation of completely green products and processes is an enormous challenge. Because the number of chemical synthesis pathways is enormous, there is no one systematic, fail-safe method for ensuring that the chemistry being implemented is green. Indeed, it is more nearly correct to inquire if a proposed chemical manufacturing process is simply "greener" than other alternatives. Thus, green chemistry recognizes the importance of incremental improvements. Anastas and Warner (1998) formulated have a set of principles to define and guide the scope of green chemistry.

1) *Prevent waste*
   It is better to prevent waste than to treat or clean it up waste after it has been created. In the past, the Earth and its rivers, oceans, atmosphere, and soil have been considered infinite sinks for the discharge of pollution and waste. The command and control philosophy implies that it is acceptable to generate waste, as long as the waste is safely treated and disposed or stored. Green Chemistry's highest principle is that waste should not be created at all. From an economic standpoint, waste is a cost that requires paying twice: the first cost is for purchase of raw materials and the second is for management or disposal of the waste.
2) *Employ the atom economy when practical*
   Synthetic methods should be designed to maximize the incorporation of all materials used in the process within the final product. Thus, it is not sufficient to achieve even a 100% yield of the desired product if the synthesis also produces by-products. Rather, it is desirable to incorporate all of the atoms of the raw materials into the products. The atom economy, as discussed by Trost (1991), and atom utilization, a term coined by Sheldon (2000), embody the same concept with slightly different definitions. The atom economy is defined as the formula weight of the desired product divided by the stoichiometrically weighted formula weights of all reactants. The degree of atom utilization is the quotient of the formula weight of the desired product and the formula weight of all products and by-products. In situations where the exact composition of all by-products is difficult to determine, the atom economy is the more practical one to use. Sheldon (2000) also defines the E factor, which is the mass ratio of waste to desired product. E factors for bulk commodity chemicals are typically less than 5, and often less than 1. By contrast, fine chemicals and pharmaceuticals typically generate large amounts of waste per unit of product, with E factors perhaps exceeding 100. Neither the atom efficiency nor the E factor provides any quantitative measure of the potential for human harm of the waste products.
3) *Use the least hazardous chemical syntheses*
   Wherever practical, synthetic methods should be designed to use and generate substances that possess little or no toxicity to human health and the environment. This principle promotes the use of less toxic reagents and intermediates, and the generation of less toxic by-products. The use of less toxic feedstocks drives the development of, for instance, renewable raw materials. Better catalysts and chemical reactor designs are also critical to developing less hazardous syntheses.
4) *Design safer chemicals*
   Chemical products should be designed to achieve their desired function while minimizing their toxicity. This principle requires matching the desired function of a given chemical compound with its chemical structure. It also requires the ability to predict beforehand the possible toxic effects of a given chemical substance. Thus, the research and development of methods for establishing structure–property relationships promote this principle. In making the decision to develop and market a particular product, industry should go beyond considerations of the traditional performance properties (vapor pressure, color, stability, etc.) and add as performance metric such favorable environmental properties as reduction in toxicity and carcinogenicity.
5) *Use safer solvents and auxiliaries*
   The use of auxiliary substances, solvents, separation agents, and other additives should be made unnecessary

wherever possible and innocuous if they must be used. Safer solvents such as water, alcohols, supercritical carbon dioxide, or biodegradable surfactant solutions are preferable to chlorinated solvents. Separation processes like liquid extraction require the use of mass separation agents, but such alternatives as reactive separations through membrane reactors (see Chapter 8) may provide a more environmentally acceptable alternative.

6) *Design for energy efficiency*
The energy requirements of the various chemical processes should be recognized for their environmental and economic impacts, which should be minimized. If possible, synthetic methods should be conducted at ambient temperature and pressure. The chemical industry has substantially reduced its use of energy in manufacturing, starting with the oil embargo and energy shortages of the 1970s. Heat integration and other engineering tools are well developed in the chemical industry and can be applied within a given process. However, it is still true that a substantial portion of energy usage in manufacturing can be attributed to the chemical industry. Thus, advances that can fundamentally alter the required energy usage in manufacturing a product are a key aspect of green chemistry.

7) *Design for water optimization and integration*
Among the various species in a chemical and allied manufacturing process, water is critical and is widely used across several industries. Water is a relatively inexpensive and useful solvent which is easy to handle and presents minimal toxicity and safety concerns. In a typical chemical process, water may play several roles including the following:

- It may act as a solvent for reaction (such as in single or multiple phase reactions).
- It may be a solvent for separations (such as in decanters and liquid–liquid extraction systems).
- It may be used for heat removal operations (such as a noncontact stream in condensers).
- It may be used in postproduction operations (such as in washing operations wherein, the final product may be washed to remove undesirable solvents, by-products, and unreacted raw materials).
- It may be used in environmental systems (such as in scrubbers that handle gaseous emissions from vents and process units).
- It may be used in safety systems (such as to provide seals, heels, and in pump seal systems).
- It may be used to generate utilities in the process (such as steam generation).

Wastewater reduction and water conservation are becoming increasingly more important issues in process industries. More stringent environmental regulations, concerns over long-term health effects on humans and nature, and the future availability of "clean" water resources are just a few of the factors that are driving efforts toward improvements in water conservation and wastewater reduction in manufacturing processes. As these issues continue to receive intense government scrutiny and to raise concern among community-organized environmental groups, industries that fail to address the problems to the satisfaction of such outside stakeholders may find present operations threatened and the sustainability of future operations in doubt. These critical concerns have refocused efforts over the past decade toward identifying cost-effective wastewater reduction and water conservation process designs, involving direct recycle and reuse of water that can be implemented within a variety of manufacturing process industries. Unsustainable strain on freshwater sources globally could be reduced if industries reuse their wastewater after advanced physico-chemical-biological treatments.

8) *Use renewable feedstocks*
Raw material, or feedstock, should be renewable rather than depletable, whenever technically and economically practical. The process of manufacturing chemicals from agricultural feedstocks is known as Chemurgy. Fuels derived from biomass may include such cultivated ones as ethanol from corn or fuels from waste biomass. Natural polymers such as chitin and cellulose are receiving attention for their use in manufacturing engineered materials. Of particular interest is the use of CO_2 as a raw material for C_1 chemistry.

9) *Reduce or avoid the use of derivatives*
The unnecessary use of derivatives in blocking groups, for protection–deprotection, and the temporary modification of physical–chemical processes should be minimized or avoided when possible, because such steps require additional materials, increased solvent usage, and greater energy requirements. This principle is related to principles 2 and 5.

10) *Use catalytic reagents*
Catalytic reagents, employed as selectively as possible, are superior to stoichiometric reagents. The use of catalysts, whether heterogeneous or heterogeneous, is essential to realizing principles 2 and 3. Anastas et al. (2001) referred to the use of catalysis as a foundational pillar of green chemistry because catalysts can help achieve a number of green chemistry's goals, including decreased material usage, increased atom efficiency, and reduced energy demands.

11) *Design for degradation*
Chemical products should be designed so that at the end of their function they are broken down into innocuous

products that degrade and do not persist in the environment. This principle clearly intersects with the concept of DfE (see Section 10.1.7). Product design involves communication along a diverse chain of stakeholders: customers, sales representatives, engineers and materials scientists, and synthetic chemists and others. Green chemistry therefore requires close collaboration of chemists with these other stakeholders.

12) a) *Use real-time analysis to pollution prevention*
    Analytical methodologies need to be developed further to allow for real-time, in-process monitoring and control prior to the formation of hazardous substances. Waste production is not always the result of synthetic chemistry; sometimes it is simply the result of poor process monitoring and control. The analytical methods currently in use frequently rely on sampling the product at the end of the production line, which may be accomplished in minutes but also may last for hours. In such cases, by the time the product analysis has been completed it is typically too late to adjust the production process for optimum performance. For this reason, real-time process monitoring and control of temperature, pressure, feed rate, feed composition, and product composition are vital to green chemical engineering.

    b) *Choose inherently safer chemistry for accident prevention*
    Substances and the forms of them used in a chemical process should be chosen to minimize the potential for chemical accidents, including releases, explosions, and fires. There is some overlap of this principle with principle 5 regarding safer solvents and auxiliaries: for instance, supercritical carbon dioxide is an inherently nontoxic substitute for chlorinated solvents. Substitutes for lead compounds in paint and gasoline are also examples of principles 5 and 12(a). This principle also applies to strategies such as eliminating the bulk transport and storage of hazardous intermediates in favor of generating these intermediates on-site. While the ultimate goal of green chemistry would be to eliminate the hazardous material entirely, the possibility of a massive release of the hazardous compound is reduced or eliminated by generating it on-site instead of relying on bulk storage.

These principles address an important criterion necessary for achieving the goal of a green technology-preventing pollution on all fronts. This includes conserving materials (feedstocks, reagents, and solvent) and energy (decreased byproduct formation and increased conversion, which, in turn, minimizes the number of process steps), using improved catalysts or catalytic processes in place of noncatalytic ones, and designing safer chemicals and chemical reactions.

As a new technology surpasses initial bench-level development, the researcher should address the 12 additional principles of green chemistry, introduced by Winterton (2001), as follows:

1) Identify byproducts; quantify if possible
2) Report conversions, selectivities, and productivities
3) Establish a full mass balance for the process
4) Quantify catalysis and solvent losses
5) Investigate basic thermochemistry to identify exotherms (safety)
6) Anticipate other potential mass and energy transfer, limitations
7) Consult a chemical or process engineer
8) Consider the effect of the overall process on choice of chemistry
9) Help develop and apply sustainable measures
10) Quantify and minimize use of utilities and other inputs
11) Recognize where operator safety and waste minimization may be incompatible
12) Monitor, report, and minimize wastes emitted to air, water, and solids from experiments or process.

10.4 The Hannover Principles

10.4.1 Leadership in Energy and Environmental Design (LEED)

The Leadership in Energy and Environmental Design (LEED) program is an internationally recognized building certification program for architects, contractors, owners, and operators (see Appendix H). It provides a framework for building design, construction, operation, and maintenance to achieve reduction in environmental impacts and improvement in conditions for building occupants. To guide design of building sustainability, design criteria from the U.S. Green Building Council's (USGBC) LEED certification program will be used (USGBC 2011a).

LEED has evolved since 1998 to more accurately represent and incorporate emerging green building technologies. The pilot version, LEED New Construction (NC) v1.0, led to LEED NCv2.0, LEED NCv2.2 in 2005 (USGBC 2011b), and LEED 2009 (previously named LEED v3) in 2009. LEED v4 was introduced in November 2013. Until 31 October 2016, new projects could choose between LEED 2009 and LEED v4. New projects registering after 31 October 2016 have been required to use LEED v4.

The LEED rating system is meant for new and existing commercial, institutional, and residential buildings, and also for entire neighborhood developments. There are currently over 800 LEED-qualified buildings in the United States (USGBC 2011c). There are separate rating systems for different building types and projects.

LEED 2009 encompasses 10 rating systems for the design, construction, and operation of buildings, homes, and neighborhoods. Five overarching categories correspond to the specialties available under the LEED professional program. That suite currently consists of the following:

1) Green building design and construction
   - LEED for new construction
   - LEED for core and shell
   - LEED for schools
   - LEED for retail: new construction and major renovations
   - LEED for healthcare
2) Green interior design and construction
   - LEED for commercial interiors
   - LEED for retail: commercial interiors
3) Green building operations and maintenance
   - LEED for existing buildings: operations and maintenance
4) Green neighborhood development
   - LEED for neighborhood development
5) Green home design and construction

The LEED for homes rating system is different from LEED v3, with different point categories and thresholds that reward efficient residential design.

Each rating system listed includes five environmental categories: sustainable sites, water efficiency, energy and atmosphere, materials resources, and indoor environmental quality. Two additional non-environmental human effects categories are also included in the rating system: innovation in design and regional priority.

In the LEED rating system, each category is subdivided into "credit" subcategories, and one or more points are assigned to each credit subcategory. Total possible points; however, 10 additional points can be awarded for the innovation in design and regional priority categories. The allocation of points among the credit subcategories is based on an analysis of potential environmental impacts and human benefits relative to a set of impact indicators. These impact indicators include GHG emissions, fossil fuel use, toxins, carcinogens, air and water pollutants, and indoor environment. The assessment of potential environmental impacts uses the TRACI impact assessment model from the USEPA (see Sections 6.3.12, 6.4.3.7, and 6.4.3.8) and then applies additional weightings to each subcategory credit from the National Institute of Standards and Technology (NIST), similar to the way that weightings are applied in LCA (see Chapter 6). The decision on how many points to assign to each credit subcategory is based on a consensus process with a committee of sustainability experts.

The USGBC LEED Program certifies that building projects achieve certain levels of desired outcomes according to a numerical scale. For example, in retail applications, the following point scale applies to both design and construction phases:

10.4.1.1 Certification Level

Buildings can qualify for four levels of certification:

1) Certified: 40–49 points
2) Silver: 50–59 points
3) Gold: 60–79 points
4) Platinum: 80 points and above

The distribution of these points among the major building categories is dependent on application, but insights into the working of the LEED rating systems can be gained by closer inspection of one application: retail buildings (LEED 2010). Here the points are distributed as follows:

Sustainable sites	26 points
Water efficiency	10 points
Energy and atmosphere	35 points
Material resources	14 points
Indoor environmental quality	15 points
Total	100 points

The greatest number of points, and therefore the greatest emphasis in green building design and operation, relates to energy consumption/atmosphere protection and sustainable sites.

10.5 Sustainable Industries and Business

The approach for sustainable industries and business or industrial sustainability primarily advocates consideration of economics, environment, ecology, equity, and ethics in decisions made in business context. While the focus is on efficient "lean" manufacturing and operations that save materials, energy, and money, care must be taken to ensure that the focus is not only on financial sustainability from operations standpoint. In the broader context of general business, corporate social responsibility (CSR) is a growing concept. It frequently is indistinguishable from similar concepts such as corporate sustainable development, corporate responsibility, and corporate citizenship. Some people believe CSR to be the private sector's way of integrating the environmental, economic, and social imperatives of their activities, resembling the concept of the triple bottom line.

Sustainable business, or green business, is an enterprise to be that has no negative impact on the global or local environment, community, society, or economy – a business that strives to meet the triple bottom line. Often, sustainable businesses have progressive environmental and human rights policies. In general, business is described as green if it matches the following four criteria:

1) It incorporates principles of sustainability into each of its business decisions.
2) It supplies environmentally friendly products or services that replace demand for nongreen products and/or services.
3) It is greener than traditional competition.
4) It has made an enduring commitment to environmental principles in its business operations. A sustainable business is any organization that participates in environmentally friendly or green activities to ensure that all processes, products, and manufacturing activities adequately address current environmental concerns while maintaining a profit. In other words, it is a business that "meets the needs of the present without compromising the ability of the future generations to meet their own needs." It is the process of assessing how to design products that will take advantage of the current environmental situation and how well a company's products perform with renewable resources.

The Brundtland Report emphasized that sustainability is a three-legged stool of people, planet, and profit. Sustainable businesses with the supply chain try to balance all three through the triple-bottom-line concept – using sustainable development and sustainable distribution to affect the environment, business growth, and the society (Cooney 2009; Rennie 2008a, b; WCED 1987).

10.5.1 Eco-Efficiency

The term eco-efficiency was coined and defined by the World Business Council for Sustainable Development (WBCSD 1992, 2000) in its 1992 publication "changing course." It is based on the concept of creating more goods and services while using fewer resources and creating less waste and pollution. Eco-efficiency is achieved through the delivery of "competitively priced goods and services that satisfy human needs and bring quality of life while progressively reducing environmental impacts of goods and resource intensity throughout the entire life-cycle to a level that is compatible with the Earth's estimated carrying capacity" (Carvalho et al. 2017).

10.5.2 Sustainable Supply Chain Systems

Products such as aircraft, automobile, and computers, which involve a series of discrete manufacturing and assembly processes along the supply chain, can generate a significant ecological footprints in terms of material, energy, and land use as well as industrial wastes and emissions. Manufacturers have used a variety of design techniques to reduce the resource intensity of their manufacturing and logistical systems. These techniques include simplifying product architecture to reduce the number of distinct parts and assembly operations (which also enables easier maintenance and disassembly); utilizing recycled materials or refurbished components; avoiding substances with undesirable properties such as carcinogenicity, toxicity, flammability, ozone depletion, or environmental persistence; reducing electrical and thermal energy use through process modifications or transportation efficiency (e.g. improved product geometry for pallet loading); utilizing quality improvement and "lean" manufacturing techniques (e.g. just-in-time inventory replacement) to reduce work-in progress and scrap; and collaborating with customers and suppliers to streamline the supply chain and minimize waste (Bag et al. 2017; Clift 2003; Clift and Wright 2000).

10.5.3 Sustainable Green Economy

Everyone affects the sustainability of the marketplace and the planet in some way. Sustainable development within a business can create value for customers, investors, and the environment. A sustainable business must meet customer needs while, at the same time, treating the environment well. The green economy presents an alternative vision for growth and development in which economic growth and improvements in people's lives are generated in ways consistent with sustainable development. While there are varying visions of what the green economy encompasses, this concept has piqued the interest of policymakers and businesses alike because it presents the possibility of new opportunities for economic growth. Green investment firms are consequently attracting unprecedented interest. In the United Kingdom, for instance, the Green Investment Bank is devoted exclusively to supporting renewable domestic energy. However, the United Kingdom and Europe as a whole are falling behind the impressive pace set by developing nations in terms of green development (Das 2005f; The CERES Principles n.d.; Harvey 2011). Thus, green investment firms are creating more and more opportunities to support sustainable development practices in emerging economies. By providing micro-loans and larger investments, these firms assist small business owners in developing nations who seek business education, affordable loans, and new distribution network for their "green" products.

10.6 Six Essential Characteristics

Tueth (2010) reiterates the ideas put forward by authors such as Paul Hawken (*The Ecology of Commerce* and *Natural Capitalism*), Bill McDonough and Michael Braungart (*Cradle to Cradle: Remaking the Way We Make Things*), and Janine Benyus (*Biomimicry*) when he proposes that a mature and authentic sustainable business contains six essential characteristics.

1) *Triple top-line (TTL) value production*
   "The TTL establishes three *simultaneous requirements* of sustainable business activities – financial benefits for the company, natural world betterment, and social advantages for employees and members of the local community – with each of these three components recognized as equal in status." Whereas many businesses use the triple bottom line, "triple top line" stresses the importance of initial design and is a term attributable to Braungart and McDonough (2002).
2) *Biomimicry – nature-based knowledge and technology*
   "This biomimicry-based principal involves the conscious emulation of natural-world genius in terms of growing our food, harnessing our energy, constructing things, conducting business healing ourselves, processing information and designing our communities" (Benyus 1997).
3) *Products of service to products of consumption*
   "*Products of service* are durable goods routinely leased by the customer that are made of technical materials and are returned to the manufacturer and reprocessed into a new generation of products when they are worn out. (These products are mostly nontoxic to human and environmental health but toxic materials that are used will be kept within a closed loop type system and not be able to escape into the environment). *Products of consumption* are shorter lived items made only of biodegradable materials. They are broken down by the detritus organisms after the products lose their usefulness. (These are also nonhazardous to human or environmental health). This principal requires that we manufacture only these two types of products and necessitates the gradual but continual reductions of products of service and their replacement with products of consumption as technological advancements allow" (Braungart and McDonough 2002).
4) *Solar, wind, geothermal and ocean energy*
   "This principal advocates employing only sustainable energy technology – solar, wind, ocean and geothermal – that can meet our energy needs indefinitely without negative effects for life on earth." Other authors have referred to this as utilizing current solar income (Hawken 1999).
5) *Local-based organizations and economies*
   "This ingredient includes durable, beautiful and healthy communities with locally owned and operated businesses and locally managed nonprofit organizations, along with regional corporations and shareholders working together in a dense web of partnerships and collaborations" (Tueth 2010).
6) *Continuous improvement process*
   "Operational processes inside successful organizations include provisions for constant advancements and upgrade as the company does its business. The continuous process of monitoring, analyzing, redesigning and implementing is used to intensify TTL value production as conditions change and new opportunities emerge" (Tueth 2010).

Business sustainability also requires firms to adhere to the above referred principles of sustainable development (WCED 1987). So, for industrial development to be more sustainable, it must address important issues at the macro level, such as economic efficiency (innovation, prosperity, productivity), social equity (poverty, community, health, and wellness), human rights, and environmental accountability (climate change, land use, and biodiversity). There are number of best practices that foster business sustainability, and help organizations move along the path from laggards to leaders. These practices include the following:

- Stakeholder engagement: Organizations can learn from customers, employee, and their surroundings community. Engagement is not only spreading out messages, but understanding opposition, finding common ground, and involving stakeholders in joint decision-making.
- Environmental management systems: These systems provide the structures and processes that help embed environmental efficiency into a firm's culture and mitigate risks. The most widely recognized standard worldwide is ISO 14001, but numerous other industry-specific and country-specific standards exist.
- Reporting and disclosure: Measurement and control are at the heart of instituting sustainable practices. Not only can organizations collect and collate the information, they can also be entirely transparent with outsiders. The Global Reporting Initiative is one of many examples of well-organized reporting standards.
- LCA: Those organizations wanting to take a large leap forward should systematically analyze the environmental and social impact of the products, processes, or activities or services they use and produce through LCA, which measure more accurately impacts (Das 2005d).

10.7 Social Services

Organizations that give back to the community, whether through employees volunteering their time or through charitable donations are often considered to be socially sustainable. Organizations also can encourage education in their communities by training their employees and offering internships to younger members of the community. Practices such as these increase the education level and quality of life in the community. In order for a business to be truly sustainable, it must sustain not only the necessary environmental resources but also its social resources, including employees, customers (the community), and its reputation (Earth Charter Principles 2019; Hahn 2008).

10.8 Environmental Regulatory Law: Command and Control Market Based, and Reflexive

Environmental law and policy is critically important and endlessly fascinating. To see its importance, just imagine the United States today without federal regulation of air and water quality, pesticide use, and solid and hazardous waste disposal. No matter how far the nation has come in addressing environmental issues, there remains much to do in terms of environmental sustainability. Federal and state laws have only begun to address such important environmental issues as global climate change, point and nonpoint sources of water pollution, ambient and indoor air pollution, groundwater pollution and groundwater depletion, and fisheries collapse. Scientific uncertainty, market failures, risk tradeoffs, cognitive biases, and fundamental policy disagreements make it extremely difficult to agree on solutions to many of these remaining problems. International pollution issues such as controlling emissions of global warming gases, climate changes, ozone depletion, and transboundary waste disposal are to be integrated with their domestic counterparts. Moreover, the increasingly controversial trade and environmental debates concerning WTO, GATT, and NAFTA, as well as the international trade in hazardous chemicals issues need to be explored. Environmental law and policy already has picked the low hanging fruit. The remaining problems continue to challenge even the wisest and most patient policy experts (Salzman and Thompson 2010).

Orts (1995) lists three types of environmental regulatory laws: command-and-control, market based, and reflexive. Each of these types of regulation aims to protect the environment, but each does so by reinforcing and subordinating different ethical and economic values and relationships. Commands and control law sets standards for pollution and imposes fines and penalties for companies that exceed those standards. From a regulator's point of view, this type of regulation can be seen as following the ethical mandate to strictly enforce the public's desire for a clean environment. However, from the point of a business person this type of regulation can be oppressive and unrealistic, and violates the freedom of business managers and owners. Orts states four objections to the command and control approach to environmental law.

- "Command-and-control depends too much on administrative agencies which are vulnerable to political influence." For example, environmental protection can be severely weakened if the President appoints directors who do not care about the environment.
- "Command and control is too static." Technology changes and regulatory agencies cannot keep pace. Another problem is how to set the proper standards for pollution. How do we know what is the right amount of pollution?
- "Command and control becomes too complex and unwieldy." Regulatory agencies churn out more and more rules, until it becomes virtually impossible to keep track of them.
- "Command and control is too harsh and punitive." Instead of rewarding and encouraging good environmental behavior, it punishes bad environmental behavior.

Orts (1995) notes that there are some instances in which command and control regulations are in order. They work, for example, if we are dealing with a highly toxic material, such as mercury, that can kill human beings in very small doses. The value of protecting human life clearly outweighs our freedom to dispose of mercury in landfills or water ways. In other areas, such as controlling emissions of coal-fired utility plants, Orts notes that the command-and-control approach has given way to a market-based approach. This approach reinforces the value of autonomy and freedom, but it has limitations as well. Orts raises concerns about four issues.

1) "Pollution charges and taxes." Companies must pay for the privileges of polluting the environment. This can make markets more efficient. One problem with this approach could be how to define the proper charges and tax levels.
2) "Expanding property rights to include the natural environment." If we can define property rights for endangered environments, we can use existing controls, such as the courts, to protect the environment. Unfortunately, some environmental problems, such as air pollution, are not easily recast in terms of property rights.
3) "Tradable pollution rights." Companies can trade the pollution rights they buy on the open market, allowing

pollution to be controlled in the most efficient way. This can work for oligopolies, where the traders are few. It would be unwieldy for dry cleaners, though.

4) "Environment marketing regulation." Government or nongovernment organizations certify products as environmentally friendly. This allows consumers to buy those products that harm the environment less than other products. There are at least two problems with this. First, how do we set standards for "environmentally friendly products"? Second, this will work only if consumers are willing to select products on the basis of the environment.

Despite their problems, market-based approaches try to work within common business incentives. By allowing businesses to buy and trade pollution rights, the environment is treated more like other factors of production. It also allows managers more freedom to tailor their environmental plans to their particular production and location characteristics, and it does so without the extreme restrictions on managerial freedom of the command-and-control. Reflexive environmental law tries to work within the institutional environment not only of business, but also of the society in which business operates. Reflexive law tries to set up an institutional environment that reinforces autonomy and a clean environment.

10.9 Business Ethics

Ethics is the study of whatever is right and good for humans, while business ethics are some business actions in light of some aspect of human value that requires the evaluation of business practice. Business practice goes beyond facts to include the "ought to" of a situation (Donaldson and Werhane 1979). Business ethics (also corporate ethics) is a form of applied ethics or professional ethics that examines ethical principles and moral or ethical problems that arise in a business environment. It applies to all aspects of business conduct and is relevant to the conduct of individuals and entire organizations. Business ethics reflects the philosophy of business, one of whose aims is to determine the fundamental purposes of a company. If a company's purpose is to maximize shareholder returns, then sacrificing profits to other concerns is a violation of its fiduciary responsibility. Corporate entities are legally considered as persons in the United States and in most nations. The "corporate persons" are legally entitled to the rights and liabilities due to citizens as persons.

Governments use laws and regulations to point business behavior in what they perceive to be beneficial directions. Ethics implicitly regulates areas and details of behavior that lie beyond governmental control. The emergence of large corporations with limited relationships and sensitivity to the communities in which they operate accelerated the development of formal ethics regimes.

Very often it is held that business is not bound by any ethics other than abiding by the law. Milton Friedman is the pioneer of the view. He held that corporations have the obligation to make a profit within the framework of the legal system, nothing more. Friedman made it explicit that the duty of the business leaders is "to make as much money as possible while conforming to the basic rules of the society, both those embodied in the law and those embodied in ethical custom." Ethics for Friedman is nothing more than abiding by "customs" and "laws." The reduction of ethics to abidance to laws and customs, however, have drawn serious criticisms (Dienhart 2000; Enderle 1999; Friedman 1970; NSPE 2019).

10.9.1 The Two Traditional Issues Involved with Ethics

1) Ethical relativism (ER) – Universal values that apply to everyone or is everything relative to individual, country, company, etc. (Donaldson and Werhane 1979).
   a) Relates to cultural relativism. This presumes that different people reason about morality varies by culture, education, and religious traditions.
   b) Arguments for ER are as follows:
      i) Empirical evidence of cultural relativism.
      ii) There is no viable universal standard that can be applied to everyone.
   c) Arguments against ER are as follows:
      i) Just because finding universals is hard, that does not imply that ER is correct.
      ii) Just because a particular issues is not resolved does not imply that it is, in principle, not ever resolvable.
      iii) Taking ER to its full extent means that you can't justify any moral judgments at all.
   d) The Foreign Corrupt Practices Act (1977) attempted to legislate that what we say is moral is how we should behave. This Act was driven by the issue that you are being inconsistent with yourself if you say something is immoral yet do it anyway just because you are in a foreign country.
2) Truth telling – Can you ever justify not telling the truth? Must we communicate honestly?
   a) A vigorous defender of truth telling is Kant. Truth telling is an essential feature of right communication (Lectures on Ethics). His reasoning is as follows:
      i) We all want others to follow this when speaking to us.

ii) All societies depend on mutual bonds of honesty and truthfulness to enforce their continued existence.
iii) Lying thwarts the discovery of new truths. Knowledge growth is required for the advancement of a society.

b) Counter agreements to this are as follows:
i) Everyone understands the game of inflating claims (not lying) in advertising etc. so it is all right.
ii) You can't ever know the perfect truth around a product/service because perfect information is not available – people can't know if a company is lying or bluffing.
iii) Outside of the advertising arena: How do you conduct a business negotiation if you never "put something on the table" that gives you room to wiggle so to speak?

10.10 International Issues

While business ethics emerged as a field in the 1970s, international business ethics did not emerge until the late 1990s, looking back on the international developments of that decade. Many new practical issues arose out of the international context of business. Theoretical issues such as cultural relativity of ethical values receive more emphasis in this field. Other, older issues can be grouped here as well. Issues and subfields include the following:

- The search for universal values as a basis for international commercial behavior.
- Comparison of business ethical traditions in different countries. Also on the basis of their respective GDP and (corruption rankings).
- Comparison of business ethical traditions from various religious perspectives.
- Ethical issues arising out of international business transactions; e.g. bioprospecting and biopiracy in the pharmaceutical industry; the fair trade movement; transfer pricing.
- Issues such as globalization and cultural imperialism.
- Varying global standards – e.g. the use of child labor.
- The way in which multinationals take advantage of international differences, such as outsourcing production (e.g. clothes) and services (e.g. call centers) to low-wage countries.
- The permissibility of international commerce with pariah states.

The success of any business depends on its financial performance. Financial accounting helps the management to report and also control the business performance.

The information regarding the financial performance of the company plays an important role in enabling people to take right decision about the company. Therefore, it becomes necessary to understand how to record based on accounting conventions and concepts ensure accurate records keeping.

Foreign countries often use dumping as a competitive threat, selling products at prices lower than their normal value. This can lead to problems in domestic markets. It becomes difficult for these markets to compete with the pricing set by foreign markets. In 2009, the International Trade Commission has been researching antidumping laws. Dumping is often seen as an ethical issue, as larger companies are taking advantage of other less economically advanced companies.

10.11 Ethical Sustainability

The five E's in sustainable development are economy, environment, efficiency, equity, and ethics, as shown in Figure 10.3. Therefore, sustainable development is an ethical concept, like "justice." We need a new vision of sustainability, a vision that focuses on understanding how sustainability is a kind of relationship, a relationship between society and the environment. All human relationships, good and bad alike, involve attitude and action. The essence of sustainability is to both develop a mature ethical attitude toward nature and a mature physical relationship with nature, which involves exploiting nature in an appropriate manner.

Technology increases our ability to exploit the environment and the efficiency of exploitation. But it does not determine how we ought to exercise that ability and efficiency. In the 1970s, technology and economic incentives led to more efficient home heating and insulation. What did we do with that ability? Use less energy? No, we built bigger houses because we could heat bigger houses more affordably. The thermal efficiency is the traditional domain of engineering and much of the activities are constrained by scientific laws, including thermodynamics. Through technical ingenuity and enhancing efficiencies within the existing economic system and achieving greater efficiencies in materials and energy use are important steps toward sustainable development. Equity on the other hand ensures a better quality of life for everyone, now and generations to come.

Society is a ship whose engine is technology and rudder is ethics; history bears plenty of witness to the wrecks caused by technologies that developed ahead of ethics. There is no reason to think sustainability is any different. Technology is important, but not as important as understanding how sustainability is first and foremost a relationship: a relationship with the environment; a relationship where ethics (or lack

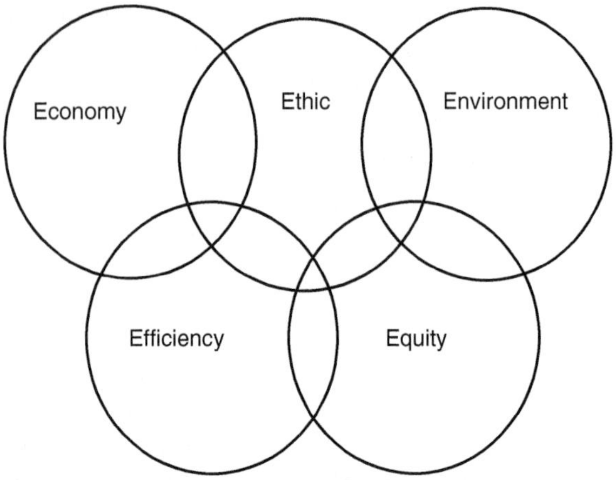

Figure 10.3 Five E's of sustainable development is an ethical concept.

of attention to ethics) is the predominant influence on how we exploit nature (Common Dreams 2010).

10.12 Social Sustainability

Sustainability issues are being expressed in scientific, environment, economic, and business terms, as well as in ethical terms of stewardship, but implementing change is a social challenge that entails, among other things, national and international law, urban planning and transport, local and individual lifestyles, and ethical consumerisms. The relationship between human rights and human development, corporate power and environmental justice, global poverty and citizen action suggests that responsible global citizenship is an inescapable element of what may at first glance seem to be simply matter of personal consumer and moral choice (Blewitt 2008).

According to the Western Australia Council of Social Services (WACOSS) (Sen 2013):

> Social sustainability occurs when the formal and informal processes; systems; structures; and relationships actively support the capacity of current and future generations to create healthy and livable communities. Socially sustainable communities are equitable, diverse, connected and democratic and provide a good quality of life.

It has the following dimensions (Anand and Sen 1996):

- *Equity* – the community provides equitable opportunities and outcomes for all its members, particularly the poorest and most vulnerable members of the community
- *Diversity* – the community promotes and encourages diversity
- *Interconnected/social cohesions* – the community provides processes, systems, and structures that promote connectedness within and outside the community at the formal, informal, and institutional level
- *Quality of life* – the community ensures that basic needs are met and fosters a good quality of life for all members at the individual, group, and community level (e.g. health, housing, education, employment, safety)
- *Democracy and governance* – the community provides democratic processes and open and accountable governance structures.
- *Maturity* – the individual accepts the responsibility of consistent growth and improvement through broader social attributes (e.g. communication styles, behavioral patterns, indirect education, and philosophical explorations)

Figure 10.4 portrays the major pathways in building sustainable development in an emerging green economy in which the social sustainability is one of the essential components, along with the economical, ethical, environmental, efficient, and equitable sustainability.

As a society, we make social investments and we have a "stock" of social and human resources. Economic development can either contribute to or deplete those social resources. Many would argue that the form of economic development championed by Thatcher and Reagan has been socially unsustainable, depleting human and social capital and resources in addition to the damage it has wrought to the natural environment. The concept of socially sustainable development, including socially sustainable urban development, has received less attention than the concept of envi-

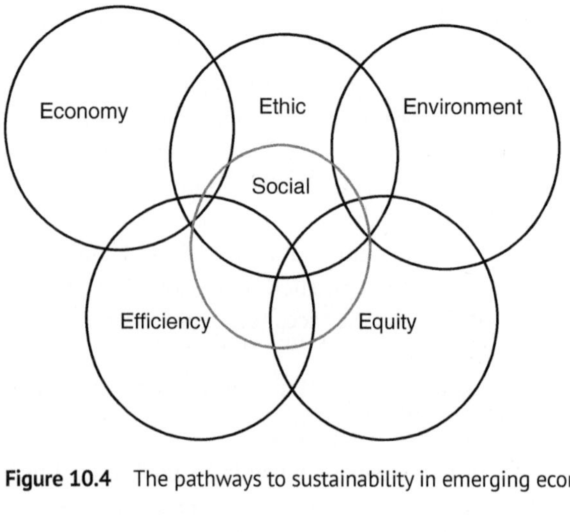

Figure 10.4 The pathways to sustainability in emerging economy.

ronmentally sustainable development (CPHA 1992). What would constitute socially sustainable development?

Some would argue that it is development that it

- meets basic needs for food, shelter, education, work, income, and safe living and working conditions;
- is equitable, ensuring that the benefits of development are distributed fairly across society;
- enhances, or at least does not impair, the physical, mental, and social well-being of the population;
- promotes education, creativity, and the development of human potential for the whole population;
- preserves our cultural and biological heritage, thus strengthening our sense of connectedness to our history and environment;
- promotes conviviality, with people living together harmoniously and in mutual support of each other;
- is democratic, promoting citizen participation and involvement, and
- is livable, linking "the form of the city's public places and city dwellers' social, emotional and physical well-being."

The systems and processes that we put in place to achieve these ends can be thought of as the "soft infrastructure" of the community, a term used by Len Duhl, Professor of Public Health and Professor of Urban Planning at the University of California at Berkeley, to describe those elements of the community that contribute to social well-being. This "soft" infrastructure includes formal human services (health, education, social services, recreation and culture, etc.) as well as the community's informal structure the web of voluntary organizations and social relationships that comprise community. Urban planning needs to integrate these elements into all its work, giving as much weight to the soft infrastructure as to the hard infrastructure if we are going to create communities that work.

10.13 Conclusions

- LCA is an effective tool to characterize environmental sustainability of a product, or a process, or an activity.
- Industrial ecology and zero discharge are key methodologies for future for many manufacturing, and chemical and allied processes. Central to an eco-industrial parks ability to exchange wastes and materials will be conversion or separation technologies capable of preparing former wastes for other uses.
- Future concern for the environment will grow, resources will become more scarce and less reliable, and demands for quality and cost competitiveness will increase. To meet these challenges a new paradigm of business and industrial organizations ask individual communities, companies, and other constituents to consider how they might work together to use resources more wisely.
- It is hoped, perhaps demanded, that the sustainable economy of this evolving century will be characterized by a conservation and decrease in the materials and energy content of products and an increase in their knowledge content.
- The challenge for businesses and governments will be to ensure that continued economic development is environmentally, ecologically, ethically, and socially sustainable. Areas in which government direction is needed include changing consumption and production pattern; appreciation for cultural diversity and traditions, development of healthy social systems; and widespread education and awareness of sustainability and sustainable development.
- Some experts have described sustainable development as the main driving force for the twenty-first century industry. Engineers, scientists, business and industry owners, lawyers, governments and policymakers, and educators all will play key roles in this endeavor.
- Sustainability and sustainable development is a journey, not an endpoint.

10.13.1 Business

The most broadly accepted criterion for corporate sustainability constitutes a firm's efficient use of natural capital. This eco-efficiency is usually calculated as the economic value added by a firm in relation to its aggregated ecological impact. This idea has been popularized by the WBCSD under the following definition: "Eco-efficiency is achieved by the delivery of competitively priced goods and services that satisfy human needs and bring quality of life, while progressively reducing ecological impacts and resource intensity throughout the life cycle to a level at least in line with the earth's carrying capacity" (DeSimone and Popoff 2000).

Similar to the eco-efficiency concept but so far less explored is the second criterion for corporate sustainability. Socio-efficiency describes the relation between a firm's value-added and its social impact. Whereas it can be assumed that most corporate impacts on the environment are negative (apart from rare exceptions such as the planting of trees) this is not true for social impacts. These can be either positive (e.g. corporate giving, creation of employment) or negative (e.g. work accidents, mobbing of employees, human rights abuses). Depending on the type of impact socio-efficiency thus either tries to minimize negative social impacts (i.e. accidents per value added) or maximize positive social impacts (i.e. donations per value added) in relation to the value added.

Both eco-efficiency and socio-efficiency are concerned primarily with increasing economic sustainability. In this process they are instrumentals, both natural and social capital aiming to benefit from win-win situations. However, as Dyllick and Hockerts (2002) point out the business case alone will not be sufficient to realize sustainable development. They point toward eco-effectiveness, socio-effectiveness, sufficiency, and eco-equity as four criteria that need to be met if sustainable development is to be reached.

CASI Global, New York, states: "CSR & Sustainability together lead to sustainable development. CSR as in corporate social responsibility is not what you do with your profits, but is the way you make profits. This means CSR is a part of every department of the company value chain and not a part of HR/independent department. Sustainability as in effects towards Human resources, Environment and Ecology has to be measured within each department of the company."

10.13.2 Corporate Sustainability

Corporate sustainability is an approach that creates long-term stakeholder value by implementing a business strategy that considers every dimension of how a business operates in the ethical, social, environmental, cultural, and economic spheres. It also formulates strategies to build a company that fosters longevity through transparency and proper employee development.

Corporate sustainability is an evolution on more traditional phrases describing ethical corporate practice. Phrases such as CSR or corporate citizenship continue to be used but are increasingly superseded by the broader term *corporate sustainability*. Unlike phrases that focus on "added-on" policies, *corporate sustainability* describes business practices built around social and environmental considerations.

The phrase is derived from two keys sources. The Brundtland Commission's Report, *Our Common Future* (WCED 1987), described sustainable development as "development that meets the needs of the present without compromising the ability of future generations to meet their own needs." This desire to grow without damaging future generations' prospects is becoming more and more central to business philosophies. Within more academic management circles, Elkington (2004) developed the concept of the Triple Bottom Line which proposed that business goals were inseparable from the societies and environments within which they operate. While short-term economic gain could be chased, a failure to account for social and environmental impacts would make those business practices unsustainable.

Measuring corporate sustainability is possible through composite indicators which aggregate environmental, social, corporate governance, and economic measures, e.g. complex performance indicator (Shaker 2015).

10.14 Strategy for Corporate Sustainability

10.14.1 Business Case for Sustainability

The challenge for many businesses in this new field is to quantify the positive impacts of sustainability. Sustainability can increase revenue, reduce energy expenses, reduce waste expenses, reduce materials and water expenses, increase employee productivity, reduce hiring and attrition expenses, and reduce strategic and operational risks (Blewitt 2014). Furthermore, sustainable business practices may attract talent and generate tax breaks (Kahle and Gurel-Atay 2013; Welford 1997).

10.14.2 Transparency

Transparency deals with the idea that by having an engaging and open environment in the company as well as the community will improve performance and increase profits. It is an open culture that promotes employee involvement in the innovation and creative processes. Reaching out to the community creates a much bigger team, is extremely cheap, and provides evaluation from all angles. Companies are looking inward and realizing changes must be made to fulfill environment needs such as energy efficiency, limiting product waste and toxicity, and designing innovative products. One way for companies to accomplish this is through open communications with stakeholders characterized by high levels of information disclosure, clarity, and accuracy (Finn 2009).

10.14.3 Stakeholder Engagement

Sustainability requires a company to look internally and externally to understand their environmental and social impacts. This requires the engagement of stakeholders to understand impacts and concerns. A business can address sustainability internally by educating employees and seeking to reduce impacts through waste reduction, energy efficiency, etc. Employee engagement can be a powerful motivator by having a philanthropy committee or a green team. As a company looks externally, stakeholders include customers, suppliers, community, and nongovernment organizations.

Problems

10.1 Where we are in terms of industrial quality environmental management (IQEM)?

10.2 Safe energy production and renewal of safety guidelines have 12 principles of green chemistry that surpass initial bench-level development. Select 6 of the 12 principles introduced by Winterton to share information on "How to do it!" as gained by individual research in the text or in classroom discussions, speakers, or projects. Please number your choices to match the list numbers. Clean, clear, concise statements limited to a paragraph for each of the principles chosen is the minimum.

10.3 List three types of environmental law (Orts) and Orts' four objections to the "command and control" approach to environmental law. In your own words, state your choice of point of view and compare it to a regulator's point of view or a business owner's point of view.

10.4 Two traditional issues involved with ethics (Donaldson and Werhane) Ethical Relativism and Truth telling when related to International Issues bring about conclusions that may have started with the learning how to make beer, the quitting of the hunter-gatherer life style, and bringing about commerce based on the value of what is created in relation to the amount of work required in business descriptions of relative wealth. Using Figures 10.2 and 10.3 *create an integrated Venn diagram that includes all components shown in both diagrams.*

10.5 (a) Define sustainability as stated in the "World Commission on Environment and Development"; (b) What are sustainable engineering design principles? (c) Design for environmental processing and manufacturing.

10.6 What are the 12 principles of green engineering?

10.7 What are the 12 principles of green chemistry?

10.8 (a) Define the Leadership in Energy and Environmental Design (LEED). (b) What are the US Green Building Council (USGBC) LEED rating system and point scale for three levels of certification? (c) Mention the range of points for each level of certification, and how the points are distributed among the USGBC's various green building design and operation criteria.

10.9 Provide definition and terms for environmental management: (a) pollution prevention; (b) source reduction; (c) in-process vs on-site vs off-site recycling; (d) waste treatment; (e) disposal; (f) direct release.

10.10 Define the sustainable supply chain systems in manufacturing.

10.11 What are some elements in international business ethic issues and ethical sustainability?

10.12 Define eco-efficiency, sustainable green economy, and its six essential characteristics.

10.13 State your points of view on strategies for corporate sustainability and corporate responsibility.

References

Abraham, M.A. (2006). *Sustainability Science and Engineering: Defining Principles*. Amsterdam: Elsevier.

Abraham, M.A. and Nguyen, N. (2003). "Green engineering: defining the principles." Results from the Sandestin conference. *Environmental Progress* 22 (4): 233–236.

Ahwahnee Principles (1991). Local Government Commission. http://www.lgc.org/ahwahnee/principles.html (accessed 14 March 2019).

AIChE, IFS (2004). American Institute of Chemical Engineers, Institute for Sustainable Development. http://www.aiche.org/sustainability (accessed 15 April 2004).

Allen, D.T. and Shonnard, D.R. (eds.) (2002). *Green Engineering: Environmentally Conscious Design of Chemical Processes*. Upper Saddle River, NJ: Prentice Hall PTR.

Allen, D.T. and Shonnard, D.R. (2012). *Sustainable Engineering: Concepts, Design, and Case Studies*. Upper Saddle River, NJ: Prentice Hall PTR.

Allenby, B.R. (2012). *The Theory and Practice of Sustainable Engineering*. Upper Saddle River, NJ: Prentice Hall PTR.

American Institute of Chemical Engineers (AIChE), Center for Waste Reduction Technology (CWRT) (2000). Collaborative Projects, Focus Area: Sustainable

Development, Development of Baseline Metrics. http://www/aiche.org/cwrt/projects/sustain.htm (accessed 5 June 2019).

Anand, S. and Sen, A.K. (1996). *Sustainable Human Development: Concepts and Priorities*, Office of Development Studies Discussion Paper, vol. 1. New York: United Nations Development Programme.

Anastas, P.T., Kirchhoff, M.M., and Williamson, T.C. (2001). Catalysis as a foundational pillar of green chemistry. *Applied Catalysis A, General* 221: 3.

Anastas, P.T. and Warner, J.C. (1998). *Green Chemistry: Theory and Practice*. New York, NY: Oxford University Press.

Anastas, P.T. and Zimmerman, J.B. (2003). Design through the 12 principles of green engineering. *Engineering Science and Technology* 37 (5): 94A–101A.

Azapagic, A. (2000). Life cycle assessment: a tool for innovation and improved environmental performance, Part VI35. In: *Science, Technology and Innovation Policy: Opportunities and Challenges for the Knowledge Economy* (eds. P. Conceição, D.V. Gibson, M.V. Heitor and S. Shariq), 519–530. Westport, CT: Quorum Books.

Bag, S., Anand, N., and Pandey, K.K. (2017). Green Supply Chain Management model for sustianble manufacturing practices. In: *Green Supply Chain Management for Sustainable Business Practices* (eds. M. Khan, M. Hussain and M. Ajmal), 153–189. Hershey, PA: IGI Global.

Bare, J.C. (2003). The tool for the reduction and assessment of chemical and other environmental impacts (TRACI), USEPA, 2003. TRACI is available on the Internet at no cost at http://epa.gov/ORD/NRMRL/std/sab/iam_traci.htm (accessed 11 May 2019).

Benyus, J.M. (1997). *Biomimicry*. New York, NY: William Morrow and Co.

Bhander, G.S., Hauschild, M., and McAloone, T. (2003). Implementing life cycle assessment in product development. *Environmental Progress* 22 (4): 255–267.

Biswas, G., Clift, R., Davis, G. et al. (1998). Ecometrics: identification, categorization and life cycle validation. *The International Journal of Life Cycle Assessment* 3: 183–190.

Blewitt, J. (2008). *Understanding Sustainable Development*. London, UK: Earthscan.

Blewitt, J. (2014). *Understanding Sustainable Development*, 2e. London, UK: Routledge.

Brandão, M., Clift, R., Canals, M.L., and Basson, L. (2010). A life-cycle approach to characterizing environmental and economic impacts of multifunctional land-use systems: an integrated assessment in the UK. *Sustainability* 2: 3747–3776.

Canadian Public Health Association (CPHA) (1992). Human and Ecosystem Heath, Ottawa.

Carvalho, H., Govindan, K., Azevedo, S.G., and Cruz-Machado, V. (2017). Modelling green and lean supply chains: an eco-efficiency perspective. *Resources, Conservation and Recycling* 120: 75–87.

Clift, R. (2003). Metrics for supply chain sustainability. *Clean Technologies and Environmental Policy* 5: 24–247.

Clift, R. and Wright, L. (2000). Relationship between environmental impacts and added value along the supply chain. *Technological Forecasting and Social Change* 65: 281–295.

Common Dreams (2010). True sustainability needs an ethical revolution. www.commondreams.org (accessed 21 May 2019).

Cooney, S. (2009). *Build a Green Small Business: Profitable Ways to Become an Ecopreneur*. New York: McGraw-Hill.

Cooper, J.S. (2003). Life cycle assessment and sustainable development indicators. *Journal of Industrial Ecology* 7 (1): 12–15.

Curran, M.A. (2003). Do bio-based products move us toward sustainability? A look at three USEPA case studies. *Environmental Progress* 22 (4): 277–292.

Curran, M.A. (ed.) (2012). *Life Cycle Assessment Handbook: A Guide for Environmentally Sustainable Product*. Salem, MA/Hoboken, NJ: Scrivener Publishing LLC/Wiley.

Curran, M.A. (ed.) (2015). *Life Cycle Assessment Student Handbook*. Salem, MA/Hoboken, NJ: Scrivener Publishing LLC/Wiley.

Das, T.K. (2002). Evaluating the life cycle environmental performance of chlorine disinfection and ultraviolet technologies. *Clean Technologies and Environmental Policy* 4: 32–43.

Das, T.K. (2005a). *Toward Zero Discharge: Innovative Methodology and Technologies for Process Pollution Prevention*. Hoboken, NJ: Wiley Interscience.

Das, T.K. (ed.) (2005b). Sustainability and sustainable development. In: *Toward Zero Discharge: Innovative Methodology and Technologies for Process Pollution Prevention*, 191–222. Hoboken, NJ: Wiley Interscience.

Das, T.K. (ed.) (2005c). Zero discharge industries. In: *Toward Zero Discharge: Innovative Methodology and Technologies for Process Pollution Prevention*, 11–32. Hoboken, NJ: Wiley Interscience.

Das, T.K. (ed.) (2005d). Fundamentals of life cycle assessment. In: *Toward Zero Discharge: Innovative Methodology and Technologies for Process Pollution Prevention*, 35–87. Hoboken, NJ: Wiley Interscience.

Das, T.K. (ed.) (2005e). Eco-industrial parks: model zero discharge communities. In: *Toward Zero Discharge: Innovative Methodology and Technologies for Process Pollution Prevention*, 252–259. Hoboken, NJ: Wiley Interscience.

Das, T.K. (ed.) (2005f). Economics of pollution prevention: toward an environmentally sustainable economy. In: *Toward Zero Discharge: Innovative Methodology and Technologies for Process Pollution Prevention*, 161–189. Hoboken, NJ: Wiley Interscience.

Das, T.K. (ed.) (2005g). Sustainable economy and the Earth. In: *Toward Zero Discharge: Innovative Methodology and Technologies for Process Pollution Prevention*, 183–188. Hoboken, NJ: Wiley Interscience.

Das, T.K., Sharma, M., and Butner, S. (2001). The concept of an eco-industrial park: a solution to industrial ecology. *Presented at the AIChE Spring Meeting Topical Conference: Industrial Ecology and Zero-Discharge Manufacturing*, Houston, TX (9 April 2001).

Das, T.K., Tsoka, C., Johns, W., and Kokossis, A. (2003). Sustainability metrics and practice: an evaluation of industrial practice. *Presented at the AIChE National Meeting Topical Conference: Sustainability and Life Cycle*, San Francisco, CA (7 November 2003).

DeSimone, L.D. and Popoff, F. (2000). *Eco-Efficiency: The Business Link to Sustainable Development*. Cambridge, MA: MIT Press.

Dewulf, J., van Langenhove, H., Mulder, J. et al. (2000). Illustrations towards quantifying the sustainability of technology. *Green Chemistry* 2: 108–114.

Dienhart, J. (2000). *Business, Institutions, and Ethics*. Oxford University Press.

Donaldson, T. and Werhane, P. (1979). *Ethical Issues in Business*. Englewood Cliffs, NJ: Prentice-Hall.

Dyllick, T. and Hockerts, K. (2002). Beyond the business case for corporate sustainability. *Business Strategy and the Environment* 11: 130–141.

Earth Charter Principles (2019). http://www.earthcharterusa.org/earth_charter.html (accessed 25 September 2019).

Elkington, J. (2004). Enter the triple bottom line. In: *The Triple Bottom Line: Does It All Add Up?* (eds. A. Henriques and J. Richardson). Earthscan.

Enderle, G. (1999). *International Business Ethics*. University of Notre Dame Press.

Finn, D. (2009). *Our Uncertain Future: Can Good Planning Create Sustainable Communities?* Urbana, IL: University of Illinois.

Friedman, M. (1970). The social responsibility of business is to increase profit. *The New York Times Magazine* (13 September 1970), pp. 1–6.

Gonzalez, M.A. and Smith, R.L. (2003). A methodology to evaluate process sustainability. *Environmental Progress* 22 (4): 269–276.

Graedel, T.E. and Allenby, B.R. (2003). *Industrial Ecology*, 2e. Upper Saddle River, NJ: Prentice Hall.

Hahn, K. (2008). What is sustainable business? http://en.verteeblog.com/what-is-sustainable-business/200847 (accessed 28 May 2019).

Harvey, F. (2011). Europe "falling behind" in green investment race. *The Guardian* (7 July 2011).

Hawken, P., Lovins, A., and Hunter Lovins, I. (1999). *Natural Capitalism: Creating the Next Industrial Revolution*. New York: Little, Brown and Company.

Hohen, P. (2000). Presentation at the OECD Business and Industry Policy Forum on *Environmental Management: Challenges for Industry* (27 September 2000).

Initiative for Global Environmental Leadership (IGEL) (2012). Special Repot: The Pathways to Sustainability in Emerging Economies, The Wharton School. The University of Pennsylvania.

International Institute for Sustainable Development (IISD), *Towards Change: The Work and Results of Mining,* Minerals and Sustainable Development, Final Report, September 2002.

International Organization of Standardization (ISO) (1997). *Environmental Management: Life Cycle Assessment – Principles and Framework*, ISO/FDIS 4040, ISO TC 207. London: BSI.

Kahle, L.R. and Gurel-Atay, E. (eds.) (2013). *Communicating Sustainability for the Green Economy*. New York: Routledge.

Leadership in Energy and Environmental Leadership (LEED) (2010). Application Guide for Multiple Buildings and on Campus Building Projects, USGBC, 11 October 2010.

Linke, P., Kokossis, A., Das, T.K., and Tsoka, C. (2003). Environmental informatics: tools and technologies to monitor trends and changes. Paper presented at the AIChE Topical Conference: Sustainability and Life Cycle, San Francisco, CA (7 November 2003).

Luttropp, C. and Lagerstedt, J. (2006). EcoDesign and The Ten Golden Rules: Generic advice for merging environmental aspects into product development. *Journal of Cleaner Production* 14 (15–16): 1396–1408.

Mathew, M.A. (2003). Green chemistry. In: *Kirk-Othmer Encyclopedia of Chemical Technology* (ed. A. Seidel). New York, NY: Wiley.

Matlack, A. (2003). Some recent trends and problems in green chemistry. *Journal of the Royal Society of Chemistry* 5: G7–G12.

McDonough, W. and Braungart, M. (2002). *Cradle to Cradle: Remaking the Way We Make Things*. New York: North Point Press.

McLennan, J.F. (2004). *Philosophy of Sustainable Design*, 4e. Ecotone Publishing.

National Society of Professional Engineers (NSPE) (2019) Code of ethics. https://www.nspe.org/resources/ethics/code-ethics (accessed 27 September 2018).

Orts, E. (1995). A reflexive model of environmental regulation. *Business Ethics Quarterly* 5 (4): 779–794.

Rennie, E. (2008a). Growing Green, Boosting the bottom line with sustainable business practices. *APICS Magazine* 18 (2).

Rennie, E. (2008b). Painting a Green Story. *APICS Extra* 3 (2).

Salzman, J. and Thompson, B. (2010). *Environmental Law and Policy*, 3e. Foundation Press.

Sen, A.K. (2013). The ends and means of sustainability. *Journal of Human Development and Capabilities* 14 (1): 6–20.

SETAC (1991). *A Technical Framework for Life Cycle Assessment*, Workshop Report from the Smugglers Notch. Vermont: The Society of Environmental Toxicology and Chemistry.

SETAC (1993). *Guidelines for Life-Cycle Assessment: A Code of Practice*. Pensacola: The Society of Environmental Toxicology and Chemistry.

Shaker, R.R. (2015). The spatial distribution of development in Europe and its underlying sustainability correlations. *Applied Geography* 63: 304–314.

Sheldon, R.A. (2000). Atom efficiency and catalysis in organic synthesis. *Pure and Applied Chemistry* 72 (7): 1233.

Shonnard, D.R., Lidner, A., Nguyen H. et al. (2007). Green engineering: integration of green chemistry, pollution prevention, and risk-based considerations, Kent, J.A. *Kent and Reigel's Handbook of Industrial Chemistry and Biotechnology*, Vol. I, 11, Springer Science and Business Media.

Sikdar, S. (2003). Journey towards sustainable development: a role for chemical engineers. *Environmental Progress* 22 (4): 227–232.

Spangenberg, J.H., Fuad-Luke, A., and Blincoe, K. (2010). Design for Sustainability (DfS): the interface of sustainable production and consumption. *Journal of Cleaner Production* 18: 1485–1493.

The Augsburg Materials Declaration Research (2002). The Augsburg materials declaration materials research. https://www.coursehero.com/file/p5k2n77/The-Augsburg-Materials-Declaration-Materials-Research-Society-2002-identifies (accessed 6 June 2019).

The CERES Principles (n.d.). Coalition for environmentally responsible economics. http://www.ceres.org/our_work/principles.htm (accessed 16 March 2019).

Trost, B.M. (1991). The atom economy: a search for synthetic efficiency. *Science* 219: 245.

Tueth, M. (2010). *Fundamentals of Sustainable Business: A Guide to the Next 100 Years*. Hackensack: World Scientific.

U.S. Green Building Council (USGBC) (2011a). *LEED 2009 for Retail: New Construction and Major Renovation*. Washington, DC: US Open Green Building Council.

U.S. Green Building Council (USGBC) (2011b). *LEED Projects and Case Studies Directory*. Washington, DC: U.S. Green Building Council http://www.usbc.org/LEED/Project/CertifiedProjectList.aspx (

U.S. Green Building Council (USGBC) (2011c). *Greenbuild is Global*. www.usgbc.org (accessed 23 August 2018).

United Nations Environment Program (UNEP) (1992). *Agenda 21*. www.unep.org (accessed 6 June 2019).

United Nations Environment Program (UNEP) (1997). *Global Environmental Outlook – 1*. www.unep.org (accessed 6 June 2019).

United Nations Environment Program (UNEP) (1999). *Global Environmental Outlook – 2000*. www.unep.org (accessed 6 June 2019).

United Nations Environmental Program–Society of Environmental Toxicology and Chemistry (UNEP-SETAC) (2000). *Letter of Intent: Best Available Practice of Life Cycle Assessment with Generic Application Dependency*. Signed by the International Council of SETAC, and UNEP, Division Technology, Industry, and Economics. http://www.uneptie.org/pc/sustain/lca/letter-of-intent.htm (accessed 6 June 2019).

USEPA (2001). The U.S. Environmental Protection Agency's Design for the Environment Program (Report), March 2001, EPA 744-F-00-020.

USEPA (2017). *History of Safer Choice and Design for the Environment*. USEPA. 19 April 2017.

Welford, R. (1997). *Hijacking Environmentalism: Corporate Responses to Sustainable Development*. London: Earthscan.

Western Governors' Association (2004). Principles for Environmental Management in the West, 3 December 2003. http://www.westgov.org/wga/policy (accessed 6 August 2019).

White, A.L. (1999). Sustainability and the accountable corporation: society's rising expectations of business. *Environment* 41 (8): 30–43.

Winterton, N. (2001). Twelve more green chemistry principles. *Green Chemistry* 6 (1): G73–G75.

World Business Council for Sustainable Development (WBCSD) (1992). *Changing Course*. Geneva, Switzerland: World Business Council for Sustainable Development.

World Business Council for Sustainable Development (WBCSD) (2000). *Eco-Efficiency: Creating More Value with Less Impact*. Geneva, Switzerland: Conches.

World Commission on Environment and Development (WCED) (1987). *Our Common Future*. Oxford: Oxford University Press.

Appendix A

Conversion Factors

Table A.1 Conversion factors.

Length
1 kilometer (km) = 1 000 meters (m); 1 meter (m) = 100 centimeter (cm)
1 centimeter (cm) = 10^{-2} m; 1 millimeter (mm) = 10^{-3} m
1 micron (μ) = 10^{-6} m; 1 nanometer (nm) = 10^{-9} m; 1 angstrom (Å) = 10^{-10} m
1 inch (in.) = 2.540 cm; 1 foot (ft) = 30.48 cm; 1 mile (mi) = 1.609 km
1 centimeter (cm) = 0.393 7 in.; 1 meter = 39.37 in.;1 kilometer (km) = 0.621 4 mile
Area
1 square meter (m^2) = 10.76 ft^2; 1 square mile (mi^2) = 640 acres
1 square foot (ft^2) = 929 cm^2; 1 acre = 43 560 ft^2
Volume
1 liter (l) = 1 000 cm^3 = 1.057 quart (qt) = 0.2642 gal = 0.035 32 ft^3
1 US barrel liquid = 0.12 m^3 = 4.24 ft^3
1 cubic meter (m^3) = 1 000 l = 35.32 ft^3
1 cubic foot (ft^3) = 7.481 US gal = 0.028 33 m^3 = 28.32 cm^3 = 28.32 l
1 US gallon (gal) = 0.13137 ft^3 = 3.785 l
Mass
1 kilogram (kg) = 2.204 6 pounds (lb) = 1 000 gram (g) = 0.001 1 ton
1 lb = 453.6 g = 7 000 gr = 0.453 6 kg = 0.000 50 ton; ton = 2 000 lb = 1.102 tonne (metric ton)
Speed
1 km/h = 0.277 8 m/s = 0.621 4 mi/h = 0.911 3 ft/s
1 mi/h = 1.467 ft/s = 1.609 km/h = 88 ft/min
Density
1 g/cm^3 = 1 000 kg/m^3 = 62.43 lb/ft^3; 1 lb/ft^3 = 0.016 02 g/cm^3 = 16.02 kg/m^3
Force
1 Newton (N) = 10^{-5} dyne = 1.0 kg-m/s^2 = 7.232 lb_m-ft/s^2 = 0.224 8 lb_f
1 pound force (lb_f) = 4.448 N = 4.448 kg-m/s^2 = 32.17 lb_m-ft/s^2
1 kg-m/s^2 = 0.224 8 lb_f = 1 N; 1 lb_m-ft/s^2 = 0.138 3 N = 1 lb_f
Energy
1 kJ = 239 cal = 737.5 ft-lb_f = 0.948 Btu = 2.778×10^{-4} kWh
1 ft-lb_f = 1.356×10^{-3} kJ = 0.323 9 cal = 1.285×10^{-3} Btu = 3.766×10^{-7}
1 calorie (cal) = 4 186 J = 3.087 ft-lb_f = 3.968×10^{-3} Btu

(*Continued*)

Industrial Environmental Management: Engineering, Science, and Policy, First Edition. Tapas K. Das.

Companion website: www.wiley.com/go/Das/IEM_1e

Table A.1 (Continued)

1 Btu (British thermal unit) = 778 ft-lb$_f$ = 1.055 kJ = 2.93×10^{-4} kWh

1 kilowatt hour (kWh) = 3.60×10^3 kJ = 8.60×10^5 cal = 3413 Btu

Power

1 Btu/h = 2.93×10^{-4} kW = 3.93×10^{-4} hp = 0.216 ft-lb$_f$/s

1 horsepower (hp) = 550 ft-lb$_f$/s = 2 545 Btu/h = 745.7 W

1 kilowatt (kW) = 1.341 hp = 737.6 ft-lb$_f$/s = 3412 Btu/h

Pressure

1 atmosphere (atm) = 101 325 N/m^2 = 14.70 lb$_f$/in.^{2} (psi) = 760 mm mercury = 406.8 in. H_2O

1 N/m^2 (Pascal) = 9.869×10^{-6} atm = 1.45×10^{-4} psi = 0.0075 mm Hg = 0.00402 in. H_2O

in. H_2O = 0.002458 atm = 0.03614 psi = 1.868 mm Hg = 249 Pa (N/m^2)

Appendix B

International Environmental Law

Following are the most commonly used abbreviations and acronyms for treaty research in international environmental law:

- Agenda 21: Program of Action for Sustainable Development (Agenda 21)
- Aarhus Convention: Convention on Access to Information, Public Participation in Decision-Making, and Access to Justice in Environmental Matters
- Basel Convention: Basel Convention on the Control of Transboundary Movements of Hazardous Wastes and Their Disposal
- Basel Ban Amendment: Protocol on Liability and Compensation for Damage Resulting from Transboundary Movements of Hazardous Wastes and Amendment of Annex B
- Biodiversity Convention: Convention on Biological Diversity (CBD)
- Bunkers Convention: International Convention on Civil Liability for Bunker Oil Pollution Damage (BUNKER): International Maritime Organization
- Cartagena Protocol: Cartagena Protocol on Biosafety to the Convention on Biological Diversity
- CCAMLR: Commission for the Conservation of Antarctic Marine Living Resources
- CCSBT: Convention for the Conservation of Southern Bluefin Tuna
- CITES: Convention on the International Trade in Endangered Species of Wild Fauna and Flora
- CLC: Convention on Civil Liability for Oil Pollution Damage
- Climate Change Convention: United Nations Framework Convention on Climate Change
- Kyoto Protocol: Kyoto Protocol to the United Nations Framework Convention on Climate Change
- LRTAP: Long-Range Transboundary Air Pollution, Geneva Convention
- Montreal Protocol: Montreal Protocol on Substances that Deplete the Ozone Layer
- NAAEC: North America Agreement on Environmental Cooperation
- OLDEPESCA: Agreement Instituting the Latin American Organization for Fisheries Development
- OSPAR Convention: Convention for the Protection of the Marine Environment of the Northeast Atlantic
- Paris Convention: Paris Convention on Third-Party Liability in the Field of Nuclear Energy
- POPs: Persistent Organic Pollutants, World Health Organization
- Rio Declaration: Rio Declaration on Environment and Development, UNCED
- Stockholm Convention: Stockholm Convention on Persistent Organic Pollutants (POPs)
- UNCCD: United Nations Convention to Combat Desertification
- UNCLOS: United Nations Convention on the Law of the Sea
- UNEP: United Nations Environmental Program
- UNFCCC: United Nations Framework Convention on Climate Change
- Vienna Convention: Convention for the Protection of the Ozone Layer
- Wetlands Convention: Convention on Wetlands of International Importance Especially as Waterfowl Habitat: UNESCO Constitution

Because of the many possible subtopics available for international environmental laws, it is helpful to have a table (see the following) of agreements and websites (accessed on 4 and 5 January 2019) that are research-based on the eight subtopics of hazardous waste, nuclear waste, ocean and marine sources, ozone and protection of the atmosphere, pollution, protection of species and wildlife, sustainable development, and trade and the environment.

Industrial Environmental Management: Engineering, Science, and Policy, First Edition. Tapas K. Das.

Companion website: www.wiley.com/go/Das/IEM_1e

Subtopics for international environmental law	Conventions and agreements	Websites for research
1. Hazardous waste	*Basel Convention (1998)*[a]	Secretariat of the Basel Convention: UN Environmental Program (http://archive.basel.int/text/documents.html)
	Protocol on Liability and Compensation for Damage Resulting from Transboundary Movements of Hazardous Wastes (1999)	"Basel Ban Amendment": Secretariat of the Basel Convention (http://www.basel.int/meeting/cop/cop5/docs/prot-e.pdf)
		United Nations Environmental
2. Nuclear waste	[b]*Convention on Nuclear Safety (1994)*	International Atomic Energy Agency Circular (http://www.iaea.org/Publications/Documents)
	Convention on Third Party Liability in the Field of Nuclear Energy ["Paris Convention"] (1960)	Reservations and Declaration to the Convention on Nuclear Safety (https://www.iaea.org/topics/nuclear-safety-and-security)
		Nuclear Energy Agency's Convention on Nuclear Third Party Liability (https://www.oecd-nea.org/law/nlparis_conv.html)
		Paris Convention Decisions, Recommendations, and Interpretations (https://www.oecd-nea.org/law/pubs/1990/156-paris-convention-decisions.pdf)
3. Ocean and marine sources	*Agreement for the Establishment of the Indo-Pacific Fisheries Council (1948)*	Australian Treaty Series: AustLII (http://www.austlii.edu.au/au/other/dfat/treaties/1949/4.html)
	Agreement Instituting the Latin American Organization for Fisheries Development (1982)	Internet Guide to Int'l Fisheries Law (http://www.oceanlaw.net/texts/oldepesca.html)
	[b]*Convention for the Conservation of Southern Bluefin Tuna (1993)*	ENTRI Treaty Locator: CIESIN (Columbia) (http://sedacc.ciesin.columbia.edu/entri/texts/salmon.north.atlantic.1982.html)
	Convention for the Prevention of Marine Pollution from Land-Based Sources (1974) and "OSPAR" Convention (1992)	Commission for the Conservation of Southern Bluefin Tuna (http://www.ccsbt.org/docs/about.html)
		OSPAR Commission Website (http://www.ospar.org/eng/html/welcome.html)
4. Ozone and protection of the atmosphere	[b]*Convention for the Protection of the Ozone Layer ["Vienna Convention"]/(1985)*	United Nations Environment Program (http://www.unep.ch/ozone/index.asp)
	[b]*Convention on Long-Range "Transboundary Air Pollution (1979)*	United Nations Economic Commission for Europe (http://www.unece.org/env/lrtap/fsulf_h1.html)
	Kyoto Protocol to the United Nations Framework Convention (1997)	UN FCCC website (http://unfccc.int/essential_background/items/2877.php)
	[b]*Montreal Protocol (1987)*	UN Environmental Program: Montreal Protocol (http://www.unep.ch/ozone/Treaties_and_Ratification/2B_montreal_protocol.asp)
	[b]*United Nations Framework Convention on Climate Change (1992)*	UN FCCC Website (http://unfccc.int/essential_background/convention/items/2627.php)

5. Pollution	*Stockholm Convention on Persistent Organic Pollutants (2001)*	Persistent Organic Pollutants Website (http://www.pops.int/default.htm)
	International Convention on Civil Liability for Bunker Oil Pollution Damage (2001)	ENTRI: Environmental Treaties and Resource Indicators (http://sedac.ciesin.columbia.edu/entri/texts/pollution.from.ships.1973.html)
	International Convention on Civil Liability for Oil Pollution Damage and 1992 Protocol (1969 and 1992)	AustLII:Australian Legal Information Institute (http://www.austlii.edu.au/au/other/dfat/treaties/2009/14.html)
		CIESIN (Columbia University)/ENTRI (http://sedac.ciesin.columbia.edu/entri/texts/civil.liability.oil.pollution.damage.1969.html)
6. Protection of species and wildlife	*Agenda 21: Program of Action for Sustainable Development (1992)*	United Nations Sustainable Development (https://sustainabledevelopment.un.org/content/documents/Agenda21.pdf)
	Cartagena Protocol (2000)	Biological Diversity Website (https://www.cbd.int/doc/legal/cartagena-protocol-en.pdf)
	Convention on Biological Diversity (1992)	Convention on Biological Diversity Website (https://www.cbd.int/convention/)
	[b]*North American Agreement on Environmental Cooperation (1993)*	Commission for Environmental Cooperation (http://www.cec.org/pubs_info_resources?law_treat_agree/naaec/index.cfm?varlan=english)
	[b]*United Nations Convention to Combat Desertification in Countries Experiencing Serious Drought (1994)*	UNCCD Website (http://www.unccd.int/convention/menu.php)
	Agenda 21: Program of Action for Sustainable development (1992)	United Nations Sustainable Development (http://www.un.org/esa/susdev/documents/agenda21/english/Agenda21.pdf)
	[b]*Convention on the Conservation of Antarctic Marine Living Resources (1980)*	Convention on the Conservation of Antarctic Marine Living Resources Secretariat (http://archive.ccamlr.org/pu/e/gen-intro.htm)
	[b]*Convention on Wetlands of International Importance especially as Waterfowl Habitat (1971)*	Ramsar Convention on the Wetlands Website (http://www.ramsar.org/index_very_key_docs.htm)
	Rio declaration on Environment and Development (1992)	UN Environment Program (http://www.unep.org/Document.multilingual/)
	Antarctic Treaty (1959) and Protocol on Environmental Protection (1991)	Committee for Environmental Protection: Antarctic Treaty (https://www.ats.aq/e/cep.htm)
7. Sustainable development		(https://sustainabledevelopment.un.org/?menu=1300)
8. Trade and the environment	*Convention on Int'l Trade in Endangered Species of Wild Fauna and Flora ["CITES"] (1973)*	CITES Official Documents (https://www.cites.org/eng/disc/text.php)
	North American Agreement on Environmental Cooperation ["NAAEC"]/(1993)	Commission for Environmental Cooperation (http://www.cec.org/about-us/about-cec)
	Aarhus Convention (1998)	United Nations Economic Commission for Europe (http://www.unece.org/env/pp/treatytext.htm)
		United Nations Economic Commission for Europe (http://www.unece.org/env/pp/treatytext.htm)

Source: Adapted from Muralikrishna and Manickam (2017).
[a] Dates referenced are the date of signature of the treaties.
[b] United States signed and ratified the treaties.

Reference

Muralikrishna, V.I. and Manickam, V. (2017). *Environmental Management: Science and Engineering for Industry*, 1e. Oxford: Butterworth-Heinemann, Elsevier.

Appendix C

Air Pollutant Emission Factors

Stationary Point and Area Sources

Table C.1 Compilation of air pollutant emission factors: stationary point and area sources.

Chapter	Title
Chapter 1	*External Combustion Sources* https://www3.epa.gov/ttn/chief/ap42/ch01/index.html
Chapter 2	*Solid Waste Disposal* https://www3.epa.gov/ttn/chief/ap42/ch02/index.html
Chapter 3	*Stationary Internal Combustion Sources* https://www3.epa.gov/ttn/chief/ap42/ch03/index.html
Chapter 4	*Evaporation Loss Sources* https://www3.epa.gov/ttn/chief/ap42/ch04/index.html
Chapter 5	*Petroleum Industry* https://www3.epa.gov/ttn/chief/ap42/ch05/index.html
Chapter 6	*Organic Chemical Process Industry* https://www3.epa.gov/ttn/chief/ap42/ch06/index.html
Chapter 7	*Liquid Storage Tanks* https://www3.epa.gov/ttn/chief/ap42/ch07/index.html
Chapter 8	*Inorganic Chemical Industry* https://www3.epa.gov/ttn/chief/ap42/ch08/index.html
Chapter 9	*Food and Agricultural Industries* https://www3.epa.gov/ttn/chief/ap42/ch09/index.html
Chapter 10	*Wood Products Industry* https://www3.epa.gov/ttn/chief/ap42/ch10/index.html
Chapter 11	*Mineral Products Industry* https://www3.epa.gov/ttn/chief/ap42/ch11/index.html
Chapter 12	*Metallurgical Industry* https://www3.epa.gov/ttn/chief/ap42/ch12/index.html
Chapter 13	*Miscellaneous Sources* https://www3.epa.gov/ttn/chief/ap42/ch13/index.html
Chapter 14	*Greenhouse Gas Biogenic Sources* https://www3.epa.gov/ttn/chief/ap42/ch14/index.html

Industrial Environmental Management: Engineering, Science, and Policy, First Edition. Tapas K. Das.

Companion website: www.wiley.com/go/Das/IEM_1e

Table C.2 List of hazardous air pollutants.

Chemical abstract service number	Chemical names
75070	Acetaldehyde
60355	Acetamide
75058	Acetonitrile
98862	Acetophenone
53963	2-Acetylaminofluorene
107028	Acrolein
79061	Acrylamide
79107	Acrylic acid
107131	Acrylonitrile
107051	Allyl chloride
92671	4-Aminobiphenyl
62533	Aniline
90040	*o*-Anisidine
1332214	Asbestos
71432	Benzene (including benzene from gasoline)
92875	Benzidine
98077	Benzotrichloride
100447	Benzyl chloride
92524	Biphenyl
117817	Bis(2-ethylhexyl)phthalate (DEHP)
542881	Bis(chloromethyl)ether
75252	Bromoform
106990	1,3-Butadiene
156627	Calcium cyanamide
105602	Caprolactam
133062	Captan
63252	Carbaryl
75150	Carbon disulfide
56235	Carbon tetrachloride
463581	Carbonyl sulfide
120809	Catechol
133904	Chloramben
57749	Chlordane
7782505	Chlorine
79118	Chloroacetic acid
532274	2-Chloroacetophenone
108907	Chlorobenzene
510156	Chlorobenzilate
67663	Chloroform
107302	Chloromethyl methyl ether
126998	Chloroprene
1319773	Cresols/Cresylic acid (isomers and mixture)
95487	*o*-Cresol
108394	*m*-Cresol
106445	*p*-Cresol
98828	Cumene
94757	2,4-D, salts and esters
3547044	DDE
334883	Diazomethane
132649	Dibenzofurans[d]
96128	1,2-Dibromo-3-chloropropane
84742	Dibutylphthalate
106467	1,4-Dichlorobenzene(p)
91941	3,3-Dichlorobenzidene
111444	Dichloroethyl ether[b] [bis(2-chloroethyl) ether]
542756	1,3-Dichloropropene
62737	Dichlorvos
111422	Diethanolamine
121697	*N,N*-Dimethylaniline
64675	Diethyl sulfate
119904	3,3-Dimethoxybenzidine
60117	Dimethyl aminoazobenzene
119937	3,3′-Dimethyl benzidine
79447	Dimethyl carbamoyl chloride
68122	Dimethyl formamide
57147	1,1-Dimethyl hydrazine
131113	Dimethyl phthalate
77781	Dimethyl sulfate
534521	4,6-Dinitro-*o*-cresol, and salts
51285	2,4-Dinitrophenol
121142	2,4-Dinitrotoluene
123911	1,4-Dioxane (1,4-diethyleneoxide)
122667	1,2-Diphenylhydrazine
106898	Epichlorohydrin (l-chloro-2,3-epoxypropane)
106887	1,2-Epoxybutane
140885	Ethyl acrylate
100414	Ethyl benzene
51796	Ethyl carbamate (urethane)
75003	Ethyl chloride (chloroethane)
106934	Ethylene dibromide (dibromoethane)
107062	Ethylene dichloride (1,2-dichloroethane)
107211	Ethylene glycol[b]
151564	Ethylene imine (aziridine)
75218	Ethylene oxide

Table C.2 (Continued)

Chemical abstract service number	Chemical names
96457	Ethylene thiourea
75343	Ethylidene dichloride (1,1-dichloroethane)
50000	Formaldehyde
76448	Heptachlor
118741	Hexachlorobenzene
87683	Hexachlorobutadiene
77474	Hexachlorocyclopentadiene
67721	Hexachloroethane
822060	Hexamethylene-1,6-diisocyanate
680319	Hexamethylphosphoramide
110543	Hexane
302012	Hydrazine
7647010	Hydrochloric acid
7664393	Hydrogen fluoride (hydrofluoric acid)
7783064	Hydrogen sulfide
123319	Hydroquinone
78591	Isophorone
58899	Lindane (all isomers)
108316	Maleic anhydride
67561	Methanol
72435	Methoxychlor
74839	Methyl bromide (bromomethane)
74873	Methyl chloride (chloromethane)
71556	Methyl chloroform (1,1,1-trichloroethane)
78933	Methyl ethyl ketone (2-butanone)
60344	Methyl hydrazine
74884	Methyl iodide (iodomethane)
108101	Methyl isobutyl ketone (hexone)
624839	Methyl isocyanate
80626	Methyl methacrylate
1634044	Methyl tert butyl ether
101144	4,4-Methylene bis(2-chloroaniline)
75092	Methylene chloride (dichloromethane)
101688	Methylene diphenyl diisocyanate
101779	4,4′-Methylenedianiline
91203	Naphthalene
98953	Nitrobenzene
92933	4-Nitrobiphenyl
100027	4-Nitrophenol
79469	2-Nitropropane
684935	*N*-Nitroso-*N*-methylurea
62759	*N*-Nitrosodimethylamine
59892	*N*-Nitrosomorpholine
56382	Parathion
82688	Pentachloronitrobenzene (quintobenzene)
87865	Pentachlorophenol
108952	Phenol
106503	*p*-Phenylenediamine
75445	Phosgene
7803512	Phosphine
7723140	Phosphorus
85449	Phthalic anhydride
1336363	Polychlorinated biphenyls (aroclors)
1120714	1,3-Propane sultone
57578	β-Propiolactone
123386	Propionaldehyde
114261	Propoxur (baygon)
78875	Propylene dichloride (1,2-dichloropropane)
75569	Propylene oxide
75558	1,2-Propylenimine (2-methyl aziridine)
91225	Quinoline
106514	Quinone
100425	Styrene
96093	Styrene oxide
1746016	2,3,7,8-Tetrachlorodibenzo-*p*-dioxin
79345	1,1,2,2-Tetrachloroethane
127184	Tetrachloroethylene (perchloroethylene)
7550450	Titanium tetrachloride
108883	Toluene
95807	2,4-Toluene diamine
584849	2,4-Toluene diisocyanate
95534	*o*-Toluidine
8001352	Toxaphene (chlorinated camphene)
120821	1,2,4-Trichlorobenzene
79005	1,1,2-Trichloroethane
79016	Trichloroethylene
95954	2,4,5-Trichlorophenol
88062	2,4,6-Trichlorophenol
121448	Triethylamine
1582098	Trifluralin
540841	2,2,4-Trimethylpentane
108054	Vinyl acetate
593602	Vinyl bromide
75014	Vinyl chloride
75354	Vinylidene chloride (1,1-dichloroethylene)

(*Continued*)

Table C.2 (Continued)

Chemical abstract service number	Chemical names
1330207	Xylenes (isomers and mixture)
95476	*o*-Xylenes
108383	*m*-Xylenes
106423	*p*-Xylenes
0	Antimony compounds
0	Arsenic compounds (inorganic including arsine)
0	Beryllium compounds
0	Cadmium compounds
0	Chromium compounds
0	Cobalt compounds
0	Coke oven emissions
0	Cyanide compounds[a]
0	Glycol ethers
0	Lead compounds
0	Manganese compounds
0	Mercury compounds
0	Fine mineral fibers[c]
0	Nickel compounds
0	Polycyclic organic matter
0	Radionuclides (including radon)[e]
0	Selenium compounds

[a] X′CN where X=H′ or any other group where a formal dissociation may occur. For example KCN or $Ca(CN)_2$.
[b] Includes mono and di-ethers of ethylene glycol, diethylene glycol, and triethylene glycol R-$(OCH_2CH_2)n$-OR′ where $n = 1, 2$, or 3.
[c] Includes mineral fiber emissions from facilities manufacturing or processing glass, rock, or slug fibers (or other mineral derived fibers) of average diameter of 1 μm or less.
[d] Includes organic compounds with more than one benzene ring and which have a boiling point greater than or equal to 100 °C.
[e] A type of atom which spontaneously undergoes radioactive decay.

Table C.3 Values of the ideal gas law constant *R* [*R* = (*PV*/*nT*)].

	Units
R	**(Pressure-volume)/(matter-temperature)**
0.08206	atm-1/gmol-K
82.06	atm-cm^3/gmol-K
62.36	mm Hg-1/gmol-K
1.314	atm-ft^3/lbmol-K
0.08314	bar-1/gmol-K
998.9	mm Hg-ft^3/lbmol-K
0.7302	atm-ft^3/lbmol-R
21.85	in. Hg-ft^3/lbmol-R
555.0	mm Hg-ft^3/lbmol-R
10.73	psia-ft^3/lbmol-R
1545	ft·lbf/lbmol-R
8.314	Pa-m^3/gmol-K
R	*(Energy)/(matter-temperature)*
1.987	cal/gmol-K
8314	J/kgmol-K

Note: 1 Pascal (Pa) = 1 N/m^2; 1 N-m = 1 Joule (J).

Table C.4 Selected properties of air.

Temperature (°F)	Specific heat at constant pressure (C_p) (Btu/lb$_m$ °F)	Absolute viscosity (lb$_m$/ft-h)	Thermal conductivity (k) (Btu/h-ft-°F)	Prandtl No. ($C_p\mu/k$) (dimensionless)	Density (ρ)[a] (lb$_m$/ft^3)
0	0.240	0.040	0.0124	0.77	0.0863
20	0.240	0.041	0.0128	0.77	0.0827
40	0.240	0.042	0.0132	0.77	0.0794
60	0.240	0.043	0.0136	0.76	0.0763
80	0.240	0.045	0.0140	0.77	0.0734
100	0.240	0.047	0.0145	0.76	0.0708
120	0.240	0.047	0.0149	0.76	0.0684
140	0.240	0.048	0.0153	0.76	0.0662
160	0.240	0.050	0.0158	0.76	0.0639
180	0.240	0.051	0.0162	0.76	0.0619
200	0.240	0.052	0.0166	0.76	0.0601
250	0.241	0.055	0.0174	0.76	0.0558
300	0.241	0.058	0.0182	0.76	0.0521
350	0.241	0.060	0.0191	0.76	0.0489
400	0.241	0.063	0.0200	0.76	0.0460
450	0.242	0.065	0.0207	0.76	0.0435
500	0.242	0.067	0.0214	0.76	0.0412
600	0.242	0.072	0.0229	0.76	0.0373
700	0.243	0.076	0.0243	0.76	0.0341
800	0.244	0.080	0.0257	0.76	0.0314
900	0.245	0.085	0.0270	0.77	0.0295
1000	0.246	0.089	0.1283	0.77	0.0275
1200	0.248	0.097	0.0308	0.78	0.0238
1400	0.251	0.105	0.0328	0.80	0.0212
1600	0.254	0.112	0.0346	0.82	0.0192
1800	0.257	0.120	0.0360	0.85	0.0175
2000	0.260	0.127	0.0370	0.83	0.0161

Source: Adapted from Danielson (1973).
[a] ρ taken at pressure of 760 mm or 29.82 in. of mercury.

Reference

Danielson, J.A. (ed.) (1973). *Air Pollution Engineering Manual*, AP-40, 2e. Washington, DC: USEPA.

Appendix D

Frequently Asked Questions and Answers

Water Quality Model, Dispersion Model and Permits

D.1 Example Chain-of-Custody form for Use by POTWs

INDUSTRIAL WASTE MONITORING PROGRAM							ANALYSIS														
SAMPLERS (Signatures)																					
Industry Code	Time	Date	Source Description	Grab/Comp	G/P																COMMENTS/ PRESERVATIVE

Relinquished by:	Date/Time	Received by:	Date/Time	Remarks:
Relinquished by:	Date/Time	Received by:	Date/Time	
Relinquished by:	Date/Time	Received by:	Date/Time	

Industrial Environmental Management: Engineering, Science, and Policy, First Edition. Tapas K. Das.

Companion website: www.wiley.com/go/Das/IEM_1e

D.2 Water Quality Models

- Cornell Mixing Zone Expert System (CORMIX): "User's Manual for CORMIX: A Hydrodynamic Mixing Zone Model and Decision Support System for Pollutant Discharges into Surface Waters", By Jirka, G.H., Donekar, R.L., and Hinton, S.W. School of Civil and Environmental Engineering, Cornell University, Ithaca, N.Y. 14853
  CORMIX1, is used to predict and analyze environmental impacts of submerged single port discharges to lakes, rivers and estuaries.
  CORMIX2, may be used to predict plume characteristics of submerged multiport discharges.
  U.S. Environmental Protection Agency
  TMDL Modeling Toolbox
  Watershed and Water Quality Modeling Technical Support Center
  U.S. EPA – NERL
  Athens, GA 30605
- A Dynamic One-Dimensional Model of Hydrodynamics and Water Quality (EPDRiv1)
- Stream Water Quality Model: QUAL2K: TMDL Development

D.3 Frequently Asked Questions

Proposed General NPDES Permit for Industrial/ Municipal Wastewaters Dischargers that Meet Objectives/Criteria at the Point of Discharge

1. *What kind of discharge or facility would be eligible for coverage under the General NPDES Permit?*
   Treatment facilities that treat domestic wastewater and meet water quality objectives/criteria at the point of discharge or "end-of-pipe."
2. *What does "at the point of discharge" mean?*
   Point of discharge means the point just prior to where the facility effluent discharge meets the receiving water. Under this General Permit, facilities are required to meet all of their effluent limits prior to discharge to the receiving water – or "at the point of discharge."
3. *What are the potential advantages of having coverage under the General NPDES Permit vs. an Individual NPDES Permit?*
   The following is a list of some of the advantages being covered under the General NPDES Permit vs. an Individual NPDES Permit:
   - Streamlined application
   - Streamlined permitting process (three to six months shorter duration)
   - Simple reenrollment under subsequent General NPDES Permit
4. *How would I apply for the General NPDES Permit?*
   The application will consist of the submittal of a Notice of Intent (NOI), discussed in the response to Question 5.
5. *What is NOI?*
   NOI includes information regarding the owner/operator/authorized representative and facility (e.g. Form). Additional data and/or information that the discharger would like the Regional Board to take into consideration can also be provided.
6. *What is the Notice of Applicability (NOA)?*
   NOA is the official document signed by the Executive Officer granting regulatory coverage under the General NPDES Permit to the discharger for the NPDES discharge. The NOA will include at minimum: facility description and applicable final effluent limitations, effluent and receiving water monitoring requirements, beneficial uses, pretreatment program requirements, and minor land discharge specifications.
7. *What would the approval process be to get coverage under the General NPDES Permit?*
   After the submittal of the NOI, Regional Board staff would review the application and perform a reasonable potential analysis of the data. Staff may request additional information in order to determine eligibility for coverage under the General NPDES Permit. Staff will develop the NOA for approval by the Executive Officer. The subject discharge would have regulatory coverage under the General NPDES Permit after the Executive Officer issues the NOA.
8. *Will the discharger have an opportunity to review a draft of the NOA?*
   Yes. The draft NOA will be provided to the discharger for review prior to the NOA being considered for approval by the Executive Officer.
9. *Will there be public review of the NOA before the NOA is approved?*
   No. The public review process will be completed during the development and approval process for the General NPDES Permit. The public will be notified by the posting of a Notice of Public Hearing and will have the opportunity to submit written comments and recommendations concerning the tentative General NPDES Permit. The Central Valley Water Board will consider all timely comments received from the public at a public hearing. At the public hearing, the Central Valley Water Board will consider testimony and whether to adopt the proposed General NPDES Permit. If the Central Valley Water Board adopts the General NPDES Permit, the Executive Officer will be

authorized to issue individual NOAs to specific NPDES discharges that fall under the coverage of the General NPDES Permit.

10. *How long would the application and approval process take?*
Since this would be a new General NPDES Permit, the complete process time is undetermined. However, staff estimates the process will take approximately three to four months from the time the applicant submits a complete NOI through issuance of the NOA.

11. *What is the term of coverage under the General NPDES Permit?*
The General NPDES Permit would be effective for a five-year term. Whenever a new discharge is issued an NOA and enrolled for coverage under the General NPDES Permit, that coverage will expire when the General NPDES Permit expires. When the General NPDES Permit is renewed, the coverage for existing enrollees will be automatically renewed for the next five-year term of the renewed General NPDES Permit, as long as the terms and conditions of the NOA are still applicable.

12. *Would this Permit apply to discharges that are effluent dominated?*
Yes. The General NPDES Permit can be used for both effluent and not-effluent dominated situations.

13. *How long is the information in an engineering study valid?*
As long as the information in the engineering study applies to the existing operations and conditions at the treatment facility, and it is useful to staff in the analysis of determining eligibility of the discharge, then the engineering study should be valid.

14. *Would a discharge that has dilution credits be eligible for the General NPDES Permit?*
No. If dilution credits are needed to meet effluent limitations, then the discharge would not be eligible for coverage under the General NPDES Permit.

15. *We do not currently meet end-of-pipe limits but are making upgrades to our facility that will enable compliance at the point of discharge – could we apply for the General NPDES Permit now or do we have to wait for the upgrades to be completed?*
A discharger whose facility upgrades are underway and will be able to meet applicable effluent limits at the point of discharge when the upgrades are complete may apply for the General NPDES Permit once it is adopted by the Central Valley Water Board. Regional Board staff would review the NOI as it is received and will develop an NOA for approval by the Executive Officer as appropriate. The existing compliance schedule would be amended to reflect that an NOA authorizes the facility's discharge is under the General NPDES Permit.

16. *Has USEPA been involved with the development of the General NPDES Permit?*
Yes. Staff will continue to request input from the USEPA during the development process.

17. *How will the General NPDES Permit work for site-specific constituents that are dependent upon hardness, pH, or temperature?*
The basis for calculating the criterion and the effluent limitations for each constituent would be described in the General NPDES Permit. The applicable site-specific details and applicable final effluent limitations for a particular facility would be contained in the NOA.

18. *If the monitoring data for hardness changes, would the corresponding effluent limits change (i.e. will floating limits be issued)?*
No, floating limits will not be issued. The General NPDES Permit would include the procedures for determining site specific limitations and the final effluent limit applicable to the discharger would be included in the NOA.

19. *When can I submit the NOI application?*
Staff will begin accepting NOIs after the Central Valley Water Board has adopted the General NPDES Permit.

20. *Where can I get information regarding what is required with the NOI?*
An application checklist will be developed listing all the necessary information for a complete NOI. Once available, the checklist will be included in the General NPDES Permit and will be posted on the Regional Board's website.

21. *If the General NPDES Permit does not work for me, can I obtain an Individual NPDES Permit?*
A discharger authorized by the General NPDES may request exclusion from the General NPDES Permit by submitting a Report of Waste Discharge for an individual permit and reasons supporting the request. If the reasons for the individual permit are adequate to support the request, an individual permit will be prepared (see also Question 27).

22. *Have groups or individuals other than dischargers been notified to take part in the development process for the General NPDES Permit?*
Yes. All NPDES permit holders and interested parties including individuals, stakeholders, and nongovernmental organizations that have expressed interest in the Central Valley Regional Water Board's NPDES program have been notified.

23. *Are the enrollees grouped under the General NPDES Permit, such as by region?*
The General NPDES Permit will only be available to dischargers in the Central Valley Region. There is no grouping of certain types of projects within the Region.

24. *Our treatment plant upgrades will not be completed before the General NPDES Permit is available. Will we miss the application window?*
No. You will not miss the application window. Please see Questions/Answers 15 and 19. You may apply for coverage under the General NPDES Permit at any time while the Order is effective.

25. *How prescriptive will the General NPDES Permit be for the operation of the UV system?*
The General NPDES Permit would include the Regional Board's standard UV specifications, which are based on the National Water Research Institute guidelines. However, just like Individual Permits, the General Permit would allow site-specific UV specifications to be included in the NOA if detailed UV specifications are contained in a Title 22 Engineering Report approved by the Division of Drinking Water.

26. *Other general orders contain more restrictive monitoring than in our Individual NPDES Permit. Will this General NPDES Permit contain restrictive monitoring?*
The goal is not to include more restrictive monitoring than what is currently included in individual orders. However, the General NPDES Permit will be developed to apply to multiple discharges, so the monitoring may be different than what is in your current Individual NPDES Permit.

27. *Will dischargers be able to choose between the General NPDES Permit and an Individual NPDES Permit for coverage of their discharge?*
Dischargers that are eligible for coverage under the General NPDES Permit have the option to be covered under the General NPDES Permit or an Individual NPDES Permit; however, the terms and conditions in the Individual NPDES Permit will likely be similar to those in the General Permit.

28. *My Individual NPDES Permit does not expire for more than 180 days. When will I be required to apply for the General NPDES Permit?*
As soon as the General NPDES Permit becomes available, you may apply for coverage for your discharge. However, you may wait to submit your application until no later than 180 days prior to the expiration date of your existing NPDES Permit, in accordance with title 23 of the California Code of Regulations.

29. *Does this General NPDES Permit apply to my facility?*
The Central Valley Water Board has identified 25 facilities that appear to be immediately eligible for the General NPDES Permit. However, there is no intent to guarantee coverage under the General NPDES Permit to the facilities on the list of 25, nor is it the intent to exclude any facility not on the list of 25 from applying for coverage under the General NPDES Permit.

30. *If a discharger has more than one discharge point, can the discharger obtain regulatory coverage under the General NPDES Permit and under an Individual NPDES Permit for the separate discharges?*
Yes. The discharger can obtain regulatory coverage under the General NPDES Permit for discharges deemed eligible for coverage and obtain regulatory coverage under an Individual NPDES Permit for discharges that are not eligible for the General NPDES Permit.

31. *Will discharges to land be covered under the General NPDES Permit?*
The General NPDES Permit will include standard land discharge specifications (e.g. no objectionable odors, no public contact, etc.). If the land discharge requires more complex or site-specific specifications to ensure the protection of water quality, the discharger may need to obtain regulatory coverage for the land discharge under a separate permit.

32. *Who is responsible for making the determination about whether a discharge of pollutants from a surface discharging system will enter waters of the United States?*
Following is a response from the response to comments document that addresses this question and also provides additional information on waters of the United States. The response indicates that the homeowner is responsible for making the determination. EPA expects that the homeowner will make the decision based on the information provided by the person who conducts the soil investigation. The response from the response to comments document is as follows:

Congress, in Section 502 of the CWA, defined "navigable waters" broadly as encompassing all "waters of the United States." EPA has issued a regulatory definition of the term "waters of the United States" at 40 CFR § 122.2. EPA's definition includes, among other things, traditional navigable waters, tributaries of traditional navigable waters, and wetlands that are adjacent to traditional navigable waters or their tributaries. The Supreme Court has determined the scope of Congress' intent to regulate "waters of the United States" in several opinions of the Court, most recently in the case of *Rapanos v. United States*, 126 S. Ct. 2208 (2006).

EPA has provided guidance to individuals and companies impacted by the *Rapanos* decision; that guidance and other materials are set forth at http://water.epa.gov/lawsregs/guidance/wetlands/CWAwaters.cfm. This guidance discusses EPA's jurisdiction over traditional navigable waters, relatively permanent non-navigable streams, non-navigable streams that are not relatively permanent, and wetlands adjacent to those waters to help EPA and the Corps of Engineers, as well as citizens, identify whether particular surface waters are "waters of the United States."

It is the responsibility of the potential discharger to determine whether or not his or her system might discharge to a water of the United States. EPA realizes, though, that the *Rapanos* guidance may be difficult for the average person to apply. As a common sense way of evaluating whether you are required to be covered by a permit, if you were to install a new or replacement surface discharging system on your land, would effluent or pollutants (even diluted ones) from your system end up in a water of the United States or a conveyance, such as a ditch, drainage pipe, channel, tunnel, conduit, discrete fissure or other means that leads to a water of the United States? In evaluating this question, consider that rain water, irrigation activities, lawn sprinkling systems and any other ways that water can carry pollutants to waters of the United States. If so, even though pollutants would not be carried to waters of the United States unless your area experienced an exceptionally wet season, you are still required to obtain coverage under a permit. Only if you are sure that your system would not discharge pollutants to a water of the United States or a conveyance that leads to a water of the United States should you forego obtaining a permit for a surface discharging system. If you do not obtain a permit, but actually discharge, you may be subject to an enforcement action under the CWA.

33. *If a complaint is received by a local health department on an existing surface discharging system, does EPA get involved in the complaint investigation process?*

EPA's involvement depends upon the definition of an existing surface discharging system. EPA considers existing surface discharging systems to be those surface discharging systems that were installed and operational prior to the 10 February 2014 effective date of EPA's NPDES general permit for new and replacement surface discharging systems in Illinois. Therefore, to the extent that a local health department receives a complaint for an existing surface discharging system as described above, those particular systems are not covered by EPA's permit, and, thus, EPA would not get involved in a complaint investigation process.

If the complaint is for a system covered under EPA's general permit, then EPA will only get involved in a complaint investigation process to the extent that the complaint involves alleged noncompliance with the NPDES general permit. If NPDES permit noncompliance is alleged, EPA would welcome the local health department's sharing of all information that it has regarding the complaint and any information the department has as the result of its own investigation.

34. *If a property is so limited in area (due to well setback, space consumed by the building and driveways) that it is not physically possible to install any subsurface system, will a soils evaluation still be required? This could also apply to a site which is completely in a floodway. In such cases, I believe that the site evaluator could attest that there is no value to a soils evaluation being done, and therefore a soils evaluation should be waived.*

The final general permit requires that any NOI submitted include a technical feasibility analysis which is comprised of both the site evaluation and soil analysis. Even though there may be situations as described where there is limited area, a technical feasibility determination, which includes a soil analysis, is required for anyone seeking coverage under the general permit.

35. *Does EPA see any difference between treated effluent discharged from an NSF Standard 40 treatment plant and an NSF Standard 350 treatment plant as far as NPDES requirements or monitoring?*

The general permit has effluent limitations that Dischargers must achieve regardless of what technology is used. Therefore, all permit requirements and monitoring are the same regardless of whether the effluent is discharged from an NSF Standard 40 treatment plant, NSF 350 treatment plant, or any other type of treatment system.

36. *If a site contains only fill material, is a soil-based system exempted from such a site, or is the suitability totally based on the soil classifier's report?*

The general permit would allow the use of a soil-based system at a site containing fill material only. As at all sites, the site containing fill material would have to be evaluated to determine whether it can support various soil-based alternatives through completing the technical feasibility analysis (soil analysis and site evaluation).

37. *With regard to the 30-day wait period, if the applicant does not hear anything from the EPA, then the applicant knows that he or she can install a surface discharging system, and the local county health department can issue a construction/installation permit. Will the EPA call a local health department once EPA has made a decision whether to grant or deny coverage if that decision is made prior to 30 days upon EPA's receipt of the NOI?*
Although it is not specified in the permit, and the final permit does not require it, EPA will send a coverage letter to the applicant and copy the applicable local county health department. EPA will notify both the applicant and the applicable county health department by telephone and/or email, and follow up with a formal letter regarding the decision. Should EPA reach a decision prior to 30 calendar days following receipt of an NOI, EPA will notify the applicant and applicable county health department upon reaching its decision.

38. *Some properties in our county are very small and the only choice is to install an aeration unit. In these particular instances, why is it necessary to go through the Notice of Intent process?*
It is the applicant's decision whether or not to apply for coverage under the general permit, but if the applicant chooses the general permit, he or she must comply with the established application/NOI process. EPA's NPDES General Permit does not allow for an abbreviated application process for instances when particular sites, upon cursory inspection, appear to be inadequate to support an alternative to a surface discharging system. An abbreviated process could result in a premature decision ruling out a site's ability to support an alternative to a surface discharging system. Requiring everyone who applies for coverage under the general permit to submit a notice of intent and the required supplemental information ensures that each site, regardless of size, soil type, or other restrictive features are consistently evaluated based upon the technical and economic criteria established in the final permit.

D.4 Answers to Frequently Asked Questions

USEPA Prevention of Significant Deterioration Permit

1 *What is PSD?*
PSD is an acronym for Prevention of Significant Deterioration rules. These rules need to be addressed when a company is adding a new source or modifying an existing source in an attainment area. The PSD rules need to be addressed for the pollutants for which the area is classified as attainment with the National Ambient Air Quality Standards (NAAQS). PSD rules are designed to keep an area with "good" air in compliance with the NAAQS. The strategy with PSD rules assumes that minor new sources and minor modifications do not significantly affect the air quality. The distinctive requirements of PSD are Best Available Control Technology (BACT), air quality analysis–modeling (allowable increments), and analysis of impacts of the project on visibility, vegetation, and soils. Under PSD, if a source is classified as a major source for any one pollutant, then a significant increase in any pollutant (even one that the source is not major for) triggers PSD review.

2 *What is PSD increment?*
PSD increments prevent the air quality in attainment areas from deteriorating to the level set by the NAAQS. A PSD increment is the maximum allowable increase in concentration that is allowed to occur above a baseline concentration for a pollutant. The ambient air quality evaluated is the ambient air quality in effect at the time the minor source baseline dates were established. An increment is defined as the maximum increase in ambient air quality that is allowed above the conditions that exist on the date the baseline dates are set. At the time increments were established, states were permitting new and modified stationary sources to consume 100% of the available NAAQS. Congress with the 1977 amendments to the federal CAA and EPA set increments at specific percentages of the NAAQS to define the maximum increment of deterioration from existing air quality. Increments limit the deterioration in the air quality and is more stringent than the NAAQS in limiting emissions increases. Exceeding (i.e. overconsuming) an increment does not result in the same requirements as exceeding a NAAQS in that it does not trigger nonattainment. Instead, upon finding an increment has been over consumed, the federal rules in 51.16(a) require the state to develop a plan to rectify the overconsumption of increment. The plan is a SIP amendment. The PSD increments are contained in 40 CFR 51.166(c) and 40 CFR 52.21(c). The federal CAA allows any increment other than an annual increment to be exceeded once per year and not trigger any requirements to address increment over consumption. The baseline concentration is defined for each pollutant and is the ambient concentration existing at the time that the first complete PSD permit application affecting the area is submitted. Significant deterioration is said to occur when

the amount of new pollution would exceed the applicable PSD increment. It is important to note, however, that the air quality cannot deteriorate beyond the concentration allowed by the applicable NAAQS, even if not all of the PSD increment is consumed. Therefore, the NAAQS is the maximum allowable concentration "ceiling."

3 *What is a minor source baseline date?*
A minor source baseline date is the date when increment consumption/expansion by nonmajor sources begins. Minor source baseline dates are set for each air quality control region established by the state under requirements of the federal CAA (Section 107 – Planning areas). To set a minor source baseline date, the PSD application has to include the pollutant (NO_x, SO_2, PM_{10}, or $PM_{2.5}$) as

- a pollutant which will be emitted at or above the major source threshold triggering the need for PSD permitting, or
- a regulated pollutant which is emitted as a result of the project subject to PSD permitting at or above the significant emission rate, and
- the PSD application is accepted as complete.

The minor source baseline date denotes the date/calendar year that a state needs to track and retain an emission inventory for the air quality control region, where the baseline date has been set. According to the federal CAA, the baseline date denotes the baseline concentration of pollutants in the air as determined by ambient monitoring date acquired by EPA and the states. After the minor source baseline date, any increase in actual emissions from both major and minor sources consumes the PSD increment for that area.

4 *What is the major source baseline date?*
The major source baseline dates for SO_2 and TSP/PM_{10} were set by congress in the 1977 amendments to the federal CAA. As directed in the law, EPA established the dates for NO_x by rule in 1988 and for $PM_{2.5}$ by rule in 2010. EPA justified that $PM_{2.5}$ was a different pollutant than TSP/PM_{10} in its rulemaking to establish the increment. Between the major source baseline date and the minor source baseline date, only major sources can consume or expand increment. After the minor source baseline date, permitting of major sources must determine how much increment has been consumed/released by the major source and how much increment remains in the area of the source.

5 *Can a minor source consume or expand increment before the minor source baseline date is triggered?*
No. Minor source changes before the minor source baseline date has been triggered have no effect on the available increment.

6 *Can a major source consume or expand increment before the minor source baseline date?*
Yes. Starting on the major source baseline date and ending on the minor source baseline date, only changes at major sources can consume or expand increment. Major sources also continue to consume and expand increment after the minor source baseline date.

7 *Can a minor source consume or expand increment after the minor source baseline date is triggered?*
Yes. A minor source in existence on the date the minor source baseline date has been set can consume or expand increment. Minor sources coming into existence after the minor source baseline date consumes increment until it closes or implements more stringent emission controls.

8 *What is a "netting exercise"?*
A netting exercise is a demonstration by the applicant which sums up emission changes which have occurred at a source over a contemporaneous time period. The resulting emission changes are then reviewed to determine if the proposed project must undergo PSD. Netting is used when a proposed project is significant by itself, but the applicant wants to avoid PSD rule applicability by taking into account other emission decreases which have occurred over the contemporaneous timeframe. The applicant must also account for other emission increases. Thus, when summing the net emissions decrease from other projects with the emissions increase from the proposed project, the overall net increase in emissions would not be significant. For PSD, netting is only required if the proposed project by itself has significant emissions. For example, a major source could have three unrelated projects of 20 tons of NO_x each; in an attainment area over a five-year period whose accumulated emissions would be significant, but netting and hence PSD review is not required because the projects were not related. Note, however, that a deliberate decision to split an otherwise "significant" project into smaller projects to avoid PSD review would be viewed as circumvention. In determining if projects are related, you need to ask two basic questions:

1) Were the projects proposed over a relatively short period of time?
2) Could the changes be considered part of a single project or business decision?

9 *What are contemporaneous emission decreases and when are they needed?*
The term emission decreases is most often used to refer to a decrease in emissions at a source which is used to counterbalance or compensate for an emissions increase in a netting exercise. For example, if a proposed project will result in an increase of 100TPY of a pollutant, but three years earlier the source removed 80TPY unit of the same pollutant, then 80TPY of decreases are available such that the contemporaneous increase from the proposed project is only 20TPY. Use of emission decreases in this manner is not "needed" but is an option available to the applicant under PSD.

10 *Does one ever need to go beyond the contemporaneous time period to consider past emissions increases?*
Yes. One cannot relax restrictions placed on a project to avoid status as a major project without considering whether the project would then have been major when originally permitted. If the project would become major with a requested relaxation in limits, appropriate PSD must be imposed as part of the issuance of a revised permit.

11 *What is a "Sham" permit?*
Sham is defined as counterfeit, untrue, or fake. A Sham permit is when a source pursues a permit limit on the potential to emit (PTE) for a proposed project in order to limit the source to minor source levels as a means of circumventing the requirements of PSD. Most often this term applies to construction situations when a company wants to expedite commencement of construction so they are willing to take what they consider temporary limits on PTE such that the proposed project is not required to undergo PSD. Therefore, the company can get their permit without any of the PSD associated delays. The company's intent in such cases would be to remove the limiting permit conditions prior to normal operation, or shortly thereafter (e.g. request a revision to the permit or appeal the permit). Another circumstance which may occur is when a major project is broken up into several smaller minor projects in order to avoid PSD requirements.

12 *What is "debottlenecking"?*
When the output of a multistep process is limited by the capacity of one unit or activity, that unit or activity is a bottleneck. Debottlenecking this step in a process can increase the capacity of the other steps both upstream and downstream. Debottlenecked emissions increase must be counted as part of a project's emissions increase. For example, if a paper cutter at the end of the printing line is replaced and the new paper cutter can handle a larger volume of paper quicker such that more printing will be performed, and emissions will increase. The cutter itself has no emissions. However, by replacing the cutter, the entire process line has been debottlenecked, and the process line's emissions were increased.

13 *What does "relied on" mean in regard to creditable emissions increases and decreases in a netting exercise and what are its implications?*
In PSD rules (40 CFR 52.21), it is stated that emissions increases and decreases are creditable to the extent that they were not previously relied on [40 CFR Part 52.21(b)(3)(iii)]. "Relied on" means that a permit was issued where the proposed project was actually required to meet PSD requirements and such decreases and increases were accounted for in any modeling or analysis of the use of available PSD increments. As per EPA guidance MEMO dated 29 December 1989, there are situations, such as when a source nets out of review, when the permitting authority does not rely on creditable emissions increases or decreases in issuing a PSD permit. For example, when a source nets out of review, no PSD permit is issued. As such, the reviewing authority has not relied on any creditable emissions increases or decreases in issuing a permit, so the emissions increases and decreases are still available for future applications.

14 *What do the terms "synthetic minor" and "natural minor" generally mean?*
A person may mean the following when using these terms, although they are not specifically defined in any known PSD rules or literature. The term "synthetic minor" is generally used to describe a source that has permit conditions which limit its potential-to-emit (PTE) to less than major source levels, but whose PTE in the absence of any permit conditions would be above major source levels. The term "natural minor" is generally used to describe a source whose PTE is less than major source levels in the absence of any permit conditions.

15 *What criteria do you use to determine if two facilities which are close by and related should be treated as the same source or separate sources for purposes of PSD?*
The PSD rules provide that facilities should be considered a single stationary source if they meet all of the following three criteria:

1) Belong to the same SIC major (two-digit) group. If the facilities could have separate SICs but a support relationship exists (e.g. 50% of the product of one is utilized by the other), then one facility is considered

a support facility and this criterion shall be considered met.

2) Are located on one or more contiguous or adjacent properties (in the same general area).
3) Are under common ownership or control. If the applicant challenges the existence of common control, it may be necessary to look at the contractual agreement between the facilities to determine if they are under common control.

In reviewing applications, one must check these factors carefully if the circumstances at the source are not straightforward. In particular, is a company trying to improperly separate a single source into multiple plants in order to avoid PSD applicability? Is a company trying to improperly claim contemporaneous decreases from another source to avoid PSD applicability?

16 *An engine manufacturing plant sprays VOC contaminated wastewater into the air to dispose of VOC. Is the activity, if new, subject to PSD?*
If the source has non-fugitive emissions greater than 250 TPY, the new emissions which are fugitive would count in determining PSD applicability. The only place fugitives are given special treatment is in determining if the source is subject to PSD review.
Reference: 40 CFR §52.21(i)(4)(vii)

17 *A major source makes a physical change which increases emissions by a "significant" amount. The project also has offsetting reductions elsewhere at the same time. In the past five years, however, there have been other increases such that the net emission increase result over five years is greater than the significant emission rate for at least one pollutant. Is the new physical change subject to PSD review?*
Yes. PSD review is triggered if

1) the proposed change, by itself, without counting any offsetting reductions, will result in a significant emissions increase and
2) when facility-wide increases and decreases over the past five years are considered, a significant net emissions increase results.

Emissions decreases are not counted in "Step 1" of the PSD applicability analysis.
Reference: 40 CFR §52.21

18 *A major source wishes to take two actions: (i) increase production at a previously PSD-permitted emission unit and (ii) build a new emission unit whose emissions will be less than the SER. Emissions of fluorides from the two actions, when added together, are greater than the SER and occur within the contemporaneous time frame. Does the physical change (new unit) trigger PSD review because of the change in actual emissions at the previously permitted units being greater than the SER?*
No. Unless the production rate of the previously permitted unit was limited in the PSD permit. An increase in production is exempt from PSD review, unless the production rate is limited in the permit. Also, the PTE of the new unit by itself must be greater than the SER to trigger review. Unless the increase in emissions from the previously permitted emission unit was a result of the new emissions unit (i.e. debottlenecking).
Reference: 40 CFR §52.21

19 *Is an iron foundry one of the 28 PSD categories?*
Yes. It is a secondary metal production plant if it uses scrap metal to produce iron, even if the metal is poured into molds.
Reference: 40 CFR §52.21(b)(1)(i)(a)

20 *Is a whiskey distillery one of the 28 categories (chemical process plants) listed in 40 CFR §52.21(b)(1)(i)(a)?*
No. A chemical process plant is any establishment in Major Group 28 of the SIC Code. Beverage distilleries are in Major Group 20.
Reference: 40 CFR §52.21(b)(1)(i)(a)

21 *A major stationary source wishes to make a physical change resulting in a 15 TPY increase in particulate matter emissions. Less than five years ago, the source had a production increase (not subject to PSD permitting) resulting in a 50 TPY increase in SO_2. There were no increases in particulate matter emissions at the source over the past five years. Is the proposed increase subject to PSD permitting?*
No. The triggering increase must be of the same pollutant as the one for which a significant increase results. Also, the proposed physical change must be greater than the SER by itself. If the PM10 emissions is 15 *TPY* or PM2.5 emissions is 10 *TPY* or more, then PSD applies.
Reference: 40 CFR §52.21

22 *A boiler at a major stationary source has been shut down for 11 years. At the time of the shutdown, extensive efforts were made to keep the boiler from deteriorating. During the shutdown period, this maintenance has continued. A recent inspection by the manufacturer shows that very little effort would be required to return the boiler to service. The air-operating permit has been allowed to expire. The owner maintains that the boiler was always intended to be used at some time in the future. Is the returning to service of the boiler subject to PSD review?*

No. Normally, a shutdown of greater than two years is considered permanent. If, however, the owner demonstrates that the shutdown was not intended to be permanent, the shutdown may be considered temporary. If the shutdown is considered temporary, a start-up would not be subject to PSD review. The "acid test" is whether the shutdown is permanent. In any case, the increase would be considered an increase in actual emissions for any future net increase calculation and for increment consumption purposes.

23 *A minor source which adds emissions of a pollutant in a major amount (100 TPY if one of the 28 listed source categories or 250 TPY if not listed) is subject to PSD review as a new major source, rather than as a modification. The netting concept is used only in the definition of major modification, and not in the definition of major stationary source. This seems to indicate that a minor source adding a major emission unit could not escape PSD review by considering previous decreases which cause the net increase to be less than the major source threshold. Is this the case?*

Yes. For example, suppose a minor source emitting 200 TPY had a decrease in actual emissions in 2010 of 50 TPY, leaving 150 TPY. In 2012, 260 TPY is proposed to be added. If the 50 TPY reduction could be used to offset the 260 TPY increase, the increase would be only 210 TPY and the source would escape review. The 50 TPY decrease cannot be used, however, so the 260 TPY increase is subject to review as a new major stationary source.

Reference: 40 CFR §52.21(b)(1)

24 *An existing source is major only because its SO_2 emissions are 120 TPY. The source proposes to add 60 TPY of particulate emissions. At the same time, the source is willing to accept a new, federally enforceable limitation which lowers its SO_2 emissions to 90 TPY. Is the proposed addition of 60 TPY of particulate subject to PSD review?*

No. Since the source will not be major after the change, the action is not subject to PSD review.

Reference: 40 CFR §52.21(b)(2)(i)

25 *The PSD baseline air quality is based on actual emissions from existing sources. Actual emissions are defined as the average emissions rate in TPY. How does Ecology interpret this in establishing short-term (24-hour, 3-hour) baseline air quality levels when air quality modeling is used?*

Baselines for 3- and 24-hour averages should be set using the maximum 3-hour average or 24-hour average emission rate of the existing source, respectively, which occurred during the period over which the annual emission rate was determined. For example, if a source's annual emission rate is determined to be 430 TPY by averaging 400 TPY in 1978 and 460 TPY in 1979, the 3-hour baseline emission rate would be the maximum 3-hour average emission rate which occurred during the period of 1978 and 1979.

For short-term PSD modeling, only the new and contemptuous equipment is modeled. Equipment with increase in utilization is not part of the short-term modeling.

26 *An ambient monitor was operated for one year (or shorter time, if representative of highest values) and then shut down. A proposed source wishes to use the data for its PSD application. Except for the time lapse, the data is representative of current air quality at the proposed site, is of good quality, and was gathered entirely in a time period less than three years before the source submits its application. Can the data be used, even though the monitor has been shut down?*

As long as all the data needed in the application are collected sequentially, and all the data are collected sometime in the previous three years, the timing requirement is satisfied. For example, suppose an agency operated an ozone monitor throughout a particular ozone season, which the agency determines to be April through September of 2010. The monitor is then shut down. This data could be used in a PSD application submitted any time before 1 April 2013, provided the data are still representative of current conditions, and all other requirements are met, such as quality assurance and monitor location.

Reference: 40 CFR §52.21(m); 45 FR 52724

27 *A minor source locates in a PSD area where the baseline has been triggered. In another nearby PSD area, the baseline is still not triggered after the minor source begins operation. The source's emissions impact this neighboring area. Do these emissions consume increment?*

No. The baseline air quality is that which actually exists in the baseline area on the baseline date, minus contributions from new major sources. Therefore, at some future baseline date for the neighboring area, the baseline air quality must include the actual contribution from the minor source. Since the emissions are in the baseline for the area, they do not consume increment. If the situation is reversed (minor source locates in an area that has not been triggered, impacts a triggered area), emissions would consume increment in the neighboring area, but not in the area where the source locates.

D.5 Dispersion Model Development and Application

Goal: Understand how to predict the time averaged concentration of contaminants emitted into the atmosphere. Real emission sources vary in complexity from a single elevated source

to complicated industrial complexes with multiple emission points.

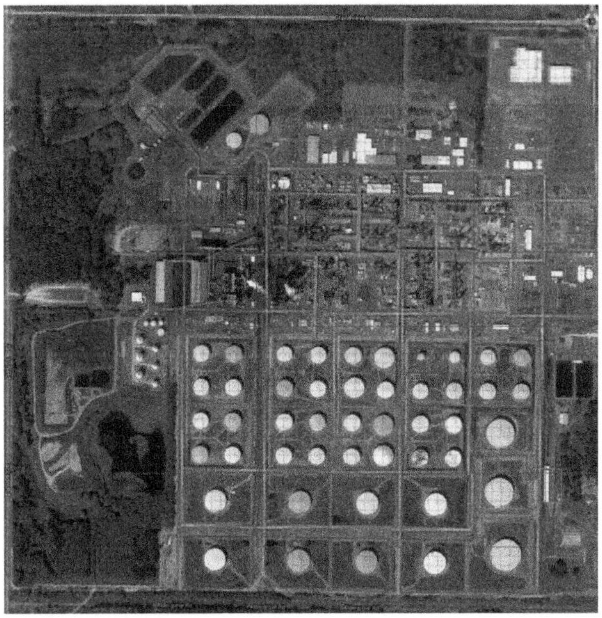

The ability to predict concentrations produced by emissions from a variety of sources is important from both forensic and predictive standpoints. Air pollution law in most industrial countries regulates contaminant concentrations and requires a determination before granting operating permits. Generally, regulations require that a computer model be run to predict the expected ground-level concentration from new or increased emissions or structural changes that will affect the dispersion of emitted contaminants. This handout will guide you through the concepts and some mathematics of dispersion models and conclude with examples using computer models to predict concentrations.

D.5.1 Model Development Is Based upon Several Assumptions

Molecular diffusion is insignificant when compared to the dispersion produced by atmospheric turbulence in both convective and stable conditions in the layer of the atmosphere between the Earth surface and an upper boundary that lies between a few tens of meters above the surface and a few thousand meters depending of meteorological conditions. This layer is called the planetary boundary layer (PBL). The PBL may be either convective (CBL), dominated by turbulent plumes, or more quiescent (stable) with much smaller fluctuations in the vertical and cross-wind directions (SBL).

Theory predicts and observations confirm that in the free atmosphere the vertical and horizontal concentration profiles have nearly Gaussian distributions about the plume centerline.

If the emission is visualized as a series of overlapping puffs, then equations for the growth of each puff are just the familiar Gaussian equation.

Let us begin with the familiar Gaussian distribution.

Gaussian or Normal Distribution

$$f(x) = \frac{1}{\sigma(2\pi)^{1/2}} \exp\left[\frac{-(x - x_0)^2}{2\sigma^2}\right] \tag{D.1}$$

- x = mean of the distribution
- σ = standard deviation

Gaussian distribution is used to model probabilities, and in this context formula is used to predict steady-state concentration at a point downstream.

Convention is to set up a Cartesian coordinate system with the x-axis extending downwind from the source, the z-axis extends in the vertical, and the y-axis in the cross-wind direction (Figure D.1).

In the plume from a continuously emitting source, maximum concentration decreases, but the width of the plume

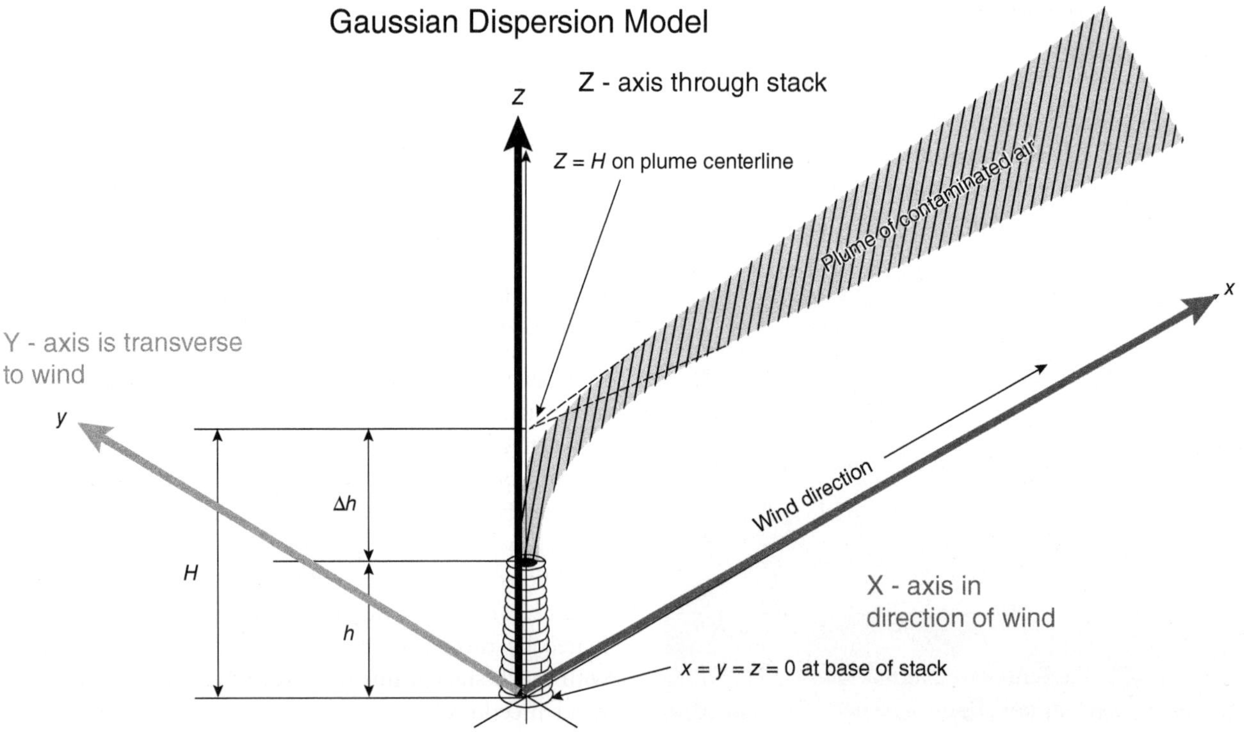

Figure D.1 Coordinate system and nomenclature for the Gaussian plume idea.

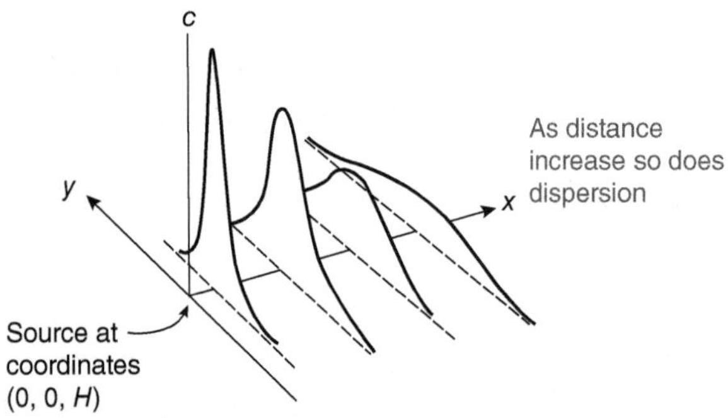

Figure D.2 Behavior of the downwind, elevated transverse concentration profiles as a function of distance downward.

increases as the plume moves downwind as shown in Figure D.2 (note: the vertical coordinate is concentration).

Observations

1) The rates of both the horizontal and vertical spreading depend on atmospheric conditions.
   a) Atmospheric conditions depend on
      i) incoming short wave radiation, e.g. sunlight
      ii) outgoing long wave radiation, mainly affected by degree of cloudiness
      iii) terrain surface roughness
      iv) terrain surface moisture
      v) wind speed
2) Buoyancy affects the plume centerline to produce a change in the height depending on the difference in temperature between the plume and the surrounding atmosphere.

The idealized concentration at any point downwind of an elevated source at height H is given by "emission rate of pollutant, wind speed, distance, and plume standard deviations in the horizontal and vertical directions."

Point Source at Elevation H

- Assumes no interference or limitation to dispersion in any direction

$$C(x,y,z)=\frac{Q}{2\pi u\sigma_y\sigma_z}\exp\left[\frac{-(y-y_0)^2}{2\sigma_y^2}\right]\exp\left[\frac{-(z-z_0)^2}{2\sigma_z^2}\right] \tag{D.2}$$

x_0 and z_0 are locations of centerline of plume
y_0 is taken as the base of the stack
z_0 is H
Q = emission strength of source (mass/time) – g/s
u = average wind speed through the plume – m/s
C = concentration – g/m^3 (notice this is not ppm)
σ_y and σ_z are horizontal and vertical standard deviations in meter

However, permitting decisions are based on the concentration at breathing height which must consider the effect of the ground surface on the plume dispersion. Because the deposition velocity of gases and most particulates of concern are small compared with the vertical velocities of atmospheric turbulence, the Earth surface behaves as a reflecting surface which modifies the dispersion equation. This effect is modeled as a virtual (imaginary) source shown in Figure D.3.

Point Source at Elevation *H* with Reflection

$$C(x,y,z)=\frac{Q}{2\pi u\sigma_y\sigma_z}\left[\exp-\left(\frac{y^2}{2\sigma_y^2}\right)\right]\left\{\exp\left[\frac{-(z-H)^2}{2\sigma_z^2}\right]+\exp\left[\frac{-(z+H)^2}{2\sigma_z^2}\right]\right\} \quad \text{(D.3)}$$

Recalling that the Gaussian distribution may be defined by two parameters, its maximum value and a measure of its spread, we may calculate the concentration at any location by estimating them. Obviously, the maximum value is related to the emission rate and the spread depends on atmospheric parameters including time. Wikipedia has a good review: https://en.wikipedia.org/wiki/Outline_of_air_pollution_dispersion.

Gaussian Dispersion Model

- For localized point sources – stacks
- General appearance
- Plume exits at height, h_s
  Rises an additional distance, Δh
  - buoyancy of hot gases
  - called plume rise
  - reaches distance where buoyancy and upward momentum cease
- Exit velocity, V_s
- Plume appears as a point source emitted at height $H = h_s + \Delta h$
- Emission rate Q (g/s)
  Assume wind blows in x direction at speed u
  - u is independent of time, elevation, or location (not really true)

In addition to point sources (stacks) important sources of pollutants of concern look more like lines (e.g. roads), areas (e.g. settling ponds, unpaved parking lots), or volumes (pollutants are emitted within the influence of a large building). Equations for line and area sources are derived by integrating the previous equation in one or two dimensions. Volume sources have initial dimensions which affect the dispersion equation.

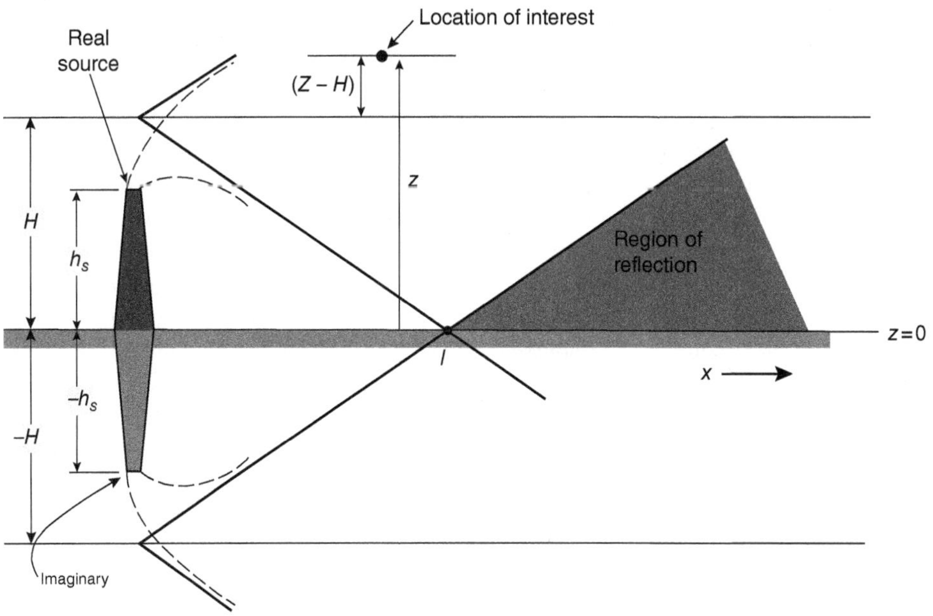

Use of an imaginary source to decribe mathematically gaseous reflection at the surface of the earth.

Figure D.3 Use of an imaginary source to describe mathematically gaseous reflection at the surface of the Earth.

D.5.2 Information Required for Modeling the Dispersion of Emissions

- Characterize the type of source
  Point, line, area, volume
- Characterize the emissions temperature
  Vertically directed exit velocity
  Initial dimension, e.g. stack diameter
  Emission rate of pollutant of concern
- Characterize the immediate area around the source
  Isolated source|nearby buildings
  Flat|Significant nearby terrain
  Land use classification, e.g. agricultural, forest, desert
- Characterize the meteorology wind speed and wind direction
  Vertical temperature gradient, i.e., lapse rate
- Existing (background) ambient concentrations of the pollutant of concern
- Develop a plan for predicting the ground-level concentrations produced by the source, often called a modeling protocol
  Obtain protocol approval from the appropriate entities

The currently accepted steady-state dispersion model for evaluating the ground-level concentrations in the United States is AERMOD. The following is taken from the AERMOD Model Formulation and Evaluation report, EPA-454/ R-17-001, May 2017.

In the following AERMIC stands for the **A**MS/**E**PA **R**egulatory **M**odel **I**mprovement **C**ommittee comprised of American Meteorological Society (AMS) and US Environmental Protection Agency (EPA) scientists.

> The AERMOD modeling system consists of two preprocessors and the dispersion model. The AERMIC meteorological preprocessor (AERMET) provides AERMOD with the meteorological information it needs to characterize the PBL. The AERMIC terrain pre- processor (AERMAP) both characterizes the terrain, and generates receptor grids for the dispersion model (AERMOD).
>
> AERMET uses meteorological data and surface characteristics to calculate boundary layer parameters (e.g. mixing height, friction velocity, etc.) needed by AERMOD. This data, whether measured off-site or on-site, must be representative of the meteorology in the modeling domain. AERMAP uses gridded terrain data for the modeling area to calculate a representative terrain-influence height associated with each receptor location. The gridded data is supplied to AERMAP in the format of the Digital Elevation Model (DEM) data. The terrain preprocessor can also be used to compute elevations for both discrete receptors and receptor grids.
>
> AERMOD is a steady-state plume model. In the stable boundary layer (SBL), it assumes the concentration distribution to be Gaussian in both the vertical and horizontal. In the convective boundary layer (CBL), the horizontal distribution is also assumed to be Gaussian, but the vertical distribution is described with a bi-Gaussian probability density function (pdf). Additionally, in the CBL, AERMOD treats "plume lofting," whereby a portion of plume mass, released from a buoyant source, rises to and remains near the top of the boundary layer before becoming mixed into the CBL. AERMOD also tracks any plume mass that penetrates into the elevated stable layer, and then allows it to re-enter the boundary layer when and if appropriate. For sources in both the CBL and the SBL AERMOD treats the enhancement of lateral dispersion resulting from plume meander.
>
> Using a relatively simple approach, AERMOD incorporates current concepts about flow and dispersion in complex terrain. Where appropriate, the plume is modeled as either impacting and/or following the terrain. This approach has been designed to be physically realistic and simple to implement while avoiding the need to distinguish among simple, intermediate and complex terrain, as required by other regulatory models. As a result, AERMOD removes the need for defining complex terrain regimes.

D.5.3 Screening Modeling

EPA's currently accepted screening dispersion model is AERSCREEN. In keeping with the screening approach of the past four decades, AERSCREEN calculates the expected maximum ground-level concentration for multiple averaging periods (1-hour to annual) for a single emission point. It uses the primary components of the AERMOD modeling system but uses an additional program, MAKEMET, to create the meteorological conditions that have been shown to produce the maximum ground-level concentration from the source and land classification/land use characteristics. A quick round of screening modeling is often all that is required to demonstrate compliance with air quality standards.

The MAKEMET program has a long complex history, dating back over 10 years. Suffice it to say that the version of MAKEMET supplied with AERSCREEN operates in the following manner. As mentioned earlier, MAKEMET generates a matrix of meteorological conditions for application of AERMOD in a screening mode and output the results in the form of AERMOD-ready surface and profile meteorological data files. The matrix is generated based on looping through a range of wind speeds, cloud covers, ambient temperatures, solar elevation angles, and convective velocity scales (w^* for convective conditions only) for user-specified surface characteristics (Z_0, B_0, r). For stable cases, the mechanical mixing height (Z_m) is calculated based on the friction velocity, u^*. A loop through Z_m factors (multiplied times the initial value calculated form u^*) is also included to account for smoothing of Z_m that occurs with refined AERMET data. Stable transition cases with solar angle greater than zero but less than the critical solar angle (ACRIT) are also included in the matrix.

AERSCREEN is an interactive wrapper for the AERMOD modeling system and computes input values according to user supplied values in response to prompts printed to the screen.

Before running AERSCREEN, users should consult and become familiar with the following:

AERMOD User's Guide (USEPA 2016a)
AERMAP User's Guide (USEPA 2016b)
AERMET User's Guide (for surface characteristics tables) (USEPA 2016c)
BPIPPRM User's Guide (USEPA 2004) – relates to building effects on dispersion
AERSURFACE User's Guide (USEPA 2008)
AERMOD Implementation Guide (USEPA 2016e)

Although the next two references are specific to the United States, there is useful information that may be important to dispersion modeling in general.

Guideline on Air Quality Models (Appendix W) (USEPA 2016d)
Screening Procedures for Estimating the Air Quality Impact of Stationary Sources (USEPA 1992)

A link to AERSCREEN (Version 16216) code including the latest MAKEMET code (Version 16216) is provided below along with documentation and test cases. The MAKEMET program interfaces with AERSCREEN to generate a site-specific matrix of screening meteorological conditions based on user inputs for input into AERMOD.

AERSCREEN code and executables (v16216) (ZIP)(759 K)
MAKEMET code and executables (v16216) (ZIP)(355 K)

Please note that before using AERSCREEN, users must have the latest *AERMOD executable* (18081) as well as the *AERMAP* terrain preprocessor and *BPIPPRM* executables. These programs can be downloaded from the *AERMOD Modeling System* section.

D.5.4 AERSCREEN Examples

Figure D.4 was produced in response for an analysis of emissions from an anaerobic digester used to produce methane for power generation. Analysis showed that the methane also contained sulfur compounds. It provides an example of the effect of stack height on ground level

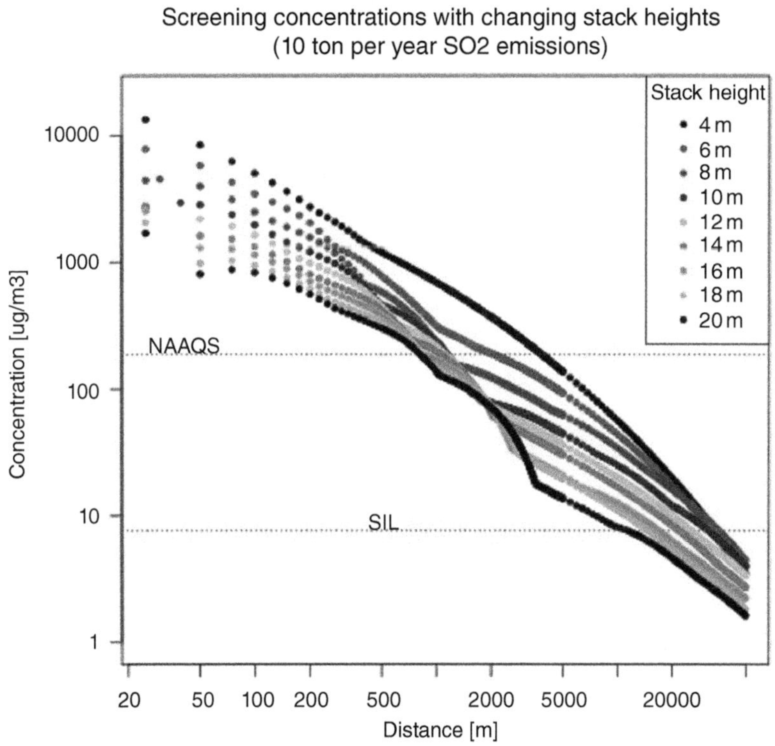

Figure D.4 SO_2 concentration with changing heights.

concentration. It also demonstrates that a very small, 10 TPY source of SO_2 with a short stack may exceed the US standard to several thousand meters from the stack. The Air Quality Impact Analysis (AQIA) starts with preliminary modeling for each pollutant to determine whether an applicant can forego detailed analysis and preconstruction monitoring. If the projected ambient concentration increase for a given pollutant is below the PSD Significant Impact Levels (SILs) for each averaging period, no further analysis of the ambient impact is required for that pollutant. For those pollutants with averaging periods that have impacts greater than the SIL, a full impact analysis (taking into account other increment consuming sources) is used to demonstrate compliance with PSD increments and NAAQS (national ambient air quality standard). Many SO_2 sources emit thousands of tons but may have stacks as high as 300 m. AERSCREEN can quickly provide an estimate of the maximum ground-level impact and distances of significant contribution to the ambient concentration. Note that the concentration axis is logarithmic.

Figure D.5 shows the effect that land use, e.g. water, forest, grassland, has on ground-level concentration for the same 10 TPY source. It is also a good demonstration of the rule of thumb that the maximum ground-level concentration occurs at approximately ten times the stack height from the source, e.g. for this 10 m stack height the maximum ground-level concentrations occurs about 100 m downwind.

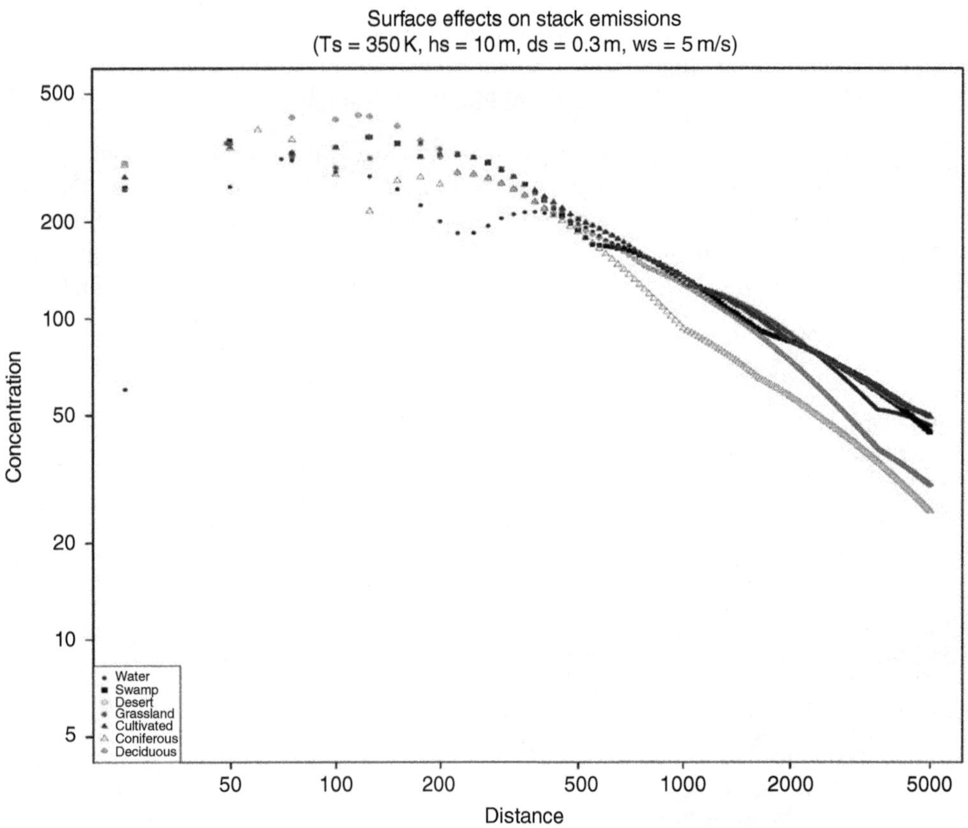

Figure D.5 Effects on stack emissions and land use.

D.6 Hazardous Waste Manifest

Please print or type. *(Form designed for use on elite (12-pitch) typewriter.)* *Form Approved. OMB No. 2000-0404. Expires 7-31-86*

UNIFORM HAZARDOUS WASTE MANIFEST

1. Generator's US EPA ID No. | Manifest Document No.

2. Page 1 of

Information in the shaded areas is not required by Federal law.

3. Generator's Name and Mailing Address

4. Generator's Phone ( )

A. State Manifest Document Number

. State Generator's ID

5. Transporter 1 Company Name

6. US EPA ID Number

C. State Transporter's ID

D. Transporter's Phone

7. Transporter 2 Company Name

8. US EPA ID Number

E. State Transporter's ID

F. Transporter's Phone

9. Designated Facility Name and Site Address

10. US EPA ID Number

G. State Facility's ID

H. Facility's Phone

GENERATOR

11. US DOT Description *(Including Proper Shipping Name, Hazard Class, and ID Number)*	12. Containers No.	12. Containers Type	13. Total Quantity	14. Unit Wt/Vol	I. Waste No.
a.					
b.					
c.					
d.					

J. Additional Descriptions for Materials Listed Above

K. Handling Codes for Wastes Listed Above

15. Special Handling Instructions and Additional Information

16. GENERATOR'S CERTIFICATION: I hereby declare that the contents of this consignment are fully and accurately described above by proper shipping name and are classed, packed, marked, and labeled, and are in all respects in proper condition for transport by highway according to applicable international and national government regulations.

Unless I am a small quantity generator who has been exempted by statute or regulation from the duty to make a waste minimization certification under Section 3002(b) of RCRA, I also certify that I have a program in place to reduce the volume and toxicity of waste generated to the degree I have determined to be economically practicable and I have selected the method of treatment, storage, or disposal currently available to me which minimizes the present and future threat to human health and the environment.

Printed/Typed Name | Signature | *Month Day Year*

TRANSPORTER

17. Transporter 1 Acknowledgement of Receipt of Materials

Printed/Typed Name | Signature | *Month Day Year*

18. Transporter 2 Acknowledgement of Receipt of Materials

Printed/Typed Name | Signature | *Month Day Year*

FACILITY

19. Discrepancy Indication Space

20. Facility Owner or Operator: Certification of receipt of hazardous materials covered by this manifest except as noted in Item 19.

Printed/Typed Name | Signature | *Month Day Year*

EPA Form 8700 22 (Rev. 4 85) Previous edition is obsolete.

References

- **Max/Min Temperatures**
  - http://www.wrcc.dri.edu/summary/Climsmwa.html
- **Population**
  - http://www.ofm.wa.gov/pop/april1/poptrends.pdf
- **AERSCREEN User's Guide**
  - http://www3.epa.gov/ttn/scram/dispersion_screening.htm
- **Background Concentration**
  - http://lar.wsu.edu/nw-airquest/lookup.html
- **National Land Cover Database**
  - http://www.mrlc.gov/viewerjs
- **National Elevation Database**
  - http://viewer.nationalmap.gov/basic
  - Also includes NLCD access
- **EPA** Support Center for Regulatory Atmospheric Modeling (SCRAM)
  - https://www.epa.gov/scram/air-quality-dispersion-modelingscreening-models

USEPA (1992). *Screening Procedures for Estimating the Air Quality Impact of Stationary Sources*. Research Triangle Park, NC: USEPA.

USEPA (2004). *User's Guide to the Building Profile Input Program*, EPA-454/R-93-038. Research Triangle Park, NC: USEPA.

USEPA (2008). *AERSURFACE User's Guide*, EPA-454/B-08-001. Research Triangle Park, NC: USEPA.

USEPA (2016a). *User's Guide for the AMS/EPA Regulatory Model – AERMOD*, EPA-454/B-16-011. Research Triangle Park, NC: USEPA.

USEPA (2016b). *User's Guide for the AERMOD Terrain Preprocessor (AERMAP)*, EPA454/B-16-012. Research Triangle Park, NC: USEPA.

USEPA (2016c). *User's Guide for the AERMOD Meteorological Preprocessor (AERMET)*, EPA-454/B-16-010. Research Triangle Park, NC: USEPA.

USEPA (2016d). *Guideline on Air Quality Models (Appendix W of 40 CFR Part 51)*. Research Triangle Park, NC: USEPA.

USEPA (2016e). *AERMOD Implementation Guide*, EPA-454/B-16013. Research Triangle Park, NC: USEPA.

USEPA (2016f). *AERSCREEN User's Guide*, EPA-454/B-16-004. Research Triangle Park, NC: USEPA.

Appendix E

Industrial Hygiene Outlines

Table E.1 Highlight and brief explanation about industrial hygiene types and techniques used in chemical and allied industries.

a) *Inherently safer workplace – eliminate or reduce hazard*
   - Eliminate chemical entirely
   - Reduce chemical inventories, including raw materials, intermediates, and products
   - Replace chemical with less hazardous chemical
   - Decrease temperature and pressure of chemical
   - Reduce pipeline size to reduce hold-up inventory
b) *Enclosures – enclose room or equipment and place under negative pressure*
   - Enclose hazardous operations such as sample points
   - Seal room, sewers, ventilation, and the like
   - Use analyzers and instruments to observe inside equipment
   - Shield high-temperature surfaces
c) *Local ventilation – contain and exhaust hazardous substances*
   - Use properly designed hoods
   - Use hoods for charging and discharging
   - Use ventilation at drumming station
   - Use local exhaust at sample points
   - Keep exhaust systems under negative pressure
d) *Dilution ventilation – design ventilation systems to control low-level toxics*
   - Design locker rooms with good ventilation and special areas or enclosures for contaminated clothing
   - Design ventilation to isolate operations from rooms and offices
e) *Wet methods – use wet methods to minimize contaminations with duds*
   - Clean vessels chemically vs. sandblasting
   - Use water sprays for cleaning
   - Clean areas frequently
   - Use water sprays to shield trenches or pump seals
f) *Good housekeeping – keep toxicants and dusts contained*
   - Use dikes around tanks and pumps
   - Provide water and steam connections for area washing
   - Provide lines for flushing and cleaning
   - Provide well-designed sewer system with emergency contaminant
g) *Personal protection – a last line of defense*
   - Use safety glasses and face shields
   - Use aprons, arm shields, and space suits
   - Wear appropriate respirators; airline respirators are required when oxygen concentration is less than 19.5%

Source: Crowl and Louvar (2011).

Industrial Environmental Management: Engineering, Science, and Policy, First Edition. Tapas K. Das.

Companion website: www.wiley.com/go/Das/IEM_1e

Reference

Crowl, D.A. and Louvar, J.F. (2011). *Chemical Process Safety*, 3e. Upper Saddle River, NJ: Prentice Hall.

Appendix F

Environmental Cost-Benefit

The cost-benefit analysis and economic input–output model are available at the website www.eiolca.net. An *economic input–output life cycle assessment*, or *EIOLCA,* involves the use of aggregate sector-level data to quantify the amount of environmental impact that can be directly attributed to each sector of the economy and how much each sector purchases from other sectors in producing its output. Combining such data sets can enable accounting for long chains (e.g. building an automobile requires energy, but producing energy requires vehicles, and building those vehicles requires energy, etc.), which somewhat alleviates the scoping problem of traditional life cycle assessments. EIOLCA analysis traces out the various economic transactions, resource requirements, and environmental emissions (including all the various manufacturing, transportation, mining, and related requirements) required for producing a particular product or service.

EIOLCA relies on sector-level averages that may or may not be representative of the specific subset of the sector relevant to a particular product. To the extent that the good or service of interest is representative of a sector, EIOLCA can provide very fast estimates of full supply chain implications for that good or service.

Researchers at the Green Design Institute of Carnegie Mellon University began developing a web-based tool for performing an EIOLCA in the 1990s. The underlying software traces out the various economic transactions, resource requirements, and environmental emissions associated with the production of a particular product or service. The model captures all the various manufacturing, transportation, mining, and related requirements to produce a product or service. For example, one might wish to trace out the implications of purchasing \$46 000 of reinforcing steel and \$104 000 of concrete for a kilometer of roadway pavement. Environmental implications of these purchases can be estimated using EIOLCA. The current (2002) model is based upon the Department of Commerce's 428 sector industry input–output model of the US economy.

In 2018, VitalMetrics Group, an environmental consultancy, developed a web-based Spend Analysis Tool for quantifying the environmental impacts associated with an organization's entire upstream supply chain. It is compliant with the approach for quantifying spend-based impacts defined in the Greenhouse Gas Protocol Corporate Value Chain Accounting and Reporting Standard. The tool utilizes the Comprehensive Environmental Data Archive (CEDA), a peer-reviewed EIOLCA database with a base year of 2014. CEDA represents 389 industrial sectors, the commodities and the linkages between them, and over 2700 environmental exchanges arising from them, including extraction of various natural resources, water consumption, land use, and emissions to air, water, and soil.

Industrial Environmental Management: Engineering, Science, and Policy, First Edition. Tapas K. Das.

Companion website: www.wiley.com/go/Das/IEM_1e

Appendix G

Resource Recovery

Waste-To-Energy Facility, City of Spokane, Washington, USA

G.1 Description of the Facility

The source recovery and waste-to-energy facility (WTE) located in Spokane, Washington, can process 800 T of municipal solid waste (MSW) daily: there are two boilers rated at 400 T/day each. The facility uses annually about 255 000 T of refuse. There is one steam turbine generators rated 13.8 kV, 30 MVA, 25.8 MW at a power factor of 0.86. The facility's power generation contract varies seasonally from \$38 to \$48 MWh (gross ~150 000 MW/year) (C. Averyt, personal communication). MSW is typically dumped directly into the refuse pit by truck. The overhead refuse cranes then move the refuse from the pit to each boiler's charging hopper. MSW is mixed and sorted in the pit by the crane operator. The ash residue is produced at a rate of about 6000 T/month. The average refuse-derived-fuel composition is given in Table G.1. A schematic diagram of the resource recovery is shown in Figure G.1.

G.1.1 Technology

The Refuse-Fired Boiler Combustion System is designed to introduce refuse into furnace at a controlled rate and provide the necessary over-fire and under-fire combustion air for proper combustion of the refuse and combustible gases. It is also designed to transport the ash inert residue to the Ash Handling System for disposal and to remove the soot and ash particles from the heat transfer surfaces. The combustion air handling portion of the Refuse-Fired Boiler Combustion System contains the necessary equipment and controls to supply and regulate the quantity of air being supplied to the boiler furnace for the combustion of refuse. It also contains the necessary equipment and controls to remove the combustion gases from the boiler furnace. The air supply side of the combustion system consists of primary and secondary forced draft fans, steam coil air heater, volume control dampers, and associated ductwork. The gas removal side consists of induced draft fans and associated ductwork. The City of Spokane WTE facility is equipped with two identical boilers, each boiler having a separate spray dryer absorber (SDA) and a separate fabric filter (FF). The lime preparation, dilution water, ash removal, and atomizing air systems are common to both the boilers.

The primary flow path in the air pollution control system starts at the flue gas discharge from the boiler economizer. The flue gas leaves the boiler containing a variety of contaminants, primarily sulfur dioxide (SO_2), hydraulic acid (HCl), and particulate ash. This contaminated gas flows through the ductwork to the inlet of the SDA. The gas enters the top of the SDA and flows downward through a distribution network of chevrons. Chevrons are used to distribute the gas evenly across the SDA cross section, where it comes in contact with atomized lime slurry. The lime slurry absorbs most of the acid gases. The clean gas flows down to the bottom of the SDA vessel where it turns and flows up and out the side of the SDA vessel.

To eliminate the need of stocking large quantities of anhydrous ammonia, a urea hydrolyzer is used for the generation of NH_3 vapor for NO_x control. This process uses a 40–50% by weight urea solution to produce ammonia. The urea solution is fed to the hydrolyzer with a diaphragm type metering pump driven by a variable frequency drive motor. The urea solution is pumped at a variable flow rate to maintain a constant level in the hydrolyzer.

The flue gas contains a significant amount of dust (ash and lime) particles as it leaves the SDA. This gas enters the FF where it is distributed into individual modules (or compartments). The FF is designed with six compartments. Five compartments are necessary for 100% load, while six compartments may be off line for maintenance. The pulse jet uses pulse of compressed air to clean the bags. The cleaning cycle can be started either by a pressure drop signal or a time cycle. Pressure drop signals are usual method.

A total of four Aldora Technologies, Inc. supplied continuous monitoring systems (CEMS) are used to measure emissions from two municipal waste combustor (MWC)

Industrial Environmental Management: Engineering, Science, and Policy, First Edition. Tapas K. Das.

Companion website: www.wiley.com/go/Das/IEM_1e

Table G.1 Waste composition: detailed results.

Material	Est. percent	± (%)	Material	Est. percent	± (%)
Paper packaging	11.2		Paper products	14.6	
Newspaper	0.3	0.5	Newspaper	1.3	0.4
Card board/Kraft packaging	5.1	2.0	Cardboard/Kraft	0.0	0.0
Other ground wood packaging	0.3	0.2	Magazines	1.2	0.6
Mixed/low-grade packaging	3.3	0.9	High-grade papers	1.4	0.6
Compostable packaging	0.1	0.1	Other ground wood	0.3	0.3
R/C paper packaging	2.0	1.3	Mixed low grade	2.2	0.7
Plastic packaging	7.7		Compostable products	7.1	4.3
#1 PET bottles	0.7	0.2	Paper processing sludge	0.0	0.0
#2 PET non-bottles	0.4	0.1	R/C paper	1.2	1.7
#3 HDPE natural bottles	0.3	0.1	Plastic products	3.3	
#2 HDPE colored bottles	0.5	0.2	#1 PET	0.0	0.0
#2 HDPE jars and tubs	0.2	0.1	#2 HDPE	0.0	0.0
#3 PVC packaging	0.0	0.0	#3 PVC	0.0	0.0
#4 LDPE packaging	0.0	0.0	#4 LDPE	0.0	0.0
#5 PP packaging	0.5	0.1	#5 PP	0.0	0.0
#6 PS packaging	0.5	0.2	#6 PS	0.0	0.0
#7 Others packaging	0.7	0.1	#7 Other	1.3	0.6
Merchandise packaging film	0.7	0.1	Garbage bags	1.1	0.2
Nonindustrial/packaging film	2.7	0.8	Film products	0.2	0.1
Industrial/packaging film	0.2	0.2	R/C products	0.7	0.5
R/C packaging	0.2	0.1	Consumer products	4.0	
Glass	1.3		Televisions – CRT	0.0	0.0
Clear containers	0.3	0.1	Televisions – LCD	0.0	0.0
Green containers	0.3	0.3	VCRs, DVDs, DVRs	0.0	0.0
Brown containers	0.4	0.2	Computer monitors – CRT	0.0	0.0
Plate glass	0.0	0.0	Computer monitors – LCD	0.0	0.0
Glassware	0.1	0.1	Computers	0.0	0.0
R/C glass	0.2	0.2	Computer peripherals	0.4	0.3
Metal	6.4		Audio equipment	0.0	0.0
Alum. beverage cans	0.5	0.1	Gaming equipment	0.0	0.0
Alum. foil/containers	0.2	0.1	Other consumer electronics	0.1	0.2
Other aluminum	0.1	0.1	Textiles – organic	1.6	0.8
Other nonferrous	0.0	0.0	Textiles – synthetic	1.6	0.6
Food cans – tinned	0.7	0.2	Shoes, purses, belts	0.2	0.2
Food cans – coated	0.2	0.1	Tires and rubber	0.0	0.0
White goods	0.0	0.0	Furniture	0.0	0.0
Other ferrous	2.4	1.7	Mattresses	0.0	0.0
R/C metals	2.3	1.3	R/C consumer	0.1	0.2
Organics	24.0		Hazardous wastes	3.5	
Food – vegetative	7.1	1.1	Pesticides/herbicides	0.0	0.0
Food-nonvegetative	3.8	1.0	Mercury vapor lighting	0.0	0.0
Leaves and grass	9.6	3.2	Compact fluorescents	0.0	0.0

Table G.1 (Continued)

Material	Est. percent	± (%)	Material	Est. percent	± (%)
Prunings	0.4	0.5	Fluorescent tubes	0.0	0.0
Animal manure	2.3	1.4	Asbestos	0.0	0.0
Animal carcasses	0.0	0.0	Latex paint	0.9	1.0
Crop residues	0.0	0.0	Solvent-based glues	0.0	0.0
Fruit waste	0.0	0.0	Latex-based glues	0.0	0.0
R/C organics	0.8	0.7	Oil-based paint and solvent	0.0	0.0
Wood wastes	8.2		Caustic cleaners	0.0	0.0
Treated wood	0.1	0.1	Dry-cell batteries	0.0	0.0
Painted wood	3.0	1.9	Wet-cell batteries	0.0	0.0
Dimensional lumber	2.0	0.6	Gasoline/kerosene	0.0	0.0
Engineered wood	2.1	1.0	Motor oil	0.0	0.0
Pallets and crates	0.3	0.5	Antifreeze	0.0	0.0
Other untreated wood	0.1	0.2	Other vehicle fluids	0.0	0.0
Wood by-products	0.2	0.3	Oil filters	0.0	0.0
R/C wood	0.4	0.5	Explosives	0.0	0.0
			Medical wastes	0.1	0.1
Construction materials	15.8		Pharmaceuticals/vitamins	0.1	0.1
Natural wood	1.9	3.1	Disposable diapers	2.3	0.7
Insulation	0.1	0.1	Other cleaners and soaps	0.0	0.0
Asphalt paving	0.0	0.0	Other hazardous	0.0	0.0
Concrete	0.0	0.0	Other nonhazardous	0.0	0.0
Drywall	0.3	0.2	Residues	0.0	
Carpet	3.4	3.6	Ash	0.0	0.0
Carpet padding	0.3	0.4	Dust	0.0	0.0
Soil, rocks, sand	8.0	2.8	Fines	0.0	0.0
Asphalt roofing	0.2	0.1	Sludge/Special industrial	0.0	0.0
Plastic flooring	0.2	0.2			
Ceramics and brick	0.8	0.5	Totals	100.0	
R/C demo	0.4	0.3	Sample count	12	

Confidence intervals are calculated at 90% confidence level. Percentages for material types may not total 100% due to rounding.

units. One CEMS is installed for each spray SDA inlet and each FF outlet. Each CEMS consists of a Perkin Elmer MCS 100E multicomponent analyzer with integral zirconium oxide oxygen analyzer and an Aldora Technologies supplied hot/wet sample transport and conditioning system. The SDA inlet and FF outlet CEMS from each MWC unit are connected to a single Environmental Systems Corporation (ESC) 8816 data logger. The two ESC data loggers send all CEMS data to a central UNIX-based, ESC supplied data acquisition system located in the plant control room. Each FF outlet CEMS measures nitrogen oxides (NO_x), carbon monoxide (CO), oxygen (O_2), controlled sulfur dioxide (SO_2), and moisture (H_2O) concentrations. Emissions are measured on a wet basis and corrected to a dry basis using the moisture measurements from the MCS 100E.

The stream from the boiler super heater outlet is piped to the 900 psig steam manifold, which is a section of pipe 10 in. in diameter and serves as the central receiving and distribution for steam at 900 psig and 830 °F. Each boiler's bottom ash/shifting system collects bottom ash, shifting from the two boilers, conditions it, and discharges it onto the main feed belt. Each boiler's fly-ash system collects ash from the boiler generating bank hoppers, the SDA, and from the FFs.

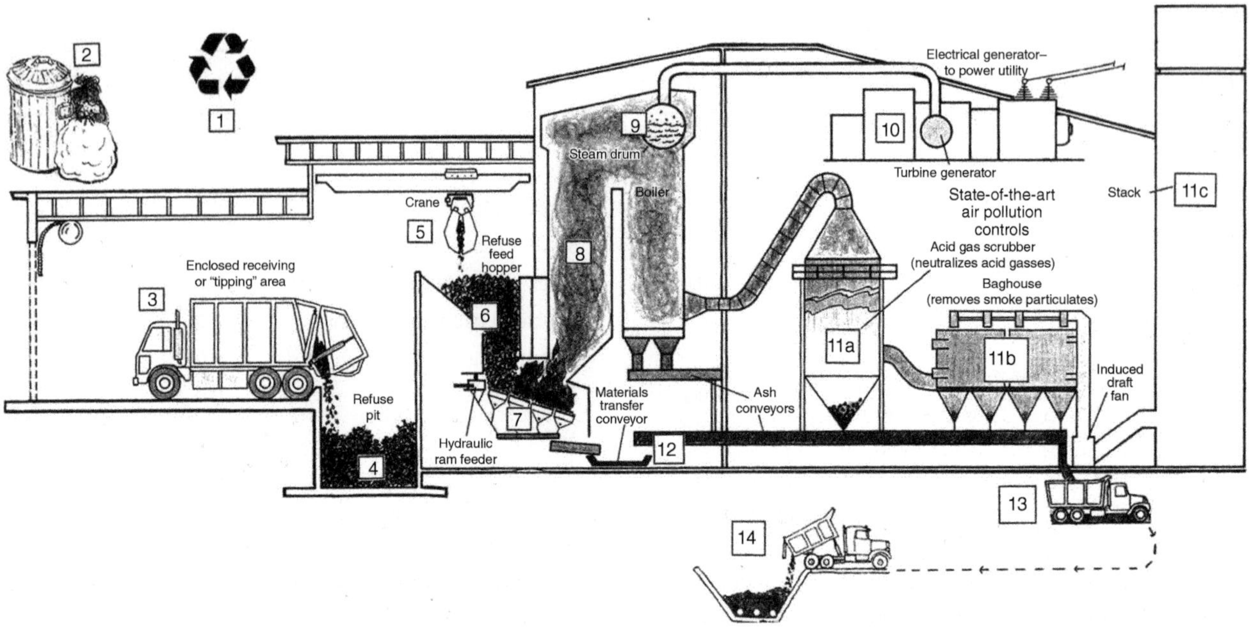

1. Reuse and recycle municipal solid waste
2. Waste collection
3. Enclosed recycling or tipping area
4. Refuse pit
5. Crane
6. Refuse feed hopper
7. Hydraulic ram feeder
8. Boiler/Incinerator
9. Steam drum
10. Turbine generator/electric power to utility
11a. Ash collection
11b. Baghouse for fly-ash collection
11c. Stack
12. Materials/Ash transfer conveyors
13. Induced draft fan blow gases through stack
14. Ash recovery, transport, and landfill

Figure G.1 Schematic of a municipal solid waste-to-energy facility in the city of Spokane, Washington, USA.

Appendix H

The Hannover Principles

Developed by William McDonough and Michael Braungart (2002), the Hannover Principles were among the first to comprehensively address the fundamental ideas of sustainability and the built environment, recognizing our interdependence with nature and proposing a new relationship that includes our responsibilities to protect it. The Principles encourage all of us – you, your organization, your suppliers and customers – to link long-term sustainable considerations with ethical responsibility, and to reestablish the integral relationship between natural processes and human activity. When you make decisions in your organization, remember these essential principles:

- Recognize interdependence. Simply put: everything you do personally, in your organization and through your work interacts with and depends upon the natural world, at every scale, both locally and across the globe.
- Eliminate the concept of waste. Are you considering the full, life-cycle consequences of what you create or buy?
- Understand the limitations of design. Treat nature as a model, not as an inconvenience to be evaded or controlled.

1) Insist on rights of humanity and nature to coexist in a healthy, supportive, diverse and sustainable condition.
2) Recognize interdependence. The elements of human design interact with and depend upon the natural world, with broad and diverse implications at every scale. Expand design considerations to recognizing even distant effects.
3) Respect relationships between spirit and matter. Consider all aspects of human settlement, including community, dwelling, industry, and trade in terms of existing and evolving connections between spiritual and material consciousness.
4) Accept responsibility for the consequences of design decisions upon human well-being, the viability of natural systems, and their right to coexist.
5) Create safe objects of long-term value. Do not burden future generations with requirements for maintenance or vigilant administration of potential danger due to the careless creation of products, processes, or standards.
6) Eliminate the concept of waste. Evaluate and optimize the full life-cycle of products and processes, to approach the state of natural systems, in which there is no waste.
7) Rely on natural energy flows. Human designs should, like the living world, derive their creative forces from perpetual solar income. Incorporate this energy efficiently and safely for responsible use.
8) Understand the limitations of design. No human creation lasts forever and design does not solve all problems. Those who create and plan should practice humility in the face of nature. Treat nature as a model and mentor, not as an inconvenience to be evaded or controlled.
9) Seek constant improvement by the sharing of knowledge. Encourage direct and open communication between colleagues, patrons, manufacturers, and users to link long-term sustainable considerations with ethical responsibility, and reestablish the integral relationship between natural processes and human activity.

The Hannover Principles should be seen as a living document committed to the transformation and growth in the understanding of our interdependence with nature, so that they may adapt as our knowledge of the world evolves.

Source: http://www.virginia.edu/arch/pub/hannover_list.html; http://policyworks.gov/org/main/mp/gsa/sd2.html

Industrial Environmental Management: Engineering, Science, and Policy, First Edition. Tapas K. Das.

Companion website: www.wiley.com/go/Das/IEM_1e

Table H.1 Factors for constructing variables control charts.

	$\bar{x}$ Chart			$\bar{R}$ Chart		
	Factors for control limits			Factors for control limits		
n^a	A_1	A_2	d_2	D_3	D_4	n
2	3.760	1.880	1.128	0	3.267	2
3	2.394	1.023	1.693	0	2.575	3
4	1880	0.729	2.059	0	2.282	4
5	1.596	0.577	2.326	0	2.115	5
6	1.410	0.483	2.534	0	2.004	6
7	1.277	0.419	2.704	0.076	1.924	7
8	1.175	0.373	2.847	0.136	1.864	8
9	1.094	0.337	2.970	0.184	1.816	9
10	1.028	0.308	3.078	0.223	1.777	10
11	0.973	0.285	3.173	0.256	1.744	11
12	0.925	0.266	3.258	0.284	1.716	12
13	0.884	0.249	3.336	0.308	1.692	13
14	0.848	0.235	3.407	0.329	1.671	14
15	0.816	0.223	3.472	0.348	1.652	15
16	0.788	0.212	3.532	0.364	1.636	16
17	0.762	0.203	3.588	0.379	1.621	17
18	0.738	0.194	3.640	0.392	1.608	18
19	0.717	0.187	3.689	0.404	1.596	19
20	0.697	0.180	3.735	0.414	1.586	20
21	0.679	0.173	3.778	0.425	1.575	21
22	0.662	0.167	3.819	0.434	1.566	22
23	0.647	0.162	3.858	0.443	1.557	23
24	0.632	0.157	3.895	0.452	1.548	24
25	0.619	0.153	3.931	0.459	1.541	25

[a] $n > 25$: $A_1 = \frac{\sigma}{\sqrt{n}}$ where n = number of observations in sample.

Table H.2 Process sigma conversion table.

Yield %	DPMO	Sigma level
6.68000	933200	0.000
8.45500	915450	0.125
10.56000	894400	0.250
13.03000	869700	0.375
15.87000	841300	0.500
19.08000	809200	0.625
22.66000	773400	0.750
26.59500	734050	0.875
30.85000	691500	1.000
35.43500	645650	1.125
40.13000	598700	1.250
45.02500	549750	1.375
50.00000	500000	1.500
54.97500	450250	1.625
59.87000	401300	1.750
64.56500	354350	1.875
69.15000	308500	2.000
73.40500	265950	2.125
77.34000	226600	2.250
80.92000	190800	2.375
84.13000	158700	2.500
86.97000	130300	2.625
89.44000	105600	2.750
91.54500	84550	2.875
93.32000	66800	3.000
94.79000	52100	3.125
95.99000	40100	3.250
96.96000	30400	3.375
97.73000	22700	3.500
98.32000	16800	3.625
98.78000	12200	3.750
99.12000	8800	3.875
99.38000	6200	4.000
99.56500	4350	4.125
99.70000	3000	4.250
99.79500	2050	4.375
99.87000	1300	4.500
99.91000	900	4.625
99.94000	600	4.750
99.96000	400	4.875
99.97700	230	5.000
99.98200	180	5.125
99.98700	130	5.250
99.99200	80	5.375
99.99700	30	5.500
99.99767	23	5.625
99.99833	17	5.750
99.99900	10	5.875
99.99966	3	6.000

DPMO – defects per million opportunities.

Reference

McDonough, W. and Braungart, M. (2002). *Cradle to Cradle: Remaking the Way We Make Things*. North Point Press.

Appendix I

Environmental Goals and Business Goals Are Not Two Distinct Goal Sets

They are integral to each other each sets goals depend upon the other of Earth's longevity. Research gains inroads into "use applications and variable cross-media impacts and positive benefits become in outcome based monitoring that balances input to outgo." Environment and business benefit through keeping product users using products that are cleanly made and reproduced to maintain a healthy environment and the environment benefits by being properly used and energized so that it is wholesome and reproduces rather that becoming extinct. Research progresses in gaining knowledge of how to do more with less and this makes a win-win situation that positively engages all concerned from inquiry to pilot to creation of backing in industrial applications. Careful evaluations include alternative methodology for containment of toxicity, pollution, and resource conservation. Constant and consistent evaluation and monitoring of health and resource retention is wise and imperative.

Global compliance considers not only those who are product users, sources of product materials, but those who design, test, process, and package products. There is the consideration of product viability, value, and depth of application over time.

Testing of design theoretically as well as in action function and pilots are imperative because need is the frequent but not the only driver of product creation. Extrapolation that reveals other needs or interactions that follow or extend the uses that start with the originally identified need then more testing is required. If need becomes a red herring driver then tests lose their ability to balance the integral process and the important ability to have strong responsible action be achieved. Planned obsolescence rarely generates value with longevity. Products that take on inflexible aspects or have no competition to encourage continued positive growth become causes of waste and are self-destructive.

Assembly of products requires knowledge of the philosophy of future needs as well as present needs and how the future helps direct the present. Understanding what drives change and its creative limits also includes what and where support for change comes from and how time is a significant limiting factor. If the answer to a need is created after a need is revealed there is potential for much waste and loss of viable access to environmental support.

Industrial Environmental Management: Engineering, Science, and Policy, First Edition. Tapas K. Das.

Companion website: www.wiley.com/go/Das/IEM_1e

Appendix J

Sample Codes of Ethics and Guidelines

J.1 National Society of Professional Engineers

J.1.1 Preamble

Engineering is an important and learned profession. As members of this profession, engineers are expected to exhibit the highest standards of honesty and integrity. Engineering has a direct and vital impact on the quality of life for all people. Accordingly, the services provided by engineers require honesty, impartiality, fairness, and equity, and must be dedicated to the protection of the public health, safety, and welfare. Engineers must perform under a standard of professional behavior that requires adherence to the highest principles of ethical conduct (NSPE 2018).

J.1.1.1 Fundamental Canons

Engineers, in the fulfillment of their professional duties, shall:

1. Hold paramount the safety, health, and welfare of the public.
2. Perform services only in areas of their competence.
3. Issue public statements only in an objective and truthful manner.
4. Act for each employer or client as faithful agents or trustees.
5. Avoid deceptive acts.
6. Conduct themselves honorably, responsibly, ethically, and lawfully so as to enhance the honor, reputation, and usefulness of the profession.

J.1.1.2 Rules of Practice

1. Engineers shall hold paramount the safety, health, and welfare of the public.
   a. If engineers' judgment is overruled under circumstances that endanger life or property, they shall notify their employer or client and such other authority as may be appropriate.
   b. Engineers shall approve only those engineering documents that are in conformity with applicable standards.
   c. Engineers shall not reveal facts, data, or information without the prior consent of the client or employer except as authorized or required by law or this Code.
   d. Engineers shall not permit the use of their name or associate in business ventures with any person or firm that they believe is engaged in fraudulent or dishonest enterprise.
   e. Engineers shall not aid or abet the unlawful practice of engineering by a person or firm.
   f. Engineers having knowledge of any alleged violation of this Code shall report thereon to appropriate professional bodies and, when relevant, also to public authorities, and cooperate with the proper authorities in furnishing such information or assistance as may be required.
2. Engineers shall perform services only in the areas of their competence.
   a. Engineers shall undertake assignments only when qualified by education or experience in the specific technical fields involved.
   b. Engineers shall not affix their signatures to any plans or documents dealing with subject matter in which they lack competence, nor to any plan or document not prepared under their direction and control.
   c. Engineers may accept assignments and assume responsibility for coordination of an entire project and sign and seal the engineering documents for the entire project, provided that each technical segment is signed and sealed only by the qualified engineers who prepared the segment.
3. Engineers shall issue public statements only in an objective and truthful manner.
   a. Engineers shall be objective and truthful in professional reports, statements, or testimony. They shall include all relevant and pertinent information in

Industrial Environmental Management: Engineering, Science, and Policy, First Edition. Tapas K. Das.

Companion website: www.wiley.com/go/Das/IEM_1e

such reports, statements, or testimony, which should bear the date indicating when it was current.

b. Engineers may express publicly technical opinions that are founded upon knowledge of the facts and competence in the subject matter.

c. Engineers shall issue no statements, criticisms, or arguments on technical matters that are inspired or paid for by interested parties, unless they have prefaced their comments by explicitly identifying the interested parties on whose behalf they are speaking, and by revealing the existence of any interest the engineers may have in the matters.

4. Engineers shall act for each employer or client as faithful agents or trustees.
   a. Engineers shall disclose all known or potential conflicts of interest that could influence or appear to influence their judgment or the quality of their services.
   b. Engineers shall not accept compensation, financial or otherwise, from more than one party for services on the same project, or for services pertaining to the same project, unless the circumstances are fully disclosed and agreed to by all interested parties.
   c. Engineers shall not solicit or accept financial or other valuable consideration, directly or indirectly, from outside agents in connection with the work for which they are responsible.
   d. Engineers in public service as members, advisors, or employees of a governmental or quasi-governmental body or department shall not participate in decisions with respect to services solicited or provided by them or their organizations in private or public engineering practice.
   e. Engineers shall not solicit or accept a contract from a governmental body on which a principal or officer of their organization serves as a member.
5. Engineers shall avoid deceptive acts.
   a. Engineers shall not falsify their qualifications or permit misrepresentation of their or their associates' qualifications. They shall not misrepresent or exaggerate their responsibility in or for the subject matter of prior assignments. Brochures or other presentations incident to the solicitation of employment shall not misrepresent pertinent facts concerning employers, employees, associates, joint venturers, or past accomplishments.
   b. Engineers shall not offer, give, solicit, or receive, either directly or indirectly, any contribution to influence the award of a contract by public authority, or which may be reasonably construed by the public as having the effect or intent of influencing the awarding of a contract. They shall not offer any gift or other valuable consideration in order to secure work. They shall not pay a commission, percentage, or brokerage fee in order to secure work, except to a bona fide employee or bona fide established commercial or marketing agencies retained by them.

J.1.1.3 Professional Obligations

1. Engineers shall be guided in all their relations by the highest standards of honesty and integrity.
   a. Engineers shall acknowledge their errors and shall not distort or alter the facts.
   b. Engineers shall advise their clients or employers when they believe a project will not be successful.
   c. Engineers shall not accept outside employment to the detriment of their regular work or interest. Before accepting any outside engineering employment, they will notify their employers.
   d. Engineers shall not attempt to attract an engineer from another employer by false or misleading pretenses.
   e. Engineers shall not promote their own interest at the expense of the dignity and integrity of the profession.
2. Engineers shall at all times strive to serve the public interest.
   a. Engineers are encouraged to participate in civic affairs; career guidance for youths; and work for the advancement of the safety, health, and well-being of their community.
   b. Engineers shall not complete, sign, or seal plans and/or specifications that are not in conformity with applicable engineering standards. If the client or employer insists on such unprofessional conduct, they shall notify the proper authorities and withdraw from further service on the project.
   c. Engineers are encouraged to extend public knowledge and appreciation of engineering and its achievements.
   d. Engineers are encouraged to adhere to the principles of sustainable development[1] in order to protect the environment for future generations.
   e. Engineers shall continue their professional development throughout their careers and should keep current in their specialty fields by engaging in professional practice, participating in continuing education courses, reading in the technical literature, and attending professional meetings and seminars.

1 "Sustainable development" is the challenge of meeting human needs for natural resources, industrial products, energy, food, transportation, shelter, and effective waste management while conserving and protecting environmental quality and the natural resource base essential for future development.

3. Engineers shall avoid all conduct or practice that deceives the public.
   a. Engineers shall avoid the use of statements containing a material misrepresentation of fact or omitting a material fact.
   b. Consistent with the foregoing, engineers may advertise for recruitment of personnel.
   c. Consistent with the foregoing, engineers may prepare articles for the lay or technical press, but such articles shall not imply credit to the author for work performed by others.
4. Engineers shall not disclose, without consent, confidential information concerning the business affairs or technical processes of any present or former client or employer, or public body on which they serve.
   a. Engineers shall not, without the consent of all interested parties, promote or arrange for new employment or practice in connection with a specific project for which the engineer has gained particular and specialized knowledge.
   b. Engineers shall not, without the consent of all interested parties, participate in or represent an adversary interest in connection with a specific project or proceeding in which the engineer has gained particular specialized knowledge on behalf of a former client or employer.
5. Engineers shall not be influenced in their professional duties by conflicting interests.
   a. Engineers shall not accept financial or other considerations, including free engineering designs, from material or equipment suppliers for specifying their product.
   b. Engineers shall not accept commissions or allowances, directly or indirectly, from contractors or other parties dealing with clients or employers of the engineer in connection with work for which the engineer is responsible.
6. Engineers shall not attempt to obtain employment or advancement or professional engagements by untruthfully criticizing other engineers, or by other improper or questionable methods.
   a. Engineers shall not request, propose, or accept a commission on a contingent basis under circumstances in which their judgment may be compromised.
   b. Engineers in salaried positions shall accept part-time engineering work only to the extent consistent with policies of the employer and in accordance with ethical considerations.
   c. Engineers shall not, without consent, use equipment, supplies, laboratory, or office facilities of an employer to carry on outside private practice.
7. Engineers shall not attempt to injure, maliciously or falsely, directly or indirectly, the professional reputation, prospects, practice, or employment of other engineers. Engineers who believe others are guilty of unethical or illegal practice shall present such information to the proper authority for action.
   a. Engineers in private practice shall not review the work of another engineer for the same client, except with the knowledge of such engineer, or unless the connection of such engineer with the work has been terminated.
   b. Engineers in governmental, industrial, or educational employ are entitled to review and evaluate the work of other engineers when so required by their employment duties.
   c. Engineers in sales or industrial employ are entitled to make engineering comparisons of represented products with products of other suppliers.
8. Engineers shall accept personal responsibility for their professional activities, provided, however, that engineers may seek indemnification for services arising out of their practice for other than gross negligence, where the engineer's interests cannot otherwise be protected.
   a. Engineers shall conform with state registration laws in the practice of engineering.
   b. Engineers shall not use association with a nonengineer, a corporation, or partnership as a "cloak" for unethical acts.
9. Engineers shall give credit for engineering work to those to whom credit is due, and will recognize the proprietary interests of others.
   a. Engineers shall, whenever possible, name the person or persons who may be individually responsible for designs, inventions, writings, or other accomplishments.
   b. Engineers using designs supplied by a client recognize that the designs remain the property of the client and may not be duplicated by the engineer for others without express permission.
   c. Engineers, before undertaking work for others in connection with which the engineer may make improvements, plans, designs, inventions, or other records that may justify copyrights or patents, should enter into a positive agreement regarding ownership.
   d. Engineers' designs, data, records, and notes referring exclusively to an employer's work are the employer's property. The employer should indemnify the engineer for use of the information for any purpose other than the original purpose.

—As Revised July 2018

> By order of the United States District Court for the District of Columbia, former Section 11(c) of the NSPE Code of Ethics prohibiting competitive bidding, and all policy statements, opinions, rulings or other guidelines interpreting its scope, have been rescinded as unlawfully interfering with the legal right of engineers, protected under the antitrust laws, to provide price information to prospective clients; accordingly, nothing contained in the NSPE Code of Ethics, policy statements, opinions, rulings or other guidelines prohibits the submission of price quotations or competitive bids for engineering services at any time or in any amount.
>
> Statement by NSPE Executive Committee

In order to correct misunderstandings which have been indicated in some instances since the issuance of the Supreme Court decision and the entry of the Final Judgment, it is noted that in its decision of 25 April 1978, the Supreme Court of the United States declared: "The Sherman Act does not require competitive bidding."

It is further noted that as made clear in the Supreme Court decision:

1. Engineers and firms may individually refuse to bid for engineering services.
2. Clients are not required to seek bids for engineering services.
3. Federal, state, and local laws governing procedures to procure engineering services are not affected, and remain in full force and effect.
4. State societies and local chapters are free to actively and aggressively seek legislation for professional selection and negotiation procedures by public agencies.
5. State registration board rules of professional conduct, including rules prohibiting competitive bidding for engineering services, are not affected and remain in full force and effect. State registration boards with authority to adopt rules of professional conduct may adopt rules governing procedures to obtain engineering services.
6. As noted by the Supreme Court, "nothing in the judgment prevents NSPE and its members from attempting to influence governmental action ..."

Note: In regard to the question of application of the Code to Corporations vis-à-vis real persons, business form or type should not negate nor influence conformance of individuals to the Code. The Code deals with professional services, which services must be performed by real persons. Real persons in turn establish and implement policies within business structures. The Code is clearly written to apply to the Engineer, and it is incumbent on members of NSPE to endeavor to live up to its provisions. This applies to all pertinent sections of the Code.

Source: National Society of Professional Engineers. Publication date revised: July 2018.

Reference

National Society of Professional Engineers (NSPE) (2018). Publication date revised: July 2018.

Index

Industrial Environmental Management: Engineering, Science, and Policy, First Edition. Tapas K. Das.

Companion website: www.wiley.com/go/Das/IEM_1e

d

e

f

k

l

m

q

r